Springer Collected Works in Mathematics

More information about this series at http://www.springer.com/series/11104

David Mumford

Selected Papers II

On Algebraic Geometry, Including Correspondence with Grothendieck

Editors
Amnon Neeman
Ching-Li Chai
Takahiro Shiota

Reprint of the 2010 Edition

Springer

Author
David Mumford
Brown University
Providence, RI, USA

Editors
Amnon Neeman
School of Mathematical Sciences
Australian National University
Canberra, ACT, Australia

Ching-Li Chai
Department of Mathematics
University of Pennsylvania
Philadelphia, PA, USA

Takahiro Shiota
Faculty of Science
Kyoto University
Kyoto, Japan

ISSN 2194-9875
Springer Collected Works in Mathematics
ISBN 978-1-4939-9588-2

This Springer imprint is published by the registered company Springer Science+Business Media, LLC, part of Springer Nature
The registered company address is: 233 Spring Street, New York, NY 10013, U.S.A.

David Mumford

Selected Papers, Volume II

On Algebraic Geometry, including Correspondence with Grothendieck

Ching-Li Chai
Amnon Neeman
Takahiro Shiota
Editors

Springer

David Mumford
Division of Applied Mathematics
Brown University
182 George Street
Providence, RI 02912
USA
david_mumford@brown.edu

Editors

Ching-Li Chai
Department of Mathematics
University of Pennsylvania
209 S. 33rd street
Philadelphia, PA 19104-6395
USA
chai@math.upenn.edu

Amnon Neeman
School of Mathematical Sciences
Australian National University
Canberra, ACT 0200
Australia
amnon.neeman@anu.edu.au

Takahiro Shiota
Department of Mathematics
Kyoto University
Kitashirakawa-Oiwake-cho
Sakyo-ku 606-8502, Kyoto
Japan
shiota@math.kyoto-u.ac.jp

ISBN 978-0-387-72491-1
Springer New York Dordrecht Heidelberg London

Library of Congress Control Number: 2009931334

Printed on acid-free paper

Springer is part of Springer Science+Business Media (www.springer.com)

Preface

It is a tremendous pleasure to thank the editors Ching-Li Chai, Amnon Neeman and Takahiro Shiota for putting this second volume of my algebro-geometric work together. The first volume contained my papers on classification of varieties and moduli spaces. At the time this was put together, it was not anticipated to go further, but Chai, Neeman and Shiota, together with Springer have proposed to complete this by publishing all my other papers in algebraic geometry plus the surviving correspondence between me and Alexander Grothendieck. The editors have assembled permissions, edited and typeset several unpublished manuscripts and edited and commented Grothendieck's letters to me (and a few of mine to him). This involved a great deal of new typesetting and proofreading and, in addition to the editors, I want to thank the large team of graduate students, of Yogananda, Chai and Neeman who carried this out. The editors have also worked hard disambiguating the references in Grothendieck's letters—with little help from me because of my terrible memory. I will always be in their debt because their work has made a readable and hopefully useful tool out of a jumble of material in my files.

I also want to thank Jean Malgoire who has given permission on behalf of Grothendieck for the publication of his letters. For me, personally, Grothendieck's letters were priceless and enabled me to understand many of his ideas in their raw form before they were generalized too far and embedded in the daunting machinery of his "Éléments". It was an unequaled pleasure to have known him. He started the movement which has added a vast and highly productive level of abstraction to algebraic geometry (and many related fields), an approach which is still growing and deepening today. I consider him the greatest genius with whom I have had the opportunity to interact and I extend my heartfelt best wishes to him on this, the year of his 80th birthday.

Although I am not active in algebraic geometry, I have been watching some of the spectacular developments in recent years. It is wonderful to see how the long attack on birational geometry and the canonical ring in higher dimensions has paid off. When I was in the field, it seemed as if there might be layer after layer of complexity here, counter-examples galore. But no: a wonderful order has now appeared. Likewise, the structure and topology of the moduli spaces of curves is being tamed (at least asymptotically). Perhaps resolution of singularities in characteristic p is near solution. The Hodge conjecture still remains a mystery, though both positive and negative hints have been found. As I said in my preface to volume I, this beautiful story is not finished!

David Mumford
Tenants Harbor
August 2008

Contents

* The footnotes in Part II and in the newly typeset articles [63b], [u64a], [u64b], [u67a], [u67b] are due to the editors, except those which explicitly say otherwise. At the end of this volume are Notes concerning other articles in Part I, as well as a few articles in [SP1].

Labels like "[SP1]," defined in the Algebraic Geometry part of the Bibliography (pp. xiii–xviii), as well as the following acronyms are used in those notes:

GB = Le groupe de Brauer

TDTE = Technique de descente et théorèmes d'existence en géométrie algébrique (or Techniques de construction et théorèmes d'existence en géométrie algébrique)

List of Scanned Images

Editors' notes

The volume contains (1) reproductions of Mumford's published papers, (2) four unpublished papers, and (3) mathematical correspondence between Grothendieck and Mumford, plus several letters from Grothendieck to other mathematicians.

We tried to provide the documents with a minimum of intrusion. In the case of published articles the originals were reproduced; the Notes at the end of this volume list misprints, a few mathematical errors, and a smattering of hints concerning relevant later developments. Article [63b] is a special case: the published version was a Japanese translation of Mumford's manuscript, and here we included the typeset original. The unpublished papers and letters were newly typeset; we corrected obvious errors silently, and used footnotes for comments and additional information.

As for the correspondence, we had to decide which letters to include and how faithful to the original we wished to be. Adopting the view that this is primarily a mathematical document, we reproduced the letters that contain interesting mathematics, and tried to make the mathematics as accessible to the reader as possible. We modernized the notation, adopting what has become standard in the last decades. In the case of letters from Grothendieck, we asked ourselves how much we may, and should, alter his English. In the end we decided to correct the spelling, add the occasional missing pronoun or article, adjust minor grammatical errors, and improve the punctuation, but to refrain from changing the wording, or correcting whole phrases. The changes are made silently, without comments.

We would like to thank W. Messing, F. Oort and S.S. Shatz for always being willing to offer us counsel on difficult decisions and marginal cases. Since we did not always follow their advice, we alone are responsible for our mistakes; but it was always good to have someone to turn to when we were uncertain what to do.

We would like to thank M. Künzer for his help in reading Grothendieck's manuscripts. To illustrate Künzer's talent we reproduced an original page of the letter of 18 October, 1962; the reader is invited to try to untangle what it says on line −6 of the page. We asked quite a few people before we turned to Künzer, and no one else was even close.

Many people helped us by contributing illuminating suggestions for footnotes to the correspondence, and to a lesser extent, by pointing out mistakes in the published papers and checking the errata against the originals. We are indebted to them all. We especially wish to express our gratitude to M. Artin, J.-L. Colliot-Thélène, P. Deligne, O. Gabber, P.A. Griffiths, R.M. Hain, L. Illusie, N.M. Katz, S.L. Kleiman, F.F. Knudsen, W.E. Lang, C. Liedtke, R.K. Lazarsfeld, A. Mayer, W. Messing, T. Oda, M. Raynaud, J.-P. Serre, J. Stix and Y.-S. Tai.

The bulk of the typesetting was looked after by C.S. Yogananda at S.J. College of Engineering, Mysore, and by team Sriranga at Sriranga Digital Software Technologies, Srirangapatna. Our thanks for their monumental effort. Yogananda also gave us valuable LaTeX tips. We are also grateful to A. Auel, D. Fithian, S. Gupta, A. Holschbach, A. Obus and P. Pandit, who typeset a few handwritten letters each. G.S.D. Stevenson was the first to proofread the entire document and standardize the mathematical notation; his help was invaluable.

August 2008 C.-L. Chai, A. Neeman and T. Shiota

Acknowledgement

Concerning Part I of this volume, the author, editors and publisher would like to thank D.A. Bayer [93], P. Deligne [69e], F.F. Knudsen [76b], A.T. Lascu [75c], D.I. Lieberman [75b], F. Oort [68a], K. Suominen [72d], J. Tate [78d] and P. van Moerbeke [79b] for their permissions to publish their respective joint work with Mumford, P. Deligne for his permission to use the photograph in [78d], and Lucas and Robin Kempf for their permission to publish their father's appendix to [70] and his photograph in [02]. We would also like to thank the following individuals and bodies for their permissions:

Accademia Nazionale dei Lincei .... [73b]
American Association for the Advancement of Science *Science* .... [78d]
American Mathematical Society .... [75b], [76a]
Proceedings of the American Mathematical Society .... [71b]
Notices of the American Mathematical Society .... [86]
Annals of Mathematics .... [62b]
Cambridge University Press
Mathematical Proceedings of the Cambridge Philosophical Society .... [75c]
Centro Internazionale Matematico Estivo .... [70]
The Hebrew University Magnes Press *J. Analyse Math.* .... [67d]
l'Institut des Hautes Études Scientifiques
Publications Mathématiques de l'IHÉS .... [61a], [69e]
Institut Mittag-Leffler *Acta Mathematica* .... [79b]
Istituto Nazionale di Alta Matematica "Francesco Severi" .... [93]
The Johns Hopkins University Press
American Journal of Mathematics .... [64], [65a], [84a], [02]
Kinokuniya Book Store .... [78a]
The Mathematical Society of Japan *Sûgaku* .... [63b]
Mathematica Scandinavica .... [76b]
Oxford University Press India .... [69b]
Carol Parikh .... [90]
Rice University .... [73c]
Marie Schilling .... [65b]
Springer-Verlag *Inventiones Mathematicae* .... [68a]
Tata Institute of Fundamental Research .... [78b], [78c]
Wolters-Noordhoff bv .... [72d]

In this context we also wish to thank E. Arbarello, A. Del Centina, E.D. Farjoun, B. Hassett, J.J. Morava, R. Sujatha and A. Vistoli, who helped us reach copyright holders.

Concerning Part II of the volume, we express our warm thanks to A. Grothendieck and his correspondents, M. Artin, R. Hartshorne, J. Murre and J. Tate, who allowed some of their letters to be published here. We also thank J. Malgoire, who, having power-of-attorney for Grothendieck, communicated to us the permission.

Bibliography of David B. Mumford

Algebraic Geometry

Books

[GIT] *Geometric Invariant Theory*. Ergebnisse der Mathematik und ihrer Grenzgebiete, Neue Folge, Band 34. Springer-Verlag, 1965, vi+145 pp. Second enlarged edition (with J. Fogarty). 1982, xii+220 pp. Third enlarged edition (with F. Kirwan and J. Fogarty). 1994, xiv+292 pp.

[CAS] *Lectures on Curves on an Algebraic Surface* (with a section by G.M. Bergman). Annals of Mathematics Studies, No. 59. Princeton University Press, Princeton, N.J., 1966, xi+200 pp.

[RB] *Introduction to Algebraic Geometry: preliminary version of the first 3 chapters*. Mimeographed notes from the Harvard Mathematics Department and bound in red. Reprinted as *The Red Book of Varieties and Schemes*. Lecture Notes in Mathematics, 1358, Springer-Verlag, 1988, vi+309 pp; second expanded edition, including the Ziwet Lectures at the University of Michigan (1974) on curves and their Jacobians (with contributions by Enrico Arbarello). 1999, x+306 pp.

[AV] *Abelian Varieties*. Tata Institute of Fundamental Research Studies in Mathematics, No. 5. Published for the Tata Institute of Fundamental Research, Bombay; Oxford University Press, London, 1970, viii+242 pp. Second edition (with Appendices by C.P. Ramanujam and Y. Manin). 1974, x+279 pp. Corrected reprint of the second edition. Hindustan Book Agency, New Delhi (international distribution by the American Mathematical Society), 2008, xii+263 pp.

[AS] Appendices to *Algebraic Surfaces* by O. Zariski, Second supplemented edition, on pages 71–74, 88–91, 118–128, 147–155, 197–206, 229–231, 238–239 (with S.S. Abhyankar and J. Lipman). Ergebnisse der Mathematik und ihrer Grenzgebiete, Band 61. Springer-Verlag, 1971, xi+270 pp.

[TE] *Toroidal Embeddings I* (with G. Kempf, F.F. Knudsen and B. Saint-Donat). Lecture Notes in Mathematics, Vol. 339. Springer-Verlag, 1973, viii+209 pp.

[CJ] *Curves and their Jacobians*. The University of Michigan Press, Ann Arbor, Mich., 1975, vi+104 pp. Reprinted as an appendix (pp. 225–304) in the second expanded edition of [RB].

[SC] *Smooth Compactification of Locally Symmetric Varieties* (with A. Ash, M. Rapoport and Y.-S. Tai). Lie Groups: History, Frontiers and Applications, Vol. IV, Math. Sci. Press, Brookline, Mass., 1975, iv+335 pp.

[AG1] *Algebraic Geometry I. Complex Projective Varieties*. Grundlehren der Mathematischen Wissenschaften, No. 221. Springer-Verlag, 1976, x+186 pp. Reprinted in the "Classics in Mathematics" series, Springer-Verlag, 1995, x+186 pp.

[Θ1] *Tata Lectures on Theta I*, containing Introduction and motivation: theta functions in one variable, Basic results on theta functions in several variables (with the assistance of C. Musili, M. Nori, E. Previato and M. Stillman). Progress in Mathematics, 28. Birkhäuser Boston, Inc., Boston, Mass, 1983, xiii+235 pp. Republished in the Modern Birkhäuser Classics series, 2007.

[Θ2] *Tata Lectures on Theta II*, Jacobian theta functions and differential equations (with the collaboration of C. Musili, M. Nori, E. Previato, M. Stillman and H. Umemura). Progress in Mathematics, 43. Birkhäuser Boston, Inc., Boston, Mass, 1984. xiv+272 pp. Republished in the Modern Birkhäuser Classics series, 2007.

[Θ3] *Tata Lectures on Theta III* (with the collaboration of M. Nori and P. Norman). Progress in Mathematics, 97. Birkhäuser Boston, Inc., Boston, Mass, 1991, viii+202 pp. Republished in the Modern Birkhäuser Classics series, 2007.

[SP1] *Selected Papers on the Classification of Varieties and Moduli Spaces*. Springer-Verlag, 2004, xiii+795 pp.

Papers

[61a] The topology of normal singularities of an algebraic surface and a criterion for simplicity. *Publ. Math. de l'I.H.É.S.* No. 9, 1961, 5–22.

[61b] Pathologies of modular algebraic surfaces. *Amer. J. Math.* **83**, 1961, 339–342.

[61c] An elementary theorem in geometric invariant theory. *Bull. Amer. Math. Soc.* **67**, 1961, 483–487.

[62a] Further pathologies in algebraic geometry. *Amer. J. Math.* **84**, 1962, 642–648.

[62b] The canonical ring of an algebraic surface. An appendix to a paper by Oscar Zariski, "The theorem of Riemann–Roch for high multiples of an effective divisor on an algebraic surface, *Ann. of Math.* (2) **76**, 1962, 560–615", *Ann. of Math.* (2) **76**, 1962, 612–615.

[63a] Projective invariants of projective structures and applications. In *Proc. Internat. Congr. Mathematicians (Stockholm, 1962)*, Inst. Mittag-Leffler, Djursholm, 1963, 526–530.

[63b] Some aspects of the problem of moduli (translated into Japanese by Y. Akizuki). *Sûgaku* **15**, 1963/1964, 155–157.

[64] Two fundamental theorems on deformations of polarized varieties (with T. Matsusaka). *Amer. J. Math.* **86**, 1964, 668–684.
Correction to: "Two fundamental theorems on deformations of polarized varieties". *Amer. J. Math.* **91**, 1969, 851.

[65a] A remark on Mordell's conjecture. *Amer. J. Math.* **87**, 1965, 1007–1016.

[65b] Picard groups of moduli problems. In *Arithmetical Algebraic Geometry*, (Proc. Conf. Purdue Univ., 1963), Harper & Row, New York, 1965, 33–81.

[66a] On the equations defining abelian varieties. I. *Invent. Math.* **1**, 1966, 287–354.

[66b] Families of abelian varieties. In *Algebraic Groups and Discontinuous Subgroups* (Proc. Sympos. Pure Math., Vol. IX, Boulder, Colo., 1965), Amer. Math. Soc., Providence, R.I., 1966, 347–351.

[67a] On the equations defining abelian varieties. II. *Invent. Math.* **3**, 1967, 75–135.

[67b] On the equations defining abelian varieties. III. *Invent. Math.* **3**, 1967, 215–244.

[67c] Pathologies. III. *Amer. J. Math.* **89**, 1967, 94–104.

[67d] Abelian quotients of the Teichmüller modular group. *J. Analyse Math.* **18**, 1967, 227–244.

[68a] Deformations and liftings of finite, commutative group schemes (with F. Oort). *Invent. Math.* **5**, 1968, 317–334.

[68b] Periods of a moduli space of bundles on curves (with P. Newstead). *Amer. J. Math.* **90**, 1968, 1200–1208.

[69a] Enriques' classification of surfaces in char *p*. I. In *Global Analysis* (Papers in Honor of K. Kodaira), Univ. Tokyo Press, Tokyo, 1969, 325–339.

[69b] Bi-extensions of formal groups. In *Algebraic Geometry* (Internat. Colloq., Tata Inst. Fund. Res., Bombay, 1968), Oxford Univ. Press, London, 1969, 307–322.

[69c] A note of Shimura's paper "Discontinuous groups and abelian varieties". *Math. Ann.* **181**, 1969, 345–351.

[69d] Rational equivalence of 0-cycles on surfaces. *J. Math. Kyoto Univ.* **9**, 1969, 195–204.

[69e] The irreducibility of the space of curves of a given genus (with P. Deligne). *Publ. Math. de l'I.H.É.S.* No. 36, 1969, 75–109.

[70] Varieties defined by quadratic equations (with an appendix by G. Kempf). In *Questions on Algebraic Varieties* (C.I.M.E., III Ciclo, Varenna, 1969), Edizioni Cremonese, Rome, 1970, 29–100.

[71a] Theta characteristics of an algebraic curve. *Ann. Sci. École Norm. Sup.* (4) **4**, 1971, 181–192.

[71b] A remark on Mahler's compactness theorem. *Proc. Amer. Math. Soc.* **28**, 1971, 289–294.

[71c] The structure of the moduli spaces of curves and abelian varieties. In *Actes du Congrès International des Mathématiciens* (Nice, 1970), Gauthier-Villars, Paris, 1971, Tome 1, 457–465.

[72a] An analytic construction of degenerating curves over complete local rings. *Compositio Math.* **24**, 1972, 129–174.

[72b] An analytic construction of degenerating abelian varieties over complete rings. *Compositio Math.* **24**, 1972, 239–272.

[72c] Some elementary examples of unirational varieties which are not rational (with M. Artin). *Proc. London Math. Soc.* (3) **25**, 1972, 75–95.

[72d] Introduction to the theory of moduli (with K. Suominen). In *Algebraic geometry, Oslo 1970* (Proc. Fifth Nordic Summer-School in Math.), Wolters-Noordhoff, Groningen, 1972, 171–222.

[73a] A rank 2 vector bundle on $\mathbb{P}^4$ with 15,000 symmetries (with G. Horrocks). *Topology* **12**, 1973, 63–81.

[73b] An example of a unirational 3-fold which is not rational. In *Atti del Convegno Internazionale di Geometria a Celebrazione del Contenario della Nascita di Federigo Enriques*, Accademia Nazionale dei Lincei, Rome, 1973, p. 99.

[73c] A remark on the paper of M. Schlessinger. In *Complex Analysis, 1972* (Proc. Conf., Rice Univ., Houston, Tex., 1972), Vol. I: Geometry of Singularities. Rice Univ. Studies 59, no. 1, 1973, 113–117.

[73d] Introduction to Part I of *Oscar Zariski: Collected Papers, Volume I, Foundations of Algebraic Geometry and Resolution of Singularities.* The MIT Press, Cambridge, Massachusetts and London, England, 1973, 3–6.

[73e] Introduction to Part II of *Oscar Zariski: Collected Papers, Volume II, Holomorphic Functions and Linear Systems.* The MIT Press, Cambridge, Massachusetts and London, England, 1973, 175–181.

[74] Prym varieties. I. In *Contributions to Analysis* (a collection of papers dedicated to Lipman Bers), Academic Press, New York, 1974, 325–350.

[75a] A new approach to compactifying locally symmetric varieties. In *Discrete Subgroups of Lie Groups and Applications to Moduli* (Internat. Colloq., Bombay, 1973), Oxford Univ. Press, Bombay, 1975, 211–224.

[75b] Matsusaka's big theorem (with D. Lieberman). In *Algebraic Geometry* (Proc. Sympos. Pure Math., Vol. XXIX, Humboldt State Univ., Arcata, Calif., 1974), Amer. Math. Soc., Providence, R.I., 1975, 513–530.

[75c] The self-intersection formula and the "formule-clef" (with A.T. Lascu and D.B. Scott). *Math. Proc. Cambridge Philos. Soc.* **78**, 1975, 117–123.

[75d] Pathologies IV. *Amer. J. Math.* **97**, no. 3, 1975, 847–849.

[76a] Hilbert's fourteenth problem—the finite generation of subrings such as rings of invariants. In *Mathematical Developments Arising from Hilbert Problems* (Proc. Sympos. Pure Math., Vol. XXVIII, Northern Illinois Univ., De Kalb, Ill., 1974), Amer. Math. Soc., Providence, R.I., 1976, 431–444.

[76b] The projectivity of the moduli space of stable curves. I: Preliminaries on "det" and "Div" (with F.F. Knudsen). *Math. Scand.* **39**, no. 1, 1976, 19–55.

[76c] Enriques' classification of surfaces in char. p. III (with E. Bombieri). *Invent. Math.* **35**, 1976, 197–232.

[77a] Enriques' classification of surfaces in char. *p*. II (with E. Bombieri). In *Complex Analysis and Algebraic Geometry*, Iwanami Shoten, Tokyo, 1977, 23–42.

[77b] Stability of projective varieties. Lectures given at the "Institut des Hautes Études Scientifiques", Bures-sur-Yvette, March-April 1976. Monographie de l'Enseignement Mathématique, No. 24. *L'Enseignement Mathématique* **23**, 1977, 39–110.

[77c] Hirzebruch's proportionality theorem in the noncompact case. *Invent. Math.* **42**, 1977, 239–272.

[78a] An algebro-geometric construction of commuting operators and of solutions to the Toda lattice equation, Korteweg de Vries equation and related non-linear equations. In *Proceedings of the International Symposium on Algebraic Geometry* (Kyoto Univ., Kyoto, 1977), Kinokuniya Book Store, Tokyo, 1978, pp. 115–153.

[78b] The work of C.P. Ramanujam in algebraic geometry. In *C.P. Ramanujam—a Tribute*, Tata Inst. Fund. Res. Studies in Math., 8, Springer-Verlag, 1978, 8–10.

[78c] Some footnotes to the work of C.P. Ramanujam. *ibid.*, 247–262.

[78d] Fields medals. IV. An instinct for the key idea (with J. Tate). *Science* **202**, no. 4369, 1978, 737–739.

[79a] An algebraic surface with K ample, $(K^2) = 9$, $p_g = q = 0$. *Amer. J. Math.* **101**, no. 1, 1979, 233–244.

[79b] The spectrum of difference operators and algebraic curves (with P. van Moerbeke). *Acta Math.* **143**, no. 1–2, 1979, 93–154.

[81] Singularities from the point of view of global moduli problems (abstract). In *Leopoldina-Symposion: Singularitäten—Singularities* [Leopoldina Symposium: Singularities] held in Thüringen, May 3–7, 1978. Nova Acta Leopoldina, n.F., Bd. 52, nr. 240, Johann Ambrosius Barth, Leipzig, 1981, p. 161.

[82] On the Kodaira dimension of the moduli space of curves (with J. Harris, with an appendix by W. Fulton). *Invent. Math.* **67**, no. 1, 1982, 23–88.

[83a] On the Kodaira dimension of the Siegel modular variety. In *Algebraic Geometry—Open Problems* (Ravello, 1982), Lecture Notes in Math., 997, Springer-Verlag, 1983, 348–375,

[83b] Towards an enumerative geometry of the moduli space of curves. In *Arithmetic and Geometry* Vol. II, Progr. Math., 36, Birkhäuser Boston, Boston, Mass, 1983, 271–328.

[84a] Proof of the convexity theorem. An appendix to a paper by Linda Ness, "A stratification of the null cone via moment map, Amer. J. Math. 106, 1984, pp. 1281-1329" *Amer. J. Math.* **106**, 1984, 1326–1329.

[86] Oscar Zariski: 1899–1986. *Notices Amer. Math. Soc.* **33**, no. 6, 1986, 891–894.

[88] Oscar Zariski and his work. A joint AMS–MAA lecture presented in Atlanta, Georgia, January 1988. AMS–MAA Joint Lecture Series. American Mathematical Society, Providence, R.I., 1988. 1 video cassette (NTSC; VHS) (60 min.)

[90] A foreword for non-mathematicians. In *The Unreal Life of Oscar Zariski* by Carol Parikh, Academic Press, Inc., Boston, Mass., 1990, pp. xv–xxvii; Springer Verlag, 2008, pp. xiii–xxii.

[93] What can be computed in algebraic geometry? (with D. Bayer). In *Computational Algebraic Geometry and Commutative Algebra* (Cortona, 1991), Sympos. Math., XXXIV, Cambridge Univ. Press, Cambridge, U.K., 1993, pp. 1–48.

[02] In memoriam: George R. Kempf, 1944–2002. *Amer. J. Math.* **124**, no. 6, 2002, iii–iv.

[u64a] The boundary of moduli schemes. Mimeographed notes, Summer Institute in Algebraic Geometry, Amer. Math. Soc., Woods Hole, 1964, 8 pp.

[u64b] Further comments on boundary points. Mimeographed notes, Summer Institute in Algebraic Geometry, Amer. Math. Soc., Woods Hole, 1964, 7 pp.

[u67a] Abstract theta functions (mimeographed notes by H. Pittie). Advanced Science Seminar in Algebraic Geometry, Sponsored by the National Science Foundation, Bowdoin College, Summer 1967, 15 pp.

[u67b] Degeneration of algebraic theta functions. Handwritten manuscript, 22 pp.

Vision

Books

1. *Filtering, Segmentation and Depth* (with Mark Nitzberg and T. Shiota). Lecture Notes in Computer Science **662**, Springer-Verlag, 1993.
2. *Two- and Three-dimensional Patterns of the Face* (with Peter Giblin, Gaile Gordon, Peter Hallinan and Alan Yuille). AK Peters, 1999.
3. *Pattern Theory: The Stochastic Analysis of Real World Signals* (with A. Desolneux). in final stages of preparation.

Papers

1. The Representation of Shape (with Andy Latto and Jayant Shah). In *Proceedings of the 1984 IEEE Workshop on Computer Vision*, 1984, pp. 183–191.
2. Boundary Detection by Minimizing Functionals I (with Jayant Shah). In *Image Understanding 1989*, ed. by Shimon Ullman & Whitman Richards, Ablex Press, 1990, pp. 19–43. Preliminary version in *1985 IEEE Conference on Computer Vision and Pattern Recognition* (CVPR), 1985, pp. 22–26.
3. The Problem of Robust Shape Descriptors. In *Proc. 1st IEEE Int. Conf. Comp. Vision* (ICCV), 1987, pp. 602–606.
4. Optimal Approximations of Piecewise Smooth Functions and Associated Variational Problems (with Jayant Shah). *Comm. Pure Appl. Math.*, 1989, **42**, pp. 577–685.

5. The 2.1*D* Sketch (with Mark Nitzberg). In *Proc. 3rd IEEE Int. Conf. Comp. Vision* (ICCV), 1990, pp. 138–144.
6. Parametrizing Exemplars of Categories. *J. Cognitive Neuroscience*, 1991, **3**, pp. 87–88.
7. Mathematical Theories of Shape: do they model perception?. In *Proc. Conference 1570, Soc. Photo-optical & Ind. Engineers*, 1991, pp. 2–10.
8. Texture Segmentation by Minimizing Vector-Valued Energy Functionals: the coupled membrane model (with Tai Sing Lee and Alan Yuille). *Proc. European Conf. Comp. Vision, 1992*, Lecture Notes in Computer Science 588, Springer-Verlag, 1992, pp. 165–173.
9. A Bayesian Treatment of the Stereo Correspondence Problem Using Half-Occluded Regions (with Peter Belhumeur). *Proc. IEEE Conf. Comp. Vision and Pattern Recognition, 1992* (CVPR), pp. 506–512.
10. Elastica and Computer Vision. In *Algebraic Geometry and its Applications* (West Lafayette, IN, 1990), 491–506, Springer-Verlag, 1994.
11. Commentary on Grenander & Miller "Representations of Knowledge in Complex Systems". *Proc. Royal Stat. Soc.*, 1994.
12. Pattern Theory: A Unifying Perspective. In *First European Congress of Mathematics: Paris, July 6–10, 1992, Vol. 1. Invited lectures* (Progress in mathematics, Vol. 119), Birkhäuser, 1994, pp. 187–224. Revised version in *Perception as Bayesian Inference*, ed. D. Knill and W. Richards, Cambridge Univ. Press, 1996, pp. 25–62; also in *Fields Medallists' lectures*, World Sci. Ser. 20th Century Math., Vol. 5, World Sci. Publ., River Edge, NJ, 1997, pp. 226–261; 2nd edition: *ibid.*, Vol. 9, 2003, pp. 234–269.
13. Chordal completions of planar graphs (with Fan Chung). *J. of Combinatorics*, **62**, 1994, pp. 96–106.
14. Bayesian Rationale for the Variational Formulation. In *Geometry-Driven Diffusion in Computer Vision*, B.M. ter Haar Romeny editor, Kluwer Academic, 1994, pp. 135–146.
15. The Statistical Description of Visual Signals, *International Congress of Industrial and Applied Math. 1995* ed. K. Kirshgassner, O. Mahrenholtz & R. Mennicken, Akademie Verlag, 1996.
16. Review of *Variational Methods in image segmentation* by J-M Morel & S. Solimini. *Bull. Amer. Math. Soc.*, **33**, 1996, pp. 211–216.
17. Learning Generic Prior Models for Visual Computation (with Song Chun Zhu). In *Proc. Conf. Comp. Vision and Pattern Rec.*, Comp Sci Press, 1997, 463–469
18. FRAME: Filters, Random Fields and Maximum Entropy (with Song Chun Zhu and Yingnian Wu). *Int. J. Comp. Vis.*, **27**, 1998, pp. 107–126.
19. Minimax Entropy Principle and Its Application to Texture Modeling (with Song Chun Zhu and Yingnian Wu). *Neural Computation*, **9**, 1997, pp. 1627–60.
20. GRADE: Gibbs Reaction and Diffusion Equations (with Song Chun Zhu). In *Proc. 6th Int. Conf. Comp. Vision*, IEEE Computer Society, 1998, pp. 847–854.
21. Prior Learning and Gibbs Reaction-Diffusion (with Song Chun Zhu). *IEEE Trans. Patt. Anal. and Mach. Int.*, **19**, 1997, pp. 1236–50.

22. The Statistics of Natural Images and Models (with Jinggang Huang). *Proc. IEEE Conf. Comp. Vision and Pattern Rec.*, Comp Sci Press, 1999, pp. 541–547.
23. Statistics of range images (with Jinggang Huang and Ann Lee). *Proc. IEEE Conf. Comp. Vision and Pattern Rec.*, Comp Sci Press, 2000, pp. 324–331.
24. Stochastic Models for Generic Images (with Basilis Gidas). *Quarterly Appl. Math.*, **59**, 2001, pp. 85–111.
25. Occlusion models for natural images: A statistical study of a scale-invariant dead-leaves model (with Ann Lee and Jinggang Huang). *Int. J. Computer Vision*, **41**, 2001, pp. 35–59.
26. Surface evolution under curvature flow (with Conglin Lu and Yan Cao). Special Issue on Partial Differential equations in Image Proc. Comp. Vision and Comp. Graphics, *Journal of Visual Communication and Image Representation*, 2002, pp. 65–81.
27. Geometric Structure Estimation of Axially Symmetric Pots from Small Fragments (with Yan Cao). In *Proc. of Int. Conf. on Signal Processing, Pattern Recognition, and Applications*, Crete, 2002.
28. Pattern Theory: The Mathematics of Perception. In *Proceedings of ICM 2002, Beijing, Vol. I*, Higher Education Press, Beijing, 2002, pp. 401–422.
29. The Nonlinear Statistics of High-contrast Patches in Natural Images (with Ann Lee and Kim Pedersen). *Int. J. Comp. Vision*, **54**, 2003, pp. 83–103.
30. Riemannian Geometries on Spaces of Plane Curves (with Peter Michor). *J. of the European Math. Society*, **8**, 2006, pp. 1–48.
31. Vanishing geodesic distance on spaces of submanifolds and diffeomorphisms (with Peter Michor). *Documenta Mathematica*, **10**, 2005, 217–245.
32. 2D-Shape Analysis using Conformal Mapping (with Eitan Sharon). *Int. J. of Comp. Vision*, **70**, 2006, pp. 55–75; preliminary version in *Proc. IEEE Conf. Comp. Vision and Patt. Rec.*, 2004, pp. 350–357.
33. Stuff It! Review of *Introduction to Circle Packing: The Theory of Discrete Analytic Functions* by Kenneth Stephenson. *The American Scientist*, **94**, 2006.
34. Empirical Statistics and Stochastic Models for Visual Signals. In *Brain and Systems: New Directions in Statistical Signal Processing*, ed. by S. Haykin, J. Principe, T. Sejnowski, and J. McWhirter, MIT Press, 2006.
35. An overview of the Riemannian metrics on spaces of curves using the Hamiltonian approach (with Peter Michor). *Applied and Computational Harmonic Analysis*, **23**, 2007, pp. 74–113.
36. A Stochastic Grammar of Images (with Song Chun Zhu). *Foundations and Trends in Computer Graphics and Vision*, **2**, 2007, pp. 259–362.
37. A metric on shape space with explicit geodesics (with Laurent Younes, Peter Michor and Jayant Shah). Atti Accad. Naz. Lincei Cl. Sci. Fis. Mat. Natur. Rend. Lincei (9) Mat. Appl. **19**, 2008, pp. 25–57.

Biology and Psychology of Vision

1. Discriminating Figure from Ground: the role of edge detection and region growing (with Steve Kosslyn, Lynn Hillger and Richard Herrnstein). *Proc. Nat. Acad. Sci.*, 1987, **84**, pp. 7354–7358.
2. Teaching Pigeons an Abstract Relational Rule: Insideness (with Richard Herrnstein, William Vaughan and Steve Kosslyn). *Perception and Psychophysics*, 1989, **46**, pp. 56–64.
3. On the Computational Architecture of the Neocortex, I: The role of the thalamo-cortical loop. *Biological Cybernetics*, 1991, **65**, pp. 135–145; II: The role of cortico-cortical loops. *Biological Cybernetics*, 1992, **66**, pp. 241–251.
4. Neuronal Architectures for Pattern-theoretic Problems. In *Large Scale Neuronal Theories of the Brain*, MIT Press, 1994, pp. 125–152.
5. Thalamus. In *The Handbook of Brain Theory and Neural Networks*, M. Arbib editor, MIT Press, 1995, pp. 981–984.
6. Neural correlates of boundary and medial axis representations in primate striate cortex (with Tai Sing Lee, K. Zipser and Peter Schiller). ARVO abstract, 1995. *Investigative Ophthamology & Visual Science*, **36**, 1995, 477–477.
7. Issues in the mathematical modeling of cortical functioning and thought. In *The Legacy of Norbert Wiener: A Centennial Symposium*, ed. D. Jerison et al, Amer. Math. Society, 1997, pp. 235–260.
8. Visual Search and Shape from Shading Modulate Contextual Processing in Macaque Early Visual Cortices (with Tai Sing Lee, R. Romero, A. Tobias and T. Moore). *Neuroscience Abstract*, 1997.
9. The Role of V1 in Shape Representation (with Tai Sing Lee, Song Chun Zhu and Victor Lamme). *Computational Neuroscience*, ed. Bower, Plenum Press, 1997.
10. The Role of the Primary Visual Cortex in Higher Level Vision (with Tai Sing Lee, R. Romero and Victor Lamme). *Vision Research*, **38**, 1998, 2429–2454.
11. Thalamus. In *MIT Encyclopedia of the Cognitive Sciences*, ed. R.A. Wilson and F.C. Keil, MIT Press, 1999, pp. 835–837.
12. Neural activity in early visual cortex reflects behavioral experience and higher-order perceptual saliency (with Tai Sing Lee, C. Yang and R. Romero). *Nature Neuroscience*, **5**, 2002, 589–597.
13. Hierarchical Bayesian Inference in the Visual Cortex (with Tai Sing Lee). *Journal of the Optical Society of America*, **20**, 2003, pp. 1434–1448.
14. Modeling and Decoding Motor Cortical Activity using a Switching Kalman Filter (with Wei Wu, Michael Black, Y. Gao, Elie Bienenstock and John Donoghue). *IEEE Trans. on Biomed. Eng.*, **51**, 2004, pp. 933–942.
15. Movement Direction Decoding using Fast oscillation in Local Field Potential and Neural Firing (with Wei Wu, W. Truccolo, M. Saleh and John Donoghue). *13th Computational Neuroscience Meeting*, 2004.
16. Minds must Unite: It's time for experimentalists to stop ignoring computational modelers (with David Donoho and Bruno Olshausen). 'Opinion' section, *The Scientist*, **19**, June 6, 2005, pp. 18–19.

Other Work

1. Contributor to *Multi-variable Calculus*, The Calculus Consortium based at Harvard, Wiley, 1995.
2. Calculus Reform—For the Millions, *Notices Amer. Math. Soc.*, May 1997, pp. 559–563.
3. Trends in the Profession of Mathematics. *Mitt. Dtsch. Math.-Ver.*, 1998, no. 2, pp. 25–29
4. The Dawning of the Age of Stochasticity. In *Mathematics: Frontiers and Perspectives*, edited by V. Arnold, M. Atiyah, P. Lax and B. Mazur, AMS, 2000, pp. 197–218. Also in *Mathematics towards the third millennium (Rome, 1999)*, Atti Accad. Naz. Lincei Cl. Sci. Fis. Mat. Natur. Rend. Lincei (9) Mat. Appl. 2000, Special Issue, pp. 107–125.
5. *Indra's Pearls* (with Caroline Series and David Wright). Cambridge University Press, 2002.
6. Mathematics in the Near East, some personal observations, *Notice of the AMS*, **52**, May 2005, pp. 526–530.
7. Mathematics Belongs in a Liberal Education, *The Arts and Humanities in Higher Education*, **5**, 2006, pp. 21–32.
8. Henri's Crystal Ball (with Phil Davis). *Notices of the AMS*, **55**, 2008, pp. 458–466.

Part I
Articles

THE TOPOLOGY OF NORMAL SINGULARITIES OF AN ALGEBRAIC SURFACE AND A CRITERION FOR SIMPLICITY

By DAVID MUMFORD

Let a variety V^n be embedded in complex projective space of dimension m. Let $P \in V$. About P, choose a ball U of small radius ε, in some affine metric $ds^2 = \Sigma dx_j^2 + \Sigma dy_j^2$, $z_j = x_j + iy_j$ affine coordinates. Let B be its boundary and $M = B \cap V$. Then M is a real complex of dimension $2n-1$, and a manifold if P is an isolated singularity. The topology of M together with its embedding in B ($=$ a $2m-1$-sphere) reflects the nature of the point P in V. The simplest case and the only one to be studied so far, to the author's knowledge, is where $n=1, m=2$, i.e. a plane curve (see [3], [14]). Then M is a disjoint union of a finite number of circles, knotted and linked in a 3-sphere. There is one circle for each branch of V at P, the intersection number of each pair of branches is the linking number of the corresponding circles, and the knots formed by each circle are compound toroidal, their canonical decomposition reflecting exactly the decomposition of each branch via infinitely near points.

The next interesting case is $n=2, m=3$. One would hope to find knots of a 3-sphere in a 5-sphere in this case; this would come about if P were an isolated singularity whose normalization was non-singular. Unfortunately, isolated non-normal points do not occur on hyper-surfaces in any Cohen-MacCaulay varieties. What happens, however, if the normalization of P is non-singular, is that M is the image of a 3-sphere mapped into a 5-sphere by a map which (i) identifies several circles, and (ii) annihilates a ray of tangent vectors at every point of another set of circles. In many cases the second does not occur, and we have an immersion of the 3-sphere in the 5-sphere. It would be quite interesting to know Smale's invariant in $\pi_3(V_{3,5})$ in this case (see [10]).

From the standpoint of the theory of algebraic surfaces, the really interesting case is that of a singular point on a *normal* algebraic surface, and m arbitrary. M is then by no means generally S^3 and consequently its own topology reflects the singularity P! In this paper, we shall consider this case, first giving a partial construction of $\pi_1(M)$ in terms of a resolution of the singular point P; secondly we shall sketch the connexion between $H_1(M)$ and the algebraic nature of P. Finally and principally, we shall demonstrate the following theorem, conjectured by Abhyankar:

Theorem. — $\pi_1(M) = (e)$ if and only if P is a simple point of F (a locally normal 1
surface); and F topologically a manifold at P implies $\pi_1(M) = (e)$.

1. — ANALYSIS OF M AND PARTIAL CALCULATION OF $\pi_1(M)$

A normal point P in F is given. A finite sequence of quadratic transformations plus normalizations leads to a non-singular surface F′ dominating F [15]. The inverse image of P on F′ is the union of a finite set of curves $E_1, E_2, \ldots, E_n$. By further quadratic transformations if necessary we may assume that all E_i are non-singular, and, if $i \neq j$, and $E_i \cap E_j \neq \emptyset$, then that E_i and E_j intersect normally in exactly one point, which does not lie on any other E_k. This will be a great technical convenience.

We note at this point the following fundamental fact about E_i : the intersection matrix $S = ((E_i . E_j))$ is negative definite. (This could also be proven by Hodge's Index Theorem.)

Proof. — Let H_1 and H_2 be two hyperplane sections of F, H_1 through P, and H_2 not (and also not through any other singular points of F). Let $(f) = H_1 - H_2$. Let H_1' be the proper transform of H_1 on F′, and H_2' the total transform of H_2. Then $H_2' \equiv H_1' + \Sigma m_i E_i$, where $m_i > 0$, all i (here m_i is positive since $m_i = \mathrm{ord}_{E_i}(f)$, f a function that is regular and zero at P on F, and moreover P is the center of the valuation of E_i on F).

Let $S' = ((m_i E_i . m_j E_j)) = M.S.M$, where M is the diagonal matrix with $M_{ii} = m_i$. To prove S′ is negative definite is equivalent with the desired assertion. Now
2 note (*a*) $S_{ij} \geq 0$, if $i \neq j$, (*b*) $\sum_i S'_{ij} = \sum_i (m_i E_i . m_j E_j) = -(H_1' . m_j E_j) \leq 0$, all j. For any symmetric matrix S′, these two facts imply negative indefiniteness. To get definiteness, look closer: we know also (*c*) $\sum_i S'_{ij} < 0$, for some j (since H_1' passes through some E_j), and (*d*) we cannot split $(1, 2, \ldots, n) = (i_1, i_2, \ldots, i_k) \cup (j_1, j_2, \ldots, j_{n-k})$ disjointly so that $S'_{i_a j_b} = 0$, any a, b (since $\cup E_i$ is connected by Zariski's main theorem [16]). Now these together give definiteness: Say

$$0 = \sum_{ij} \alpha_i \alpha_j S'_{ij} = \Sigma \alpha_i^2 S'_{ii} + 2 \sum_{i<j} \alpha_i \alpha_j S'_{ij}$$
$$= \sum_j \Big(\sum_i S'_{ij} \Big) \alpha_j^2 - \sum_{i<j} S'_{ij} (\alpha_i - \alpha_j)^2$$

where α_i are real. Then by (*c*), some $\alpha_j = 0$, and by (*d*), $\alpha_i = \alpha_j$, all i, j.

Our first step is a close analysis of the structure of M. We have defined it informally in the introduction in terms of an affine metric (depending apparently on the choice of this metric). Here we shall give a more general definition, and show that all these manifolds coincide, by virtue of having identical constructions by patching maps.

In the introduction, M is a level manifold of the positive C^∞ fcn.

$$\rho^2 = |Z_1|^2 + \ldots + |Z_n|^2,$$

(Z_i affine coordinates near $P \in F$). Now notice that M may also be defined as the level manifolds of ρ^2 on the non-singular F′ (ρ^2 being canonically identified to a fcn. on F′). It is as a "tubular neighborhood" of $\cup E_i \subset F'$ that we wish to discuss M. Now the general problem, given a complex $K \subset E^n$, Euclidean n-space, to define a tubular neighborhood, has been attacked by topologists in several ways although it does not appear to have been treated definitively as yet. J. H. C. Whitehead [13], when K is a subcomplex in a triangulation of E^n, has defined it as the boundary of the star of K in the second barycentric subdivision of the given triangulation. I am informed that Thom [11] has considered it more from our point of view: for a suitably restricted class of positive C^∞ fcns. f such that $f(P)=0$ if and only if $P\in K$, define the tubular neighborhood of K to be the level manifolds $f=\varepsilon$, small ε. The catch is how to suitably restrict f; here the archtype for f^{-1} may be thought of as the potential distribution due to a uniform charge on K. In our case, as we have no wish to find the topological ultimate, we shall merely formulate a convenient, and convincingly broad class of such f, which includes the p^2 of the introduction.

Let us say that a positive C^∞ real fcn. f on F′ such that $f(P)=0$ iff $P\in E_i$, is *admissible* if

1) $\forall P\in E_i-\bigcup_{j\neq i} E_j$, if $Z=0$ is a local equation for E_i near P, $f=|Z|^{2n_i}.g$, where g is C^∞ and neither 0 nor ∞ near P.

2) If $P_{ij}=E_i\cap E_j$, and $Z=0$, $W=0$ are local equations for E_i, E_j respectively then $f=|Z|^{2n_i}.|W|^{2n_j}.g$, where g is C^∞ and neither 0 nor ∞ near P_{ij}.

The following proposition is left to the reader.

Proposition: (i) If F″ dominates F′, and f is admissible for $\bigcup E_i$ on F′, and $g: F''\to F'$ is the canonical map, then $f\circ g$ is admissible for $g^{-1}(\bigcup E_i)$ on F′.

(ii) For a suitable F″ dominating F′, p^2 is an admissible map for $g^{-1}(\bigcup E_i)$.

Let me say, however, that in (ii), the point is to take F″ high enough so that the linear system of zeroes of the functions $(\Sigma\alpha_i Z_i)$ less its fixed components, has no base points.

What we must now show is that there is a unique manifold M such that, if f is any admissible fcn., M is homeomorphic to $\{P\,|\,f(P)=\varepsilon\}$ for all sufficiently small ε. Fix a fcn. f to be considered. Notice that at each of the points P_{ij}, there exist real C^∞ coordinates X_{ij}, Y_{ij}, U_{ij}, V_{ij}, such that

$$f=(X_{ij}^2+Y_{ij}^2)^{n_i}(U_{ij}^2+V_{ij}^2)^{n_j}\alpha_{ij},$$

α_{ij} *a constant*, valid in some neighborhood U given by

$$X_{ij}^2+Y_{ij}^2<1$$
$$U_{ij}^2+V_{ij}^2<1.$$

Assume E_i is $X_{ij}=Y_{ij}=0$, and E_j is $U_{ij}=V_{ij}=0$.

Our first trick consists of choosing a C^∞ metric $(ds)^2$ (depending on f), such that within

$$U'=\begin{Bmatrix}X_{ij}^2+Y_{ij}^2<1/2\\ U_{ij}^2+V_{ij}^2<1/2\end{Bmatrix},$$

$$ds^2=dX_{ij}^2+dY_{ij}^2+dU_{ij}^2+dV_{ij}^2.$$

Such a metric exists, e.g. by averaging a Hodge metric with these Euclidean metrics by some partition of unity. Now let

$$\begin{array}{ccc} N_i & & S_i \\ \downarrow{\scriptstyle \pi_i} & \text{and} & \downarrow{\scriptstyle \psi_i} \\ E_i & & E_i \end{array}$$

be the normal 2-plane bundle to E_i and normal S^1-bundle to E_i in F′ respectively. Consider the map $(\exp)_i$: $N_i \to F'$ obtained by mapping N_i into F along geodesics perpendicular to E_i. Let $f_i = f\circ(\exp)_i$. Now for every point $Q \in E_i - \bigcup_{j\neq i} E_j$, there is a neighborhood W of $Q \in E_i$, and an ε_0 such that if $\varepsilon < \varepsilon_0$, the locus $f_i(P) = \varepsilon$, $\pi_i(P) \in W$ cuts once each ray in $\pi_i^{-1}(W)$ (because f_i^{1/n_i} is a well-defined pos. C^∞ fcn. vanishing on the zero cross-section, with non-degenerate Hessian in normal directions; this is the standard situation of Morse theory, see [9]). Consequently, for any $W \subset E_i$ open, such that $E_j \cap W = \emptyset$, $j \neq i$, there is an ε_0 such that if $\varepsilon < \varepsilon_0$, the locus $f(P) = \varepsilon$ canonically contains a homeomorphic image of $\psi_i^{-1}(W)$ (recall $(\exp)_i$ is a local homeomorphism near the zero-section of N_i). Therefore, we see that the manifold M for which we are seeking a definition independent of f, is to be put together out of pieces of S_i; we need only seek its structure near P_{ij}. Let us therefore look in U′. Let us fix neighborhoods U_{ij} of $P_{ij} \in E_i$ and U_{ji} of $P_{ij} \in E_j$ by $(U_{ij}^2 + V_{ij}^2) < 1/4$ and $(X_{ij}^2 + Y_{ij}^2) < 1/4$ respectively. Let $E_k^* = E_k - \bigcup_{j\neq k} U_{kj}$ for all k. Now choose $\varepsilon_0 < \alpha_{i,j}/8^{n_i+n_j}$ and so that if $\varepsilon < \varepsilon_0$, $f(P) = \varepsilon$ contains $\psi_i^{-1}(E_i^*)$ and $\psi_j^{-1}(E_j^*)$ canonically. Then in the local coordinates in U′ about P_{ij}, $\psi_i^{-1}(\partial E_i^*) \subset \{P \mid f(P) = \varepsilon\}$ equals

$$\left\{(X_{ij}, Y_{ij}, U_{ij}, V_{ij}) \mid U_{ij}^2 + V_{ij}^2 = 1/4,\ X_{ij}^2 + Y_{ij}^2 = \left(\frac{4^{n_j}\varepsilon}{\alpha_{ij}}\right)^{1/n_i}\right\}$$

and $\psi_j^{-1}(\partial E_j^*) \subset \{P \mid f(P) = \varepsilon\}$ equals

$$\left\{(X_{ij}, Y_{ij}, U_{ij}, V_{ij}) \mid X_{ij}^2 + Y_{ij}^2 = 1/4,\ U_{ij}^2 + V_{ij}^2 = \left(\frac{4^{n_i}\varepsilon}{\alpha_{ij}}\right)^{1/n_j}\right\}$$

(because of the Euclidean character of the metric ds^2 near P_{ij}, $\exp_i$ takes the simplest possible form!). Note $\left(\frac{4^{n_j}\varepsilon}{\alpha_{ij}}\right)^{1/n_i} < 1/8$. Therefore, we see that $\psi_i^{-1}(E_i^*)$ and $\psi_j^{-1}(E_j^*)$ are patched by a standard "plumbing fixture":

$$\{(x, y, u, v) \mid (x^2+y^2) \leq 1/4,\ (u^2+v^2) \leq 1/4,\ (x^2+y^2)^n . (u^2+v^2)^m = \varepsilon < 1/8^{n+m}\}$$

where n and m are integers.

One sees immediately that this is simply $S^1 \times S^1 \times [0, 1]$, and if we set $M_i^* = \psi_i^{-1}(E_i^*)$, then it simply attaches ∂M_i^* to ∂M_j^*. Moreover, what is this attaching? There is a coordinate system on both ∂M_i^* and ∂M_j^* via

$$\left(\frac{X_{ij}}{\sqrt{X_{ij}^2+Y_{ij}^2}}, \frac{Y_{ij}}{\sqrt{X_{ij}^2+Y_{ij}^2}}\right) = \xi \in S^1 \text{ (in the usual embedding in } E^2)$$
$$\left(\frac{U_{ij}}{\sqrt{U_{ij}^2+V_{ij}^2}}, \frac{V_{ij}}{\sqrt{U_{ij}^2+V_{ij}^2}}\right) = \eta \in S^1 \text{ (in the usual embedding in } E^2)$$

and relative to these coordinates, the attaching is readily seen to be the identity. To complete the invariant topological description of M, we need only to show that the cycles $\{(\xi, \eta_0) \mid \xi \in S^1, \eta_0 \text{ fixed}\}$ and $\{(\xi_0, \eta) \mid \xi_0 \text{ fixed}, \eta \in S^1\}$ are invariantly determined (since an identification of 2 tori is determined up to isotopy by an identification of a basis of 1-cycles). But on M_i^* for instance, the 1st one is just the fibre of S_i over a point of E_i, and the 2nd is the loop ∂E_i^* lifted to S_i so that it is contractible in $\psi_i^{-1}(U_{ij})$; similarly on M_j^*, but *vice versa*.

This determines M uniquely. We have essentially found, moreover, not only M but also for any fixed f, maps

$$\varphi : M \to \cup E_i$$
$$\psi : \{P \mid 0 < f(P) \leqslant \varepsilon\} \to M$$

where ψ induces a homeomorphism of any $\{P \mid f(P) = \varepsilon' \leqslant \varepsilon\}$ onto M. Namely, define φ on M_i^* by ψ_i: projection into E_i, and in U′ near P_{ij}, define it as follows (fig. 1):

$$\begin{aligned}\varphi((X_{ij}, Y_{ij}, U_{ij}, V_{ij})) &= (0, 0, U_{ij}, V_{ij}) \in E_i \quad \text{if } U_{ij}^2+V_{ij}^2 \geq 1/4 \\ &= (0, 0, pU_{ij}, pV_{ij}) \in E_i \quad \text{if } X_{ij}^2+Y_{ij}^2 \leqslant U_{ij}^2+V_{ij}^2 \leqslant 1/4 \\ &= (p'X_{ij}, p'Y_{ij}, 0, 0) \in E_j \quad \text{if } U_{ij}^2+V_{ij}^2 \leqslant X_{ij}^2+Y_{ij}^2 \leqslant 1/4 \\ &= (X_{ij}, Y_{ij}, 0, 0) \in E_j \quad \text{if } X_{ij}^2+Y_{ij}^2 \geqslant 1/4,\end{aligned}$$

where
$$p = \tau(X_{ij}^2+Y_{ij}^2, U_{ij}^2+V_{ij}^2)$$
$$p' = \tau(U_{ij}^2+V_{ij}^2, X_{ij}^2+Y_{ij}^2)$$

and where
$$\tau(\alpha, \beta) = \frac{\beta - \alpha}{1 - 4\alpha}.$$

As for ψ, away from P_{ij}, define ψ by first $(\exp)_i^{-1}$, then the projection of N_i — (0-section) to S_i, and then the identification of S_i into M; near P_{ij}, define it by identifying those points whose ξ and η coordinates are equal, and that have the same image in $E_i \cup E_j$ under the map φ.

Note that φ induces a map $\varphi_*: \pi_1(M) \to \pi_1(\cup E_i)$, which is onto as all the "fibres" are connected [1]. In order not to be lost in a morass of confusion, we shall now restrict ourselves to computing only H_1 in general, and π_1 only if $\pi_1(\cup E_i) = (e)$. Note thats this last is equivalent to (*a*) E_i connected together as a tree (i.e. it never happens $E_1 \cap E_2 \neq \emptyset, E_2 \cap E_3 \neq \emptyset, \ldots, E_{k-1} \cap E_k \neq \emptyset, E_k \cap E_1 \neq \emptyset$ and $k > 2$ for some ordering of the E_i's), (*b*) all E_i are rational curves.

First, to compute $H_1(M)$, start with $H_i(\cup E_i)$. Let $\cup E_i$, as a graph, be p-connected,

(1) M is, of course, not a fibre space in the usual sense. However, the map φ_* in question is onto for any simplicial map such that the inverse image of every point is connected.

i.e. there exist some $P_1, \ldots, P_p$ such that if these points are deleted from $\cup E_i$, then $\cup E_i$ becomes a tree, but this does not happen for fewer P_i. Choose such P_i, and to $\cup E_i - \cup P_i$, for each P_i, add two points P_i' and P_i'', one to each E_j to which P_i belonged. The result, T, is, up to homotopy type, simply the wedge of the (closed) surfaces E_i (1). $\cup E_i$ is itself obtained from T by identifying the p pairs of points P_i', P_i''; therefore up to homotopy

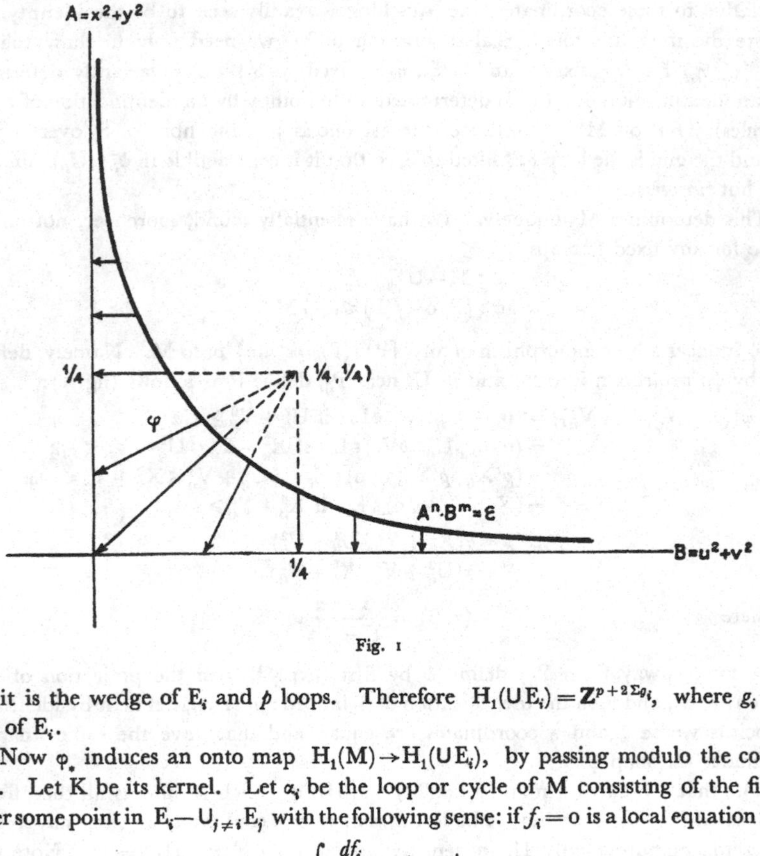

Fig. 1

type, it is the wedge of E_i and p loops. Therefore $H_1(\cup E_i) = \mathbf{Z}^{p+2\Sigma g_i}$, where g_i is the genus of E_i.

Now φ_* induces an onto map $H_1(M) \to H_1(\cup E_i)$, by passing modulo the commutators. Let K be its kernel. Let α_i be the loop or cycle of M consisting of the fibre of M over some point in $E_i - \cup_{j \neq i} E_j$ with the following sense: if $f_i = 0$ is a local equation for E_i,

$$\int_{\alpha_i} \frac{df_i}{f_i} = +2\pi i$$

or equivalently α_i as a loop about the origin of a fibre of the normal bundle N_i to E_i should have positive sense in its canonical orientation. I claim α_i generate K, and their relations are exactly $\Sigma(E_i \cdot E_j)\alpha_j = 0$, $i = 1, \ldots, n$.

(1) For example, proceeding surface by surface in any order, we may deform the complex $\cup E_i$ so that all the E_j which meet some one E_i meet it at the same point.

Proof. — First introduce the auxiliary cycles β_{ij} on $\varphi^{-1}(E_i)=M_i$, whenever $E_i \cap E_j = (P_{ij}) \neq \emptyset$. Namely, move the cycle α_j along the fibres until it lies on $\varphi^{-1}(P_{ij}) \subset M_i$, and there call it β_{ij}. By my construction of the patching of M_i and M_j, we know that β_{ij} is what I called η, while α_i is ξ. Now compute the subgroup K_i of $H_1(M_i)$ defined by

$$\begin{array}{ccccccccc} 0 & \to & K_i & \longrightarrow & H_1(M_i) & \to & H_1(E_i) & \to & 0 \\ & & \downarrow & & \downarrow & & \downarrow & & \\ 0 & \to & K & \longrightarrow & H_1(M) & \to & H_1(\cup E_i) & \to & 0. \end{array}$$

As above, let U_{ij} be a small disc on E_i about P_{ij}, and $E_i^* = E_i - \cup U_{ij}$, and $M_i^* = \varphi^{-1}(E_i^*)$. Then M_i^* is a deformation retract of M_i, and is, on the one hand canonically the restriction of the bundle S_i to E_i^*, and on the other hand uncanonically homeomorphic to $S^1 \times E_i^*$. In this last description, α_i is canonically identified to $S^1 \times (\text{point})$, while β_{ij} are identified to $(\text{point}) \times \partial(U_{ij})$ only up to adding a multiple of α_i. Therefore we see that K_i is generated by α_i, β_{ij}, with one relation (1)

$$\sum_j \beta_{ij} + N\alpha_i = 0, \text{ some } N.$$

To evaluate N, note that β_{ij} considered as cycles in S_i are locally contractible (i.e. in the neighborhood of $\varphi^{-1}(P_{ij})$ described by my plumbing fixture). It is well known that when the oriented fundamental 2-cycle of E_i is lifted to S_i, its boundary is $(E_i^2)\alpha_i$. Therefore, this same lifting in M_i^* will have boundary $\sum_j \beta_{ij} + (E_i^2)\alpha_i$. Now by the Mayer-Vietoris sequence, $H_1(M)$ is generated by $H_1(M_i)$, hence K is by K_i, and has extra relations imposed by the identification of cycles on $M_i \cap M_j$. Since $H_1(M_i \cap M_j)$ is generated by β_{ij} and β_{ji}, these relations are implicit in our choice of generators.

As a consequence of our result, since $\det(E_i \cdot E_j) = \mu \neq 0$, K is a finite group of order μ, and is the torsion subgroup of $H_1(M)$.

Now consider the case E_i rational, and $\cup E_i$ tree-like. We shall compute $\pi_1(M)$, using $\pi_1(M_i)$ as building blocks. In order to keep these various groups, with their respective base points, under control, it is necessary to define a skeleton of basic paths leading throughout E_i. Let $Q_i \in E_i - \bigcup_{j \neq i} E_j$ be chosen as base point in E_i. On E_i, choose a path l_i as illustrated in Diagram II touching on each $P_{ij} \in E_i$. Lift all the l_i together into M by a map s, so that $\varphi(s(l_i)) = l_i$, and so that at $\varphi^{-1}(P_{ij})$, $s(l_i) \cap s(l_j) \neq \emptyset$. Choose, e.g. $s(Q_1)$ as base point for all of M. Let $G = \cup l_i$. Now the lifting s enables us to give the following recipe for paths α_i:

1. Go along $s(G)$ from $s(Q_1)$ to a point P in M_i.
2. Go once around the fibre of M_i through P in the canonical direction explained above.
3. Go back to $s(Q_1)$ along $s(G)$.

(1) In the map $H_1(E_i^*) \to H_1(E_i)$, the kernel is generated by $\{\partial(U_{ij})\}$ with the single relation $\Sigma_{j \neq i} \partial(U_{ij}) = \partial(\text{fundamental 2-cycle of } E_i^*) \sim 0$.

This is clearly independent of the choice of P.

Our result can now be stated: firstly, the α_i generate π_1; secondly, their only relations are (*a*) α_i and α_j commute if $E_i \cap E_j \neq \emptyset$, (*b*) if $k_i = (E_i^2)$, and $E_{j_1}, E_{j_2}, \ldots, E_{j_m}$ are those E_j intersecting E_i, written in the order in which they intersect l_i, then

$$e = \alpha_{j_1}\alpha_{j_2}, \ldots, \alpha_{j_m}\alpha_i^{k_i}.$$

To prove this, we use the following theorem of Van Kampen (see [8], p. 30): if X and Y are subcomplexes of a complex Z, and $Z = X \cup Y$, while $X \cap Y$ is connected, then $\pi_1(Z)$ is the free product of $\pi_1(X)$ and $\pi_1(Y)$ modulo amalgamation of the sub-

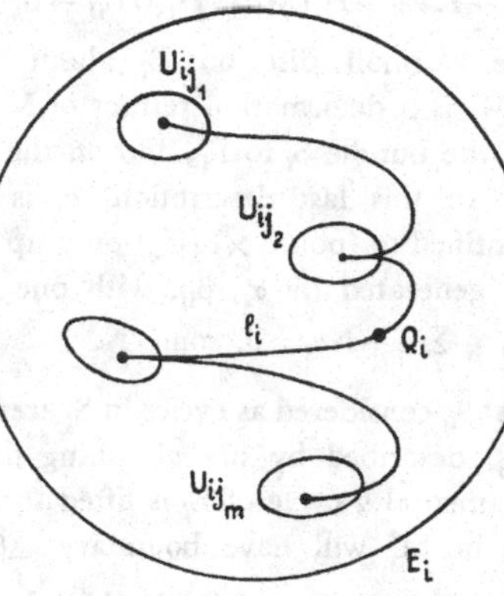

Fig. 2

groups $\pi_1(X \cap Y)$. Now since E_i is tree-like, M can be gotten from the M_i by successively joining on a new M_i with *connected* intersection with the part so far built up. Let $\pi_1(M_i)$ be mapped into $\pi_1(M)$ by mapping a loop in M_i with base point $s(Q_i)$ to one in M with base point $s(Q_1)$ by simply tagging on to both ends of it the section of $s(G)$ joining these two points. Then $\pi_1(M)$ is simply the free product of the $\pi_1(M_i)$ with amalgamation of the loops in $M_i \cap M_j$. Now recalling the structure of M_i^*, we have an exact sequence that splits:

$$0 \to \pi_1(S^1) \to \pi_1(M_i^*) \overset{\pi}{\to} \pi_1(E_i^*) \to 0$$

(S^1 the fibre of M_i, a 1-sphere). The path α_i is clearly a generator of $\pi_1(S^1)$ here, and hence in the center of $\pi_1(M_i^*)$.

Now the important thing to notice is that if E_i meets E_j, then α_j in $\pi_1(M_j)$ can be moved by modifying the point P on $s(G)$ where α_j detours around the fibre S^1; in particular, it may do this at $s(l_i) \cap s(l_j)$. In that position the loop α_j may be regarded canonically as in $\pi_1(M_i)$. Under the identification of $\pi_1(M_i)$ to $\pi_1(M_i^*)$ and the projection π of this group onto $\pi_1(E_i^*)$, what happens to the loop α_j? Recalling the patching map on the boundaries of M_i^* and M_j^* which was examined above, we see that this path proceeds along G from Q_i to near P_{ij}, then circles around the boundary of U_{ij}

in a positively oriented direction, then returns along G to Q_i. Referring again to our diagram, we see the relation $e=\pi(\alpha_{j_1}).\pi(\alpha_{j_2}).\ldots.\pi(\alpha_{j_m})$. Now it is well-known that these loops $\pi(\alpha_{j_k})$ generate the fundamental group of the m-times punctured sphere, and that this is the unique relation. Consequently, looking at the above exact sequence, it is clear that $\alpha_i, \alpha_{j_1}, \ldots, \alpha_{j_m}$ (when distorted into M_i as indicated above) generate $\pi_1(M_i)$. Moreover, the only relations among these generators are, therefore, that α_i and α_{j_k} commute, and $\alpha_{j_1}\ldots.\alpha_{j_m}\in\pi_1(S^1)$, i.e. $=\alpha_i^N$. But, using our results on $H_1(M)$, $N=-(E_i^2)$.

It follows that α_i generate $\pi_1(M)$ with relations (a) and (b), and that the only additional relations are those coming from the amalgamation of $\pi_1(M_i\cap M_j)=\mathbf{Z}+\mathbf{Z}$. But α_i and α_j are generators here, and as loops in M_i and M_j, these have already been identified. Hence we are through, Q.E.D.

II. — ALGEBRO-GEOMETRIC SIGNIFICANCE OF $H_1(M)$

(a) Local Analytic Picard Varieties and Unique Factorization.

We shall study in this section two questions of algebro-geometric interest in the solution of which the topological structure of M, in particular its homological structure, is reflected. The first of these is the problem of the local Picard Variety at $P\in F$. Generally speaking, this, as a group, should be the group of *local* divisors at P modulo local linear equivalence to zero. (We shall be more precise below.) However, if by divisor one refers to an algebraic divisor and by local one means in the sense of the Zariski topology, one sees by example that the resulting group has little significance: it is not local enough. Ideally, one should mean by an irreducible local divisor a minimal prime ideal in the formal completion of the local ring of the point in question. However, I have been unable to establish the structure of the resulting Picard group. A compromise between these two groups is possible over the complex numbers. Take as divisors *analytic* divisors, and the usual complex topology to interpret local. There results a local *analytic* Picard variety that is quite accessible. In this section, we shall first analyze the group of local *analytic* divisors near $\cup E_i$ modulo local linear equivalence and then consider the singular point P. Here by local analytic divisors we mean formal sums of irreducible analytic divisors defined in a neighborhood of $\cup E_i$ (including the divisors E_i themselves). Such a sum, $\Sigma n_i D_i$, is said to be locally linearly equivalent to zero if there exists a neighborhood U of $\cup E_i$ where all D_i are defined and a meromorphic function f on U such that $(f)=\Sigma n_i(D_i\cap U)$. This quotient we shall call the local analytic Picard Variety at $\cup E_i$, or Pic ($\cup E_i$).

Denote by Ω the sheaf of germs of holomorphic functions on F'; by $\Omega^*\subset\Omega$ the sheaf of germs of non-zero holomorphic functions. One has the usual exact sequence:

$$0\to\mathbf{Z}\to\Omega\xrightarrow{\exp(2\pi i x)}\Omega^*\to 0$$

where $\mathbf{Z}$ is the constant sheaf of integers. Let $\pi: F'\to F$ be the regular projection from the non-singular surface F' to the singular F.

Proposition. — $\mathrm{Pic}(\cup E_i) \simeq (R^1\pi)(\Omega^*)_P$.

Proof. — Define $\mathrm{Pic}(\cup E_i) \to (R^1\pi)(\Omega^*)_P$, by associating to $\Sigma n_i D_i$, defined in $U \supset \cup E_i$, the following 1-cocycle: assume $P \in V, \pi^{-1}(V) \subset U$, assume f_j is a local equation for $\Sigma n_i D_i$ in V_j, $\{V_j\}$ a covering of V, then $\{f_{j_1}/f_{j_2}\} \in H^1(\{V_j\}, \Omega^*)$ induces an $\alpha \in H^1(\pi^{-1}(V), \Omega^*)$, hence an $\alpha' \in (R^1\pi)(\Omega^*)_V$, hence an $\alpha'' \in (R^1\pi)(\Omega^*)_P$. It is well known that $\alpha \in H^1(\pi^{-1}(V), \Omega^*)$ is uniquely determined by $\Sigma n_i D_i$, hence so is α''.

To see that $\Sigma n_i D_i \to \alpha''$ is $1-1$, say $\alpha'' = 0$. Therefore $\exists V' \subset V$ say, $\mathrm{Res}_{V'} \alpha' = 0$, i.e. $\mathrm{Res}_{\pi^{-1}(V')}(\alpha) = 0$. Therefore the covering $\{V_j \cap \pi^{-1}(V')\} = \{V'_j\}$ has a refinement $\{V''_k\}$ such that there exist non-zero functions g_k on V''_k such that $g_{k_1}/g_{k_2} = f_{\tau k_1}/f_{\tau k_2}$ (for some map τ from the indices of $\{V''_k\}$ to those of $\{V_j\}$ such that $V''_k \subset V'_{\tau k}$). Therefore $f = \dfrac{f_{\tau k}}{g_k}$ defines a function throughout $\pi^{-1}(V')$ such that $(f) = \Sigma n_i D_i$.

To see that $\Sigma n_i D_i \to \alpha''$ is onto $(R^1\pi)(\Omega^*)_P$, let $\beta'' \in (R^1\pi)(\Omega^*)_P$ be represented by $\beta \in H^1(\pi^{-1}(V), \Omega^*)$ and let this define the line bundle L over $\pi^{-1}(V)$ in the usual way. Let $\mathscr{J}$ be the sheaf of germs of cross-sections of L: a coherent sheaf. Now by a result of Grauert and Remmert (cf. Borel-Serre [2], p. 104), $(R^0\pi)(\mathscr{J})$ is coherent on F. But $(R^0\pi)(\mathscr{J})$ is not the zero sheaf on F (at all points $Q \neq P$, $\mathscr{J}_Q \simeq (R^0\pi)(\mathscr{J})_Q$), hence there exists some element $S \in (R^0\pi)(\mathscr{J})_P$, $S \neq 0$. S corresponds to a section in $\mathscr{J}_{\pi^{-1}(V')}$, for some open $V' \ni P$, $V' \subset V$. Therefore, the line bundle $L \mid \pi^{-1}(V)$ has a section S. But if β is represented by a cocycle f_{ij} with respect to a covering $\{V_i\}$ of V, then S is given by a set of holomorphic functions f_i on V_i such that $f_j = f_i(f_{ij})$. It follows that $f_i = 0$ define a divisor which is represented by β.

A. Grothendieck has posed the problem, for any proper map $f: V_1 \to V_2$ (onto), to define a relative Picard Variety of the map f. It seems clear, in the classical case, that if Ω^* is the sheaf of holomorphic units on V_1, $(R^1 f)(\Omega^*)$ is the logical choice although no nice properties have been established in general so far as the writer knows. In our case, $(R^1 f)(\Omega^*)_Q$, for $Q \neq P$, is simply (1), but at P, we have seen it to be $\mathrm{Pic}(\cup E_i)$. We now wish to show that in our case, $(R^1 f)(\Omega^*)_P$ is an analytic group variety. This is seen by the exact sequence for derived functors:

$$0 \to (R^0\pi)(\mathbf{Z}) \to (R^0\pi)(\Omega) \xrightarrow{\varphi} (R^0\pi)(\Omega^*) \to$$
$$\to (R^1\pi)(\mathbf{Z}) \xrightarrow{\chi} (R^1\pi)(\Omega) \to (R^1\pi)(\Omega^*) \xrightarrow{\psi}$$
$$\to (R^2\pi)(\mathbf{Z}) \to \dots$$

(i) Note first that if $x \in (R^0\pi)(\Omega^*)_P$, then x is a non-zero function on $\pi^{-1}(V)$, $P \in V$, and necessarily constant on $\cup E_i$ which is connected and compact, therefore, at least on some $\pi^{-1}(V')$, $P \in V' \subset V$, $x = \exp(2\pi i y)$, y a holomorphic function on $\pi^{-1}(V')$, hence $x = \varphi(y)$, $y \in (R^0\pi)(\Omega)_P$.

(ii) Note secondly that $(R^i\pi)(\mathbf{Z})_p \simeq H^i(\cup E_i, \mathbf{Z})$, since for $P \in V$, V small, $\pi^{-1}(V)$ is contractible to $\cup E_i$.

(iii) Note thirdly that if $i > 0$, $(R^i\pi)(\Omega)_Q = (0)$ for $Q \neq P$, and being a coherent sheaf, for $Q = P$ must be a finite dimensional vector space over $\mathbf{C}$.

(iv) Note fourthly that if $\gamma \in H^2(\cup E_i, \mathbf{Z}) \simeq (R^2\pi)(\mathbf{Z})_P$, there exists $\alpha \in (R^1\pi)(\Omega^*)_P$ such that $\psi\alpha = \gamma$. To show this, note that $H^2(\cup E_i, \mathbf{Z}) \simeq \mathbf{Z}^n$, ($n$ = number of irreducible curves in $\cup E_i$) with generators γ_i whose value on the 2-cycle E_j is δ_{ij}; it is enough to verify it for the generators γ_i. But let D_i be an irreducible analytic curve through $Q \in E_i - \bigcup_{j \neq i} E_i$, with a simple point at Q, and tangent transversal to that of E_i at Q. If $D_i \to \alpha_i \in (R^1\pi)(\Omega^*)$, I claim $\psi\alpha_i = \gamma_i$. This is left to the reader. Therefore, we obtain

$$0 \to H^1(\cup E_i, \mathbf{Z}) \xrightarrow{\chi} (R^1\pi)(\Omega)_P \to \operatorname{Pic}(\cup E_i) \to H^2(\cup E_i, \mathbf{Z}) \to 0$$
$$(R^1\pi)(\Omega)_P \simeq \mathbf{C}^N, \text{ some } N.$$

(v) Note lastly that χ maps $H^1(\cup E_i, \mathbf{Z})$ into a *closed* subgroup of $(R^1\pi)(\Omega)_P$, hence the connected component of $\operatorname{Pic}(\cup E_i)$ is an analytic group. If this were false, there would be a *real* sum of elements of $H^1(\cup E_i, \mathbf{Z})$ that was zero without having to be, i.e. $\{\alpha_{ij}\} \in H^1(\pi^{-1}(V), \mathbf{R})$ (with respect to some covering $\{U_i\}$) such that $\{\alpha_{ij}\} \sim 0$ in the sheaf Ω (in some $\pi^{-1}(V')$, $V' \subset V$). In other words, $\alpha_{ij} = f_i - f_j$, f_i holomorphic in U_i. But let p_i be a real, C^∞ function on U_i such that $\alpha_{ij} = p_i - p_j$ (Poincaré's lemma). Then $f_i - p_i = F$, $df_i = \omega$ and $dp_i = \eta$, are defined all over $\pi^{-1}(V')$, $\omega - \eta = dF$. I claim actually all the periods of η are zero (which implies $\eta = df$, and $\{\alpha_{ij}\} \sim 0$ in $H^1(\cup E_i, \mathbf{R})$ and we are through). First of all, the periods of η equal those of ω. Look at its periods on the 1-cycles of any E_i: since η is real, all the periods of the holomorphic differential ω are also real. But it is wellknown that then all the periods of ω must be identically zero, and therefore ω reduces to *zero* on paths in E_i. Since this is true for all i, ω has no periods along *any* path in $\cup E_i$, and since $\pi^{-1}(V')$ is contractible to $\cup E_i$, ω has no periods at all. Therefore neither does η and we are through.

There is another way of looking at $\operatorname{Pic}(\cup E_i)$. Namely, let $\mathfrak{o}$ be the local ring of (convergent) holomorphic functions at P, i.e. $(R^0\pi)(\Omega)_P$ (by the theorem of Riemann, cf. the report of Behnke and Grauert ([1], p. 18)). Now every divisor D' in $\pi^{-1}(V')$, except for the E_i's, defines a divisor D in V', hence a minimal prime ideal $\mathfrak{p}$ in $\mathfrak{o}$. Let us 3
set Pic(P) equal to the group of ideal classes in $\mathfrak{o}$: i.e. to the semi-group of pure rank 1 ideals $\mathfrak{a}$ of $\mathfrak{o}$, modulo the principal ideals [1]. Then the association of D to $\mathfrak{p}$ defines a map from $\operatorname{Pic}(\cup E_i) \to \operatorname{Pic}(P)$, (if we define the image of each E_i to be (1), the identity). This is quite clear once one sees that every meromorphic function f in $\pi^{-1}(V)$ is a quotient

---

(1) The composition law is the "Kronecker" product treated so elegantly by Hermann Weyl [12], cf. chapter 2, namely:

$$(\mathfrak{a}, \mathfrak{b}) \to \text{rank 1 component of } \mathfrak{a}.\mathfrak{b}$$
$$= \bigcup_{n=1}^{\infty} (\mathfrak{a}.\mathfrak{b}) : \mathfrak{m}^n$$

where $\mathfrak{m}$ = maximal ideal of $\mathfrak{o}$
(:) = residual quotient operation.

of two holomorphic functions in some $\pi^{-1}(V')$, $V' \subset V$: but given f, consider the coherent sheaf $\mathcal{J}$ given by $\{g \mid (fg)$ is a positive divisor$\}$. $(R^0\pi)(\mathcal{J})$ is coherent, hence there exists $g_1 \in (R^0\pi)(\mathcal{J})_P$, and if $fg_1 = g_2$, then $f = g_2/g_1$ is the desired decomposition. Now the map $\text{Pic}(\cup E_i) \rightarrow \text{Pic}(P)$ is onto as every minimal prime ideal $\mathfrak{p} \subset \mathfrak{o}$ defines some divisor through P. Its kernel is immediately seen to be generated by the E_i themselves. Hence we see

Proposition:
$$\frac{\text{Pic}(\cup E_i)}{\{\Sigma n_i E_i\}} \simeq \text{Pic}(P)$$

Corollary. — We have
$$0 \rightarrow H^1(\cup E_i, \mathbf{Z}) \rightarrow (R^1\pi)(\Omega)_P \xrightarrow{\varphi} \text{Pic}(P) \xrightarrow{\psi} H_1(M)_0 \rightarrow 0$$

where $H_1(M)_0 =$ torsion subgroup of $H_1(M)$ and ψ associates to the divisor D through P, the 1-cycle $D \cap M$.

Proof of Corollary: Note that $\Sigma n_i E_i$ is never in the image of $(R^1\pi)(\Omega)_P$ since that would require $(\Sigma n_i E_i, E_j) = 0$ for all j. To see the exactness at ψ, note that the co-kernel of φ is obtained by associating to a divisor $\Sigma n_i D_i$ (where we may assume $E_i \cap E_j \cap (\bigcup_l \text{Supp } D_l) = \emptyset$, all $i \neq j$) the formal sum
$$\sum_k \Big(\sum_i n_i D_i \cdot E_k\Big)\gamma_k \text{ modulo } \Big\{\sum_k (E_i \cdot E_k)\gamma_k\Big\},$$

the γ_k as in (iv) above. But ψ is given by associating to $\Sigma n_i D_i$, the element
$$\sum_k (\Sigma n_i D_i \cdot E_k)\alpha_k,$$

in terms of our basis for $H_1(M)_0$ in (I); but by our enumeration of the relations on the α_k we see γ_k can be interchanged with α_k.

Do these results have purely algebraic counterparts? First, note that it is hopeless to expect that the ideal structure of $\mathfrak{o}_0$ (= algebraic local ring of P on F) will reflect the homology of the singularity so well. This is seen in the following example: Take a non-singular cubic curve E in the projective plane, and let $P_1, \ldots, P_{15}$ be points on E in general position except that on E the divisor $\Sigma_1^{15} P_i \equiv 5 \times$ (plane section). Blow up every point P_i to a divisor E_i, and call F′ the resulting surface. On F′, the proper transform E′ of E is exceptional: it is shrunk by the linear system of quintics through the P_i. Then $E_i - E_j$ as a divisor in Pic(E′) is in the component of the identity, but as an algebraic divisor is not algebraically locally equivalent to zero: in fact F′ is regular, hence algebraic and linear equivalence are the same, but since $\text{Tr}_{E'}(E_i - E_j) \not\equiv 0$, $E_i - E_j$ is not locally linearly equivalent to zero.

4 However, I conjecture that the ideal class group of $\mathfrak{o}^*$ (= completion of $\mathfrak{o}_0$ and $\mathfrak{o}$) is *identical* to that of $\mathfrak{o}$, and that sums of formal branches through $\cup E_i$ modulo holomorphic linear equivalence (in the sense of Zariski [17]) gives $\text{Pic}(\cup E_i)$. If this is so, it should
5 give $\text{Pic}(\cup E_i)$ an *algebraic* structure, which would be a decided improvement on our results. At present, I am unable to prove these statements.

240

(b) Intersection Theory on Normal Surfaces.

We consider here the problem of defining, for divisors A, B through P on F, *(a)* total transforms A′, B′ on F′, and *(b)* intersection multiplicities $i(A.B;P)$. This problem has been posed by Samuel (see [7]) and considered by J. E. Reeve [19]. In this case, I suggest the following as a canonical solution:

a) To define $A' = A'_0 + \Sigma r_i E_i$, where A'_0 is the proper transform of A, require

$$(A'.E_i) = 0, \quad i = 1, 2, \ldots, n,$$

or

$$(A'_0.E_i) + \sum_j r_j (E_j.E_i) = 0, \quad i = 1, 2, \ldots, n.$$

Since $\det(E_i.E_j) = \mu \neq 0$, this has a unique solution.

b) To define $i(A.B;P)$, set it equal to

$$\begin{aligned}(A'.B') &\text{ over } P\\ &= \sum_{P' \text{over} P} [i(A'_0.B'_0;P') + \Sigma r_i i(E_i.B'_0;P')]\\ &= \sum_{P' \text{over} P} [i(A'_0.B'_0;P') + \Sigma s_i i(A'_0.E_i;P')]\end{aligned}$$

where

$$A' = A'_0 + \Sigma r_i E_i; \quad B' = B'_0 + \Sigma s_i E_i.$$

We note the following properties:

(i) $A = (f)_F$, then $A' = (f)_{F'}$; hence $A \equiv B$ implies $A' \equiv B$

Proof. — For $((f)_{F'}.E_i) = 0$.

(ii) A effective, then all r_i are positive.

Proof. — Say some $r_i \leq 0$. Say also $r_i/m_i \leq r_j/m_j$, all j, where the m_j are the same as in the proof of negative definiteness. Then we see:

$$\begin{aligned}0 &\geq \sum_j r_j (E_j.E_i) = \sum_j r_j/m_j \, (m_j E_j.E_i),\\ &\geq r_i/m_i \sum_j (m_j E_j.E_i) \geq 0.\end{aligned}$$

Therefore, if $E_i \cap E_j \neq \emptyset$, $r_i/m_i = r_j/m_j$ and $r_j \leq 0$. As $\cup E_i$ is connected, this gives ultimately $r_i/m_i = R$, independent of i. But then also $(\Sigma m_j E_j.E_i) = 0$, all i, which contradicts property *(c)* in the proof just referred to.

(iii) $i(A.B;P)$ is symmetric and distributive.

(iv) A and B effective, then $i(A.B;P)$ is greater than 0.

(v) $i(A.B;P)$ independent of the choice of F′.

Proof. — To show this, it suffices, since any two non-singular models are dominated by a third, see Zariski [15], to compare F′ with F″ gotten by blowing up some point P′ over P. But let A′, B′ be the total transforms of A, B on F′, and A″, B″ those on F″, and let T be the map from F″ to F′. Then with respect to T, A″ is the total transform of A' on F'', and B'' that of B'. In that case it is well-known that, for any point set S in F' (including all the points of any common components of A', B'), $(A'.B')_S = (A''.B'')_{T^{-1}(S)}$.

(vi) A' is integral if and only if $\Sigma(A'_0.E_i)\alpha_i = 0$ in $H_1(M)$.

Proof. — $\Sigma(A'_0.E_i)\alpha_i = 0$ if and only if there are integers k_j such that

$$(A'_0.E_i) = \Sigma k_j(E_j.E_i),$$

i.e. if the relation $\Sigma(A'_0.E_i)\alpha_i = 0$ is an integral sum of the relations defining $H_1(M)$. But this is equivalent to $(A'_0 + \Sigma k_j E_j . E_i) = 0$ for all i, i.e. $A' = A'_0 + \Sigma k_j E_j$, k_j integral. Q.E.D.

The element $\Sigma(A'_0.E_i)\alpha_i$ has this simple interpretation: if M is chosen near enough to P, it represents the 1-cycle $A \cap M$. We see that this is again the fundamental map: (Group of Local Divisors at P) $\to H_1(M)$ considered in the final corollary of part (*a*). By the results of part (*a*), moreover, we can interpret (vi) as saying: A' is integral if and only if A is locally analytically equivalent to zero (i.e. A is in the connected component of $\mathrm{Pic}(P)$). Essentially, our definition of intersection multiplicity on a normal surface is the unique linear theory that has the correct limiting properties for divisors that can be analytically deformed off the singular points.

III. — THE CASE $\pi_1(M) = (e)$

We shall prove the following theorem, stronger than that announced above:

Theorem. — Let F be a non-singular surface, and E_i, $i = 1, 2, \ldots, n$, a connected collection of non-singular curves on F, such that $E_i \cap E_j$ is empty, or consists of one point on a transversal intersection, and $E_i \cap E_j \cap E_k$ is always empty. Let M be a tubular neighborhood of $\bigcup E_i$, as defined in section I. If (*a*) $\pi_1(M) = (e)$, and (*b*) $((E_i.E_j))$ is negative definite, then $\bigcup E_i$ is exceptional of first kind, i.e. is the total transform of some simple point on a surface dominated by F and birational to it.

Proof. — As above, $\pi_1(M) = (e)$ implies that all E_i are rational, and connected together as a tree. Now suppose that $\bigcup E_i$ is not exceptional of first kind. Assume that among all collections of E_i with all the properties of the theorem, there is no collection not exceptional with *fewer* curves E_i. As a consequence, no E_i of our collection has the two properties (*a*) $(E_i^2) = -1$, (*b*) E_i intersects at most two other E_j. For if it did, one could shrink E_i by Castelnuovo's criterion, preserving all the properties required (that the negative definiteness is preserved is clear as follows: the self-intersection of a cycle of the E_j's on the blown down surface equals the self-intersection of its total transform on F which must be negative). We allow the case where there is only one E_i. Now the central fact on which this proof is based is the following group-theoretic proposition:

Proposition. — Let G_i, $i = 1, 2, 3$, be non-trivial groups, and a_i an element of G_i. Then denoting the free product of A and B by $A * B$, it follows $G_1 * G_2 * G_3/\text{modulo } (a_1 a_2 a_3 = e)$ is non-trivial.

Proof. — First of all, if $\infty \geq n_1, n_2, n_3 > 1$, then $Z_{n_1} * Z_{n_2} * Z_{n_3}/(a_1 a_2 a_3 = e)$ is non-trivial, where Z_k denotes the integers modulo k, and each a_i is a generator. For, as a matter of fact, these are well-known groups easily constructed as follows: choose a triangle with angles π/n_1, π/n_2, and π/n_3 (modular if some $n_i = \infty$), in one of the three standard planes. Reflections in the three sides of the triangle generate a group of motions of the plane, and the group we seek is the subgroup, of index 2, of the orientation preserving motions in this group. Secondly, reduce the general statement to this case by means of:

($\#$) If n = order of a_1 in G_1, and a_1 is identified to a generator of $Z_n \subseteq G_1$, then $G_1 * G_2 * G_3/(a_1 a_2 a_3 = e)$ trivial $\Rightarrow Z_n * G_2 * G_3/(a_1 a_2 a_3 = e)$ trivial.

To show this, let $H = G_2 * G_3/((a_2 a_3)^n = e)$, and note that H is isomorphic to $Z_n * G_2 * G_3/(a_1 a_2 a_3 = e)$. Let n' be the order of a_1 in H. Then $G_1 * G_2 * G_3/(a_1 a_2 a_3 = e)$ is the free product of $G_1/(a_1^{n'} = e)$ and H with amalgamation of the subgroups generated by $a_2 a_3$ and a_1^{-1}. But by O. Schreier's construction of amalgamated free products (see [5], p. 29) this is trivial only if H is, hence ($\#$). Now the proposition is trivial if any $a_i = e$; hence let n_i = order $(a_i) > 1$. By ($\#$) iterated, $G_1 * G_2 * G_3/(a_1 a_2 a_3 = e)$ trivial implies $Z_{n_1} * Z_{n_2} * Z_{n_3}/(a_1 a_2 a_3 = e)$ trivial, which is absurd. Q.E.D.

Returning to the theorem, we wish to show the absurdity of $\pi_1(M) = (e)$, while no E_i is such that (a) $(E_i^2) = -1$, and (b) E_i meets at most two other E_j. There are two cases to consider: either *some* E_i meets three or more other E_j; or every E_i meets at most two other E_j (this includes the case of only one E_i).

Case 1. — Let E_1 meet $E_2, \ldots, E_m$, where m is at least 4. For $i = 2, 3, \ldots, m$, let T_i be the set of E_j's (besides E_1) such that E_j is connected to E_i by a series of E_k other than E_1. The T_i's are disjoint. Let M_i be the manifold bounding a neighborhood of T_i as above. Let $G_i = \pi_1(M_i)$, and $G = \pi_1(M)$/modulo $\alpha_1 = e$, where α_1 represents, as in (I), the loop about E_1. Then by the results of (I),

$$G = G_2 * G_3, \ldots, * G_m/(\alpha_2 \alpha_3 \ldots \alpha_m = e),$$

if the G_i are ordered suitably, and α_i in G_i represents a loop about E_i. Now $m \geq 4$, and $\pi_1(M) = (e)$, hence $G = (e)$, hence by the above theorem, there exists an i (say $i = 2$) such that $G_2 = \pi_1(M_2) = (e)$. By the induction assumption, the tree of curves T_2 is exceptional of first kind. Therefore, by Zariski's theorem on the factorization of anti-regular transformations on non-singular surfaces (see [18]), some E_j in T_2 enjoys the properties (a) and (b) with respect to T_2. Then E_j would also enjoy them in $\cup E_i$ (which is impossible) *unless* $E_j = E_2$, in which case E_j could meet only two other E_k (say E_{m+1}, E_{m+2}) in T_2, but would meet *three* other E_k in $\cup E_i$. Pursuing this further, apply the same reasoning to the curve E_2 which meets exactly three other E_k. Again, either some curve shrinks, or else either E_1, E_{m+1}, or E_{m+2} has in any case property (a), i.e. self-intersection -1. But then compute $((E_2 + E_i)^2)$ ($i = 1, m+1$, or $m+2$ according as which E_i has property (a)), and we get 0, contradicting negative definiteness of the intersection matrix.

Case 2. — It remains to consider the case where no E_i intersects more than two others. Then the E_i are arranged as follows:

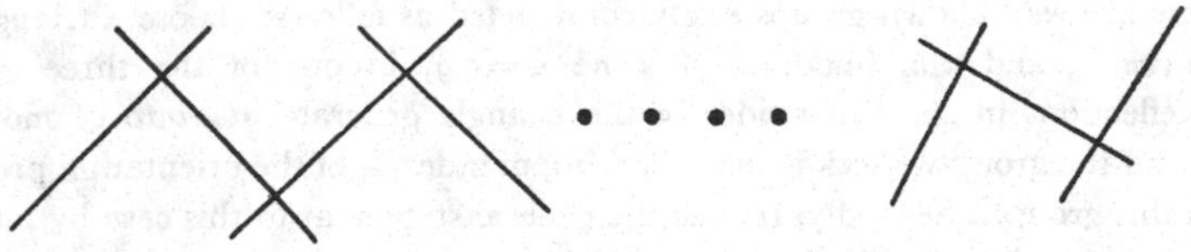

Fig. 3

In this case, it is immediate that π_1 is commutative, hence $=H_1$. It is given (in additive notation) by the equations:

$$\begin{array}{ll} k_1\alpha_1 - \alpha_2 \ldots\ldots\ldots & =0 \\ -\alpha_1 + k_2\alpha_2 - \alpha_3 \ldots\ldots & =0 \\ -\alpha_2 + k_3\alpha_3 \ldots\ldots & =0 \\ \ldots\ldots\ldots\ldots\ldots & \\ -\alpha_{n-1} + k_n\alpha_n & =0, \end{array}$$

where $k_i = -(E_i^2)$. Assume all $k_i \geq 2$, and prove

$$\mu = \det \begin{pmatrix} k_1 & -1 & 0 & 0 & \ldots & 0 \\ -1 & k_2 & -1 & 0 & \ldots & 0 \\ 0 & -1 & k_3 & -1 & \ldots & 0 \\ \ldots & & & & & \\ 0 & \ldots & & 0 & -1 & k_n \end{pmatrix} > 1,$$

hence the equations have a solution mod μ. To show this, use induction on n, using the stronger induction hypothesis $k_1 > 1$, $k_2, \ldots, k_n \geq 2$, allowing k_i to be rational. Then note the identity:

$$\det \begin{pmatrix} k_1 & -1 & 0 & \ldots & 0 \\ -1 & k_2 & -1 & \ldots & 0 \\ \ldots & & & & \\ 0 & \ldots & & -1 & k_n \end{pmatrix} = k_1 \det \begin{pmatrix} (k_2 - 1/k_1) & -1 & \ldots & 0 \\ -1 & k_3 & \ldots & 0 \\ \ldots & & & \\ 0 & \ldots & -1 & k_n \end{pmatrix}$$

This completes the proof of our theorem.

Corollary. — P a normal point of an algebraic surface F. If F has a neighborhood U homeomorphic to a 4-cell, P is a simple point of F.

Proof. — Let W be the intersection of an affine ball about P with F, as considered in the introduction, and so small that its boundary M lifted to a non-singular model F′ dominating F qualifies as a tubular neighborhood of the total transform of P. It suffices to show that $\pi_1(M) = (e)$, in view of the theorem just proven. Let U′ be a 4-cell-neighborhood of P contained in W, and let W′ be an affine ball about P contained in U′. We have constructed in section I a continuous map ψ from $U' - (P)$ to M that induces the canonical identification of M as the boundary of W' to M (as the boundary of W). Therefore if γ is any path in M, regard γ as a path in the boundary of W'; as a path in $U'-(P)$ (which is homotopic to a 3-sphere) it can be contracted to a point; but then ψ maps this homotopy to contraction of γ as a path in M. Q.E.D.

IV. — AN EXAMPLE

It is instructive to note that there exist singular points P, for which $H_1(M)=(0)$, while, of course, $\pi_1(M)\neq(e)$. Take P to be the origin of the equation $0=x^p+y^q+z^n$, where p, q, and n are pairwise relatively prime. Look at the equation as $-(z)^n=x^p+y^q$; this shows that M is an n-fold cyclic covering of the 3-sphere $|x|^2+|y|^2=1$, x,y complex, branched along the points $x^p+y^q=0$, i.e. along a torus knot, K, in S^3. Therefore M is a manifold of the type considered by M. Seifert [20], p. 222; he shows $H_1(M)=(0)$.

The singular point $0=x^2+y^3+z^5$ is of particular interest as illustrating the possibility of a singular point on a surface whose local analytic Picard Variety is trivial contrary to a conjecture of Auslander. To show $\text{Pic}(P)$ $(P=(0,0,0))$, is trivial amounts to showing $(R^1\pi)(\Omega)_P=(0)$, where $\pi\colon F'\to F$ is the map from a non-singular model to $0=x^2+y^3+z^5$ (since we know $H_1(M)=(0)$ already). Let us choose a slightly better global surface F (our statement being local, we are free to choose a different model of $k(F)$ outside a neighborhood of P): namely take F_0 to be the double plane with sextic branch locus $B: u(u^2y^3+z^5)$, where u, y, z are homogeneous coordinates. F_0 has two singularities: one is over $y=z=0$ and this is P; the other is over $u=z=0$ — call it Q. Let F_1 be the result of resolving Q alone, and F_2 be the non-singular surface obtained by resolving P and Q. Let $\pi: F_2\to F_1$. We must show $(R^1\pi)(\Omega_{F_2})_P\simeq(0)$. But since $(R^1\pi)(\Omega_{F_2})$ is (0) outside of P, it is equivalent to show $H^0(F_1,(R^1\pi)(\Omega_{F_2}))=(0)$. First of all, note that F_2 is birational to P^2: indeed $0=x^2+y^3+z^5$ is uniformized by the substitution:

$$x=1/u^3v^5(u+v)^7,\ y=-1/u^2v^3(u+v)^5,\ z=-1/uv^2(u+v)^3.$$

Therefore $0=H^1(F_2,\Omega_{F_2})=H^2(F_2,\Omega_{F_2})$. Now consider the Spectral Sequence of Composite Functors:

$$H^i(F_1,(R^j\pi)(\Omega_{F_2}))\underset{i}{\Rightarrow}H^k(F_2,\Omega_{F_2}).$$

Noting that $(R^0\pi)(\Omega_{F_2})=\Omega_{F_1}$, it follows:

a) $H^1(F_1,\Omega_{F_1})=(0)$
b) $d_2^{0,1}\colon H^0(F_1,(R^1\pi)(\Omega_{F_2}))\to H^2(F_1,\Omega_{F_1})$
is $1-1$, onto.

Therefore, it suffices to show $H^2(F_1,\Omega_{F_1})=(0)$, or $0\geqslant p_a(F_1)$ $(=\dim H^2-\dim H^1)$. Now unfortunately $p_a(F_0)=1$, since, in general, if G is a double plane with branch locus of order $2m$, $p_a(G)=(m-1)(m-2)/2$ (none of the singularities of G being resolved, of course) (1). To compute $p_a(F_1)$, embed F_0 in a family of double planes $F_{0,\alpha}$, where the branch locus B_α for $F_{0,\alpha}$ is

$$u(u^2y^3+z^5+\alpha u^4 z).$$

Now $F_{0,\alpha}$ have singularities over $u=z=0$ of identical type for all α, hence one may resolve these, and obtain a family of surfaces $F_{1,\alpha}$ containing F_1. But since B_α, for general α, has no singularity except $u=z=0$, the general $F_{1,\alpha}$ is non-singular. Now by the invariance of p_a [21], $p_a(F_1)=p_a(F_{1,\alpha})\leqslant \dim H^2(F_{1,\alpha},\Omega)=\dim H^0(F_{1,\alpha},\Omega(K))$, K the canonical class on $F_{1,\alpha}$. But if ω is the double *quadratic* differential (i.e. of type $A(dx\wedge dy)^2$ locally) on P^2 with poles exactly at B_α, one can readily compute $(f_\alpha^*\omega)$, where $f_\alpha\colon F_{1,\alpha}\to P^2$; it turns out strictly negative, and as it represents 2 K, it follows

$$p_g(F_{1,\alpha})=\dim H^0(F_{1,\alpha},\Omega(K))=0.$$

For details on the behaviour of p_a of double planes, which include our result as a particular case, see the works of Enriques and Campedelli cited in [4], p. 203-4, and the doctoral thesis of M. Artin [Harvard, 1960].

REFERENCES

[1] H. Behnke and H. Grauert, Analysis in non-compact Spaces, in *Analytic Functions*, Princeton, 1960.
[2] A. Borel and J.-P. Serre, Le Théorème de Riemann-Roch, *Bull. Soc. Math. France*, *86*, 1958.
[3] K. Brauner, Klassifikation der Singularitäten Algebroiden Kurven, *Abh. Math. Semin. Hamburg*, *6*, 1928.
[4] F. Enriques, *Le Superficie Algebriche*, Zanichelli, Bologna, 1949.
[5] A. Kurosch, *Theory of Groups*, vol. 2, Chelsea, N.Y.
[6] T. Matsusaka, On Algebraic Familes of Positive Divisors, *J. Math. Soc. Japan*, *5*, 1953.
[7] P. Samuel, Multiplicités de certaines composantes singulières, *Ill. J. Math.*, *3*, 1959.
[8] H. Seifert and W. Threlfall, *Lehrbuch der Topologie*, Teubner, 1934.
[9] H. Seifert and W. Threlfall, *Variationsrechnung im Grossen*, Chelsea, 1951.
[10] S. Smale, The Classification of Immersions of Spheres in Euclidean Space, *Annals Math.*, *69*, 1959.
[11] R. Thom, *Forthcoming paper on the differential structures on cells and spheres.*
[12] H. Weyl, *Algebraic Theory of Numbers*, Princeton, 1940.
[13] J. H. C. Whitehead, Simplicial Spaces, Nuclei, and *m*-Groups, *Proc. London Math. Soc.*, *45*, 1939.
[14] O. Zariski, On the Topology of Algebroid Singularities, *Am. J. Math.*, *54*, 1932.
[15] O. Zariski, The Reduction of Singularities of an Algebraic Surface, *Annals Math.*, *40*, 1939.
[16] O. Zariski, Foundations of a General Theory of Birational Correspondences, *Trans. Am. Math. Soc.*, *53*, 1943.
[17] O. Zariski, Theory and Applications of Holomorphic Functions on an Algebraic Variety of Arbitrary Characteristic, *Memoirs of the Am. Math. Soc.*, 1951.
[18] O. Zariski, Introduction to the Problem of Minimal Models, *J. Math. Soc. Japan*, 1958.
[19] J. E. Reeve, A Note on Fractional Intersection Multiplicities, *Rend. Circolo Mat. Palermo*, *1*, 1958.
[20] H. Seifert, Topologie Dreidimensionaler Gefaserter Räume, *Acta Math.*, *60*, 1932.
[21] J. Igusa, Arithmetic Genera of Normal Varieties in an Algebraic Family, *Proc. Nat. Acad. Sci.*, *41*, 1955.

Reçu le 20 mai 1960.
Révisé le 15 février 1961.

(1) This may be seen by means of a suitable resolution of $(R^0 f)(\Omega_G)$, $f\colon G\to P^2$ being its double covering. It is, however, classical: cf. [4], p. 180-2 using the formula:

$$4p_a=n+P-3\pi-k/3-2 \text{ where } n=2,\ k=0,$$
$$\pi=m-1, \text{ and}$$
$$P=(2m-1)(2m-2)/2=p_a(\text{Branch Locus}).$$

APPENDIX

THE CANONICAL RING OF AN ALGEBRAIC SURFACE

BY DAVID MUMFORD

In this appendix we wish to examine how the general theory developed by O. Zariski applies to the canonical divisor class. To be precise, suppose F is a non-singular algebraic surface over an algebraically closed field k, which

(a) is not birationally equivalent to a ruled surface, and

(b) is minimal [11].

Moreover let K be the canonical divisor class. We set

$$R = \bigoplus_{n=0}^{\infty} H^0(\mathfrak{o}_F(nK)),$$

and we call R the canonical ring of the surface ($R = R^*[K]$ in Zariski's notation). There are three essentially different cases to consider according as (K^2) is negative, zero, or positive. We assert:

---

[9] The assumption that $\dim|hD| > 0$ for some h is necessary. Thus, it is possible to have a prime cycle E such that $(E^2) = 0$ (and which is therefore arithmetically effective, whence $\mathcal{E} = 0$) and such that $\dim|nE| = 0$ or all n (whence $B_n = nE$, and B_n is not bounded). To obtain such a cycle E, we use the construction of §2, with the following modifications: we take for E' a generic plane section of F', we take for $\mathfrak{A}$ the divisor class determined on E' by $|E'|$ (i.e., we take for h the integer 0) and we determine $P'_1, P'_2, \cdots, P'_m$ (where $m = (E'^2)$) by the condition (4). Then it is immediate that the proper transform E of E' satisfies the desired conditions.

D. Mumford, *Selected Papers*, Vol. II, .

THEOREM. (i) *Under the above hypotheses,* $(K^2) < 0$ *is impossible.*

(ii) *If* $(K^2) = 0$, *then for some* n *either* $nK \equiv 0$, *or* $|nK|$ *is a linear system without base points, composite with a pencil. Therefore* R *is a finitely generated ring of dimension* 1 *or* 2.

1 (iii) *If* $(K^2) > 0$, *then for sufficiently large* n, $|nK|$ *is a linear system without base points. Therefore* R *is a finitely generated ring of dimension* 3. *Moreover* $s(nK) = \dim H^1(\mathfrak{o}_F(nK)) = 0$ *for sufficiently large* n.

Only the proof of (iii) will be given in this Appendix, since the proof of (i) and (ii) is rather long and will be published elsewhere. If the characteristic is 0, the latter proof depends chiefly upon Enriques' theorem: *if* F *is a relatively minimal non-singular algebraic surface, and* $|nK|$ *is empty for all* n, *then* F *is ruled.* The first complete proof of this in characteristic 0 (and of its refinement: if $|12K|$ is empty, then F is ruled) was obtained several years ago by K. Kodaira (unpublished). In characteristic p, new difficulties arise, but Enriques' result and parts (i) and (ii) of the theorem can still be proved.

We shall now establish (iii). Notice first that by the Riemann-Roch theorem, either $|nK|$ or $|-nK|$ is non-empty for large n. The latter case is impossible. For suppose q is the irregularity of F (= dimension of the Picard variety), and ρ is the base number (= rank of the Neron-Severi group). Then by Noether's formula for $p_a(F)$ and by Igusa's inequality,[10] we see that

2 $$12(p_a(F) + 1) = (K^2) - \deg(c_2) > 2 - 4q + \rho .$$

But since $p_g(F) = 0$, it follows that $p_a(F) = -q$.[11] Therefore:

$$8(1 - q) \geqq \rho - 1 .$$

But if $q = 0$, then F is rational, by Castelnuovo's criterion[12] which we have excluded; and if $q = 1$, then the Albanese map is a regular map onto a curve, and $\rho \geqq 2$. Therefore this last inequality cannot be fulfilled.

Therefore $|nK|$ is at least non-empty for sufficiently large n. Let D be any irreducible curve on F. Suppose $(D \cdot K) \leqq 0$. Then by Hodge's index theorem, since $(K^2) > 0$, it follows that $(D^2) < 0$. But also $-2 \leqq 2p_a(D) - 2 = (D^2) + (D \cdot K)$. Therefore $p_a(D) = 0$; i.e., D is a non-singular rational curve, and (D^2) equals -1 or -2. In the first case, D would be

---

[10] See J. I. Igusa, *Betti and Picard numbers of abstract algebraic surfaces*, Proc. Nat. Acad. Sci. U.S.A., 46 (1960), p. 724

[11] See Y. Nakai, *On the characteristic linear systems of algebraic families*, Ill. J. Math., 1 (1957), p. 552.

[12] See O. Zariski, *On Castelnuovo's criterion of rationality* $p_a = p_2 = 0$ *of an algebraic* surface, Ill. J. of Math., 2 (1958), p. 303.

exceptional of the first kind and F would not be minimal [11]. Therefore, we conclude:

(*) *If D is an irreducible curve, and* $(D \cdot K) \leqq 0$, *then D is a non-singular rational curve,* $(D^2) = -2$, *and* $(D \cdot K) = 0$.

Notice that there can be at most a finite number of such irreducible curves D. In fact, by the Riemann-Roch theorem, there is an m such that $\dim |mK| \geqq 2$. Then every curve D such that $(D \cdot K) = 0$ is either a fixed component of the linear system $|mK|$ or else is disjoint from every divisor of $|mK|$. In either case, there is only a finite set of such irreducible curves.

Let $E_1, E_2, \cdots, E_n$ be the set of all irreducible curves D such that $(D \cdot K) = 0$. Then by a very beautiful theorem of M. Artin,[13] which is the central point of this proof, there is a normal surface F^*, and a regular birational map $f: F \to F^*$, with the following five properties:

(i) f is biregular on $F - \bigcup E_i$,

(ii) f maps each E_i to one point,

(iii) the canonical divisor K^* on F^* is a Cartier divisor,

(iv) $f^{-1}(K^*) = K$,

(v) $p_a(F) = p_a(F^*)$.

By (iv), the linear systems $|nK|$ (on F), and $|nK^*|$ (on F^*) are canonically isomorphic. The proof that for sufficiently large n, $|nK^*|$ has no base points proceeds in three steps:

Step I. For all sufficiently large n, $|nK|$ (and hence $|nK^*|$) is non-empty. This is a corollary of the Riemann-Roch theorem.

Step II. For all sufficiently large n, $|nK^*|$ has no fixed components. For let k and l be relatively prime integers such that $|kK|$ and $|lK|$ are non-empty. Then by Theorem 9.1 above, for all sufficiently large n, the only fixed components of $|nkK|$ and $|nlK|$ are the irreducible curves E_i. But since all sufficiently large integers are of the form $nk + n'l$, for "sufficiently" large n and n', it follows that the only fixed components of $|nK|$ for sufficiently large n are the curves E_i. Hence by (ii) and (iv), the corresponding linear system $|nK^*|$ has no fixed components.

Step III. For sufficiently large n, $|nK|$ has no base points at all. The proof of this depends on a slight extension of Theorem 6.2 above. Namely, notice that this theorem, together with the proof of that theorem (§ 6), are equally valid whenever (in the notation of that theorem) V is a normal surface, and D is a Cartier divisor. Now let k and l be relatively prime integers such that $|kK^*|$ and $|lK^*|$ have no fixed components. By (iii),

3 [13] See M. Artin, *Some numerical criteria for contractability of curves on an algebraic surface*, Amer. J. Math., forthcoming, Th. (2.7).

and this extension of Theorem 6.2, for all sufficiently large n, the linear systems $|nkK^*|$ and $|nlK^*|$ have no base points. Hence just as before, for all sufficiently large n, $|nK^*|$ (and hence $|nK|$) has no base points.

The result on the superabundance follows from (v) and Theorem 6.5 above, once one observes that for sufficiently large n, the linear system $|nK|$ must define a regular map of F into projective space, with image F^*.

Finally, one sees that R is finitely generated as follows: Let k and l be relatively prime integers such that $|kK|$ and $|lK|$ have no base points. Then by Theorem 6.5, the rings $R^*[kK]$, and $R^*[lK]$ are finitely generated. But these two rings together generate a subring of R that contains all but a finite number of its homogeneous components. As each component of R is a finite dimensional vector space, R itself is therefore finitely generated. q.e.d.

HARVARD UNIVERSITY

REFERENCES

1. A. GROTHENDIECK, *Sur une note de Mattuck-Tate*, J. für reine und angew. Math., 20 (1958), 208-215.
2. W. V. D. HODGE, *Note on the theory of the base for curves on an algebraic surface*, J. London Math. Soc., 12 (1937), 58-63.
3. Y. NAKAI, *Non-degenerate divisors on an algebraic surface*, J. Science Hiroshima Univ., 24 (1960).
4. D. REES, *On a problem of Zariski*, I. J. of Math., 2 (1958), 145-149.
4′. M. ROSENLICHT, *Equivalence relations on algebraic curves*, Ann. of Math., 56 (1952), 169-191.
5. P. SAMUEL, *Sur les anneaux gradués*, C. R. Acad. Cien. Brasil (1958).
6. O. ZARISKI and P. SAMUEL, Commutative Algebra, volumes 1 and 2 (1958 and 1960), D. van Nostrand Company, Princeton.
7. O. ZARISKI, Algebraic Surfaces, Ergebnisse der Mathematik und ihrer Grenzgebiete, 3 (1935), n. 5.
8. ———, *Proof of a theorem of Bertini*, Trans. Amer. Math. Soc., 50 (1941), 48-70.
9. ———, *Complete linear systems on normal varieties and a generalization of a lemma of Enriques-Severi*, Ann. of Math., 55 (1952), 552-592.
10. ———, *Interprétations algébro-géometriques du quatorzième problème de Hilbert*, Bull. Soc. Math., 78 (1954), 155-168.
11. ———, Introduction to the problem of minimal models in the theory of algebraic surfaces, Publications of the Mathematical Society of Japan 4 (1958).

Article [63b]
Some Aspects of the Problem of Moduli

David Mumford

Abstract of the Lecture given at the Annual Meeting of the Mathematical Society of Japan on May 24, 1963
Sûgaku *15, 1963/1964, 155–157.*

I. The first aspect which I wish to discuss is the question of how to make precise the heuristic concept of moduli. For example, suppose one is concerned with curves of genus g: then, for every algebraically closed field Ω, let $\mathfrak{M}_g(\Omega)$ be the set of curves of genus g, defined over Ω, up to isomorphism. Since the moduli scheme M_g is to classify curves, one asks at least that there be given an isomorphism between the set of Ω-rational points of M_g and $\mathfrak{M}_g(\Omega)$. This obviously does not determine M_g, however. A stronger demand is to ask for a collection of isomorphisms between the set of R-valued points of M_g, and the set of curves of genus g over R, for *every commutative ring* R; here a curve over R means a scheme, simple and proper over Spec(R), whose geometric fibres are curves of genus g. Moreover, these isomorphisms should be functorial in R. Then, in fact, this determines M_g, if it exists. An essentially equivalent demand is to ask that there exists a "Universal Family" of curves over M_g itself. Such an M_g I call a fine moduli scheme; unfortunately, it does not exist unless the classificational problem is slightly modified (via a "higher level structure"). For higher dimensions, to find suitable modifications to "eliminate the automorphisms" is an interesting problem.

In any case, one can compromise for a coarse moduli scheme: here one merely asks for some collection of maps, from the sets of curves over R to the sets of R-valued points of M_g, which are (i) functorial in R, (ii) isomorphisms when R is an algebraically closed field. Finally, to determine M_g completely, one should ask that it satisfy a universal mapping property with respect to all other solutions of the first two demands.

II. The next aspect we consider is that of the qualitative properties of the sought-for moduli scheme: especially, whether it is a true scheme, or only a pre-scheme; and whether it is of finite type over the integers. But, in fact, examples due to Kodaira, Nagata, Nishi, and others indicate the absence of both of these properties in the general case of classifying higher dimensional varieties. To remedy this difficulty, the simplest solution seems to be to modify the problem: instead of classifying varieties, one seeks to classify polarized varieties. By a polarized variety, we mean a variety V

D. Mumford, *Selected Papers*, Vol. II,

together with a Cartier divisor class D, determined up to algebraic equivalence and torsion, such that nD is induced by a projective embedding of V, if $n \gg 0$.

For this classificational problem, Matsusaka and the writer have shown that the moduli scheme should be a true scheme, if the varieties are assumed nonsingular, and not birationally ruled. Moreover, note that a Hilbert polynomial $P(n) = \chi(\mathcal{O}_V(nD))$ can be attached to any polarized variety, and that it remains constant in flat families of such polarized varieties. Then we have also shown that, for nonsingular *surfaces*, the moduli scheme of polarized surfaces with fixed Hilbert Polynomial should be of finite type. Whether the same is true in dimension 3 is a very intriguing question. Another difficult problem is to ascertain how essential is the role of the nonsingularity assumption in these matters. In the complex analytic case, nonsingular families recommend themselves as being differentiably trivial, so that they can be visualized as families of complex structures on a fixed manifold. In the algebraic case, however, there seems to be no compelling reason for thinking that this is a reasonable assumption.

III. Beyond the qualitative problems already discussed, there looms the big question of whether, although possessing all good local and global properties, the moduli scheme may fail to exist for more subtle reasons. One may put the problem this way: the "moduli scheme" may be formally described as the quotient of a scheme by some topologically beautiful equivalence relation but it may be impossible to give a scheme realizing this quotient. For instance, it is sometimes impossible to "blow down" certain subvarieties, or to "divide" some variety by the action of some group. In this case, there would be only an open subset U of *stable* polarized varieties which could be realized as a scheme.

This problem appears to be closely connected with the local projective differential geometry of embedded varieties $V \subset \mathbb{P}_n$. To illustrate, suppose V is a nonsingular curve, and that the embedding is determined by a complete linear system on V of high degree. Then the Weierstrass gap theorem, and the Frenet–Serret equations give a very explicit picture of this embedded curve. This enables us to do two things: In the first place, you can look at the set of $x \in V$ where the Frenet–Serret equations break down. I call these points of *Hyper-Osculation*, and with convenient multiplicities, they can be added together to give a divisor on V. This possesses two key properties: (i) as V and the embedding vary continuously, this divisor varies continuously, (ii) the maximum multiplicity with which any x occurs in this divisor is bounded by g^2 ($g =$ the genus of V). This being so, projective invariants of V can be constructed in a highly explicit fashion out of determinants in the coordinates of these points. This is tantamount to constructing the moduli scheme for curves. In the second place, the very explicit expression of V gives directly information on the Chow form of V: especially on the monomials which occur in the Chow form with nonzero coefficient and are extremal in the convex hull of all monomials of fixed degree with this property. This, too, leads to projective invariants of space curves, hence to moduli. In this connexion, the difficulty in the surface case appears to be lack of very much information on the local projective differential geometry of surfaces in $\mathbb{P}_n$.

IV. Riemann originally asked for $3g-3$ complex numbers, called moduli, to be attached to each curve over the complex numbers. One interpretation of this assertion is to ask, not only for a construction of M_g but for a projective embedding $M_g \subset \mathbb{P}_N$. This leads to the fourth aspect: to study the Picard group of M_g. One interesting point in this connexion is that it is possible to define the Picard group of the moduli problem itself without reference to the moduli scheme.[1] Namely, by a line bundle on the moduli problem we shall mean a collection of line bundles, one on each scheme S for each family of curves over S; plus, for each morphism between families, a corresponding morphism between line bundles. Heuristically, such line bundles arise from attaching *canonically* one dimensional vector spaces over Ω to each curve over Ω.

I can prove that the group of line bundles on the moduli problem, i.e., the Picard group, is finitely generated; and that, up to torsion, there is exactly a subgroup isomorphic to $\mathbb{Z}$ of line bundles which extend to line bundles on the whole moduli problem of principally polarized abelian varieties (via the Jacobian). I conjecture that the group itself is $\mathbb{Z}$,[2] but in this connexion I can give only some curious relations.[3] For example, to any curve C, we can attach two 1-dimensional vector spaces: a) $\bigwedge^g \mathrm{H}^0(C,\Omega_C)$, where Ω_C is the sheaf of differentials on C, and b) $\bigwedge^{3g-3} \mathrm{H}^0(C,(\Omega_C)^2)$, where $(\Omega_C)^2$ is the sheaf of quadratic differentials. These extend naturally to line bundles $\mathscr{L}_1$ and $\mathscr{L}_2$ on the whole moduli problem. Then, up to torsion:

$$\mathscr{L}_2 \cong (\mathscr{L}_1)^{13}.$$

---

[1] See [65b].

[2] This conjecture, that the image of the natural map $\mathrm{Pic}(\mathscr{A}_g) \to \mathrm{Pic}(\mathscr{M}_g)$ is isomorphic to $\mathbb{Z}$ for $g \geq 3$, where $\mathscr{A}_g$ is the moduli problem for g-dimensional principally polarized abelian varieties over the same base field as $\mathscr{M}_g$, is true when the base field is $\mathbb{C}$, or any algebraically closed field of characteristic 0. In fact $\mathrm{Pic}(\mathscr{M}_g)$ itself is isomorphic to $\mathbb{Z}$ in characteristic 0 when $g \geq 3$; see the notes for [67d] in this volume.

[3] The displayed relation between the Hodge line bundle $\mathscr{L}_1$ and the canonical bundle $\mathscr{L}_2$ on the moduli problem was proved using the Grothendieck–Riemann–Roch theorem. This is the first time Mumford applied the Grothendieck–Riemann–Roch theorem to obtain a relation between two tautological classes on the moduli problem $\mathscr{M}_g$.

TWO FUNDAMENTAL THEOREMS ON DEFORMATIONS OF POLARIZED VARIETIES.

By T. Matsusaka and D. Mumford.[1]

Introduction. In contrast to the theory of moduli of curves, the global theory of moduli of higher dimensional varieties—with the exception of Abelian varieties—is largely unexplored. The work of the authors and of others[2] has begun at least to clarify the problem, and to pose some plausible conjectures. One thing that is clear, however, is that there is a complexity here of a higher order of magnitude from that encountered for curves. The purpose of the present article is to present two results of a qualitative nature that limit the degree of possible complexity of various sought for varieties or scheme of moduli. The first result of ours asserts that two non-singular projective varieties with polarizations, which are isomorphic as polarized varieties, remain isomorphic after specializations over a discrete valuation-ring, whenever they remain non-singular polarized varieties and at least one of them is non-ruled (cf. Th. 2). The second asserts that a set of non-singular polarized surfaces, which are deformations of each other, can be realized as an algebraic family (i.e. a finite union of an irreducible algebraic family) of non-singular projective surfaces in a projective space, if their ranks are bounded; and, in fact, the set of non-singular surfaces with non-degenerate divisors with a given Hilbert polynomial and of any characteristic can be realized as an algebraic family over the ring of integers. From this, it can be shown that the variety of moduli of such surfaces, which are not ruled, is a finite union of Q-varieties, which will be discussed in a near future.

In Chapter I, we shall settle the first result we mentioned. In Chapter II, we give an estimation for $l(X)$ when X is a non-degenerate divisor on a projective variety. Our second main theorem will be settled in Chapter III, as well as in Chapter IV, under slightly different technique. In the first three Chapters, essentially the terminonolgy and conventions of Weil's book [18] are followed. In Chapter IV, because of the nature of the technique which

Received October 25, 1963.

[1] This work was done while the first named author was supported by the N. S. F. and the second named author was supported by the Sloan Foundation, and the Army Research Office (Durham).

[2] Cf. [5], [6], [8], [9], [10], [11].

are followed, essentially Grothendieck's terminology and conventions in [2] are followed. However, in order to keep the uniformity, the word *"ample"* (resp. *"non-degenerate") is used for "very ample"* (resp. *"ample"*) in the sense of Grothendieck.

By a *specialization* of a variety or a cycle, we understand a reduction of such over a discrete valuation-ring (cf. [17]). For the theorem of Riemann-Roch in general, we follow quite often the sheaf-theoretic terminology which can be found in [15] and [22]. Let V be a normal variety and M a finitely generated module of functions on V. When $Y = \inf_{g \in M}(\mathrm{div}(g))$, the set $\Lambda(M)$ of V-divisors $\mathrm{div}(g) - Y$, $g \in M$, is called the *reduced linear system* determined by M. When F is any positive V-divisor, $\Lambda(M) + F$ is called a *linear system.* Assume that V is complete. When X is a V-divisor, the set $L(X)$ of functions g on V such that $\mathrm{div}(g) + X \succ 0$ forms a finite dimensional vector space ([18], App. 1, Th. 3). We denote by $\Lambda(X)$ the set of positive V-divisors which are linearly equivalent to X, and call it the *complete linear system* determined by X. We denote by $|X|$ the support of X. We have $\Lambda(X) = \Lambda(L(X)) + F$, where $F = X + \inf_{g \in L(X)}(\mathrm{div}(g))$. We denote by $l(X)$ the dimension of $L(X)$. When V is a projective variety, we denote by $\mathfrak{o}_V$ the sheaf of local rings on V, the defining sheaf of functions on a scheme V. If X is a Cartier divisor on V, we denote by $\mathfrak{L}(X)$ the corresponding invertible sheaf. With this sheaf theoretic notations, $H^0(V, \mathfrak{L}(X)) = L(X)$ when V is normal. Moreover, when V is a non-singular projective surface, $H^2(V, \mathfrak{L}(X))$ is isomorphic to the dual of $H^0(V, \mathfrak{L}(K(V) - X))$ and $\dim H^1(V, \mathfrak{L}(X)) = s(X)$ is the *superabundance* of X. When there is no danger of confusion, we write $H^i(\mathfrak{L}(X))$ for $H^i(V, \mathfrak{L}(X))$.

Chapter I.

THEOREM 1.[3] *Let V be a complete abstract variety, W an abstract variety and T a birational correspondence between V and W. Let $\mathfrak{o}$ be a discrete valuation-ring with the quotient field k, such that V, W and T are defined over k. Let (V', W', T') be a specialization of (V, W, T) over $\mathfrak{o}$ and assume that V', W' are abstract varieties and that V' is complete. When W' is not a ruled variety, there is a component T'' of T' with the coefficient 1 in T' such that T'' is a birational correspondence between V' and W' and that $\mathrm{pr}_i(T' - T'') = 0$ for $i = 1, 2$.*

Proof. From the compatibility of specializations with the operation of algebraic projection (cf. [17]), we see that T' has a component T'' with the

[3] This theorem was pionted out to us by M. Artin.

following properties: (a) $\mathrm{pr}_2 T'' = W'$; (b) the coefficient of T'' in T' is 1; (c) $\mathrm{pr}_2(T' - T'') = 0$. Let $\mathfrak{p}$ be the maximal ideal of $\mathfrak{o}$ and κ the residue field of $\mathfrak{o}$ with respect to $\mathfrak{p}$. Let (x') be a generic point of a representative of W' over κ. Then, there is a representative (x) of a generic point of W over k such that (x') is a specialization of (x) over $\mathfrak{o}$, over

$$(V, W, T) \xrightarrow{\mathfrak{o}} (V', W', T')$$

(cf. [17], Th. 7). Let R_v be the specialization-ring of the specialization $(x) \xrightarrow{\mathfrak{o}} (x')$ in $k(x)$. Then R_v is a discrete valuation ring of $k(x)$ (cf. [17], Prop. 5 and Th. 15). Hence, it determines a valuation v of $k(x)$. Let $Q \times (x)$ be a generic point of T over k. Since V and V' are complete, there is at least one representative (y) of Q such that the coordinates y_i of (y) are in R_v, that (y') is a representative of Q' if $(Q \times (x), (y)) \xrightarrow{\mathfrak{o}} (Q' \times (x'), y'))$ and that $Q' \times (x')$ is contained in $|T'|$. When that is so, $Q' \times (x')$ is contained in T''; in fact, it is a generic point of T'' since $\mathrm{pr}_2\colon T'' \to W'$ is birational and (x') is a generic point of W' over $\bar{\kappa}$. It follows that Q' is a generic point of the projection A of T'' on V' over $\bar{\kappa}$. Let R be the specialization ring of $(y) \xrightarrow{\mathfrak{o}} (y')$. Then, the valuation v is a prime divisor of R in the sense of Abhyankar, and W' is a ruled variety over A unless $A = V'$ (cf. [1], Prop. 3). Therefore, $A = V'$. When that is so, T'' is a birational correspondence between V' and W', which can be seen easily, using the compatibility of specializations with the operation of intersection-product (cf. [17]).

THEOREM 2. *Let $\mathfrak{o}$ be a discrete valuation-ring with the quotient field k; let V and W be non-singular projective varieties, defined over k, and T the graph of an isomorphism, defined over k, between V and W. Let X (resp. Y) be a non-degenerate divisor on V (resp. W), both rational over k, such that $Y = T(X)$. Let $(V, W, X, Y, T) \xrightarrow{\mathfrak{o}} (V', W', X', Y', T')$ and assume that V', W' are non-singular and that X' (resp. Y') is also non-degenerate on V' (resp. W'). Then T' is the graph of an isomorphism between V' and W', if one of the V', W' is not ruled.*

Proof. By Theorem 1, we have $T' = T'' + T^*$, where T'' is a birational correspondence between V' and W', and $\mathrm{pr}_l T^* = 0$ for $l = 1, 2$. Let $F_1, \cdots, F_t$ be the projections of the components of T^* on V'. Note that none of the F_i is 0-dimensional: for if F_i were 0-dimensional, the corresponding component of the n-dimensional cycle T^* would have to be of the form $F_i \times W'$, and this contradicts $\mathrm{pr}_2 T^* = 0$. If X'_m is a divisor in $\Lambda(mX')$, then T' and $X'_m \times W'$ intersect properly if and only if $|X'_m| \not\supset F_i$ for any i. Let U be the set of such divisors X'_m. For every such X'_m, $T'(X'_m)$ is defined. The Chow-variety of U, i.e. the set of Chow-points of members of U, is an open subset of that of $\Lambda(T'(X'_m))$. When U is not empty, the mapping $X'_m \to T'(X'_m)$ defines, as is well-known, an injection of U into $\Lambda(T'(X'_m))$; and, as a matter of fact, defines an injective linear rational map of the Chow variety of U into that of $\Lambda(T'(X'_m))$ cf. [18], Chap. IX, Th. 3 and [18], Chap. VIII, Th. 4). Now assume that (a) $T'(X'_m) \sim mY'$ and (b) $l(mX') = l(mY')$ for large m.

Suppose that P' is a point of V' and let $\Lambda(mX')_{P'}$ be the linear subsystem of divisors which pass through P'. For sufficiently large m, mX' is ample, hence P' is the only base point of $\Lambda(mX')_{P'}$, hence $\Lambda(mX')_{P'} \cap U$ is not empty. Then the set A of divisors $T'(Z')$, $Z' \in \Lambda(mX')_{P'} \cap U$, consists of divisors passing through every point Q' such that $P' \times Q' \in |T'|$. If there were more than one such Q', the closure of the Chow-variety of A is at least of co-dimension 2 in that of $\Lambda(mY')$, since mY' is ample for sufficiently large m. On the other hand, its co-dimension has to be 1 as the closure of the image of the Chow-variety of $\Lambda(mX')_{P'} \cap U$ by the injective rational map, since $l(mX') = l(mY')$. Hence $T^* = 0$ and T' is single-valued on the points of V', hence everywhere regular by Zariski's Main Theorem. Similarly T'^{-1} is everywhere regular.

To prove (a) and (b), note that $p_a(mX) = p_a(mY)$ for all integers m. Hence $p_a(V') = p_a(V) = p_a(W) = p_a(W')$; and $p_a(mX') = p_a(mX) = p_a(mY) = p_a(mY')$ for all integers m (cf. [13]). It follows that $l(mX) = l(mY) = l(mX') = l(mY')$ for large positive integer m by the theorem of Riemann-Roch (cf. [21]). Thus (b) is satisfied. Now let C and D be the supports of the Chow-varieties of $\Lambda(mX)$, $\Lambda(mY)$. Since the linear equivalence is preserved under specializations (cf. [17]), C', D' will be the supports of the Chow-varieties of $\Lambda(mX')$, $\Lambda(mY')$ for large m, if $(C, D) \xrightarrow{\mathfrak{o}} (C', D')$. Then (a) follows from the compatibility of specializations with the operation of intersection-product and from the invariance of linear equivalence under specializations.

Let V and V' be two complete non-singular polarized varieties (cf. [20]), k a field of definition of V and $\mathfrak{o}$ a discrete valuation ring with the quotient field k. Let X be a polar divisor of W and W, W' the underlying varieties of V, V'. If $(W, X) \xrightarrow{\mathfrak{o}} (W', X')$ and X' is a polar divisor of V', we shall say that V' is a specialization of V over $\mathfrak{o}$. With this definition, de have the following corollary.

COROLLARY 1. *Let $\boldsymbol{V}$ and $\boldsymbol{W}$ be two varieties over a discrete valuation ring $\mathfrak{o}$ (i.e. $\mathfrak{p}$-variety in the sense of Shimura; a scheme in the sense of Grothendieck). Let generic fibres V, W of $\boldsymbol{V}$, $\boldsymbol{W}$ be non-singular projective varieties, defined over the quotient field k of $\mathfrak{o}$. Let the special fibres V', W' be non-singular projective varieties. Assume that V, W, V', W' are underlying varieties of polarized varieties $\tilde{V}$, $\tilde{W}$, $\tilde{V}'$, $\tilde{W}'$ and that $(V, W) \xrightarrow{\mathfrak{o}} (V', W')$ can be extended to $(\tilde{V}, \tilde{W}) \xrightarrow{\mathfrak{o}} (\tilde{V}', \tilde{W}')$. Then, when there is an isomorphism $\bar{f}$ between $\tilde{V}$ and $\tilde{W}$ over k, $\bar{f}$ can be extended to an isomorphism $\boldsymbol{f}$ of $\boldsymbol{V}$ and $\boldsymbol{W}$, if W' is not ruled. Moreover, the graph of $\bar{f}$ specializes to the graph of an isomorphism f' between $\tilde{V}'$ and $\tilde{W}'$ over $\mathfrak{o}$.*

COROLLARY 2. *Let V be a projective, non-ruled, non-singular variety with a structure of polarization and G the connected component, containing the identity, of the group of automorphisms of V. Then G is an Abelian variety.*

Proof. The group of automorphisms of V is an algebraic group (cf. [8]). If G is not complete, the graph of an automorphism, corresponding to a suitable element of G, can be specialized, over some field of definition of V, to a $V \times V$-cycle which is not the graph of an automorphism. This is impossible by Theorem 2. Hence G is complete and is an Abelian variety by the theorem of Chevalley (cf. [19], Th. 5).

Chapter II.

Let V^n be a normal projective variety and X a non-degenerate divisor on V. $L(mX)$ defines a projective embedding f_m of V for large m by the definition. Let W^r be a simple subvariety of V and k a common field of definition for W and V, over which X is rational. Then $L(mX)$ has a basis over k (cf. [18], Ch. IX, Cor. 1 of Th. 8). Let $A_1, \cdots, A_r$ be independent generic divisors of $\Lambda(mX)$ over k. Then every component of $W \cap A_1 \cap \cdots \cap A_r$ is simple on V and on W (cf. [18], Ch. V, Th. 1). We set $[W \cdot X^{(r)}] = (1/m^r)\deg(W \cdot A_1 \cdots A_r)$ and $X^{(n)} = [V \cdot X^{(n)}]$. Then $[W \cdot X^{(r)}]$ does not depend upon the choice of independent generic divisors $A_1, \cdots, A_r$. Moreover, it does not depend upon the choice of m, as long as it is sufficiently large, and is a positive integer (cf. Bezout's theorem).

THEOREM 3. *Let V^n be a normal projective variety and X a non-degenerate divisor on V. Then $l(X) \leqq X^{(n)} + n$.*

Proof. If $l(X) \leqq 1$, there is nothing to prove. If $n = 1$, our theorem is an immediate consequence of the theorem of Riemann-Roch. Therefore, we assume that $l(X) > 1$ and that $n > 1$. Let X_0 be a generic divisor of the complete linear system $\Lambda(X)$. Then $X_0 = \sum_1^d Y_i + F$, where F is the fixed component of $\Lambda(X)$ and $Y_i \neq Y_j$ for $i \neq j$, since $\Lambda(X)$ is complete (cf. [18], Ch. IX, Cor. of Th. 15). If $d > 1$, the Y_i are generic divisors of one and the same pencil on V by the theorem of Bertini (cf. [18], Ch. IX, Th. 17). Hence $\dim \Lambda(X) = l(X) - 1 \leqq d$. On the other hand, we get $d \leqq X^{(n)}$ by computing $X^{(n)} = [(\sum_1^d Y_i + F) \cdot X^{(n-1)}]$. Therefore, our theorem is true in this case also.

Assume now that $d = 1$. Then $X_0 = Z + F$, where Z is an absolutely irreducible subvariety of V. Let K be an algebraically closed field, containing k, over which Z and F are rational, and (Z^*, α) a normalization of Z over K. Let m_0 be a positive integer such that mX is ample for $m \geqq m_0$ and X_m a generic divisor of $\Lambda(mX)$ over K for such m. We note that every component of $X_m \cap Z$ is simple both on V and Z and is proper on V. We contend that: (a) When $g \in L(X_{m+1} - X_m)$, $g \to g^* = g \circ \alpha^{-1}$ is a homomorphism of $L(X_{m+1} - X_m)$ into $L(X^*)$, where $X^* = \alpha(Z \cdot (X_{m+1} - X_m))$; (b) The kernel of the above homomorphism is a vector space of dimension 1; (c) X^* is non-degenerate on Z^* and $X^{*(n-1)} \leqq X^{(n)}$. Our theorem will be an immediate consequence of (a), (b), (c). For, we have $l(X^*) \leqq X^{*(n-1)} + (n-1)$ by the induction hypothesis, hence $l(X^*) \leqq X^{(n)} + (n-1)$ by (c), and $l(X) = l(X_{m+1} - X_m) \leqq l(X^*) + 1$ by (a) and (b).

To prove (a), we may assume that $g^* \neq 0$. We first remark the following two facts: (i) If U is a subvariety of Z of co-dimension 1, which is simple both on V and Z, and g' is the function induced on Z by g, then the coefficient of U in $\operatorname{div}(g) \cdot Z$, that of U in $\operatorname{div}(g')$ and that of $\alpha(U)$ in $\operatorname{div}(g^*)$ all coincide; (ii) If W^* is a component of $\operatorname{div}(g^*)_\infty = g^{*-1}(\infty)$, its geometric image W by α^{-1} is a component of $X_{m+1} \cap Z$, and is simple both on V and Z. In fact, U has the same coefficient a in $\operatorname{div}(g) \cdot Z$ as in $\operatorname{div}(g')$ (cf. [18]-IX, Th. 3). Since g^* can be written as $g' \circ \alpha^{-1}$, and since α is biregular along U, it follows that the coefficient of $\alpha(U)$ in $\operatorname{div}(g^*)$ is also a. As for (ii), g^* is not finite along W^* (i. e. at a generic point of W^* over a field of definition of W^*, containing K), and hence, g is not also finite along W. Consequently, $W \subset |g^{-1}(\infty)| = |X_{m+1}|$ and W is a component of $Z \cap X_{m+1}$.

Now let U be a component of $X_m \cap Z$ or of $X_{m+1} \cap Z$. Since

$$\operatorname{div}(g) \cdot Z + Z \cdot (X_{m+1} - X_m) \succ 0,$$

it follows that the coefficient of $\alpha(U)$ in $\operatorname{div}(g^*) + X^*$ is non-negative by (i). Therefore, if $\operatorname{div}(g^*) + X^*$ has a component W^* of negative coefficient, it is a component of $\operatorname{div}(g^*)_\infty = g^{*-1}(\infty)$, which is impossible by (ii).

To prove (b), let g be a function in $L(X_{m+1} - X_m)$ such that $g^* = 0$. Then $\operatorname{div}(g) = W - (X_{m+1} - X_m)$, where W is a positive V-divisor such that Z is a component of it. Since $W \sim X_{m+1} - X_m \sim X$, it follows that $W = Z + F$ and that g is uniquely determined up to a constant factor. (b) is thereby proved.

To prove (c), choose a positive integer $r \geqq m_0$ and identify $\Lambda(rX)$ with the linear system of hyperplane sections of V by means of the embedding f_r. Let s be another large positive integer. Then $sX_r \sim srX_{m+1} - srX_m$ and sX_r, srX_m, srX_{m+1} are sections of V by hypersurfaces of degrees s, sm, $s(m+1)$ respectively, since the linear system of hypersurface sections of a normal projective variety is complete when the degree of hypersurfaces is large enough (Zariski's normalization theorem). Consequently, $sX_r \cdot Z$, $srX_m \cdot Z$, $srX_{m+1} \cdot Z$ are also sections of Z by hypersurfaces of degrees s, sm, $s(m+1)$. When s is chosen large enough so that α is determined by homogeneous functions of homogenity s, $\alpha(sX_r \cdot Z)$, $\alpha(srX_m \cdot Z)$, $\alpha(srX_{m+1})$ are hypersurface sections of Z^* by hypersurfaces of degrees 1, m, $m+1$ respectively. Hence

$$\alpha(sX_r \cdot Z) \sim sr\alpha(Z \cdot (X_{m+1} - X_m)).$$

Thus, $X^* = \alpha(Z \cdot (X_{m+1} - X_m))$ is non-degenerate. $\Lambda(rX)$ is the linear system of hyperplane sections of V. Hence

$$X^{(n)} = (1/r^{n-1})\deg(X) = (1/r^{n-1})\deg(Z+F) \geqq (1/r^{n-1})\deg(Z).$$

$\Lambda(srX^*)$ is the linear system of hyperplane sections of Z^*. Hence $X^{*(n-1)} = (1/(sr)^{n-1})\deg(Z^*)$. But $\deg(Z^*) = s^{n-1}\deg(Z)$ as is well-known and easy to see. (c) is thus proved.

Remark 1. Let $\mathfrak{L}$ be a non-degenerate invertible sheaf (ample invertible sheaf in the sense of Grothendieck) on a projctive variety V. Let d be the leading coefficient of $\chi(\mathfrak{L}^m)$. Then it is easy to deduce that $\dim H^0(\mathfrak{L}) \leqq d + \dim V$ from our theorem. In fact, when (V^*, β) is a normalization of V and X a Cartier divisor on V^* determined by $\mathfrak{L}$, then X is non-degenerate and $d = X^{(n)}$ if $\dim V^* = n$.

Remark 2. In our theorem, we assumed that X is non-degenerate. Assume now that V is a non-singular projective surface and X a V-divisor such that $X^{(2)} > 0$ and $[Y \cdot X] > 0$ for all positive V-divisors Y. Then, we can prove directly that $l(X) \leqq X^{(2)} + 2$. In fact, we may assume, as in the proof of our theorem, that a generic divisor X_0 of $\Lambda(X)$ is of the form $Z + F$, where Z is an irreducible curve. Let Y be a V-divisor such that $Y \sim X$ and that $|Y|$ contains neither Z nor any singular point of Z. Then $[X \cdot Z] = [Y \cdot Z] \leqq X^{(2)}$ and $L(X)$ and $L(Y)$ are isomorphic. As in the proof of our theorem, $L(Y)$ induces on Z a module M' of functions on Z; $\alpha^{-1}(M')$ is then a submodule of $M^* = L(\alpha^{-1}(Z \cdot Y))$ and the kernel of the homomorphism $L(Y) \to M^*$ is a vector space of dimension 1. Hence, we have our inequality by the theorem of Riemann-Roch. Our divisor X is in fact a non-degenerate divisor on V according to [12], and our theorem is available according to this. But using this remark and our Theorem 4, we recover this result.

Chapter III.

Let V be a non-singular projective surface and X, Y two divisors. There is a V-divisor X' such that $X' \sim X$ and that X' and Y intersect properly on V. We denote by $X_\wedge Y$ the intersection-product $X' \cdot Y$, and by $[X \cdot Y]$ the degree of $X_\wedge Y$. When $X = Y$, $[X \cdot X]$ is denoted by $X^{(2)}$. We denote by $K(V)$ a canonical divisor on V and set $p_a(X) = (1/2)[X \cdot (X + K(V))] + 1$. When X is irreducible, $(X + K(V))_\wedge X$ is a canonical divisor $K(X)$ of X and $\deg(K(X)) = 2p_a(X) - 2$ (cf. [16]). When $X = \sum_i a_i X_i$, we have

$$
(1) \qquad \begin{aligned} p_a(X) &= \sum_i a_i p_a(X_i) + \sum_i (1/2) a_i(a_i - 1) X_i^{(2)} \\ &\quad + \sum_{i,j;\, i \neq j} (1/2) a_i a_j [X_i, X_j] - \sum_i a_i - 1. \end{aligned}
$$

According to the theorem of Riemann-Roch on V, we have

$$
l(X) - s(X) + l(K(V) - X) = X^{(2)} - p_a(X) + p_a(V) + 2.
$$

1. Denote by Σ the set of pairs (V, X) of a projective non-singular surface V and a V-divisor X satisfying the following conditions.

(I) $[X \cdot Y] > 0$ whenever Y is a positive V-divisor;

(II) $0 < X^{(2)} < c_1$;

(III) $|p_a(X)| < c_2$;

(IV) $|p_a(V)| < c_3$.

In order to simplify further discussions, we assume that the constants c_i $(i > 3)$ which will be introduced are *positive integers, satisfying* $c_i > c_{i-1}$ and depending *only upon* c_1, c_2, c_3.

14

LEMMA 1. *There are constants* c_4, c_5 *such that* $|[X\cdot K(V)]| < c_4$ *and that* $p_a(mX) > 0$, $l(K(V) - mX) = 0$, $m^2X^{(2)} - p_a(mX) + p_a(V) + 1 > 0$ *whenever* $(V, X) \in \Sigma$ *and* $m > c_5$.

This is an easy consequence of (I), (II), (III), (IV), the theorem of Riemann-Roch and of the formula (1).

LEMMA 2. *Let* (V, X) *be a member of* Σ *and* $T = \sum_1^t a_iY_i$ *the reduced expression for a member* T *of* $\Lambda(2c_5X)$. *Then, there are constants* c_6, c_7, c_8 *and* c_9 *with the following properties:*

(i) $\sum_1^t a_i < c_6$;

(ii) $|[Y_i\cdot Y_j]| < c_7$;

(iii) $|[K(V)\cdot Y_i]| < c_8; 0 \leqq p_a(Y_i) < c_8$;

(iv) The multiplicity of any point on Y_i is at most c_9.

Proof. (i) is a consequence of $\sum_1^t a_i \leqq \sum_1^t a_i[Y_i\cdot X] = [T\cdot X] \leqq 2c_1c_5$. (ii) and (iii) follow from the three inequalities:

(A) $a_iY_i^{(2)} + \sum_{i\neq j} a_j[Y_i\cdot Y_j] = [Y_i\cdot 2c_5X] \leqq \sum a_i[Y_i\cdot 2c_5X] \leqq 4c_5{}^2c_1$.

(B) $-2 \leqq 2p_a(Y_i) - 2 = [Y_i\cdot(Y_i + K(V))]$.

(C) $\sum a_i[K(V)\cdot Y_i] \leqq 2c_4c_5$.

In fact, (A) gives an upper bound for every $Y_i^{(2)}$. Hence, (B) gives a lower bound for every $[K(V)\cdot Y_i]$. Then (C) gives an upper bound for every $[K(V)\cdot Y_i]$ and (iii) is proved. Returning to (B), we obtain a lower bound for every $Y_i^{(2)}$, and using this, (A) gives upper bounds for all $[Y_i\cdot Y_j]$. This gives (ii), since $[Y_i\cdot Y_j] \geqq 0$ if $i \neq j$. Finally, the arithmetic genus of Y_i is bounded by (ii) and (iii). If the r_{ij} are the multiplicities of the singular points x_{ij} of Y_i, an inequality of Noether (cf. [4]) states

$$\sum_j r_{ij}(r_{ij} + 1)/2 + p_a(Y_i^*) \leqq p_a(Y_i),$$

where Y_i^* is a non-singular model of Y_i. This gives (iv).

The following lemma is an easy consequence of the generalized Riemann-Roch theorem for curves.

LEMMA 3. *Let* W *be a non-singular surface in a projective space and* Y *a divisor on* W. *Let* C *be an irreducible curve on* W *such that* $[Y\cdot C] > 2p_a(C) - 2$. *Then* $H^1(\mathfrak{L}(Y)/\mathfrak{L}(Y - C)) = 0$.

COROLLARY. *Using the same assumptions and notations of our lemma,* $s(Y-C)=s(Y)$ *if and only if*

$$0\to H^0(\mathfrak{L}(Y-C))\to H^0(\mathfrak{L}(Y))\to H^0(\mathfrak{L}(Y)/\mathfrak{L}(Y-C))\to 0.$$

Proof. This is an immediate consequence of our lemma and of an exact sequence $0\to\mathfrak{L}(Y-C)\to\mathfrak{L}(Y)\to\mathfrak{L}(Y)/\mathfrak{L}(Y-C)\to 0$.

LEMMA 4. *Let* $T=\sum_i a_iY_i$ *be a positive divisor on a non-singular projective variety* W, $\Lambda(A)$ *a complete linear system on* W *and assume that* $[(A-T')\cdot Y_i]>2p_a(Y_i)-2$ *for all* i *and for all* T' *such that* $0\prec T'\prec T$. *Then we have* $H^1(\mathfrak{L}(A)/\mathfrak{L}(A-T))=0$.

Proof. If $\sum_i a_i=1$, our lemma follows from Lemma 3. Assume that our lemma has been proved for those positive W-divisors $T''=\sum_i a''_iY_i$ with $\sum_i a''_i<\sum_i a_i$. Set $T'=\sum_i a'_iY_i$ with $a_1-1=a'_1$, $a_i=a'_i$ for $i\geqq 2$. In the exact cohomology sequence of an exact sequence

$$0\to\mathfrak{L}(A-T')/\mathfrak{L}(A-T)\to\mathfrak{L}(A)/\mathfrak{L}(A-T)\to\mathfrak{L}(A)/\mathfrak{L}(A-T')\to 0,$$

we have $H^1(\mathfrak{L}(A)/\mathfrak{L}(A-T'))=0$ by the induction assumption, and $H^1(\mathfrak{L}(A-T')/\mathfrak{L}(A-T))=0$ by our assumption and Lemma 3. Hence we get $H^1(\mathfrak{L}(A)/\mathfrak{L}(A-T))=0$.

COROLLARY 1. *Let* (V,X) *be a member of* Σ *and* $T=\sum_i a_iY_i$ *the reduced expression for a member of* $\Lambda(2c_5X)$. *Set* $T'=\sum_i a'_iY_i$, $U=\sum_i a''_iY_i$ *with* $0\leqq a'_i$, $a''_i\leqq a_i$. *Then, there is a constant* c_{10} *such that*

$$H^1(\mathfrak{L}(2mc_5X-U)/\mathfrak{L}(2mc_5X-U-T'))=0$$

for $m\geqq c_{10}$.

Proof. This follows at once from our lemma, (1) and from Lemma 2.

COROLLARY 2. *There is a constant* c_{11} *such that*

$$c_{11}\geqq s(2mc_5X)\geqq s(2(m+1)c_5X-B)\geqq s(2(m+1)c_5X)$$

whenever $(V,X)\in\Sigma$, $l(2c_5X-B)\geqq 1$ *and* $m\geqq c_{10}$.

Proof. This is an easy consequence of Theorem 3, Corollary 1 above and of Lemma 2. (cf. Remark 2.)

COROLLARY 3. *There is a constant* c_{12} *such that* $\Lambda(2mc_5X)$ *is irreducible (i.e. contains an irreducible curve) whenever* $(V,X)\in\Sigma$ *and* $m>c_{12}$.

Proof. Let T be a member of $\Lambda(2c_5X)$. If Y is a fixed component of $\Lambda(mT)$, we have $l(mT)=l(mT-Y)$. Then $m[T\cdot Y]=s(mT-Y)-s(mT)+p_a(Y)-1$ by the theorem of Riemann-Roch, which leads to a contradiction if $m>\max(c_{10},2c_{11}+c_8-1)=c_{11}'$ by the above Corollary 1, (I) and by (iii) of Lemma 2. If $\Lambda(mT)$ is composed of a pencil for $m>c_{11}'$, a generic divisor of $\Lambda(mT)$ can be written as $\sum_1^t T_i$, where the T_i belong to one and the same pencil by the theorem of Bertini. Clearly, we have $\dim\Lambda(mT)\leqq t$ and $[T_i\cdot T_j]\geqq 1$ by (I). Hence $(mT)^{(2)}\geqq t^2$ and $2mc_5\cdot c_1^{\frac{1}{2}}\geqq t\geqq\dim\Lambda(mT)$. On the other hand,

$$\dim\Lambda(mT)\leqq(2m^2c_5^2c_1-mc_5c_4)-c_3$$

by (I), Lemma 1 and by the theorem of Riemann-Roch. Our corollary now follows from this easily.

2. When Λ and Λ' are two linear systems on a complete normal variety, the smallest linear system Λ'' containing the divisors $X+X'$, $X\in\Lambda$, $X'\in\Lambda'$, is called the *minimum sum* of Λ and Λ'. Then the following lemma is easy to prove.

LEMMA 5. *Let $\Lambda(C)$ be a non-empty complete linear system on a complete normal variety W. Assume that $\Lambda(C)$ has no base point and that the minimum sum of $\Lambda(C)$ and $\Lambda(mC)$ is complete. Let h_i be a non-degenerate map of W into a projective space determined by $\Lambda(iC)$ for $i=m,m+1$. Then there is an isomorphism α between images W_m, W_{m+1} of W by h_m, h_{m+1} such that $h_{m+1}=\alpha\circ h_m$.*

In the following three lemmas, denote by C an irreducible curve on a non-singular projective surface V and $\mathfrak{R}_C$ the intersection of local rings of C at the singular points of C. Using only those functions of C which are in $\mathfrak{R}_C$, we can define linear systems as in the case of normal varieties. Throughout this chapter, *linear systems on curves lying on V are understood in this sense.* By the *degree* of a linear system on C, we understand the degree of a generic divisor of the linear system. The Riemann-Roch theorem on C then states $l(\mathfrak{m})=\deg(\mathfrak{m})-p_a(C)+1+l(K(C)-\mathfrak{m})$ for a C-divisor $\mathfrak{m}$ (cf. [13], [16]). The following lemma is known as a lemma of Castelnuovo when C is non-singular, which can be proved in the same way as in the ordinary case.

LEMMA 6. *Let Λ' be a linear system on C without base point and Λ a complete non-special linear system on C. Let $\mathfrak{n}'$ be a generic divisor of Λ' and assume that $\Lambda-\mathfrak{n}'$ is non-special and is of degree equal to* $\deg(\Lambda)-\deg(\mathfrak{n}')$. *Then the minimum sum of Λ and Λ' is complete.*

Actually, it is enough to know a special case of this lemma, under an additional assumption that $\deg(\Lambda) - \deg(\mathfrak{n}') \geqq 2p_a(C)$, which makes a proof very easy.

Let Λ be a linear system $\Lambda(M) + F$, where M is a finitely generated module of functions on V. Let C^* be the largest non-singular open subset of C and denote by $*$ the restriction of a C-chain (i.e. a zero-cycle on V whose support is contained in C) to C^*. Assume that C and F intersect properly on V and that every g in M induces a function g' in $\mathfrak{R}_C$ on C. Denote by M' the set of such functions g' and by Λ' the set of C-divisors $(X \cdot C)^* + (C \cdot F)^*$, by taking for X all divisors from $\Lambda(M)$ such that X and C intersect properly on V. Then Λ' is a linear system on C, whose reduced part is determined by M'. We denote Λ' by $\mathrm{Tr}_C \Lambda$ and call it the linear system on C induced by Λ. When Λ is a complete linear system $\Lambda(X)$ whose fixed component F satisfies our requirement, it always induces a linear system on C, since there is a V-divisor Z such that $X \sim Z$ and that the support of Z does not contain C and the singular points of C.

LEMMA 7. *Assume that $C^{(2)} > 0$ and that $\Lambda(C)$ has no base point. If $s(mC)$ is a constant for all positive integers m, the minimum sum of $\Lambda(C)$ and $\Lambda(mC)$ is complete for $m > [C \cdot K(V)] + 4$.*

Proof. Let C' be a V-divisor such that $C' \sim C$ and that $|C'|$ contains neither C nor the singular points of C. We have $[C \cdot (mC)] - 2p_a(C) > 0$ when m satisfies our condition. Hence $\mathrm{Tr}_C \Lambda(mC')$ is complete by Corollary of Lemma 3. By our assumption, $\mathrm{Tr}_C \Lambda(C')$ has no base point. Hence the minimum sum of $\mathrm{Tr}_C \Lambda(mC')$ and $\mathrm{Tr}_C \Lambda(C')$ is complete for such m by Lemma 6. Thus, the minimum sum of $\Lambda(mC')$ and $\Lambda(C')$ induces on C a complete linear system. Let M be the module generated by $f \cdot g$ with $f \in L(mC')$, $g \in L(C')$. Then M induces on C the module $L((m+1)C' \cdot C)$. When h is a function in $L((m+1)C')$, inducing 0 on C, we have $\mathrm{div}(h) = C + H - (m+1)C'$ with $H \succ 0$. Hence h is in M (cf. [18], Chap. IX, Cor. 2 of Th. 8). Our lemma follows from this at once.

It is not true in general that a complete linear system $\Lambda(\mathfrak{m})$ on C contains a divisor of the same degree as $\mathfrak{m}$, unless $\mathfrak{m}$ is positive. But we have the following.

LEMMA 8. *When $\deg(\mathfrak{m}) \geqq 2p_a(C)$, $\Lambda(\mathfrak{m})$ contains a divisor of the same degree as $\mathfrak{m}$.*

Proof. Let $\mathfrak{a}$ be a positive C-divisor such that $\deg(\mathfrak{m}) = \deg(\mathfrak{a}) = m$. Set $p_a(C) = t$ and let k be a field of definition for C over which $\mathfrak{m}$ and $\mathfrak{a}$

are rational. Then, there are two generic divisors $\mathfrak{p}$ and $\mathfrak{q}$ of degrees t over k and a unit f in $\mathfrak{R}_C$ such that $\mathrm{div}(f)=(\mathfrak{m}-\mathfrak{a})+(\mathfrak{p}-\mathfrak{q})$ (cf. [14], Lemma 3). Since $\mathfrak{a}$ is positive, a generic divisor of $\Lambda(\mathfrak{a})$ over k has the degree m; moreover, we have $l(\mathfrak{a})\geqq t+1$. Therefore, there is a unit g in $\mathfrak{R}_C$ and a generic divisor $\mathfrak{a}'=\mathfrak{p}+\mathfrak{b}$ of $\Lambda(\mathfrak{a})$ of degree m such that $\mathrm{div}(g)=\mathfrak{a}-\mathfrak{a}'$. Then $\mathrm{div}(f\cdot g)=\mathfrak{m}-\mathfrak{q}-\mathfrak{b}$ and our lemma is proved.

3. THEOREM 4. *There is a constant c_{13} such that mX is ample for all $(V,X)\in\Sigma$, whenever $m>c_{13}$.*

Proof. In order to prove our theorem, it is enough to prove that m_0X is ample for all (V,X) in Σ, where m_0 is a constant, depending only on c_1, c_2, c_3. In fact, the sum of two ample divisors is also ample; moreover, if Y is an ample divisor on a non-singular projective variety W and $B=B'-B''$, $B'>0$, $B''>0$, is a W-divisor, then, whenever

$$d\geqq \deg(B')\cdot(\deg(W)-2)+\deg(V)+\deg(B'').$$

$dY+B$ is ample (cf. [18], Chap. IX, Cor., Th. 13). Our theorem follows from these two facts and from Lemma 1 as an easy exercise.

Let T be a member of $\Lambda(2c_5X)$ and Z an irreducible member of $\Lambda(c_{12}T)$ (cf. Cor. 3 of Lemma 4). Then we have

(a) $[rZ\cdot(2tZ+iT)]>2p_a(rZ)$ *for* $0<r\leqq t$ *and for all* $i\geqq 0$. In fact, this follows from (II), Lemma 1 and from the formula $2p_a(D)-2=[D\cdot(D+K(V))]$.

(b) *There is a constant d, depending only on c_1, c_2, c_3, such that $\Lambda(mdZ)$ has no base point for $m\geqq 1$.* In fact, there is an integer a such that $3\leqq a\leqq c_{11}+3$ and that $s(aZ-Z)=s(aZ)$ (cf. Cor. 2 of Lemma 4). Set $r=t=1$ in (a). Then we see that $\mathrm{Tr}_Z\,\Lambda(aZ)$ is complete by Corollary of Lemma 3; moreover, it is of degree $>2p_a(Z)$ since $\mathfrak{L}(aZ)/\mathfrak{L}(aZ-Z)$ is isomorphic to $\mathfrak{L}((aZ)_\wedge Z)$. When that is so, $\Lambda(aZ)$ has no base point by Lemma 8.

Now let E be an irreducible member of $\Lambda(dZ)$ (cf. Cor. 3 of Lemma 4). By Corollary 2 of Lemma 4, there is an integer b such that $3\leqq b\leqq c_{11}+3$ and that $s(bE)=s(bE-E)$. $\mathrm{Tr}_E\,\Lambda(E)$ has no base point by (b). By (a) and Corollary of Lemma 3, $\mathrm{Tr}_E\,\Lambda(bE)$ is complete and is of degree $>2p_a(E)$. Moreover, $\mathrm{Tr}_E\,\Lambda(E)$ and $\mathrm{Tr}_E\,\Lambda(bE)$ satisfy the conditions of Lemma 6 by (a) and (b). Therefore, the minimum sum of them is complete, which implies that $\mathrm{Tr}_E\,\Lambda(bE+E)$ is complete and $s(bE+E)=s(bE)$

(cf. Cor. of Lemma 3). Repeating this, we see that $s(c_{11}E+3E+mE)$ is a constant for all $m\geqq 1$. By (I) and (b), a non-degenerate map h_m of V into a projective space, determined by $\Lambda(m(c_{11}+3)E)$, is a morphism and has no fundamental curve on V. Then it is a projective embedding for large m, which is an easy consequence of the possibility of projective normalization in an algebraic extension of the function field (cf. [7], Chap. IV, Prop. 8). Then h_m is already a projective embedding if

$$m>[c_{11}+3)E\cdot K(V)]+r\geqq 2(c_{11}+3)dc_{12}c_5c_4+4$$

by Lemma 5 and Lemma 7. Our theorem is thereby proved.

Chapter IV.

So far, we have restricted our technique to the use of irreducible curves on the surface. Since the generalized Riemann-Roch theorem is available for such curves, it was easy to see, for instance, whether some linear systems on such curves are free from base points, etc. On the other hand, if we generalize a criterion of ampleness to reducible curves by means of the theory of schemes, we can simplify the latter part of Chapter III to some extent. Let V be a projective variety. Denote by $\mathfrak{O}_V$ the sheaf of local rings on V. If D is a Cartier divisor on V, we mean by the associated subscheme $\mathfrak{D}$ the subscheme of the scheme V (a) whose underlying space is the support of D and (b) whose sheaf $\mathfrak{O}_{\mathfrak{D}}$ is defined at a point x of D to be $\mathfrak{O}_{x,V}/(f)$ for any local equation f of the divisor D, where $\mathfrak{O}_{x,V}$ is the local ring of V at x. We further denote by $\mathfrak{m}_{x,V}$ the maximal ideal of $\mathfrak{O}_{x,V}$. In the following proposition, we discuss a criterion of ampleness on a positive 1-cycle on V. The first half of the proposition has been settled essentially in Lemma 4, and we give only a brief account of the proof for it in the sheaf theoretic terminology.

PROPOSITION. *Let $T=\sum_1^t a_iY_i$ be a positive divisor on a non-singular projective surface W. Let τ be the subscheme of the scheme W, associated to T, and $\mathfrak{M}$ an invertible sheaf on τ. Let d_i be the degree of the Cartier divisor class on Y_i defined by $\mathfrak{M}\otimes\mathfrak{O}_{Y_i}$. If $d_i>[(K(W)+T')\cdot Y_i]$ for all divisors T' such that $0\prec T'\prec T$, and for all i such that $1\leqq i\leqq t$, it follows that $H^1(\mathfrak{M})=0$. Moreover, if $d_i>[(K(W)+T')\cdot Y_i]+2\max_{x\in Y_i}$ (multiplicity of x on Y_i), then $\mathfrak{M}$ is ample on τ.*

Proof. We have $\mathfrak{O}_W/\mathfrak{L}(-T)\cong\mathfrak{O}_\tau$, $\mathfrak{O}_W/\mathfrak{L}(-T')\cong\mathfrak{O}_{\tau'}$ from the definitions of $\mathfrak{O}_\tau$, $\mathfrak{O}_{\tau'}$, where τ' is the subscheme of W, associated to T'. Moreover, we have $\mathfrak{L}(-T')/\mathfrak{L}(-T)=\mathfrak{L}(-T')\otimes\mathfrak{O}_{Y_1}$ when we use the same notations as in the proof of Lemma 4. Hence, we have the exact sequence:

$$0\to\mathfrak{L}(-T')\otimes\mathfrak{O}_{Y_1}\to\mathfrak{O}_T\to\mathfrak{O}_{T'}\to 0.$$

Tensoring the above with $\mathfrak{M}$, we get

$$H^1(\mathfrak{M}\otimes\mathfrak{L}(-T')\otimes\mathfrak{O}_{Y_1})\to H^1(\mathfrak{M})\to H^1(\mathfrak{M}\otimes\mathfrak{O}_{T'})\to 0$$

instead of

$$\begin{aligned}H^1(\mathfrak{L}(A-T')/\mathfrak{L}(A-T))&\to H^1(\mathfrak{L}(A)/\mathfrak{L}(A-T))\to\\ &\to H^1(\mathfrak{L}(A)/\mathfrak{L}(A-T'))\to 0\end{aligned}$$

in the proof of Lemma 4. Computing the degree of the Cartier divisor class on Y_1 determined by $\mathfrak{M}\otimes\mathfrak{L}(-T')\otimes\mathfrak{O}_{Y_1}$, we get $H^1(\mathfrak{M})=0$ as in the proof of Lemma 4.

For an invertible sheaf $\mathfrak{L}$ on a complete algebraic scheme W to be ample, it is necessary and sufficient that:

(i) *the sections of $\mathfrak{L}$ separate points,*

(*) (ii) *for any point x on W, if we identify the stalk $\mathfrak{L}_x$ with $\mathfrak{O}_{x,W}$, the sections of $\mathfrak{L}$ which are zero at x span $\mathfrak{m}_{x,W}/\mathfrak{m}_{x,W}^2$.*

For every point x in the support $|T|$ of T, let D_x be a positive W-divisor such that $i(D_x\cdot Y_i,x;V)$ gives the multiplicity of x on Y_i for $1\leqq i\leqq t$. Moreover, for every pair of distinct points x, y in $|T|$, define an invertible sheaf $\mathfrak{M}_{x,y}$ on τ as follows:

$$\begin{aligned}\mathfrak{M}_{x,y}&=\mathfrak{M}\otimes\mathfrak{L}(-D_x)\ \text{ in } |T|-y-(|T|\cap|D_x|-x),\\ &=\mathfrak{M}\otimes\mathfrak{L}(-D_y)\ \text{ in } |T|-x-(|T|\cap|D_y|-y),\\ &=\mathfrak{M}\ \text{ elsewhere.}\end{aligned}$$

Similarly, for every x in $|T|$, define $\mathfrak{M}_{x,x}$ as follows:

$$\begin{aligned}\mathfrak{M}_{x,x}&=\mathfrak{M}\otimes\mathfrak{L}(-2D_x)\ \text{ in } |T|-(|T|\cap|D_x|-x),\\ &=\mathfrak{M}\ \text{ elsewhere.}\end{aligned}$$

The first half of the proposition implies $H^1(\mathfrak{M}_{x,y})=0$ for all the pairs (x,y), because the degree of the restriction of $\mathfrak{M}_{x,y}$ to Y_i is at least $d_i-2\max_{z\in Y_i}$ (multiplicity of z on Y_i). Therefore, we have the exact sequence

$$0\to H^0(\mathfrak{M}_{x,y})\to H^0(\mathfrak{M})\to H^0(\mathfrak{M}/\mathfrak{M}_{x,y})\to H^1(\mathfrak{M}_{x,y})=0$$

for all pairs (x,y). Now suppose $x \neq y$. Then $H^0(\mathfrak{M}/\mathfrak{M}_{x,y})$, whose support is the union of two points x and y, contains a section which is 0 at x and a unit at y. Hence $H^0(\mathfrak{M})$ contains a section with the same property. Then the sections of $\mathfrak{M}$ separates points. Let $f=0$ be a local equation of D_x at x, f being an element of $\mathfrak{O}_{x,W}$. Then f induces an element f' in $\mathfrak{O}_{x,\tau}$, and we have $(\mathfrak{M}/\mathfrak{M}_{x,x})_x = \mathfrak{O}_{x,\tau}/f'^2$. But this maps surjectively to the ring $\mathfrak{O}_{x,\tau}/\mathfrak{m}_{x,\tau}^2$. Hence, if we identify $\mathfrak{M}_x$ to $\mathfrak{O}_{x,\tau}$, the sections of $\mathfrak{M}$ which are zero at x span $\mathfrak{m}_{x,\tau}/\mathfrak{m}_{x,\tau}^2$.

Using this proposition, we recover Corollary 1 and Corollary 2 of Lemma 4. Moreover, Corollary 1 can be expressed as $H^1(\mathfrak{L}(mX-U)\otimes\mathfrak{O}_\tau) = 0$ for $m \geqq c_{10}$, where τ is the subscheme of V, corresponding to a member T of $\Lambda(2c_5X)$. Furthermore, we see that $\mathfrak{L}(mX)\otimes\mathfrak{O}_\tau$ is ample on τ for $m \geqq c_{10}$ when c_{10} is chosen suitably. Then, by Corollary 2 of Lemma 4, one can find an integer r such that $s(c_{10}T+rT) = s(c_{10}T+rT+T)$ and that $0 \leqq r \leqq c_{11}$. Set $m = 2c_5(c_{10}+r+1)$. Since $H^1(\mathfrak{L}(mX)\otimes\mathfrak{O}_\tau) = (0)$ and $s(mX) = s(mX-T)$, it follows that the restriction map $H^0(\mathfrak{L}(mX)) \to H^0(\mathfrak{L}(mX)\otimes\mathfrak{O}_\tau)$ is surjective. Moreover, $\mathfrak{L}(mX)\otimes\mathfrak{O}_\tau$ is ample on τ. To prove that $\mathfrak{L}(mX)$ is ample, we again use the criterion (*) cited above. Let x and y be two given points on V. Let T_1 and T_2 be members of $\Lambda(c_5X)$ such that T_1 goes through x and that T_2 goes through y (cf. Lemma 1). When we set $T = T_1 + T_2$, x and y are in the support of τ. Since $\mathfrak{L}(mX)\otimes\mathfrak{O}_\tau$ is ample, there is a section of this sheaf which is zero at x and not zero at y if $x \neq y$. Lifting this to a section of $\mathfrak{L}(mX)$, the same is true of $\mathfrak{L}(mX)$. If $x=y$, set $T=2T_1$. The restriction of functions from V to τ induces an isomorphism between $\mathfrak{m}_{x,V}/\mathfrak{m}_{x,V}^2$ and $\mathfrak{m}_{x,\tau}/\mathfrak{m}_{x,\tau}^2$. When we identify the stalk of $\mathfrak{L}(mX)\otimes\mathfrak{O}_\tau$ at x with $\mathfrak{O}_{x,\tau}$, the sections of $\mathfrak{L}(mX)\otimes\mathfrak{O}_\tau$ which vanish at x span $\mathfrak{m}_{x,\tau}/\mathfrak{m}_{x,\tau}^2$. Lifting these sections, we see that the same is true of $\mathfrak{L}(mX)$ and $\mathfrak{m}_{x,V}/\mathfrak{m}_{x,V}^2$. Hence $\mathfrak{L}(mX)$ is ample for $m = 2c_5(c_1 + r + 1)$, where r is a certain integer such that $0 \leqq r \leqq c_{11}$. Since the sum of two ample divisors is ample, we see that $\mathfrak{L}(dX)$ is ample for a suitable d, which depends only on c_1, c_2, c_3. Then we get our Theorem 4 again.

REFERENCES.

[1] S. Abhyankar, "On the valuations centered in a local domain," *American Journal of Mathematics*, vol. 78 (1956), pp. 321-348.

[2] A. Grothendieck, "Éléments de géometrie algébriques, I, II, III," *Publications Mathematiques Institut des Hautes Études Scientifiques*, Paris.

[3] H. Hironaka, "A note on algebraic geometry over ground rings," *Illinois Journal of Mathematics*, vol. 2 (1958), pp. 355-366.

[4] ———, "On the arithmetic genera and the effective genera of algebraic curves," *Memoirs of the College of Science, University of Kyoto*, vol. 30 (1957), pp. 177-195.

[5] K. Kodaira and D. C. Spencer, "On deformations of complex analytic structures, I, II," *Annals of Mathematics*, vol. 67 (1958), pp. 328-466.

[6] M. Kuranishi, "On the locally complete families of complex analytic structures," *Annals of Mathematics*, vol. 75 (1962), pp. 536-577.

[7] S. Lang, *Introduction to Algebraic Geometry*, Interscience Tracts, No. 5, 1958.

[8] T. Matsusaka, "Polarized varieties, fields of moduli and generalized Kummer varieties of polarized Abelian varieties," *American Journal of Mathematics*, vol. 80 (1958), pp. 45-82.

[9] ———, "Q-varieties," *to appear.*

[10] D. Mumford, "Geometric invariant theory," *to appear.*

[11] ———, "Further pathologies in algebraic geometry," *American Journal of Mathematics*, vol. 84 (1962), pp. 642-647.

[12] Y. Nakai, "Non-degenerate divisors on an algebraic surface," *Journal of Science of the Hiroshima University*, vol. 24 (1960), pp. 1-6.

[13] M. Rosenlicht, "Equivalence relations on algebraic curves," *Annals of Mathematics*, vol. 56 (1952), pp. 169-191.

[14] ———, "Generalized Jacobian varieties," *Annals of Mathematics*, vol. 59 (1954), pp. 505-530.

[15] J.-P. Serre, "Faisceaux algébriques coherents," *Annals of Mathematics*, vol. 61 (1955), pp. 197-278.

[16] ———, *Groupes algébriques et corps de classes*, Actualités Scientifiques et Industrielles, No. 1264, Paris.

[17] G. Shimura, "Reduction of algebraic varieties with respect to a discrete valuation of the basic field," *American Journal of Mathematics*, vol. 77 (1955), pp. 134-176.

[18] A. Weil, *Foundations of algebraic geometry*, revised edition, American Mathematical Society Colloquium Publications, No. 29 (1960).

[19] ———, *Variétiés Abéliennes et Courbes algébriques*, Actualités Scientifiques et Industrielles, No. 1064, Paris.

[20] ———, "On the theory of complex multiplications," *Proceedings, International Symposium on Algebraic Number Theory*, Nikko, Japan, 1955.

[21] O. Zariski, "Complete linear systems on normal varieties and a generalization of a lemma of Enriques-Severi," *Annals of Mathematics*, vol. 55 (1952), pp. 552-592.

[22] ———, "Scientific report on the second summer institute, III," *Bulletin of the American Mathematical Society*, vol. 62 (1956), pp. 117-141.

ERRATA.

Correction to "Two Fundamental Theorems on Deformations of Polarized Varieties," by T. Matsusaka and D. Mumford, *Am. J. Math.*, vol. 86 (1964), pp. 668-684.

The following one sentence is missing towards the end of the statement of the Lemma 6:

"Assume further that Λ' is contained in the complete linear system $\Lambda(\mathfrak{n}')$ determined by $\mathfrak{n}'$ and $\Lambda = \Lambda(\mathfrak{a})$ where $\mathfrak{a}$ is a positive C-divisor."

BRANDEIS UNIVERSITY AND HARVARD UNIVERSITY.

A REMARK ON MORDELL'S CONJECTURE.

By David Mumford.*

It is somewhat surprising that the systematic evaluation of the heights of rational points on a curve and on its jacobian variety and particularly of their relation to each other should yield any new information. Nonetheless this appears to be the case and the result is described in this article. Although the main theorem is not even a special case of the very fascinating conjecture of Mordell, still it is an estimate that already reveals that rational points on curves of genus at least 2 are much harder to come by than on curves of genus 0 or 1. It is a quantitative limitation on the heights of such points which is well-known to be false in the case of genus 0 or 1. Incidentally, there is a good explanation why an estimate of this type can be obtained so cheaply, whereas Mordell's conjecture itself could not: namely, results obtained by our methods will more or less automatically apply to the analogous "function field" case [where the ground field is a function field in one variable over a finite field, rather than an algebraic number field]. And in this case, unless further restrictions are imposed, there are curves of any genus with an infinite number of rational points whose heights increase exactly at the rate which we will find.

Let k be an algebraic number field of finite degree over $\boldsymbol{Q}$. Let C be a non-singular projective curve over k of genus g at least 2. Mordell's conjecture asserts that the set of k-rational points on C is finite. Now suppose that a projective embedding of C is fixed, allowing us to talk of the heights, $ht(x)$, of k-rational points of x. Then my result is this:

Theorem. *There are real constants a and b, $a > 0$, such that if the countable set of k-rational points of C is ordered by increasing height—call the points $x_1, x_2, \cdots$—then*

$$ht(x_i) \geqq e^{ai+b}.$$

Because of the well-known properties of heights, this result is not affected by changing the projective embedding of C. An example of the theorem is given by Fermat's curve:

Received February 25, 1965.

* This research was partially supported by the AMS 1964 Summer Institute in Algebraic Geometry and NSF-GP3512.

COROLLARY. *Let $(\alpha_i, \beta_i, \gamma_i)$ be an infinite set of distinct positive integral solutions of the equation*

$$X^n + Y^n = Z^n$$

such that α_i, β_i, γ_i have no common factors and such that $\{\gamma_i\}$ is an increasing sequence. Assume $n \geqq 4$. There are real constants a and b, $a > 0$, such that

$$\gamma_i \geqq e^{(e^{ai+b})}.$$

A final word: that the proof of the theorem appears in as natural and simple a form as it does is due to the collaboration of John Tate; that it appears in print, needless to say, is not.

1. The theory of heights. We fix an algebraic number field k, of finite degree over $\boldsymbol{Q}$. The main result of the "classical" Theory of Weil (cf. [1] and [4]) is the construction of a set of functions as follows:
Given: a scheme X, projective over k, and an element $\delta \in Pic(X)$.
Constuct: a real-valued function on the set of k-rational points X_k, written

1 $$h_\delta(x),\ x \in V_k$$

In fact, h_δ is not constructed precisely, but only the class of all functions, differing from one member of this class by a *bounded* function is constructed. This construction has the following properties (where $O(x)$ denotes a bounded function of x):

a) If $f\colon X \to Y$ is a k-morphism of schemes X and Y as above, and if $\delta \in Pic(Y)$, then

$$h_\delta(f(x)) = h_{f^*(\delta)}(x) + O(x)$$

b) If $\delta_1, \delta_2 \in Pic(X)$, for X as above, then

$$h_{\delta_1+\delta_2}(x) = h_{\delta_1}(x) + h_{\delta_2}(x) + O(x)$$

c) If D is an effective Cartier divisor on the projective scheme X, and if D defines the element $\delta \in Pic(X)$, then there is a real constant K such that

$$h_\delta(x) \geqq K, \text{ all } x \in X - \text{Support } (D).$$

d) If $\delta \in Pic(X)$ is ample, then for all constants K, the set of points $x \in X_k$ such that $h_\delta(x) \leqq K$ is finite.

The lack of a really definite height function is one of the most awkward aspects of this theory. In case X is assumed to be an abelian variety, this defest has been remedied by Néron and Tate (cf. [2], [3], [4½]). The simplest way to state their result is this:

THEOREM. *Let X be an abelian variety, and let $\delta \in Pic(X)$. Then the class of functions h_δ on X contains a "quadratic" function on X, i.e., a function f satisfying the identity:*

$$f(x+y+z)-f(x+y)-f(x+z)-f(y+z)$$
$$+f(x)+f(y)+f(z)-f(0)=0.$$

One checks immediately that a real-valued bounded quadratic function is constant. Therefore, if we put the two requirements on h_δ that (1) it is quadratic, and (2) it is 0 at the identity point e, then we obtain a completely well-defined height function. Moreover, we get the important Corollary:

COROLLARY. 1) *If X is an abelian variety, and $\delta_1, \delta_2 \in Pic(X)$, then the normalized height functions on X satisfy:*

$$h_{\delta_1+\delta_2}(x)=h_{\delta_1}(x)+h_{\delta_2}(x), \text{ all } x \in X_k.$$

2) *If $f\colon X \to Y$ is any morphism of abelian varieties, and $\delta \in Pic(Y)$, then*

$$h_{f^*\delta}(x)=h_\delta(f(x))-h_\delta(f(e)),$$

all $x \in X_k$. In particular, if f is a homomorphism (i.e., takes the identity to the identity), then

$$h_{f^*\delta}(x)=h_\delta(f(x)).$$

2. The set-up derived from a curve. We shall assume given a non-singular projective curve C, over k, with genus $g \geqq 1$. The purpose of this section is to give a thorough account of the auxiliary varieties associated to C, the canonical divisor classes that they carry, and their universal properties. For the sake of simplicity, we also assume that a base point $x_0 \in C_k$ has been chosen once and for all; and that all other schemes X occurring in the discussion have base points p_X. (The base points on abelian varieties will be assumed to be their identity points). - A general concept which is central to the discussion is the following:

Definition. Let X and Y be connected algebraic schemes over k. A *divisorial correspondence* on $X \times Y$ is an element $\delta \in Pic(X \times Y)$ which is 0
2 restricted to either of the subschemes $X \times \{p_Y\}$ of $\{p_X\} \times Y$.

First of all, let J be the connected component of the identity of the

Picard scheme of C: i.e., the so-called "Jacobian variety" of C. It is an abelian variety of dimension g. Moreover, J is characterized by the existence of a canonical divisorial correspondence

$$\delta_1 \in Pic(C \times J)$$

which has the universal mapping property (cf. [5] and [6]):

$$(*) \qquad \begin{cases} \text{For all connected algebraic schemes } X, \text{ and all} \\ \text{divisorial correspondences } \eta \text{ on } C \times X, \text{ there is} \\ \text{a unique morphism } f\colon X \to J \text{ such that} \\ \qquad (1_C \times f)^*(\delta_1) = \eta. \end{cases}$$

Secondly, on the non-singular surface $C \times C$ the Weil divisor

$$\Delta - C \times \{x_0\} - \{x_0\} \times C$$

defines an element $\Delta^* \in Pic(C \times C)$ which is clearly a divisorial correspondence. By the $UMP(*)$, there is a unique morphism

$$\phi\colon C \to J$$

such that $\Delta^* = (1_C \times \phi)^*(\delta_1)$.

Thirdly, let $\hat{J}$ be the connected component of the identity of the Picard
3 scheme of J: i.e., the dual abelian variety. J is characterized by the existence of a canonical divisorial correspondence

$$\delta_2 \in Pic(J \times \hat{J})$$

which has the universal mapping property:

$$(**) \qquad \begin{cases} \text{For all connected algebraic schemes } X, \text{ and all} \\ \text{divisorial correspondences } \eta \text{ on } J \times X, \text{ there is} \\ \text{a unique morphism } f\colon X \to \hat{J} \text{ such that} \\ \qquad (1_J \times f)^*(\delta_2) = \eta. \end{cases}$$

Fourthly, the morphism ϕ dualizes to a morphism $\hat{\phi}\colon \hat{J} \to J$. Namely,
4 apply the Universal mapping property (*) with $X = J$, $\eta = (\phi \times 1_{\hat{J}})^*(\delta_2)$. This means that we get a diagram:

$$(***) \qquad \begin{array}{ccc} C \times \hat{J} & \xrightarrow{\phi \times 1_{\hat{J}}} & J \times \hat{J} \\ \big\downarrow {\scriptstyle 1_C \times \hat{\phi}} & & \\ C \times J & & \end{array}$$

such that δ_1 and δ_2 induce the same correspondence on $C \times \hat{J}$.

Fifthly, recall the general construction by which divisor classes η on abelian varieties X define homomorphisms from X to its dual $\hat{X}$. There are three maps from $X \times X$ to X—the group law μ and the two projections p_1 and p_2. Then one checks that for any $\eta \in Pic(X)$, the divisor class

$$\mu^*(\eta) - p_1^*(\eta) - p_2^*(\eta)$$

is a divisorial correspondence on $X \times X$. Therefore, by definition of $\hat{X}$, there is a unique morphism $f\colon X \to \hat{X}$ such that

$$\mu^*\eta - p_1^*\eta - p_2^*\eta = (1_X \times f)^* \begin{bmatrix} \text{canonical class} \\ \text{on } X \times \hat{X} \end{bmatrix}.$$

We will denote f by $\Lambda(\eta)$. Recall that Λ is itself a homomorphism: $\Lambda(\eta_1 \pm \eta_2) = \Lambda(\eta_1) \pm \Lambda(\eta_2)$. In terms of this definition, the central result concerning jacobians is the following (due to Weil [7]).

THEOREM. *$\exists$ an ample divisor Θ on J such that*

$$\hat{\phi} = -\Lambda(\Theta)^{-1}.$$

In fact, recall that Θ is nothing but the sum of the subset $\phi(C)$ in J with itself (with respect to the group law in J) $(g-1)$ times. For reference we write the meaning of this Theorem out as follows:

$$\begin{cases} \psi = -\hat{\phi}^{-1} \\ \text{class of } \underbrace{\mu^*\Theta - p_1^*\Theta - p_2^*\Theta}_{\text{call this } \theta} = (1_J \times \psi)^*(\delta_2). \end{cases}$$

The net result of all this is the following: suppose we identify J with $\hat{J}$ via the isomorphism ψ, or $\Lambda(\Theta)$. Then we have defined the canonical divisor classes:

On $C \times C$: Δ^*

On $C \times J$: δ_1

On $J \times J$: $\theta =$ class of $\mu^*\Theta - p_1^*\Theta - p_2^*\Theta$

$= \delta_2$ via our identifications

These are related by the equations

(a) $\Delta^* = (1_C \times \phi)^*(\delta_1)$

(b) $\delta_1 = -(\phi \times 1_J)^*(\theta)$.

hence

(c) $\Delta^* = -(\phi\times\phi)^*(\theta)$.

Proof. (a) has been pointed out before, and (c) follows from (a) and (b). As for (b), first use the fact that $\Lambda(-\Theta) = -\Lambda(\Theta) = -\psi$. Therefore

$$-\theta = (1_J\times(-\psi))^*\delta_2 = (1_J\times\hat{\phi}^{-1})^*\delta_2.$$

Hence

$$-(1_J\times\hat{\phi})^*\theta = \delta_2$$

and finally:

$$\begin{aligned}(1_C\times\hat{\phi})^*\delta_1 &= (\phi\times 1_{\hat{J}})^*\delta_2 \qquad \text{(This is (***))}\\ &= -(\phi\times 1_{\hat{J}})^*(1_J\times\hat{\phi})^*\theta\\ &= -(\phi\times\hat{\phi})^*\theta\\ &= -(1_C\times\hat{\phi})^*(\phi\times 1_{\hat{J}})^*\theta.\end{aligned}$$

Since $1_C\times\hat{\phi}$ is an isomorphism, (b) follows. Q. E. D.

3. The basic estimates. Once again, we consider a curve C over a number field k, as above. Now we will use the maps obtained in § 2 to obtain properties of the height functions introduced in § 1. The most important height function is $h_\theta(x,y)$ defined for $x,y\in J_k$.

PROPOSITION 1. *$h_\theta(x,y)$ is a symmetric, bilinear form on $J_k\times J_k$. Moreover it is positive definite on J_k/mod torsion.*

Proof. Let $f_1\colon J\to J\times J$ be the homomorphism mapping x to $x\times e$, and let f_2 map x to $e\times x$. Since θ is a divisorial correspondence, $f_1{}^*\theta = f_2{}^*\theta = 0$. Therefore

$$\begin{aligned}h_\theta(x,e) &= h_\theta(f_1(x)) = h_{f_1^*\theta}(x) = 0,\\ h_\theta(e,x) &= h_\theta(f_2(x)) = h_{f_2^*\theta}(x) = 0.\end{aligned}$$

But this means that h_θ is a quadratic function on the product of two groups which is 0 on both factors alone. It is easy to check that this implies that h_θ is bilinear.

Let $\xi\colon J\times J\to J\times J$ be the morphism mapping $x\times y$ to $y\times x$. Then clearly $\xi^*\theta = \theta$, hence

$$h_\theta(x,y) = h_\theta(\xi(y,x)) = h_{\xi^*\theta}(y,x) = h_\theta(y,x).$$

To evaluate $h_\theta(x,x)$, let $\Delta\colon J\to J\times J$ be the diagonal morphism, and let $\lambda_2\colon J\to J$ be multiplication by 2. Then

$$\begin{aligned} h_\theta(x,x) &= h_\theta(\Delta(x)) \\ &= h_{\Delta^*\theta}(x) \\ &= h_{\Delta^*(\mu^*\Theta - p_1^*\Theta - p_2^*\Theta)}(x) \\ &= h_{\lambda_2^*\Theta}(x) - 2h_\Theta(x). \end{aligned}$$

since $\lambda_2 = \mu \circ \Delta$, $1_J = p_i \circ \Delta$. On the other hand, if D is any divisor on J, let D' be the divisor obtained by reflecting D in the origin. Then $\lambda_2^*(D)$ is in the same divisor class as $3D + D'$. Therefore,

$$\begin{aligned} h_\theta(x,x) &= h_\Theta(x) + h_{\Theta'}(x) \\ &= h_\Theta(x) + h_\Theta(-x) \end{aligned}$$

I claim that if this is not positive, then x must be a torsion point on J. Namely, assume $h_\theta(x,x) \leqq 0$. Then for all integers n,

$$\begin{aligned} h_\Theta(nx) + h_\Theta(-nx) &= h_\theta(nx,nx) \\ &= n^2 h_\theta(x,x) \\ &\leqq 0, \end{aligned}$$

hence either $h_\Theta(nx) \leqq 0$ or $h_\Theta(-nx) \leqq 0$. This means that if x is not a torsion point, there are an infinite number of distinct points x_i such that $h_\Theta(x_i) \leqq 0$. Since Θ is ample, this contradicts property (d) of heights. Q. E. D.

By the Mordell-Weil theorem, J_k is a finitely generated abelian group. In particular

$$X = J_k \otimes \mathbf{R}$$

is a finite-dimensional real vector space. Moreover, h_θ makes it into a Euclidean space: we will abbreviate the norm $h_\theta(x,y)$ to $\langle x,y\rangle$. The inner product $\langle x,y\rangle$ can be used to compute other heights too:

PROPOSITION 2. *Let $\eta \in Pic(C)$ be a divisor class of degree 0. Then there is a unique point $\tilde{\eta} \in J_k$ such that η equals the restriction of δ_1 to $C \times \{\tilde{\eta}\}$, and*

$$\langle \phi x, \tilde{\eta}\rangle = -h_\eta(x) + O(x), \text{ all } x \in C_k.$$

Proof. The first assertion is part of the definition of the jacobian J of C. The second is an immediate consequence of (b), § 2:

$$\begin{aligned} \langle \phi x, \tilde{\eta}\rangle &= h_\theta(\phi x, \tilde{\eta}) \\ &= h_{(\phi \times 1_J)^*\theta}(x, \tilde{\eta}) + O(x) \\ &= -h_{\delta_1}(x, \tilde{\eta}) + O(x) \\ &= -h_\eta(x) + O(x). \end{aligned}$$ Q. E. D.

PROPOSITION 3. $\langle \phi x, \phi y \rangle = - h_{\Delta^*}(x, y) + O(x, y)$.

Proof. This follows from (c), § 2. Q. E. D.

COROLLARY 1. *There is a constant K such that for $x, y \in C_k$, $x \neq y$,*

$$\langle \phi x, \phi y \rangle \leqq h_{x_0}(x) + h_{x_0}(y) + K.$$

Proof. Recall that $\Delta^* = \Delta - (x_0) \times C - C \times (x_0)$. Apply property (c), § 1 of heights to $h_\Delta(x, y)$; note that the divisor $(x_0) \times C$ (resp. $C \times (x_0)$) is of the form $p_1{}^*(x\ )$ (resp. $p_2{}^*(x_0)$); hence $h_{(x_0)\times C}(x, y)$ equals $h_{x_0}(x)$ to within a bounded function and $h_{C\times(x_0)}(x, y)$ equals $h_{x_0}(y)$ to within a bounded function. Q. E. D.

COROLLARY 2. *There is a divisor class $\kappa \in Pic(C)$ of degree 0 such that for $x \in C_k$,*

$$\langle \phi x, \phi x \rangle = 2g h_{x_0}(x) + h_\kappa(x) + O(x).$$

Proof. The self-intersection number (Δ^2) of the diagonal on $C \times C$ is well-known to be $2 - 2g$. Therefore the divisor class on Δ obtained by restricting the class of Δ^* has degree $-2g$. Let

$$f: C \to C \times C$$

be the diagonal map. Then there is a divisor class $\kappa \in Pic(C)$ of degree 0 such that

$$f^*(\Delta^*) = -(2g x_0 + \kappa).$$

Therefore

$$\begin{aligned} \langle \phi x, \phi x \rangle &= - h_{\Delta^*}(f(x)) + O(x) \\ &= - h_{f^*(\Delta^*)}(x) + O(x) \\ &= 2g h_{x_0}(x) + h_\kappa(x) + O(x). \end{aligned}$$

Q. E. D.

Putting Proposition 2 and Corollary 1 and 2 together, we obtain the basic estimate:

There is a constant K, and an element $\tilde{\kappa} \in J_k$ such that if $x, y \in C_k$, $x \neq y$, then

$$\langle \phi x, \phi y \rangle \leqq 1/2g \{ \langle \phi x, \phi x \rangle + \langle \phi x, \tilde{\kappa} \rangle + \langle \phi y, \phi y \rangle + \langle \phi y, \tilde{\kappa} \rangle \} + K.$$

4. A packing argument. From here on, we have only to make some elementary observations about Euclidean geometry. First of all, define a new map:

$$C_k \xrightarrow{\ \psi\ } X$$

via $\psi(x) = \phi(x) + \frac{\bar{\kappa}}{2g-2}$. One checks easily that ψ has the property:

$$\begin{cases} \text{There is a constant } K_2 \text{ such that if } x, y \in C_k,\ x \neq y, \text{ then} \\ \langle \psi x, \psi y \rangle \leqq 1/g \left[\dfrac{\langle \psi x, \psi x \rangle + \langle \psi y, \psi y \rangle}{2} \right] + K_2. \end{cases}$$

Let $\| z \| = \sqrt{\langle z, z \rangle}$, let $f(s) = 1/2(s + 1/s)$, and let

$$\cos(u, v) = \langle u, v \rangle / \| u \| \cdot \| v \|$$

be the cosine of the angle between points $u, v \in X$ in the given norm. Then we can rewrite the above formula as:

$$\cos(\psi x, \psi y) \leqq \frac{1}{g} f\left(\frac{\| \psi x \|}{\| \psi y \|}\right) + \frac{K_2}{\| \psi x \| \cdot \| \psi y \|}$$

Now arrange the countable set of points C_k in a sequence so that

$$\| \psi x_1 \| \leqq \| \psi x_2 \| \leqq \cdots.$$

Note that as $\| \psi x \| \sim \sqrt{2g h_{x_0}(x)}$, (Cor. 2, § 3), it follows that $\| \psi x_i \| \to +\infty$ as $i \to \infty$. The following "packing" lemma is well-known:

Lemma. *There is an integer N such that if $A_1, \cdots, A_N$ are any non-zero elements of X, then for some pair of integers $1 \leqq i, j \leqq N$,*

$$\cos(A_i, A_j) \geqq \tfrac{2}{3}.$$

Corollary. *If $g \geqq 2$ and $\| \psi x_n \| > \sqrt{12 K_2}$, then $\| \psi x_{n+N} \| \geqq \frac{5}{3} \| \psi x_n \|$.*

Proof. If not, whenever $n \leqq i \leqq j \leqq n + N$ then $1 \leqq \| \psi x_j \| / \| \psi x_i \| < \frac{5}{3}$. Hence

$$1 \leqq f\left(\frac{\| \psi x_j \|}{\| \psi x_i \|}\right) < 7/6,$$

and

$$\cos(\psi x_j, \psi x_i) < 7/6g + \frac{K_2}{\| \psi x_i \| \cdot \| \psi x_j \|} < \tfrac{2}{3}.$$

This contradicts the lemma. Q. E. D.

Corollary. *If $g \geqq 2$, then there are real constants a and b, $a > 0$, such that*

$$\| \psi x_n \| \geqq e^{an+b}.$$

It is now easy to argue backwards and show that $\| \phi x_n \|$, and $ht_{x_0}(x_n)$, and finally $ht_\delta(x_n)$—for any $\delta \in Pic(C)$ of positive degree—also increase exponentially. This will be left to the reader.

REFERENCES.

I. On the theory of heights:

[1] A. Weil, "Arithmetic on algebraic varieties," *Annals of Mathematics*, vol. 53 (1951), p. 412.

[2] S. Lang, "Les formes bilinéaires de Néron et Tate," *Séminaire Bourbaki*, Expose 274 (1964).

[3] A. Néron, "Hauteurs sur les variétés abéliennes" (to appear).

[4] S. Lang, *Diophantine Geometry*, Interscience-Wiley, N. Y., 1962.

[4½] J. Manin, "The Tate height of points on an abelian variety, its variants and applications," *Izvestia Academii Nauk*, vol. 28 (1964), p. 1363.

II. On the theory of Picard schemes and abelian varieties:

[5] S. Lang, *Abelian varieties*, Interscience-Wiley, N. Y., 1959.

[6] A. Grothendieck, *Fondéments de la géométrie algébrique*, Collected Bourbaki talks, Paris, 1962.

[7] A. Weil, *Variétés Abéliennes et Courbes Algébriques*, Hermann & Cie, Paris, 1948.

Picard Groups of Moduli Problems

David Mumford

The purpose of this lecture is to describe a single specific calculation which gives a modern formulation of an old fact. However, I want to devote a large part of this lecture to the explanation of the machinery which has been developed to give a new and, I think, enlightening setting to a whole group of old questions.

Severi, for one, raised the question: look at maximal families of (irreducible) space curves—is the parameter space of such families
1 rational [10]? A more intrinsic question is whether the moduli variety for nonsingular curves of genus g is rational; in other words, look at the parameter space of the universal family of nonsingular curves of genus g and ask whether this is rational; † this question

† Actually, there is no such family. But if $g \geq 3$, then almost all such curves admit no automorphisms, and there is a universal family of the automorphism-free nonsingular curves.

may be very difficult. However, it can be approximated by any number of weaker questions: is this space unirational, or is it regular in the sense that its function field does not admit everywhere regular differential forms (cf. [5], Chapter 7, §2)? Or, still weaker, is the Picard variety of its function field trivial (cf. [6], Chapter 6, §1)? One of the principal results of our theory is that the last statement is true in characteristic 0. In the same line, can we determine various cohomology groups of this moduli variety?

Now all these questions, especially the last two, suffer from a certain vagueness because of our uncertainty about

1. Whether to look only at birational invariants of the function field,
2. Or, if we want to look at invariants of a definite model, which model to select (since there is no universal family of nonsingular curves),
3. If we settle for the usual moduli variety (i. e., the *coarse* one, cf. [7]), it has singularities (cf. [9]) and is not compact.

If we want an answer which has some pretense of being a basic fact, or of being more than idle, we certainly need to start with the *correct* variety, that is, the one which is most relevant to the set of all nonsingular curves with whatever structure is contained therein. Now the real clue here, I contend, is that we must not ask for the cohomology or the Picard group simply of a variety; there is a much better object, which is much more intrinsically related to the moduli problem and which possesses equally (a) a function field, (b) a Picard group, and (c) both étale and coherent cohomology theories. The invariants of this object—call it X—are the basic pieces of information.

In the first section, I want to describe the whole class of objects of which our X is an example. These objects, "topologies," were discovered by Grothendieck, and are the basic concept on which his theory of étale cohomology is constructed. In fact, it was chiefly in order to better understand this important concept that I made the calculations described in this paper. In the second section, I want to describe the étale topology proper, and its relation to the Zariski and the classical topology. In the third section, I want to introduce the topologies relevant to the problem of moduli. All this is nothing but definitions, and I hope that they possess enough intrinsic symmetry and interest to make the reader bear with their mounting abstractness. In the fourth section, I try to alleviate this abstractness by giving the full gory details of the topology relevant to the computations described later. In the fifth section, I describe precisely in two different ways the Picard groups associated to the moduli problem. In the last two sections, for $g = 1$, we give two separate computations of this group.

§1. TOPOLOGIES

In the classical definition of a topology, we start with a basic set X, the space, and we are given a collection A of subsets of X, called the open subsets. Suppose we try to eliminate the set X from our description and develop the theory from A alone: then we will have to endow A with extra structure to compensate for the loss of X. First of all, we make A into a category $\mathbf{A}$ by defining:

$$\begin{aligned} \mathrm{Hom}(U, V) &= \text{set with one element } f_{U,V}, \text{ if } U \subset V \\ &= \text{empty set, if } U \not\subset V \\ &\quad (\text{all } U, V \in A). \end{aligned}$$

Notice that the operation of intersecting two open sets U, V can be defined in terms of this category:

1.1.

$U \cap V$ is the product of U and V in $\mathbf{A}$, that is, it fits into a diagram

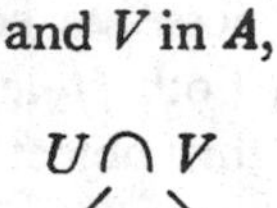

and has the universal mapping property: for all $W \in A$, and for all maps f, g as below, there is a unique h making the diagram commute:

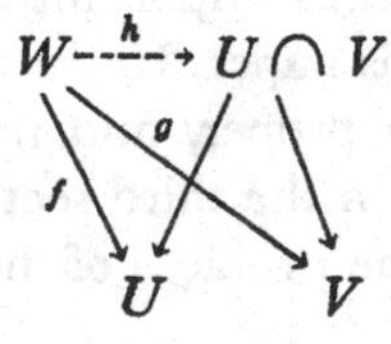

Similarly, arbitrary unions of open sets can be defined as *sums* in the category A:

1.2.
If $U = \bigcup_{i \in I} U_i$, then with respect to the inclusions

$$\begin{array}{l} U_1 \\ \quad \searrow \\ U_2 \rightarrow U \\ \quad \vdots \end{array}$$

U is the categorical sum of the U_i's.

Moreover, the whole space X as an object of A—but not as a set—can be recovered as the final object of A; X is the unique element Y of A such that for all other $U \in A$, there is one and only one map from U to Y.

Now suppose that we want to define the concept of a sheaf $\mathfrak{F}$ (of sets) on X purely in terms of A. This goes as follows: first of all we must have a presheaf. This will be a collection of sets $\mathfrak{F}(U)$, one for each $U \in A$; and a collection of restriction maps, that is, if $U \subset V$, or if, equivalently, there is an element $f_{U,V} \in \mathrm{Hom}(U, V)$, then we must have a map

$$\mathrm{res}_{U,V} : \mathfrak{F}(V) \longrightarrow \mathfrak{F}(U).$$

This is nothing more than a contravariant functor $\mathfrak{F}$ from A to the category (Sets). In order to be a sheaf, it must have an additional property:

1.3.
If U_α is a *covering* of U, that is, each U_α is contained in U and

$$\bigcup_\alpha U_\alpha = U,$$

then an element x of $\mathfrak{F}(U)$ is determined by its restrictions to the subsets U_α; and every set of elements $x_\alpha \in \mathfrak{F}(U_\alpha)$, such that x_α and x_β always have the same restriction to $U_\alpha \cap U_\beta$, come from such an x.

To define sheaves, it is now clear that we may as well start with any category $\mathcal{C}$, instead of A, and call its objects the open sets, provided that:

a. If U, V are open sets, $\mathrm{Hom}(U, V)$ contains at most one element.
b. Finite products and arbitrary sums of objects in $\mathcal{C}$ exist; $\mathcal{C}$ has a final object X.

Also this turns out to be essential:

c. $$V \cap [\bigcup_{i \in I} U_i] = \bigcup_{i \in I} (V \cap U_i)$$

where $\cap$, $\cup$ denote products and sums.

Then sheaves are simply contravariant functors $\mathfrak{F}$ from $\mathcal{C}$ to (Sets) such that, whenever $U = \bigcup_{i \in I} U_i$, the following diagram of sets is exact:

$$\mathfrak{F}(U) \rightarrow \prod_{i \in I} \mathfrak{F}(U_i) \rightrightarrows \prod_{i,j \in I} \mathfrak{F}(U_i \cap U_j).$$

Moreover, the "global sections" $\Gamma(\mathfrak{F})$ of a sheaf $\mathfrak{F}$ are nothing but the elements of the set $\mathfrak{F}(X)$. If we look at sheaves of abelian groups instead of sheaves of Sets, then we can define the higher cohomology groups as well as Γ $(= H^0)$. Namely, we verify in the standard way:

a. The category of abelian sheaves is an abelian category with lots of injective objects.
b. Γ is a left-exact functor from this category to the category (abelian groups).

Hence, as usual, if $\mathfrak{F}$ is an abelian sheaf, put $H^i(\mathfrak{F})$ $(i \geq 0)$ equal to the ith derived functor of Γ (cf. [4], §3.2; [1], Ch. 2, §2).

So far, the theory is essentially trivial: it is nothing more than an exercise in avoiding the explicit mention of points. Grothendieck's fantastic idea is to enlarge the set of possibilities by dropping the assumption that $\mathrm{Hom}(U, V)$ contains at most one element; for example, open sets may even have nontrivial automorphisms. Notice first of all that then it is no longer sufficient to say simply that open sets U_α cover the open set U: it will be necessary to specify particular maps

$$p_\alpha : U_\alpha \rightarrow U$$

with respect to which U is covered by the U_α's. Moreover, it is generally not enough to say that the U_α's cover U only when U is the categorical sum of the U_α's: usually there are other collections of maps $\{p_\alpha\}$ which we will want to call coverings. The concept which emerges from these ideas is the following:

Definition. A "topology" T is a category $\mathcal{C}$ whose objects are called open sets and a set of "coverings." Each covering is a set of morphisms in $\mathcal{C}$, where all the morphisms have the same image; that is, it is a set of the form:

$$\{U_\alpha \xrightarrow{p_\alpha} U\}.$$

The axioms are:

a. Fibred products† $U_1 \times_V U_2$ of objects in $\mathcal{C}$ exist.

b. $\{U' \xrightarrow{p} U\}$ is a covering if p is an isomorphism; if $\{U_\alpha \xrightarrow{p_\alpha} U\}$ is a covering and if, for all α,

$$\{U_{\alpha,\beta} \xrightarrow{q_{\alpha,\beta}} U_\alpha\}$$

is a covering, then the whole collection

$$\{U_{\alpha,\beta} \xrightarrow{p_\alpha \circ q_{\alpha,\beta}} U\}$$

is a covering.

c. To generalize property (c) under 1.3, if $\{U_\alpha \xrightarrow{p_\alpha} U\}$ is a covering, and $V \longrightarrow U$ is any morphism, then

$$\{V \times_U U_\alpha \xrightarrow{q_\alpha} V\}$$

is a covering (q_α being the projection of the fibre product on its first factor).

† In any category, given morphisms $p : X \rightarrow Z$ and $q : Y \rightarrow Z$, a fibre product is a commutative diagram:

$$\begin{array}{ccc} & W & \\ {}^{u}\swarrow & & \searrow^{v} \\ X & & Y \\ {}_{p}\searrow & & \swarrow_{q} \\ & Z & \end{array}$$

such that for all objects W' and morphisms $u' : W' \rightarrow X$, $v' : W' \rightarrow Y$ such that $p \circ u' = q \circ v'$, there is a unique morphism $t : W' \rightarrow W$ such that $u' = u \circ t$, $v' = v \circ t$. This object W is usually written

$$X \times_Z Y$$

and referred to alone as the fibre product of X and Y over Z.

In general, we want to assume that $\mathcal{C}$ possesses a final object X, but this is not necessary. We want to generalize the concept of a sheaf to an arbitrary topology:

Definition. A sheaf (of sets) on T is a contravariant functor $\mathfrak{F}$ from $\mathcal{C}$ to the category (Sets) such that, for all coverings $U_\alpha \xrightarrow{p_\alpha} U$ in T, the following diagram of sets is exact:

$$\mathfrak{F}(U) \to \prod_\alpha \mathfrak{F}(U_\alpha) \rightrightarrows \prod_{\alpha,\beta} \mathfrak{F}(U_\alpha \times_U U_\beta)$$

(the arrows being the usual maps given by the functor $\mathfrak{F}$, contravariant to p_α and to the projections of $U_\alpha \times_U U_\beta$ to U_α and to U_β).

Exactly as before, each sheaf $\mathfrak{F}$ of abelian groups has a group of global sections:

$$\Gamma(\mathfrak{F}) = \mathfrak{F}(X)$$

(X the final object) and hence, by the method of derived functors, higher cohomology groups $H^i(T, \mathfrak{F})$.

A topology in the classical sense gives a topology in an obvious way. To give the theory some content, consider the following example:

Let a group π act freely and discontinuously on a topological space X; that is, for all $x \in X$, there is an open neighborhood U of x such that $U \cap U^\sigma = \emptyset$ for all $\sigma \in \pi$, $\sigma \neq e$. For every set S and action of π on S, we can construct the topological space $\mathfrak{S} = (X \times S)/\pi$ (endowing S with the discrete topology). With two π sets S and T and a π-linear map $f : S \to T$, we obtain a local homeomorphism

$$\begin{array}{ccc} (X \times S)/\pi & \xrightarrow{\tilde{f}} & (X \times T)/\pi \\ \| & & \| \\ \mathfrak{S} & & \mathfrak{T} \end{array}$$

that makes $\mathfrak{S}$ into a covering space of $\mathfrak{T}$. Let the category $\mathcal{C}$ consist in the set of such spaces $\mathfrak{S}$ and such maps $\tilde{f}$; let the coverings consist of maps $\tilde{f}_\alpha : \mathfrak{S}_\alpha \to \mathfrak{T}$ such that, equivalently, $\mathfrak{T} = \bigcup_\alpha \tilde{f}_\alpha(\mathfrak{S}_\alpha)$ or $T = \bigcup_\alpha f_\alpha(S_\alpha)$. The final object in this topology is the topological space X/π, since every other open set has a unique projection

$$\mathfrak{S} = (X \times S)/\pi \to X/\pi$$

in the category. In other words, what has happened is that the open sets are no longer subsets of X/π; they are covering spaces of X/π.

If X is simply connected and connected, then X is just the universal covering space of X/π, and the topology consists in fact in *all* covering spaces $\mathfrak{S}$ of X/π, and all continuous maps $\mathfrak{S} \to \mathfrak{T}$ making the following diagram commute:

$$\begin{array}{ccc} \mathfrak{S} & \xrightarrow{f} & \mathfrak{T} \\ & \searrow \quad \swarrow & \\ & X/\pi & \end{array}$$

On the other hand, this topology is actually independent of X: we may as well "call" the π-sets S themselves the open sets, and call the π-linear maps $f : S \to T$ the morphisms. Then the space X corresponds to the π-set π, (say with left multiplication as the action of π on itself), and the final object X/π corresponds to the π-set $\{0\}$, with trivial action of π. We shall call this topology T_π.

In this form, it is easy to give an explicit description of a sheaf $\mathfrak{F}$ on the topology. Let π, considered only as a set with the *group* π acting on the left, be denoted $\langle\pi\rangle$. Then the right action of π on $\langle\pi\rangle$ makes π into a group of automorphisms of the π-set $\langle\pi\rangle$. But the group of automorphisms of $\langle\pi\rangle$ obviously acts on the set $\mathfrak{F}(\langle\pi\rangle)$ for every sheaf $\mathfrak{F}$. Let $M = \mathfrak{F}(\langle\pi\rangle)$. Then M itself becomes a π-set. I claim that $\mathfrak{F}$ is canonically determined by the π-set M.

a. Let S be a π-set on which π acts transitively. Then there is a π-linear surjection

$$\langle\pi\rangle \xrightarrow{p} S$$

making $\langle\pi\rangle$ into a covering of S. By applying the sheaf axiom to this covering, we check that $\mathfrak{F}(S)$ is isomorphic to the subset M^h of M of elements, left fixed by $h \subset \pi$, where h is the stabilizer of $p(e)$.

b. If S is any π-set, then S is the disjoint union of π-subsets S_α on which π acts transitively. If i_α is the inclusion of S_α in S, apply the sheaf axiom to the covering

$$\{S_\alpha \xrightarrow{i_\alpha} S\}.$$

We check that, via $\mathfrak{F}(i_\alpha)$,

$$\mathfrak{F}(S) \cong \prod_\alpha \mathfrak{F}(S_\alpha).$$

Conversely, given the π-set M, the isomorphisms in (a) and (b) define a sheaf $\mathfrak{F}$: hence to give a sheaf (of sets) $\mathfrak{F}$ in this topology and to give a π-set M are one and the same thing. In particular, a sheaf of abelian groups $\mathfrak{F}$ is the same thing as a π-module M. As the π-set $\{0\}$ is the final object in T_π, we find by means of (a) that the global sections $\Gamma(\mathfrak{F})$ of the sheaf $\mathfrak{F}$ are just the invariant elements M^π of M. Now it is well known that the category of π-modules is an abelian category, and that

$$M \to M^\pi$$

is a left-exact functor on this category. Its derived functors are known as the cohomology groups of π with coefficients in M:

$$H^i(\pi, M)$$

(cf. [8], §10.6). Therefore, we find:

$$H^i(T_\pi, \mathfrak{F}) \cong H^i(\pi, M).$$

One final set of concepts: if T_1 and T_2 are two topologies with final object, a *continuous map* F from T_1 to T_2 consists in a functor from the category of open sets of T_2 to the category of open sets of T_1 such that:

a. It takes the final object to the final object.
b. It takes fibre products in T_2 to fibre products in T_1.
c. It takes coverings in T_2 to coverings in T_1.

For the sake of tradition, if U is an open set in T_2, we let $F^{-1}(U)$ denote the open set in T_1 associated to U by this functor; in other words, requirement (b) means:

$$F^{-1}(U \underset{U}{\times} U_2) \cong (F^{-1}(U_1) \underset{F^{-1}(U)}{\times} F^{-1}(U_2)).$$

If F is a continuous map, then F induces a map F_* from sheaves on T_1 to sheaves on T_2: let $\mathfrak{F}$ be a sheaf on T_1. Define

$$F_*(\mathfrak{F})(U) = \mathfrak{F}(F^{-1}(U))$$

for all open sets U in T_2. This is clearly a sheaf. By standard techniques (cf. [4] and [1], Ch. 2, §4), we find that there is a canonical homomorphism:

$$H^i(T_2, F_*(\mathfrak{F})) \to H^i(T_1, \mathfrak{F}).$$

Moreover, let U be an open set in a topology T. Then "U with its induced topology" is a topology T_U defined as follows:

a. Its open sets are morphisms $V \to U$ in T.
b. Its morphisms are commutative diagrams:

$$\begin{array}{ccc} V_1 & \longrightarrow & V_2 \\ & \searrow \quad \swarrow & \\ & U & \end{array}$$

c. A set of morphisms

$$\begin{array}{ccc} V_\alpha & \longrightarrow & V \\ & \searrow \quad \swarrow & \\ & U & \end{array}$$

is a covering, if the set of morphisms $\{V_\alpha \to V\}$ is a covering in T.

Then there is a canonical continuous "inclusion" map:

$$i : T_U \to T,$$

that is, to the open set V in T, associate the open set $i^{-1}(V)$ which is the projection:

$$p_2 : V \times U \to U.$$

§2. ÉTALE AND CLASSICAL TOPOLOGIES

From now on, we will be talking about schemes. For the sake of simplicity, we will work over an algebraically closed field k, and all schemes will be assumed separated and of finite type over k, *without further mention.*

Definition. Let $f : X \to Y$ be a morphism. Then if, for all closed points $y \in Y$, $f^{-1}(y)$ is a finite set and for all $x \in f^{-1}(y)$, the induced homomorphism

$$f^* : \mathfrak{o}_y \to \mathfrak{o}_x$$

gives rise to an isomorphism of the completions of these rings

$$\hat{f}^* : \hat{\mathfrak{o}}_y \xrightarrow{\sim} \hat{\mathfrak{o}}_x,$$

then f is *étale.*

As an exercise, the reader might prove that this is equivalent to assuming:

a. f is flat, that is, for all $x \in f^{-1}(y)$, $\mathfrak{o}_x$ is a flat $\mathfrak{o}_y$-module,
b. The scheme-theoretic fibre $f^{-1}(y)$ is a reduced finite set, that is, $f^{-1}(y)$ is a finite set, and for all $x \in f^{-1}(y)$, $\mathfrak{m}_x = f^*(\mathfrak{m}_y) \cdot \mathfrak{o}_x$.

Clearly, "étale" is the scheme-theoretic analog of "local homeomorphism" for topological spaces. Now let X be a scheme.

Definition. The étale topology $X_{\text{ét}}$ of X consists of

a. The category whose objects are étale morphisms $p : U \to X$, and whose morphisms are arbitrary X-morphisms; in other words, given $U \xrightarrow{p} X$, $V \xrightarrow{q} X$, then $\operatorname{Hom}(p, q)$ is the set of commutative diagrams

$$\begin{array}{ccc} U & \xrightarrow{f} & V \\ & {\scriptstyle p}\searrow \quad \swarrow{\scriptstyle q} & \\ & X & \end{array}$$

(For simplicity, we shall refer to the objects of this category as schemes U, the étale morphism p to X being understood).
b. The coverings consist in arbitrary sets of morphisms $\{U_\alpha \xrightarrow{p_\alpha} U\}$ provided that

$$U = \bigcup_\alpha p_\alpha(U_\alpha).$$

Let X_{Zar} be the Zariski topology on X: its category consists in the open *subsets* of X and the inclusion maps between them; and a set of inclusion maps $p_\alpha : U_\alpha \subset U$ is said to be a covering if

$$U = \bigcup_\alpha U_\alpha.$$

These two topologies are related by a continuous map

$$\sigma : X_{\text{ét}} \to X_{\text{Zar}}.$$

Namely, if $U \subset X$ is an open subset, the inclusion morphism i of U in X is obviously étale, so that i is an open set in $X_{\text{ét}}$. The reader should check to see that the map from U to i extends to a functor from the category of X_{Zar} to the category of $X_{\text{ét}}$, which takes coverings to coverings and fibre products to fibre products.

If $k = \mathbf{C}$, the field of complex numbers, we can compare $X_{\text{ét}}$ with the classical topology too. The set of closed points of X forms an analytic set $X_{\mathbf{C}}$, and has an underlying topology inherited from the usual topology on $\mathbf{C}$: call this X_{cx}. Unfortunately, there is no continuous map in either direction between $X_{\text{ét}}$ and X_{cx}. However, there is a third topology related to both: let open sets consist in analytic sets U and holomorphic maps

$$f : U \to X_{\mathbf{C}},$$

which are local homeomorphisms—as usual, coverings are just sets of maps

$$\begin{array}{ccc} U_\alpha & \xrightarrow{f_\alpha} & U \\ & \searrow \quad \swarrow & \\ & X_{\mathbf{C}} & \end{array}$$

such that $U = \bigcup_\alpha f_\alpha(U_\alpha)$. Call this topology X^*_{cx}. Then there are continuous maps

$$\begin{array}{ccc} & X^*_{\mathrm{cx}} & \\ {}^{a}\swarrow & & \searrow{}^{b} \\ X_{\mathrm{cx}} & & X_{\text{ét}}, \end{array}$$

since

a. An open set in X_{cx} is an "open" subset $U \subset X_{\mathbf{C}}$; and this defines the inclusion map

$$i : U \to X_{\mathbf{C}},$$

which is a holomorphic local homeomorphism.

b. An open set in $X_{\text{ét}}$ is an étale morphism $f : U \to X$ of a scheme U to the scheme X; and this defines the holomorphic local homeomorphism

$$f_{\mathbf{C}} : U_{\mathbf{C}} \to X_{\mathbf{C}}$$

of the corresponding analytic sets.

On the other hand, although a is not an isomorphism of topologies, it is very nearly one in the following sense:

Definition. Let $f : T_1 \to T_2$ be a continuous map of topologies. f is an *equivalence* of T_1 and T_2 if

a. The functor f^{-1} from the category of open sets of T_2 to that of T_1 is fully faithful,

b. Every open set U in T_1 admits a covering in T_1 of the form $\{f^{-1}(V_\alpha) \xrightarrow{g_\alpha} U\}$, with suitable open sets V_α in T_2,

c. A collection of maps $\{V_\alpha \xrightarrow{g_\alpha} V\}$ in T_2 is a covering, if the collection of maps $\{f^{-1}(V_\alpha) \xrightarrow{f^{-1}(g_\alpha)} f^{-1}(V)\}$ in T_1 is a covering.

We leave it to the reader to check several simple points: a is an equivalence of topologies; if $f : T_1 \to T_2$ is an equivalence of topologies, f_* defines an equivalence between the category of sheaves on T_1 and the category of sheaves on T_2; hence if $\mathfrak{F}$ is a sheaf of abelian groups on T_1, the canonical homomorphism:

$$H^i(T_1, \mathfrak{F}) \xrightarrow{\sim} H^i(T_2, f_*\mathfrak{F}),$$

is an isomorphism. In fact, there is no significant difference between equivalent topologies. For this reason, we often speak of "the continuous map" from X_{cx} to $X_{\text{ét}}$, although strictly speaking this does not exist. Finally, there is a very nice result of M. Artin: let $\mathbf{Z}/n$ denote the sheaf on X^*_{cx}

$$\mathbf{Z}/n(U) = \bigoplus_{\substack{\text{connected components of} \\ U \text{ in complex topology}}} \mathbf{Z}/n;$$

(this is the same as the sheaf associated to the presheaf which simply assigns the group $\mathbf{Z}/n$ to every open set U.) We shall denote $b_*(\mathbf{Z}/n)$ simply by $\mathbf{Z}/n$; since the connected components of a scheme U in its complex and in its Zariski topologies are the same, we have:

$$b_*(\mathbf{Z}/n)(U) = \bigoplus_{\substack{\text{connected components} \\ \text{of } U \text{ in Zariski topology}}} \mathbf{Z}/n.$$

If X is nonsingular, M. Artin has proven that the canonical homomorphism

$$H^i(X_{\text{ét}}, \mathbf{Z}/n) \to H^i(X^*_{\mathrm{cx}}, \mathbf{Z}/n) \xrightarrow{\sim} H^i(X_{\mathrm{cx}}, \mathbf{Z}/n)$$

is an isomorphism. This result assures us that, at least for nonsingular varieties, the étale topology, defined purely in terms of schemes, captures much of the topological information contained in the a priori finer complex topology.

To complete this comparison of the topologies associated to a scheme X, we must mention three other topologies, defined over any k, which are interesting. The idea behind these topologies is to enlarge the category as much as you want, but to keep the coverings relatively limited. In all of them, an open set is an *arbitrary* morphism

$$f : U \longrightarrow X,$$

and a map between two open sets f_1 and f_2 is a commutative diagram:

$$\begin{array}{ccc} U_1 & \longrightarrow & U_2 \\ & {\scriptstyle f_1}\searrow \quad \swarrow{\scriptstyle f_2} & \\ & X & \end{array}$$

The restriction on the coverings involves new classes of morphisms, defined as follows:

Definition. A morphism $f : X \longrightarrow Y$ is *flat* if for all $x \in X$, the local ring $\mathfrak{o}_x$ is a flat module over $\mathfrak{o}_{f(x)}$. Moreover, f is *smooth* if it is flat and if the scheme-theoretic fibres of f are nonsingular varieties (not necessarily connected).

To understand smoothness better, the reader might check that it is equivalent to requiring, for all $x \in X$, that the completion $\hat{\mathfrak{o}}_x$ is isomorphic, as $\hat{\mathfrak{o}}_{f(x)}$-algebra, to

$$\hat{\mathfrak{o}}_{f(x)}[[X_1, \ldots, X_n]]$$

for some n.

For the purposes of §3, it is very important to know that smooth morphisms are also characterized by the following property (cf. [3], exposé 3, Theorem 3.1).

2.1.

Let A be a finite-dimensional commutative local k-algebra, and let $I \subset A$ be an ideal. Let a commutative diagram of solid arrows be given:

$$\begin{array}{ccc} & \mathrm{Spec}\,(A/I) & \\ \swarrow & & \searrow \\ \mathrm{Spec}\,(A) & \dashrightarrow & X \\ \searrow & & \swarrow \\ & Y & \end{array}$$

Then there exists a morphism denoted by the dotted arrow filling in the commutative diagram.

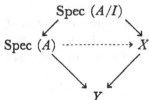

This should be understood as a kind of "homotopy lifting property," so that smooth morphisms are somewhat analogous to fibre spaces.

We can now define three topologies,

$$X^*_{\text{ét}},\ X^*_{\text{smooth}},\ \text{and } X^*_{\text{flat}},$$

by defining a covering as a collection of morphisms $\{U_\alpha \xrightarrow[f_\alpha]{} U\}$ such that $U = \bigcup_\alpha f_\alpha(U_\alpha)$ and f_α is étale; f_α is smooth; f_α is flat, respectively. You can check to see that all our topologies are related by continuous maps as follows:

$$X^*_{\text{flat}} \longrightarrow X^*_{\text{smooth}} \longrightarrow X^*_{\text{ét}} \longrightarrow X_{\text{ét}} \longrightarrow X_{\text{Zar}}.$$

The important fact about these maps is that, in particular, they set up isomorphisms between the cohomology of X^*_{smooth}, $X^*_{\text{ét}}$ and $X_{\text{ét}}$. Therefore, as far as cohomology is concerned, any one of these three topologies is just as good as the others.

§3. MODULI TOPOLOGIES

For this entire section, fix a nonnegative integer g. We first recall the basis of the moduli problem for curves of genus g:

2 *Definition.* A "curve" (over k, of genus g) is a connected, reduced, one-dimensional scheme X, such that

$$\dim H^1(X, \mathfrak{o}_X) = g.$$

Definition. A "family of curves" over S (or, parametrized by S) is a flat, projective morphism of schemes

$$\pi : \mathfrak{X} \to S,$$

whose fibres over all closed points are curves.

Definition. A "morphism" F of one family $\pi_1 : \mathfrak{X}_1 \to S_1$ to another $\pi_2 : \mathfrak{X}_2 \to S_2$ is a diagram of morphisms of schemes:

$$\begin{array}{ccc} \mathfrak{X}_1 & \longrightarrow & \mathfrak{X}_2 \\ {\scriptstyle \pi_1}\downarrow & & \downarrow{\scriptstyle \pi_2} \\ S_1 & \longrightarrow & S_2 \end{array}$$

making $\mathfrak{X}_1$ into the fibre product of S_1 and $\mathfrak{X}_2$ over S_2. F is smooth/flat/étale if the morphism from S_1 to S_2 is smooth/flat/étale.

Definition. Given a family of curves $\pi : \mathfrak{X} \to S$ and a morphism $g : T \to S$, the "induced family of curves" over T is the projection:

$$p_2 : \mathfrak{X} \times_S T \to T.$$

The most natural problem is to seek a universal family of curves, that is, one such that every other one is induced from it by a unique morphism of the parameter spaces. As indicated in the introduction, the usual compromises made in order that this existence problem has a solution are exactly what we want to avoid now. Instead, we want to define a topology; in the ideal case, if a universal family of curves had existed, this would be one of the standard topologies on the universal parameter space. Inasmuch as such a family does not exist (unless stringent conditions on the curves in our families are adopted), this topology is a new object.

Definition (Provisional Form). The moduli topologies $\mathfrak{M}^*_{\text{ét}}$, $\mathfrak{M}^*_{\text{smooth}}$, and $\mathfrak{M}^*_{\text{flat}}$ are as follows:

a. Their open sets are families of curves.
b. Morphisms between open sets are morphisms between families.
c. A collection of such morphisms

$$\begin{array}{ccc} \mathfrak{X}_\alpha & \longrightarrow & \mathfrak{X} \\ {\scriptstyle \pi_\alpha}\downarrow & & \downarrow{\scriptstyle \pi} \\ S_\alpha & \xrightarrow[g_\alpha]{} & S \end{array}$$

is called a covering, if $S = \bigcup_\alpha g_\alpha(S_\alpha)$ and if each g_α is étale, smooth, or flat, respectively.

The first thing to check is that this is a topology and, in particular, that fibre products exist in our category. However, unlike the examples considered in §2, there is not necessarily a final object in our category. Such a final object would be a universal family of curves. A second point is that, if $\pi : \mathfrak{X} \to S$ is any family of curves, the topology induced on the open set π is equivalent to the topology $S^*_{\text{ét}}$, S^*_{smooth}, or S^*_{flat} on S.

A less trivial fact is that absolute products exist in our category. Let $\pi_i : \mathfrak{X}_i \to S_i$ $(i = 1, 2)$ be two families of curves: I shall sketch the construction of the product family. First, over the scheme

$S_1 \times S_2$, we have two induced families of curves,

$$\begin{array}{ccc} \mathfrak{X}_1 \times S_2 & & S_1 \times \mathfrak{X}_2 \\ & \searrow \quad \swarrow & \\ & S_1 \times S_2. & \end{array}$$

Now suppose $\mathfrak{Y} \to T$ is a third family of curves, and that the following morphisms are morphisms of families:

$$\begin{array}{ccccc} & & \mathfrak{Y} & & \\ & \swarrow & \downarrow & \searrow & \\ \mathfrak{X}_1 & & \downarrow & & \mathfrak{X}_2 \\ \downarrow & & T & & \downarrow \\ \downarrow & \swarrow & & \searrow & \downarrow \\ S_1 & & & & S_2 \end{array} \qquad \text{(a)}$$

To have such morphisms is obviously equivalent to having (1) a morphism $T \to S_1 \times S_2$, and (2) isomorphisms *over* T of the three families of curves:

3 $$\begin{array}{ccccc} [\mathfrak{Y}] & & [(\mathfrak{X}_1 \times S_2) \underset{(S_1 \times S_2)}{\times} T] & & [(S_2 \times \mathfrak{X}_2) \underset{(S_1 \times S_2)}{\times} T] \\ & \searrow & \downarrow & \swarrow & \\ & & T & & \end{array} \qquad \text{(b)}$$

But now we must digress for a minute; consider, á la Grothendieck, the following class of universal mapping properties which can be used to define auxiliary schemes. Let S be a scheme, and let

$$\begin{array}{ccc} X_1 & & X_2 \\ & \searrow \quad \swarrow & \\ & S & \end{array}$$

be two morphisms. Look at all pairs (T, Φ) consisting of schemes T over S (i. e., with given morphism to S) and isomorphisms over T:

$$\begin{array}{ccc} X_1 \times_S T & \underset{\Phi}{\xrightarrow{\sim}} & X_2 \times_S T. \\ & \searrow \quad \swarrow & \\ & T & \end{array}$$

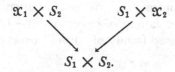

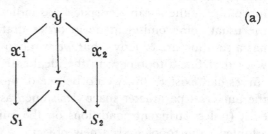

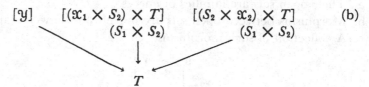

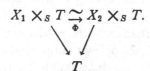

If there is one such pair (T, Φ) such that for every other such pair (T', Φ'), there is a unique morphism (over S)

$$f : T' \to T$$

making the following diagram commute:

$$\begin{array}{ccc} X_1 \times_S T' & \xrightarrow{\Phi'} & X_2 \times_S T' \\ \downarrow{\scriptstyle 1_{X_1} \times f} & & \downarrow{\scriptstyle 1_{X_2} \times f} \\ X_1 \times_S T & \xrightarrow{\Phi} & X_2 \times_S T \end{array}$$

then (T, Φ) is uniquely determined up to canonical isomorphism by this property. And T is denoted

$$\mathrm{Isom}_S(X_1, X_2).$$

Now returning to our families of curves, suppose that the scheme

$$I = \mathrm{Isom}_{S_1 \times S_2}(\mathfrak{X}_1 \times S_2, S_1 \times \mathfrak{X}_2)$$

exists. Then, in situation (a), we get not only a canonical morphism from T to $S_1 \times S_2$ but even a canonical morphism from T to I. Now over I, the two families of curves induced from $\mathfrak{X}_1/S_1$ and $\mathfrak{X}_2/S_2$ are canonically identified: call this family $\mathfrak{X}/I$. Then the situation (a) is obviously equivalent with a morphism from the family Y/T to the family $\mathfrak{X}/I$. In other words, $\mathfrak{X}/I$ is the only possible product of the families $\mathfrak{X}_1/S_1$ and $\mathfrak{X}_2/S_2$. Fortunately, I does exist in our case. This is a consequence of a general result of Grothendieck's (cf. [2], exposé 221), and we will pass over this point completely.

Definition. The product of the families $\pi_i : \mathfrak{X}_i \to S_i$ $(i = 1, 2)$ will be denoted:

$$\pi : (\mathfrak{X}_1, \mathfrak{X}_2) \to \mathrm{Isom}(\pi_1, \pi_2).$$

Since products do exist in the common category of the topologies $\mathfrak{M}^*$, there is no reason not to add a final object M to this category in a perfectly formal way. In order to enlarge the topology, though, we have to define the coverings of the final object M. The point is, however, that if $\pi : \mathfrak{X} \to S$ is part of a covering of M, and if $\pi' : \mathfrak{X}' \to S'$ is any other family of curves whatsoever, then the morphism from the product family

$$(\mathfrak{X}, \mathfrak{X}') \to \mathrm{Isom}(\pi, \pi')$$

to the family π' must be part of a covering of π'. In particular, the projection from $\mathrm{Isom}(\pi, \pi')$ to S' must be étale, smooth, or flat according to the case involved. This leads to:

Definition. A family of curves $\pi : \mathfrak{X} \to S$ is *étale*, *smooth*, or *flat over M* if, for all other families $\pi' : \mathfrak{X}' \to S'$, the projection from $\mathrm{Isom}(\pi, \pi')$ to S' is étale, smooth, or flat.

If we use criterion 2.1 for smoothness given at the end of §2 (p. 46), the condition that π is smooth over M can be reformulated. In fact, after unwinding all the definitions by various universal mapping properties, this condition comes out as follows.

3.1.

Let A be a finite-dimensional commutative local k-algebra, and let $I \subset A$ be an ideal. Suppose we are given a diagram of solid arrows:

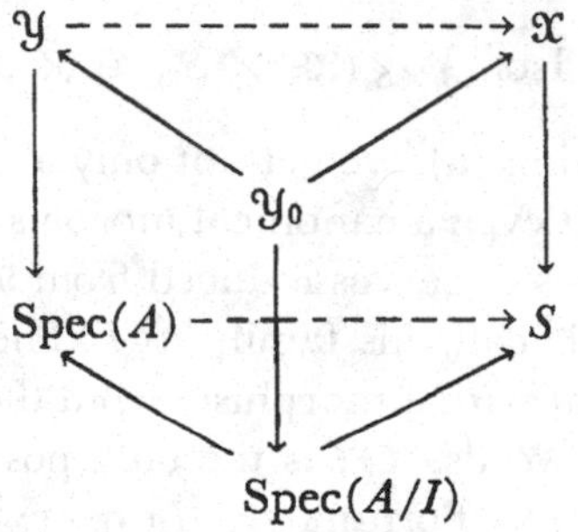

where $\mathfrak{Y}/\mathrm{Spec}(A)$ and $\mathfrak{Y}_0/\mathrm{Spec}(A/I)$ are families of curves, and where the two solid squares are morphisms of families of curves. Then there should be a morphism of families denoted by the dotted arrows filling in the commutative diagram.

Such families have been considered already: compare, especially, the thesis of M. Schlessinger. The important thing is that plenty of such families exist. In particular, if C is any curve over k, we certainly want C to be part of such a family. This can be proven by the method of "linear rigidifications" (cf. [7], §5.2 and §7.2). A fortiori, plenty of families π flat over M exist too.

With families étale over M, it is another matter. In fact, unless we stick to curves C without global vector fields (i. e., everywhere finite derivations), such families do not exist. Let us analyze what it means for $\pi : \mathfrak{X} \to S$ to be étale over M. Let C be any curve over k, and let

$$p : C \to \mathrm{Spec}(k)$$

be the trivial family given by C. Then if π is étale over M, $\mathrm{Isom}(\pi, p)$ must be étale over $\mathrm{Spec}(k)$; that is, $\mathrm{Isom}(\pi, p)$ must consist in a finite set of reduced points. But the points of $\mathrm{Isom}(\pi, p)$ represent isomorphisms of C with the fibres $\pi^{-1}(s)$ of the family π. Therefore, if π is étale over M, the following is satisfied.

3.2.

For all curves C over k, C occurs only a finite number of times in the family $\pi : \mathfrak{X} \to S$, that is C is only isomorphic to a finite number of curves $\pi^{-1}(s)$. Moreover, if C occurs at all in π, the group of automorphisms of C is finite.

Now conversely, the smoothness of π and (3.2) (i.e., 3.1 and 3.2), guarantee the étaleness of π. To see this, let $\pi' : \mathfrak{X}' \to S'$ be any other family of curves. Assume (3.1) and (3.2). Then we know that $\mathrm{Isom}(\pi, \pi')$ is smooth over S'; for it also to be étale over S' means simply that $\mathrm{Isom}(\pi, \pi')$ has only a finite number of closed points over every closed point $s' \in S'$. But let C be the curve $\pi'^{-1}(s')$. There is an isomorphism between the set of closed points of $\mathrm{Isom}(\pi, \pi')$ over s' and the set of isomorphisms of C with the curves $\pi^{-1}(s)$ (s a closed point of S). Therefore, the finiteness of this set follows from 3.2.

Definition. A family $\pi : \mathfrak{X} \to S$ of curves satisfying (3.1) and (3.2) will be called a "modular" family of curves.

Modular families have the following very nice property. Let $\pi_i : \mathfrak{X}_i \to S_i$ be two modular families of curves, and suppose the curve C occurs in π_1 over the point $s_1 \in S_1$ and in π_2 over the point $s_2 \in S_2$, that is,

$$\pi_1^{-1}(s_1) \cong C \cong \pi_2^{-1}(s_2).$$

I claim that S_1 and S_2 are formally isomorphic at the points s_1, s_2; in other words, the complete local rings $\hat{o}_{s_1}$ and $\hat{o}_{s_2}$ are isomorphic. To see this, fix an isomorphism τ of $\pi_1^{-1}(s_1)$ and $\pi_2^{-1}(s_2)$. Then τ determines a point

$$t \in \mathrm{Isom}(\pi_1, \pi_2)$$

lying over s_1 and s_2. But since both π_1 and π_2 are modular, $\mathrm{Isom}(\pi_1, \pi_2)$

is étale over both S_1 and S_2. Therefore, via the projections, we get isomorphisms:

$$\hat{o}_{s_1} \cong \hat{o}_t \cong \hat{o}_{s_2}.$$

More precisely, two modular families containing the same curve are related by an étale correspondence at the point where this curve occurs. As a consequence, for example, either all or none of such families are nonsingular, and they all have the same dimension.

Another very important property of modular families is that any morphism between two such families is necessarily étale. Assume that

$$\begin{array}{ccc} \mathcal{Y} & \longrightarrow & \mathcal{X} \\ {\scriptstyle\varpi}\downarrow & & \downarrow{\scriptstyle\pi} \\ T & \longrightarrow & S \end{array}$$

is a morphism of modular families. This morphism defines an isomorphism of $\mathcal{Y}/T$ and the family induced by $\mathcal{X}/S$ over T: so it defines a morphism of T to Isom(ϖ, π) by the universal mapping property defining Isom. We get the diagram

4

$$\begin{array}{ccccc} & & \mathrm{Isom}(\varpi, \pi) & & \\ & {\scriptstyle\sigma}\nearrow\swarrow{\scriptstyle p_1} & & \searrow{\scriptstyle p_2} & \\ T & & \longrightarrow & & S \end{array}$$

where $p_1 \circ \sigma = 1_T$, and $p_2 \circ \sigma = g$. Since p_1 is étale, a section, such as σ, of p_1 defines an isomorphism of T with an open component

$$I_0 \subset \mathrm{Isom}(\varpi, \pi).$$

Since p_2 is étale, the restriction of p_2 to I_0 is étale; hence g is étale.

We now return to our topologies.

Definition (*Final Form*). The moduli topologies $\mathfrak{M}^*_{\text{smooth}}$, and $\mathfrak{M}^*_{\text{flat}}$ are as follows:

a. Their open sets are families of curves, and a final object M.
b. Their morphisms are morphisms of families of curves, and projections onto the final object M,
c. A collection of such morphisms with image a family $\pi : \mathcal{X} \rightarrow S$ is called a covering exactly as before; a collection of projections of families $\pi_\alpha : \mathcal{X}_\alpha \rightarrow S_\alpha$ onto the final object M is called a covering

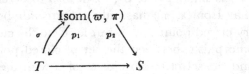

of M if: (1) each family π_α is smooth, or flat over M, and (2) every curve C occurs in one of the families π_α.

It is clear that a topology $\mathfrak{M}^*_{\text{ét}}$ could be defined in the same way, but then the final object M would have no coverings at all. This is because some curves have an infinite group of automorphisms, and hence do not occur in any modular families. One result is that sheaves on this topology would not be sufficiently restricted; the topology is too loosely tied together and would not be useful.

However, suppose that we happen to be interested only in nonsingular curves. This is perhaps short sighted, but never mind. By considering only families of nonsingular curves, we can modify $\mathfrak{M}^*_{\text{smooth}}$, for example, and get a smaller topology. Now if the genus g is at least 2, it is well known that such nonsingular curves have only a finite group of automorphisms. It is to be expected that they belong to modular families, and indeed this is the case. Therefore, we can define an étale moduli topology by looking only at nonsingular curves and modular families. We make the definition in analogy to the scheme topology $X_{\text{ét}}$ rather than $X^*_{\text{ét}}$:

Definition. The moduli topology $\mathfrak{M}_{\text{ét}}$ is as follows:

a. Open sets are modular families of nonsingular curves, and a final object M.
b. Morphisms are morphisms of families of curves, and projections onto the final object M.
c. A collection of morphisms:

$$\begin{array}{ccc} \mathfrak{X}_\alpha & \longrightarrow & \mathfrak{X} \\ \downarrow & & \downarrow \\ S_\alpha & \xrightarrow[g_\alpha]{} & S \end{array}$$

is a covering if $S = \bigcup_\alpha g_\alpha(S_\alpha)$; a collection of projections of families $\pi_\alpha : \mathfrak{X}_\alpha \rightarrow S_\alpha$ onto the final object M is a covering if every curve C occurs in one of the families π_α.

In the rest of this paper, this is the topology we will be interested in; therefore, we will refer to it simply as $\mathfrak{M}$, *rather than* $\mathfrak{M}_{\text{ét}}$.

It is, I think, a very important topology. At a future occasion, I hope to give some deeper results about this topology and compute some of its cohomology groups. For the present, I just want to mention a few nice facts about it:

a. The induced topology on an open set $\pi : \mathfrak{X} \to S$ is equivalent to the étale topology $S_{\text{ét}}$ on S.
b. If $\pi : \mathfrak{X} \to S$ is an open set, S is a nonsingular $3g - 3$-dimensional variety.
c. The so-called "higher level moduli schemes" form (for $n \geq 3$, n prime to the characteristic of k) modular families

$$\pi_n : \mathfrak{X}_n \to S_n$$

each of which is, by itself, a covering over M. Moreover, $\text{Isom}(\pi_n, \pi_n)$ is a finite Galois covering of S_n.

§4. THE ELLIPTIC TOPOLOGY

The last topology that I want to define is the one which we shall study closely in §§6 and 7. It is essentially the topology $\mathfrak{M}_{\text{ét}}$ in the case $g = 1$, except that certain modifications are necessary to extend the definition given in §3 when $g \geq 2$. With this topology everything can be made very explicit, and hopefully the abstractness of all our definitions will be enlivened by this case. This topology is the classical proving ground for all notions of moduli, and, as such, it is found in various forms in hundreds of places.

The difficulty in using the definitions of §3 when $g = 1$ is that a nonsingular curve of genus 1 admits a structure of a group scheme, and therefore it has an infinite group of automorphisms. But by a minor modification, we can make everything go through. The key is to consider not curves, but *pointed* curves, that is, curves with a distinguished base point.

Definition. A nonsingular pointed curve of genus 1 is an "elliptic curve."

Definition. A "family of pointed curves" is a family of curves $\pi : \mathfrak{X} \to S$ with a given section $\varepsilon : S \to \mathfrak{X}$ (i. e., $\pi \circ \varepsilon = 1_s$). If $g = 1$, and the curves are nonsingular, this is called a "family of elliptic curves".

We can define a modular family of elliptic curves just as before.

Since modular families of elliptic curves do exist, it makes sense to state the next definition.

Definition. The topology $\mathfrak{M}$ is as follows:

a. Its open sets are modular families of elliptic curves, and a final object M,
b. Its morphisms are (étale) morphisms of families of elliptic curves, and projections of every open set to M.
c. Coverings of a family $\pi : \mathfrak{X} \longrightarrow S$ are collections of morphisms of families:

$$\begin{array}{ccc} \mathfrak{X}_\alpha & \longrightarrow & \mathfrak{X} \\ {\scriptstyle \pi_\alpha}\downarrow & & \downarrow{\scriptstyle \pi} \\ S_\alpha & \xrightarrow{f_\alpha} & S \end{array}$$

such that $S = \bigcup_\alpha f_\alpha(S_\alpha)$; coverings of M are collections of projections of families π_α to M, provided that every elliptic curve occurs in one of the families π_α.

We want to describe this topology explicitly. First, we shall outline the basic facts about elliptic curves, and then indicate step by step, without complete proofs, how this leads to our final description. We shall assume from now on that the characteristic of k is not 2 or 3, so as to simplify the situation.

The basic fact is that elliptic curves are exactly the curves obtained as double coverings of the line ramified in four distinct points. Therefore, they are the curves C described by equations

$$y^2 = (x - \alpha_1)(x - \alpha_2)(x - \alpha_3)(x - \alpha_4).$$

Since the group of automorphisms acts transitively on the curve C, we can assume that the distinguished point e on C is the point $x = \alpha_4$, $y = 0$. By a projective transformation in the coordinate x, we can put α_4 at ∞, and the equation becomes:

$$y^2 = (x - \alpha_1')(x - \alpha_2')(x - \alpha_3'),$$

where e is now the unique point of this curve over $x = \infty$.

In the language of schemes, the conclusion is that every elliptic curve is isomorphic as *pointed* curve to the subscheme of $\mathbf{P}_2$ defined by homogeneous ideal

$$\mathfrak{A} = (X_2^2X_0 - (X_1 - \alpha_1'X_0)(X_1 - \alpha_2'X_0)(X_1 - \alpha_3'X_0))$$

together with the distinguished point

$$X_0 = 0;\ X_1 = 0;\ X_2 \neq 0.$$

It can be shown that this representation is essentially unique; in fact, the triple $(\alpha_1', \alpha_2', \alpha_3')$ is uniquely determined by the curve up to permutations and to affine substitutions of the form

$$\beta_i' = A\alpha_i' + B.$$

It follows easily from this that elliptic curves are *classified* by the number:

$$j = -64\left[\frac{(\lambda-2)\cdot(2\lambda-1)\cdot(\lambda+1)}{\lambda\cdot(\lambda-1)}\right]^2 \qquad (1)$$

where $\lambda = \dfrac{\alpha_3' - \alpha_1'}{\alpha_2' - \alpha_1'}$.

Why is this? First, λ determines the triple $(\alpha_1', \alpha_2', \alpha_3')$ up to affine transformations. And, if we permute the α_i''s, λ is transformed into one of six numbers:

5 $$\lambda,\ 1-\lambda,\ \frac{1}{\lambda},\ \frac{\lambda-1}{\gamma},\ \frac{1}{1-\lambda},\ \frac{\lambda}{\lambda-1}.$$

Also, the values $\lambda = 0$ and $\lambda = 1$ are excluded, since the three numbers α_1', α_2', and α_3' are distinct. It can be checked that j is unchanged by any of these substitutions in λ, and, conversely, that only λ's related by these substitutions give the same j. The factor -64 arose historically, and turns out to be crucial if we specialize to characteristic 2. In characteristics other than 2, it is obviously harmless!

How about automorphisms of elliptic curves as *pointed* curves? Every elliptic curve C obviously possesses the automorphism

$$x \longrightarrow x$$
$$y \longrightarrow -y$$

corresponding to its being a double covering of the x-line. We will call this the *inversion* ρ of C. A very important fact is that if $\pi : \mathfrak{X} \longrightarrow S$, $\varepsilon : S \longrightarrow \mathfrak{X}$ is any family of elliptic curves, then the inversions of all the fibres piece together to an automorphism $P : \mathfrak{X} \longrightarrow \mathfrak{X}$, of the family π; we will also call this the inversion of π. A related fact is that ρ commutes with any other automorphism α of C. Since $k(x)$ is the field of functions on C fixed by the inversion, such an α will take $k(x)$ into itself; that is, it will be given by a projective transformation in x. Also since α leaves e fixed, it leaves $x = \infty$ fixed; and it must permute the other three branch points $\alpha_1', \alpha_2', \alpha_3'$. It is now an elementary result that such an α occurs only in two cases:

a. $j = 0$; $\lambda = 2, \frac{1}{2}$, or -1; $\alpha_1', \alpha_2', \alpha_3'$ of the form $\beta, \beta + \gamma, \beta + 2\gamma$.
b. $j = 12^3$; $\lambda = -\omega$ or $-\omega^2$ (ω a cube root of 1); $\alpha_1', \alpha_2', \alpha_3'$ of the form $\beta + \gamma, \beta + \omega\gamma, \beta + \omega^2\gamma$.

Now normalizing the first case by choosing $\alpha_1' = -1$, $\alpha_2' = 0$, $\alpha_3' = 1$, we find that C possesses the automorphism σ of order 4:

$$x^\sigma = -x$$
$$y^\sigma = i \cdot y$$

such that σ^2 is the inversion. Normalizing the second case by choosing $\alpha_1' = 1$, $\alpha_2' = \omega$, $\alpha_3' = \omega^2$, we find that C possesses the automorphism τ of order 6:

$$x^\tau = \omega \cdot x$$
$$y^\tau = -y$$

such that τ^3 is the inversion. These are the only automorphisms.

Now, what about modular families. Since only one parameter j is involved, it is natural to expect that modular families are always parametrized by nonsingular curves S. This is true. The most natural thing would be to look for a modular family parametrized by j itself. The following is an example of such a family:

$$y^2 = x^3 + A \cdot (x + 1)$$

where
$$A = \frac{27}{4} \cdot \frac{12^3 - j}{j}.$$

We check that if $j \neq 0, 12^3$, then A is finite and the roots of $x^3 + A(x + 1)$ are all distinct—so we have an elliptic curve. And we can compute its j-invariant in an elementary way, and it is the j we had at the start.

In the language of schemes, let

$$\mathbf{A}_j = \operatorname{Spec} k[j]$$
$$S = \mathbf{A}_j - (0, 12^3),$$

and let $\mathfrak{X}$ be the closed subscheme of $\mathbf{P}_2 \times S$ defined by the vanishing of the section

$$X_2^2 \cdot X_0 - X_1^3 - \frac{27}{4} \cdot \frac{12^3 - j}{j} \cdot (X_1 X_0^2 + X_0^3)$$

of the sheaf $\mathit{o}(3)$. Let ϵ be the morphism

$$S \simeq (0, 0, 1) \times S \subset \mathfrak{X}.$$

Then a rigorous analysis of the infinitesimal deformations of an elliptic curve shows that this is a modular family.

Can we extend this family π to cover the points $j = 0$ and 12^3? For the value $j = 0$, A is infinite; and for $j = 12^3$, our equation degenerates. But even a priori it is clear that there has to be trouble. If π is a modular family, then $\mathrm{Isom}(\pi, \pi)$ must be étale over S. Now for each closed point $t \in S$, the closed points of $\mathrm{Isom}(\pi, \pi)$ over t stand for: (a) closed points $t' \in S$ such that $\pi^{-1}(t)$ and $\pi^{-1}(t')$ are isomorphic, plus (b) isomorphisms of $\pi^{-1}(t)$ and $\pi^{-1}(t')$. If S is an open set in the j-line, $\pi^{-1}(t)$ and $\pi^{-1}(t')$ can never be isomorphic unless $t = t'$. Therefore, the number of points in $\mathrm{Isom}(\pi, \pi)$ over t equals the order of the group of automorphisms of $\pi^{-1}(t)$. For $j \neq 0, 12^3$, this is 2, so $\mathrm{Isom}(\pi, \pi)$ is a double covering of S; and $\mathrm{Isom}(\pi, \pi)$ could not have four or six points over $j = 0$ or $j = 12^3$. The real problem here is that j is not the "right" parameter at $j = 0$ and 12^3. At $j = 0$, $\sqrt{j}$ or something analytically equivalent is needed; at $j = 12^3$, $\sqrt[3]{j - 12^3}$ is needed. This works out as follows. Let $\pi : \mathfrak{X} \to S$ be *any* modular family. In particular, S is a nonsingular curve. Suppose we define a function on the closed points of S by assigning to the point $s \in S$ the j-invariant of the curve $\pi^{-1}(s)$. It can be proven that this function is a morphism:

$$S \xrightarrow{j} \mathbf{A}_j.$$

We can then prove the following.

4.1.

Each component of S dominates $\mathbf{A}_j$ and the ramification index of j at a closed point $s \in S$ is 1, 2, or 3 according to whether $j(s) \neq 0$ and 12^3, $j(s) = 0$, or $j(s) = 12^3$.

We now want to return to the problem of giving an explicit description of the topology $\mathfrak{M}$. The morphism j is one invariant which we can attach to the family $\pi : \mathfrak{X} \to S$. Unfortunately, a given j may correspond to more than one family π. A second invariant is needed. The key is to use more strongly the particular modular family over $\mathbf{A}_j - (0, 12^3)$ which we have constructed. With this as a reference point, so to speak, we will get the second invariant. Let $\pi_0 : \mathfrak{X}_0 \to S_0$ denote this one family. We use the diagram:

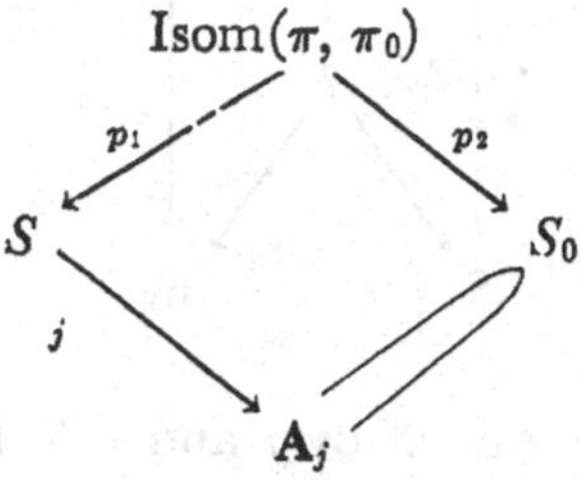

The first thing to notice is that this is commutative: let t be a closed point of $\text{Isom}(\pi, \pi_0)$. If $s = p_1(t)$ and $s_0 = p_2(t)$, then t represents an isomorphism of $\pi^{-1}(s)$ and $\pi_0^{-1}(s_0)$. Therefore, $\pi^{-1}(s)$ and $\pi_0^{-1}(s_0)$ have the same j-invariant, that is, $p_1(t)$ and $p_2(t)$ have the same image in $\mathbf{A}_j$.

Now what is $\text{Isom}(\pi, \pi_0)$? Over a closed point $s \in S$, its points represent isomorphisms of $\pi^{-1}(s)$ with curves $\pi_0^{-1}(s_0)$, $s_0 \in S_0$. In other words, $\text{Isom}(\pi, \pi_0)$ has no points over s if $j(s) = 0$ or 12^3; two points otherwise. $\text{Isom}(\pi, \pi_0)$ is a double étale covering of the open set:

$$j^{-1}(S_0) \subset S.$$

This covering extends uniquely to a covering I of all of S (not necessarily étale!)†. The covering I/S is the second invariant. I claim that j and I/S determine the modular family π uniquely.

Indication of Proof. The first step is to check that there is at most

† By a double covering T/S, I mean a second nonsingular curve T, and a finite, flat, surjective morphism $f : T \to S$ of degree 2, étale over an open dense subset $S' \subset S$. Now either Isom is the disjoint union of two copies of $j^{-1}(S_0)$; and then I is the disjoint union of two copies of S; or Isom is the normalization of $j^{-1}(S_0)$ in a quadratic extension of its function field, and then I is the normalization of S in this field.

one family $\mathfrak{X}/S$ extending the restriction of this family to the open subset $j^{-1}(S_0)$. After this, we may assume $j(S) \subset S_0$. Let $\mathfrak{Y}$ be the given family of elliptic curves over $I = \mathrm{Isom}(\pi, \pi_0)$. Then we have a diagram of morphisms of families:

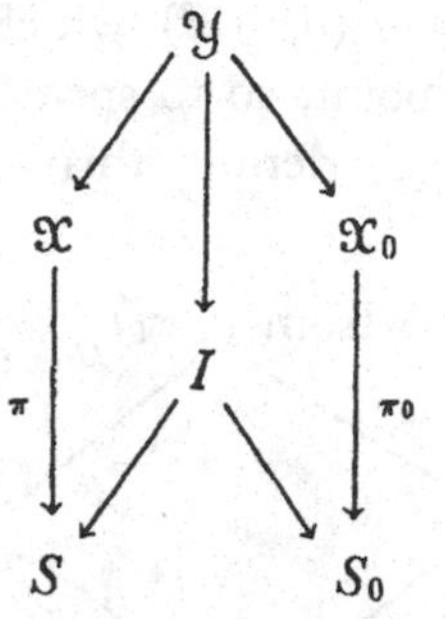

The family $\mathfrak{Y}/I$ is determined by j and I/S, because it is just the family induced over I by the base extension

$$I \to S \xrightarrow{j} S_0$$

from the standard family $\mathfrak{X}_0$. On the other hand, $\mathfrak{Y}$ is also induced from $\mathfrak{X}$ via the double étale covering I/S. Therefore, $\mathfrak{Y}$ is a double étale covering of $\mathfrak{X}$. We could recover $\mathfrak{X}$ from $\mathfrak{Y}$ if we knew the involution ι of $\mathfrak{Y}$ interchanging the two sheets of this covering. But let P_0 be the inversion of the family π_0: this is an involution of $\mathfrak{X}_0$ over S_0. Let $\bar{\iota}$ be the involution of I corresponding to the covering I/S: this is an automorphism of I over S_0 too. Since the diagram sets up an identification

$$\mathfrak{Y} = I \times_{S_0} \mathfrak{X}_0,$$

$\bar{\iota}$ and P_0 induce an involution $\bar{\iota} \times P_0$ of $\mathfrak{Y}$. We check that $\iota = \bar{\iota} \times P_0$.

Q.E.D.

The next question is whether there are any restrictions on j and I/S for these to come from a modular family. Besides the restriction (4.1) on j mentioned above, it turns out that the following is the only further restriction.

4.2.

I is ramified over all points s of S where $j(s) = 0$ or 12^3.

Turning all this around, we can make it into a second definition of the topology $\mathfrak{M}$:

Definition. The topology $\mathfrak{M}$ is as follows:

a. Its open sets are morphisms j of nonsingular curves S to $\mathbf{A}_j$ satisfying restriction (4.1), plus double coverings I/S satisfying restriction (4.2); and a final object M.

b. Its morphisms are commutative diagrams:

$$\begin{array}{ccc} I_1 & \longrightarrow & I_2 \\ \downarrow & & \downarrow \\ S_1 & \longrightarrow & S_2 \\ & {\scriptstyle j_1}\searrow \quad \swarrow{\scriptstyle j_2} & \\ & \mathbf{A}_j & \end{array}$$

making I_1 into the fibre product $S_1 \times_{S_2} I_2$; and projections of every open set onto M,

c. Coverings of $(j, I/S)$ are collections of morphisms

$$\begin{array}{ccc} I_\alpha & \longrightarrow & I \\ \downarrow & & \downarrow \\ S_\alpha & \xrightarrow{f_\alpha} & S \end{array}$$

such that $S = \bigcup_\alpha f_\alpha(S_\alpha)$; coverings of M are collections of projections of open sets $(j_\alpha, I_\alpha/S_\alpha)$ onto M such that $\mathbf{A}_j = \bigcup_\alpha j_\alpha(S_\alpha)$.

Note that, because of restriction (4.1), given a morphism of open sets:

$$\begin{array}{ccc} I_1 & \longrightarrow & I_2 \\ \downarrow & & \downarrow \\ S_1 & \xrightarrow{f} & S_2 \end{array}$$

the morphism f is necessarily étale.

Let us work out (absolute) products in these terms to see how it all fits together. Say $(j_1, I_1/S_1)$ and $(j_2, I_2/S_2)$ are two open sets. Suppose we want to map a third open set $(j, I/S)$ to both:

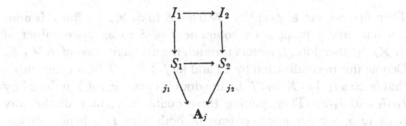

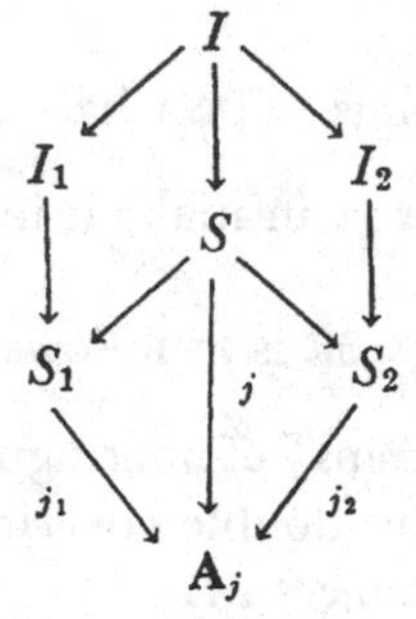

Then first we get a morphism f_1 from S to $S_1 \times_{\mathbf{A}} S_2$. But S is non-singular, and f maps each component of S to an open subset of $S_1 \times_{\mathbf{A}} S_2$; therefore, f_1 factors through the normalization of $S_1 \times_{\mathbf{A}} S_2$. Denote this normalization by T, and let $f_2 : S \to T$ be the morphism that factors f_1. Let I_1' and I_2' be the double coverings of T induced by I_1/S_1 and I_2/S_2. Then pulling these double coverings all the way back to S, we get isomorphisms of both with I/S, hence an isomorphism between them. Exactly as in §3, we get a factorization of f_2 via $S \xrightarrow{f_3} \mathrm{Isom}_T(I_1', I_2')$. But what is this Isom? At points where I_1' and I_2' are unramified, it is just the "quotient" double covering; that is, if I_1' is defined by extracting $\sqrt{f_1}$, and I_2' by $\sqrt{f_2}$, then Isom is the double covering given by $\sqrt{f_1/f_2}$. Since I_1' and I_2' are ramified over exactly the same points of T, this "quotient" covering extends to an étale double covering I_3' over all of T. It turns out that I_3' is a closed subscheme of $\mathrm{Isom}_T(I_1', I_2')$ and f_3 factors via

$$S \xrightarrow{f_4} I_3'.$$

This I_3' is the S of the product open set. Over I_3', I_1 and I_2 can be canonically identified to the I of the product open set.

§5. THE PICARD GROUPS

Now we come to the Picard groups, which are one of the interesting invariants of our topologies. There are two quite different ways to define these groups. One is a direct method going back to the moduli problem itself; the other is a cohomological method using our topologies. We will first explain the direct method:

Fix, as before, the genus g.

Definition. An "invertible sheaf" L (on the moduli problem itself) consists in two sets of data:

a. For all families of nonsingular curves (of genus g) $\pi : \mathfrak{X} \to S$, an invertible sheaf $L(\pi)$ on S.
b. For all morphisms F between such families:

$$\begin{array}{ccc} \mathfrak{X}_1 & \longrightarrow & \mathfrak{X}_2 \\ {\scriptstyle \pi_1}\big\downarrow & & \big\downarrow{\scriptstyle \pi_2} \\ S_1 & \xrightarrow{f} & S_2 \end{array}$$

an isomorphism $L(F)$ of $L(\pi_1)$ and $f^*(L(\pi_2))$.†

The second set of data should satisfy a compatibility condition with respect to composition of morphisms:

Let

$$\begin{array}{ccccc} \mathfrak{X}_1 & \longrightarrow & \mathfrak{X}_2 & \longrightarrow & \mathfrak{X}_3 \\ {\scriptstyle \pi_1}\big\downarrow & & \big\downarrow{\scriptstyle \pi_2} & & \big\downarrow{\scriptstyle \pi_3} \\ S_1 & \xrightarrow[f]{} & S_2 & \xrightarrow[g]{} & S_3 \end{array}$$

be a composition of the morphism F from π_1 to π_2, and G from π_2 to π_3. Then the diagram:

$$\begin{array}{ccccc} & & f^*(L(\pi_2)) & & \\ & \overset{L(F)}{\nearrow} & & \overset{f^*(L(G))}{\searrow} & \\ L(\pi_1) & & & & f^*(g^*(L(\pi_3))) \\ & \underset{L(G \circ F)}{\searrow} & & \Vert & \\ & & (g \circ f)^*(L(\pi_3)) & & \end{array}$$

should commute.

This definition has an obvious translation into the language of fibred categories, which is left to the reader who has a taste for that approach. Loosely speaking, an invertible sheaf is simply a pro-

† Note that the morphism F is the whole diagram, while f is simply the morphism from S_1 to S_2. In the sequel, we will denote morphisms of families by capital letters and the component morphism of base spaces by the same small letters.

cedure for attaching canonically a one-dimensional vector space to every nonsingular curve of genus g: Start with L as above. If C/k is such a curve, let

$$\pi : C \to \operatorname{Spec}(k)$$

be the projection. Then $L(\pi)$ is a one-dimensional vector space (over k) attached to C. Conversely, if this procedure is "canonical" enough, then given a family $\pi : \mathfrak{X} \to S$, the one-dimensional vector spaces attached to the curves $\pi^{-1}(s)\,(s \in S_k)$ should form a line bundle over S; and its sections then form an invertible sheaf $L(\pi)$.

Example. Given any $\pi : \mathfrak{X} \to S$ as above, let

$$E(\pi) = R^1\pi_*(o_X).$$

This is known to be a locally free sheaf on S of rank g. Let

$$L(\pi) = \Lambda^g\{R^1\pi_*(o_X)\}.$$

This is an invertible sheaf on S. Moreover, for all morphisms of families:

$$\begin{array}{ccc} \mathfrak{X}_1 & \longrightarrow & \mathfrak{X}_2 \\ {\scriptstyle \pi_1}\downarrow & & \downarrow{\scriptstyle \pi_2} \\ S_1 & \xrightarrow{f} & S_2 \end{array}$$

there is a canonical identification of $E(\pi_1)$ and $f^*(E(\pi_2))$, hence of $L(\pi_1)$ and $f^*(L(\pi_2))$. This is, therefore, an invertible sheaf on the moduli problem. It corresponds to attaching to each curve C the one-dimensional vector space

$$\Lambda^g\{H^1(C, o_C)\}.$$

It is clear what is meant by an isomorphism of two invertible sheaves.

Definition. The set of isomorphism classes of such invertible sheaves is called $\operatorname{Pic}(\mathfrak{M})$.

As usual, $\operatorname{Pic}(\mathfrak{M})$ is an abelian group. Given L and M, two invertible sheaves, define $L \otimes M$ by

$$(L \otimes M)(\pi) = L(\pi) \otimes M(\pi)$$
$$(L \otimes M)(F) = L(F) \otimes M(F).$$

This induces the product on the set of isomorphism classes $\operatorname{Pic}(\mathfrak{M})$.

Now we give the second definition of Pic($\mathfrak{M}$). Recall that, by definition, a scheme X is a particular type of topological space, together with a given sheaf of rings $\boldsymbol{o}_X$. Now that we have generalized the concept of a topological space, it is clear that an important type of object to look at will be a topology T, together with a given sheaf of rings $\boldsymbol{o}_T$. This combination is known as a "site." For example, if X is a scheme it is not only the Zariski topology which comes with the sheaf of rings $\boldsymbol{o}_X$. Recall the five topologies on X and the continuous maps:

$$X^*_{\text{flat}} \rightarrow X^*_{\text{smooth}} \rightarrow X^*_{\text{ét}} \rightarrow X_{\text{ét}} \rightarrow X_{\text{Zar}}.$$

Let $\pi : U \rightarrow X$ be a morphism, that is, an open set in X^*_{flat}. Then define† a sheaf $\boldsymbol{o}$ on X_{flat} by

$$\boldsymbol{o}(U \rightarrow X) = \Gamma(U, \boldsymbol{o}_U).$$

By taking direct images, this also defines a sheaf $\boldsymbol{o}$ on X^*_{smooth}, $X^*_{\text{ét}}$, $X_{\text{ét}}$ and X_{Zar}; on X_{Zar} this is just the original sheaf $\boldsymbol{o}_X$. Thus each of these topologies is a site.

What is more important now is that the topologies $\mathfrak{M}$ are also sites. Let $\pi : \mathfrak{X} \rightarrow S$ be an open set in $\mathfrak{M}$, that is, a modular family of nonsingular curves. Let

$$\boldsymbol{o}(\mathfrak{X} \xrightarrow{\pi} S) = \Gamma(S, \boldsymbol{o}_S).$$

This defines a sheaf of rings $\boldsymbol{o}$ on $\mathfrak{M}$, except for the ring $\boldsymbol{o}(M)$: this is simply determined by the sheaf axiom. Fix a covering of $\mathfrak{M}$ by open sets $\{\mathfrak{X}_\alpha \xrightarrow{\pi_\alpha} S_\alpha\}$. Let the product of π_α and π_β be the open set

$$\mathfrak{X}_{\alpha,\beta} \xrightarrow{\pi_{\alpha,\beta}} S_{\alpha,\beta}.$$

Then $\boldsymbol{o}(M)$ is the kernel of the usual homomorphism

$$\prod_\alpha \boldsymbol{o}(\mathfrak{X}_\alpha \rightarrow S_\alpha) \rightarrow \prod_{\alpha,\beta} \boldsymbol{o}(\mathfrak{X}_{\alpha,\beta} \rightarrow S_{\alpha,\beta}).$$

In fact, if $g \geq 3$, it is known that $\boldsymbol{o}(M)$ is just k. In any case, this defines $\boldsymbol{o}$, and it brings $\mathfrak{M}$ into a familiar context: we can now develop a theory of coherent sheaves, and their cohomology on $\mathfrak{M}$,

† This is not obviously a sheaf; it is so as a consequence of the theory of descent (cf. [3], exposé 8).

as well as a general theory of (étale) cohomology. Moreover, in addition to $\boldsymbol{o}$ we get the two auxiliary sheaves:

a. $\boldsymbol{o}^*$, defined by $\boldsymbol{o}^*(\mathfrak{X} \xrightarrow{\pi} S)$ = group of units in $\boldsymbol{o}(\mathfrak{X} \xrightarrow{\pi} S)$.
b. $\boldsymbol{K}$, defined as the sheaf associated to the presheaf $\tilde{\boldsymbol{K}}(\mathfrak{X} \xrightarrow{\pi} S)$ = total quotient ring of $\boldsymbol{o}(\mathfrak{X} \xrightarrow{\pi} S)$.

The ring of global sections of $\boldsymbol{K}$ is, so to speak, the function field of the moduli problem. Now the second definition of Pic($\mathfrak{M}$) is simply the cohomology group:

$$H^1(\mathfrak{M}, \boldsymbol{o}^*).$$

Sketch of Proof of Isomorphism. The first thing to do is to set up a map between these groups. The map goes like this: let L be an invertible sheaf on the moduli problem. Then we will associate to L an element:

$$\lambda \in H^1(\mathfrak{M}, \boldsymbol{o}^*).$$

First choose any collection of families $\pi_\alpha : \mathfrak{X}_\alpha \to S_\alpha$ which is a covering of the final object M. Then $L(\pi_\alpha)$ is an invertible sheaf on S_α. By replacing S_α with a suitable set of (Zariski) open subsets and replacing π_α by the set of induced families over these subsets, we can assume that for each α there is an isomorphism:

$$L(\pi_\alpha) \underset{\phi_\alpha}{\xrightarrow{\sim}} \boldsymbol{o}_{S_\alpha}$$

For each α, choose such an isomorphism. For all α, β, let

$$\pi_{\alpha,\beta} : (\mathfrak{X}_\alpha, \mathfrak{X}_\beta) \to \mathrm{Isom}(\pi_\alpha, \pi_\beta) = I_{\alpha,\beta}$$

be the product of the families π_α and π_β. Let p_1 and p_2 denote the projections of $\mathrm{Isom}(\pi_\alpha, \pi_\beta)$ onto S_α and S_β. By definition of an invertible sheaf, we are given isomorphisms of $p_1^*(L(\pi_\alpha))$ and $p_2^*(L(\pi_\beta))$ with $L(\pi_{\alpha,\beta})$. Now look at the composite isomorphism:

$$\begin{aligned} \boldsymbol{o}_{I_{\alpha,\beta}} &= p_1^*(\boldsymbol{o}_{S_\alpha}) \\ &\xrightarrow{\sim} p_1^*(L(\pi_\alpha)) && \text{via } \phi_\alpha \\ &\cong L(\pi_{\alpha,\beta}) \\ &\cong p_2^*(L(\pi_\beta)) \\ &\xleftarrow{\sim} p_2^*(\boldsymbol{o}_{S_\beta}) && \text{via } \phi_\beta \\ &= \boldsymbol{o}_{I_{\alpha,\beta}}. \end{aligned}$$

This isomorphism is set up by multiplication by a unit:

$$\sigma_{\alpha,\beta} \in \Gamma(I_{\alpha,\beta}, o^*_{I_{\alpha,\beta}}) = o^*(\pi_{\alpha,\beta}).$$

I claim that, for the covering $\{\pi_\alpha\}$ of $\mathfrak{M}$, $\{\sigma_{\alpha,\beta}\}$ forms a 1-Czech cocycle in the sheaf o^*. This is checked using the compatibility property for the invertible sheaf L (cf., last part of the definition of an invertible sheaf). Then this cocycle induces an element λ_1 in the first Czech cohomology group for this covering, hence an element λ_2 of $H^1(\mathfrak{M}, o^*)$.

Now suppose the isomorphisms ϕ_α are varied? The only possible change is to replace ϕ_α by $\phi'_\alpha = \sigma_\alpha \cdot \phi_\alpha$, where σ_α means multiplication by the unit:

$$\sigma_\alpha \in \Gamma(S_\alpha, o^*_{S_\alpha}) = o^*(\pi_\alpha)$$

But then $\sigma_{\alpha,\beta}$ is replaced by the homologous cocycle:

$$\sigma'_{\alpha,\beta} = p_1^*(\sigma_\alpha) \cdot p_2^*(\sigma_\beta^{-1}) \cdot \sigma_{\alpha,\beta}.$$

Therefore even λ_1 is unaltered. Now suppose the covering $\{\pi_\alpha\}$ is changed. Any two coverings are dominated by a finer covering, so we can assume that the new covering is finer. It is immediate that the new λ_1 is just the element of the new Czech cohomology group induced by the old λ_1 under restriction. Therefore, λ_2 is unaltered.

This defines a map from $\text{Pic}_1(\mathfrak{M})$ (the first group) to $\text{Pic}_2(\mathfrak{M})$ (the second group). To show that this is a surjective, we first use the fact that (for any sheaf F),

$$H^1(\mathfrak{M}, F) = \varinjlim_{\text{coverings } \mathfrak{A}} H^1(\mathfrak{A}, F)$$

where $H^1(\mathfrak{A}, -)$ denotes the first Czech cohomology group for the covering $\mathfrak{A}$. Now suppose $\lambda_2 \in H^1(\mathfrak{M}, o^*)$ is given. Then λ_2 is induced by a $\lambda_1 \in H^1(\mathfrak{A}, o^*)$ for some covering $\mathfrak{A}$. And λ_1 is defined by some cocycle $\{\sigma_{\alpha,\beta}\}$ in o^*, (if $\mathfrak{A}$ is the covering $\pi_\alpha : \mathfrak{X}_\alpha \to S_\alpha$). Now suppose $\pi : \mathfrak{X} \to S$ is any modular family of nonsingular curves. Then

$$\{I_\alpha = \text{Isom}(\pi, \pi_\alpha) \to S\}$$

is an étale covering of S. Moreover, via the natural projection

$$I_\alpha \times_S I_\beta \to \mathrm{Isom}(\pi_\alpha, \pi_\beta),$$

the cocycle $\{\sigma_{\alpha,\beta}\}$ induces a cocycle $\{\tau_{\alpha,\beta}\}$ for the covering $\{I_\alpha \to S\}$ of S and the sheaf o_S^*. We then require a theorem of Grothendieck:

Theorem 90 (Hilbert-Grothendieck). Let $\{U_\alpha \xrightarrow{q_\alpha} X\}$ be a flat covering of X; for all α, let L_α be an invertible sheaf on U_α; and for all α, β, let $\phi_{\alpha,\beta}$ be an isomorphism on $U_\alpha \times_X U_\beta$ of the sheaves $p_1^*(L_\alpha)$ and $p_2^*(L_\beta)$. Assume an obvious compatibility of isomorphisms on $U_\alpha \times_X U_\beta \times_X U_\gamma$ (for all α, β, γ). Then there is an invertible sheaf L on X, and for all α, isomorphisms ψ_α on U_α of L_α and $q_\alpha^*(L)$ such that, on $U_\alpha \times_X U_\beta$, the diagram:

$$\begin{array}{ccc} p_1^*(L_\alpha) & \xrightarrow[\phi_{\alpha,\beta}]{\sim} & p_2^*(L_\beta) \\ {\scriptstyle p_1^*(\psi_\alpha)}\downarrow\wr & & \wr\downarrow{\scriptstyle p_2^*(\psi_\alpha)} \\ p_1^*(q_\alpha^*(L)) & = & p_2^*(q_\beta^*(L)) \end{array}$$

commutes. Moreover, L and ψ_α are uniquely determined, up to canonical isomorphisms. (cf. [3], exposé 8, Theorem 1.1).

There is a shorthand which is used in connection with this theorem: given the L_α, the isomorphisms $\{\phi_{\alpha,\beta}\}$ are called "descent data" for $\{L_\alpha\}$. The L obtained is said to be gotten by "descending" the sheaves L_α to X (that is, reversing the process, the L_α are gotten by lifting L to U_α).

Apply this theorem with $U_\alpha = I_\alpha$, $X = S$, $L_\alpha = o_{I_\alpha}$, and $\phi_{\alpha,\beta}$ given by $\sigma_{\alpha,\beta}$. The L constructed is to be our $L(\pi)$. We leave it to the reader to construct the isomorphisms $L(F)$ required for an invertible sheaf; and to check that this L does induce λ_2 when the process is reversed.

Finally, why is the map injective? If λ_2 were 0, then for a suitable covering λ_1 would be 0, and for suitable choices of the ϕ_α's, the cocycle $\sigma_{\alpha,\beta}$ itself would come out 1. The question is then, if $\sigma_{\alpha,\beta} = 1$ for all α, β show that $L = o$ (the trivial invertible sheaf). What we need to do is to construct, for every family $\pi : \mathfrak{X} \to S$, an isomorphism

$$\psi(\pi) : L(\pi) \xrightarrow{\sim} o_S,$$

such that, for every morphism F of families:

$$\begin{array}{ccc} \mathfrak{X}_1 & \longrightarrow & \mathfrak{X}_2 \\ \downarrow{\scriptstyle \pi_1} & & \downarrow{\scriptstyle \pi_2} \\ S_1 & \xrightarrow{f} & S_2 \end{array}$$

the diagram:

$$\begin{array}{ccc} L(\pi_1) & \xrightarrow[\sim]{L(F)} & f^*(L(\pi_2)) \\ \wr\downarrow{\scriptstyle \psi(\pi_1)} & & \wr\downarrow{\scriptstyle f^*(\psi(\pi_2))} \\ \boldsymbol{o}_{S_1} & = & f^*(\boldsymbol{o}_{S_2}) \end{array}$$

commutes. Exactly as before, we use the induced étale covering

$$\{I_\alpha = \mathrm{Isom}(\pi, \pi_\alpha) \xrightarrow{q_\alpha} S\}.$$

The family of curves $\mathcal{Y}_\alpha$ over I_α induced via q_α from $\mathfrak{X}/S$ is canonically isomorphic to the family induced from $\mathfrak{X}_\alpha/S_\alpha$. But we are given an isomorphism of $L(\pi_\alpha)$ and $\boldsymbol{o}_{S_\alpha}$. This induces an isomorphism of $L(\mathcal{Y}_\alpha/I_\alpha)$ and $\boldsymbol{o}_{I_\alpha}$; hence an isomorphism

$$q_\alpha^*(L(\pi)) \underset{\psi_\alpha}{\overset{\sim}{\rightarrow}} \boldsymbol{o}_{I_\alpha}.$$

The fact that $\sigma_{\alpha,\beta} = 1$ can be easily seen to imply that the diagram of sheaves on $I_\alpha \times_S I_\beta$:

$$\begin{array}{ccc} p_1^*(q_\alpha^*(L(\pi))) & \xrightarrow[\sim]{p_1^*(\psi_\alpha)} & p_1^*(\boldsymbol{o}_{I_\alpha}) \\ \| & & \| \\ p_2^*(q_\beta^*(L(\pi))) & \xrightarrow[p_2^*(\psi_\beta)]{\sim} & p_2^*(\boldsymbol{o}_{I_\beta}) \end{array}$$

commutes. In order words, both L and $\boldsymbol{o}_S$ satisfy the conclusions of Theorem 90 for the setup $U_\alpha = I_\alpha$, $X = S$, $L_\alpha = \boldsymbol{o}_{I_\alpha}$, and $\phi_{\alpha,\beta} = 1$. Therefore, the uniqueness half of that theorem states that there is a canonical isomorphism of L and $\boldsymbol{o}_S$. This is to be $\psi(\pi)$. We omit the rest of the details.

§6. COMPUTATIONS: DIRECT METHOD

We return to the case $g = 1$, and its topology $\mathcal{M}$. In this section, for $\mathrm{char}(k) \neq 2, 3$, we shall give a direct computation of $\mathrm{Pic}(\mathcal{M})$. In the next section, for $k = \mathbf{C}$, we shall give a transcendental computation of this same group.

Let L be an invertible sheaf on the moduli problem. First of all, let us try to extract some numerical invariants directly from L. Start with a family of curves $\pi : \mathcal{X} \to S$. Any such family has one nontrivial automorphism: the inversion ρ of order 2. By definition of an invertible sheaf, the morphism of families:

$$\begin{array}{ccc} \mathcal{X} & \xrightarrow{\rho} & \mathcal{X} \\ {\scriptstyle\pi}\downarrow & & \downarrow{\scriptstyle\pi} \\ S & \xrightarrow[1_S]{} & S \end{array}$$

induces an automorphism $L(\rho)$ of $L(\pi)$. Since ρ has order 2, so does $L(\rho)$. But $L(\rho)$, as any automorphism of an invertible sheaf, is given by multiplication by an element $\alpha \in \Gamma(S, \mathfrak{o}_S^*)$. Therefore, $\alpha^2 = 1$: hence on each connected component S_i of S, α equals $+1$ or -1. In particular, suppose $S = \mathrm{Spec}(k)$, and $\mathcal{X} = C$ is an elliptic curve. Then we have defined a number:

$$\alpha(C) = \pm 1.$$

Moreover, if $\pi : \mathcal{X} \to S$ is any family, then the fact that the inversion ρ for π induces the inversion on each fibre $\pi^{-1}(s)$ of the family implies that the function $\alpha \in \Gamma(S, \mathfrak{o}_S^*)$ has value $\alpha(\pi^{-1}(s))$ at the point $s \in S$. This shows that α is a "continuous" function of C; that is, if we have a family $\pi : \mathcal{X} \to S$ with connected base S, then α is constant on the set of curves $\pi^{-1}(s)$ occuring as fibres in the family π. Actually this shows that α is constant on all curves; either $\alpha(C) = +1$ for all C, or $\alpha(C) = -1$. Namely, it is easy to exhibit a family π with connected base S, such that every C occurs in π. For example, take the family of all nonsingular cubic curves; or take the modular family of cubic curves

$$y^2 = x(x - 1)(x - \lambda),$$

where $\lambda \neq 0, 1, \infty$. Therefore, in fact, we have defined one number $\alpha(L)$ equal to ± 1. And, quite clearly, this gives a homomorphism

$$\mathrm{Pic}(\mathfrak{M}) \xrightarrow{\alpha} \mathbf{Z}/2.$$

In fact, this same method goes further. After all, there are two elliptic curves with bigger groups of automorphisms. Let C_A be the curve with a group of automorphisms of order 4 (i. e., $j = 0$); let C_B be the curve with a group of order 6 (i. e., $j = 12^3$). Pick generators σ and τ of $\mathrm{Aut}(C_A)$ and $\mathrm{Aut}(C_B)$. Note that σ^2 is the inversion of C_A and τ^3 is the inversion of C_B. Let

$$\pi_A : C_A \longrightarrow \mathrm{Spec}(k)$$
$$\pi_B : C_B \longrightarrow \mathrm{Spec}(k)$$

be the trivial families. Then L gives us one-dimensional vector spaces $L(\pi_A)$ and $L(\pi_B)$, *and* L gives us an action of $\mathrm{Aut}(C_A)$ on $L(\pi_A)$ and of $\mathrm{Aut}(C_B)$ on $L(\pi_B)$. In particular, σ acts on $L(\pi_A)$ by multiplication by a fourth root of 1: call it $L(\sigma)$; and τ acts on $L(\pi_B)$ by multiplication of a sixth root of 1: call it $L(\tau)$. Clearly,

$$L(\sigma)^2 = \alpha(L)$$
$$L(\tau)^3 = \alpha(L).$$

If we also fix a primitive twelfth root of 1, ζ, then we can determine uniquely an integer β mod 12 by the equations:

$$\zeta^{6\beta} = \alpha(L); \ \zeta^{3\beta} = L(\sigma); \ \zeta^{2\beta} = L(\tau).$$

Then this associates an invariant $\beta(L) \in \mathbf{Z}/12$ to each invertible sheaf L. It is easy to see that this is a homomorphism:

$$\mathrm{Pic}(\mathfrak{M}) \xrightarrow{\beta} \mathbf{Z}/12.$$

Actually, β is not quite as nice as α, in that to define β we had to make three arbitrary choices, namely, σ, τ, and ζ. Our next step is to simultaneously make β more canonical and to prove that β is surjective. Recall the invertible sheaf Λ on $\mathfrak{M}$ given as an example in §3:

$$\Lambda(\mathfrak{X} \xrightarrow{\pi} S) = R^1\pi_*(o_{\mathfrak{X}})$$

(with the obvious compatibility morphisms for each morphism of families). The interesting fact is that $\beta(\Lambda)$ is a generator of $\mathbf{Z}/12$. To verify this, all we have to check is that $\Lambda(\sigma)$ [resp. $\Lambda(\tau)$] is a *primitive* fourth root (resp. a sixth root) of 1. But this means simply that $\mathrm{Aut}(C_A)$ [resp. $\mathrm{Aut}(C_B)$] acts faithfully on $\Lambda(\pi_\alpha)$ [resp. $\Lambda(\pi_B)$]. Now by definition:

$$\Lambda(\pi_A) = H^1(C_A, o_{C_A})$$
$$\Lambda(\pi_B) = H^1(C_B, o_{C_B}).$$

We could say, at this point, that it is a classical fact that these actions are faithful. But this is not hard to check:

Proof of Faithfulness.

a. By Serre duality, for any curve C, $H^1(C, o_C)$ is canonically dual to the vector space of regular differentials on C.

b. If C is the elliptic curve:

$$y^2 = x^3 + Ax + B,$$

then the differential dx/y is regular, and is a basis of the space of such differentials.

c. C_A is the curve

$$y^2 = x^3 - x = x(x+1)(x-1)$$

and σ may be taken to be

$$x \longmapsto -x$$
$$y \longmapsto iy.$$

Then $dx/y \longmapsto i(dx/y)$, so the action of $\mathrm{Aut}(C_A)$ is faithful.

d. C_B is the curve

$$y^2 = x^3 - 1 = (x-1)(x-\omega)(x-\omega^2)$$

and τ may be taken to be

$$x \longmapsto \omega \cdot x$$
$$y \longmapsto -y.$$

Then $dx/y \longmapsto -\omega(dx/y)$, so the action of $\mathrm{Aut}(C_B)$ is faithful.

Q.E.D.

Therefore β is indeed surjective. But also β can be normalized by the requirement:

$$\beta(\Lambda) \equiv 1 \pmod{12}.$$

Then β becomes completely canonical.

The last step is that β is injective, completing the proof of:

Main Theorem. If char(k) $\neq$ 2, 3, and if g = 1, then there is a canonical isomorphism

$$\mathrm{Pic}(\mathfrak{M}) \cong \mathbf{Z}/12.$$

Sketch of Rest of Proof. Let L be an invertible sheaf on the moduli problem such that $\beta(L) \equiv 0 \pmod{12}$. Then all automorphisms of all elliptic curves C induce trivial automorphisms of the corresponding vector spaces $L(C/\mathrm{Spec}(k))$. We must set up consistent isomorphisms of all the invertible sheaves $L(\pi)$ with the sheaves $\boldsymbol{o}_S$. But say $\pi : \mathfrak{X} \to S$ is a modular family of curves containing every elliptic curve as a fibre. Then according to the results of §5, it is sufficient to set up an isomorphism ϕ of this one $L(\pi)$ and $\boldsymbol{o}_S$ provided that the compatibility property written out in §5 is satisfied.

Look at the diagram:

$$\begin{array}{c} \mathrm{Isom}(\pi, \pi) \\ \downarrow f \\ \{\text{Normalization of } S \times_{\mathbf{A}} S\} \\ \downarrow g \\ S \times_{\mathbf{A}_j} S \\ p_1 \downarrow \; \downarrow p_2 \\ S \end{array}$$

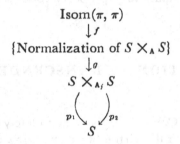

(cf. §4, last part). Recall that f is an étale double covering. Let $q_i = p_i \circ g \circ f$. By definition of an invertible sheaf, we are given an isomorphism ψ of $q_1^*(L)$ and $q_2^*(L)$. We can use the fact that $\beta(L) \equiv 0$ to show that there is actually an isomorphism ψ_0 of $p_1^*(L)$ and $p_2^*(L)$ which induces ψ via $f^* \circ g^*$. Set theoretically, we see this is as follows: let t and t' be two closed points of $\mathrm{Isom}(\pi, \pi)$ over the same point $\bar{t}$ of $S \times_{\mathbf{A}} S$. Let $\bar{L}_1$ and $\bar{L}_2$ be the one-dimensional vector spaces obtained by restricting the invertible sheaves $p_1^*(L)$ and $p_2^*(L)$ to the one-point subscheme $\{\bar{t}\}$. If $s_i = p_i(\bar{t})$, then $\bar{L}_i = L(\pi^{-1}(s_i))$. Moreover, t and t' define two isomorphisms τ and τ' of $\pi^{-1}(s_1)$ and $\pi^{-1}(s_2)$. By hypothesis, $L(\tau' \circ \tau^{-1})$ is the identity! Therefore, $L(\tau) = L(\tau')$. But $L(\tau)$ and $L(\tau')$ are just the isomorphisms of $\bar{L}_1$ and $\bar{L}_2$ given by looking at the action of ψ at the points t and t'. Therefore, ψ induces a *unique* isomorphism ψ_0 of $\bar{L}_1$ and $\bar{L}_2$ at $\bar{t}$. One must still check that this isomorphism ψ_0 is given by functions in the local rings of $S \times_{\mathbf{A}} S$ (this scheme is not normal, so this is not obvious). We omit this technical point.

Now the compatibility property of ψ shows immediately that ψ_0 is descent data for the invertible sheaf $L(\pi)$ on S with respect to the morphism $j : S \to \mathbf{A}_j$. Also j is clearly a flat covering of $\mathbf{A}_j$. Therefore, we can apply Theorem 90 of §5! In other words, we find an invertible sheaf L_0 on $\mathbf{A}_j$, and an isomorphism ϕ of $L(\pi)$ with $j^*(L_0)$ such that the following commutes:

$$\begin{array}{ccc} p_1^*(L(\pi)) & \xrightarrow[\sim]{\psi} & p_2^*(L(\pi)) \\ {\scriptstyle p_1^*(\phi)}\downarrow\wr & & \downarrow\wr{\scriptstyle p_2^*(\phi)} \\ p_1^*(j^*(L_0)) & = & p_2^*(j^*(L_0)) \end{array} \tag{2}$$

But now every invertible sheaf on the affine line is trivial, that is, $L_0 \cong \mathfrak{o}_{\mathbf{A}}$. Use this isomorphism to set up an isomorphism of L with $\mathfrak{o}_S$. Finally, the compatibility property follows immediately from (2). Q.E.D.

§7. COMPUTATIONS: TRANSCENDENTAL METHOD

Now assume $k = \mathbf{C}$. We shall give a completely different approach to Pic($\mathfrak{M}$) which has the virtue of generalizing to higher genus in various ways. This approach is based on:

Definition. An "analytic family of elliptic curves" is a morphism $\pi : \mathfrak{X} \to S$ of analytic spaces, which is proper and flat, plus a section $\varepsilon : S \to \mathfrak{X}$ of π, such that the fibres of π are elliptic curves.

We can now define a modular analytic family in two ways: either by the same properties used to define an (algebraic) modular family;† or else by defining the j-morphism from the base S to the complex j-plane and requiring that

a. S is a nonsingular one-dimensional complex space.
b. j is open.
c. j has ramification index 1, 2, 3 at $x \in S$ according to $j(x) \neq 0, 12^3$, $j(x) = 0$, or $j(x) = 12^3$.

† The lifting property goes over verbatim. But instead of asking that each elliptic curve only occur a finite number of times in a modular family $\pi : \mathfrak{X} \to S$, we should ask that it occur only over the points of a discrete subset $\Delta \subset S$.

A morphism of families is defined exactly as before, using analytic maps rather than algebraic ones.

Definition. The topology $\mathfrak{M}_{cx}$ is as follows:

a. Its open sets are analytic modular families of elliptic curves $\pi : \mathfrak{X} \to S$, and a final object M,
b. Its morphisms and coverings are exactly as in $\mathfrak{M}$.

We check to see that products exist in this topology and that they have exactly the same interpretation as before. Moreover, we get a continuous map of topologies:

$$\mathfrak{M}_{cx} \xrightarrow{\alpha} \mathfrak{M},$$

just as, in §2, we found a continuous map from the complex topology to the étale topology on a scheme. For all integers n, define a sheaf $\mathbf{Z}/n$ on $\mathfrak{M}_{cx}$ by

$$\mathbf{Z}/n[\mathfrak{X} \xrightarrow{\pi} S] = \bigoplus_{\substack{\text{topological components} \\ S_\alpha \text{ of } S}} \mathbf{Z}/n$$

The direct image $\alpha_*(\mathbf{Z}/n)$ of this sheaf is simply the "same" sheaf:

$$\mathbf{Z}/n[\mathfrak{X} \xrightarrow{\pi} S] = \bigoplus_{\substack{\text{topological components} \\ \text{(in Zariski topology) of } S}} \mathbf{Z}/n.$$

An immediate extension of Artin's result tells us that the canonical homomorphism

$$H^i(\mathfrak{M}, \mathbf{Z}/n) \longrightarrow H^i(\mathfrak{M}_{cx}, \mathbf{Z}/n)$$

is an isomorphism.†

This gives us a transcendental approach to the cohomology groups $H^i(\mathfrak{M}, \mathbf{Z}/n)$. These are related to the Picard group by virtue of the standard exact sequences of sheaves:

$$0 \longrightarrow \mathbf{Z}/n \longrightarrow \mathfrak{o}^* \xrightarrow{n} \mathfrak{o}^* \longrightarrow 0,$$

where n indicates the homomorphism $f \mapsto f^n$ (cf. [1], p. 102). The

† The stronger form in which Artin gave his result is that if $g : X_{cx} \to X_{\acute{e}t}$ is the canonical map, then:

$$R^i g_*(\mathbf{Z}/n) = (0), \qquad i > 0.$$

This gives $R^i\alpha_*(\mathbf{Z}/n) = (0)$, $(i > 0)$ as a corollary because $\mathfrak{M}$ (respectively, $\mathfrak{M}_{cx}$) induces on an open set $\mathfrak{X} \xrightarrow{\pi} S$ the topology $S_{\acute{e}t}$ (respectively, S_{cx}).

cohomology sequence tells us:

$$0 \to H^0(\mathfrak{M}, \mathbf{Z}/n) \to H^0(\mathfrak{M}, \boldsymbol{o}^*) \xrightarrow{n} H^0(\mathfrak{M}, \boldsymbol{o}^*) \to H^1(\mathfrak{M}, \mathbf{Z}/n) \to \mathrm{Pic}(\mathfrak{M}) \to \mathrm{Pic}(\mathfrak{M}) \to H^2(\mathfrak{M}, \mathbf{Z}/n).$$

Via these sequences, we can work out the structure of Pic($\mathfrak{M}$), given that of $H^i(\mathfrak{M}, \mathbf{Z}/n)$. This is because we can prove by general arguments that

a. $H^0(\mathfrak{M}, \boldsymbol{o}^*)$ has the subgroup $\mathbf{C}^*$ of constant functions, with factor group isomorphic to $\mathbf{Z}^a$.
b. Pic($\mathfrak{M}$) has a subgroup $\mathrm{Pic}^\tau(\mathfrak{M})$ of the type $\mathbf{R}^b/\mathbf{Z}^c$, where the lattice $\mathbf{Z}^c$ spans $\mathbf{R}^b$ (it need not be discrete), with finitely generated factor group.

Corollary. If there is a prime p such that $H^1(\mathfrak{M}, \mathbf{Z}/p) = (0)$, then $H^0(\mathfrak{M}, \boldsymbol{o}^*) = \mathbf{C}^*$, and Pic($\mathfrak{M}$) is finitely generated.

Corollary. If there is a prime p such that $H^1(\mathfrak{M}, \mathbf{Z}/p) = H^2(\mathfrak{M}, \mathbf{Z}/p) = (0)$, then Pic($\mathfrak{M}$) is a finite group, and

$$\mathrm{Pic}(\mathfrak{M}) = \varinjlim H^1(\mathfrak{M}, \mathbf{Z}/n),$$

where the limit is taken with respect to the ordering:

$$n_1 \geq n_2 \quad \text{if} \quad n_2 \mid n_1,$$

and the maps

$$\mathbf{Z}/n_2 \xrightarrow{n_1/n_2} \mathbf{Z}/n_1.$$

(The proofs are obvious.)

We now go on to consider the topology $\mathfrak{M}_{\mathbf{cx}}$. The point is that there is one open set in $\mathfrak{M}_{\mathbf{cx}}$ which is very well known:

Let $\mathfrak{H} = \{z \in \mathbf{C} \mid \mathrm{Im}(z) > 0\}$.

Let $\mathbf{Z} \oplus \mathbf{Z}$ act on the analytic space $\mathbf{C} \times \mathfrak{H}$ so that the generators act by:

$$(x, z) \mapsto (x + 1, z)$$
$$(x, z) \mapsto (x + z, z).$$

Let $\mathfrak{X} = (\mathbf{C} \times \mathfrak{H}/\mathbf{Z} \oplus \mathbf{Z})$.

Let $\pi : \mathfrak{X} \to \mathfrak{H}$ be induced by $p_2 : \mathbf{C} \times \mathfrak{H} \to \mathfrak{H}$.

Let $\epsilon : \mathfrak{H} \to \mathfrak{X}$ be induced by the section $\mathfrak{H} \xrightarrow{\sim} (0) \times \mathfrak{H} \subset \mathbf{C} \times \mathfrak{H}$.

Then π (and ϵ) define a modular analytic family of elliptic curves. Moreover, every elliptic curve occurs in π, so it is a covering of M.

Let $\Gamma = SL(2; \mathbf{Z})$

$$= \text{group of integral } 2 \times 2 \text{ matrices } \begin{pmatrix} a & b \\ c & d \end{pmatrix}$$

such that $ad - bc = 1$.

Let Γ act on $\mathfrak{H}$ via

$$\begin{pmatrix} a & b \\ c & d \end{pmatrix} \times (z) \longmapsto \left(\frac{az+b}{cz+d}\right)$$

Call this morphism $\tau_0 : \Gamma \times \mathfrak{H} \to \mathfrak{H}$.

Let Γ act on $\mathbf{C} \times \mathfrak{H}$ via

$$\begin{pmatrix} a & b \\ c & d \end{pmatrix} \times (x, z) \longmapsto \left(\frac{x}{cz+d}, \frac{az+b}{cz+d}\right)$$

Then we check that the action of Γ normalizes the action of $\mathbf{Z} \oplus \mathbf{Z}$, hence it induces an action of Γ on $\mathfrak{X}$. Denote by $\tau : \Gamma \times \mathfrak{X} \to \mathfrak{X}$ the morphism giving this action. Clearly, π and ϵ commute with this action of Γ, so that we have made Γ into a group of automorphisms of the family $\mathfrak{X}/\mathfrak{H}$. This action of Γ has the following interpretation: via the diagram

$$\begin{array}{ccccc} & & \Gamma \times \mathfrak{X} & & \\ & \swarrow^{\tau} & \downarrow{\scriptstyle 1_\Gamma \times \pi} & \searrow^{p_2} & \\ \mathfrak{X} & & & & \mathfrak{X} \\ \downarrow{\scriptstyle \pi} & & \Gamma \times \mathfrak{H} & & \downarrow{\scriptstyle \pi} \\ & \swarrow^{\tau_0} & & \searrow^{p_2} & \\ \mathfrak{H} & & & & \mathfrak{H} \end{array} \tag{3}$$

the family of elliptic curves $\Gamma \times \mathfrak{X}/\Gamma \times \mathfrak{H}$ is made into the product of $\mathfrak{X}/\mathfrak{H}$ with itself; in particular,

$$\Gamma \times \mathfrak{H} = \text{Isom}(\pi, \pi).$$

The effect of this is to make a connection between the topology $\mathfrak{M}_{ex}$ and the topology T_Γ of the discrete group Γ (cf. §1). We recall:

Definition. The topology T_Γ is as follows:

a. Its open sets are Γ-sets S, that is, sets plus action of Γ.
b. Its morphisms are Γ-linear maps between Γ-sets.
c. Its coverings are collections $S_\alpha \xrightarrow{p_\alpha} S$ of morphisms such that

$$S = \bigcup_\alpha p_\alpha(S_\alpha).$$

For our purpose, we need a slight modification of this topology.

Definition. Let Γ_n be the subgroup of Γ of matrices such that

$$\begin{pmatrix} a & b \\ c & d \end{pmatrix} \equiv \begin{pmatrix} 1 & 0 \\ 0 & 1 \end{pmatrix} \pmod{n}$$

Definition. Let T'_Γ be the following topology:

a. Its open sets are Γ-sets S such that, for all $x \in S$, the subgroup of Γ of elements leaving x fixed is contained in Γ_3; and a final object M.
b. Its morphisms are Γ-linear maps of Γ-sets, and projections of Γ-sets to M.
c. Its covering are collections $S_\alpha \xrightarrow{p_\alpha} S$ of morphisms such that

$$S = \bigcup_\alpha p_\alpha(S_\alpha);$$

and any collections of morphisms to M.

There is a continuous map:

$$T_\Gamma \xrightarrow{\beta} T'_\Gamma$$

such that β^{-1} of a Γ-set S is S; and $\beta^{-1}(M)$ is the Γ-set $\{e\}$ with one element. It is easy to check that β_* sets up an equivalence between the category of abelian sheaves on T_Γ and the category of abelian sheaves on T'_Γ. Therefore, cohomologically T_Γ and T'_Γ are identical. In fact, as we saw in §1, these categories of sheaves are equivalent to the category of Γ-modules (where the group of global sections of a sheaf is equal to the subgroup of Γ-invariants of the corresponding module). Therefore, the cohomology of T_Γ and T'_Γ is also the same as the cohomology of the group Γ.

Finally, there is a continuous map

$$\mathfrak{M}_{cx} \xrightarrow{\gamma} T'_{\Gamma}$$

which is as follows:

Definition. Let S be a Γ-set in T'_{Γ}; give S the discrete topology. Then $\gamma^{-1}(S)$ is the family:

$$\begin{array}{c} \mathfrak{X} \times S/\Gamma \\ \downarrow \\ \mathfrak{H} \times S/\Gamma \end{array}$$

where Γ acts on $\mathfrak{X} \times S$ and $\mathfrak{H} \times S$ by a product of τ and τ_0 with the given action on S.

This makes sense only provided that Γ acts freely on $\mathfrak{H} \times S$ (hence on $\mathfrak{X} \times S$). But if an element $a \in \Gamma$ leaves fixed some element of S, then by definition of T'_{Γ}, $a \in \Gamma_3$; then it is easy to check that a acts on $\mathfrak{H}$ without fixed points. Therefore, the action is free. Of course, $\gamma^{-1}(M)$ is to be M. The key point to check is that fibre products in T'_{Γ} go into fibre products in $\mathfrak{M}_{cx}$. We omit the proof, except to say that this fact follows readily from the fact that diagram (3) makes $\Gamma \times \mathfrak{X}/\Gamma \times \mathfrak{H}$ into the product of $\mathfrak{X}/\mathfrak{H}$ with itself.

Recapitulating, we have unwound the structure of M by the following continuous maps:

$$\begin{array}{ccc} & & \mathfrak{M} \\ & \overset{\alpha}{\nearrow} & \\ \mathfrak{M}_{cx} & \overset{\gamma}{\searrow} & \\ & & T'_{\Gamma}. \\ & \overset{\beta}{\nearrow} & \\ T_{\Gamma} & & \end{array}$$

The final step is to prove that, via γ, we get an isomorphism:

$$H^i(T'_{\Gamma}, \mathbf{Z}/n) \xrightarrow{\sim} H^i(\mathfrak{M}_{cx}, \mathbf{Z}/n)$$

This follows from the Leray spectral sequence, once we know that:

$$R^i\gamma_*(\mathbf{Z}/n) = (0), \qquad i > 0;$$

and this is equivalent to knowing that $\mathbf{Z}/n$ has no higher cohomology in the induced topology on the open set $\mathfrak{X}/\mathfrak{H}$ in $\mathfrak{M}_{cx}$. But this is just the classical topology on $\mathfrak{H}$; and since $\mathfrak{H}$ is homeomorphic to a cell, $\mathbf{Z}/n$ has no higher cohomology in this topology.

Corollary. There are canonical isomorphisms:

$$H^i(\mathfrak{M}, \mathbf{Z}/n) = H^i(\Gamma, \mathbf{Z}/n) \quad \text{for all } i.$$

Now it is well known that

a. $H^i(\Gamma, M) = (0)$, $i \geq 2$, for any Γ-module M which is p-torsion, $p \neq 2, 3$,
b. $H^1(\Gamma, M) = \mathrm{Hom}(\mathbf{Z}/12, M)$ for any abelian group M with trivial Γ action.

Putting all the results of this section together, we have proven again that $\mathrm{Pic}(\mathfrak{M}) \cong \mathbf{Z}/12$.

REFERENCES

1. Artin, M., *Grothendieck Topologies*, Mimeographed notes, Harvard University, Cambridge, Mass., 1962.
2. Grothendieck, A., Fondements de la géométrie algébrique, *Collected Bourbaki Talks*, mimeographed by the Secrétariat Mathématique, Paris, 1962.
3. Grothendieck, A., *Séminaire de géométrie algébrique*, Institut des hautes études scientifique, Paris, 1960–61.
4. Grothendieck, A., Sur quelques points d'algèbre homologique, *Tôhoku Math. Jour.*, **9**(1957), 119.
5. Lang, S., *Introduction to algebraic geometry*, Interscience, N.Y., 1958.
6. Lang, S., *Abelian varieties*, Interscience, N.Y., 1959.
7. Mumford, D., *Geometric invariant theory*, Springer-Verlag, Berlin-Göttingen-Heidelberg, 1965.
8. Northcott, D., *An introduction to homological algebra*, Cambridge University Press, Cambridge, 1960.
9. Rauch, H., The singularities of the moduli space, *Bull. Amer. Math. Soc.*, **68**(1962), 390.
10. Severi, F., La géométrie algébrique italienne (esp. p. 37), in *Colloque de géométrie algébrique*, Georges Thone, Liége, and Masson, Paris, 1949.

ABELIAN QUOTIENTS
OF THE TEICHMÜLLER MODULAR GROUP*

By

DAVID MUMFORD

in Cambridge, Mass., U.S.A.

1

Let D be the Teichmüller space of genus g: it is a $3g$-3-dimensional complex analytic manifold isomorphic to a bounded and contractible domain of holomorphy in $\mathbf{C}^{3g-3}$. Let Γ be the Teichmüller modular group: it is a discrete group acting discontinuously on D. This is the analytic side of the moduli problem.

On the other hand, let $M = D/\Gamma$: this analytic space has a canonical structure of algebraic variety. In fact, it turns out to be a Zariski-open subset of a projective variety: a so-called *quasi-projective* variety. Therefore, for all subgroups $\Gamma' \subset \Gamma$ of finite index, the analytic spaces $M(\Gamma') = D/\Gamma'$, being coverings of M, are also quasi-projective algebraic varieties. This follows from the

Generalized Riemann existence theorem.** *If X is any normal algebraic variety, Y any normal analytic space, and $f: Y \to X$ is a proper holomorphic map with finite fibres, and if there is a Zariski-open set $U \subset X$ such that $f^{-1}(U)$ is dense in Y and* res $f: f^{-1}(U) \to U$ *is unramified, then Y has one and only one structure of algebraic variety making f into a morphism.*

Thus from the algebraic standpoint, one has an inverse system of quasi-projective varieties:

---

* This work was supported by a Sloan Foundation Grant.

** In this form, the theorem is due to J. P. Serre and M. Artin. It can be reduced to the comparison theorem of GAGA [9], using either the paper of Grauert-Remmert [2], or resolution of singularities and elementary arguments; or, by methods of Artin and Grothendieck it can be reduced to the 1-dimensional case where it is classical.

D. Mumford, *Selected Papers*, Vol. II,

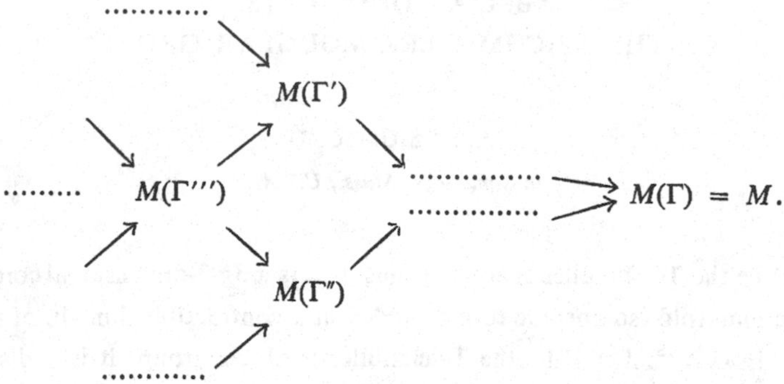

Moreover all sufficiently small $\Gamma' \subset \Gamma$ of finite index act freely on D, hence all $M(\Gamma')$ sufficiently "far up" in this inverse system are non-singular varieties, with D as their common universal covering space. The purpose of this note is to prove two closely related results:

Theorem 1. *If* $[\Gamma,\Gamma]$ *is the commutator subgroup of* Γ, *then* $\Gamma/[\Gamma,\Gamma]$ *is a finite cyclic group, whose order divides* 10. (?!)

Theorem 2. *The Albanese variety of M is trivial, i.e. there are no non-trivial rational maps from M to an abelian variety.**

Moreover, in the terminology of [6], these results are also equivalent to:

Theorem. *The Picard group of the moduli problem is a finitely generated abelian group isomorphic to* $H^2(\Gamma, \mathbf{Z})$.

(Cf. §7 of [6]). Analogy with the many calculations that have been made for arithmetic groups acting on symmetric domains, as well as the general feeling that M should be quantitatively similar to projective space, leads me
2 to conjecture that the rank of $H^2(\Gamma, \mathbf{Z})$ is *one*.

I want to thank Professor Magnus for a very informative letter acquainting me with the literature on the Teichmüller modular group and explaining what seems to be really proven, and what is not too clear. I also want to

* It is equivalent to consider morphisms or rational maps from M to abelian varieties 0, since M is locally a quotient of a non-singular variety by a finite group.

thank Professors Maskit and Grothendieck for helpful comments and for checking my computations.

§1. The connection between Theorem 1, 2.

Recall that if V is any algebraic variety, and $x \in V$ is a base point, then among all morphisms:

$$\phi: V \to A$$

$$\phi(x) = 0$$

where A is an abelian variety, O its origin, there is a "universal" one, i.e. one such morphism:

$$\phi_0: V \to A_0,$$

such that if $\psi: V \to B$ is any other, there is a unique morphism $\chi: A_0 \to B$ (which is necessarily a homomorphism) such that $\psi = \chi \circ \phi_0$. A is called the *Albanese* variety of V. We want to describe the Albanese variety of M.

Theorem 3. *Let V be a non-singular projective variety and let*

$$f: V \to V_0$$

be a birational morphism of V onto a normal projective variety V_0. Let $Z \subset V$ be a Zariski closed subset such that

$$\operatorname{codim} f(Z) \geqq 2.$$

Then

$$H_1(V - Z, Q) \xrightarrow{\sim} H_1(V, Q).$$

Proof. Since $H_1(V - Z, \mathbf{Z})$ and $H_1(V, \mathbf{Z})$ are finitely generated abelian groups, it suffices to show that for almost all primes p,

$$H_1(V - Z, \mathbf{Z}/p\mathbf{Z}) \xrightarrow{\sim} H_1(V, \mathbf{Z}/p\mathbf{Z}).$$

But, for any reasonable topological space X,

$$H_1(X, Z/pZ) \cong \pi_1(X)/\pi_1(X)^p \cdot [\pi_1(X), \pi_1(X)].$$

Therefore the isomorphism we need amounts to showing:

(a) all p-cyclic unramified coverings of $V-Z$ extend to p-cyclic unramified coverings of V

(b) a p-cyclic unramified covering of V that splits over $V-Z$ (i.e., restricted to $V-Z$, it is isomorphic to the disjoint union of p copies of $V-Z$), also splits over V.

However, p-cyclic unramified topological coverings of V and $V-Z$ are canonically algebraic varieties in view of the generalized Riemann Existence Theorem already quoted. Therefore if V is normal, connected coverings of V are obtained by taking the normalization of V in suitable finite extension fields L of the function field $\mathbb{C}(V)$. But such coverings do not split over *any* Zariski-open subset of V. Therefore (b) holds for all p.

To prove (a), we need a preliminary step:

Lemma (Matsumura). *If $D_1, \cdots, D_n$ are the components of Z of codimension 1, then the fundamental classes of the D's are independent in $H^2(V, Z)$.*

Proof. Suppose there was a relation

$$\sum_{i=1}^{n} n_i \cdot \text{class}(D_i) = 0.$$

Assume $n_i \neq 0$, if $1 \leqq i \leqq k$, and $n_i = 0$, $i > k$. Let $Z^* = f(D_1 \cup \cdots \cup D_k)$. Then so long as $\dim(Z^*) > 0$, cutting V_0 by a sufficiently general hyperplane H, we can replace V_0 by $V_0 \cap H$, V by $f^{-1}(V_0 \cap H)$, and Z by $f^{-1}(f(Z) \cap H)$, and obtain the same pathology with a V of lower dimension. If $\dim Z^* = 0$, then so long as $\dim(V) \geqq 3$ we can cut V by a sufficiently general hyperplane H, replace V by $V \cap H$, V_0 by the normalization of $f(V \cap H)$ and Z by the exceptional locus of

$$\text{res} f \colon V \cap H \to f(V \cap H)$$

(which still includes $D_1 \cap H, \cdots, D_k \cap H$ since they are collapsed to point in V_0). Finally, we obtain a situation contradicting the lemma, with $\dim V = \dim V_0 = 2$. Then by [7], p. 6, the intersection matrix (D_i, D_j) is negative definite, hence the D's must be homologically independent. Q.E.D.

To complete the proof of (a), we use Kummer Theory. If

$$\pi: W' \to V - Z$$

is an unramified algebraic p-cyclic extension, it is obtained by normalizing $V - Z$ in the extension of $\mathbf{C}(V)$ given by $\sqrt[p]{f}$, some $f \in \mathbf{C}(V)$. Define W to be the normalization of V in this extension field. Let $D_1, \cdots, D_n$ be the divisor components of Z, and let

$$(f) = \sum_{i=1}^{n} n_i D_i + \sum_{i=1}^{m} m_i E_i .$$

Since $\sqrt[p]{f}$ defines an unramified covering outside of Z, it follows that $p \mid m_i$ all i. Therefore in $H^2(V, \mathbf{Z})$,

$$0 = \operatorname{class}((f)) = \Sigma n_i \cdot \operatorname{class}(D_i) + p \cdot \operatorname{class}\left(\Sigma \frac{m_i}{p} E_i\right).$$

But since by the lemma, class (D_i), $1 \leq i \leq n$, in $H^2(V, \mathbf{Z})$ are independent, for almost all primes p this relation implies that p divides all the n_i as well. In that case

$$(f) = p \cdot D$$

for some divisor D on V, and $\sqrt[p]{f}$ defines an everywhere unramified covering of V, i.e. W/V is unramified. Q.E.D.

Corollary 1. *Let V be a non-singular algebraic variety. Assume that V is isomorphic to $V^* - Z^*$ where V^* is a projective variety and* $\operatorname{codim} Z^* \geq 2$. *Let*

$$\phi: V \to A$$

be the Albanese morphism of V. Then

$$\phi_*: H_1(V, Q) \xrightarrow{\sim} H_1(A, Q).$$

Proof. Let V' be a desingularization of V^* obtained by blowing up only points of Z^*, and let

$$f: V' \to V$$

be the birational morphism. Let $Z' = f^{-1}(Z^*)$. Then $V' - Z' \cong V$. The morphism ϕ extends to a morphism ϕ':

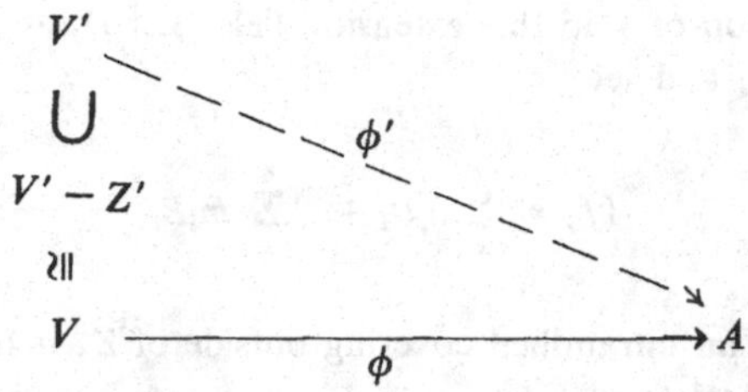

since V' is non-singular ([3], p. 20), and ϕ' is the Albanese morphism for V'. Then it is well-known# that

$$\phi': H_1(V', Q) \xrightarrow{\sim} H_1(A, Q).$$

This, plus the Theorem applied to $f: V' \to V$, and Z', imply the Corollary. Q.E.D.

This Corollary does not apply directly to the moduli variety M since M is not non-singular. However, if $g \geqq 3$ M does meet the other condition of the Corollary. In fact, let M_A denote the moduli variety for g-dimensional principally polarized abelian varieties, and let M_A^* denote the Satake compactification of M_A. Let

$$\theta: M \to M_A$$

This is proven, for example by constructing A as

$$H_1(V', \mathbf{R})/\text{Image}\ [H_1(V', \mathbf{Z})]$$

and constructing ϕ' by integrating the differentials $H^0(V', \Omega^1_{V'})$ cf. [10], p. 82.

be the morphism induced by associating the Jacobian variety and its Θ-polarization to each curve of genus g. In [8], A. Mayer and I determined the set $\overline{\theta(M)} - \theta(M)$ (closure in M_A^*). It turns out to be what you expect — g-dimensional products of lower dimensional jacobians in M_A itself, as Matsusaka and Hoyt had shown; h-dimensional products of lower dimensional jacobians ($h < g$), in $M_A^* - M_A$. In particular, for each type of product decomposition we get a locally closed subset of $\overline{\theta(M)} - \theta(M)$. There is no piece of dimension $3g-4$ (codimension 1). The only piece of dimension $3g-5$ is the locus corresponding to products

$$J_1 \times J_{g-1}$$

J_1, J_{g-1} being 1 and $(g-1)$-dimensional jacobians respectively. The other pieces, such as:

a) $J_i \times J_j$, $i+j=g$, $1<i$, $j<g$

b) J_{g-1} (giving a point in $(M_A^* - M_A) \cap \overline{\theta(M)})$ have codimension at least 3.

To make the Corollary apply, choose a normal subgroup $\Gamma' \subset \Gamma$ of finite index that acts freely on the Teichmüller space D. Let $M' = D/\Gamma'$ and let $\pi: M' \to M$ be the canonical morphism. Then M' is non-singular *and* compactifiable in codimension 2 (take the normalization of $\overline{\theta(M)}$ in the function field of M'). Therefore, we get

$$(\Gamma'/[\Gamma',\Gamma']) \otimes Q \cong H_1(M', Q) \cong H_1(A', Q)$$

if $\phi': M' \to A'$ is the Albanese morphism. Furthermore, the finite group (Γ/Γ') acts on $\Gamma'/[\Gamma',\Gamma'] \otimes Q$, on M', and hence on A'.

(I). The canonical homomorphim

$$(\Gamma'/\Gamma',\Gamma']) \otimes Q \to (\Gamma/[\Gamma,\Gamma]) \otimes Q,$$

is surjective and its kernel is generated by elements

3 $$x - x^{\gamma},\ \gamma \in \Gamma/\Gamma',\ x \in (\Gamma'/[\Gamma',\Gamma']) \otimes Q.$$

(This is well-known; for example, use the Lyndon Spectral Sequence, [4], p. 354, formula (10.6)).

(II). If $\phi: M \to A$ is the Albanese of M, there is a canonical diagram:

$$\begin{array}{ccc} M' & \xrightarrow{\phi'} & A' \\ \pi \downarrow & & \downarrow \chi \\ M & \xrightarrow[\phi]{} & A \end{array}$$

for some homomorphism χ. χ: $A' \to A$ can be characterized as the universal homomorphism $\psi: A' \to B$, B abelian, such that $\psi(x^\gamma) = \psi(x)$, all $x \in A'$, $\gamma \in \Gamma/\Gamma'$. Therefore

$$A \cong A'/[\text{Subgroup generated by points } x^\gamma - x,\ x \in A',\ \gamma \in \Gamma/\Gamma'].$$

Using this, it is easy to see that

$$H_1(A,Q) \cong H_1(A',Q)/[\text{Subgroup generated by points } x^\gamma - x,\ x \in H_1, \gamma \in \Gamma/\Gamma'].$$

Putting all this together, we finally conclude:

Corollary 2. *If $\phi: M \to A$ is the Albanese morphism of the moduli variety M and $g \geqq 3$, then there is a canonical isomorphism:*

$$(\Gamma/[\Gamma,\Gamma]) \otimes Q \xrightarrow{\sim} H_1(A,Q).$$

§2. Dehn's presentation of Γ.

Everything that follows depends on the fundamental paper [1] of Dehn. Let F be a fixed differentiable surface of genus g. We shall picture F as follows:

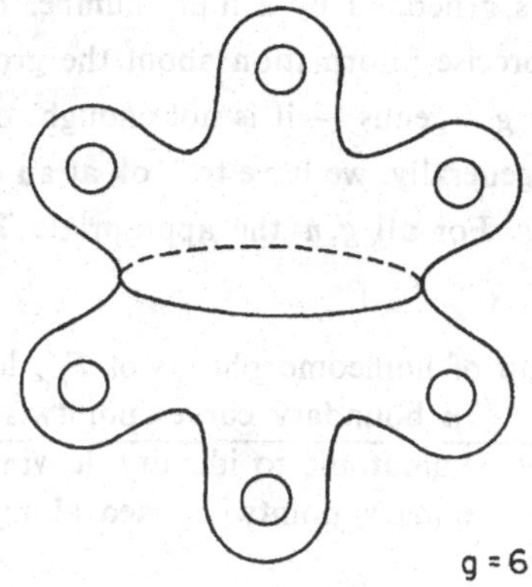

Figure 1

The first basic fact is that:

$$\Gamma \cong \frac{\{\text{Group of orientation preserving homeomorphisms of } F\}}{\{\text{Those homotopic to identity}\}}$$

$$\cong \frac{\{\text{Group of automorphisms of } \pi_1(F, p), \det = +1\}}{\{\text{Inner Automorphisms}\}}.$$

Here the determinant of an automorphism $\alpha: \pi_1(F) \to \pi_1(F)$ refers to the determinant of the induced map on $\pi_1/[\pi_1, \pi_1]$, which is a lattice of rank $2g$. To obtain generators of Γ, Dehn introduced particular homeomorphism classes of F into itself that he called "*Schraubungen*" — we shall call them "*screw maps*". Let γ be a simple closed curve on F: Draw a small collar around γ, i.e. choose a continuous injective orientation-preserving map $f: [0,1] \times S^1 \to F$ such that $\gamma = f[(\frac{1}{2}) \times S^1]$. The screw map S_γ associated to γ is defined by

$$\begin{cases} S_\gamma(x) = x \quad \text{if } x \notin \text{Im}(f). \\ S_\gamma(f(\alpha, \theta)) = f(\alpha, \theta + 2\pi\alpha). \end{cases}$$

In other words, if γ_α is the curve $f[(\alpha) \times S^1]$, we rotate γ_α through an angle $2\pi\alpha$, ranging from 0 on γ_0, to π on γ itself, to 0 again on γ_1*. The main theorem

* Note that this screw map depends, modulo homotopy, only on γ and the orientation of F, and *not* on an orientation for γ.

of his paper is that Γ is generated by a finite number of these screw maps (p. 203). To get more precise information about the group Γ — we should have called it Γ_g, where g = genus — it is not enough to refer to the groups Γ_h, where $h < g$. More generally, we have to look at an oriented surface $F_{g,n}$ of genus g *with n holes*. For all g, n the appropriate Teichmüller modular group $\Gamma_{g,n}$ is:

$$\Gamma_{g,n} = \left\{ \frac{\text{Group of homeomorphisms of } F_{g,n} \text{ leaving the } n \text{ boundary curves pointwise fixed}}{\text{Those homotopic to identity, leaving the boundary pointwise fixed along the way}} \right\}.$$

Note that whenever we choose a continuous, injective orientation preserving map $f: F_{h,n} \to F$, we obtain a map $f_*: \Gamma_{h,n} \to \Gamma$, since homeomorphisms of $F_{h,n}$ extend uniquely to homeomorphisms of F which are the identity on $F - f(F_{h,n})$.

Dehn's main result is that $\Gamma_{g,n}$ is generated by the screw maps along the set of curves β_k and γ_{ij}:

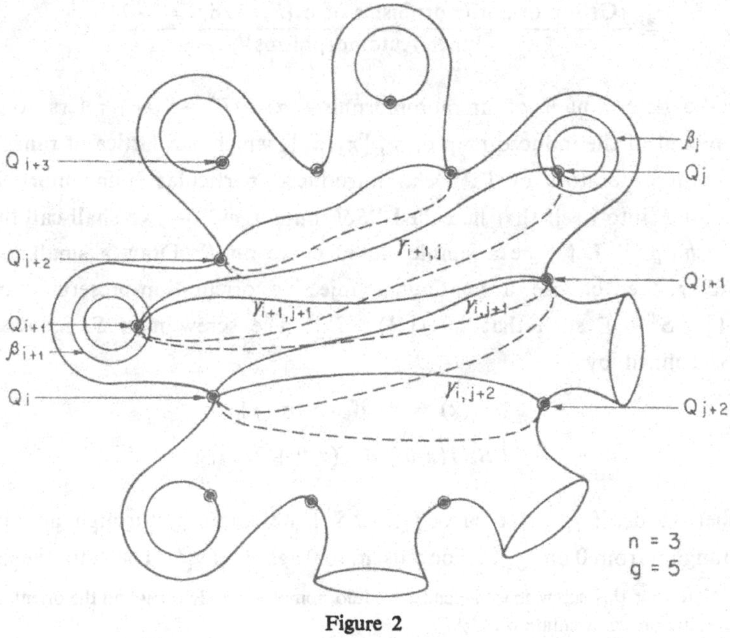

Figure 2

This result can easily be improved. On our surface F, consider the following piece homeomorphic to $F_{2,1}$:

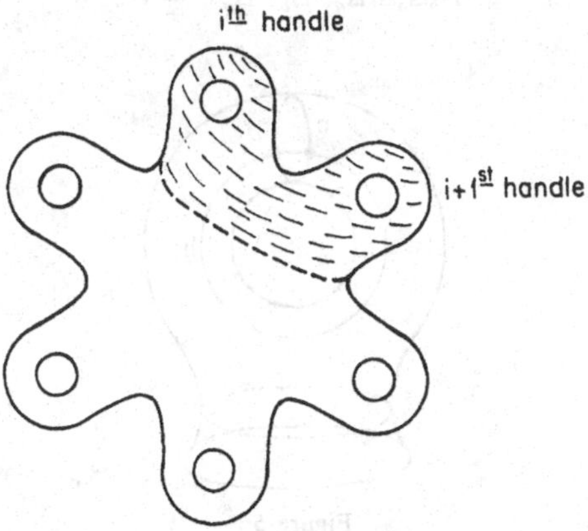

Figure 3

For all i, $1 \leqq i \leqq g$, this defines $f_{i,*}: \Gamma_{2,1} \rightarrow \Gamma$.

Theorem 4. *Γ is generated by the images of $\Gamma_{2,1}$ under $f_{1,*}, \cdots, f_{g,*}$.*

Before beginning to prove this, let me list what is known about the groups $\Gamma_{g,n}$ for low g and n:

(a) $\Gamma_{0,0} = \Gamma_{0,1} = (e)$

(b) $\Gamma_{0,2} = \boldsymbol{Z}$: $F_{0,2}$ is just a collar, and $\Gamma_{0,2}$ is a free group on the screw map along its midline.

(c) $\Gamma_{0,3} = \boldsymbol{Z} \oplus \boldsymbol{Z} \oplus \boldsymbol{Z}$, with generators given by screw maps along curves which are the 3 boundary curves, pulled slightly in:

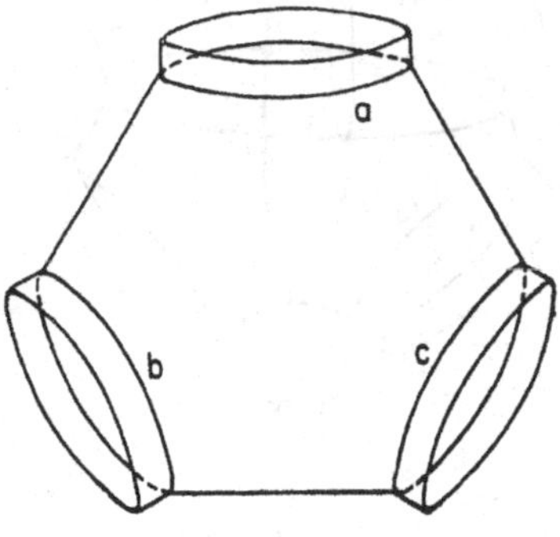

Figure 4

(d) $\Gamma_{1,1}$ = free group on screw maps Δ_a, Δ_b along a, b, modulo the relation

$$\Delta_b\Delta_a\Delta_b\Delta_a^{-1}\Delta_b^{-1}\Delta_a^{-1} = e$$

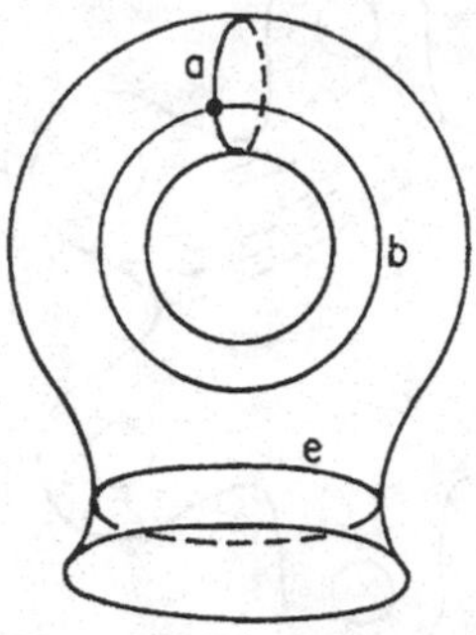

Figure 5

Moreover, $\Delta_e = (\Delta_a\Delta_b\Delta_a)^4$. (Dehn, pp. 156 and 172).

(e) $\Gamma_{0,4}$ is generated by screw maps Δ_a, Δ_b, Δ_{r_1}, Δ_{r_2}, Δ_{r_3}, Δ_{r_4} where the Δ_{r_i}'s are in the center. It appears that there are no other relations, i.e. $\Gamma_{0,4} \cong Z^4 \oplus$ (free group on 2 elements).

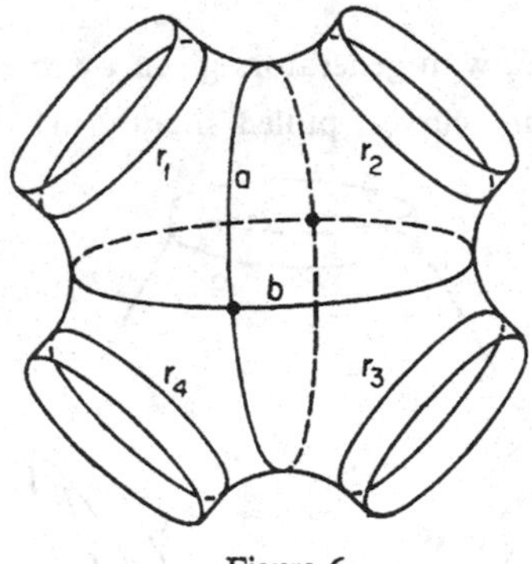

Figure 6

(f) $\Gamma_{1,2}$ is generated by screw maps Δ_a, Δ_b, Δ_c, Δ_e and Δ_f:

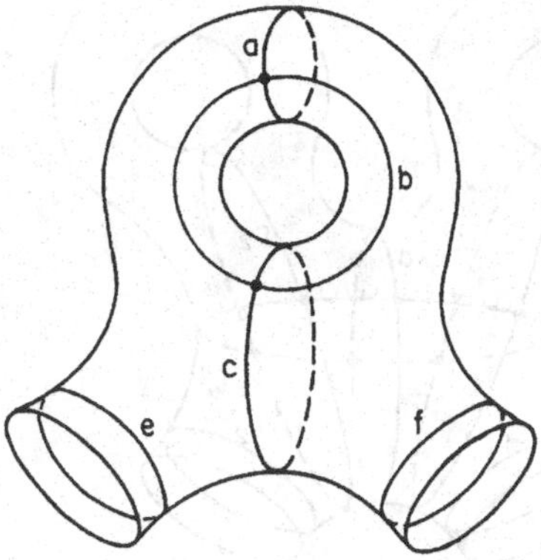

Figure 7

(Dehn, p. 188). The relations among these were worked out by Magnus [5], in a somewhat different form. To begin with, one has the obvious relations:

$$\Delta_e, \Delta_f \in \text{Center}$$
$$\Delta_a\Delta_c = \Delta_c\Delta_a.$$

Much subtler is the following*

$$\Delta_e\Delta_f = (\Delta_a\Delta_b)^2\Delta_c\Delta_b(\Delta_a\Delta_b)^2\Delta_b\Delta_c\Delta_b\Delta_c^{-1}.$$

There are still further relations that we will not list. $\Gamma_{1,2}$ is the first really complicated group and it is related to Artin's Braid group.

Proof of Theorem 4. Because of (f), the mapping class group $\Gamma_{1,2}$ is generated by Δ_a, Δ_b, Δ_c and *either* Δ_e or Δ_f, whichever is convenient. We use this to prove:

Lemma. *$\Gamma_{2,2}$ is generated by*

(a) $\Gamma_{2,1}$

(b) Δ_{c_1} *or* Δ_{c_2}, *whichever you want*

(c) Δ_{e_1} *or* Δ_{e_2} *whichever you want* (cf. Figure 8).

* This can be readily derived from the formula

$$[\Theta^{-1}u\Theta, \Theta^{-1}a\Theta, \Theta^{-1}b\Theta] = s^{-4}\rho\tau^{-1}\rho^{-1}\tau$$

on p. 638 of Magnus' article.

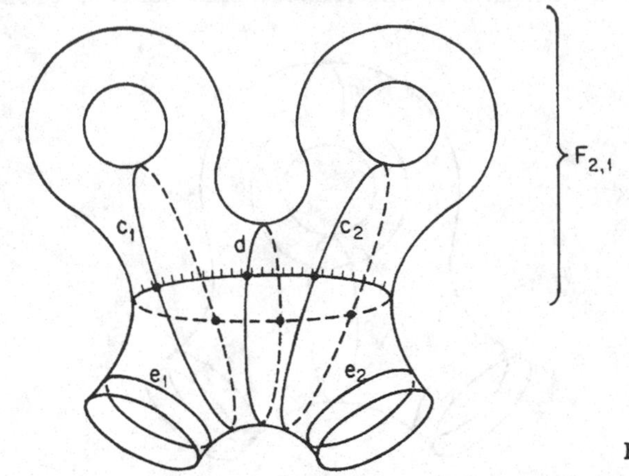

Figure 8

Proof. According to Dehn's result, $\Gamma_{2,2}$ is generated by $\Gamma_{2,1}$, Δ_{c_1}, Δ_{c2}, Δ_d, Δ_{e_1} and Δ_{e_2}. But first of all Δ_{c_1} and Δ_{c_2} are conjugate with respect to a suitable element of $\Gamma_{2,1}$ (Dehn, §10, (1), p. 200; it is not hard to work this case out explicitly). Secondly, suppose you cut the surface open along d: then $F_{2,2} = F'_{1,2} \cup F''_{12}$. Applying the results of (f) to each piece, it follows that

1) $\Delta_{e_1} \cdot \Delta_d \in$ group generated by $\Delta_{c_1}, \Gamma_{2,1}$
2) $\Delta_d \cdot \Delta_{e_2} \in$ group generated by $\Delta_{c_2}, \Gamma_{2,1}$. Q.E.D.

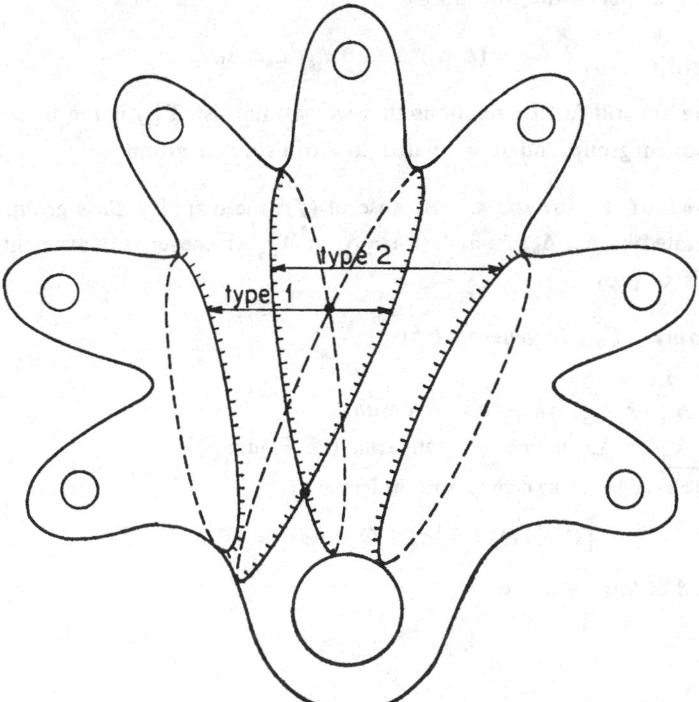

Figure 9

Apply this lemma to the $\Gamma_{2,2}$'s of types 1, 2 in Figure 9. It follows that any of the screw maps $\Delta_{\gamma_{ij}}$ of Fig. 2 can be written successively as products involving various elements of the $f_{k,*}(\Gamma_{2,1})$'s and screw maps $\Delta_{\gamma_{i',j'}}$ where $|i'-j'|<|i-j|$. Loosely speaking, modulo terms in the $f_{k,*}(\Gamma_{2,1})$'s, one replace $\Delta_{\gamma_{ij}}$ by $\Delta_{\gamma_{i',j'}}$'s with $\gamma_{i',j'}$ wrapping round a smaller part of F. This evetually eliminates the Δ_γ's altogether. Q.E.D.

Corollary 1. *Γ is generated by the four homeomorphisms Δ_a, Δ_b, Δ_d and R* (cf. Fig. 10).

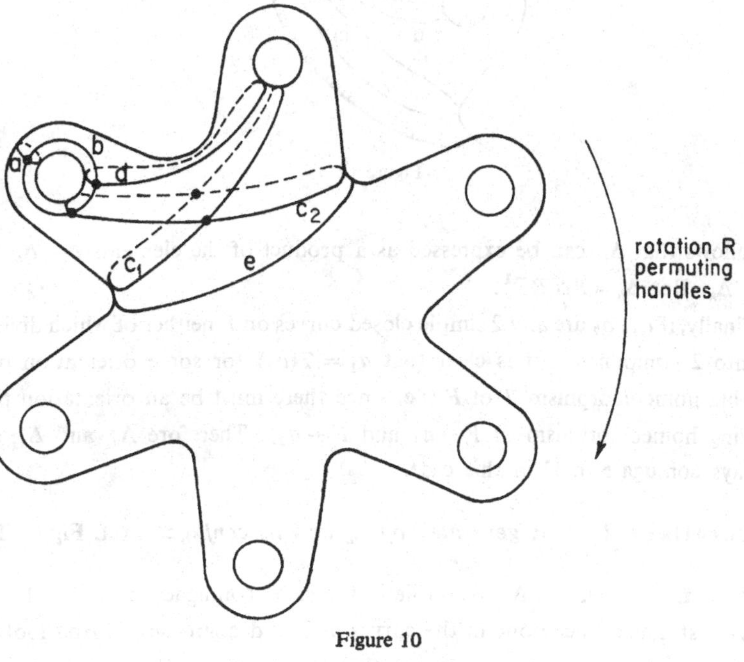

Figure 10

Proof. According to the Theorem, R plus Δ_a, Δ_b, Δ_{c_1}, Δ_{c_2}, Δ_d and Δ_e certainly suffice. On the other hand, cutting out an $F_{1,2}$ as follows:

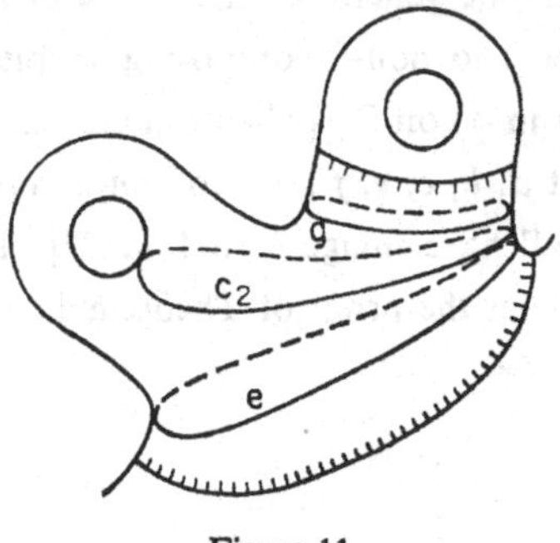

Figure 11

it follows from (f) that Δ_e can be expressed as a product of the elements Δ_a, Δ_b, Δ_{c_2} and Δ_g; and Δ_g is a product of the elements $R\Delta_a R^{-1}$, $R\Delta_b R^{-1}$. This gets rid of Δ_e. Secondly, cutting out an $F_{1,2}$ like this:

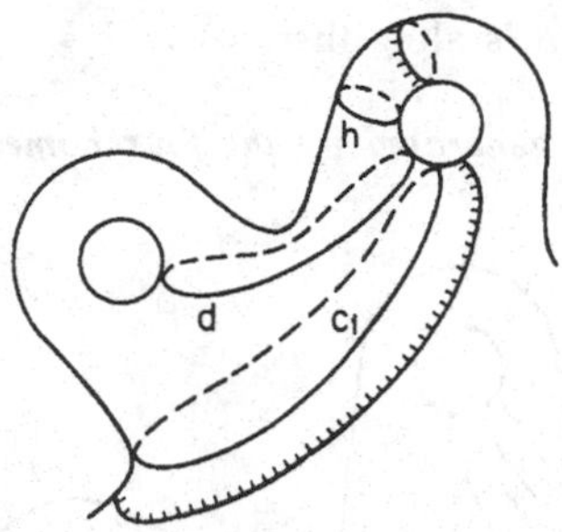

Figure 12

it follows that Δ_{c_1} can be expressed as a product of the elements Δ_a, Δ_b, Δ_d and Δ_h. But $\Delta_h = R\Delta_a R^{-1}$. Q.E.D.

Finally, if a_1, a_2 are any 2 simple closed curves on F neither of which divides F into 2 components, it is clear that $a_1 = T(a_2)$ for some orientation preserving homeomorphism T of F (i.e. since there must be an orientation preserving homeomorphism of $F - a_1$ and $F - a_2$). Therefore Δ_{a_1} and Δ_{a_2} are always conjugate in Γ in this case.

Corollary 2. *Γ is generated by Δ_a and its conjugates* (cf. Figure 10).

Proof. In fact, Δ_a, Δ_b, Δ_d all lie in the same conjugacy class by the remark just made, since none of the curves a, b, d decomposes F. And looking at the proof of Corollary 1, we see that we generated Γ by conjugates of Δ_a, Δ_b, Δ_d with respect to powers of R. Q.E.D.

Therefore, $\Gamma/[\Gamma,\Gamma]$ has one generator, the image of the conjugacy class δ of screw maps with respect to non-decomposing simple closed curves. Now refer back to the relation in (f) on $\Gamma_{1,2}$. We can embed $F_{1,2}$ in any surface F of genus $g (g \geqq 2)$ so that a, b, c, e, f go into *non-decomposing* curves. Then Δ_a, Δ_b, Δ_c, Δ_e and Δ_f all have image δ in $\Gamma/[\Gamma,\Gamma]$, and the relation says that $10\delta = 0$. This completes the proof of Theorem 1.

Theorem 2 follows from Theorem 1 by the results of §1 if $g \geqq 3$. If $g = 2$, Theorem 1 is known from the results of Igusa [11].

Actually, when $g = 2$, it is not hard to check that $\Gamma/[\Gamma,\Gamma]$ is exactly $\mathbf{Z}/10\mathbf{Z}$. Closely related to this is the fact that the Picard group Pic($\mathcal{M}$) of the moduli problem of genus 2 is also $\mathbf{Z}/10\mathbf{Z}$.

Proof. The methods in [6], §7 carry over without modification to prove that

$$\{\text{Torsion subgroup of Pic}(\mathcal{M})\} \cong \text{Hom}(\Gamma/[\Gamma,\Gamma], \mathbf{Q}/\mathbf{Z}).$$

(cf. 1st Corollary, p. 77; Corollary, p. 81, [6]). On the other hand, if M is the moduli space of genus 2, then, *modulo torsion*, Pic(M) and Pic($\mathcal{M}$) are isomorphic. Igusa showed ([11], p. 638) that there is a finite morphism $\pi: \mathbf{A}^3 \to M$. So if $d = \text{degree}(\pi)$, and D is any Cartier divisor on M, then

$$\begin{aligned} d.D &= \pi(\pi^{-1}(D)) \\ &= \pi((f)), \text{ some } f \in \Gamma(A^3, o_A) \\ &= (Nmf). \end{aligned}$$

Therefore Pic(M) is torsion and Pic($\mathcal{M}$) and $\Gamma/[\Gamma,\Gamma]$ are dual finite abelian groups. But we can easily check that 2 and 5 divide the order of Pic($\mathcal{M}$). To do this, use the element $\delta \in \text{Pic}(\mathcal{M})$ gotten by attaching to each curve C of genus 2 the one-dimensional vector space:

$$\delta(C) = \Lambda^2[H^0(C, \Omega_C^1)]$$

(Ω^1 = sheaf of 1-forms on C; cf. [6], §5). Taking C_1 to be the curve $y^2 = x^5 - 1$, the automorphisms

$$\begin{cases} x \to \eta x \\ y \to y \\ \eta^5 = 1, \ \eta \neq 1 \end{cases}$$

of C_1 all act non-trivially on $\delta(C_1)$. Therefore $5 \mid \text{order}(\delta)$. Taking C_2 to be the curve $y^2 = x \cdot (x^4 - 1)$, the automorphisms

$$x \to \eta^2 \cdot x$$

$$y \to \eta \cdot x$$

$$\eta^8 = 1, \eta^4 \neq 1$$

all act non-trivially on $\delta(C_2)$, although their squares act trivially. Therefore $2 \mid \text{order}(\delta)$. Q.E.D.

BIBLIOGRAPHY

1. M. Dehn, Die Gruppe der Abbildungsklassen, *Acta Math.*, **69** (1938), p. 135.
2. H. Grauert and R. Remmert, Komplexe Räume, *Math. Ann.*, **136** (1958), p. 245.
3. S. Lang, Abelian varieties, Wiley-Interscience, N. Y., 1959.
4. S. MacLane, Homology, Springer-Verlag, Berlin-Göttingen-Heidelberg, 1963.
5. W. Magnus, Über Automorphismen von Fundamentalgruppen berandeter Flächen, *Math Ann.*, **109** (1934), p.617.
6. D. Mumford, Picard groups of moduli problems, *Proc. Symp. Arith. Alg. Geom.*, Harper and Row, N.Y., 1966.
7. D. Mumford, The topology of normal singularities of an algebraic surface, *Publ. Inst. Hautes Et. Sci.*, No. 9, Paris, 1960.

4 8. D. Mumford, article in Supplementary notes to the Woods Hole Summer Institute of the AMS, 1964.

9. J.-P. Serre, Géométrie analytique et géométrie algébrique, *Ann. Inst. Fourier*, **6** (1956), p. 1.
10. A. Weil, Variétés kählériennes, Hermann, Paris, 1958.
11. J.-I. Igusa, Arithmetic variety of moduli: for genus 2, *Annals. of Math.*, **72** (1960), p. 612.

HARVARD UNIVERSITY
CAMBRIDGE, MASS., U.S.A.

(Received May 3, 1966)

Inventiones math. 5, 317–334 (1968)

Deformations and Liftings of Finite, Commutative Group Schemes*

FRANS OORT (Amsterdam) and DAVID MUMFORD (Cambridge, Mass.)

1. Introduction

Consider the following problems:

(A) Given a field k, a finite k-group scheme N_0, and a ring R with a surjective ringhomomorphism $R \to k$. Does there exist a finite, flat R-group scheme N such that $N_0 \cong N \otimes_R k$? (If so, we say that N_0 is obtained from N by reduction mod $\mathfrak{m}$, where $\mathfrak{m} = \operatorname{Ker}(R \to k)$, or, we say that N is a lifting of N_0 to R.)

(B) Given a field k (of characteristic $p>0$), and a finite k-group scheme N_0. Does there exist a ring R (integral domain of characteristic zero) with a reduction $R \to k$, and a finite, flat R-group scheme N such that $N_0 \cong N \otimes_R k$?

The answers to (A) and to the weaker question (B) are negative in general. However if in (B) moreover is given that N_0 is a *commutative* finite group scheme, the answer is affirmative; it is the aim of this paper to give a proof of this fact via deformation theory of finite group schemes in characteristic $p>0$. As a byproduct we obtain a proof for the fact that any finite, local group scheme can be embedded into a formal Lie group with coefficients in the same field, on the same number of parameters.

Example (−A). Let k be a field of characteristic $p>0$ (e.g. the prime field $k=\mathbf{F}_p$), and let R be a ring with a reduction $R \to k = R/\mathfrak{m}$, such that $p \cdot 1 \notin \mathfrak{m}^2$ (an "unramified" situation) (e.g. $R = W_\infty(k)$, so $W_\infty(\mathbf{F}_p) = \mathbf{Z}_p$, the ring of p-adic integers, or $R = W_\infty(k)/p^2$). Let $N_0 = \alpha_{p,k}$, i.e. $N_0 = \operatorname{Spec}(k[\tau])$, $\tau^p = 0$, and the group law is defined by $s_0\colon E_0 \to E_0 \otimes_k E_0$, $E_0 = k[\tau]$, with $s_0(\tau) = \tau \otimes 1 + 1 \otimes \tau$; we claim that in this case the answer to problem (A) is negative. Suppose R to be local (localize if necessary), and suppose N as indicated could be found; then $N = \operatorname{Spec}(E)$, $E = R[\sigma]$, where $\sigma^p = a_1 \sigma + \cdots + a_{p-1} \sigma^{p-1}$ with $a_i \in \mathfrak{m}$; the group law would be given by some ringhomomorphism $s\colon E \to E \otimes_R E$, so

$$s(\sigma) = \sigma \otimes 1 + 1 \otimes \sigma + \sum b_{ij} \sigma^i \otimes \sigma^j, \qquad b_{ij} \in \mathfrak{m};$$

* This work was partially supported by NSF grant GP 3512.

as $(s\,\sigma)^p = s(\sigma^p)$, we obtain:

$$p\cdot(\sigma\otimes\sigma^{p-1}+\cdots+\sigma^{p-1}\otimes\sigma)\equiv 0 \qquad (\text{mod}\ \mathfrak{m}^2\cdot E\otimes E),$$

which is a contradiction.

Remark. In the previous situation, by a result of Tate (cf. [13]), we know that α_p can be lifted to R (e.g. R is a complete local ring) if and only if $p\in R$ admits a factorization $p=a\,b$, with $a\in\mathfrak{m}$, and $b\in\mathfrak{m}$.

Example (−B). Let R be an integral domain of characteristic zero, and let $N=\text{Spec}(E)$ be a finite R-group scheme such that E is a free R-module of rank p^2 (where p is a prime number). Then N is commutative. This can be seen as follows: let L be an algebraic closure of the field of fractions of R; we know that $N\otimes_R L$ is reduced (cf. [1], footnote on p. 109; cf. [9], lecture 25, theorem 1; cf. [11]), so by group theory it follows that $N\otimes L$, and hence that N is commutative. This shows that any non-commutative group scheme of rank p^2 cannot be lifted to characteristic zero. It is easy to give an example: take the kernel of the Frobenius homomorphism of a suitable non-commutative linear group. For example, let N_0 be given by: k is a field of characteristic p, and for any k-algebra B,

$$N_0(B)=\left\{\text{the multiplicative group of matrices } \begin{pmatrix}\alpha & \beta\\ 0 & 1\end{pmatrix},\ \alpha\in B,\ \beta\in B,\ \alpha^p=1,\ \beta^p=0\right\};$$

so $N_0=\text{Spec}(E_0)$, $E_0=k[\tau,\rho]$ with $\tau^p=1$, $\rho^p=0$, with $s_0(\tau)=\tau\otimes\tau$ and $s_0(\rho)=\rho\otimes 1+\tau\otimes\rho$.

2. Liftings of Deformations

The first example makes it clear that in order to lift a finite (local, unipotent) group scheme to characteristic zero, in general one has to allow ramification at p; but it is difficult to obtain directly from N_0 the information "how much ramification" is needed. Therefore we solve the problem B in the commutative case via deformation theory in characteristic $p>0$. The following lemma is a special case of a general principle: that specializations of liftable "objects" are liftable.

Lemma (2.1). *Assume we are given rings: $A\subset K\xleftarrow{\pi} R$, where R is a characteristic zero local domain, $\pi\colon R\to R/\mathfrak{m}=K$ its residue class map, and A a subring of K, and that we are given finite free group schemes over these rings*

$$\begin{array}{ccccc} N_0 & \longleftarrow & M_0 & \longrightarrow & M\\ \downarrow & & \downarrow & & \downarrow\\ \text{Spec}(A) & \longleftarrow & \text{Spec}(K) & \longrightarrow & \text{Spec}(R), \end{array}$$

where $M_0 \cong N_0 \otimes_A K \cong M \otimes_R K$. *Write* $R' = \{x \in R \mid \pi(x) \in A\}$; *there is a finite free group scheme* $N \to \mathrm{Spec}(R')$ *such that* $N_0 \cong N \otimes_{R'} A$ *and* $M \cong N \otimes_{R'} R$.

Proof. Let $N_0 = \mathrm{Spec}(E_0)$, $M_0 = \mathrm{Spec}(F_0)$, $M = \mathrm{Spec}(F)$. Then $F_0 \cong E_0 \otimes_A K \cong F \otimes_R K$. Identify E_0 with the corresponding subset of F_0, and identify F_0 with the corresponding quotient of F, so $E_0 \subset F_0 \xleftarrow{\pi'} F$. Each of these three is a free module of rank d, say, over either A, K or R, and has the structure of a bialgebra. Let $E = \{x \in F \mid \pi'(x) \in E_0\}$, and choose a basis $\{b_1, \ldots, b_d\}$ of E_0 over k; let $a_i \in F$ satisfy $\pi'(a_i) = b_i$; one checks easily that E is a free R'-module with basis $\{a_1, \ldots, a_d\}$. Moreover, one can also check

i) that the identity 1 of F is in E,

ii) E is closed under multiplication in the ring F,

iii) the comultiplication $F \to F \otimes_R F$ carries E in $E \otimes_{R'} E$,

iv) the augmentation $F \to R$ carries E in R',

v) the inverse $F \to F$ carries E to E.

Therefore $N = \mathrm{Spec}(E)$ is a finite free group scheme over R' with all the required properties.

Actually, what we need:

Corollary (2.2). *Let* $A = k$ *be a field, and let* N_0 *be a finite k-group scheme; this group scheme can be lifted to characteristic zero if and only if for some field extension* $k \subset K$ *(or for every field extension* $k \subset K$*),* $N_0 \otimes_k K$ *can be lifted to characteristic zero.*

The "if" part follows from (2.1). The "only if" part for example is an easy consequence of the place extension theorem (cf. EGA 0_{III}, 10.3.1).

Corollary (2.3). *Let* $k \leftarrow A \hookrightarrow K$ *be ringhomomorphisms, and let* $N_0 = \mathrm{Spec}(E_0)$ *be a finite free A-group scheme such that* $N_0 \otimes_A K$ *can be lifted to characteristic zero. Then* $N_0 \otimes_A k$ *can be lifted to characteristic zero.*

If $N_0 \cong N \otimes_{R'} A$, then $N \otimes_A k \cong N \otimes_{R'} A \otimes_A k \cong N \otimes_{R'} k$.

3. Moduli of Rigidified Local Group Schemes

It is clear that in general the moduli functor for finite group schemes is not representable.

Example. Let $\mathrm{char}(k) = p > 0$, take $B = k[T]$, and define a B-bialgebra by $E = B[\tau]$ with $\tau^p = T\tau$ and $s(\tau) = \tau \otimes 1 + 1 \otimes \tau$; for any field $K \supset k$ and for any $t \in \mathrm{Spec}(B)(K)$ with $t \neq 0$ (i.e. for any k-algebra homomorphism φ: $B \to K$ such that $\varphi(T) \neq 0$) E_t is the bialgebra of a reduced

22*

group scheme, isomorphic to $\mathbf{Z}/p$ in case K is algebraically closed, while E_0 is the bialgebra of the group scheme α_p.

However by an obvious rigidification of the underlying scheme of the group schemes we can obtain a moduli space. In order to see that any finite group scheme admits a nice deformation we would like to know that this moduli space is irreducible. It is easy to see it is connected, and by imposing extra conditions we can actually obtain a variety.

First we recall the following fact, due to Dieudonné and Cartier. Let N be a finite local k-group scheme, where k is a *perfect* field; $N = \mathrm{Spec}(E)$. Then there exist integers $v_1, \ldots, v_m$ and an isomorphism

$$E \cong k[X_1, \ldots, X_m]/(X_1^{p\exp(v_1)}, \ldots, X_m^{p\exp(v_m)})$$

(cf. SGAD, Exp. VII$_B$, 5.4; we are writing $p\exp(a) = p^a$ for typographical reasons); in this case we say that E admits a *truncation type* $v = (v_1, \ldots, v_m)$.

By the way, the following example shows that in general a finite local group scheme over an imperfect field does not admit a truncation type: let $a \in k$, $a \notin k^p$, $E = k[X, Y]/(X^{p^2}, X^p - aY)$, and $s(X) = X \otimes 1 + 1 \otimes X$, $s(Y) = Y \otimes 1 + 1 \otimes Y$.

Notation. Let $\alpha = (\alpha_1, \ldots, \alpha_m)$ be a set of non-negative integers; we write X^α for

$$X^\alpha = X_1^{\alpha_1} \times \cdots \times X_m^{\alpha_m}$$

(with $X_i^0 = 1$), and we denote by $|\alpha| = \alpha_1 + \cdots + \alpha_m$.

Definition. Let p be a prime number, $v = (v_1, \ldots, v_m)$ a set of positive integers, and $\mu = X^\alpha$ a monomial in m variables, where $\alpha = (\alpha_1, \ldots, \alpha_m)$. We say that μ satisfies the condition $(P\,v)_i$ for $1 \leqq i \leqq m$, if there exists an index j such that

$$\alpha_j \cdot p^{v_i} \geqq p^{v_j}$$

or, equivalently $(X^\alpha)^{p\exp(v_i)}$ is in the ideal generated by $X_1^{p\exp(v_1)}, \ldots, X_m^{p\exp(v_m)}$. We say that a polynomial in $X_1, \ldots, X_m$ satisfies $(P\,v)_i$ if is can be written as a sum of monomials which all satisfy condition $(P\,v)_i$. We say that a polynomial in the variables $X_j \otimes X_k$, $1 \leqq j \leqq m$, $1 \leqq k \leqq m$, satisfies condition $(P\,v)_i$ if it can be written as a sum

$$\sum_t \mu_{1t} \otimes \mu_{2t}$$

where μ_{1t} and μ_{2t} are monomials such that for each index t either μ_{1t} or μ_{2t} satisfies $(P\,v)_i$. Analogous definition for a polynomial in the variables $X_j \otimes X_k \otimes X_l$.

Remark. Let B be an *integral* domain of characteristic p, and let $N = \mathrm{Spec}(E)$ be a finite B-group scheme, $E = B[\tau_1, \ldots, \tau_m]$ with $\tau_i^{p\exp(v_i)} = 0$,

$1 \leqq i \leqq m$; the comultiplication is denoted by $s\colon E \to E \otimes E$. As s is a ringhomomorphism it follows that $(s\,\tau_i)^{p\exp(\nu_i)}=0$, so $s(\tau_i)$ is a polynomial in $\tau_j \otimes \tau_k$ which satisfies condition $(P\,\nu)_i$. The same for the polynomials $\gamma(\tau_i)$, where $\gamma\colon E \to E$ defines the inverse.

We fix k, a field of characteristic $p>0$, and $\nu=(\nu_1, \ldots, \nu_m)$, a set of positive integers; $\mathbf{C}=\mathbf{C}_k$ denotes the category of k-algebras. Define a functor $\Sigma_\nu=\Sigma\colon \mathbf{C} \to \mathbf{Ens}$ by:

$\Sigma(B)=\{$all cocommutative B-bialgebra structures on $B[\tau_1, \ldots, \tau_m]=E$, such that $s(\tau_i)$ are polynomials satisfying condition $(P\,\nu)_i$ for $1 \leqq i \leqq m\}$,

where $\tau_i^{p\exp(\nu_i)}=0$ for $1 \leqq i \leqq m$, and where the augmentation ideal of E is generated by $\tau_1, \ldots, \tau_m$. Note that a B-bialgebra F can correspond to various elements of $\Sigma(B)$, as there may exist several isomorphisms $F \cong B[\tau_1, \ldots, \tau_m]$.

Theorem (3.1). *We fix k, and $\nu=(\nu_1, \ldots, \nu_m)$ as before; the functor $\Sigma\colon \mathbf{C} \to \mathbf{Ens}$ is represented by a k-algebra U, and there exists an integer n such that $U \cong k[T_1, \ldots, T_n]$.*

It is easy to see that Σ is representable; however the first step of the proof will be more complicated as we want to obtain information for late use.

Proof, first step: Σ is representable. Consider all combinations $(i,\ \alpha=(\alpha_1, \ldots, \alpha_m),\ \beta=(\beta_1, \ldots, \beta_m))$ such that $1 \leqq i \leqq m$, $0 \leqq \alpha_j < p\exp(\nu_j)$, $0 \leqq \beta_j < p\exp(\nu_j)$, and such that the monomial $\tau^\alpha \otimes \tau^\beta$ satisfies condition $(P\,\nu)_i$ (i.e. either $(\tau^\alpha)^{p\exp(\nu_i)}=0$, or $(\tau^\beta)^{p\exp(\nu_i)}=0$), and such that $|\alpha|>0$ and $|\beta|>0$; let $A=k[\ldots, Y_{i,\alpha,\beta}, \ldots]$, and let $F=A[\tau_1, \ldots, \tau_m]$ with $\tau_i^{p\exp(\nu_i)}=0$, $1 \leqq i \leqq m$. Then we are given an A-algebra homomorphism

$$s\colon F \to F \otimes_A F$$

by

$$s(\tau_i)=\tau_i \otimes 1 + 1 \otimes \tau_i + \sum_{\alpha,\beta} Y_{i,\alpha,\beta}\, \tau^\alpha \otimes \tau^\beta$$

(s is a ringhomomorphism because of the conditions $(P\,\nu)_i$, but this is not the point where these conditions are used essentially). Let μ_1, $\mu_2, \ldots$ be all non-zero monomials of the form $\tau^\alpha \otimes \tau^\beta \otimes \tau^\gamma$; we write $\Gamma s=(s \otimes 1)\cdot s-(1 \otimes s)\cdot s$, and

$$(\Gamma s)(\tau_i)=\sum_j H_{ij}\,\mu_j, \qquad 1 \leqq i \leqq m,$$

with $H_{ij} \in A$; let $\mathfrak{p} \subset A$ be the ideal generated by these polynomials, and by the symmetry relations:

$$\mathfrak{p}=(\ldots, H_{ij}, \ldots, \ldots, Y_{i,\alpha,\beta}-Y_{i,\beta,\alpha}, \ldots)\cdot A.$$

We define $U=A/\mathfrak{p}$, and $E=U[\tau_1,\dots,\tau_m]$. It is clear that s induces a coassociative comultiplication

$$s\colon E\to E\otimes_U E,$$

defined by

$$s(\tau_i)=\tau_i\otimes 1+1\otimes\tau_i+\sum_{\alpha,\beta} y_{i,\alpha,\beta}\,\tau^\alpha\otimes\tau^\beta,$$

where $y_{i,\alpha,\beta}=Y_{i,\alpha,\beta} \bmod \mathfrak{p}$. Clearly the pair (U,E) represents the functor $\Omega_{m,\nu}=\Omega$ defined by:

$\Omega(B)=\{$all cocommutative coassociative B-algebra homomorphisms $s\colon E\to E\otimes_B E$, where $E=B[\tau_1,\dots,\tau_m]$, such that $s(x)\equiv x\otimes 1+1\otimes x \pmod{\mathfrak{a}\otimes\mathfrak{a}}$, $\mathfrak{a}=(\tau_1,\dots,\tau_m)\cdot E$, and such that $s(\tau_i)$ satisfies condition $(P\nu)_i$ for $1\leqq i\leqq m\}$.

The following lemma asserts that $\Sigma(B)\twoheadrightarrow\Omega(B)$:

Lemma (3.2). *Let B be a ring in which $p\cdot 1=0$, let $E=B[\tau_1,\dots,\tau_m]$ with $\tau_i^{p\exp(\nu_i)}=0$, $1\leqq i\leqq m$, and with augmentation ideal $\mathfrak{a}=(\tau_1,\dots,\tau_m)\cdot E$. Let $s\colon E\to E\otimes_B E$ be a B-algebra homomorphism such that*

$$s(x)\equiv x\otimes 1+1\otimes x \qquad (\mathrm{mod}\ \mathfrak{a}\otimes\mathfrak{a})$$

for all $x\in\mathfrak{a}$ (i.e. the augmentation is a left- and a right-coidentity), and such that $s(\tau_i)$ satisfies condition $(P\nu)_i$ for $1\leqq i\leqq m$. Then there exists a unique B-algebra homomorphism $\gamma\colon E\to E$ such that $m(\gamma\otimes 1)\,s(x)=0$ for all $x\in\mathfrak{a}$ (where $m\colon E\otimes_B E\to E$ is the multiplication).

Proof. We define $\gamma_1(\tau_i)=-\tau_i$; thus we have defined a B-algebra homomorphism $\gamma_1\colon E\to E$ having the property

$$m(\gamma_1\otimes 1)\,s(x)\in\mathfrak{a}^2 \qquad \text{for all } x\in\mathfrak{a},$$

and it is unique modulo $\mathfrak{a}^2$ among all having this property. Suppose for some $N\geqq 1$ there is given a B-algebra homomorphism $\gamma_N\colon E\to E$ such that

$$m(\gamma_N\otimes 1)\,s(x)=\rho_N(x)\in\mathfrak{a}^{N+1} \qquad \text{for all } x\in\mathfrak{a},$$

and such that $\gamma_N(\tau_i)$ satisfies condition $(P\nu)_i$ for $1\leqq i\leqq m$. It is easy to see that $\rho_N(\tau_i)$ satisfies condition $(P\nu)_i$; thus

$$\gamma_{N+1}(\tau_i)=\gamma_N(\tau_i)-\rho_N(\tau_i), \qquad 1\leqq i\leqq m,$$

defines a B-algebra homomorphism $\gamma_{N+1}\colon E\to E$; it is clear that

$$m(\gamma_{N+1}\otimes 1)\,s(\tau_i)\in\mathfrak{a}^{N+2} \qquad \text{for } 1\leqq i\leqq m,$$

and it is readily verified that if γ' also has the property $m(\gamma'\otimes 1)\,s(x)\in\mathfrak{a}^{N+2}$ for all $x\in\mathfrak{a}$, and $\gamma'(\tau_i)-\gamma_{N+1}(\tau_i)\in\mathfrak{a}^{N+1}$ for all i, then $\gamma'(x)\equiv\gamma_{N+1}(x)$ (mod

$\mathfrak{a}^{N+2}$) for all $x \in \mathfrak{a}$. Thus the construction of γ and its uniqueness follow by induction as $\mathfrak{a}^{|\gamma|}=0$.

Thus the ring U and the bialgebra structure on E represent the functor $\Sigma \cong \Omega$, and the first step of the proof is concluded. Let $W=\mathrm{Spec}(U)$; consider the point $0 \in W(k)$ defined by $y_{i,\alpha,\beta} \mapsto 0$, i.e. $s(\tau_i)=\tau_i \otimes 1 + 1 \otimes \tau_i$ and $\gamma(\tau_i)=-\tau_i$; that is the point corresponding to the rigidified group scheme $\alpha_{p\exp(v_1)} \times \cdots \times \alpha_{p\exp(v_m)}$.

The crucial part of the proof of the theorem is: $0 \in W(k)$ is a *non-singular* point of W (note that this is false if W were the moduli space of all rigidified group schemes, say of a fixed rank, not necessarily local; note that this is also false if W were the moduli space of all rigidified local group schemes, not all the v_i equal, and not imposing the extra conditions $(P\,v)_i$). This we can show in two ways. It can be deduced from results of Lazard about formal group laws; this will be done in the next section. We could also have used the group-cohomology as described in SGAD, Exp. III, especially p. III. 42/43, Theorem 3.5 (also cf. [8]), and using a result of G. Efroymson, which says that $H^3_{\mathrm{symm}}(N, \mathbf{G}_a)=0$ (trivial action of the commutative finite group scheme N on the additive linear group $\mathbf{G}_a$) (proved in his Harvard thesis, 1966, later generalized into a structure theorem about the cohomology ring $H^{\cdot}(N, \mathbf{G}_a)$, not yet published).

4. Finite Group Schemes and Buds

First we recall some definitions and results to be found in a paper by Lazard, cf. [5]. Let m and r be positive integers, R a ring (commutative, and $1 \in R$), and

$$f\colon R[X_1, \ldots, X_m]=E \to E \otimes_R E$$

an R-algebra homomorphism; we say that f defines an r-bud ("r-bourgeon") on m parameters, with coefficients in R if (we write $(f \otimes 1)\cdot f - (1 \otimes f)\cdot f = \Gamma f$):

$$(\Gamma f)(X_i) \equiv 0 \qquad (\text{mod degree } r+1) \text{ for } 1 \leqq i \leqq m$$

(degree means total degree in the variables $X_1 \otimes 1, \ldots, 1 \otimes X_m$); f and g define the same r-bud if and only if $f(X_i) \equiv g(X_i)$ (mod degree $r+1$) for $1 \leqq i \leqq m$ (cf. [5], p. 381, Definition 13.1); a system $f_1, f_2, \ldots$ such that f_r is an r-bud on m parameters, and such that f_r and f_{r+1} define the same r-bud is called a formal Lie group on m parameters. We write

$$\Lambda_{m,r}(R)=\Lambda(R)=\{\text{all cocommutative } r\text{-buds ("}r\text{-bourgeons abéliens") on } m \text{ parameters with coefficients in } R\};$$

clearly we have thus obtained a covariant functor $\Lambda_{m,r}$ defined on the category of commutative rings with identity; if $f \in \Lambda_{m,r}(E)$ and $\varphi\colon E \to R$

is a ring homomorphism we write $(\Lambda\varphi)(f)\in\Lambda_{m,r}(R)$ for the r-bud over R obtained from f, applying φ. Lazard has proved:

(i) (cf. [5], pp. 394–399, and previous pages). Let

$$N(m,r)=N=m\left(\binom{r+m}{m}-m-1\right);$$

there exists a universal

$$F_r\in\Lambda_{m,r}(A_r),\qquad A_r=\mathbf{Z}[T_1,\ldots,T_{N(m,r)}],$$

i.e. (A_r,F_r) represents the functor $\Lambda_{m,r}$, or: the map

$$\mathrm{RHom}(A_r,\mathrm{R})\twoheadrightarrow\Lambda_{m,r}(R)$$

defined by $\varphi\mapsto(\Lambda\varphi)(F_r)$ is bijective for every R.

(ii) The natural restriction map $\Lambda_{m,r+1}(R)\to\Lambda_{m,r}(R)$ is surjective if R is without integral torsion (cf. [5], p. 396, Lemma 15.2), hence, by (i), this map is surjective for every R; it corresponds to the inclusion map

$$A_r=\mathbf{Z}[T_1,\ldots,T_{N(m,r)}]\hookrightarrow A_{r+1}=\mathbf{Z}[T_1,\ldots,T_{N(m,r+1)}],$$

such that $F_r\in\Lambda_r(A_r)\subset\Lambda_r(A_{r+1})$ and $F_{r+1}\in\Lambda_{r+1}(A_{r+1})$ define the same r-bud.

(iii) Suppose f_r and f_{r+1} define the same r-bud on m parameters with coefficients in R; $(\Lambda\varphi_r)(F_r)=f_r$ and $(\Lambda\varphi_{r+1})(F_{r+1})=f_{r+1}$; then the diagram

$$\begin{array}{ccc} A_r & \hookrightarrow & A_{r+1} \\ & {\scriptstyle\varphi_r}\searrow & \downarrow{\scriptstyle\varphi_{r+1}} \\ & & R \end{array}$$

commutes. Hence

$$A=\bigcup A_r=\mathbf{Z}[T_1,T_2,\ldots]$$

represents the functor of all formal Lie groups on m parameters (cf. [5], p. 397, Theorem 15.1); in particular, any r-bud on m parameters can be extended to a formal Lie group on m parameters with coefficients in the same ring.

Suppose we fix k, a field of characteristic $p>0$, a positive integer m, and positive integers $v_1,\ldots,v_m$. We choose an integer r so that

$$r\geqq 3\cdot\sum_{i=1}^{m}(p\exp(v_i)-1).$$

We consider only rings R containing k, in particular $p\cdot 1=0$ in R. We restrict the functor Λ to the category of k-algebras; for such rings we define a functor Δ by:

$$\Delta_{m,r,v}=\Delta\subset\Lambda_{m,r}$$

$\Delta(R)=\{f\in\Lambda_{m,r}(R)$ such that $f(X_i)$ satisfies condition $(Pv)_i$ for $1\leqq i\leqq m\}$.

For $f \in \Delta(R)$, we define $\rho(f)$ by

$$\rho(f)(\tau_i) = f(X_i) \bmod (X_1^{p\exp(v_1)}, \ldots, X_m^{p\exp(v_m)});$$

because of the conditions $(P\,v)_i$ we thus obtain an R-algebra homomorphism (!)

$$\rho(f)\colon E \to E \otimes_R E, \quad E = R[\tau_1, \ldots, \tau_m],$$

where $\tau_i^{p\exp(v_i)} = 0$, $1 \leqq i \leqq m$, and because of the choice of r it follows that

$$(\Gamma s)(\tau_i) = 0, \qquad 1 \leqq i \leqq m,$$

so $\rho(f) \in \Omega(R)$ (in the notation introduced in Section 3). So we have the following morphisms of functors (defined on k-algebras):

$$\Sigma \cong \Omega_{m,v} = \Omega \leftarrow \Delta_{m,r,v} \subset \Lambda_{m,r}.$$

Proposition (4.1). *We fix k, m, $v_1, \ldots, v_m$, and $r \geqq 3 \cdot \sum (p\exp(v_i) - 1)$ as before. The functors*

$$\Lambda, \Delta, \Omega\colon \mathbf{C} \to \mathbf{Ens}$$

are representable, say by L, D, and W. The schemes D and W (and also L) are isomorphic to affine spaces over k. In suitable coordinates the morphism $\rho\colon D \to W$ is given by a projection

$$D \cong \operatorname{Spec}(k[T_1, \ldots, T_n, T_1', \ldots, T_m']) \to \operatorname{Spec}(k[T_1, \ldots, T_n]) \cong W;$$

in particular, for every $R \supset k$ the map $\rho\colon D(R) \to W(R)$ is surjective.

In order to deduce these facts from Lazard's results, we need the following tools:

Lemma (4.2). *Let*

$$f(X_i) = \sum_{\alpha, \beta} a_{i,\alpha,\beta} X^\alpha \otimes X^\beta$$

be polynomials with coefficients in a ring R with $p \cdot 1 = 0$, such that $f(X_i)$ satisfies condition $(P\,v)_i$, $1 \leqq i \leqq m$; then $(f \otimes 1) f(X_i)$, and also $(1 \otimes f)$ $f(X_i)$, can be written as a sum of monomials satisfying condition $(P\,v)_i$.

Proof.

$$(f \otimes 1) f(X_i) = \sum_{\alpha,\beta} a_{i,\alpha,\beta} \{\prod_j [\sum a_{j,\gamma,\delta} X^\gamma \otimes X^\delta]^{\alpha_j}\} \otimes X^\beta = \sum_{\alpha,\beta} a_{i,\alpha,\beta} Q_{i,\alpha,\beta}.$$

It suffices to consider each $Q_{i,\alpha,\beta}$ separately; either X^β satisfies condition $(P\,v)_i$, and we are done, or there exists an index e such that $\alpha_e \cdot p\exp(v_i) \geqq p\exp(v_e)$, so $p\exp(n + v_i) \geqq p\exp(v_e)$ with $\alpha_e \geqq p^n$, and $n \geqq 0$; in that case

$$\begin{aligned} Q_{i,\alpha,\beta} &= \{[\sum a_{e,\gamma,\delta} X^\gamma \otimes X^\delta]^{p^n} \times (\cdots)\} \otimes X^\beta \\ &= \{\{\sum [a_{e,\gamma,\delta} X^\gamma \otimes X^\delta]^{p^n}\} \times (\cdots)\} \otimes X^\beta; \end{aligned}$$

for each (e, γ, δ) there exists an index d such that $\gamma_d \cdot p\exp(v_e) \geqq p\exp(v_d)$, or $\delta_d \cdot p\exp(v_e) \geqq p\exp(v_d)$, hence

$$p^n \cdot \gamma_d \cdot p\exp(v_i) \geqq \gamma_d \cdot p\exp(v_e) \geqq p\exp(v_d),$$

or the same with δ_d, and $(Q_{i,\alpha,\beta})^{p\exp(v_i)}$ is divisable by $(X_d \otimes 1 \otimes 1)^{p\exp(v_d)}$, respectively divisable by $(1 \otimes X_\alpha \otimes 1)^{p\exp(\gamma_d)}$, and the lemma is proved.

Lemma (4.3). *Let R be a ring, M an ideal in R, and $b \in R$ so that $M \cdot b = 0$. Let $E = R[X_1, \ldots, X_m]$, and $g: E \to E \otimes E$ so that*

$$g(X_i) \equiv X_i \otimes 1 + 1 \otimes X_i \qquad (\text{mod } M \cdot E \otimes E).$$

Let $P = b\,X^\alpha \otimes X^\beta$ be a monomial such that X^α and X^β do not satisfy condition $(P\,v)_i$ (for some fixed index i); then $(g \otimes 1)(P)$, and also $(1 \otimes g)(P)$, can be written as a sum of monomials none of which satisfy condition $(P\,v)_i$.

Proof.

$$(g \otimes 1)(P) = b \cdot g(X^\alpha) \otimes X^\beta = b \cdot \{\prod_j (X_j \otimes 1 + 1 \otimes X_j)^{\alpha_j}\} \otimes X^\beta$$

as $M \cdot b = 0$, and the lemma is proved.

Let k be a field, W a k-algebraic scheme, and $w \in W(k)$. The following statements are known to be equivalent:

(i) w is a non-singular point on W;

(ii) the local ring $\mathcal{O}$ of w on W is a regular local ring, i.e. its completion $\hat{\mathcal{O}}$ is a formal power series ring $\hat{\mathcal{O}} \cong k[[e_1, \ldots, e_n]]$;

(iii) (Grothendieck's criterion, cf. SGA, III.3.1 and II.5.10) for every local artinian k-algebra R, maximal ideal M, and any ideal $I \subset R$ so that $M \cdot I = 0$, the map $W(R)_w \to W(R/I)_w$ is surjective (we write $W(R)_w$ for the set of morphisms $W \to \operatorname{Spec}(R)$ with $(W \to \operatorname{Spec}(R) \to \operatorname{Spec}(k)) = w$).

Lemma (4.4). *Let $\rho: D \to W$ be a morphism of k-algebraic schemes, and $d \in D(k)$ a non-singular point on D; suppose the tangential map*

$$\rho_*: t_{D,d} \to t_{W,\rho(d)}$$

to be surjective. Then $\rho(d) = w \in W(k)$ is a non-singular point on W.

Proof. Let $e_1, \ldots, e_n \in \mathcal{O}_{W,w}$ be choosen in such a way that their residues modulo $\mathfrak{m}^2$ form a k-base for $\mathfrak{m}/\mathfrak{m}^2$, where $\mathfrak{m}$ is the maximal ideal of $\mathcal{O}_{W,w}$. We obtain:

$$k[[e_1, \ldots, e_n]] \xrightarrow{\ \pi\ } \hat{\mathcal{O}}_{W,w} \xrightarrow{\ \varphi\ } \hat{\mathcal{O}}_{D,d};$$

as the tangential map ρ_* is surjective, the images of the e_i's are linearly independent modulo the square of the maximal ideal of $\mathcal{O}_{D,d}$; as d is a non-singular point this implies that the composition $\varphi \cdot \pi$ is injective; thus π is injective (and it is also surjective), so $\hat{\mathcal{O}}_{W,w}$ is a formal power series ring, hence $w \in W(k)$ is a non-singular point, and the lemma is proved.

Elimination Lemma (4.5). *Let $A = k[T_1, \ldots, T_N]$, and $H_1, \ldots, H_d \in A$. Suppose given positive integers $w(T_1), \ldots, w(T_N)$ such that $H_1, \ldots, H_d$ are homogeneous polynomials in the weighed variables $T_1, \ldots, T_n$ (i.e. we write $w(\prod T_{n_i}) = \sum w(T_{n_i})$; if μ_1 and μ_2 are monomials occuring with non-zero coefficients in some H_j, then $w(\mu_1) = w(\mu_2)$). Suppose $H_1(0) = 0 = H_2(0) = \cdots = H_d(0)$, such that 0 is a non-singular point of $V = \operatorname{Spec}(A/(H_1, \ldots, H_d)A)$. Then we can renumber the variables, and we can choose $0 \leqq n \leqq N$ so that*

$$A/(H_1, \ldots, H_d)A \cong k[T_1, \ldots, T_n].$$

Proof. Suppose $(H_1, \ldots, H_d)A \neq 0$ (otherwise the conclusion is obvious); in that case at least one of these polynomials has a linear term: if not, we would have

$$(H_1, \ldots, H_d)A \subset (T_1^2, \ldots, T_i T_j, \ldots, T_N^2)A = \mathfrak{b},$$

so

$$\operatorname{Spec}(A/\mathfrak{b}) \subset V \subsetneqq \mathbf{A}_k^N = \operatorname{Spec}(k[T_1, \ldots, T_N]),$$

a contradiction with the fact that $0 \in V(k)$ is non-singular. So let

$$H_d = c\,T_N + G, \qquad c \in k,\ c \neq 0$$

so that T_N does not appear in the linear term of G (renumber the variables and the polynomials if necessary); as $w(T_i)$ are positive integers for all i, it follows that $G \in k[T_1, \ldots, T_{N-1}]$. We write

$$G_i = H_i\left(T_1, \ldots, T_{N-1}, -\frac{1}{c}\,G(T_1, \ldots, T_{N-1})\right), \qquad 1 \leqq i < d,$$

and clearly

$$A/(H_1, \ldots, H_d) \cong k[T_1, \ldots, T_{N-1}]/(G_1, \ldots, G_{d-1})$$

(the variable T_N is eliminated); moreover it is clear that the polynomials $G_1, \ldots, G_{d-1}$ are homogeneous in the weighed variables $T_1, \ldots, T_{N-1}$; thus the lemma is proved by induction on d.

Proof of Proposition (4.1). We proved that Ω is represented by W in Section 3, by the results of Lazard we know Λ is representable, and it is easy to see that Δ is representable (cf. below). The point $0 \in D(k)$ is defined by $f \in \Delta(k)$, $f(X_i) = X_i \otimes 1 + 1 \otimes X_i$; first we show that this is a non-singular point on D. Let R be a local artinian k-algebra, with maximal ideal M, and let $I \subset R$ be an ideal such that $M \cdot I = 0$; we write

$R'=R/I$. By Grothendieck's criterion it suffices to show that

$$D(R)_0 \to D(R')_0$$

is a surjective map. Thus given $f' \in \Delta(R')_0 = D(R')_0$, we would like to construct $f \in \Delta(R)_0$ so that $f' \equiv f(\mathrm{mod}(I \cdot E \otimes E))$ (where $E = k[X_1, \ldots, X_m]$); by the result of Lazard we know that Λ is represented by a non-singular scheme (in fact affine space of dimension $N(m, r)$), so for $f' \in \Delta(R')_0 \subset \Lambda(R')_0$ there exists a $g \in \Lambda(R)_0$ so that

$$f' \equiv g \qquad (\mathrm{mod}\, I \cdot E \otimes E).$$

We know that

$$g(X_i) \equiv X_i \otimes 1 + 1 \otimes X_i \qquad (\mathrm{mod}\, M \cdot E \otimes E),$$

as we work in the point $0 \in D(k) \subset L(k)$; we write

$$g(X_i) = f(X_i) + c(X_i),$$

where $c(X_i)$ consists of monomials none of which satisfy condition $(P\nu)_i$, and $f(X_i)$ consists of monomials which satisfy condition $(P\nu)_i$. We claim that

$$(\Gamma f) \equiv 0 \qquad (\mathrm{mod\ degree}\ r+1),$$

i.e. $f \in \Delta(R)_0$; in fact let

$$f(X_i) = X_i \otimes 1 + 1 \otimes X_i + \sum_{\alpha, \beta} a_{i,\alpha,\beta} X^\alpha \otimes X^\beta,$$

$$c(X_i) = \sum_{\alpha, \beta} b_{i,\alpha,\beta} X^\alpha \otimes X^\beta;$$

then $a_{i,\alpha,\beta} \in M$ and $b_{i,\alpha,\beta} \in I$. Using $M \cdot I = 0$, we obtain:

$$(g \otimes 1)\, g(X_i) = [(f \otimes 1)\, f(X_i)] + [\sum_{\alpha,\beta} b_{i,\alpha,\beta} X^\alpha \otimes X^\beta \otimes 1 + (g \otimes 1)(\sum_{\alpha,\beta} b_{i,\alpha,\beta} X^\alpha \otimes X^\beta)].$$

By (4.2) the first term in square brackets can be written as a sum of monomials all satisfying condition $(P\nu)_i$; by (4.3) the second term can be written as a sum of monomials none of which satisfy condition $(P\nu)_i$. Thus the equation $(\Gamma g)(X_i) \equiv 0 (\mathrm{mod\ degree}\ r+1)$ proves, by sorting out all $(P\nu)_i$-monomials, that

$$(\Gamma f)(X_i) \equiv 0 \qquad (\mathrm{mod\ degree}\ r+1),$$

thus $f \in \Delta(R)_0$, and we have proved that $0 \in D(k)$ is a nonsingular point on D.

Next we show that $0 \in W$ is a non-singular point on W. Let $R = k[\varepsilon]$, with $\varepsilon^2 = 0$. We know that $t_{D,0} = \Delta(k[\varepsilon])_0$, hence by (4.4) it suffices to show that

$$\rho_*\colon \Delta(k[\varepsilon])_0 \to \Omega(k[\varepsilon])_0$$

is a surjective map. Hence we are given

$$s\colon E \to E \otimes E, \quad E = R[\tau_1, \ldots, \tau_m],$$

with

$$s(\tau_i) = \tau_i \otimes 1 + 1 \otimes \tau_i + \varepsilon \cdot \sum c_{i,\alpha,\beta}\, \tau^\alpha \otimes \tau^\beta, \qquad c_{i,\alpha,\beta} \in k,$$

satisfying $(P\, v)_i$ and $(\Gamma\, s) = 0$, and we have to construct an r-bud f satisfying again the conditions $(P\, v)_i$ extending s. We choose

$$f(X_i) = X_i \otimes 1 + 1 \otimes X_i + \varepsilon \cdot \sum c_{i,\alpha,\beta}\, X^\alpha \otimes X^\beta;$$

as $\varepsilon^2 = 0$, we obtain

$$\begin{aligned}(f \otimes 1)\, f(X_i) = {} & X_i \otimes 1 \otimes 1 + 1 \otimes X_i \otimes 1 + 1 \otimes 1 \otimes X_i \\ & + \varepsilon \cdot \sum c_{i,\alpha,\beta}\, X^\alpha \otimes X^\beta \otimes 1 \\ & + \varepsilon \cdot \sum c_{i,\alpha,\beta} \Big\{\prod_j (X_j \otimes 1 + 1 \otimes X_j)^{\alpha_j}\Big\} \otimes X^\beta;\end{aligned}$$

in each of these terms the exponent of X_j is smaller than $p \exp(v_j)$, thus $\Gamma s = 0$ proves that $(\Gamma f)(X_i) = 0$. Thus $f \in \Delta(R)_0$, and certainly $\rho(f) = s$, and we have shown the tangential map ρ_* to be surjective; as $0 \in D$ is a non-singular point we conclude by (4.4) that $0 \in W$ is non-singular.

Now we prove that D and W are isomorphic to affine spaces over k. Let Δ' be the set of pairs (α, β) with $\alpha = (\alpha_1, \ldots, \alpha_m)$, $\beta = (\beta_1, \ldots, \beta_m)$ so that $1 \leqq |\alpha|$ and $1 \leqq |\beta|$ and $|\alpha| + |\beta| \leqq r$; let Δ'' be the set of triples (α, β, γ) with $1 \leqq |\alpha|$, $1 \leqq |\beta|$, $1 \leqq |\gamma|$, and $|\alpha| + |\beta| + |\gamma| \leqq r$. Let Ω' be the set of pairs (α, β) with $1 \leqq |\alpha|$ and $0 \leqq \alpha_j < p \exp(v_j)$ for $1 \leqq j \leqq m$, and $1 \leqq |\beta|$ and $0 \leqq \beta_k < p \exp(v_k)$ for $1 \leqq k \leqq m$; let Ω'' be the set of triples (α, β, γ) with $1 \leqq |\alpha|$ and $0 \leqq \alpha_j < p \exp(v_j)$, etc. Consider

$$F(X_i) = X_i \otimes 1 + 1 \otimes X_i + \sum_{\alpha,\beta} T_{i,\alpha,\beta}\, X^\alpha \otimes X^\beta,$$

summation taken over all $(\alpha, \beta) \in \Delta'$, respectively summation taken over all $(\alpha, \beta) \in \Omega'$; we write $k[\Delta']$, resp. $k[\Omega']$, for the polynomial ring $k[\ldots, T_{i,\alpha,\beta}, \ldots]$, $1 \leqq i \leqq m$ and $(\alpha, \beta) \in \Delta'$, resp. $1 \leqq i \leqq m$ and $(\alpha, \beta) \in \Omega'$. We define polynomials $H_{i,\alpha,\beta,\gamma} \in k[\Delta']$, resp. $H_{i,\alpha,\beta,\gamma} \in k[\Omega']$ by

$$(\Gamma F)(X_i) = \sum_{\alpha,\beta,\gamma} H_{i,\alpha,\beta,\gamma}\, X^\alpha \otimes X^\beta \otimes X^\gamma.$$

Clearly the scheme D, resp. W, is defined by the equations

$T_{i,\alpha,\beta} = T_{i,\beta,\alpha}$,	all $1 \leqq i \leqq m$ and $(\alpha, \beta) \in \Delta'$, resp. $(\alpha, \beta) \in \Omega'$;
$T_{i,\alpha,\beta} = 0$	if $X^\alpha \otimes X^\beta$ does not satisfy $(P\, v)_i$;
$H_{i,\alpha,\beta,\gamma} = 0$,	all $1 \leqq i \leqq m$, and $(\alpha, \beta, \gamma) \in \Delta''$, resp. $(\alpha, \beta, \gamma) \in \Omega''$.

Consider $(F\otimes 1)F(X_i)$; part of this has the form

$$\sum T_{i,\alpha,\beta}\{\prod_j (X_j\otimes 1+1\otimes X_j+\sum T_{j,\gamma,\delta}X^\gamma\otimes X^\delta)^{\alpha_j}\}\otimes X^\beta;$$

each term of this sum is of the form

$$T_{i,\alpha,\beta}\cdot\prod_{1\leq t\leq|\alpha|}(T_{?,\lambda_t,\mu_t}X^{\lambda_t}\otimes X^{\mu_t})\otimes X^\beta$$

(where the question mark indicates some integer, $1\leq ?\leq m$, and where $T_{?,1,0}=1=T_{?,0,1}$); the monomial in the T's obtained thus has weight

$$|\alpha|+|\beta|-1+\sum_t(|\lambda_t|+|\mu_t|-1)=a,$$

while the corresponding term in the X's has total degree

$$\sum_t|\lambda_t|+\sum_t|\mu_t|+|\beta|=a+1;$$

so each term in the polynomial $H_{i,\alpha,\beta,\gamma}$ has weight $|\alpha|+|\beta|+|\gamma|-1$.

Thus both D and W are defined by homogeneous equations in the weighed variables $T_{i,\alpha,\beta}$ and as $0\in D(k)$, resp. $0\in W(k)$ are non-singular points we deduce from the elimination lemma that both D and W are isomorphic to affine space over k. This finishes the proof of the first statement of (4.1). Hence Theorem (3.1) is proved, as we have seen (3.2) that $\Sigma\cong\Omega$.

Let $\mathfrak{a}\subset k[\Delta']$, respectively $\mathfrak{b}\subset k[\Omega']$ be the ideal defining D, resp. W. Renaming the variables we obtain: $k[\Omega']=k[T_1,\dots,T_N]$ and $k[\Delta']=k[T_1,\dots,T_N,T_{N+1},\dots,T_{N+M}]$. We have proved already that there exists a number n, with $0\leq n\leq N$, so that

$$\begin{array}{ccc} k[T_1,\dots,T_n] & \hookrightarrow & k[T_1,\dots,T_N] \\ & \searrow\wr & \downarrow \\ & & k[T_1,\dots,T_N]/\mathfrak{b}. \end{array}$$

The morphism $\rho\colon D\to W$ comes from the ringhomomorphism φ:

$$\begin{array}{ccc} k[T_1,\dots,T_N] & \hookrightarrow & k[T_1,\dots,T_{N+M}] \\ \downarrow & & \downarrow \\ k[T_1,\dots,T_n]\cong U=k[T_1,\dots,T_N]/\mathfrak{b} & \xrightarrow{\varphi} & k[T_1,\dots,T_{N+M}]/\mathfrak{a}=B, \\ \operatorname{Spec}(U)=W & \xleftarrow{\rho} & D=\operatorname{Spec}(B), \qquad \rho={}^a\varphi; \end{array}$$

we are done if we can prove that if we apply the elimination lemma to $\mathfrak{a}\subset k[T_1,\dots,T_{N+M}]$, none of the variables $T_1,\dots,T_n$ is eliminated: in that case

$$k[T_1,\dots,T_n]\cong U\to B\cong k[T_1,\dots,T_n,T_{N+1},\dots,T_{N+m}]$$

for some m with $0 \leqq m \leqq M$ (renumber the variables if necessary); of course in that case every R-point of W comes from an R-point of D. So we have to show: if $T_{i,\alpha,\beta}$ with $(\alpha,\beta)\in\Omega'$ appears in the linear term of some $H_{j,\gamma,\delta,\varepsilon}$, with $(\gamma,\delta,\varepsilon)\in\Delta''$, then $(\gamma,\delta,\varepsilon)\in\Omega''$; but this is clear: computing $(\Gamma F)(X_i)$ we obtain:

$$\sum T_{i,\alpha,\beta} X^\alpha \otimes X^\beta \otimes 1 - \sum T_{i,\alpha,\beta} 1 \otimes X^\alpha \otimes X^\beta$$
$$+\sum T_{i,\alpha,\beta}\{\prod_j (X_j \otimes 1 + 1 \otimes X_j + \sum T_{j,\gamma,\delta} X^\gamma \otimes X^\delta)^{\alpha_j}\} \otimes X^\beta$$
$$-\sum T_{i,\alpha,\beta} X^\alpha \otimes \{\prod_j (X_j \otimes 1 + 1 \otimes X_j + \sum T_{j,\gamma,\delta} X^\gamma \otimes X^\delta)^{\beta_j}\};$$

so "$T_{i,\alpha,\beta}$ appears in the linear term of $H_{j,\gamma,\delta,\varepsilon}$" and $(\alpha,\beta)\in\Omega'$ imply that $(\gamma,\delta,\varepsilon)\in\Omega''$. Thus we have shown that the variables $T_{n+1},\dots,T_N$ can be expressed in the variables $T_1,\dots,T_n$, that $T_{N+1},\dots,T_{N+M}$ depend on $T_1,\dots,T_n$, $T_{N+1},\dots,T_{N+m}$, and that the variables $T_1,\dots,T_n$ cannot be eliminated. Thus the proof of the proposition is concluded.

Remark. The multiplicative semi-group scheme $\mathbf{A}_1^\times = \operatorname{Spec}(k[T])$ acts on $k[\Delta']$ and on $k[\Omega']$ (use the weights of the variables). Under this action D and W are stable, as their defining equations are homogeneous in weight. In this way we originally proved W to be connected; as $D-\{0\}/\mathbf{G}_m$ and $W-\{0\}/\mathbf{G}_m$ are projective schemes, it easily follows that $\rho\colon D(k)\to W(k)$ is surjective in case k is an algebraically closed field.

Remark. One could ask for the dimension of W. It is easy to compute directly the equations for the tangent space at $W(k)$. However we do not see a formula expressing dim W in terms of m and $(\nu_1,\dots,\nu_m)$.

Remark. Let V be the k-algebraic scheme such that for every $B\supset k$, $V(B)=\{$all commutative B-bialgebra structures on $B[\tau_1,\dots,\tau_m]=E\}$; then $V_{\text{red}}=W$, and $V=W$ if and only if $\nu_1=\dots=\nu_m$.

5. Conclusions

Corollary (5.1). *Let k be a field of characteristic $p>0$, and let N be a finite commutative k-group scheme; N can be lifted to characteristic zero (in the sense of problem* (B) *of Section 1).*

Proof. By (2.2) it suffices to show the result for some $K\supset k$; so we can suppose k to be an algebraically closed field. Then $N=N_{\text{loc}}\times N_{\text{sep}}$ (cf. CGS, 2.14). As a reduced finite group scheme over an algebraically closed field corresponds uniquely to a finite group (cf. CGS, 2.16), it is clear that any separable group scheme can be lifted to characteristic zero (we know $N_{\text{sep}}=\operatorname{Spec}(k\times\dots\times k)$, take any characteristic zero domain R with a reduction $R\to k$, choose $M=\operatorname{Spec}(R\times\dots\times R)$, etc.). As k is supposed to be algebraically closed, hence perfect, N_{loc} admits a truncation type $\nu=(\nu_1,\dots,\nu_m)$, hence by (3.1) there exists a point

$w \in W(k)$, where W is an irreducible, smooth k-algebraic scheme, and a finite, free group scheme $M \to W$, such that $N_{loc} \cong M_w$ (i.e. the fibre of M at the point w is isomorphic, as a group scheme, with N_{loc}). Next we note there exists a point $u \in W(k)$ such that

$$\mu_{p\exp(v_1)} \times \cdots \times \mu_{p\exp(v_m)} \cong M_u;$$

thus the fibre of the morphism $M^D \to W$ over the point $u \in W(k)$ is reduced (by D we denote the dualizing functor associating with each finite flat commutative group scheme its linear, or: Cartier, dual; e.g. compare CGS, p. 3). Let L be an algebraic closure of the field of fractions of U, where $W = \operatorname{Spec}(U)$. It follows that the group scheme M_L^D is reduced, so M_L^D can be lifted to characteristic zero by what is said before, so M_L can be lifted to characteristic zero as D commutes with base extension, so by (2.3) it follows that $M \otimes_U k \cong M_w \cong N_{loc}$ can be lifted to characteristic zero, and the corollary is proved.

Question. Let R_0 be a local, artinian ring, and let N_0 be a finite flat, commutative R_0-group scheme. Can we lift N_0 to characteristic zero? In case the rank of N_0 is prime we can, cf. [13]. However it seems that the methods developed above do not work if R_0 is not a field.

Corollary (5.2). *Let R be a ring in which $p \cdot 1 = 0$, and let $N = \operatorname{Spec}(E)$ be a commutative R-group scheme such that E admits a truncation type $E \cong R[\tau_1, \ldots, \tau_m]$, $\tau_i^{p\exp(v_i)} = 0$, $1 \leqq i \leqq m$ (e.g. N is any finite, commutative, local group scheme over a perfect field $k = R$). There exists a commutative formal Lie group on m parameters with coefficients in R, having N as a subgroup scheme (i.e. there exists a commutative formal group*

$$f\colon R[\![X_1, \ldots, X_m]\!] \to R[\![X_1, \ldots, X_m, Y_1, \ldots, Y_m]\!]$$

inducing the given comultiplication on $R[\tau_1, \ldots, \tau_m]$).

Proof. We take $k = \mathbf{F}_p \subset R$; the R-bialgebra E with its truncation type defines a point $e \in W(R)$. We choose a big integer r; by (4.1) there exists a point $d \in D(R)$ such that $\rho(d) = e$; by the results of Lazard (cf. the beginning of section 4) any commutative r-bud on m parameters $e \in D(R) = \Delta_{m,r}(R) \subset \Lambda_{m,r}(R)$ can be extended to a formal Lie group on the same number of parameters, with coefficients in the same ring. Thus the corollary is proved.

Example (constructed by M. Hazewinkel). There exist non-commutative finite local group schemes on m parameters which cannot be embedded into a formal Lie group on m parameters. Let $\operatorname{char}(k) = p$, n and m are positive integers, and $a, b \in k$. We define

$$E = k[\tau]/(\tau^{p\exp(n+m)}),$$

$$s(\tau) = \tau \otimes 1 + 1 \otimes \tau + a\,\tau^{p^n} \otimes \tau^{p^m} + b\,\tau^{p^m} \otimes \tau^{p^n}.$$

The s thus defined is associative; it is not cocommutative if we choose $n \neq m$ and $a \neq b$; in that case we have a local bialgebra on one parameter, which cannot be extended to a formal Lie group on one parameter if k is a field, because every one-parameter formal Lie group over k is commutative, cf. [6], and [7], Theorem 1, p. 253.

Remark. By different methods it was proved that any finite commutative group scheme over any field k can be embedded into an irreducible smooth k-algebraic group scheme G (cf. CGS, 15.4; cf. [12], in that case we can even take for k a complete local noetherian ring); however in general the dimension of G is much bigger than the number of parameters of N (suppose N to be local); in fact, if the rank of N is p^d, and k is algebraically closed, an imbedding of N into a d-dimensional group variety was constructed. In general a local finite, commutative group scheme on m parameters cannot be embedded into a group variety of dimension m (i.e. N being fixed, none of the formal Lie groups constructed in 5.2 need to be algebraizable), as is shown by the following

Example. Let k be a perfect field of characteristic p, and let N be the k-group scheme having as Dieudonné-module $W_\infty(k)[F,V]/(V-F^2, F^i)$, with $i \geqq 3$; this is a local group scheme on one parameter; it has rank p^i, the rank of $\mathrm{Ker}(p \cdot 1_N)$ is p^3 and the rank of $\mathrm{Ker}(V_N)$ is p^2. If G is an abelian variety of dimension one, the rank of $\mathrm{Ker}(p \cdot 1_G)$ is p^2, so $N \subset G$ is excluded. As $0 \neq \mathrm{Ker}(V_N)$, the case $N \subset \mathbf{G}_m$ is not possible. As $\mathrm{Ker}(V_N) \neq N$, we cannot embed N into a one-dimensional unipotent group-variety G (because any one-dimensional unipotent group variety is killed by V). Thus the N we have choosen cannot be embedded into a one-dimensional group variety.

Remark. Let $v_1 \leqq v_2 \leqq \cdots \leqq v_m$, $\mu_1 \leqq \mu_2 \leqq \cdots \leqq \mu_m$, with $\mu_i \geqq v_i$ for $1 \leqq i \leqq m$, and $v_j - v_i \geqq \mu_j - \mu_i$ for $1 \leqq i < j \leqq m$; using the methods exposed above, one can show that any $s \in \Omega_v(R)$ can be extended to an element $t \in \Omega_\mu(R)$; taking $\mu_1 = a = \mu_2 = \cdots = \mu_m$, and letting a grow, we obtain again (5.2).

References

1. Cartier, P.: Groupes algébriques et groupes formels. Coll. CBRM, Brussels 1962, 87–111.
2. Demazure, M., et A. Grothendieck: Schémas en groupes. Sém. géom. algébrique, IHES, 1963–1964. Referred to as SGAD.
3. Grothendieck, A., et J. Dieudonné: Éléments de géométrie algébrique. Publ. Math., IHES. Referred to as EGA.
4. Grothendieck, A.: Séminaire de géométrie algébrique. IHES, 1960. Referred to as SGA.
5. Lazard, M.: Lois de groupes et analyseurs. Ann. Sc. Éc. norm. sup. **72**, 299–400 (1955).
6. — La non-existence des groupes de Lie formels non abéliens à un paramètre. C. R. Acad. Sci. **239**, 942–945 (1954).

7. Lazard, M.: Sur les groupes de Lie formels à un paramètre. Bull. Soc. Math. France **83**, 251–274 (1955).
8. Lubin, J., and J. Tate: Formal moduli for one-parameter Lie groups. Bull. Soc. Math. France **94**, 49–60 (1966).
9. Mumford, D.: Lectures on curves on an algebraic surface (lecture notes Harvard University, 1964). Princeton Math. Notes 59.
10. Oort, F.: Commutative group schemes. Lecture Notes in Math. 15. Berlin-Heidelberg-New York: Springer 1966. Referred to as CGS.
11. – Algebraic group schemes in characteristic zero are reduced. Inv. Math. **2**, 79–80 (1966).
12. – Embedding of finite group schemes into abelian schemes. Mimeographed notes from the advanced science seminar in algebraic geometry, Bowdoin college, summer 1967.
13. Tate, J., and F. Oort: Finite group schemes of prime rank (to appear).

Frans Oort
Mathematisch Instituut
Nieuwe Achtergracht 121
Amsterdam
The Netherlands

David Mumford
Department Mathematics
Harvard University
2 Divinity Avenue
Cambridge, Mass., USA

(Received February 22, 1968)

BI-EXTENSIONS OF FORMAL GROUPS

By DAVID MUMFORD

In the Colloquium itself, I announced that all abelian varieties can be lifted to characteristic zero. The proof of this, as sketched
1 there, is roughly as follows.

(i) It suffices to prove that every char p abelian variety is a specialization of a char p abelian variety with multiplicative formal group (an "ordinary" abelian variety), since Serre (unpublished) has shown that these admit liftings.

(ii) A preliminary reduction of the problem was made to abelian varieties X such that the invariant

$$\alpha(X) = \dim_k \operatorname{Hom}(\alpha_p, X)$$

is 1.

(iii) A method was found to construct deformations of a polarized abelian variety from deformations of its polarized Dieudonné module.

(iv) Finally, some simple deformations of polarized Dieudonné modules were constructed to establish the result.

However, it seems premature to give this proof here, since the basic method used in (iii) promises to give much fuller information on the local structure of the formal moduli space of a polarized abelian variety, and this would make my *ad hoc* method obsolete. I want instead to give some basic information on the main new technical tool which is used in (iii).

2 **1. Cartier's result.** In the note [1], Cartier has announced a module-theoretic classification of formal groups over arbitrary ground-rings R. We require only the special case where $p = 0$ in R, which is foreshadowed in Dieudonné's original paper [2], before the category men got a hold of it, modifying the technique until the restriction "R = perfect field" came to seem essential.

Definition. *Let R be a ring of characteristic p. Let $W(R)$ be the ring of Witt vectors over R, and let*

$$(a_0, a_1, a_2, \ldots)^{\sigma} = (a_0^p, a_1^p, a_2^p, \ldots),$$
$$(a_0, a_1, a_2, \ldots)^{t} = (0, a_0, a_1, \ldots).$$

3 *Then A_R will denote the ring*

$$W(R)[[V]][F]$$

modulo the relations:

(a) $FV = p$,

(b) $VaF = a^t$,

(c) $Fa = a^{\sigma}F$,

(d) $aV = Va^{\sigma}$,

for all $a \in W(R)$.

Theorem (Dieudonné-Cartier). *There is a covariant equivalence of categories between*

(A) *the category of commutative formal groups Φ over R, and*

(B) *the category of left A_R-modules M such that*

(a) $\bigcap_i V^i M = (0)$,

(b) $Vm = 0 \Rightarrow m = 0$, *all $m \in M$,*

(c) *M/VM is a free R-module of finite rank.*

The correspondence between these 2 categories can be set up as follows. Recall first that a formal group Φ/R (by which we mean a set of n power series $\phi_i(x_1, \ldots, x_n;\ y_1, \ldots,\ y_n)$, $1 \leqslant i \leqslant n$, satisfying the usual identities, c.f. Lazard [3]) defines a covariant functor F_Φ from R-algebras S to groups : i.e. $\forall\ S/R$,

$$F_\Phi(S) = \{ (a_1, \ldots, a_n) \mid a_i \in S,\ a_i \text{ nilpotent} \}$$

where

$$(a_1, \ldots, a_n).(b_1, \ldots, b_n) = (\phi_1(a_1, \ldots, a_n; b_1, \ldots, b_n), \ldots, \phi_n(a_1, \ldots, a_n; b_1, \ldots, b_n)).$$

N. B. In what follows, we will often call the functor F_Φ instead of the power series Φ the formal group, for simplicity.

Let $\hat{\mathrm{W}}$ be the functor

$$\begin{cases} \hat{\mathrm{W}}(S) = \{ (a_0, a_1, \ldots) \mid a_i \in S,\, a_i \text{ nilpotent, almost all } a_i = 0\}, \\ \text{gp law} = \text{Witt vector addition.} \end{cases}$$

Then we attach to the commutative formal group Φ the set

$$M = \mathrm{Hom}_{\mathrm{gp.\ functors}/R}(\hat{\mathrm{W}}, F_\Phi),$$

and since $A_R \cong \mathrm{Hom}\,(\hat{\mathrm{W}}, \hat{\mathrm{W}})^0$, we can endow M with the structure of left A_R-module. Conversely, to go in the other direction, first note that any A_R-module M as in the theorem can be resolved:

$$(*) \qquad 0 \longrightarrow A_R^n \xrightarrow{\ \beta\ } A_R^n \xrightarrow{\ \alpha\ } M \longrightarrow 0.$$

In fact, choose $m_1, \ldots, m_n \in M$ whose images mod VM are a basis of M/VM as R-module. Define

$$\alpha(P_1, \ldots, P_n) = \sum_{i=1}^{n} P_i\, m_i.$$

It is easy to check that Fm_i can be expanded in the form $\sum_{j=1}^{n} Q_{ij}(V)\, m_j$, Q_{ij} a power series in V with coefficients in $W(R)$. Define

$$\beta(P_1, \ldots, P_n) = \left(\sum_{i=1}^{n} P_i \cdot Q_{i1} - \delta_{i1} F, \ldots, \sum_{i=1}^{n} P_i \cdot Q_{in} - \delta_{in} F \right).$$

It is not hard to check that (*) is exact. Then β defines a monomorphism of group functors $\beta^*\colon (\hat{\mathrm{W}})^n \to (\hat{\mathrm{W}})^n$, and let F be the quotient functor $(\hat{\mathrm{W}})^n / \beta^*(\hat{\mathrm{W}})^n$. Then F is isomorphic to F_Φ for one and—up to canonical isomorphism—only one formal group Φ.

Moreover, we get a resolution of the functor F_Φ :

$$0 \longrightarrow (\hat{\mathrm{W}})^n \xrightarrow{\ \beta^*\ } (\hat{\mathrm{W}})^n \longrightarrow F_\Phi \longrightarrow 0.$$

When R is a perfect field, the above correspondence can be extended to an analogous correspondence between p-divisible groups over R and $W(R)[F, V]$-modules of suitable type (c.f. [4], [5]).

However, it does not seem likely at present that such an extension exists for non-perfect R's. This is a key point.

4 **2. Bi-extensions of abelian groups.** Let A,B,C be 3 abelian groups. A bi-extension of $B \times C$ by A will denote a set G on which A acts freely, together with a map

$$G \xrightarrow{\pi} B \times C$$

making $B \times C$ into the quotient G/A, together with 2 laws of composition:

$$+_1 : G \underset{B}{\times} G \to G \quad ; \quad +_2 : G \underset{C}{\times} G \to G$$

$$\text{def} \Big\| \qquad\qquad\qquad \text{def} \Big\|$$

$$\{(g_1, g_2) \mid \pi(g_1), \pi(g_2) \text{ have same } B\text{-component}\} \qquad \{(g_1, g_2) \mid \pi(g_1), \pi(g_2) \text{ have some } C\text{-component}\}$$

These are subject to the requirements:

(i) for all $b \in B$, $G_b' = \pi^{-1}(b \times C)$ is an abelian group under $+_1$, π is a surjective homomorphism of G_b' onto C, and via the action of A on G_b', A is isomorphic to the kernel of π;

(ii) for all $c \in C$, $G_c^2 = \pi^{-1}(B \times c)$ is an abelian group under $+_2$, π is a surjective homomorphism of G_c^2 onto B, and via the action of A on G_c^2, A is isomorphic to the kernel of π;

(iii) given $x, y, u, v \in G$ such that

$$\pi(x) = (b_1, c_1)$$

$$\pi(y) = (b_1, c_2)$$

$$\pi(u) = (b_2, c_1)$$

$$\pi(v) = (b_2, c_2),$$

then

$$(x +_1 y) +_2 (u +_1 v) = (x +_2 u) +_1 (y +_2 v).$$

This may seem like rather a mess, but please consider the motivating example: let X be an abelian variety over an algebraically closed field k, let $\hat{X}$ be its dual, and let P be the universal, or Poincaré, line bundle on over $X \times \hat{X}$. Then P_k, the underlying set of closed points of P, is a bi-extension of $X_k \times \hat{X}_k$ by k^* !

Notice that if G is a bi-extension of $B \times C$ by A, then $\pi^{-1}(B \times 0)$ splits canonically into $A \times B$, and $\pi^{-1}(0 \times C)$ splits canonically into $A \times C$. In fact, we can lift B to $\pi^{-1}(B \times 0)$ by mapping $b \in B$ to the element of G which is the identity in $\pi^{-1}(b \times C)$; and we can lift C to $\pi^{-1}(0 \times C)$ by mapping $c \in C$ to the element of G which is the identity in $\pi^{-1}(B \times c)$.

Bi-extensions can be conveniently described by co-cycles: choose a (set-theoretic) section

$$G \underset{\pi}{\overset{s}{\rightleftarrows}} B \times C$$

Via s and the action of A on G, we construct an isomorphism

$$G \simeq A \times B \times C$$

such that the action of A on G corresponds to the action of A on $A \times B \times C$ which is just addition of A-components, leaving the B- and C-components fixed. Then $+_1$ and $+_2$ go over into laws of composition on $A \times B \times C$ given by:

$$(a, b, c) +_1 (a', b, c') = (a + a' + \phi(b; c, c'), b, c + c')$$
$$(a, b, c) +_2 (a', b', c) = (a + a' + \psi(b, b'; c), b + b', c).$$

For $+_1$, $+_2$ to be abelian group laws, we need:

(a) $$\phi(b; c + c', c'') + \phi(b; c, c') = \phi(b; c, c' + c'') + \phi(b; c', c'')$$
$$\phi(b; c, c') = \phi(b; c', c);$$

(b) $$\psi(b + b', b''; c) + \psi(b, b'; c) = \psi(b, b' + b''; c) + \psi(b', b''; c)$$
$$\psi(b, b'; c) = \psi(b', b; c).$$

The final restriction comes out as:

(c)
$$\phi(b+b';c,c') - \phi(b;c,c') - \phi(b';c,c')$$
$$= \psi(b,b';c+c') - \psi(b,b';c) - \psi(b,b';c').$$

What are the co-boundaries? If you alter s by adding to it a map $\rho : B \times C \to A$, then you check that the new ϕ', ψ' are related to the old ones by

$$\phi'(b;c,c') - \phi(b;c,c') = \rho(b,c+c') - \rho(b,c) - \rho(b,c')$$
$$\psi'(b,b';c) - \psi(b,b';c) = \rho(b+b',c) - \rho(b,c) - \rho(b',c).$$

Using this explicit expression by co-cycles and co-boundaries, it is clear that the set of all bi-extensions of $B \times C$ by A forms itself an abelian group, which we will denote

$$\text{Bi-ext}\,(B \times C,\ A).$$

It is also clear, either from the definition or via co-cycles, that Bi-ext is a covariant functor in A, and a contravariant functor in B and C.

3. Bi-extensions of group-functors.

DEFINITION. *If F,G,H are 3 covariant functors from the category of R-algebras to the category of abelian groups, a bi-extension of $G \times H$ by F is a fourth functor K such that for every R-algebra S, $K(S)$ is a bi-extension of $G(S) \times H(S)$ by $F(S)$ and for every R-homomorphism $S_1 \to S_2$, the map $K(S_1) \to K(S_2)$ is a homomorphism of bi-extensions (in the obvious sense). In particular, if F,G,H are formal groups, this gives us a bi-extension of formal groups.*

If F,G,H are formal groups, it is easy again to compute the bi-extensions K by power series co-cycles. In fact, one merely has to check that:

(i) there is a functorial section

$$K \underset{\pi}{\overset{s}{\rightleftarrows}} G \times H$$

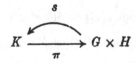

(this follows using the "smoothness" of the functor F, i.e. $F(S) \to F(S/I)$ is surjective if I is a nilpotent ideal);

(ii) any morphism of functors from one product of formal groups to another such product is given explicitly by a set of power series over R in the appropriate variables.

In fact, we will be exclusively interested in the case where $F = \hat{\mathbf{G}}_m$ is the formal multiplicative group; that is

$$\hat{\mathbf{G}}_m(S) = \left\{ \begin{array}{l} \text{Units in } S \text{ of form } 1+x,\ x \text{ nilpotent,} \\ \text{composed via multiplication.} \end{array} \right\}$$

Then if G and H are formal groups in variables $x_1, \ldots, x_n$ and $y_1, \ldots, y_m$, a bi-extension of $G \times H$ by $\hat{\mathbf{G}}_m$ is given by 2 power series $\sigma(x_1, \ldots, x_n; y_1, \ldots, y_m, y_1', \ldots, y_m')$, $\tau(x_1, \ldots, x_n, x_1', \ldots, x_n'; y_1, \ldots, y_m)$ with constant terms 1 such that—abbreviating n-tuples and m-tuples:

$$\sigma(x; \Phi(y, y'), y'').\sigma(x; y, y') = \sigma(x; y, \Phi(y', y'')).\sigma(x; y', y'')$$
$$\sigma(x; y, y') = \sigma(x; y', y)$$
$$\tau(\Psi(x, x'), x'', y).\tau(x, x'; y) = \tau(x, \Psi(x', x''); y).\tau(x', x''; y)$$
$$\tau(x, x'; y) = \tau(x', x; y)$$
$$\sigma(\Psi(x, x'); y, y').\sigma(x; y, y')^{-1}.\sigma(x'; y, y')^{-1} = \tau(x, x'; \Phi(y, y')). \tau(x, x'; y)^{-1}. \tau(x, x'; y')^{-1},$$

5 if Φ, Ψ are the group laws of G and H respectively.

We want one slightly non-trivial fact about general bi-extensions. This result gives essentially the method for computing Bi-ext's via resolutions.

Proposition 1. *Let E, G, G' be abelian group functors as above. Suppose*

$$0 \longrightarrow F_1 \longrightarrow F_0 \longrightarrow G \longrightarrow 0$$
$$0 \longrightarrow F_1' \longrightarrow F_0' \longrightarrow G' \longrightarrow 0$$

are 2 exact sequences of such functors. Then

$$\operatorname{Ker}\{\text{Bi-ext}\,(G \times G', E) \longrightarrow \text{Bi-ext}\,(F_0 \times F_0', E)\}$$

$$\cong \frac{\{(f,g) \mid f\colon F_0 \times F_1' \to E \text{ and } g : F_1 \times F_0' \to E \text{ bi-homomorphisms } \operatorname{res} f = \operatorname{res} g \text{ on } F_1 \times F_1'\}}{\{(f,g) \mid \exists h\colon F_0 \times F_0' \to E \text{ bi-homomorphism}, f \text{ and } g \text{ restrictions of } h\}}$$

The proof goes along these lines: let H be a bi-extension of $G \times G'$ by E. If it lies in the above kernel, then the induced bi-extension of $F_0 \times F_0'$ is trivial:

$$H \underset{(G\times G')}{\times} (F_0 \times F_0') \cong E \times F_0 \times F_0'.$$

Consider the equivalence relation on the functor $E \times F_0 \times F_0'$ induced by the mapping of it onto H. It comes out that there are maps $f\colon F_0 \times F_1' \to E$, $g\colon F_1 \times F_0' \to E$ such that this equivalence relation is generated by

$$(a,b,c) \sim (a + f(b,\bar{c}), b, c + \bar{c}), \; a \in E(S), b \in F_0(S)$$
$$c \in F_0'(S), \bar{c} \in F_1'(S). \quad (1)$$

and

$$(a,b,c) \sim (a + g(\bar{b},c), b + \bar{b}, c), \quad a \in E(S), b \in F_0(S)$$
$$\bar{b} \in F_1(S), c \in F_0'(S). \quad (2)$$

Moreover, f and g have to be bi-homomorphisms with $\operatorname{res} f = \operatorname{res} g$ on $F_1 \times F_1'$. Conversely, given such f and g, define the functor H to be the quotient of $E \times F_0 \times F_0'$ by the above equivalence relation. H turns out to be a bi-extension. Finally, the triviality of H can be seen to be equivalent to f and g being the restrictions of a bi-homomorphism $h\colon F_0 \times F_0' \to E$.

4. Bi-extensions of $\hat{W}$.

PROPOSITION 2. Bi-ext $(\hat{W} \times \hat{W}, \hat{G}_m) = (0)$.

PROOF. Consider functors F from (R-algebras) to (abelian groups) which are isomorphic as set functors to D^I, where

$$D^I(S) = \{(a_i) \mid a_i \in S, \text{ all } i \in I, a_i \text{ nilpotent, almost all } a_i = 0\}$$

and where I is an indexing set which is either finite or countably infinite. Note that all our functors are of this type. Then I claim that for all R of char p, all such F, there is a canonical retraction p_F:

$$\underset{\text{set-functors}}{\operatorname{Hom}}(\hat{\mathrm{W}}, F) \underset{\text{inclusion}}{\overset{p_F}{\rightleftarrows}} \underset{\text{gp-functors}}{\operatorname{Hom}}(\hat{\mathrm{W}}, F)$$

which is functorial both with respect to (1) any homomorphism $F \to G$, and (2) base changes $R_1 \to R_2$.

The construction of p_F is based on Theorem 1 of Cartier's note [1]. Let $\hat{\mathrm{W}}^*$ be the full Witt group functor (i.e. based on all positive integers, rather than powers of p), and let $i: D \to \hat{\mathrm{W}}^*$ be the canonical inclusion used in [1]. Then Theorem 1 asserts that for all formal groups F, every morphism $\phi: D \to F$ extends uniquely to a homomorphism $u: \hat{\mathrm{W}}^* \to F$.

$$\begin{array}{ccc} D & \xrightarrow{\ \phi\ } & F \\ {\scriptstyle i}\downarrow & \nearrow{\scriptstyle u} & \\ \hat{\mathrm{W}}^* & & \end{array}$$

Cartier informs me that this theorem extends to all F's of our type. On the other hand, $\hat{\mathrm{W}}$, over a ring of char p, is a direct summand of $\hat{\mathrm{W}}^*$:

$$\hat{\mathrm{W}}^* \underset{\pi}{\overset{j}{\rightleftarrows}} \hat{\mathrm{W}}.$$

Construct p_F as follows: given $f: \hat{\mathrm{W}} \to F$, let $\phi =$ res to D of $f \circ \pi$; let $u =$ extension of ϕ to a homomorphism u; let $p_F(f) = u \circ j$.

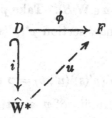

Now let F be a bi-extension of $\hat{\mathrm{W}} \times \hat{\mathrm{W}}$ by $\hat{\mathrm{G}}_m$. For every R-algebra S and every $a \in \hat{\mathrm{W}}(S)$, let F'_a (resp F''_a) denote the fibre functor of F over $\{a\} \times \hat{\mathrm{W}}$ (resp $\hat{\mathrm{W}} \times \{a\}$) (i.e. $F_a(T) = \{b \in F(T) \mid 1^{\text{st}}$ (resp 2nd) component of $\pi(b)$ is induced by a via $S \to T\}$). Then F'_a and F''_a are group functors of the good type extending $\hat{\mathbf{W}}$ by $\hat{\mathbf{G}}_m$ over ground ring S. Now since $\hat{\mathbf{G}}_m$ is smooth, one can choose a section s to π:

$$F \underset{\pi}{\overset{s}{\rightleftarrows}} \hat{\mathbf{W}} \times \hat{\mathbf{W}}$$

s restricts to morphisms $s_a : \hat{\mathbf{W}}/S \to F'_a$, for all $a \in \hat{\mathbf{W}}(S)$. Take $p_{F'_a}(s_a)$. As a varies, these fit together into a new section $p'(s)$ to π. But $p'(s)$ is now a homomorphism with respect to addition into the 2nd variable, i.e.

$$p'(s)(u, v) +_1 p'(s)(u, v') = p'(s)(u, v + v'). \qquad (*)'$$

Now switch the 2 factors: $p'(s)$ restricts to morphism

$p'(s)_a : \hat{\mathbf{W}}/S \to F''_a$, for all $a \in \hat{\mathbf{W}}(S)$. Take $p_{F''_a}(p'(s)_a)$. As a varies, these fit together into a new section $p''(p'(s))$ to π.

Then this satisfies :

$$p''(p'(s))(u, v) +_2 p''(p'(s))(u', v) = p''(p'(s))(u + u', v). \qquad (*)''$$

But now, using the functoriality of p, and the property of bi-extensions linking $+_1$ and $+_2$, it falls out that $p''(p'(s))$ still has property $(*)'$ enjoyed by $p'(s)$! So $p''(p'(s))$ preserves both group laws and splits the extension F. Q.E.D.

Definition. *A_R will denote the ring $W(R)[[F, V]]$ modulo the relations*

(a) $FV = p$

(b) $VaF = a^t$

(c) $Fa = a^\sigma F$

(d) $aV = Va^\sigma$, *all* $a \in W(R)$.

Every element in this ring can be expanded uniquely in the form:

$$P = a_0 + \sum_{i=1}^{\infty} V^i a_i + \sum_{i=1}^{\infty} a_{-i} F^i.$$

For every such P, let

$$P^* = a_0 + \sum_{i=1}^{\infty} a_i F^i + \sum_{i=1}^{\infty} V^i a_{-i}.$$

Then * is an anti-automorphism of $\bar{A}_k$ of order 2. We shall consider $\bar{A}_R$ as an $A_R \times A_R$-module via

$$(P, Q).x = P.x.Q^*. \qquad (*)$$

PROPOSITION 3. Bi-hom$_R$ $(\hat{\mathbf{W}} \times \hat{\mathbf{W}}, \hat{\mathbf{G}}_m) \simeq \bar{A}_R$.

Moreover, since $A_R = \mathrm{Hom}_R(\hat{\mathbf{W}}, \hat{\mathbf{W}})^0$, *the left-hand side is an* $A_R \times A_R$*-module; under the above isomorphism, this structure corresponds to the* $A_R \times A_R$*-module structure on* $\bar{A}_R$ *defined by* (*).

PROOF. Cartier [1] has shown that for all R, the Artin-Hasse exponential defines isomorphisms

$$\mathrm{Hom}_R(\hat{\mathbf{W}}, \hat{\mathbf{G}}_m) \simeq \mathbf{W}(R)$$

where $\mathbf{W}$ is the full Witt functor

$$\begin{cases} \mathbf{W}(R) = \{(a_0, a_1, \ldots) \mid a_i \in R\} \\ \text{group law} = \text{addition of Witt vectors.} \end{cases}$$

Therefore,

$$\text{Bi-Hom}_R(\hat{\mathbf{W}} \times \hat{\mathbf{W}}, \hat{\mathbf{G}}_m) \simeq \mathrm{Hom}_R(\hat{\mathbf{W}}, \mathbf{W}).$$

Define a homomorphism

$$\bar{A}_R \xrightarrow{\phi} \mathrm{Hom}_R(\hat{\mathbf{W}}, \mathbf{W})$$

by $$P \longrightarrow \text{the map } [b \longmapsto P(b)].$$

Here $P(b)$ means that V and F operate on Witt vectors in the usual way: note that the doubly infinite series P operates on b since b has only a finite number of components and all are nilpotent, whereas $P(b)$ is allowed to have all components non-zero.

Let

$$\hat{\mathbf{W}}_n(R) = \{(a_0, a_1, \ldots) \mid a_i^{p^n} = 0, \text{ all } i; \text{ almost all } a_i = 0\}.$$

Notice that

$$\mathrm{Hom}_R(\hat{\mathbf{W}}, \mathbf{W}) \simeq \varprojlim_n \mathrm{Hom}_R(\hat{\mathbf{W}}_n, \mathbf{W}),$$

and that ϕ factors through maps

$$\bar{A}_R/\bar{A}_R.F^n \xrightarrow{\phi_n} \text{Hom}_R(\hat{\mathbf{W}}_n, \mathbf{W}).$$

It suffices to show that ϕ_n is an isomorphism for all n. But for $n=1$, $\bar{A}_R/\bar{A}_R.F \cong R[[V]]$, while

$$\text{Hom}_R(\hat{\mathbf{W}}_1, \mathbf{W}) \cong \text{Hom}_{p\text{-Lie algebras}}(\text{Lie}(\hat{\mathbf{W}}), \text{Lie}(\mathbf{W})).$$

Also Lie $(\hat{\mathbf{W}})$ is the free R-module on generators $\hat{e}_0, \hat{e}_1, \hat{e}_2, \ldots$ with $\hat{e}_i^{(p)} = \hat{e}_{i+1}$; and Lie $(\mathbf{W})$ is the R-module of all expressions $\sum_{i=0}^{\infty} a_i e_i$, $a_i \in R$, with same p^{th} power map. Moreover $\sum_{i=0}^{\infty} V^i a_i \in R[[V]]$ goes via ϕ_1 to the lie algebra map taking $\hat{e}_0$ to $\sum_{i=0}^{\infty} a_i e_i$. Thus ϕ_1 is an isomorphism. Now use induction on n, and the exact sequences

$$0 \longrightarrow \hat{\mathbf{W}}_{n-1} \longrightarrow \hat{\mathbf{W}}_n \xrightarrow{F^{n-1}} \hat{\mathbf{W}}_1 \longrightarrow 0.$$

This leads to the diagram:

$$\begin{array}{ccccccc}
0 \longrightarrow & \text{Hom}_R(\hat{\mathbf{W}}_1, \mathbf{W}) & \xrightarrow{\circ F^{n-1}} & \text{Hom}_R(\hat{\mathbf{W}}_n, \mathbf{W}) & \longrightarrow & \text{Hom}_R(\hat{\mathbf{W}}_{n-1}, \mathbf{W}) & \\
& \uparrow \phi_1 & & \uparrow \phi_n & & \uparrow \phi_{n-1} & \\
0 \longrightarrow & \bar{A}_R/\bar{A}_R.F & \xrightarrow{\times F^{n-1}} & \bar{A}_R/\bar{A}_R.F^n & \longrightarrow & \bar{A}_R/\bar{A}_R.F^{n-1} & \longrightarrow 0.
\end{array}$$

The bottom line is easily seen to be exact, so if ϕ_1 and ϕ_{n-1} are isomorphisms, the diagram implies that ϕ_n is an epimorphism.

Q.E.D.

Corollary. *Let F_1 and F_2 be group functors isomorphic to $(\hat{\mathbf{W}})^{n_i}$ for some n_1, n_2. Let $M_i = \text{Hom}_R(\hat{\mathbf{W}}, F_i)$ be the corresponding finitely generated, free A_R-module. Then there is a $1-1$ correspondence between bi-homomorphisms*

$$B : F_1 \times F_2 \to \hat{\mathbf{G}}_m$$

and maps

$$\beta: M_1 \times M_2 \to \bar{A}_R,$$

bi-linear in the following sense:

6 $$\beta(Pm, Qn) = P.(m,n).Q^*$$

(all $m \in M_1$, $n \in M_2$, $P, Q \in A_R$).

5. Applications. Putting Propositions 1, 2 and 3 together, we conclude the following

COROLLARY.

(a) *Let Φ, Ψ be formal groups over R.*

(b) *Let M, N be the corresponding Dieudonné modules.*

(c) *Let*

$$0 \longrightarrow F_1 \longrightarrow F_0 \longrightarrow M \longrightarrow 0$$

$$0 \longrightarrow G_1 \longrightarrow G_0 \longrightarrow N \longrightarrow 0$$

be resolutions of M and N by finitely generated, free A_R-modules. Then the group $\text{Bi-ext}_R(\Phi \times \Psi, \hat{\mathbf{G}}_m)$ *of bi-extensions of formal groups can be computed as the set of pairs of bi-linear maps:*

$$\beta: F_0 \times G_1 \to \bar{A}_R,$$

$$\gamma: F_1 \times G_0 \to \bar{A}_R,$$

such that $\beta = \gamma$ on $F_1 \times G_1$, taken modulo restrictions of bi-linear maps $\alpha: F_0 \times G_0 \to \bar{A}_R$.

In another direction, bi-extensions can be linked to p-divisible groups, as defined by Tate [6].

PROPOSITION 4. *Let F and F' be formal groups over a* char p *ring R. Assume that the subgroups G_n (resp G_n') $=$* Ker*(p^n in F (resp F')) form p-divisible groups over R (i.e. F and F' are "equi-dimensional", or of "finite height"). Then there is a 1–1 correspondence between (1) bi-extensions of $F \times F'$ by $\hat{\mathbf{G}}_m$ and (2) sets of bi-homomorphisms $\beta_n: G_n \times G_n' \to \mu_{p^n}$, such that for all $x \in G_{n+1}(S)$, $y \in G_{n+1}'(S)$,*

$$\beta_n(px, py) = \beta_{n+1}(x, y)^p.$$

Proof. We will use descent theory and existence of quotients by finite, flat equivalence relations: c.f. Raynaud's article in the same volume as Tate's talk [6]. Starting with the β_n's, let L_n be the quotient functor in the flat topology of $\hat{\mathbf{G}}_m \times G_n \times G'_{2n}$ by the equivalence relation:

$$(\lambda, x, y) \sim (\lambda.\beta_n(x, b), x, y + b)$$

where $\lambda \in \hat{\mathbf{G}}_m(S)$ $x \in G_n(S)$, $y \in G'_{2n}(S)$, $b \in G'_n(S)$. Then L_n is a bi-extension of $G_n \times G'_n$ by $\hat{\mathbf{G}}_m$. Moreover, L_n is a subfunctor of L_{n+1}, so if we let L be the direct limit of the functor L_n, then L is a bi-extension of $F \times F'$ by $\hat{\mathbf{G}}_m$.

Conversely, if we start with L, let L_n be the restriction of L over $G_n \times G'_n$. In the diagram

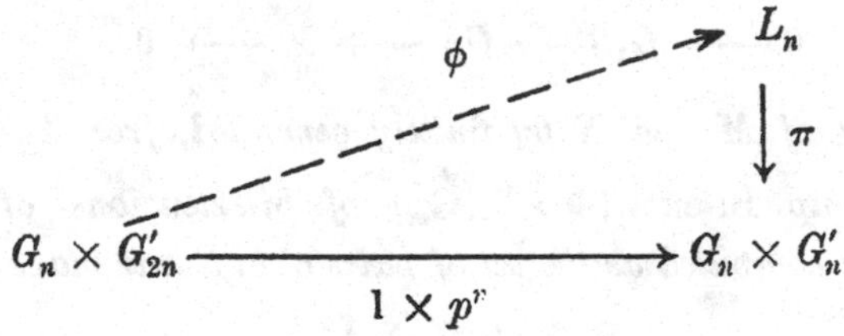

I want to define a canonical map ϕ which is a homomorphism in both variables, i.e. which splits the induced bi-extension over $G_n \times G'_{2n}$. Suppose $x \in G_n(S)$, $y \in G'_n(S)$ for some R-algebra S. Choose $z_1 \in L(S)$ such that $\pi(z_1) = (x,y)$. If we add z_1 to itself p^n times in the 1st variable, we obtain a point:

$$[p^n]_{+_1}(z_1) = z_2$$
$$\pi(z_2) = (0,y).$$

But $\pi^{-1}((0) \times F')$ is canonically isomorphic to $\hat{\mathbf{G}}_m \times (0) \times F'$, so $z_2 = (\lambda, 0, y)$, some $\lambda \in \hat{\mathbf{G}}_m(S)$. Now choose a finite flat S-algebra S' such that $\lambda = \mu^{p^n}$ for some $\mu \in \hat{\mathbf{G}}_m(S')$. Letting z_1 also denote the element of $L(S')$ induced by z_1, define $z'_1 = \mu^{-1}.z_1$. This is a new point of L over (x, y), which now satisfies $[p^n]_{+1}(z'_1) = (1,0,y)$. Now add z'_1 to itself p^n times in the 2nd variable. This gives a point

$$[p^n]_{+_2}(z_1') = z_3' \in L_n^*(S'),$$
$$\pi(z_3') = (x, p^n y).$$

Clearly, z_3' is independent of the choice of μ, so by descent theory, z_3' must be induced by a unique element $z_3 \in L_n(S)$. Define $\phi(x, y) = z_3$. It is easy to check that ϕ is a homomorphism in both variables.

We can use ϕ to set up a fibre product diagram:

$$\begin{array}{ccc} \hat{\mathbf{G}}_m \times G_n \times G'_{2n} & \xrightarrow{\ \alpha\ } & L_n \\ \downarrow{\scriptstyle \pi} & & \downarrow{\scriptstyle \pi} \\ G_n \times G'_{2n} & \xrightarrow[(1\times p^n)]{} & G_n \times G'_n \end{array}$$

where α is a homomorphism of bi-extensions. Since p^n is faithfully flat, so is α, and L_n is therefore the quotient of $\hat{\mathbf{G}}_m \times G_n \times G'_{2n}$ by a suitable flat equivalence relation. For every $x \in G_n(S)$, $y \in G'_{2n}(S)$, $b \in G'_n(S)$ and $\lambda \in \hat{\mathbf{G}}_m(S)$, there is a unique element $\beta_n(x, y, b, \lambda) \in \hat{\mathbf{G}}_m(S)$ such that

$$\alpha((\lambda, x, y)) = \alpha((\lambda . \beta_n(x, y, b, \lambda), x, y + b)$$

and this function β_n describes the equivalence relation. Using the fact that α is a homomorphism of bi-extensions, we deduce

(1) that β_n does not depend on λ,

(2) $\beta_n(x, y, b).\beta_n(x, y+b, b') = \beta_n(x, y, b + b')$ (via associativity of equivalence relation),

(3) $\beta_n(x, y, b).\beta_n(x', y, b) = \beta_n(x+x', y, b)$ (α preserves $+_1$),

(4) $\beta_n(x, y, b).\beta_n(x, y', b') = \beta_n(x, y+y', b+b')$ (α preserves $+_2$).

By (4) and (2) with $b = y' = 0$,

$$\beta_n(x, y, 0).\beta_n(x, 0, b') = \beta_n(x, y, b') = \beta_n(x, y, 0).\beta_n(x, y, b'),$$

hence β_n is independent of y too. Then (3) and (4) show that β_n is a bi-homomorphism, so L_n is constructed from a β_n as required.

We leave it to the reader to check that if we start from a set of β_n's, and construct a bi-extension L, then the above procedure leads you back to these same β_n's. Q.E.D.

I think that with these results, bi-extensions can be applied to the problem of determining the local structure of the moduli space of polarized abelian varieties.

REFERENCES

1. P. Cartier : Modules associés à un groupe formel commutatif, *Comptes Rendus Acad. France*, Series A, 265 (1967), 129.

2. J. Dieudonné : Lie groups and Lie hyperalgebras over a field of char. p, *Amer. J. Math.* 77 (1955), p. 203.

3. M. Lazard : Lois de groupes et analyseurs, *Ann. Ecoles Normales Sup.* 72 (1955), p. 299.

4. Y. Manin : Theory of commutative formal groups over fields of finite characteristic, *Usp. Math. Nauk*, 18 (1963), p. 3 (English transl. : *Russian Math. Surveys*, 18, p. 1).

5. T. Oda : *Abelian Varieties and Dieudonné Modules*, Thesis, Harvard University, 1967.

6. J. Tate : p-divisible groups, in *Local Fields*, Springer-Verlag, 1967.

Tata Institute of Fundamental Research
Bombay

THE IRREDUCIBILITY OF THE SPACE OF CURVES OF GIVEN GENUS

by P. DELIGNE and D. MUMFORD (1)

Fix an algebraically closed field k. Let M_g be the moduli space of curves of genus g over k. The main result of this note is that M_g is irreducible for every k. Of course, whether or not M_g is irreducible depends only on the characteristic of k. When the characteristic is 0, we can assume that $k=\mathbf{C}$, and then the result is classical. A simple proof appears in Enriques-Chisini [E, vol. 3, chap. 3], based on analyzing the totality of coverings of $\mathbf{P}^1$ of degree n, with a fixed number d of ordinary branch points. This method has been extended to char. p by William Fulton [F], using specializations from char. 0 to char. p *provided that* $p>2g+1$. Unfortunately, attempts to extend this method to all p seem to get stuck on difficult questions of wild ramification. Nowadays, the Teichmüller theory gives a thoroughly analytic but very profound insight into this irreducibility when $k=\mathbf{C}$. Our approach however is closest to Severi's incomplete proof ([Se], Anhang F; the error is on pp. 344-345 and seems to be quite basic) and follows a suggestion of Grothendieck for using the result in char. 0 to deduce the result in char. p. The basis of both Severi's and Grothendieck's ideas is to construct families of curves X, some *singular*, with $p_a(X)=g$, over *non-singular* parameter spaces, which in some sense contain enough singular curves to link together any two components that M_g might have.

The essential thing that makes this method work now is a recent " stable reduction theorem " for abelian varieties. This result was first proved independently in char. 0 by Grothendieck, using methods of etale cohomology (private correspondence with J. Tate), and by Mumford, applying the easy half of Theorem (2.5), to go from curves to abelian varieties (cf. [M_2]). Grothendieck has recently strengthened his method so that it applies in all characteristics (SGA 7, 1968). Mumford has also given a proof using theta functions in char. $\neq 2$. The result is this:

Stable Reduction Theorem. — Let R *be a discrete valuation ring with quotient field* K. *Let* A *be an abelian variety over* K. *Then there exists a finite algebraic extension* L *of* K *such*

---

(1) The first author wishes to thank the Institut des Hautes Études scientifiques, Bures-sur-Yvette, for support in this research and N. Katz, for his invaluable assistance in the preparation of this manuscript; the second author wishes to thank the Tata Institute of Fundamental Research, Bombay, and the Institut des Hautes Études scientifiques.

D. Mumford, *Selected Papers*, Vol. II,

that, if R_L = *integral closure of* R *in* L, *and if* $\mathcal{A}_L$ *is the Néron model of* $A \times_K L$ *over* R_L, *then the closed fibre* $A_{L,s}$ *of* $\mathcal{A}_L$ *has no unipotent radical.*

We shall give two related proofs of our main result. One of these is quite elementary, and follows by quite standard techniques once the Stable Reduction Theorem for abelian varieties is applied, in § 2, to deduce an analogous stable reduction theorem for curves. The other proof is more powerful, and is based on the use of a larger category than the category of schemes, and on proving for the objects of this category many of the standard theorems for schemes, especially the Enriques-Zariski connectedness theorem (EGA 3, (4.3)). Unfortunately, this larger category is not quite a category — it is a simple type of *2-category*; in fact, if X, Y are objects, then Hom(X, Y) is itself a category, but one in which all morphisms are isomorphisms. The objects of this 2-category we call *algebraic stacks* (1). The moduli space M_g is just the " underlying coarse variety " of a more fundamental object, the *moduli stack* $\mathcal{M}_g$ studied in [M3]. Full details on the basic properties and theorems for algebraic stacks will be given elsewhere. In this paper, we will only give definitions and state without proof the general theorems which we apply. Using the method of algebraic stacks, we can prove not only the irreducibility of M_g itself, but of all *higher level* moduli spaces of curves too (cf. § 5 below).

§ 1. Stable curves and their moduli.

The key definition of the whole paper is this:

Definition (1.1). — *Let* S *be any scheme. Let* $g \geq 2$. *A* stable curve of genus g *over* S *is a proper flat morphism* $\pi: C \rightarrow S$ *whose geometric fibres are reduced, connected, 1-dimensional schemes* C_s *such that:*

(i) C_s *has only ordinary double points;*

(ii) *if* E *is a non-singular rational component of* C_s, *then* E *meets the other components of* C_s *in more than* 2 *points;*

(iii) $\dim H^1(\mathcal{O}_{C_s}) = g$.

We will study in this section three aspects of the theory of stable curves: their pluri-canonical linear systems, their deformations, and their automorphisms.

Suppose $\pi: C \rightarrow S$ is a stable curve. Since π is flat and its geometric fibres are local complete intersections, the morphism π is locally a complete intersection (i.e., locally, C is isomorphic as S-scheme to $V(f_1, \ldots, f_{n-1}) \subset \mathbf{A}^n \times U$, where $U \subset S$ is open, and $f_1, \ldots, f_{n-1} \in \Gamma(\mathcal{O}_{\mathbf{A}^n \times U})$ are a regular sequence). Therefore, by the theory of duality of coherent sheaves [H], there is a canonical invertible sheaf $\omega_{C/S}$ on C — the unique non-zero cohomology group of the complex of sheaves $f^!(\mathcal{O}_S)$. We need to know the following facts about $\omega_{C/S}$:

a) for all morphisms $f: T \rightarrow S$, $\omega_{C \times_S T/T}$ is canonically isomorphic to $f^*(\omega_{C/S})$;

(1) A slightly less general category of objects, called *algebraic spaces*, has been introduced and studied very deeply by M. Artin [A1] and D. Knutson [K]. The idea of enlarging the category of varieties for the study of moduli spaces is due originally, we believe, to A. Weil.

b) if $S=\mathrm{Spec}(k)$, k algebraically closed, let $f: C'\to C$ be the normalization of C, $x_1, \dots, x_n, y_1, \dots, y_n$ the points of C′ such that the $z_i=f(x_i)=f(y_i)$, $1\leq i\leq n$, are the double points of C. Then $\omega_{C/S}$ is the sheaf of 1-forms η on C′ regular except for simple poles at the x's and y's and with $\mathrm{Res}_{x_i}(\eta)+\mathrm{Res}_{y_i}(\eta)=0$;

c) if $S=\mathrm{Spec}(k)$, and $\mathscr{F}$ is a coherent sheaf on C, then

$$\mathrm{Hom}(H^1(C, \mathscr{F}), k) \cong \mathrm{Hom}_{\mathcal{O}_C}(\mathscr{F}, \omega_{C/S}).$$

Theorem (**1.2**). — *If $g\geq 2$ and C is a stable curve of genus g over an algebraically closed field k, then $H^1(C, \omega_{C/k}^{\otimes n})=(0)$ if $n\geq 2$, and $\omega_{C/k}^{\otimes n}$ is very ample if $n\geq 3$.*

Proof. — Since C is stable, of genus $g\geq 2$, every irreducible component E of C either 1) has (arithmetic) genus ≥ 2 itself, 2) has genus 1, but meets other components of C in at least one point, or 3) is non-singular, rational and meets other components of C in at least three points. But by *b)* above, $\omega_{C/k}\otimes\mathcal{O}_E$ is isomorphic to $\omega_{E/k}(\sum_i Q_i)$, where $\{Q_i\}$ are the points where E meets the rest of C. Since the degree of $\omega_{E/k}$ is $2g_E-2$, it follows that in any of the cases 1, 2 or 3, $\omega_{C/k}\otimes\mathcal{O}_E$ has positive degree. This shows immediately that $\omega_{C/k}$ is ample on each component E of C, hence is ample.

Next, by *c)* above, $H^1(\omega_{C/k}^{\otimes n})$ is dual to $H^0(\omega_{C/k}^{\otimes 1-n})$. Since $\omega_{C/k}\otimes\mathcal{O}_E$ has positive degree, $\omega_{C/k}^{\otimes 1-n}\otimes\mathcal{O}_E$ has no sections for any E, any $n\geq 2$; therefore $H^0(\omega_{C/k}^{\otimes 1-n})=(0)$ if $n\geq 2$, and so $H^1(\omega_{C/k}^{\otimes n})=(0)$ if $n\geq 2$.

To prove that an invertible sheaf $\mathscr{L}$ on any scheme C, proper over k, is very ample, it suffices to show

a) for all closed points $x\neq y$

$$H^0(C, \mathscr{L})\to H^0(C, (\mathscr{L}\otimes k(x))\oplus(\mathscr{L}\otimes k(y)))$$

is surjective,

b) for all closed points x,

$$H^0(C, \mathscr{L})\to H^0(C, \mathscr{L}\otimes\mathcal{O}_x/\mathfrak{m}_x^2)$$

is surjective.

Using the exact sequence of cohomology, these both follow if $H^1(C, \mathfrak{m}_x.\mathfrak{m}_y.\mathscr{L})=(0)$ for all closed points $x, y\in C$. In our case, $\mathscr{L}=\omega_C^{\otimes n}$, $n\geq 3$, so if we use duality, we must show:

$$(*)\qquad \mathrm{Hom}(\mathfrak{m}_x.\mathfrak{m}_y, \omega_X^{\otimes -n})=(0), \quad \text{if } n\geq 2.$$

If x is a non-singular point, $\mathfrak{m}_x$ is an invertible sheaf. If x is a double point, let $\pi: C'\to C$ be the result of blowing up x, and let $x_1, x_2\in C'$ be the two points in $\pi^{-1}(x)$. Then it is easy to check that for any invertible sheaf $\mathscr{L}$ on C:

$$\mathrm{Hom}(\mathfrak{m}_x, \mathscr{L}) \cong H^0(C', \pi^*\mathscr{L})$$
$$\mathrm{Hom}(\mathfrak{m}_x^2, \mathscr{L}) \cong H^0(C', \pi^*\mathscr{L}(x_1+x_2)).$$

Therefore, we have 3 cases of $(*)$ to check:

Case 1. — x, y non-singular points of C, then $H^0(\omega_C^{-n}(x+y))=(0)$, if $n\geq 2$.

Case 2. — x double point of C, $\pi : C' \to C$ blowing up x, $\{x_1, x_2\} = \pi^{-1}(x)$, and y a non-singular point of C. Then

$$H^0(\pi^*(\omega_C)^{-n}(y)) = (0), \quad H^0(\pi^*(\omega_C)^{-1}(x_1+x_2)) = (0), \quad \text{if } n \geq 2.$$

Case 3. — x, y double points of C, $\pi : C' \to C$ blowing up x *and* y. Then

$$H^0(\pi^*(\omega_C)^{-n}) = (0) \quad \text{if } n \geq 2.$$

Now since the degree of ω_C^{-n} $(n \geq 2)$ on all components E of C is less than or equal to -2, all of this is clear, except in those cases where $n = 2$, the degree of ω_C on some E is 1, and in which two poles are allowed on E. This occurs if:

(i) case 1, $p_a(E) = 1$, E meets $C - E$ at only one point, $x, y \in E$.
(ii) case 1, $p_a(E) = 0$, E meets $C - E$ at only three points, $x, y \in E$.
(iii) case 2, E a rational curve with one double point meeting $C - E$ at one point, $x =$ double point of E.

But in all these cases, C has components besides E and a section in the H^0 in question must definitely vanish on all these other components. So at the points where E meets $C - E$, the section has extra zeroes. Since the sheaf in question has degree 0 on E, the section is zero on E too. Q.E.D.

Corollary. — *Let* $\pi : C \to S$ *be any stable curve of genus* $g \geq 2$. *Then* $\omega_{C/S}^{\otimes n}$ *is relatively very ample if* $n \geq 3$ *and* $\pi_*(\omega_{C/S}^{\otimes n})$ *is a locally free sheaf on* S *of rank* $(2n-1)(g-1)$.

Proof. — In fact, since for all $s \in S$, $H^1(\omega_{C/S}^{\otimes n} \otimes \mathcal{O}_{C_s}) = (0)$, it follows from [EGA, chap. 3, § 7], that $\pi_*(\omega_{C/S}^{\otimes n})$ is locally free and that $\pi_*(\omega_{C/S}^{\otimes n}) \otimes k(s) \simeq H^0(\omega_{C/S}^{\otimes n} \otimes \mathcal{O}_{C_s})$. Therefore the corollary follows. Q.E.D.

Taking $n = 3$, it follows that every stable curve C/S can be realized as a family of curves in $\mathbf{P}^{5g-6}$ with Hilbert polynomial:

$$P_g(n) = (6n-1)(g-1).$$

Following standard arguments ([M_1], p. 99), it is easy to prove that there is a subscheme

$$H_g \subset \mathbf{Hilb}_{\mathbf{P}^{5g-6}}^{P_g}$$

of " all " tri-canonically embedded stable curves. (**Hilb** is the Hilbert scheme *over* **Z**.) To be precise, there is an isomorphism of functors:

$$\mathrm{Hom}(S, H_g) \cong \left\{ \begin{array}{l} \text{set of stable curves } \pi : C \to S, \text{ plus isomorphisms:} \\ \qquad \mathbf{P}(\pi_*(\omega_{C/S}^{\otimes 3})) \cong \mathbf{P}^{5g-6} \times S \\ \text{(modulo isomorphism)} \end{array} \right\}$$

We will denote by $Z_g \subset H_g \times \mathbf{P}^{5g-6}$ the universal tri-canonically embedded stable curve. The functor of stable curves itself is the sheafification of the quotient of functors: $\underline{H}_g/\underline{PGL}(5g-6)$.

We now consider the deformation theory of stable curves. Let k be any ground field. The deformation theory of X's smooth over k can be found in [SGA, 60-61]; for singular S's, the theory has been worked out in [Sc]. We shall indicate here the results of this theory for a scheme X which is

(i) one-dimensional;

(ii) generically smooth over k;

(iii) locally a complete intersection.

The advantages of this case are two-fold: first, the " cotangent complex " of Grothendieck, Lichtenbaum and Schlessinger reduces, in view of (ii) and (iii), to the single coherent sheaf $\Omega_{X/k}$, the Kähler differentials. Secondly, we have:

Lemma (**1.3**). — $\mathrm{Ext}^2(\Omega_{X/k}, \mathcal{O}_X) = (0)$.

Proof. — Use the spectral sequence:

$$\mathrm{H}^p(X, \mathcal{E}xt^q(\Omega_{X/k}, \mathcal{O}_X)) \Rightarrow \mathrm{Ext}^{p+q}(\Omega_{X/k}, \mathcal{O}_X).$$

Then (i) $\mathrm{H}^2(X, \mathcal{E}xt^0) = (0)$ since $\dim X = 1$. (ii) Since $\Omega_{X/k}$ is locally free except at a finite number of points, $\mathcal{E}xt^1(\Omega_{X/k}, \mathcal{O}_X)$ has 0-dimensional support, hence $\mathrm{H}^1(X, \mathcal{E}xt^1) = (0)$. (iii) Locally, if we embed $X \subset \mathbf{A}^n$, then $\Omega_{X/k}$ has a free resolution of length 2:

$$0 \to \mathcal{I}/\mathcal{I}^2 \to \Omega_{\mathbf{A}^n} \otimes \mathcal{O}_X \to \Omega_{X/k} \to 0$$

where $\mathcal{I}$ = sheaf of ideals defining X. Therefore $\mathcal{E}xt^2 = (0)$. Q.E.D.

In Schlessinger's theory, the significance of Lemma (1.3) is that all obstructions vanish, i.e., deformations of X over base schemes $\mathrm{Spec}(A/\mathfrak{I})$ (A = local Artin ring with residue field k) can always be embedded in deformations over Spec(A). Moreover, the theory says that there is a canonical one-one correspondence between $\mathrm{Ext}^1(\Omega_{X/k}, \mathcal{O}_X)$ and the first order deformations of X, i.e., proper, flat morphisms p, and isomorphisms α as follows:

$$\begin{array}{ccccc} X_1 & \supset & X_1 \times_{\mathrm{Spec}\, k[\varepsilon]} \mathrm{Spec}\, k & \xleftarrow[\alpha]{\sim} & X \\ \downarrow{\scriptstyle p} & & \downarrow & & \downarrow \\ \mathrm{Spec}\, k[\varepsilon]/\varepsilon^2 & \supset & \mathrm{Spec}\, k & = & \mathrm{Spec}\, k. \end{array}$$

Since the obstructions vanish, there is a *versal formal deformation* $\mathcal{X}$ of X over the base scheme

$$\mathcal{M} = \mathrm{Spec}\, \mathfrak{o}_k[[t_1, \ldots, t_N]],$$

where $\mathfrak{o}_k = k$ if the char. is 0, or the complete regular local ring, max. ideal $p.\mathfrak{o}_k$, residue field k, if char.$(k) = p$, unique (by Cohen's structure theorem)

and $N = \dim_k \mathrm{Ext}^1(\Omega_X, \mathcal{O}_C)$.

This means that $\mathscr{X}$ is a formal scheme, proper and flat over $\mathscr{M}$, with fibre X over Spec(k) and the two properties:

a) Every deformation of X *is induced from* $\mathscr{X}$, i.e., if A is a local Artin $\mathfrak{o}_k$-algebra with residue field k, and $p : Y \rightarrow \mathrm{Spec}(A)$ is proper and flat with fibre X over Spec(k), then there is a commutative diagram:

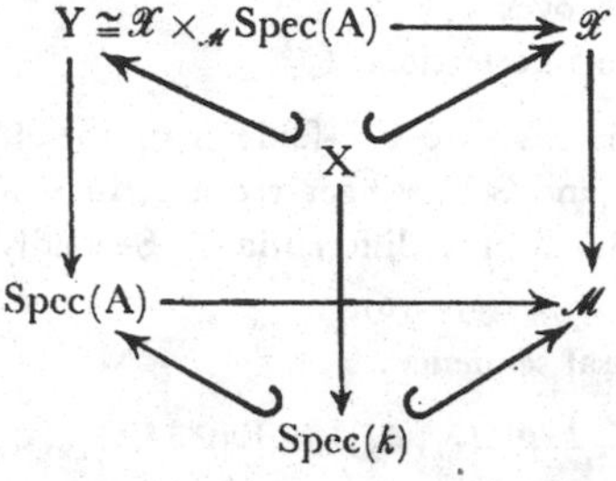

b) If $A = k[\varepsilon]/(\varepsilon^2)$, the above morphism f is uniquely determined by the diagram. This implies that the tangent space to $\mathscr{M} \times_{\mathfrak{o}_k} k$ at its closed point is canonically isomorphic to $\mathrm{Ext}^1(\Omega_{X/k}, \mathcal{O}_X)$.

In case $\mathrm{Ext}^0(\Omega_{X/k}, \mathcal{O}_X) = (0)$, $\mathscr{X}/\mathscr{M}$ is, in fact, *universal*: i.e., in property *a)*, f is *always* unique, which means that the functor represented by $\mathscr{M}$ in the category of artin, local $\mathfrak{o}_k$-algebras is isomorphic to the functor of deformations Y/A of X. This fortunately holds for stable curves:

Lemma (**1.4**). — *If* X *is a stable curve,* $\mathrm{Ext}^0(\Omega_{X/k}, \mathcal{O}_X) = (0)$.

Proof. — We may assume that k is algebraically closed. Now a homomorphism from Ω_X to $\mathcal{O}_X$ is given by an everywhere regular vector field D on X. Such a vector field is given, in turn, by a regular vector field D′ on the normalization X′ of X which vanishes at all points of X′ lying over the double points of X. In particular, D′ and hence D vanishes identically on all components E of X whose normalization E′ has genus ≥ 2. There remain the following possibilities for E:

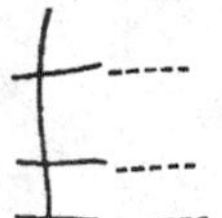

E non-singular rational

E′ rational, E one double pt.

E′ rational, E$\geq$2 double pts.

E non-singular elliptic

E′ elliptic, E$\geq$1 double points

In all cases where E′ is rational, note that D′ has to have at least 3 zeroes; and where E′ is elliptic, D′ has to have at least one zero. So D′ vanishes on all components E′. This proves that $\mathrm{Ext}^0(\Omega_{X/k}, \mathcal{O}_X)=(0)$. Q.E.D.

Schlessinger's theory also allows us to trace what happens to the singularities of X in this deformation $\mathcal{M}$. For each closed point $x \in X$, he studies deformations of the complete local ring $\hat{\mathcal{O}}_{x,X}$ alone, i.e., flat A-algebras $\mathcal{O}$ plus isomorphisms:

$$\mathcal{O}\otimes_A k \cong \hat{\mathcal{O}}_{x,X}.$$

This is a functor of A exactly as before. Since $\mathrm{Ext}^2_{\hat{\mathcal{O}}_x}(\Omega_{\hat{\mathcal{O}}_x/k}, \hat{\mathcal{O}}_x)=(0)$, there are no obstructions to extending deformations. First order deformations with $A=k[\varepsilon]/(\varepsilon^2)$ are classified by $\mathrm{Ext}^1_{\hat{\mathcal{O}}_x}(\Omega_{\hat{\mathcal{O}}_x/k}, \mathcal{O}_x)$. And there is a versal formal deformation $\hat{\mathcal{O}}$, which is a complete local ring and a flat $\mathfrak{o}_k[[t_1, \ldots, t_N]]$-algebra, $N=\dim_k \mathrm{Ext}^1_{\hat{\mathcal{O}}_x}(\Omega_{\hat{\mathcal{O}}_x/k}, \hat{\mathcal{O}}_x)$, such that

$$\hat{\mathcal{O}}/(p, t_1, \ldots, t_N)=\hat{\mathcal{O}}_{x,X}.$$

Whenever X is smooth over k at x, $N=0$, and the theory is uninteresting. The first non-trivial example is a k-rational ordinary double point with k-rational tangent lines:

$$\hat{\mathcal{O}}_{x,X}\cong k[[u, v]]/(u.v).$$

Then $N=1$, and

$$\hat{\mathcal{O}}\cong \mathfrak{o}_k[[u, v, t_1]]/(uv-t_1).$$

In other words, if $\mathcal{O}$ is any deformation of $\hat{\mathcal{O}}_{x,X}$ and u, v are lifted suitably into $\mathcal{O}$, then $u.v=w\in A$ and $\mathcal{O}$ is induced from $\hat{\mathcal{O}}$ via the homomorphism $\mathfrak{o}_k[[t_1]]\to A$ taking t_1 to w. This is easy to prove.

Finally, Schlessinger's theory connects global deformations to local ones. Let $\mathcal{X}/\mathcal{M}_{gl}$ with $\mathcal{M}_{gl}=\mathrm{Spec}(A)$ be the versal global deformation, let $x_1, \ldots, x_k \in X$ be the points where X is not smooth over k, and let $\hat{\mathcal{O}}_i$ as A_i-algebra be the versal deformation of the local ring $\hat{\mathcal{O}}_{x_i,X}$. Let $\mathcal{M}_{lo}=\mathrm{Spec}(A_1\hat{\otimes}_{\mathfrak{o}_k}\ldots\hat{\otimes}_{\mathfrak{o}_k}A_k)$. Then we may consider the local rings

$$\hat{\mathcal{O}}_{x_i,\mathcal{X}}$$

of $\mathcal{X}$ at x_i. There are $\mathfrak{o}_k$-homomorphisms $\varphi_i: A_i\to A$ such that $\hat{\mathcal{O}}_{x_i,\mathcal{X}}\cong\hat{\mathcal{O}}_i\hat{\otimes}_{A_i}A$. Dualizing, we obtain a morphism

$$\Phi=\prod_i \mathrm{Spec}(\varphi_i): \mathcal{M}_{gl}\to\mathcal{M}_{lo}$$

which describes exactly how the various singularities of X behave in the versal deformation $\mathcal{X}$. The final fact that we need is:

Proposition (1.5). — $\Phi: \mathcal{M}_{gl}\to\mathcal{M}_{lo}$ *is formally smooth*, i.e., *there are isomorphisms*

$$\mathcal{M}_{gl}\cong \mathrm{Spec}\ \mathfrak{o}_k[[t_1, \ldots, t_{N+M}]]$$
$$\mathcal{M}_{lo}\cong \mathrm{Spec}\ \mathfrak{o}_k[[t_1, \ldots, t_N]]$$

such that $\Phi^*(t_i)=t_i$, $1\leq i\leq N$.

Proof. — In view of the functorial significance of $\mathcal{M}_{gl}$ and $\mathcal{M}_{10}$, this follows if we prove that the natural map:

$$(*) \qquad \mathrm{Ext}^1_{\mathcal{O}_X}(\Omega_{X/k}, \mathcal{O}_X) \to \prod_{i=1}^{k} \mathrm{Ext}^1_{\hat{\mathcal{O}}_{x_i}}(\hat{\Omega}_{x_i}, \hat{\mathcal{O}}_{x_i})$$

is surjective; $(*)$ is the induced map $d\Phi$ on the tangent spaces to $\mathcal{M}_{gl}$ and $\mathcal{M}_{10}$. Since Ω_X is an invertible $\mathcal{O}_X$-Module outside the x_i's, it follows that:

$$\prod_{i=1}^{k} \mathrm{Ext}^1_{\hat{\mathcal{O}}_{x_i}}(\hat{\Omega}_{x_i}, \hat{\mathcal{O}}_{x_i}) \cong \prod_{i=1}^{k} \mathrm{Ext}_{\mathcal{O}_{x_i}}(\Omega_{x_i}, \mathcal{O}_{x_i})$$
$$\cong \mathrm{H}^0(X, \mathcal{E}xt^1_{\mathcal{O}_x}(\Omega_X, \mathcal{O}_X)).$$

Therefore, $(*)$ is surjective by the spectral sequence used in Lemma (1.3) and the fact that $\mathrm{H}^2(X, \mathcal{E}xt^0(\Omega_X, \mathcal{O}_X)) = (0)$. Q.E.D.

In particular, suppose C is a stable curve over an algebraically closed ground field k, and let $x_1, \ldots, x_k$ be the double points of C. Let $N = \dim \mathrm{Ext}^1(\Omega_X, \mathcal{O}_X)$. Then C has a universal formal deformation $\mathcal{C}/\mathcal{M}$ where $\mathcal{M} = \mathrm{Spec}\, \mathfrak{o}_k[[t_1, \ldots, t_N]]$. Note that since in this case, the invertible sheaf $\omega_{\mathcal{C}/\mathcal{M}}$ is relatively ample, $\mathcal{C}$ is not only a formal scheme over $\mathcal{M}$, but also the formal completion of a unique scheme proper and flat over $\mathcal{M}$, which we will also denote by $\mathcal{C}$. $\mathcal{C}$ is clearly a stable curve over $\mathcal{M}$. Now each double point x_i has *one* modulus (cf. our example above) so the versal deformation space of the rings $\mathcal{O}_{x_i, C}$ is $\mathfrak{o}_k[[t'_1, \ldots, t'_k]]$. By Proposition (1.5), we may identify t_i with t'_i, and we conclude that for suitable u_i, v_i:

$$\hat{\mathcal{O}}_{x_i, \mathcal{C}} \cong \mathfrak{o}_k[[u_i, v_i, t_1, \ldots, t_N]]/(u_i v_i - t_i).$$

In particular, $t_i = 0$ is the locus in $\mathcal{M}$ where " x_i remains a double point ".

The relation between the formal moduli space $\mathcal{M}$ of C and the local structure of H_g at a point x with $\varkappa(x) = k$ corresponding to some tri-canonical model of C is exactly the same as in the case of non-singular curves ([M$_1$], chap. 5, § 2). Let $\hat{\mathcal{O}}_x$ be the completion of the local ring $\mathcal{O}_{x, \mathrm{H}_g}$, and let $\mathrm{T} = \mathrm{Spec}(\hat{\mathcal{O}}_x)$. Let x denote the closed point of T too. The universal family of stable curves $Z_g \subset \mathrm{H}_g \times \mathbf{P}^{5g-6}$ induces a family $Z' \subset \mathrm{T} \times \mathbf{P}^{5g-6}$, whose fibre Z'_x over x is C. Then there is a unique morphism $f: \mathrm{T} \to \mathcal{M}$ such that $Z' \cong \mathcal{C} \times_{\mathcal{M}} \mathrm{T}$, with this isomorphism restricting to the identity on the fibres over x, both of which are C. I claim that via f, T is formally smooth over $\mathcal{M}$, i.e., $\hat{\mathcal{O}}_x \cong \mathfrak{o}_k[[t_1, \ldots, t_N, t_{N+1}, \ldots, t_M]]$. In fact, by choosing an isomorphism $\mathbf{P}(\pi_*(\omega^{\otimes 3}_{\mathcal{C}/\mathcal{M}})) \cong \mathbf{P}^{5g-6} \times \mathcal{M}$, we obtain a tri-canonical embedding $\mathcal{C} \subset \mathbf{P}^{5g-6} \times \mathcal{M}$ of $\mathcal{C}$, hence a morphism $s: \mathcal{M} \to \mathrm{H}_g$ such that $\mathcal{C}$, with this embedding, is the pull-back of Z_g. Then s factors through T and $s: \mathcal{M} \to \mathrm{T}$ is a section of f. On the other hand, consider the action of $\mathbf{PGL}(5g-6)$ on H_g. Let S_x be the stabilizer of the k-valued point x. Then S_x is finite and reduced. Because if it were not, S_x would have a non-trivial tangent space at the origin, i.e., there would be a $k[\varepsilon]/(\varepsilon^2)$-valued point of $\mathbf{PGL}(5g-6)$ centered at the identity, which maps the embedded stable curve $C \subset \mathbf{P}^{5g-6}$ corresponding to x into itself. But this action is given by an everywhere regular derivation on C, and we have seen that all such vanish. This means that this $k[\varepsilon]/(\varepsilon^2)$-valued automorphism is the identity at all points of C and, since C is connected and spans $\mathbf{P}^{5g-6}$, the automorphism is the identity everywhere. Thus S_x is finite and reduced. It follows that the action of $\mathbf{PGL}(5g-6)$ on T is *formally free*, and hence that T is formally a principal fibre bundle over $\mathscr{M}$ with group $\mathbf{PGL}(5g-6)$. Therefore T is formally smooth over $\mathscr{M}$ as required.

Putting this together with what we know about $\mathscr{M}$, we conclude the following:

Let k be any algebraically closed field,
let $H'_g = H_g \times \mathrm{Spec}(\mathfrak{o}_k)$, $Z'_g = Z_g \times \mathrm{Spec}(\mathfrak{o}_k)$,
let $x \in H'_g$ be a closed point,
let $C \subset Z'_g$ be the stable curve over x,
let $x_1, \ldots, x_k \in C$ be its double points.

Then

Theorem (**1.6**). — *There are isomorphisms*

$$\hat{\mathcal{O}}_{x, H'_g} \cong \mathfrak{o}_k[[t_1, \ldots, t_N]]$$
$$\hat{\mathcal{O}}_{x_i, Z'_g} \cong \mathfrak{o}_k[[u_i, v_i, t_1, \ldots, t_N]]/(u_i v_i - t_i).$$

Corollary (**1.7**). — H_g *is smooth over* $\mathbf{Z}$. *In particular, for all algebraically closed fields* k, $H_g \times \mathrm{Spec}(k)$ *is a disjoint union of a finite number of non-singular algebraic varieties over* k.

Let

$$H^0_g = \{x \in H_g \mid \text{the corresponding stable curve } (Z_g)_x \text{ is non-singular}\}.$$
$$S = \{x \in Z_g \mid \text{the projection } \pi : Z_g \to H_g \text{ is } \textit{not} \text{ smooth at } x\}.$$

Definition (**1.8**). — *Let* $p : X \to Y$ *be a smooth morphism of finite type, with* Y *a noetherian scheme, and let* $D \subset X$ *be a relative Cartier divisor. Then* D has normal crossings relative to Y *if for all* $x \in D$, *the local equation* $d = 0$ *of* D *decomposes in the strict completion* (1) $\tilde{\mathcal{O}}_{x,X}$ *of* $\mathcal{O}_{x,X}$ *as* $d = d_1 \ldots d_k$, *where* $d_1, \ldots, d_k$ *are linearly independent in* $\tilde{\mathfrak{m}}_{x,X}/\tilde{\mathfrak{m}}^2_{x,X} + \mathfrak{m}_{y,Y}.\tilde{\mathcal{O}}_{x,X}$, *with* $y = p(x)$.

Corollary (**1.9**). — $H^0_g = H_g - S^*$, *where* S^* *is a divisor with normal crossings relative to* $\mathbf{Z}$. Z_g *and* S *are smooth over* $\mathbf{Z}$, *and the projection* $p : S \to S^*$ *is finite and an isomorphism at all points where* S^* *is smooth over* $\mathbf{Z}$, i.e., S *is the normalization of* S^*.

Proof. — In the notation of Theorem (1.6), S^* is defined in $\hat{\mathcal{O}}_{x, H'_g}$ by the local equation $t_1 \ldots t_k = 0$. And

$$\hat{\mathcal{O}}_{x_i, Z'_g} \cong \mathfrak{o}_k[[u_i, v_i, t_1, \ldots, t_{i-1}, t_{i+1}, \ldots, t_N]]$$
$$\hat{\mathcal{O}}_{x_i, S} \cong \hat{\mathcal{O}}_{x_i, Z'_g}/(u_i, v_i)$$
$$\cong \mathfrak{o}_k[[t_1, \ldots, t_{i-1}, t_{i+1}, \ldots, t_N]].$$

Q.E.D.

(1) The complete local ring, formally etale over $\mathcal{O}_{x,X}$ with residue field the separable closure of $\mathcal{O}_{x,X}/\mathfrak{m}_{x,X}$.

Next we take up the isomorphisms and automorphisms of stable curves. Suppose $p: X \to S$, $q: Y \to S$ are two stable curves:

Definition (**1.10**). — $\underline{\mathrm{Isom}}_S(X, Y)$ *is the functor on* (Sch/S) *associating to each* S*-scheme* S′ *the set of* S′*-isomorphisms between* $X \times_S S'$ *and* $Y \times_S S'$. *If* $X = Y$, *we denote* $\underline{\mathrm{Isom}}_S(X, X)$ *by* $\underline{\mathrm{Aut}}_S(X)$.

Since both X and Y have the canonical polarizations $\omega_{X/S}$, $\omega_{Y/S}$ respectively, any isomorphism $f: X \to Y$ must satisfy $f^*(\omega_{Y/S}) \cong \omega_{X/S}$. Therefore, by Grothendieck's results on the representability of the Hilbert scheme and related functors [Gr_1], we conclude that $\underline{\mathrm{Isom}}_S(X, Y)$ is represented by a scheme $\mathbf{Isom}_S(X, Y)$, quasi-projective over S. Concerning this scheme, we have:

Theorem (**1.11**). — $\mathbf{Isom}_S(X, Y)$ *is finite and unramified over* S.

Proof. — To check that $\mathbf{Isom}_S(X, Y)$ is unramified, we may take S to be the spectrum of an algebraically closed field k, in which case $\mathbf{Isom}_S(X, Y)$ is either empty or isomorphic to $\mathbf{Aut}_k(X)$. A point of $\mathbf{Aut}_k(X)$ with values in $k[\varepsilon]/(\varepsilon^2)$ with image the identity may be identified with a vector field on X. By Lemma (1.4), stable curves have no non-zero vector fields. This proves that $\mathbf{Isom}_S(X, Y)$ is unramified over S, and since it is also of finite type over S, it is quasi-finite over S. It remains to check that $\mathbf{Isom}_S(X, Y)$ is proper over S.

Locally over S, X and Y are the pull-backs of the universal tri-canonically embedded stable curve by some morphisms from S to H_g, so that it suffices to prove the properness of $\mathbf{Isom}_S(X, Y)$ in the " universal " case where $S = H_g \times H_g$, X and Y being the two inverse images of the universal curve on H_g. In that case, the open subset of $\mathbf{Isom}_S(X, Y)$ corresponding to smooth curves is dense, so that the Theorem follows from the valuative criterion of properness which holds by:

Lemma (**1.12**). — *Let* X *and* Y *be two stable curves over a discrete valuation ring* R *with algebraically closed residue field. Denote by* η *and* s *the generic and closed points of* Spec(R), *and assume that the generic fibres* X_η *and* Y_η *of* X *and* Y *are smooth. Then any isomorphism* φ_η *between* X_η *and* Y_η *extends to an isomorphism* φ *between* X *and* Y.

(*A posteriori*, it follows from Theorem (1.11) that the lemma holds for any valuation ring R and without assuming X_η or Y_η smooth.)

Proof. — Another way to put the lemma is that if we start with a smooth curve X_η of genus $g \geq 2$ over the quotient field K of R, there is, up to canonical isomorphism, at most one stable curve X over R with X_η as its generic fibre. We shall deduce this from the analogous uniqueness assertion for minimal models ([L] and [Š]): given a smooth curve X_η of genus $g \geq 1$ over K, there is, up to canonical isomorphism, at most one regular 2-dimensional scheme X, proper and flat over R, with X_η as its generic fibre, without exceptional curves of the first kind in X_s.

Let z denote a generator of the maximal ideal of R and consider the affine plane curve C_n over R given by:

$$xy = z^n.$$

Let $\tilde{C}_n$ denote the scheme obtained by: 1) blowing up the maximal ideal at the unique singularity of C_n; 2) blowing up the maximal ideal at the unique singularity of this scheme..., and so on $\left[\frac{n}{2}\right]$ times. It is easy to check that $\tilde{C}_n$ is a regular scheme whose special fibre is the same as that of C_n except that the singular point is replaced by a sequence of $n-1$ projective lines as follows:

Now suppose x is a singular point of the stable curve X over R. At x, X is formally isomorphic as scheme over R to one of the schemes C_n, so we may blow up X the same way we blew up C_n. If we do this for all singular points of X, we get a regular scheme $\tilde{X}$ with generic fibre X_η. In addition, any non-singular rational component of $\tilde{X}_s$ is linked to the other irreducible components by at least two points, hence it is not exceptional of first kind. Therefore $\tilde{X}$ is *the* minimal model of X_η. Note finally that C_n is a normal scheme, hence so is X; therefore X is the unique normal scheme obtained from $\tilde{X}$ by contracting all non-singular rational components of $\tilde{X}_s$ linked to the other irreducible components by exactly two points. This proves that X is essentially unique. Q.E.D.

Another important fact about the automorphisms of stable curves is:

Theorem (**1.13**). — *Let k be an algebraically closed field and* X *a stable curve over k. Let* $\mathbf{Pic}^0(X)$ *denote the group of invertible sheaves on* X *of degree* 0 *on each component. Then the map (of ordinary groups):*

$$\mathrm{Aut}_k(X) \to \mathrm{Aut}_k(\mathbf{Pic}^0(X))$$

is injective.

Proof. — Let φ be an automorphism of X inducing the identity on $\mathbf{Pic}^0 X$.

Lemma (**1.14**). — *If* X *is smooth, then φ is the identity.*

Proof. — If not, by the Lefschetz-Weil fixed point formula, the number n of fixed points of φ, counted with their multiplicities, is

$$n = 1 - \mathrm{Tr}(\varphi, T_l(\mathbf{Pic}^0(X))) + 1 = 2 - 2g < 0$$

which is absurd. Q.E.D.

Lemma (**1.15**). — *If* X *is irreducible, then φ is the identity.*

Proof. — Let φ' be the action of φ on the normalization X′ of X. Each singular point of X, together with an ordering of its 2 inverse images in X′, defines a distinct morphism from $\mathbf{G}_m$ to $\mathbf{Pic}^0(X)$, so that the inverse image S of the singular locus of X is pointwise fixed by φ'. One has either

a) genus$(X') \geq 2$: then conclude by Lemma (1.14);

b) genus$(X') = 1$, $|S| \geq 2$, then φ' is a translation on X′ leaving a point fixed, and so φ' is the identity;

c) X′ is the projective line, $|S| \geq 4$ and φ' is a projectivity leaving more than three points fixed, so is the identity. Q.E.D.

Let Γ be the following (unoriented) graph:

(i) The set of vertices of Γ is the set Γ^0 of irreducible components of X,

(ii) the set of edges of Γ is the set Γ^1 of the singular points of X which lie on two distinct irreducible components,

(iii) an edge $x \in \Gamma^1$ has for extremities the irreducible components on which x lies.

Lemma **(1.16)**. — *If φ induces the identity on Γ, then φ is the identity.*

Proof. — If X_1 is an irreducible component of X, then $\varphi(X_1) = X_1$ and φ leaves fixed the points of intersection of X_1 with the other components. In addition, $\mathbf{Pic}^0(X)$ maps onto $\mathbf{Pic}^0(X_1)$ so that φ acts trivially on $\mathbf{Pic}^0(X_1)$. Either:

a) genus $(X_1) \geq 2$ and $\varphi | X_1$ is the identity by Lemma (1.15);

b) genus $(X_1) = 1$, φ acts by a translation and leaves a point fixed, so is the identity on X_1;

c) X_1 is the projective line and φ leaves fixed at least three points, so is the identity on X_1. Q.E.D.

Lemma **(1.17)**. — (i) *Any edge in Γ has distinct extremities.*

(ii) *Any vertex which is the extremity of* 0, 1 *or* 2 *edges is fixed by φ.*

(iii) *φ acts trivially on* $H^1(\Gamma, \mathbf{Z})$.

Proof. — It is easy to check that the subgroup of $\mathbf{Pic}^0(X)$ corresponding to invertible sheaves whose restriction to each irreducible component of X is trivial is canonically isomorphic to

$$H^1(\Gamma, \mathbf{Z}) \otimes \mathbf{G}_m.$$

This implies (iii), and (i) is trivial.

The morphism from $\mathbf{Pic}^0(X)$ to the product $\prod_i \mathbf{Pic}^0(X_i)$, extended over the irreducible components of X, is surjective, so that if $\mathbf{Pic}^0(X_i) \neq \{e\}$, then $\varphi(X_i) = X_i$. This is the case, unless X_i is a projective line, linked to the other components in at least three points. Q.E.D.

We prove now that if an automorphism φ of any finite graph Γ has the properties stated in Lemma (1.17), it is the identity. Make induction on the sum of the number of vertices and edges of Γ. If Γ has an isolated point x, then $\varphi(x) = x$ so let $\Gamma^* = \Gamma - \{x\}$. Then $\varphi =$ identity on Γ^* by induction, so $\varphi =$ identity on Γ too. If Γ has an extremity x, then $\varphi(x) = x$, and again let Γ^* be Γ minus x and the edge abutting at x. Then Γ^* has all the properties Γ has, so $\varphi =$ identity on Γ^*, hence $\varphi =$ identity on Γ. If Γ has a vertex x on which only 2 edges abut, we have one of the two cases:

(1)

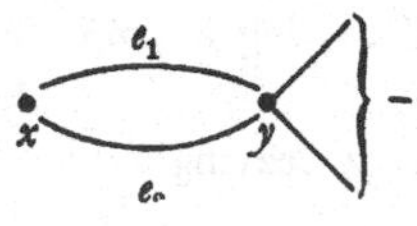

(2)

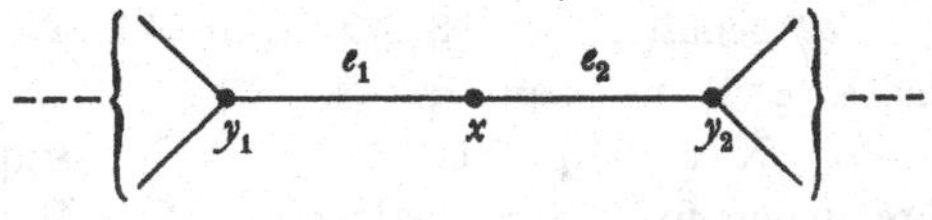

In the first case, $\varphi(y)=y$, and let Γ^* be Γ minus x, e_1 and e_2. Then $\varphi=$ identity on Γ^* and $\varphi(e_i)=e_i$ too, since if φ reverses the e_i's, this contradicts (iii). In the second case, let Γ^* be Γ minus x, and with e_1 and e_2 identified:

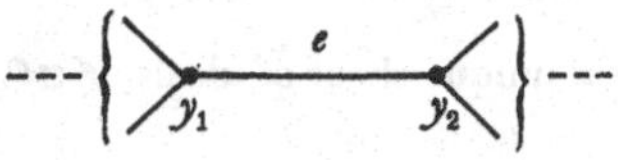

Then $\varphi=$ identity on Γ^*, so $\varphi=$ identity on Γ. Next, say Γ has an edge e with extremities x and y such that $\varphi(x)=x$, $\varphi(y)=y$, $\varphi(e)=e$. Let Γ^* be Γ minus e. Then $\varphi=$ identity on Γ^*, so $\varphi=$ identity on Γ. If none of these reductions are possible, we must be in a situation where *a)* every vertex is the abutment of at least three edges and *b)* no edge is left fixed. It is easily seen that the first Betti number b_1 of any of the connected components of Γ is at least 2. Let

$$n_0=\text{number of fixed vertices}$$
$$n_1=\text{number of edges reversed by } \varphi.$$

Then, unless $\Gamma=\emptyset$, the Lefschetz fixed point formula reads:

$$n_0+n_1=b_0-b_1<0,$$

which is impossible. Q.E.D.

§ 2. Degenerations of curves and their jacobians.

We consider the situation:

K = discretely-valued field;
R = integers in K, $k=\mathrm{R}/\mathfrak{M}=$ residue field (assumed algebraically closed);
S = Spec(R), η and s its generic and closed points respectively;
C = a curve, smooth, geometrically irreducible and proper over K, of genus $g\geq 2$;
J = the jacobian variety of C;
$\mathcal{J}=$ the Néron model of J over R (cf. [N]);
$\mathcal{J}^0\subset\mathcal{J}$ the open subgroup scheme with $\mathcal{J}^0_s=$ identity component of $\mathcal{J}_s$;
$\mathscr{C}=$ the minimal model of C over R.

A word about the existence and uniqueness of $\mathscr{C}$ is needed. We recall that $\mathscr{C}$ is to be a regular scheme, flat and proper over R, with generic fibre $\mathscr{C}_\eta=\mathrm{C}$ such that for any other regular scheme $\mathscr{C}'$, flat over R, with generic fibre $\mathscr{C}'_\eta=\mathrm{C}$, the birational map $\mathscr{C}'\to\mathscr{C}$ is a morphism. Šafarevich in [Š] and Lichtenbaum [L] have proven that such a $\mathscr{C}$ (which is obviously unique) exists, provided that there is some regular $\mathscr{C}'$, proper and flat over R, with generic fibre C. And, in fact, that $\mathscr{C}$ is projective over R. To construct such a $\mathscr{C}'$, proceed as follows: first let $\mathscr{C}''$ be any scheme, projective and flat over R with generic fibre C. Let $\hat{\mathrm{R}}$ be the completion of R, and let $\hat{\mathscr{C}}''=\mathscr{C}''\times_{\mathrm{Spec\,R}}\mathrm{Spec}\,\hat{\mathrm{R}}$. Then $\hat{\mathscr{C}}''$ is an excellent surface, so by [Ab] and by unpublished results of Hironaka, there is a sheaf of ideals $\hat{\mathscr{I}}$ with support in the singular locus of $\hat{\mathscr{C}}''$ such that blowing up $\hat{\mathscr{I}}$ leads to a regular surface $\hat{\mathscr{C}}'$. But then

$$\hat{\mathscr{I}} \supset \mathfrak{M}^n . \mathcal{O}_{\hat{\mathscr{C}}''}$$

for some n, so $\hat{\mathscr{I}}$ is induced by a unique sheaf of ideals $\mathscr{I} \subset \mathcal{O}_{\mathscr{C}''}$. Let $\mathscr{C}'$ be obtained by blowing up $\mathscr{I}$. Then

$$\hat{\mathscr{C}}' \cong \mathscr{C}' \times_{\operatorname{Spec} R} \operatorname{Spec} \hat{R}$$

so $\mathscr{C}'$ is regular. $\mathscr{C}'$ is also projective over R, hence a projective $\mathscr{C}$ exists.

Definition (**2.1**). — J *has* stable reduction *if* $\mathscr{J}_s$ *has no unipotent radical.*

Definition (**2.2**). — C *has* stable reduction in sense 1 *if* $\mathscr{C}_s$ *is reduced and has only ordinary double points.* C *has* stable reduction in sense 2 *if there is a stable curve* $\mathscr{C}'$ *over* R *with generic fibre* $\mathscr{C}'_\eta = C$.

Note that if a stable $\mathscr{C}$ exists, then by Theorem (1.11) it is unique.

Proposition (**2.3**). — *The two senses of stable reduction for* C *are equivalent.*

Proof. — Say a stable $\mathscr{C}'$ exists. Blowing up the singularities of $\mathscr{C}'$ as in Lemma (1.12), we obtain the minimal model $\mathscr{C}$ of C and it is seen that $\mathscr{C}_s$ is reduced with only ordinary double points. Conversely, suppose the minimal model $\mathscr{C}$ has this property. Let $E_1, \ldots, E_n$ be the non-singular rational components of $\mathscr{C}_s$ which meet the other components in only two points. Then the E_i's divide into several chains of the type:

unless the entire fibre consisted of E_i's and has the type:

$n \geq 2$
$\mathscr{C}_s =$ loop of E_i's
$(E_i^2)_{\text{on}\,\mathscr{C}} = -2$.

But in this case genus(C) = genus($\mathscr{C}_s$) = 1, which contradicts our assumption. Now, according to Theorem (27.1) of Lipman [Li] (generalizing a result of Artin [A_2], which works for surfaces of finite type over a field) any set of k non-singular rational curves connected in a chain as above on a regular surface X, with self-intersection 2 on X, can be blown down to a rational double point P of type A_k on a normal surface X_0. Reversing the process and blowing up a rational double point of type A_k, it is easy to see that non-singular branches γ on X, crossing transversally only the first or the last rational curve in the chain:

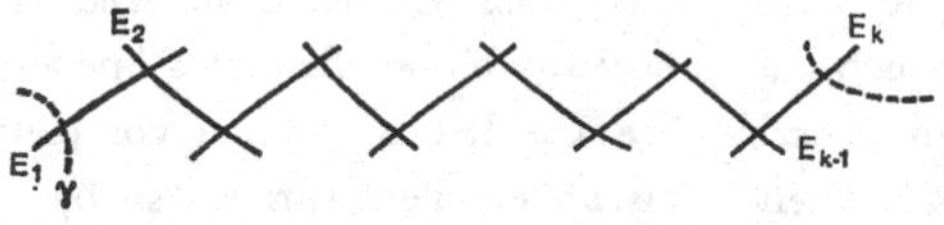

are still non-singular branches when mapped to X_0; and if γ_1, γ_2 intersect E_1 or E_k in distinct points, then they cross transversally on X_0. Therefore, suppose we blow down all the chains of E_i's on $\mathscr{C}$. Let $\mathscr{C}'$ be the normal surface so obtained. Then if one of these chains fits into $\mathscr{C}_s$ like this:

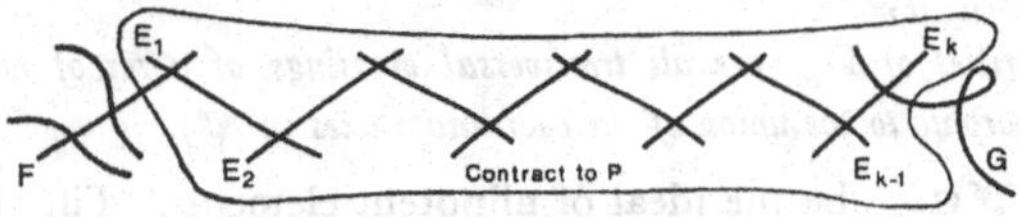

then on $\mathscr{C}'_s$, the images of F and G still have only ordinary double points, each has one non-singular branch through the singular point P, and these branches cross transversally. Therefore $\mathscr{C}'$ is a stable curve over R. Q.E.D.

We are now ready to prove the key result on which our proof of irreducibility depends:

Theorem (2.4). — *J has stable reduction if and only if C has stable reduction.*

Proof. — The connection between $\mathscr{C}$ and $\mathscr{J}$ is based on the following result of Raynaud [R]:

Theorem (2.5). — *If $\mathscr{C}$ and $\mathscr{J}$ are as above, and the greatest common denominator d of the multiplicities of the components of $\mathscr{C}_s$ is* 1, *then $\mathscr{J}^0$ represents the functor* $\underline{\mathrm{Pic}}^0(\mathscr{C}/S)$.

(This result is not stated as such in [R]. It comes out like this, in the terminology of is paper:

a) Condition (N) is verified and $p_*(\mathcal{O}_{\mathscr{C}})=\mathcal{O}_S$, so

b) $\mathscr{C}$ is cohomologically flat over S in dim. 0 by Theorem 4;

c) therefore $\underline{\mathrm{Pic}}^0$ is representable and separated over S by Theorem 3;

d) since $E=\overline{\{0\}}$ and $Q=P/E$, we find $P^0=Q^0$, and since $\mathscr{C}$ is 1-dimensional over S, P^0 and Q^0 are smooth over S; therefore $R^0=Q^0$ and $\tilde{R}=R$.

e) Then by Theorem 5, $\mathbf{Pic}^0$ is the identity component of the Néron model of $\mathbf{Pic}^0(\mathscr{C}_\eta)=J$.)

Now assume that C has stable reduction. Then $\mathscr{C}_s$ is reduced so $d=1$. By Theorem (2.5), $\mathscr{J}^0=\mathbf{Pic}^0(\mathscr{C}/S)$. Therefore $\mathscr{J}^0_s=\mathbf{Pic}^0(\mathscr{C}_s/k)$. Since $\mathscr{C}_s$ is reduced with only ordinary double points, its generalized jacobian $\mathbf{Pic}^0(\mathscr{C}_s/k)$ is an extension of an abelian variety by a torus, i.e., has no unipotent radical. Therefore J has stable reduction.

The converse is more difficult. Assume J has stable reduction. We first prove that C has stable reduction under the additional hypothesis that C has a K-rational point [1]. In this case, $\mathscr{C}$ has an R-rational point, and since $\mathscr{C}$ is regular, sections of $\mathscr{C}$ over R pass through components of $\mathscr{C}_s$ of multiplicity one. Therefore $d=1$, and

---

[1] This is in fact the only case which will be needed in our application to questions of irreducibility.

Theorem (2.5) applies. In particular, $\mathbf{Pic}^0(\mathscr{C}_s/k)=\mathscr{J}_s^0$ so $\mathbf{Pic}^0(\mathscr{C}_s/k)$ has no unipotent radical. We apply:

Lemma **(2.6)**. — *Let* D *be a complete* 1*-dimensional scheme over* k *such that* $H^0(\mathscr{O}_D)\simeq k$, *and such that the generalized jacobian of* D *has no unipotent radical. Then*

(i) $\chi(\mathscr{O}_D)=\chi(\mathscr{O}_{D_{red}})$;

(ii) *the singularities of* D_{red} *are all transversal crossings of a set of non-singular branches* (i.e., *analytically isomorphic to the union of the coordinate axes in* $\mathbf{A}^n$).

Proof. — Let $\mathscr{I}\subset\mathscr{O}_D$ be the ideal of nilpotent elements. Filtering $\mathscr{I}$ by a chain of ideals $\mathscr{I}_k$ such that $\mathscr{I}.\mathscr{I}_k\subset\mathscr{I}_{k+1}$, and using the exact sequence:

$$0\to\mathscr{I}_k/\mathscr{I}_{k+1}\xrightarrow[a\mapsto 1+a]{}(\mathscr{O}_D/\mathscr{I}_{k+1})^*\to(\mathscr{O}_D/\mathscr{I}_k)^*\to 0$$

it is easy to deduce that $\mathbf{Pic}^0(D/k)$ is an extension of $\mathbf{Pic}^0(D_{red}/k)$ by a unipotent group. Therefore, since by assumption $\mathbf{Pic}^0(D/k)$ has no unipotent subgroups,

$$\mathbf{Pic}^0(D/k)\cong\mathbf{Pic}^0(D_{red}/k).$$

Since $H^1(\mathscr{O}_D)$, resp. $H^1(\mathscr{O}_{D_{red}})$, is naturally isomorphic to the Zariski tangent space to $\mathbf{Pic}^0(D/k)$, resp. $\mathbf{Pic}^0(D_{red}/k)$, it follows that

$$H^1(\mathscr{O}_D)\cong H^1(\mathscr{O}_{D_{red}})$$

hence (i) is proven. Let $\pi:C\to D_{red}$ be the normalization of D_{red} and let D^* be the local ringed space which, as topological space is D, and whose structure sheaf is given by:

$$\Gamma(U,\mathscr{O}_{D^*})=\{f\in\Gamma(U,\pi_*(\mathscr{O}_C))\mid x_1,x_2\in\pi^{-1}(U),f(x_1)=f(x_2)\text{ if }\pi(x_1)=\pi(x_2)\}$$

It is easy to check that D^* is a 1-dimensional scheme whose singularities are all transversal crossings of a set of non-singular branches and that π factors:

$$C\overset{\pi'}{\to}D^*\overset{\pi''}{\to}D_{red}.$$

I claim that $\pi'':D^*\to D_{red}$ is an isomorphism. Filter $\mathscr{O}_{D^*}/\mathscr{O}_{D_{red}}$ so as to obtain a chain of coherent $\mathscr{O}_{D_{red}}$-algebras:

$$\mathscr{O}_{D^*}=\mathscr{O}^{(0)}\supset\mathscr{O}^{(1)}\supset\ldots\supset\mathscr{O}^{(k)}=\mathscr{O}_{D_{red}}$$

such that $l(\mathscr{O}^{(n)}/\mathscr{O}^{(n+1)})=1$. Equivalently, this factors π'':

$$D^*=D_0\to D_1\to\ldots\to D_k=D_{red}$$

where $\mathscr{O}^{(h)}\cong\mathscr{O}_{D_n}$, and all arrows are homeomorphisms. If $\{x_n\}=\mathrm{Supp}(\mathscr{O}^{(n)}/\mathscr{O}^{(n+1)})$, then we get an exact sequence:

$$0\to\mathscr{O}^*_{D_{n+1}}\to\mathscr{O}^*_{D_n}\to k_{x_n}\to 0$$

(k_y denotes the residue field at y, as sheaf on D), and hence:

$$0\to k\to H^1(\mathscr{O}^*_{D_{n+1}})\to H^1(\mathscr{O}^*_{D_n})\to 0.$$

It follows easily that $\mathbf{Pic}^0(D_{n+1}/k)$ is an extension of $\mathbf{Pic}^0(D_n/k)$ by $\mathbf{G}_a$. But $\mathbf{Pic}^0(D_{red}/k)$ has no unipotent subgroups, and this can only happen if $D_{red}=D^*$. Q.E.D. for lemma.

We apply the lemma to $\mathscr{C}_s$. According to Lichtenbaum [L] and Šafarevich [Š], there is a divisor K on $\mathscr{C}$ such that for all positive divisors D lying over the closed point of Spec(R), we have:

$$\chi(\mathcal{O}_D) = -\frac{(D.(D+kK))}{2}.$$

Let $E_1, \ldots, E_n$ be the components of $\mathscr{C}_s$, $d_1, \ldots, d_n$ their multiplicities. Then conclusion (i) of the lemma implies that:

$$(\sum_i d_i E_i . (\sum_i d_i E_i + K)) = (\sum_i E_i . (\sum_i E_i + K)).$$

But $((\sum_i d_i E_i).E_k) = 0$, all k, since $\sum_i d_i E_i$ is the divisor of a function $\pi \in R$ if (π) = maximal ideal of R. Therefore:

$$(*) \qquad ((\sum_i (d_i - 1).E_i).K) = (\sum_i E_i . \sum_i E_i).$$

Note that at least one d_i equals 1 since $\mathscr{C}$ has a section over Spec(R) and every section must pass through a component of $\mathscr{C}_s$ of multiplicity 1. Moreover the intersection matrix $(E_i.E_j)$ is negative indefinite, with one-dimensional degenerate subspace generated by $\sum_i d_i E_i$, hence *if some* $d_i > 1$, it follows that $(\sum_i E_i . \sum_i E_i) < 0$. Therefore, by (*), $(E_{i_0}.K) < 0$ for some i_0. Then we have:

a) $(E_{i_0}.K) < 0$;

b) $(E_{i_0}.E_{i_0}) < 0$;

c) $(E_{i_0}.(E_{i_0}+K)) = -2\chi(\mathcal{O}_{E_{i_0}}) \geq -2$;

hence in fact $(E_{i_0}.E_{i_0}) = (E_{i_0}.K) = -1$ so E_{i_0} is an exceptional curve of the first kind. This contradicts our assumption that on $\mathscr{C}$ all possible curves have been blown down. Therefore $d_i = 1$, *all i*.

This proves that $\mathscr{C}_s$ is reduced. By conclusion (ii) of the lemma, plus the fact that the dimension of its Zariski tangent-space is everywhere one or two (since $\mathscr{C}_s$ lies on a regular surface $\mathscr{C}$), we deduce that $\mathscr{C}_s$ has only double points. This proves that C has stable reduction in the case that $C(K) \neq \emptyset$.

In the general case, C will acquire a rational point in a finite extension K′ of K. Let S′ be the spectrum of the localisation at some maximal ideal of the integral closure of R in K′; S′ is the spectrum of a valuation ring and is faithfully flat over S.

We put $S'' = S' \times_S S'$ and denote by C′ and C″ the inverse images of C on $S'_\eta = \mathrm{Spec}(K')$ and S''_η respectively.

$$\begin{array}{ccccc} C'' & \rightrightarrows & C' & \longrightarrow & C \\ \downarrow & & \downarrow & & \downarrow \\ S'' & \underset{pr_2}{\overset{pr_1}{\rightrightarrows}} & S' & \xrightarrow{p} & S \end{array}$$

Let $\overline{C}'$ be the stable curve on S' having C' as generic fibre. The restriction, C', of $\overline{C}'$ to S'_η carries a descent datum with respect to p, i.e. an isomorphism, φ_η, between the restrictions of $pr_1^*(\overline{C}')$ and $pr_2^*(\overline{C}')$ to S''_η. It remains only to extend the isomorphism φ_η to an isomorphism, φ, between the $pr_i^*(\overline{C}')$. Then φ will be a descent datum for $\overline{C}'$ with respect to p. As $\overline{C}'$ is canonically polarized (1.2), this descent datum will be effective, and so define a stable curve, $\overline{C}$, over S with generic fibre C.

Because $\mathscr{J}^0_s$ has no unipotent radical, the inverse image, $p^*\mathscr{J}^0$, of $\mathscr{J}^0$ on S' is the identity component of the Néron Model of the jacobian of C', so that, defining $q = p \circ pr_1 = p \circ pr_2$, one has

$$\mathbf{Pic}^0(\overline{C}'/S') = p^*\mathscr{J}^0$$
$$\mathbf{Pic}^0(pr_1^*\overline{C}') = \mathbf{Pic}^0(pr_2^*\overline{C}') = q^*\mathscr{J}^0$$

We denote by T the closed subscheme of **Isom**(S'', $pr_1^*(C')$, $pr_2^*(\overline{C}')$) corresponding to those isomorphisms which induce (via the preceeding identifications) the identity on the inverse image of $\mathscr{J}^0$.

By (1.11), T is finite and unramified over S'', and by (1.13) T is radicial over S''. We conclude that the morphism from T to S'' identifies T with a closed subscheme X of S''. As X contains S''_η, which is schematically dense in S'', we have that X is S'' and T "is" the desired section, φ, of **Isom**(S'', $pr_1^*(\overline{C}')$, $pr_2^*(\overline{C}')$) over S'' which extends φ_η. Q.E.D.

Combining Proposition (2.3), Theorem (2.4), and the stable reduction theorem for abelian varieties quoted in the introduction, we obtain the most important consequence:

Corollary **(2.7)**. — *Let* R *be a discrete valuation ring with quotient field* K. *Let* C *be a smooth geometrically irreducible curve over* K *of genus* $g \geq 2$. *Then there exists a finite algebraic extension* L *of* K *and a stable curve* $\mathscr{C}_L$ *over* R_L, *the integral closure of* R *in* L, *with generic fibre* $\mathscr{C}_{L,\eta} \cong C \times_K L$.

§ 3. Elementary derivation of the theorem.

Let k be an algebraically closed field of char. $p \neq 0$. We use the notation of § 1, except that we will now denote by H_g the product $H_g \times \mathrm{Spec}(k)$ of the previous H_g with Spec(k): it is a disjoint union of non-singular varieties $H_{g,1}, \ldots, H_{g,n}$ over k and is the subscheme of $\mathbf{Hilb}^{P_g}_{\mathbf{P}^{5g-6}/k}$ of tri-canonical stable curves. Similarly, H^0_g is the open dense subset of H_g of tri-canonical non-singular curves. By the results of [M1], we know that a coarse geometric quotient

$$M^0_g = H^0_g/\mathbf{PGL}(5g-6)$$

exists, that it is a disjoint union of normal varieties over k and is the coarse moduli space for non-singular curves of genus g. Let $S^* = H_g - H^0_g$. Then everything decomposes into the same set of components.

Let

$$H_{g,i} = \text{components of } H_g$$

then $$H^0_{g,i} = H_{g,i} \cap H^0_g = \text{components of } H^0_g$$

and $$M^0_{g,i} = H^0_{g,i}/\mathbf{PGL}(5g-6) = \text{components of } M^0_g$$

and $$S^* = \text{disjoint union of } S^*_1, \ldots, S^*_n, \quad S^*_i = H_{g,i} \cap S_i.$$

We want to prove that M^0_g, or equivalently H^0_g, or equivalently H_g is irreducible. We shall use: (i) the fact that these statements are true in char. 0; (ii) the inductive assumption that these statements are true for smaller genus.

Step I. — No component of M^0_g is complete (i.e., proper over k).

Proof. — Here we use the char. 0 result. By [M_1], there is a scheme X, quasi-projective over Spec(W(k)), W(k) the Witt vectors, whose closed fibre is M^0_g, and whose generic fibre X_η is the char. 0 coarse moduli space over the quotient field of W(k). In particular, X_η is known to be connected. Since X is quasi-projective over W(k), we can embed X as an open dense subset of a scheme $\overline{X}$ projective over W(k). $\overline{X}_\eta$ is still connected, hence by the connectedness theorem of Enriques-Zariski [EGA 3], the closed fibre $\overline{X}_0$ of $\overline{X}$ is connected. But if Y were a complete variety which is a component of M^0_g, then: *a)* Y is an open subset of M^0_g, which is an open subset of $\overline{X}_0$, and: *b)* since Y is proper/k, Y would be a closed subset of $\overline{X}_0$ too. Therefore, $Y = \overline{X}_0$, hence M^0_g is itself irreducible and complete. On the other hand, if A_g is the coarse moduli space of principally polarized g-dimensional abelian varieties, then the map associating to each curve its jacobian defines a morphism:

$$\theta : M^0_g \to A_g.$$

If M^0_g were complete, the image of θ would be closed. But it is well known that the closure of the image of θ contains all products of lower dimensional jacobians too, so it is not closed. Q.E.D.

Step II. — No component of H_g consists entirely of non-singular curves, i.e., $S^*_i \neq \emptyset$ for all i.

Proof. — Here we combine Step I with the result of § 2. Take any i. Let $T = \text{Spec } k[[t]]$. Since $M^0_{g,i}$ is not complete, there is a morphism φ of the generic point T_η of T into $M^0_{g,i}$ which does *not* extend to a morphism of T into $M^0_{g,i}$. Now we replace T by its normalization T′ in a finite algebraic extension and let $\varphi' : T'_\eta \to M^0_{g,i}$ be the induced morphism, which still does not extend to a morphism from T′ to $M^0_{g,i}$. By the results of § 2, if T′ is chosen suitably there exists a stable curve $\pi : C' \to T'$ over T′ whose generic fibre C'_η is a non-singular curve corresponding to the morphism $\varphi' : T'_\eta \to M^0_g$ via the functorial properties of the moduli space. Since T′ is the spectrum of a local ring, we can choose an isomorphism

$$\mathbf{P}(\pi_*(\omega^{\otimes 3}_{C'/T'})) \cong \mathbf{P}^{5g-6} \times T'$$

and get a tri-canonical embedding $C' \subset \mathbf{P}^{5g-6} \times T'$. C′, with this embedding, is then induced from the universal tri-canonically embedded stable curve by a morphism $\psi : T' \to H_g$. Since the generic fibre of C' is C'_η, we get a commutative diagram:

$$\begin{array}{ccc} T' & \xrightarrow{\psi} & H_g \\ \cup & & \cup \\ T'_\eta & \xrightarrow{\mathrm{res}(\psi)} & H^0_g \\ & \searrow^{\varphi'} & \downarrow \\ M^0_{g,i} & \subset & M^0_g \end{array}$$

If $\mathrm{Im}(\psi) \subset H^0_g$, then φ' would extend to a morphism from T' to M^0_g, hence from T' to $M^0_{g,i}$, and this is a contradiction. Moreover, $\psi(T_\eta)$ must be a point of $H^0_{g,i}$ since its image in M^0_g is in $M^0_{g,i}$. Therefore the image x of the closed point of T' by ψ is in the closure of $H^0_{g,i}$, i.e., in $H_{g,i}$, but not in $H^0_{g,i}$ itself. Q.E.D.

Step III. — S^* is connected.

Proof. — This will follow using only the induction assumption of irreducibility for lower genera. Let $Z \subset H_g \times \mathbf{P}^{5g-6}$ be the universal tri-canonically embedded stable curve. Let $S \subset Z$ be the set of points where Z is not smooth over H_g. As we proved in § 1, S is non-singular and is the normalization of S^*. In particular, this shows that if $x \in S^*$, then the corresponding curve Z_x has exactly one double point if and only if x is a non-singular point of S^*. Stable curves C of genus g with exactly one double point belong to one of the following types:

type o:  C irreducible
normalization C' of C has genus $g-1$.

type k:  C has two non-singular components C_1, C_2

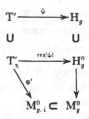

$1 \leq k \leq \left[\frac{g}{2}\right]$ genus$(C_1) = k$
genus$(C_2) = g-k$

If $0 \leq k \leq \left[\frac{g}{2}\right]$ let $S^*(k) = \{x \in S^* \mid Z_x \text{ has one double point and is of type } k\}$. Then the open dense subset of S^* of non-singular points is the disjoint union of open subsets $S^*(0), \ldots, S^*\left(\left[\frac{g}{2}\right]\right)$. We first check:

($*$) Each set $S^*(k)$ is irreducible.

Proof of ($*$). — Take the case $k=0$: the cases $1 \leq k \leq \left[\frac{g}{2}\right]$ are similar. Let

$$T = \{(x_1, x_2) \mid x_1 \neq x_2 \text{ and } \pi(x_1) = \pi(x_2) \in H^0_{g-1}\} \subset Z_{g-1} \times_{H_{g-1}} Z_{g-1}.$$

T is smooth with irreducible fibres over H^0_{g-1}, hence T is irreducible. Consider the correspondence relating $S^*(0)$ and T:

$$W=\left\{\begin{array}{l}\text{set of pairs } x\in S^*(o),\ \{x_1, x_2\}\in T \text{ such that if } y=\pi(x_1)=\pi(x_2),\\ \text{then there exists a birational morphism}\\ \qquad f: (Z_{g-1})_y \to (Z_g)_x\\ \text{such that } f(x_1)=f(x_2).\end{array}\right\}$$

It is easy to check that W is Zariski-closed. Moreover, for any $\{x_1, x_2\}\in T$, $W\cap(S^*(o)\times\{x_1, x_2\})$ is an orbit in $S^*(o)$ under $\mathbf{PGL}(5g-6)$: so these are non-empty irreducible subsets all of the same dimension. It follows that W itself is irreducible. But the projection from W to $S^*(o)$ is surjective, so $S^*(o)$ is irreducible. Q.E.D. *for* (*).

Now for any k, $1\leq k\leq\left[\frac{g}{2}\right]$, choose any stable curve $C(k)$ of the type:

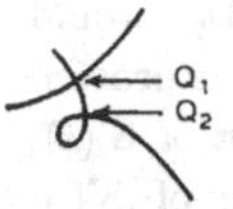

one non-singular component $C(k)'$ of genus k.
one component $C(k)''$ with one double point, normalization of genus $g-k-1$.

Let $P(k)$ be a point of H_g such that $Z_{P(k)}\cong C(k)$. Step III will be completed if we prove:

(**) $P(k)$ is in the closure of $S^*(o)$ and of $S^*(k)$, hence $S^*(o)$, $S^*(k)$ both lie in the same topological component of S^*.

Proof of (**). — Let $T=\operatorname{Spec} k[[t]]$. Using the fact that S^* has two branches through $P(k)$, one for each of the double points of $C(k)$, we see that there exist two morphisms

$$f_1, f_2 : T\to S^*$$
$$f_1(T_s)=f_2(T_s)=P(k), \qquad T_s=\text{closed point of } T$$

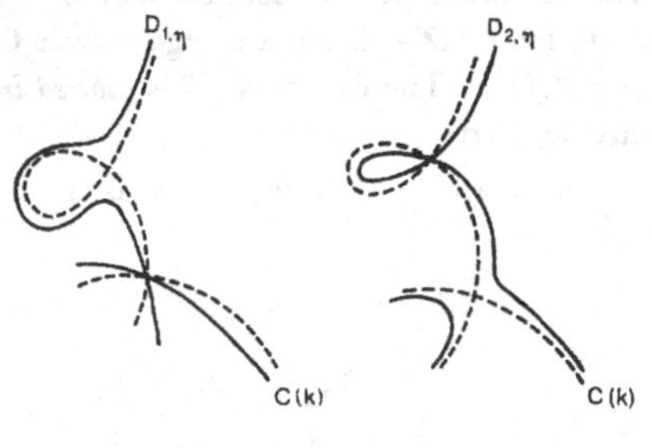

such that if $\pi_1 : D_1\to T$, $\pi_2 : D_2\to T$ are the two stable curves over T induced by f_1 and f_2, then: *a)* the closed fibres $D_{1,s}$, $D_{2,s}$ are $C(k)$; *b)* there are sections $s_1 : T\to D_1$, $s_2 : T\to D_2$ whose images are non-smooth points of π_1, π_2 and such that $s_1(T_s)$ and $s_2(T_s)$ are the two double points Q_1 and Q_2 of $C(k)$ respectively, and: *c)* the generic fibres $D_{1,\eta}$, $D_{2,\eta}$ have only one double point (cf. figure). I claim:

A) $D_{1,\eta}$ is of type (k);

B) $D_{2,\eta}$ is of type (o).

To prove A), let D_1' be the result of blowing up the subscheme $s_1(T)$ of D_1. Then D_1' is still flat and proper over T and its special fibre is the special fibre of D_1 with Q_1 blown up (use the fact that formally at Q_1, D_1 is isomorphic to $k[[t, x, y]]/(x.y)$ with the section given by $x=y=0$). Therefore the special fibre of D_1' is the disjoint union of $C(k)'$, $C(k)''$, so the general fibre of D_1' is the disjoint union of two irreducible curves which specialize to $C(k)'$, $C(k)''$ respectively. Since $D_{1,\eta}$ has only one double point, $D_{1,\eta}'$ is non-singular, so $D_{1,\eta}'$ is the disjoint union of two non-singular irreducible curves which must then have the same genera as $C(k)'$ and $C(k)''$, i.e., k, $g-k$. Thus $D_{1,\eta}$ has type (k).

To prove B), it suffices to check that $D_{2,\eta}$ is geometrically irreducible. If not, $D_{2,\eta}$ would have two components meeting at the single point $s_2(T_\eta)$. Since D_2 is smooth over T at each generic point of its special fibre, distinct geometric components of $D_{2,\eta}$ have to have specializations which are distinct components of $(D_2)_s$. Then $(D_2)_s$ would have two components meeting at the point $Q_2=s_2(T_s)$. This is false, so B) is proven.

Now because of A), $f_1(T_\eta)\in S^*(k)$, hence $P(k)=f_1(T_s)$ is in the closure of $S^*(k)$. And because of B), $f_2(T_\eta)\in S^*(0)$, hence $P(k)=f_2(T_s)$ is in the closure of $S^*(0)$ too. Q.E.D. *for* (**).

This completes the proof of Step III since we now see that all irreducible components of S^* are part of the same topological component. Finally, from Steps II and III, we see that

a) S^*, being connected, is part of a single component of H_g, while

b) each component of H_g contains part of S^*. Thus H_g is irreducible, as was to be proven.

§ 4. Some results on algebraic stacks.

The proofs of the results stated in this section will be given elsewhere.

Let C be a category and let $p:\mathcal{S}\to C$ be a category over C. For each $U\in \mathrm{Ob}\, C$, we denote by $\mathcal{S}_U$ the fibre $p^{-1}(U)$. The category $\mathcal{S}$ is *fibered in groupoids* over C if the following two conditions are verified:

a) For all $\varphi: U\to V$ in C and $y\in \mathrm{Ob}\,\mathcal{S}_V$ there is a map $f: x\to y$ in $\mathcal{S}$ with $p(f)=\varphi$.

b) Given a diagram

$$\begin{array}{ccc} x & & \\ & \searrow^{f} & \\ & & z \\ & \nearrow_{g} & \\ y & & \end{array}$$

in $\mathcal{S}$, let

$$\begin{array}{ccc} U & & \\ & \searrow^{\varphi} & \\ & & W \\ & \nearrow_{\psi} & \\ V & & \end{array}$$

be its image in C. Then for all $\chi : U \to V$ such that $\varphi = \psi\chi$, there is a unique $h : x \to y$ such that $f = g.h$ and $p(h) = \chi$.

Condition *b)* implies that the $f : x \to y$ whose existence is asserted in *a)* is unique up to canonical isomorphism.

Assume that for each $\varphi : U \to V$ in C and each $Y \in \text{Ob}\, \mathscr{S}_V$, such an $f : x \to y$ has been chosen. This x will be written as $\varphi^* y$. Then, φ^* " is " a functor from $\mathscr{S}_V$ to $\mathscr{S}_U$ and if $\varphi\psi$ is a composite morphism in C, the functors $(\varphi\psi)^*$ and $\psi^*\varphi^*$ are canonically isomorphic.

We propose the terminology " stack " for the French word " champ " of non-abelian cohomology (Giraud [G]).

Definition (4.1). — *Let* C *be a category with a Grothendieck topology. We assume products and fibre products exist in* C. *A* stack in groupoids *over* C *is a category over* $C, p : \mathscr{S} \to C$ *such that*:

(i) $\mathscr{S}$ *is fibered in groupoids over* C.

(ii) *For any* $U \in \text{Ob}\, C$ *and any objects* x, y *in* $\mathscr{S}_U$ *the functor from* C/U *to (sets) which to any* $\varphi : V \to U$ *associates* $\text{Hom}_{\mathscr{S}_V}(\varphi^* x, \varphi^* y)$ *is a sheaf.*

(iii) *If* $\varphi_i : V_i \to U$ *is a covering family in* C, *any descent datum relative to the* φ_i, *for objects in* $\mathscr{S}$, *is effective.*

For each $x \in \text{Ob}\, \mathscr{S}_U$, there are given isomorphisms between the inverse images of $x_i = \varphi_i^* x$ and $x_j = \varphi_j^* x$ over $V_{ij} = V_i \times_U V_j$, and the pull-backs of these isomorphisms on $V_{ijk} = V_i \times_U V_j \times_U V_k$ satisfy a " cocycle " condition. In (iii) it is required that reciprocally, any such " descent datum " be defined by some $x \in \mathscr{S}_U$.

In what follows, for the sake of brevity, we will use " stack " to mean " stack in groupoids ".

If $U \in \text{Ob}\, C$ and if $\mathscr{S}$ is a stack over C, the fibre $\mathscr{S}_U$ will be called the category of sections of $\mathscr{S}$ over U.

Let C be as in (4.1). The stacks over C are the objects of a 2-category [B] (stacks/C): 1-morphisms are functors from one stack to another, compatible with the projection into C; and 2-morphisms are morphisms of functors. In this 2-category every 2-morphism is an isomorphism. Products and 2-fibre products exist in this 2-category.

To each $X \in \text{Ob}\, C$ is associated the " representable " stack over C whose category of sections over U is the discrete category whose objects are the morphisms from U to X. This stack will be denoted simply X. For any stack $\mathscr{S}$, the category

$$\text{Hom}(X, \mathscr{S})$$

is canonically equivalent to the category of sections of $\mathscr{S}$ over X. Because of this, $\mathscr{S}$ is sometimes said to " classify " its sections over variable $X \in \text{Ob}\, C$. In the category $\text{Hom}(\mathscr{S}, X)$, all morphisms are identities, i.e., $\text{Hom}(\mathscr{S}, X)$ is just a set.

Let us denote also by C the 2-category having the same objects and morphisms as C, and in which the identities are the only 2-morphisms. The above construction then identifies C with a full sub-2-category of (stacks/C).

For each $S \in \text{Ob}\, C$, the category C/S satisfies the assumptions of (4.1). Any stack $\mathscr{S}_0$ over C/S (an " S-*stack* ") gives rise to a stack $\mathscr{S}$ over C; a section (φ, η) of $\mathscr{S}$ over $U \in \text{Ob}\, C$ consists of

(i) a morphism $\varphi : U \to S$;

(ii) a section η of $\mathscr{S}_0$ over (U, φ).

Definition (4.2). — *A* 1-*morphism of stacks over* C, $F : \mathscr{S}_1 \to \mathscr{S}_2$, *will be called* representable *if for any* X *in* C *and any* 1-*morphism* $x : X \to \mathscr{S}_2$, *the fibre product* $X \times_{\mathscr{S}_2} \mathscr{S}_1$ *is a representable stack.*

In down to earth terms, this means the following:

(i) for any $f : Y \to X$ in C, the category whose objects are pairs

$$\{\text{a section, } y, \text{ of } \mathscr{S}_1 \text{ over } Y; \text{ an isomorphism } F(y) \xrightarrow{\sim} f^*(x)\}$$

is equivalent to a category $S(f)$ in which all morphisms are identities;

(ii) the functor $f \mapsto \text{Ob}\, S(f)$ is representable by some $g : Z \to X$. Such a Z represents the fibre product $X \times_{\mathscr{S}_2} \mathscr{S}_1$.

Let **P** be a property of morphisms in C, stable by change of base and of a local nature on the target.

Definition (4.3). — *A representable morphism* $F : \mathscr{S}_1 \to \mathscr{S}_2$ *of stacks over* C *has property* **P** *if for any* 1-*morphism* $x : X \to \mathscr{S}_2$ *the morphism in* C *deduced by base change*: $F' : X \times_{\mathscr{S}_2} S_1 \to X$ *has that property.*

Proposition (4.4). — *Let* $\mathscr{S}$ *be a stack. The diagonal map*

$$\mathscr{S} \to \mathscr{S} \times \mathscr{S}$$

is representable if and only if for all $X, Y \in \text{Ob}\, C$ *and* 1-*morphisms* $x : X \to \mathscr{S}$, $y : Y \to \mathscr{S}$, *the fibre product* $X \times_{\mathscr{S}} Y$ *is representable.*

If $X \in \text{Ob}\, C$ and x, y are sections of $\mathscr{S}$ over X, we denote by $\underline{\text{Isom}}(X, x, y)$ the sheaf on C/X which to every Z over X associates the set of isomorphisms between the inverse images of x and y over Z. Then the object representing $\mathscr{S} \times_{(\mathscr{S} \times \mathscr{S})} X$ (the product taken with the map $(x, y) : X \to \mathscr{S} \times \mathscr{S}$) is just $\text{Isom}(X, x, y)$. If $x : X \to \mathscr{S}$, $y : Y \to \mathscr{S}$ are 1-morphisms, then the object representing $X \times_{\mathscr{S}} Y$ is just $\text{Isom}(X \times Y, p_1^* x, p_2^* y)$.

Henceforth, C *will be the category of schemes with the etale topology* (SGAD, IV, (6.3)).

Definition (4.5). — *A stack* $\mathscr{S}$ *is* quasi-separated *if the diagonal morphism from to* $\mathscr{S}$ *to* $\mathscr{S} \times \mathscr{S}$ *is representable, quasi-compact and separated.*

Definition (4.6). — *A stack* $\mathscr{S}$ *is an* algebraic stack [1] *if*

(i) $\mathscr{S} \to \mathscr{S} \times \mathscr{S}$ *is representable;*

(ii) *there exists a* 1-*morphism* $x : X \to \mathscr{S}$ *such that for all* $y : Y \to \mathscr{S}$, *the projection morphism* $X \times_{\mathscr{S}} Y \to Y$ *is surjective and etale* (i.e., x *is etale and surjective*).

---

[1] This definition is the " right " one only for quasi-separated stacks. It will however be sufficient for our purposes.

The 2-category of algebraic stacks contains the representable stacks and is stable under products and fibre products. If $\mathscr{S}$ is a quasi-separated algebraic stack, the diagonal map is unramified and quasi-affine.

Definition (**4.7**). — *An algebraic stack* $\mathscr{S}$ *is* separated *if* $\mathscr{S}\to\mathscr{S}\times\mathscr{S}$ *is proper (or, equivalently, finite). A* 1*-morphism* $f:\mathscr{S}_1\to\mathscr{S}_2$ *is* separated (resp. quasi-separated) *if for any morphism* $x:X\to\mathscr{S}_2$ *from a separated scheme* X *to* $\mathscr{S}_2$, *the fibre product* $\mathscr{S}_1\times_{\mathscr{S}_2}X$ *is separated* (resp. *quasi-separated*).

Example (**4.8**). — Let X be a scheme over S. Let G be a group scheme over S, etale, separated and of finite type over S, which operates on X. We will denote by [X/G] the S-stack whose category of sections over an S-scheme T is the category of principal homogeneous spaces (p.h.s.) E over T, with structural group G (i.e., a p.h.s. under G_T), provided with a G-morphism $\varphi:E\to X$. The principal homogeneous space $G\times X$ over X (G acting only on the first factor) plus the G-morphism $G\times X\to X$ (given by the action of G on X) is a section of [X/G] over X. The corresponding morphism $q:X\to[X/G]$ is etale and surjective, so that [X/G] is an algebraic stack. In addition, X is a principal homogeneous space over [X/G]; the stack [X/G] is representable if and only if X is a principal homogeneous space over a scheme Y, in which case

$$[X/G]\sim Y.$$

If $X=S$, then $[X/G]=[S/G]$ might be called the " classifying stack " of G over S.

Example (**4.9**). — Suppose a stack $\mathscr{S}$ has the property that in each category $\mathscr{S}_X$ the only morphisms are the identity morphisms. Then $\mathscr{S}_X$ is just a set $\mathscr{F}(X)=\mathrm{Ob}\,\mathscr{S}_X$, and this set, under pull-back, is a contravariant functor $\mathscr{F}$ in X. Conditions (ii) and (iii) of (4.1) assert that the functor $\mathscr{F}$ is a sheaf on C. Artin and Knutson [K] have defined an *algebraic space* to be a sheaf $\mathscr{F}$ such that:

(i) for any morphisms $X\to\mathscr{F}$, $Y\to\mathscr{F}$ of representable functors to $\mathscr{F}$, the fibre product $X\times_{\mathscr{F}}Y$ is representable;

(ii) there exists a morphism $X\to\mathscr{F}$, represented by surjective, etale morphisms of schemes.

This is exactly what we have called an algebraic stack in this case.

Definition (**4.10**). — *Let* $\mathscr{S}$ *be an algebraic stack. The* etale site $\mathscr{S}_{et}$ *of* $\mathscr{S}$ *is the category with objects the etale morphisms*

$$x:X\to\mathscr{S}$$

and where a morphism from (X,x) *to* (Y,y) *is a morphism of schemes* $f:X\to Y$ *plus a* 2*-morphism between the* 1*-morphism* $x:X\to\mathscr{S}$ *and* $y\cdot f:X\to\mathscr{S}$. *A collection of morphisms* $f_i:(X_i,x_i)\to(X,x)$ *is a covering family if the underlying family of morphisms of schemes is surjective.*

The site $\mathscr{S}_{et}$ is in a natural way ringed. When we speak of sheaves on $\mathscr{S}$ we mean sheaves on $\mathscr{S}_{et}$.

We now explain how many concepts from the theory of schemes may be applied to algebraic stacks.

Let $\boldsymbol{P}$ be a property of morphisms of schemes, stable by etale change of base, and of a local nature (for the etale topology) on the target.

For instance: being an open immersion with dense image, being dominant, birational...

A representable morphism of algebraic stacks $f: T_1 \to T_2$, is said to have property $\boldsymbol{P}$ if for one (and hence for every) surjective etale morphism $x: X \to T_2$, the morphism of schemes deduced by base change $f': X \times_{T_2} T \to X$ has that property.

Let $\boldsymbol{P}$ be a property of morphisms of schemes, which, at source and target, is of a local nature for the etale topology. This means that, for any family of commutative squares

$$\begin{array}{ccc} X_i & \xrightarrow{g_i} & X \\ {\scriptstyle f_i}\downarrow & & \downarrow{\scriptstyle f} \\ Y_i & \xrightarrow{h_i} & Y \end{array}$$

where the g_i (resp. h_i) are etale and cover X (resp. Y):

$$\boldsymbol{P}(f) \Leftrightarrow \forall_i \boldsymbol{P}(f_i).$$

For instance: f flat, smooth, etale, unramified, normal, locally of finite type, locally of finite presentation.

If $f: T_1 \to T_2$ is a morphism of algebraic stacks, we say f has property $\boldsymbol{P}$ if for one, then necessarily for every, commutative diagram

$$\begin{array}{ccc} X & \xrightarrow{x} & T_1 \\ {\scriptstyle f'}\downarrow & & \downarrow{\scriptstyle f} \\ Y & \xrightarrow{y} & T_2 \end{array}$$

where X and Y are schemes and x, y are etale and surjective, f' has property $\boldsymbol{P}$.

Similarly, if $\boldsymbol{P}$ is a property of schemes, of a local nature for the etale topology, an algebraic stack T will be said to have property $\boldsymbol{P}$ if for one (and hence for every) surjective etale morphism $x: X \to T$, X has property $\boldsymbol{P}$. This applies to, for instance, the properties of being regular, normal, locally noetherian, of characteristic p, reduced, Cohen-Macaulay...

An algebraic stack T will be called *quasi-compact* if there exists a surjective etale morphism $x: X \to T$ with X quasi-compact. A morphism $f: T_1 \to T_2$ of algebraic stacks will be called *quasi-compact* if for any quasi-compact scheme X over T_2, the fiber product $T_1 \times_{T_2} X$ is quasi-compact. It is enough to test the condition for a surjective family $f_i: X_i \to T_2$. We define a morphism $f: T_1 \to T_2$ to be of *finite type*, if it is quasi-compact, and locally of finite type; of *finite presentation*, if it is quasi-compact, quasi-separated, and locally of finite presentation. An algebraic stack is *noetherian*, if it is quasi-compact, quasi-separated, and locally noetherian.

The key point in what follows will be the definition of a " proper morphism " and the analogue of Chow's lemma.

A morphism $f: T_1 \to T_2$ is said to have a property $\boldsymbol{P}$, locally on T_2, if there exists a surjective etale morphism $x: X \to T_2$ such that the morphism f' deduced from f by the change of base by x has property $\boldsymbol{P}$.

Definition (**4.11**). — *A morphism* $f: T_1 \to T_2$ *is* proper *if it is separated, of finite type and if, locally over* T_2, *there exists commutative diagrams*

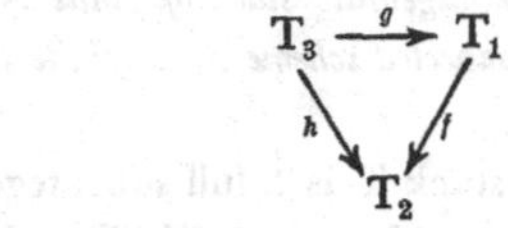

with g surjective and h representable and proper.

The following form of Chow's Lemma will be sufficient for our purposes.

Theorem (**4.12**). — *Let* S *be a noetherian scheme and f be a morphism from an etale site* T *to* S. *We assume f to be separated and of finite type. Then, there exists a commutative diagram*

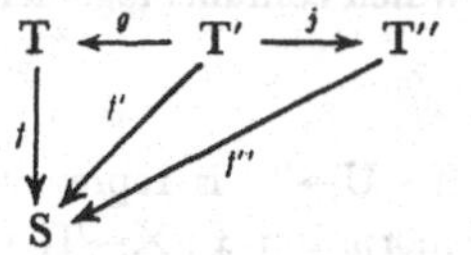

in which T′ *and* T″ *are schemes and such that*

(i) *g is proper, surjective and generically finite;*

(ii) *j is an open immersion;*

(iii) *f″ is projective.*

Using (4.12), it is easy to extend the cohomological theory of coherent sheaves to the present situation. In fact if $f: T_1 \to T_2$ is a proper morphism of noetherian algebraic stacks and if $\mathcal{F}$ is a coherent sheaf on T_1, then the $R^i f_*(\mathcal{F})$ are coherent sheaves on T_2.

However, the $R^i f_*(\mathcal{F})$ don't need to be zero for i large enough. Let S be a scheme and G a finite group. We denote by p the projection $p: [S/G] \to S$. Quasi-coherent (resp. coherent) sheaves of modules on [S/G] may (and will) be identified with quasi-coherent (resp. coherent) sheaves of modules on S, on which G acts. One has

$$R^i p_*(\mathcal{F}) \cong H^i(G, \mathcal{F}).$$

In general, the (quasi-coherent sheaf) cohomology of algebraic stacks appears as a mixture of finite group cohomology and of scheme cohomology.

The *disjoint sum* T of a family $(T_i)_{i \in I}$ of stacks is the stack a section of which over a scheme X consists of

(i) a decomposition $X = \coprod_i X_i$ of X;

(ii) a section of T_i over X_i for each i.

The *void stack* $\emptyset$ is the one represented by the void scheme.

A stack is *connected* if it is non-void and is not the disjoint sum of two non-void stacks.

Proposition (4.13). — *A locally noetherian algebraic stack is in one and only one way the disjoint sum of a family of connected algebraic stacks (called its* connected components).

We denote by $\pi_0(T)$ the set of connected components of locally noetherian algebraic stack T. If $x : X \to T$ is surjective and etale, $\pi_0(T)$ is the cokernel of the two maps

$$\pi_0(X \times_T X) \rightrightarrows \pi_0(X) \to \pi_0(T).$$

Proposition (4.14). — *Let* T *be an algebraic stack of finite type over a field* k. *Then,* T *is connected if and only if there exists a connected scheme* X, *of finite type over* k, *and a surjective morphism from* X *to* T.

An *open subset* U of an algebraic stack T is a full subcategory $U \subset T$ which is an algebraic stack, which contains together with any $t \in \mathrm{Ob}(T)$ all isomorphic t' and such that the inclusion $j : U \to T$ is representable by open immersions. The open subsets of T corresponds bijectively to the open subsets of its etale site.

For each open subset U of T, there exists one and only one full subcategory $T-U$ of T, which is an algebraic stack, which contains together with any $t \in \mathrm{Ob}(T)$ all isomorphic t' and such that

(i) $T-U$ is reduced;

(ii) the inclusion map $i : T-U \to T$ is representable by closed immersions;

(iii) for any etale surjective morphism $x : X \to T$, the inverse image of $T-U$ on X is the complement of the inverse image of U.

An algebraic stack F in T satisfying (i) and (ii) is a *closed subset* of T, and the functor $U \mapsto T-U$ is an isomorphism of the set of open and the set of closed subsets of T. If F satisfies only (ii), F_{red} satisfies (i) and (ii) so that F defines a closed subset of T.

An algebraic stack T is *irreducible* if it is not the union of two closed subsets, non void and distinct from T.

Proposition (4.15). — *A noetherian algebraic stack* T *is in one and only one way the union of irreducible closed subsets, none of which contains any other. They are called the* irreducible components *of* T. *If* U *is an open dense subset of* T, *the irreducible components of* U *are the non-void intersections of* U *with the irreducible components of* T.

Each irreducible component of T is contained in a connected component of T. Conversely:

Proposition (4.16). — *The connected components of a normal noetherian algebraic stack are irreducible.*

Theorem (4.17). — *Let* f *be a morphism of finite type from an algebraic stack* T *to a noetherian scheme* S. *For* $s \in S$, *let* $n(s)$ *be the number of connected components of the geometric fibre of* T *at* s. *Then*

(i) $n(s)$ *is a constructible function of* s;

(ii) *if* f *is proper and* flat, *then* $n(s)$ *is lower-semi-continuous;*

(iii) *if* f *is proper flat, and has geometrically normal fibres, then* $n(s)$ *is constant.*

Let $f: T \to S$ be a morphism of finite type from an algebraic stack T to a noetherian scheme S. Assume that the diagonal map $T \to T \times_S T$ is separated and quasi-compact.

Theorem (**4.18**) (Valuative criterion for separation.) — *The morphism* f *is separated if and only if, for any complete discrete valuation ring with algebraically closed residue field and any commutative diagram*

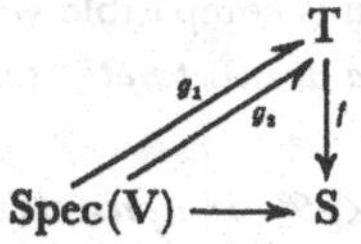

any isomorphism between the restrictions of g_1 *and* g_2 *to the generic point of* Spec(V) *can be extended to an isomorphism between* g_1 *and* g_2.

This criterion is nothing other but the valuative criterion of properness (EGA, II, 7.3.8) applied to the (representable) diagonal morphism.

Theorem (**4.19**) (Valuative criterion for properness.) — *If* f *is separated, then* f *is proper if and only if, for any discrete valuation ring* V *with field of fractions* K *and any commutative diagram*

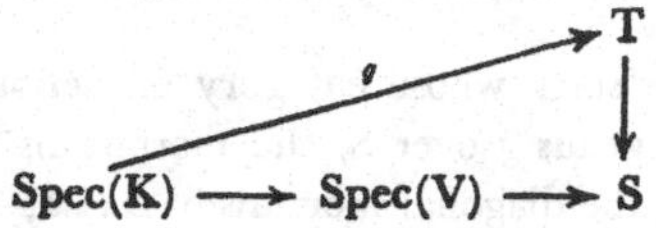

there exists a finite extension K′ *of* K *such that* g *extends to* Spec(V′), *where* V′ *is the integral closure of* V *in* K′

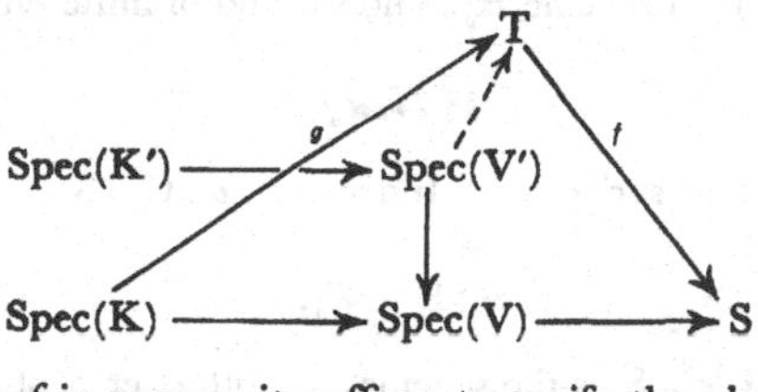

To prove a given f is proper, it suffices to verify the above criterion under the additional hypothesis that V is complete and has an algebraically closed residue field. Further, given a dense open subset U of T, it is enough to test only g's which factor through U.

Proposition (**4.20**). — *Let* $\mathscr{S}$ *be an algebraic stack. The functor which, to any algebraic stack over* $\mathscr{S}$, $f: \mathscr{E} \to \mathscr{S}$, *associates the* $\mathcal{O}_{\mathscr{S}}$ *sheaf of algebras* $f_*\mathcal{O}_{\mathscr{E}}$ *induces an equivalence of categories between:*

(i) *the category of algebraic stacks representable and affine over* $\mathscr{S}$;

(ii) *the dual of the category of quasi-coherent* $\mathcal{O}_{\mathscr{S}}$*-algebras.*

Let $\mathscr{A}$ be a quasi-coherent sheaf of $\mathcal{O}_{\mathscr{S}}$-algebras on an algebraic stack $\mathscr{S}$. For each etale morphism $x: U \rightarrow \mathscr{S}$, with U affine, let $\mathscr{A}'(U)$ be the integral closure of $\Gamma(U, \mathcal{O}_{\mathscr{S}})$ in $\mathscr{A}(U)$. By (EGA, II, 6.3.4), the $\mathscr{A}'(U)$ for variable U are the sections over U of a quasi-coherent sheaf $\mathscr{A}'$ on $\mathscr{S}$, which will be called the *integral closure* of $\mathcal{O}_{\mathscr{S}}$ in $\mathscr{A}$.

Let $f: \mathscr{E} \rightarrow \mathscr{S}$ be representable and affine. The algebraic stack which is associated by (4.20) to the integral closure of $\mathcal{O}_{\mathscr{S}}$ in $f_* \mathcal{O}_{\mathscr{E}}$ will be called the *normalization of* $\mathscr{S}$ *with respect to* $\mathscr{E}$. Its formation is compatible with any etale change of basis.

Theorem (**4.21**). — *Let* $\mathscr{S}$ *be a quasi-separated stack over a noetherian scheme* S. *Assume that*

(i) *the diagonal map* $\mathscr{S} \rightarrow \mathscr{S} \times_S \mathscr{S}$ *is representable and unramified;*

(ii) *there exists a scheme* X *of finite type over* S *and a smooth and surjective* S-*morphism from* X *to* $\mathscr{S}$.

Then, $\mathscr{S}$ *is an algebraic stack of finite type over* S.

M. Artin has developed powerful methods to relate pro-representability of a stack to the existence of etale surjective maps $x: X \rightarrow \mathscr{S}$.

§ 5. Second proof of the irreducibility theorem.

Let $\mathscr{M}_g$ $(g \geq 2)$ be the stack whose category of sections over a scheme S is the category of stable curves of genus g over S, the morphisms being the isomorphisms of schemes over S. By (1.11), the diagonal morphism $\Delta: \mathscr{M}_g \rightarrow \mathscr{M}_g \times \mathscr{M}_g$ is representable, finite and unramified.

We saw in § 1 that the stack classifying the tricanonically embedded stable curves of genus g is represented by a scheme H_g, smooth and of finite type over Spec(**Z**). The " forgetful " morphism

$$H_g \rightarrow \mathscr{M}_g$$

is representable, smooth and surjective. Indeed, if $p: C \rightarrow S$ is a stable curve over S, defining a morphism

$$\gamma: S \rightarrow \mathscr{M}_g,$$

then the fibre product $H_g \times_{\mathscr{M}_g} S$ is the scheme, smooth over S, of isomorphisms between the standard projective space of dimension $5g-6$ over S and $\mathbf{P}(p_*(\omega_{C/S}^{\otimes 3}))$. We deduce from this and (4.21) that:

Proposition (**5.1**). — $\mathscr{M}_g$ *is a separated algebraic stack of finite type over* Spec(**Z**).

Let us denote by $\mathscr{M}_g^0$ the open subset of $\mathscr{M}_g$ which " consists of " smooth curves, and by $\mathscr{Z}_g$ the " universal curve " over $\mathscr{M}_g$, the algebraic stack classifying pointed stable curves.

Theorem (**5.2**). — *The algebraic stacks* $\mathscr{M}_g$ *and* $\mathscr{Z}_g$ *are proper and smooth over* Spec(**Z**) *and the complement of* $\mathscr{M}_g^0$ *in* $\mathscr{M}_g$ *is a divisor with normal crossings relative to* Spec(**Z**).

Proof. — Let $x : \mathrm{X} \to \mathcal{M}_g$ be etale and consider the following commutative diagram of algebraic stacks of finite type over Spec($\mathbf{Z}$):

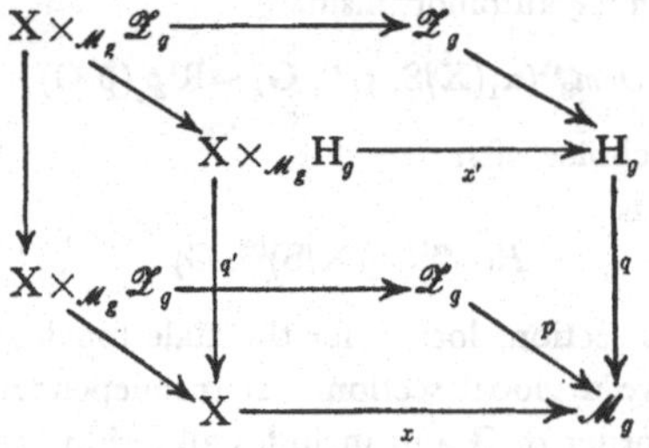

In this diagram, the four horizontal arrows are etale and the four vertical arrows are smooth and surjective. As H_g and Z_g (p. 78). are smooth over Spec($\mathbf{Z}$), so are X and $\mathrm{X}\times_{\mathcal{M}_g}\mathcal{Z}_g$. In addition, $q'^{-1}x^{-1}(\mathcal{M}_g^0)=x'^{-1}(\mathrm{H}_g^0)$ is the complement of a divisor with normal crossings relative to Spec($\mathbf{Z}$), so that the inclusion of $x^{-1}(\mathcal{M}_g)$ in X has the same property. This being true for any x, $\mathcal{M}_g$ and $\mathcal{Z}_g$ are smooth over Spec($\mathbf{Z}$) and $\mathcal{M}_g^0$ is the complement of a divisor with normal crossings relative to Spec($\mathbf{Z}$). In particular $\mathcal{M}_g^0$ is dense in $\mathcal{M}_g$.

We may now use the valuative criterion for properness in its modified form (4.19) to deduce the properness of $\mathcal{M}_g$ from the stable reduction theorem. The properness of $\mathcal{Z}_g$ then results from that of p.

(5.3) Let $p : \mathrm{X}\to\mathrm{S}$ be a stable smooth curve of genus $g\geq 2$ over S. If $k\in\mathbf{N}$ is invertible in $\mathcal{O}_\mathrm{S}$, the sheaf $\mathrm{R}^1p_*(\mathbf{Z}/k\mathbf{Z})$ on S_{et} is locally free over $\mathbf{Z}/k\mathbf{Z}$, of rank $2g$, and the cup-product is a non degenerate alternating form

$$\mathrm{R}^1p_*(\mathbf{Z}/k\mathbf{Z})\otimes\mathrm{R}^1p_*(\mathbf{Z}/k\mathbf{Z}) \to \mathrm{R}^2p_*(\mathbf{Z}/k\mathbf{Z})\sim\mu_k^{\otimes -1}.$$

Locally on S_{et}, μ_k is isomorphic to $\mathbf{Z}/k\mathbf{Z}$, and thus, locally, $\mathrm{R}^1p_*(\mathbf{Z}/k\mathbf{Z})$ is provided with a non degenerate symplectic form with values in $\mathbf{Z}/k\mathbf{Z}$, which is determined up to an unit in $\mathbf{Z}/k\mathbf{Z}$, or, as we shall say, $\mathrm{R}^1p_*(\mathbf{Z}/k\mathbf{Z})$ is provided with an homogeneous symplectic structure. The constant sheaf $(\mathbf{Z}/k\mathbf{Z})^{2g}$ will be provided with the homogeneous symplectic structure induced by the standard symplectic structure of $(\mathbf{Z}/k\mathbf{Z})^{2g}$.

Definition **(5.4)**. — *A* Jacobi structure of level *k on* X *is an isomorphism (respecting their homogeneous symplectic structures) between* $\mathrm{R}^1p_*(\mathbf{Z}/k\mathbf{Z})$ *and* $(\mathbf{Z}/k\mathbf{Z})^{2g}$.

(5.5) Given a section s of p, and a set of prime numbers $\mathbf{P}$ including all residue characteristics of S, the specialisation theorem for the fundamental group enables one to construct a pro-objet $\pi_1(\mathrm{X}/\mathrm{S}, s)^{(\mathbf{P})}$ of the category (l.c. $\mathrm{gr}^{(\mathbf{P})}$) of locally constant sheaves of finite groups of order prime to $\mathbf{P}$, with the properties

(i) $\mathit{Hom}_\mathrm{S}(\pi_1(\mathrm{X}/\mathrm{S}, s)^{(\mathbf{P})}, \mathrm{G})=\mathrm{R}^1p_*(\mathrm{X} \bmod s, p^*\mathrm{G})=\mathrm{R}^1p_*(\mathrm{Ker}(p^*\mathrm{G}\to s_*\mathrm{G}))$ functorially in $\mathrm{G}\in(\text{l.c. gr}^{(\mathbf{P})})$;

(ii) the formation of $\pi_1(\mathrm{X}/\mathrm{S}, s)^{(\mathbf{P})}$ is compatible with any change of base.

If G and H are two sheaves of groups on S_{et}, we define the sheaf of exterior morphisms from H to G, $Hom^{ext}(H, G)$, as the quotient of $Hom(H, G)$ by the action of H induced by its action on itself by inner automorphism.

The sheaf

$$Hom_S^{ext}(\pi_1(X/S, s)^{(P)}, G) \sim R^1 p_*(p^* G) \qquad (\text{for } G \in (\text{l.c. gr}^{(P)}))$$

is " independent " of the choice of s.

We shall denote it as

$$Hom_S^{ext}(\pi_1(X/S)^{(P)}, G).$$

As $p: X \to S$ admits sections locally for the etale topology, this sheaf makes sense without assuming p to have a global section. It is independent of the choice of **P**, so long as **P** is prime to the order of G and includes all residue characteristics of S.

Definition (**5.6**). — *Let* G *be a finite group of order n prime to* **P**. *A* Teichmüller structure of level G *on* X *is a surjective exterior homomorphism from* $\pi_1(X/S)$ *to* G.

The finite generation of $\pi_1(X/S)$ (SGA, 60/61, exp. 10) implies:

Lemma (**5.7**). — *The sheaf on* (Sch/S) *of the Teichmüller structures of level* G *on* X *is represented by an etale covering of* S.

We denote by ${}_G\mathcal{M}_g^0$ the stack classifying the stable smooth curves of genus g and characteristic prime to n, with a Teichmüller structure of level G.

For any algebraic stack $\mathcal{M}$, we denote by $\mathcal{M}[1/n]$ its open subset $\mathcal{M} \times \text{Spec}(\mathbf{Z}[1/n])$. Lemma (5.7) may now be rephrased:

Proposition (**5.8**). — *The " forgetful " morphism*

$${}_G\mathcal{M}_g^0 \to \mathcal{M}_g^0[1/n]$$

is representable, finite and etale.

The stack ${}_G\mathcal{M}_g^0$ thus in an algebraic stack. Let ${}_G\mathcal{M}_g$ be the normalisation of $\mathcal{M}_g[1/n]$ with respect to ${}_G\mathcal{M}_g^0$. The stack ${}_G\mathcal{M}_g$, being representable and finite over $\mathcal{M}_g[1/n]$, is proper over $\text{Spec}(\mathbf{Z}[1/n])$.

Theorem (**5.9**). — *The geometric fibres of the projection of* ${}_G\mathcal{M}_g$ *onto* $\text{Spec}(\mathbf{Z}[1/n])$ *are normal, and, fibre by fibre,* ${}_G\mathcal{M}_g^0$ *is dense in* ${}_G\mathcal{M}_g$.

Proof. — We will use Abhyankar-Artin's lemma, in its " absolute " form:

Lemma (**5.10**). — *Let* D *be a divisor with normal crossings on an excellent regular scheme* X, Y *an etale covering of* X—D *and* $\overline{Y}$ *the normalisation of* X *with respect to* Y. *Assume that the generic points of the irreducible components of* D *are all of characteristic* 0. *Then every geometric point of* X *has an etale neighbourhood* $x: X' \to X$ *such that, on* X':

(i) D *becomes a union of regular divisors* $(D_i)_{i \in I}$, D_i *of equation* $t_i = 0$.

(ii) *There exists an integer k prime to residue characteristics of* X' *such that* $\overline{Y}$ *becomes isomorphic to a disjoint union of quotients (by subgroups of* μ_k^I*) of the covering of* X' *obtained by extracting the* k^{th}*-roots of the* t_i*'s.*

If $x: X \to \mathcal{M}_g\left[\frac{1}{n}\right]$ is an etale morphism, then $X^0 = x^{-1}(\mathcal{M}_g)$ is the complement in X of a divisor with normal crossing relative to Spec(**Z**), $X_1^0 = X \times_{\mathcal{M}_g} ({}_G\mathcal{M}_g^0)$ is an etale covering of X^0 and $X_1 = S \times_{\mathcal{M}_g} ({}_G\mathcal{M}_g)$ is the normalisation of X with respect to this covering of X^0.

By the explicit local description (5.10), for any prime number l prime to n, $X_1 \times \mathrm{Spec}(\overline{\mathbf{F}}_l)$ is the normalisation of $X \times \mathrm{Spec}(\overline{\mathbf{F}}_l)$ with respect to $X_1 \times \mathrm{Spec}(\overline{\mathbf{F}}_l)$. As this is true for any modular family, we get (5.9).

Corollary **(5.11)**. — *The geometric fibres of the projection*

$$ {}_G\mathcal{M}_g^0 \to \mathrm{Spec}\left(\mathbf{Z}\left[\frac{1}{n}\right]\right) $$

all have the same number of connected components.

Proof. — By (4.17), all geometric fibres ${}_G\mathcal{M}_g \times \mathrm{Spec}(\overline{\mathbf{F}}_l)$ of ${}_G\mathcal{M}_g$ over $\mathrm{Spec}(\mathbf{Z}[1/n])$ have the same number of connected components. These connected components are irreducible (4.16). Furthermore ${}_G\mathcal{M}_g^0 \times \mathrm{Spec}(\overline{\mathbf{F}}_l)$ is dense in ${}_G\mathcal{M}_g \times \mathrm{Spec}(\overline{\mathbf{F}}_l)$ and thus has the same number of connected (or irreducible) components as ${}_G\mathcal{M}_g \times \mathrm{Spec}(\overline{\mathbf{F}}_l)$, and this number is independent of l.

(5.12) Let us denote by Π the fundamental group of an oriented closed differentiable surface S_0 of genus g. The group Π may be defined by generators and relations as follows:

(i) generators: x_i for $1 \leq i \leq 2g$;
(ii) relation: $(x_1, x_{g+1}) \ldots (x_i, x_{g+i}) \ldots (x_g, x_{2g}) = e$, where $(a, b) = aba^{-1}b^{-1}$.

Let (e_i) be the standard basis of $\mathbf{Z}^{2g}$. The morphism φ from Π to $\mathbf{Z}^{2g}$ with $\varphi(X_i) = e_i$ identifies $\Pi/(\Pi, \Pi) = H_1(\Pi, Z)$ with $\mathbf{Z}^{2g}$.

The surface S_0 is a $K(\Pi, 1)$ and thus

$$ H^2(\Pi, \mathbf{Z}) = H^2(S_0, \mathbf{Z}) = \mathbf{Z}, $$

and the cup product defines a symplectic structure on $\Pi/(\Pi, \Pi)$. This structure is identified by φ with the standard symplectic structure of $\mathbf{Z}^{2g}$.

We denote by $\mathrm{Aut}^0(\Pi)$ the subgroup of $\mathrm{Aut}(\Pi)$ which acts trivially on $H^2(\Pi, \mathbf{Z})$. Dehn has proved that each exterior automorphism of Π is induced by a diffeomorphism of S_0 onto itself and that the map induced by φ:

$$ \mathrm{Aut}^0(\Pi) \to \mathbf{Sp}_{2g}(\mathbf{Z}), $$

is surjective (see[Ma]).

Theorem **(5.13)**. — *The number of connected components of any geometric fibre of the projection of* ${}_G\mathcal{M}_g^0$ *onto* $\mathrm{Spec}(\mathbf{Z}[1/n])$ *is equal to the number of orbits of* $\mathrm{Aut}^0(\Pi)$ *in the set of exterior epimorphisms from* Π *to* G.

Proof. — By (5.13), it suffices to prove that ${}_G\mathcal{M}_g^0 \times \mathrm{Spec}(\mathbf{C})$ has the said number of connected components. As results from (4.14),

$$ \pi_0({}_G\mathcal{M}_g^0 \times \mathrm{Spec}(\mathbf{C})) = \pi_0({}_G M_g) $$

where ${}_G M_g$ is the coarse moduli scheme classifying stable smooth curves of genus g over $\mathbf{C}$ with a Teichmüller structure of level G.

Recall that a *Teichmüller curve* of genus g is a stable smooth curve C of genus g over **C** together with an exterior isomorphism φ of the transcendental fundamental group $\pi_1(C)$ with Π, which induces a symplectic isomorphism [1] between

$$H_1(C, \mathbf{Z}) \sim \pi_1(C)/(\pi_1(C), \pi_1(C))$$

and $\Pi/(\Pi, \Pi)$. By Teichmüller's theory [W], the analytic space T_g classifying Teichmüller curves of a given genus $g \geq 2$ is homeomorphic to a ball, and hence connected.

If ψ is a surjective homomorphism from Π to G, the map

$$(C, \varphi) \mapsto (C, \psi\varphi)$$

defines a morphism $t_\psi : T_g \to {}_G M_g$. Two such maps t_ψ and $t_{\psi'}$ have the same image if and only if $\psi = \psi'\sigma$ for $\sigma \in \mathrm{Aut}^0(\Pi)$, and

$${}_G M_g = \coprod_{\psi \bmod \mathrm{Aut}^0(\Pi)} t_\psi(T_g)$$

which implies (5.13).

(5.14) Let us denote by ${}_n\mathcal{M}_g^0$ the algebraic stack classifying stable smooth curves with a Jacobi structure of level n. This algebraic stack " is " a true scheme for $n \geq 3$ (by [S]).

If φ is a Jacobi structure of level n on a stable smooth curve $p : X \to S$, we define the " multiplicator " $\mu(\varphi)$ of φ by the commutative diagram

$$\begin{array}{ccc} \wedge^2(\mathbf{Z}/n\mathbf{Z})^{2g} & \longrightarrow & \mathbf{Z}/n\mathbf{Z} \\ \Big\downarrow{\scriptstyle \wedge^2\varphi} & & \Big\downarrow{\scriptstyle \mu(\varphi)} \\ \wedge^2 R^1 p_*(\mathbf{Z}/n\mathbf{Z}) & \xrightarrow{\ \wedge\ } & \mu_n^{\otimes -1} \sim R^2 p_*(\mathbf{Z}/n\mathbf{Z}) \end{array}$$

The scheme of isomorphisms between $\mathbf{Z}/n\mathbf{Z}$ and $\mu_n^{\otimes -1}$ is $\mathrm{Spec}(\mathbf{Z}[e^{2\pi i/n}, 1/n])$ thus $\varphi \to \mu(\varphi)$ defines a morphism μ from S to $\mathrm{Spec}(\mathbf{Z}[e^{2\pi i/n}, 1/n])$. This being true for any X and S, μ is induced by

$$\mu : {}_n\mathcal{M}_g^0 \to \mathrm{Spec}(\mathbf{Z}[e^{2\pi i/n}, 1/n]).$$

Theorem (5.15). — *The geometric fibres of the morphism* μ *are connected.*

Proof. — By definition, ${}_n\mathcal{M}_g^0$ is open and closed in ${}_G\mathcal{M}_g^0$ for $G = (\mathbf{Z}/n\mathbf{Z})^{2g}$. The group $\mathbf{GL}_{2g}(\mathbf{Z}/n\mathbf{Z})$ acts on G, and thus on ${}_G\mathcal{M}_g^0$. One has:

(i) the open subset ${}_n\mathcal{M}_g^0$ of ${}_G\mathcal{M}_g^0$ is stable under the subgroup $H = \mathbf{CSp}_{2g}(\mathbf{Z}/n\mathbf{Z})$ of symplectic similitudes;

(ii) ${}_G\mathcal{M}_g^0 = \coprod_{G/H} \sigma({}_n\mathcal{M}_g^0)$.

---

[1] An arbitrary isomorphism φ induces an isomorphism betwen $H_1(C, \mathbf{Z})$ and $\Pi/(\Pi, \Pi)$ which is always symplectic up to sign.

It now results from (5.11) that all geometric fibres of μ have the same number of connected components.

Consider the geometric fibre of μ at the standard complex place of $\mathrm{Spec}(\mathbf{Z}[e^{2\pi i/n}, 1/n])$. This fibre is the algebraic stack classifying complex stable smooth curves C provided with a *symplectic* isomorphism

$$H^1(C, \mathbf{Z}/n\mathbf{Z}) \xrightarrow{\sim} (\mathbf{Z}/n\mathbf{Z})^{2g}.$$

Reasoning as in (5.13), we are reduced to proving

Lemma (**5.16**). — *The homomorphism*

$$\mathrm{Aut}^0(\Pi) \to \mathbf{Sp}_{2g}(\mathbf{Z}/n\mathbf{Z})$$

is surjective.

Proof. — This results from Dehn's theory (5.12) and from the fact that the groups $\mathbf{Sp}_n$, being split semi-simple simply connected groups, are generated by their unipotent elements, so that the reduction map

$$\mathbf{Sp}_{2g}(\mathbf{Z}) \to \mathbf{Sp}_{2g}(\mathbf{Z}/n\mathbf{Z})$$

is surjective.

BIBLIOGRAPHY

[Ab] S. Abhyankar, Resolution of singularities of arithmetical surfaces, *Proc. Conf. on Arith. Alg. Geom. at Purdue.* 1963.

[A1] M. Artin, The implicit function theorem in algebraic geometry, *Proc. Colloq. Alg. Geom.*, Bombay, 1968.

[A2] M. Artin, Some numerical criteria for contractibility, *Am. J. Math.*, **84** (1962), p. 485.

[B] J. Benabou, Thèse, Paris, 1966.

[E] F. Enriques and Chisini, *Teoria geometrica delle equazioni e delle funzioni algebriche*, Bologna, 1918.

[F] W. Fulton, Hurwitz schemes and irreducibility of moduli of algebraic curves, *Ann. of Math.* (forthcoming).

[G] J. Giraud, *Cohomologie non abélienne*, University of Columbia.

[Gr] A. Grothendieck, Techniques de descente et théorèmes d'existence en géométrie algébrique, IV, *Sem. Bourbaki*, **221**, 1960-1961.

[H] R. Hartshorne, Residues and Duality, *Springer Lecture Notes*, **20**, 1966.

[K] D. Knutson, Algebraic spaces, *Thesis*, M.I.T., 1968.

[Li] J. Lipman, Rational singularities, *Publ. Math. I.H.E.S.*, n° 36 (1969).

[L] M. Lichtenbaum, Curves over discrete valuation rings, *Am. J. Math.*, **90** (1968), p. 380.

[Ma] W. Manger, Die Klassen von topologischen Abbildungen einer geschlossenen Fläche auf sich, *Math. Zeit.*, **44** (1939), p. 541.

[M1] D. Mumford, *Geometric Invariant Theory*, Springer-Verlag, 1965.

[M2] D. Mumford, *Notes on seminar on moduli problems*, Supplementary mimeographed notes printed at A.M.S. Woods Hole Summer Institute, 1964.

[M3] D. Mumford, Picard groups of moduli problems, *Proc. Conf. on Arith. Alg. Geom. at Purdue*, 1963.

[N] A. Néron, Modèles minimaux des variétés abéliennes sur les corps locaux et globaux, *Publ. Math. I.H.E.S.*, n° 21.

[R] M. Raynaud, Spécialisation du foncteur de Picard, *Comptes rendus Acad. Sci.*, **264**, p. 941 and p. 1001.

[Š] I. Šafarevich, Lectures on minimal models, *Tata Institute Lectures notes*, Bombay, 1966.

[Sc] M. Schlessinger, *Thesis*, Harvard.

[S] J.-P. Serre, *Rigidité du foncteur de Jacobi d'échelon* $n \geq 3$, app. à l'exposé 17 du séminaire Cartan 60/61.

[Se] F. Severi and Löffler, *Vorlesungen über algebraische Geometrie*, Teubner, 1924.

[W] A. Weil, Modules des surfaces de Riemann, *Sém. Bourbaki*, 168, 1957-58.

Manuscrit reçu le 21 janvier 1969.

CENTRO INTERNAZIONALE MATEMATICO ESTIVO

(C. I. M. E.)

David MUMFORD

(with an appendix by George Kempf)

VARIETIES DEFINED BY QUADRATIC EQUATIONS

Corso tenuto a Varenna dal 9 al 17 settembre 1969

VARIETIES DEFINED BY QUADRATIC EQUATIONS

by David Mumford (with an appendix by George Kempf)
(University of Harvard)

Introduction

First of all, let me fix my terminology and set-up. I will always be working over an algebraically closed ground field k. We will be concerned almost entirely with projective varieties over k (although many of our results generalize immediately to arbitrary projective schemes). By a projective variety, I will understand a topological space X all of whose points are closed, plus a sheaf $\mathcal{O}_X$ of k-valued functions on X isomorphic to some subvariety of $\mathbb{P}^n$ for some n. By a subvariety of $\mathbb{P}^n$, I will mean the subset $X \subset \mathbb{P}^n(k)$ defined by some homogeneous prime ideal $\wp \subset k[X_0,\cdots,X_n]$, with its Zariski-topology and with the sheaf $\mathcal{O}_X$ of functions from X to k induced locally by polynomials in the affine coordinates. Note that our varieties have only k-rational points — no generic points. In this, we depart slightly from the language of schemes. Note too that a projective variety can be isomorphic to many different subvarieties of $\mathbb{P}^n$. An isomorphism of X with a subvariety of $\mathbb{P}^n$ will be called an immersion of X in $\mathbb{P}^n$.

Let me begin with an elementary but somewhat startling result:

Definition: For all d, the d-ple immersion of $\mathbb{P}^n$ is the morphism:

$$s_d: \mathbb{P}^n \longrightarrow \mathbb{P}^N, \qquad N = \binom{n+d}{d} - 1$$

given by:

$$s_d(a_o,\cdots,a_n) = (a^{\alpha^{(o)}},\cdots,a^{\alpha^{(N)}})$$

where $\alpha^{(o)},\cdots,\alpha^{(N)}$ runs through the (n+1)-tuples $\alpha = (\alpha_o,\cdots,\alpha_n)$, such that $\alpha_i \geq 0$, $\Sigma\alpha_i = d$, and

$$a^\alpha = \prod_{i=0}^{n} a_i^{\alpha_i} .$$

Theorem 1: Let $X \subset \mathbb{P}^n$ be a subvariety, and let d_o be the degree of X. For all $d \geq d_o$, consider the new projective embedding:

$$X \subset \mathbb{P}^n \xrightarrow{s_d} \mathbb{P}^N .$$

Then the subvariety of $\mathbb{P}^N$ so obtained is an intersection of quadrics.*

Proof: Let r = dim X. For all linear spaces L of dimension n-r-2, disjoint from X, let H_L be the join of X and L, i.e., the locus of lines joining X and L. H_L is a hypersurface of degree $\leq d_o$. Then it is easy to see that

$$X = \bigcap_{L\cap X=\emptyset} H_L .$$

In fact, if $x \in \mathbb{P}^n - X$, let

$$\pi:\ \mathbb{P}^n - \{x\} \longrightarrow \mathbb{P}^{n-1}$$

---

* When we talk about an r-dimensional subvariety X of $\mathbb{P}^n$ being an intersection of quadrics, we never mean an intersection of only n-r quadrics (called usually a "complete intersection"). We just mean that there is a large set of quadrics Q_α, $\alpha \in S$, such that $X = \bigcap Q_\alpha$. Of course, S can be assumed finite.

be projection with center x. Then $\pi(X)$ is an r-dimensional subvariety of $\mathbb{P}^{n-1}$ so there exists a linear subspace $M \subset \mathbb{P}^{n-1}$ disjoint from $\pi(X)$ of dimension $(n-1)-r-1$. Choose L such that $\pi(L) = M$. Then $x \notin H_L$.

Thus X is an intersection of hypersurfaces of degree $\leq d_o$. Therefore, for all $d \geq d_o$, X is the intersection of those hypersurfaces of degree d that contain it. But by definition of s_d, if $H_1 \subset \mathbb{P}^n$ is a hypersurface of degree d, there is a hyperplane $H_2 \subset \mathbb{P}^N$ such that

$$H_1 = s_d^{-1}(H_2).$$

Therefore, there is a linear space $K \subset \mathbb{P}^N$ such that $X = s_d^{-1}(K)$, or

$$s_d(X) = K \cap s_d(\mathbb{P}^n).$$

To prove the theorem, it remains to check that $s_d(\mathbb{P}^n)$ is an intersection of quadrics. This follows from the remark:

For all $b_o, \cdots, b_N$,

$$(*) \quad \left[\begin{array}{l}\text{There exists } a_o, \cdots, a_n \\ \text{such that } b_i = a^{\alpha(i)}\end{array}\right] \Longleftrightarrow \left[\begin{array}{l}b_i b_j = b_k b_\ell \text{ whenever} \\ \alpha(i) + \alpha(j) = \alpha(k) + \alpha(\ell)\end{array}\right]$$

We leave this to the reader.

<u>QED</u>

I want to make 2 remarks. Suppose by the <u>rank</u> r of quadric we mean the rank of the corresponding symmetric matrix. Then the proof of this theorem shows that X is actually an intersection of quadrics of rank ≤ 4. Suppose we make the definition:

<u>Definition</u>: A subvariety $X \subset \mathbb{P}^n$ is <u>ideal-theoretically</u> an intersection of hypersurfaces $H_1, \cdots, H_m$ if set-theoretically:

$$X = H_1 \cap \cdots \cap H_m$$

and moreover, every $x \in X$ has an affine open neighborhood $U \subset \mathbb{P}^n$ such that the ideal $I(X)$ of $X \cap U \subset U$ is generated by the affine equations $f_1, \cdots, f_n$ of $H_1, \cdots, H_n$.

<u>Lemma</u>: If X is non-singular, then X is ideal-theoretically the intersections of $H_1, \cdots, H_n$ if and only if

1) $X = \bigcap_{i=1}^{n} H_i$

2) for all $x \in X$,

$$T_{x,X} = \bigcap_{i=1}^{n} T_{x,H_i}$$

(the intersection being taken in $T_{x,\mathbb{P}^n}$; here T means Zariski tangent space).

We leave the proof to the reader. Using this, we can then prove a variant of Theorem 1 to the effect that if X is non-singular, then $s_d(X)$ is ideal-theoretically an intersection of quadrics.

D. Mumford

§1. The cohomological method.

In setting up the concepts of linear systems and ampleness and in the construction of projective embeddings, we have to make a choice between 3 equivalent forrulations — that of divisor classes, of line bundles, or of invertible sheaves. It is well known that on any variety X, the group of (Cartier) divisors mod linear equivalence, the group of line bundles and the group of invertible sheaves are all canonically isomorphic. For our purposes, it is most convenient to use the sheaves:

Definition: An invertible sheaf L on X is a sheaf of $\mathcal{O}_X$-modules, locally isomorphic to $\mathcal{O}_X$ itself.

Two such sheaves L_1, L_2 can be tensored to form a 3rd $L_1 \otimes L_2$; $\mathcal{O}_X$ itself is an invertible sheaf forming a unit for this multiplication; and for any L, $L^{-1} = \mathrm{Hom}(L, \mathcal{O}_X)$ is an inverse since $L \otimes L^{-1} \cong L^{-1} \otimes L \cong \mathcal{O}_X$. The s^t of all invertible sheaves, mod isomorphisms, thus forms an abelian group, called Pic(X).

$\Gamma(L)$ or $H^0(L)$ will be the vector space of global sections of L. If $s \in \Gamma(L)$, and $x \in X$, then via an isomorphism $L|_U \cong \mathcal{O}_X|_U$ in some neighborhood U of x, we can find a value s(x); and the conditions $s(x) = 0$ or $s(x) \neq 0$ are independent of this local isomorphism.

Definition: The base points of $\Gamma(L)$ are the points $x \in X$ such that for all $s \in \Gamma(L)$, $s(x) = 0$.

If $\Gamma(L)$ is base point free, L defines a canonical morphism into projective space. Let $\mathbb{P}(\Gamma(L))$ be the projective space of hyperplanes in $\Gamma(L)$. Then define

$$\phi_L: X \longrightarrow \mathbb{P}(\Gamma(L))$$

by
$$\phi_L(x) = \{s \in \Gamma(L) \mid s(x) = 0\}.$$

This is easily checked to be a morphism. More explicitly, let $s_o, s_1, \cdots, s_n$ be a basis of $\Gamma(L)$. Define:

$$\phi_L: X \longrightarrow \mathbb{P}^n$$

by
$$\phi_L(x) = \text{pt. with homog. coord. } (s_o(x), s_1(x), \cdots, s_n(x)).$$

Definition: L is very ample if $\Gamma(L)$ is base point free and ϕ_L is an immersion (= an isomorphism of X with $\phi_L(X)$). L is ample if L^n is very ample for some $n \geq 1$.

Write $\mathbb{P}$ for $\mathbb{P}(\Gamma(L))$ and suppose L is very ample. Then the vector space $\Gamma(L)$ is canonically isomorphic to the space of homogeneous coordinate functions on the projective space $\mathbb{P}$, i.e,

$$\Gamma(L) \cong \Gamma(\mathbb{P}, \mathcal{O}_{\mathbb{P}}(1)).$$

And the kth symmetric power of $\Gamma(L)$, which we write $S^k\Gamma(L)$, is canonically isomorphic to the space of homogeneous polynomials of degree k in the homogeneous coordinates on $\mathbb{P}$, i.e.,

$$S^k\Gamma(L) \cong \Gamma(\mathbb{P}, \mathcal{O}_{\mathbb{P}}(k)).$$

Thus the vector space of homogeneous polynomials of degree k that vanish on $\phi_L(X)$ is nothing but the kernel of the canonical map:

$$S^k\Gamma(L) \longrightarrow \Gamma(L^k).$$

A strengthening of the assertion that $\phi_L(X)$ is an intersection of quadrics is that its homogeneous ideal is generated by quadrics. This is the same as asking whether the canonical map:

$$S^{k-2}\Gamma(L) \otimes \mathrm{Ker}\left[S^2\Gamma(L) \to \Gamma(L^2)\right] \longrightarrow \mathrm{Ker}\left[S^k\Gamma(L) \longrightarrow \Gamma(L^k)\right]$$

is surjective for all $k \geq 2$.

Our basic definition is this:

<u>Definition</u>: Let $\mathcal{F}, \mathcal{G}$ be coherent sheaves on X. Define $\mathcal{R}(\mathcal{F},\mathcal{G})$, $\mathcal{S}(\mathcal{F},\mathcal{G})$ as the kernel and cokernel of the canonical map α:

$$0 \longrightarrow \mathcal{R}(\mathcal{F},\mathcal{G}) \longrightarrow \Gamma(\mathcal{F})\otimes\Gamma(\mathcal{G}) \xrightarrow{\alpha} \Gamma(\mathcal{F}\otimes\mathcal{G}) \longrightarrow \mathcal{S}(\mathcal{F},\mathcal{G}) \longrightarrow 0.$$

Thus if L is a very ample invertible sheaf, $\mathcal{R}(L,L)$ is the space (a) of alternating elements of $\Gamma(L)\otimes\Gamma(L)$, and (b) of the quadratic relations holding on $\phi_L(X)$.

<u>**Definition**</u>: Let L be an ample sheaf on X. Then L is <u>normally generated</u> if

$$\Gamma(L)^{\otimes k} \longrightarrow \Gamma(L^k)$$

is surjective, all $k \geq 1$.

This is clearly equivalent to the condition $\mathcal{S}(L^i, L^j) = (0)$, $i,j \geq 1$. Note that if L is normally generated then L is necessarily very ample too! In fact, consider the 2 morphisms:

$$\begin{array}{ccc} & & \mathbb{P}(\Gamma(L^n)) \\ & \overset{\phi_{L^n}}{\nearrow} & \\ X & & \\ & \underset{\phi_L}{\searrow} & \\ & & \mathbb{P}(\Gamma(L)) \end{array}$$

The n-ple embedding of the projective space $\mathbb{P}(V)$ of hyperplanes for <u>any</u> vector space V is canonically a map

$$s_n : \mathbb{P}(V) \longrightarrow \mathbb{P}(S^n V).$$

Moreover, via the surjection

$$S^n\Gamma(L) \longrightarrow \Gamma(L^n),$$

we can identify $\mathbb{P}(\Gamma(L^n))$ canonically with a linear subspace of $\mathbb{P}(S^n\Gamma(L))$. Putting this together, we get a diagram:

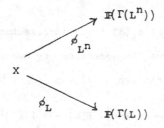

$$\begin{array}{ccccc} & & \mathbb{P}(\Gamma(L^n)) & & \\ & \overset{\phi_{L^n}}{\nearrow} & & \subset\!\!\searrow & \\ X & & & & \mathbb{P}(S^n\Gamma(L)) \\ & \underset{\phi_L}{\searrow} & & \hookrightarrow\!\!\nearrow_{s_n} & \\ & & \mathbb{P}(\Gamma(L)) & & \end{array}$$

It is easy to check that this commutes. Now for large n, L^n is very ample, hence ϕ_{L^n} is an immersion, so it follows from the diagram that ϕ_L is an immersion too, i.e., L is very ample.

Definition: Let L be a normally generated invertible sheaf. Then L is normally presented if one of the 4 equivalent conditions holds:

(A) $\mathrm{Ker}[S^2\Gamma(L) \longrightarrow \Gamma(L^2)] \otimes \Gamma(L^{k-2}) \longrightarrow \mathrm{Ker}[S^k\Gamma(L) \longrightarrow \Gamma(L^k)]$

is surjective, all $k \geq 2$

(B) $\bigoplus_{1 \leq i < j \leq n} [\mathcal{R}(L,L) \otimes \Gamma(L)^{k-2}] \longrightarrow \mathrm{Ker}[\Gamma(L)^{\otimes k} \longrightarrow \Gamma(L^k)]$

is surjective, all $k \geq 2$.

The above homomorphism maps an element $a \otimes b$ in the $(i,j)^{th}$ factor to the element of $\Gamma(L)^{\otimes k}$ whose i^{th} and j^{th} components are determined by a, and the rest by b.

(C) $\Gamma(L^{i-1}) \otimes \mathcal{R}(L,L) \otimes \Gamma(L^{j-1}) \longrightarrow \mathcal{R}(L^i,L^j)$

is surjective, if $i,j \geq 1$.

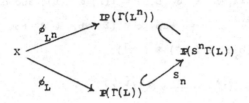

Here, if $\Sigma a_i \otimes b_i \in \mathcal{R}(L,L) \subset \Gamma(L) \otimes \Gamma(L)$, and $c \in \Gamma(L^{i-1})$, $d \in \Gamma(L^{j-1})$, then we map $c \otimes (\Sigma a_i \otimes b_i) \otimes d$ to $\Sigma(a_i c) \otimes (b_i d) \in \Gamma(L^i) \otimes \Gamma(L^j)$.

(D) $\mathcal{R}(L^i,L^j) \otimes \Gamma(L^k) \longrightarrow \mathcal{R}(L^i,L^{j+k})$

is surjective if $i,j,k \geq 1$.

It is not so obvious that all these properties are equivalent! Thus to see (A)$\Longleftrightarrow$(B), note that $\mathcal{R}(L,L) \subset \Gamma(L) \otimes \Gamma(L)$ contains the alternating tensors, so the image of

$$\bigoplus_{1\leq i<j\leq n} [\mathcal{R}(L,L) \otimes \Gamma(L)^{k-2}]$$

in $\Gamma(L)^k$ contains all the alternating tensors. So the image equals $\mathrm{Ker}(\Gamma(L)^k \longrightarrow \Gamma(L^k))$ if and only if its image in $S^k\Gamma(L)$ equals $\mathrm{Ker}(S^k\Gamma(L) \longrightarrow \Gamma(L^k))$. But its image in $S^k\Gamma(L)$ is the same as the image of the map in (A).

(C)$\Longrightarrow$(D) follows immediately using normal generation and (D)$\Longrightarrow$(C) follows by factoring the map in (C) thus:

$$\Gamma(L^{i-1}) \otimes \mathcal{R}(L,L) \otimes \Gamma(L^{j-1}) \longrightarrow \Gamma(L^{i-1}) \otimes \mathcal{R}(L,L^j) \longrightarrow \mathcal{R}(L^i,L^j).$$

Next, to prove (C)$\Longrightarrow$(B), factor $\Gamma(L)^k \longrightarrow \Gamma(L^k)$ as follows:

$$\Gamma(L) \otimes \Gamma(L)^{k-1} \xrightarrow{\text{onto}} \Gamma(L^2) \otimes \Gamma(L)^{k-2} \xrightarrow{\text{onto}} \cdots \xrightarrow{\text{onto}} \Gamma(L^k)\ .$$

To prove (B), it is enough to show that $\oplus[\mathcal{R}(L,L)\otimes\Gamma(L)^{k-2}]$ goes onto the kernel at each stage of this sequence. Thus it is enough if $\Gamma(L)^{i-1} \otimes \mathcal{R}(L,L)$ is mapped onto $\mathrm{Ker}[\Gamma(L^i) \otimes \Gamma(L) \longrightarrow \Gamma(L^{i+1})]$. This last space is $\mathcal{R}(L^i,L)$, so this ontoness is part of (C).

Finally, to prove (B) $\Longrightarrow$ (C), factor $\Gamma(L)^k \longrightarrow \Gamma(L^k)$ when $k = i+j$, as follows:

$$\Gamma(L)^{i+j} \xrightarrow[\alpha]{\text{onto}} \Gamma(L^i) \otimes \Gamma(L^j) \xrightarrow[\beta]{\text{onto}} \Gamma(L^{i+j}).$$

It follows from normal generation that we get a surjection:

$$\mathrm{Ker}(\beta\circ\alpha) \xrightarrow{\text{onto}} \mathrm{Ker}(\beta) = \mathcal{R}(L^i,L^j).$$

But $\mathrm{Ker}(\beta\,\alpha)$ is generated by $\oplus[\mathcal{R}(L,L) \otimes \Gamma(L)^{i+j-2}]$. The image of this last space in $\Gamma(L^i)\otimes\Gamma(L^j)$ is the same as the image of $\Gamma(L^{i-1}) \otimes \mathcal{R}(L,L) \otimes \Gamma(L^{j-1})$, so (C) follows.

This at least gives us a nice definition to work with! It seems easier to prove things about $\mathcal{S}$ first, and then to use these results to obtain things about $\mathcal{R}$. Our first result is:

Theorem 2 (Generalized lemma of Castelnuovo): Suppose L is an ample invertible sheaf on a variety X such that $\Gamma(L)$ has no base points. Suppose $\mathfrak{F}$ is a coherent sheaf on X such that

$$H^i(\mathfrak{F} \otimes L^{-i}) = (0), \quad i \geq 1.$$

Then (a) $H^i(\mathfrak{F} \otimes L^j) = (0)$ if $i+j \geq 0$, $i \geq 1$

and (b) $\mathcal{S}(\mathfrak{F} \otimes L^i, L) = (0)$, $i \geq 0$.

To motivate this, look at the case of Castelnuovo's original lemma: X = non-singular curve, $\mathfrak{A}$, $\mathfrak{B}$ divisors on X, $\mathfrak{J} = \mathcal{O}_X(\mathfrak{B})$, $L = \mathcal{O}_X(\mathfrak{A})$. In classical language:

$$\begin{pmatrix}\Gamma(L) \text{ has no} \\ \text{base points}\end{pmatrix} \underset{\text{def}}{\equiv} \begin{pmatrix}|\mathfrak{A}| \text{ is base point} \\ \text{free}\end{pmatrix}$$

$$\left(H^1(\mathfrak{J} \otimes L^{-1}) = (0)\right) \underset{\text{def}}{\equiv} \left(|\mathfrak{B} - \mathfrak{A}| \text{ is non-special}\right)$$

Translating the conclusion, we find:

$$\left(\mathcal{S}(\mathfrak{J}, L) = (0)\right) \underset{\text{def}}{\equiv} \left(|\mathfrak{A} + \mathfrak{B}| = \begin{matrix}\text{the minimal sum} \\ |\mathfrak{A}| + |\mathfrak{B}|\end{matrix}\right)$$

<u>Proof of Theorem 2</u>: Use induction on dim(Supp $\mathfrak{J}$). If dim(Supp $\mathfrak{J}$) = 0, then choose $s \in \Gamma(L)$ such that $s(x) \neq 0$ for all $x \in \mathrm{Supp}(\mathfrak{J})$. Then

$$\Gamma(\mathfrak{J}) \underset{k}{\otimes} (s^i \cdot k) \xrightarrow{\approx} \Gamma(\mathfrak{J} \otimes L^i)$$

is an isomorphism, so certainly

$$\Gamma(\mathfrak{J}) \underset{k}{\otimes} \Gamma(L^i) \longrightarrow \Gamma(\mathfrak{J} \otimes L^i)$$

is surjective. Therefore $\mathcal{S}(\mathfrak{J}, L^i) = (0)$. Also, all groups $H^i(\mathfrak{J} \otimes \text{anything})$, $i \geq 1$, vanish.

Now suppose we are given an $\mathfrak{J}$, and we have proven the theorem for all $\mathfrak{J}^*$'s with dim(Supp $\mathfrak{J}^*$) < dim(Supp $\mathfrak{J}$). I claim that there is an element $s \in \Gamma(L)$ sufficiently "generic" so that for every $x \in X$, if we choose an isomorphism $L|_U \cong \mathcal{O}_X|_U$ near x, so that s can be considered as a function, then s is <u>not</u> a 0-divisor in the stalk $\mathfrak{F}_x$ of $\mathfrak{F}$. To see, recall that by the Noetherian decomposition theorems, for any coherent $\mathfrak{F}$, there is a finite set of irreducible subsets $Z_1,\cdots,Z_n \subset \text{Supp}(\mathfrak{F})$ (including the components of $\text{Supp}(\mathfrak{F})$, but possibly including some "embedded components" too) such that the support of any element

$$\alpha \in \Gamma(U,\mathfrak{F})$$

is a union of some of the sets $U \cap Z_i$. For each i, not all sections $s \in \Gamma(L)$ vanish identically on Z_i. Therefore there is an element $s \in \Gamma(L)$ not identically zero on any Z_i. If $\alpha \in \Gamma(U,\mathfrak{F})$, then s must be non-zero at at least one point x of $\text{Supp}(\alpha)$, hence $\alpha \otimes s \in \Gamma(U, \mathfrak{F}\otimes L)$ is not zero near x. Thus s has the required property and the map $\mathfrak{F} \longrightarrow \mathfrak{F} \otimes L$, defined by $\alpha \longmapsto \alpha \otimes s$, is injective.

It is more convenient to use the map $\mathfrak{F} \otimes L^{-1} \longrightarrow \mathfrak{F}$, defined by $\alpha \longmapsto \alpha\otimes s$. Let $\mathfrak{F}^*$ be the cokernel. Then for all i, we have exact sequences:

$$(*)_i \qquad 0 \longrightarrow \mathfrak{F} \otimes L^{-i-1} \xrightarrow{\otimes s} \mathfrak{F} \otimes L^{-i} \longrightarrow \mathfrak{F}^* \otimes L^{-i} \longrightarrow 0.$$

Note that $\dim(\text{Supp}\ \mathfrak{F}^*) < \dim(\text{Supp}\ \mathfrak{F})$. In fact, for all i, $\otimes s$ is an isomorphism on almost all of Z_i, hence $Z_i \not\subset \text{Supp}(\mathfrak{F}^*)$. Therefore every component of Supp($\mathfrak{J}^*$) is a proper closed subset of some component of Supp($\mathfrak{J}$). By $(*)_i$, we get an exact sequence:

$$\begin{array}{ccccc} H^i(\mathfrak{J}\otimes L^{-i}) & \longrightarrow & H^i(\mathfrak{J}^*\otimes L^{-i}) & \longrightarrow & H^{i+1}(\mathfrak{J}\otimes L^{-i-1}), \quad i \geq 1 \\ \| & & & & \| \\ (0) & & & & (0) \end{array}$$

hence $H^i(\mathfrak{J}^*\otimes L^{-i}) = (0)$. Thus the hypothesis of the theorem is valid for $\mathfrak{J}^*$, so by our induction hypothesis, so is the conclusion. Going back from $\mathfrak{J}^*$ to $\mathfrak{J}$, use the exact sequence:

$$H^i(\mathfrak{J}\otimes L^{-i}) \longrightarrow H^i(\mathfrak{J}\otimes L^{-i+1}) \longrightarrow H^i(\mathfrak{J}^*\otimes L^{-i+1}).$$

The 1st group is (0) by the hypothesis on $\mathfrak{J}$; the 3rd group is (0) by the theorem for $\mathfrak{J}^*$: so the 2nd is (0). Replacing $\mathfrak{J}$ by $\mathfrak{J}\otimes L$, we continue in this way and prove by induction on i+j that

$$H^i(\mathfrak{J}\otimes L^j) = (0), \quad i+j \geq 0, \quad i \geq 1.$$

As for the $\mathcal{S}$'s, look at the diagram of solid arrows:

$$\begin{array}{ccccccccc} 0 & \longrightarrow & \Gamma(\mathfrak{J}\otimes L^{-1})\otimes\Gamma(L) & \longrightarrow & \Gamma(\mathfrak{J})\otimes\Gamma(L) & \longrightarrow & \Gamma(\mathfrak{J}^*)\otimes\Gamma(L) & \longrightarrow & 0 \\ & & \downarrow & & \downarrow & & \downarrow & & \\ 0 & \longrightarrow & \Gamma(\mathfrak{J}) & \longrightarrow & \Gamma(\mathfrak{J}\otimes L) & \longrightarrow & \Gamma(\mathfrak{J}^*\otimes L) & \longrightarrow & 0 \\ & & \downarrow & & \downarrow & & \downarrow & & \\ & & \mathcal{S}(\mathfrak{J}\otimes L^{-1},L) & \xrightarrow{\alpha} & \mathcal{S}(\mathfrak{J},L) & \longrightarrow & \mathcal{S}(\mathfrak{J}^*,L) & \longrightarrow & 0 \\ & & \downarrow & & \downarrow & & \downarrow & & \\ & & 0 & & 0 & & 0 & & \end{array}$$

It has exact rows since $H^1(\mathfrak{F}\otimes L^{-1}) = H^1(\mathfrak{F}) = (0)$, and exact columns by definition. Define the dotted arrow by $\alpha \longmapsto \alpha\otimes\tau$. Then the shaded triangle commutes, which proves that the map α is zero! Since $\mathcal{S}(\mathfrak{F}^*,L) = (0)$, it follows that $\mathcal{S}(\mathfrak{F},L) = (0)$. As we may replace $\mathfrak{F}$ by $\mathfrak{F}\otimes L^i$, $i \geq 1$, the rest of (b) follows too.

QED

A useful remark is that a close examination of this proof shows a slightly more precise result. Namely, that if $n = \dim(\mathrm{Supp}\,\mathfrak{F})$; and if $s_o,\cdots,s_n \in \Gamma(L)$ are sufficiently "generic" elements, then in fact $\Gamma(\mathfrak{F}\otimes L)$ is spanned by the images of $\Gamma(\mathfrak{F}) \otimes_k (s_i \cdot k)$, for $0 \leq i \leq n$.

<u>Theorem 3</u>: Let L be an ample invertible sheaf on an n-dimensional variety X. Suppose $\Gamma(L)$ has no base points and

$$H^i(L^j) = (0), \quad i \geq 1,\ j \geq 1.$$

Then $\mathcal{S}(L^i,L^j) = (0)$ if $i \geq n+1$, $j \geq 1$.

In particular, if $i \geq n+1$, L^i is ample with normal generation, hence very ample.

<u>Proof</u>: Apply Theorem 2 to $\mathfrak{F} = L^{n+1}$. It follows that $\mathcal{S}(L^i,L) = (0)$, if $i \geq n+1$. Explicitly $\Gamma(L^i) \otimes \Gamma(L) \longrightarrow \Gamma(L^{i+1})$ is surjective if $i \geq n+1$. Composing these maps, $\Gamma(L^i) \otimes \Gamma(L)^j \longrightarrow \Gamma(L^{i+j})$ is surjective if $i \geq n+1$. Therefore $\Gamma(L^i) \otimes \Gamma(L^j) \longrightarrow \Gamma(L^{i+j})$ is surjective too, if $i \geq n+1$.

QED

Next we want to prove similar results about $\mathcal{R}$. We need the preliminary result:

<u>6-lemma</u>: If $0 \longrightarrow \mathcal{F}_1 \longrightarrow \mathcal{F}_2 \longrightarrow \mathcal{F}_3 \longrightarrow 0$ is an exact sequence of coherent sheaves, and $\Gamma(\mathcal{F}_2) \longrightarrow \Gamma(\mathcal{F}_3)$ is surjective — e.g., if $H^1(\mathcal{F}_1) = (0)$ — then for all invertible sheaves L there is an exact sequence:

$$0 \longrightarrow \mathcal{R}(\mathcal{F}_1,L) \longrightarrow \mathcal{R}(\mathcal{F}_2,L) \longrightarrow \mathcal{R}(\mathcal{F}_3,L) \longrightarrow \mathcal{S}(\mathcal{F}_1,L) \longrightarrow \mathcal{S}(\mathcal{F}_2,L) \longrightarrow \mathcal{S}(\mathcal{F}_3,L).$$

Also, even if $\Gamma(\mathcal{F}_2) \longrightarrow \Gamma(\mathcal{F}_3)$ is not surjective, the 1st 3 terms form an exact sequence.

<u>Proof</u>: Look at the diagram of solid arrows:

$$\begin{array}{ccccccccc}
 & & 0 & & 0 & & 0 & & \\
 & & \downarrow & & \downarrow & & \downarrow & & \\
0 & \longrightarrow & \mathcal{R}(\mathcal{F}_1,L) & \dashrightarrow & \mathcal{R}(\mathcal{F}_2,L) & \dashrightarrow & \mathcal{R}(\mathcal{F}_3,L) & & \\
 & & \downarrow & & \downarrow & & \downarrow & & \\
0 & \longrightarrow & \Gamma(\mathcal{F}_1)\otimes\Gamma(L) & \longrightarrow & \Gamma(\mathcal{F}_2)\otimes\Gamma(L) & \longrightarrow & \Gamma(\mathcal{F}_3)\otimes\Gamma(L) & \longrightarrow & 0 \\
 & & \downarrow & & \downarrow & & \downarrow & & \\
0 & \longrightarrow & \Gamma(\mathcal{F}_1\otimes L) & \longrightarrow & \Gamma(\mathcal{F}_2\otimes L) & \longrightarrow & \Gamma(\mathcal{F}_3\otimes L) & & \\
 & & \downarrow & & \downarrow & & \downarrow & & \\
 & \dashrightarrow & \mathcal{S}(\mathcal{F}_1,L) & \dashrightarrow & \mathcal{S}(\mathcal{F}_2,L) & \dashrightarrow & \mathcal{S}(\mathcal{F}_3,L) & & \\
 & & \downarrow & & \downarrow & & \downarrow & & \\
 & & 0 & & 0 & & 0 & &
\end{array}$$

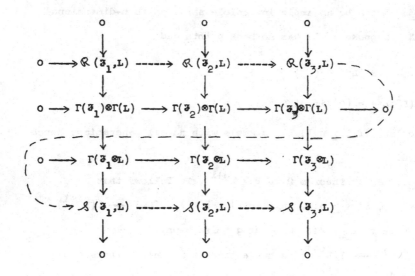

The rows and columns are exact, by hypothesis. By the so-called "serpent" argument, you get an exact sequence indicated by the dotted arrows.

QED

We apply this to prove:

Theorem 4: Let L and M be ample invertible sheaves on a projective variety X, let $\mathfrak{F}$ be a coherent sheaf on X, and assume:

i) $\Gamma(L)$, $\Gamma(M)$ have no base points,

ii) $H^{i+j-1}(\mathfrak{F}\otimes L^{-i}\otimes M^{-j}) = (0)$ if $i,j \geq 1$.

Then the natural map:

$$\mathcal{R}(\mathfrak{F},L) \otimes \Gamma(M) \longrightarrow \mathcal{R}(\mathfrak{F} \otimes M, L)$$

is surjective.

[One can also check that the hypotheses imply that

$$H^k(\mathfrak{F}\otimes L^{-i}\otimes M^{-j}) = (0) \text{ if } k \geq 1,\ i+k \geq 0,\ j+k \geq 0,\ i+j+k \geq -1.$$

Therefore the hypotheses are stable under the substitution $\mathfrak{F} \longmapsto \mathfrak{F} \otimes L$ or $\mathfrak{F} \otimes M$. However, we may as well stick to the simplest case of the theorem.]

Proof: As in Theorem 2, we use induction on $\dim(\operatorname{Supp} \mathfrak{F})$. If $\dim(\operatorname{Supp} \mathfrak{F}) = 0$, we get the diagram:

$$\begin{array}{ccccc}
0 & & 0 & & 0 \\
\downarrow & & \downarrow & & \downarrow \\
\mathcal{R}(\mathfrak{I},L)\otimes\Gamma(M) & \longrightarrow & \mathcal{R}(\mathfrak{I}\otimes M, L) & \underset{\alpha}{\cong} & \mathcal{R}(\mathfrak{I},L) \\
\downarrow & & \downarrow & & | \\
\Gamma(\mathfrak{I})\otimes\Gamma(L)\otimes\Gamma(M) & \longrightarrow & \Gamma(\mathfrak{I}\otimes M)\otimes\Gamma(L) & \underset{\beta}{\cong} & \Gamma(\mathfrak{I})\otimes\Gamma(L) \\
\downarrow & & \downarrow & & \downarrow \\
\Gamma(\mathfrak{I}\otimes L)\otimes\Gamma(M) & \longrightarrow & \Gamma(\mathfrak{I}\otimes L\otimes M) & \underset{\gamma}{\cong} & \Gamma(\mathfrak{I}\otimes L)
\end{array}$$

where the isomorphisms β and γ are obtained by choosing a section $s \in \Gamma(M)$ non-zero at all points of $\mathrm{Supp}(\mathfrak{I})$, hence an isomorphism of M and $\mathcal{O}_X$ in a neighborhood of $\mathrm{Supp}(\mathfrak{I})$. β and γ induce an isomorphism α. But in the map on the top row, if $p \in \mathcal{R}(\mathfrak{I}, L)$, then $p\otimes s \in \mathcal{R}(\mathfrak{I}, L)\otimes\Gamma(M)$ is taken to $p \in \mathcal{R}(\mathfrak{I}, L)$, so this map is surjective. This proves the theorem when $\dim(\mathrm{Supp}\ \mathfrak{I}) = 0$.

In the general case, choose a good section $s \in \Gamma(M)$ as in the proof of Theorem 2 so as to obtain an exact sequence:

$$0 \longrightarrow \mathfrak{I}\otimes M^{-1} \xrightarrow{\otimes s} \mathfrak{I} \longrightarrow \mathfrak{I}^* \longrightarrow 0$$

with $\dim(\mathrm{Supp}\ \mathfrak{I}^*) < \dim(\mathrm{Supp}\ \mathfrak{I})$. We obtain exact sequences:

$$H^{i+j-1}(\mathfrak{I}\otimes L^{-i}\otimes M^{-j}) \longrightarrow H^{i+j-1}(\mathfrak{I}^*\otimes L^{-i}\otimes M^{-j}) \longrightarrow H^{i+j}(\mathfrak{I}\otimes L^{-i}\otimes M^{-j-1}).$$

The 1st and 3rd groups are 0 by hypothesis, so the 2nd is also. This shows that $\mathfrak{I}^*$ satisfies the hypotheses of the theorem too. So by the induction hypothesis, $\mathcal{R}(\mathfrak{I}^*,L)\otimes\Gamma(M) \longrightarrow \mathcal{R}(\mathfrak{I}^*\otimes M, L)$ is surjective. Moreover, by Castelnuovo's lemma (Theorem 2), applied to

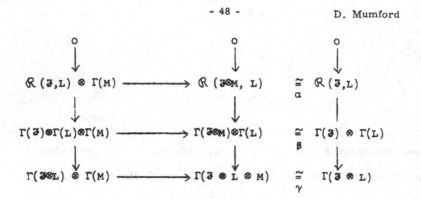

$\mathfrak{F} \otimes M^{-1}$ and L, $\mathcal{S}(\mathfrak{F} \otimes M^{-1}, L) = (0)$ and $H^1(\mathfrak{F} \otimes M^{-1}) = (0)$.

Applying the 6-lemma, we deduce that:

$$0 \longrightarrow \mathcal{R}(\mathfrak{F} \otimes M^{-1}, L) \longrightarrow \mathcal{R}(\mathfrak{F}, L) \longrightarrow \mathcal{R}(\mathfrak{F}^*, L) \longrightarrow 0$$

is exact. Now consider the diagram of solid arrows:

$$\begin{array}{ccccccccc}
0 & \longrightarrow & \mathcal{R}(\mathfrak{F} \otimes M^{-1}, L) \otimes \Gamma(M) & \longrightarrow & \mathcal{R}(\mathfrak{F}, L) \otimes \Gamma(M) & \longrightarrow & \mathcal{R}(\mathfrak{F}^*, L) \otimes \Gamma(M) & \longrightarrow & 0 \\
 & & \downarrow & \nearrow & \downarrow{\scriptstyle \beta} & & \downarrow{\scriptstyle \gamma} & & \\
0 & \longrightarrow & \mathcal{R}(\mathfrak{F}, L) & \xrightarrow{\ \alpha\ } & \mathcal{R}(\mathfrak{F} \otimes M, L) & \longrightarrow & \mathcal{R}(\mathfrak{F}^* \otimes M, L) & &
\end{array}$$

If you define the dotted arrow by $a \longmapsto a \otimes s$, it is clear that the shaded triangle commutes. Therefore $\mathrm{Im}(\alpha) \subset \mathrm{Im}(\beta)$ and using the surjectivity of γ, the surjectivity of β follows.

QED

To apply this Theorem, we need another result:

Proposition: Let $\mathfrak{F}$ be a coherent sheaf, and L, M invertible sheaves on X. If

a) $\mathcal{R}(\mathfrak{F}, L) \otimes \Gamma(M) \longrightarrow \mathcal{R}(\mathfrak{F} \otimes M, L)$ is surjective

b) $\mathcal{S}(\mathfrak{F}, L) = (0)$,

then

c) $\mathcal{R}(\mathfrak{F}, M) \otimes \Gamma(L) \longrightarrow \mathcal{R}(\mathfrak{F} \otimes L, M)$ is surjective.

Proof: Use the diagram:

$$\begin{array}{ccccccc}
 & & & & 0 & & 0 \\
 & & & & \downarrow & & \downarrow \\
 & & & & \Gamma(L) \otimes \mathcal{R}(\mathfrak{J},M) & \xrightarrow{\ \gamma\ } & \mathcal{R}(L \otimes \mathfrak{J}, M) \\
 & & & & \downarrow & & \downarrow \\
0 & \longrightarrow & \mathcal{R}(L,\mathfrak{J}) \otimes \Gamma(M) & \longrightarrow & \Gamma(L) \otimes \Gamma(\mathfrak{J}) \otimes \Gamma(M) & \xrightarrow{\ \beta\ } & \Gamma(L \otimes \mathfrak{J}) \otimes \Gamma(M) \\
 & & \downarrow{\scriptstyle \alpha} & & \downarrow & & \downarrow \\
0 & \longrightarrow & \mathcal{R}(L, \mathfrak{J} \otimes M) & \longrightarrow & \Gamma(L) \otimes \Gamma(\mathfrak{J} \otimes M) & \longrightarrow & \Gamma(L \otimes \mathfrak{J} \otimes M)
\end{array}$$

By assumption, α and β are surjective. "Chasing" the diagram, one sees quickly that γ is surjective too.

QED

Theorem 5: Let L be an ample invertible sheaf on an n-dimensional variety X. Assume:

i) $\Gamma(L)$ is base point free,

ii) $H^i(L^j) = (0)$ if $i,j \geq 1$.

Then it follows that:

$$\mathcal{R}(L^i, L^j) \otimes \Gamma(L^k) \longrightarrow \mathcal{R}(L^{i+k}, L^j)$$

is surjective, if $i \geq n+2$, $j,k \geq 1$. In particular, if $i \geq n+2$, L^i is ample with normal presentation.

Proof: By Theorem 4,

$$\mathcal{R}(L^i, L) \otimes \Gamma(L) \longrightarrow \mathcal{R}(L^{i+1}, L)$$

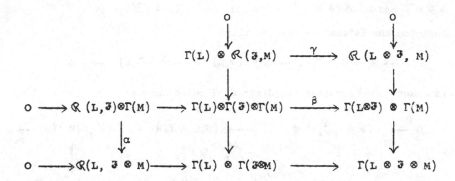

is surjective, if $i \geq n+2$. Iterating, we find that:

$$\mathcal{R}(L^i, L) \otimes \Gamma(L^j) \longrightarrow \mathcal{R}(L^{i+j}, L)$$

is surjective, if $i \geq n+2$, $j \geq 1$. Since $\mathcal{S}(L^i, L) = (0)$, $i \geq n+2$ apply the Proposition to prove that:

$$\mathcal{R}(L^i, L^j) \otimes \Gamma(L) \longrightarrow \mathcal{R}(L^{i+1}, L^j)$$

is surjective, if $i \geq n+2$, $j \geq 1$. Iterating again, we get the required assertion.

QED

§2. The case of curves.

For the whole of this section, X will be assumed to be a non-singular complete curve of genus g. We want to strengthen the results of §1 in this case. We need some more concepts and definitions. A divisor $\mathfrak{A}$ is a formal linear combination $\Sigma n_i x_i$ of points of X. For all divisors $\mathfrak{A}$, $\mathfrak{O}(\mathfrak{A})$ is the invertible sheaf of functions f which are regular except at the x_i's, and at x_i have at most an n_i-fold pole, if $n_i \geq 0$, or must have at least a $(-n_i)$-fold zero if $n_i \leq 0$. A fact that we need is that if an invertible sheaf L has a section s with zeroes exactly at $x_1, \ldots, x_k$ of multiplicities $n_1, \ldots, n_k$, then $L \cong \mathfrak{O}(\mathfrak{A})$, with $\mathfrak{A} = \Sigma n_i x_i$. If L is an invertible sheaf, $L(\mathfrak{A})$ stands for $L \otimes \mathfrak{O}(\mathfrak{A})$. Ω will be the sheaf of regular differentials on X.

Theorem 6: Let L,M be invertible sheaves on X such that $\deg L \geq 2g+1$, $\deg M \geq 2g$. Then $\mathcal{S}(L,M) = (0)$.

Proof: Let $d = \deg L$. $\mathfrak{A}$ is to be a positive divisor of degree $d-(g+1)$ which will be chosen later. Then $L(-\mathfrak{A})$ is naturally a subsheaf of L, and we get an exact sequence:

$$0 \longrightarrow L(-\mathfrak{A}) \longrightarrow L \longrightarrow L^* \longrightarrow 0$$

where $\operatorname{Supp} L^* = \operatorname{Supp} \mathfrak{A}$. The 1st requirement on $\mathfrak{A}$ is that $H^1(L(-\mathfrak{A})) = (0)$. Assuming for the moment that $\mathfrak{A}$ has this property, by the 6-lemma of §1, we get an exact sequence:

$$\mathcal{S}(L(-\mathfrak{A}),M) \longrightarrow \mathcal{S}(L,M) \longrightarrow \mathcal{S}(L^*,M)$$

But it is well known that if K is an invertible sheaf on X with $\deg K \geq 2g$, $\Gamma(K)$ has no base points. In particular, $\Gamma(M)$ has no base points, and $\mathrm{Supp}(L^*)$ is 0-dimensional. So by Castelnuovo's lemma, $\mathcal{S}(L^*,M) = (0)$.

Next, apply the Riemann-Roch theorem to $L(-\mathfrak{A})$:

$$\dim H^0(L(-\mathfrak{A})) = \deg L(-\mathfrak{A}) - (g-1) + \dim H^1(L(-\mathfrak{A}))$$

$$= 2.$$

Thus $\Gamma(L(-\mathfrak{A}))$ is a "pencil" and the 2nd requirement on $\mathfrak{A}$ is that it is base point free. Finally we want to apply Castelnuovo's lemma to deduce that $\mathcal{S}(L(-\mathfrak{A}),M) = (0)$. For this we need only that

$$H^1(M \otimes L(-\mathfrak{A})^{-1}) = H^1(M \otimes L^{-1}(\mathfrak{A})) = (0).$$

This is the 3rd requirement on $\mathfrak{A}$ Putting all this together, it will follow that $\mathcal{S}(L,M) = (0)$.

Can we find an $\mathfrak{A}$ with these 3 properties? Since $\mathfrak{A}$ consists in $d-(g+1) \geq g$ points all of which can be chosen arbitrarily, it is well known that for a suitable choice of $\mathfrak{A}$, $\mathfrak{o}(\mathfrak{A})$ will be isomorphic to any invertible sheaf K of degree $d-(g+1)$. Now the set of all invertible sheaves K of degree $d-(g+1)$ forms a projective variety J, which is exactly the Jacobian of X except that J does not have any natural base point on it to serve as the origin. It suffices to find a K such that

i) $H^1(L \otimes K^{-1}) = (0)$

ii) for all $x \in X$, $\dim H^0(L \otimes K^{-1}(-x)) = 1$

iii) $H^1(M \otimes L^{-1} \otimes K) = (0)$.

Now if (i) is false, $\dim H^0(L \otimes K^{-1}) > 2$ by Riemann-Roch, hence (ii) will be false for <u>all x</u>! Therefore it is enough to check (ii) and (iii) for all x. But by Riemann-Roch,

$$\dim H^0(L\otimes K^{-1}(-x)) > 1 \iff \dim H^1(L\otimes K^{-1}(-x)) > 0$$

$$\iff \dim H^0(\Omega\otimes L^{-1}\otimes K(x)) > 0$$

$$\iff \exists\, y_1,\dots,y_{g-2} \text{ such that } \Omega \otimes L^{-1} \otimes K(x) \cong \mathcal{O}(\Sigma y_i)$$

$$\iff \exists\, y_1,\dots,y_{g-2} \text{ such that } K \cong \Omega \otimes L^{-1}(x - \Sigma y_i).$$

We have only g-1 variable points here, so the locus of K's not satisfying (ii) has dimension at most g-1. Similarly if $\deg M = e + 2g$, we find:

$$
\begin{aligned}
H^1(M \otimes L^{-1} \otimes K) \neq (0) &\iff H^o(\Omega \otimes M^{-1} \otimes L \otimes K^{-1}) \neq (0) \\
&\iff \exists\ y_1, \ldots, y_k \text{ where} \\
&\qquad k = \deg(\Omega \otimes M^{-1} \otimes L \otimes K^{-1}) = g\text{-}1\text{-}e \\
&\qquad \text{such that} \\
&\qquad \Omega \otimes M^{-1} \otimes L \otimes K^{-1} \cong \mathcal{O}_C(\Sigma y_i) \\
&\iff \exists\ y_1, \ldots, y_k \text{ such that} \\
&\qquad K \cong \Omega \otimes M^{-1} \otimes L(-\Sigma y_i).
\end{aligned}
$$

Again there are at most g-1 variable points here, so the locus of K's not satisfying (iii) has dimension at most g-1. Since dim J = g, almost all K's do satisfy (ii) and (iii). Thus an $\mathcal{U}$ with the required properties exists.

QED

Corollary: If L is an invertible sheaf of degree $\geq$ 2g+1, then L is ample with normal generation.

If the argument in the above proof is traced through, it is not hard to show that it proves the following:

$$
\begin{aligned}
&\exists\ s_1, s_2 \in \Gamma(L) \\
&\exists\ \quad t \in \Gamma(M) \qquad \text{such that}
\end{aligned}
$$

$$
[k s_1 \otimes_k \Gamma(M) + k.s_2 \otimes_k \Gamma(M) + \Gamma(L) \otimes_k k.t] \longrightarrow \Gamma(L \otimes M)
$$

is surjective.

Our argument is essentially the same as the classical argument used to prove that if X is not hyperelliptic, then Ω is normally generated (See Hensel-Landsberg). We can paraphrase this argument in our language as follows:

We begin as before with an exact sequence:

$$0 \longrightarrow \Omega(-\mathfrak{A}) \longrightarrow \Omega \longrightarrow \Omega^* \longrightarrow 0$$

where we now assume that $\mathfrak{A}$ is a positive cycle of degree g-2. In order to apply the 6-lemma, it is not necessary that $H^1(\Omega(-\mathfrak{A})) = (0)$. In fact, it is enough if:

i) $H^1(\Omega(-\mathfrak{A})) \longrightarrow H^1(\Omega)$ is an isomorphism.

This is the 1st requirement on $\mathfrak{A}$. We then deduce as before that

$$\mathcal{S}(\Omega(-\mathfrak{A}),\Omega) \longrightarrow \mathcal{S}(\Omega,\Omega) \longrightarrow \mathcal{S}(\Omega^*,\Omega)$$

is exact. Since $\Gamma(\Omega)$ has no base points, we know that $\mathcal{S}(\Omega^*,\Omega) = (0)$ by Castelnuovo's lemma. By the Riemann-Roch theorem, it follows as before that $\Gamma(\Omega(-\mathfrak{A}))$ is a pencil and our 2nd requirement is that it is base point free. Unfortunately, we cannot apply Castelnuovo's lemma to prove $\mathcal{S}(\Omega(-\mathfrak{A}),\Omega) = (0)$, since $H^1(\Omega \otimes \Omega(-\mathfrak{A})^{-1}) = H^1(\mathfrak{O}(\mathfrak{A}))$ is never (0). We use instead a direct computation of dimensions to prove $\mathcal{S}(\Omega(-\mathfrak{A}),\Omega) = (0)$! Let s_1, s_2 be a basis of $\Gamma(\Omega(-\mathfrak{A}))$. Look at the map:

$$\underbrace{\Gamma(\Omega)s_1 \oplus \Gamma(\Omega)s_2}_{\dim = 2g} \xrightarrow{\ \alpha\ } \underbrace{\Gamma(\Omega^2(-\mathfrak{A}))}_{\dim = 2g-1}$$

(The dimension on the right is computed by the Riemann-Roch theorem.) We want α to be surjective. But the kernel will be isomorphic to the spaces of pairs $\omega_1, \omega_2 \in \Gamma(\Omega)$ such that $\omega_1 \otimes s_1 = -\omega_2 \otimes s_2$. Since s_1 and s_2 have no common zeroes, this implies that ω_1 is zero at the zeroes $\mathfrak{b}_2$ of s_2, i.e., $\omega_1 = \eta \otimes s_2$ where $\eta \in \Gamma(\Omega(-\mathfrak{b}_2))$. Then ω_2 is necessarily $-\eta \otimes s_1$, so

$$\mathrm{Ker}(\alpha) \cong \Gamma(\Omega(-\mathfrak{b}_2)).$$

Since $\mathcal{O}(\mathfrak{b}_2) \cong \Omega(-\mathfrak{A})$, it follows that

$$\begin{aligned} \dim \mathrm{Ker}(\alpha) &= \dim \Gamma(\Omega \otimes \Omega(-\mathfrak{A})^{-1}) \\ &= \dim \Gamma(\mathcal{O}(\mathfrak{A})) \\ &= \dim H^1(\Omega(-\mathfrak{A})) = 1. \end{aligned}$$

Therefore α is surjective, hence $\mathcal{S}(\Omega,\Omega) = (0)$.

Now let $\Omega(-\mathfrak{A}) = K$. K is a sheaf of degree g, and conversely every sheaf K of degree g such that $\dim \Gamma(K) \geq 2$ has the property $\dim \Gamma(\Omega \otimes K^{-1}) \geq 1$ by Riemann-Roch, hence $\Omega \otimes K^{-1} \cong \mathcal{O}(\mathfrak{A})$, some $\mathfrak{A}$, hence $K \cong \Omega(-\mathfrak{A})$, some $\mathfrak{A}$. Therefore we have proven:

<u>Theorem 7</u>: If X carries an invertible sheaf K of degree g such that $\Gamma(K)$ is a base point free pencil, then $\mathcal{S}(\Omega,\Omega) = (0)$.

The existence of such a K is not hard to show whenever X is <u>not</u> hyperelliptic, but we omit this. The proof that $\mathcal{S}(\Omega,\Omega^i) = (0)$ if $i \geq 2$, is even easier.

Theorem 6 for the vanishing of $\mathcal{S}$ is definitely the best possible unless further restrictions are placed on L and M. For example, if $L = \Omega(P+Q)$, then although L is ample and $\Gamma(L)$ has no base points, $\phi_L(P) = \phi_L(Q)$, so L is not very ample. Since L^2 is very ample, there must be sections $s \in \Gamma(L^2)$ such that $s(P) = 0$, $s(Q) \neq 0$, hence $s \notin \mathrm{Im}(\Gamma(L) \otimes \Gamma(L))$. Therefore $\mathcal{S}(L,L) \neq (0)$!

We now go on to results about $\mathcal{R}$ for curves. I don't think, unfortunately, that my results here are best possible. I shall prove:

<u>Theorem 8</u>: **Let L,M,N be invertible sheaves on X such that deg $L \geq 3g+1$, deg M, deg $N \geq 2g+2$. Then**

$$\mathcal{R}(L,M) \otimes \Gamma(N) \longrightarrow \mathcal{R}(L \otimes N,M)$$

is surjective.

From this we deduce immediately:

<u>Corollary</u>: Let L be an invertible sheaf on X such that deg $L \geq 3g+1$. Then L is normally presented.

<u>Proof of the Theorem</u>: We shall use the following lemma:

<u>Lemma</u>: For all invertible sheaves N on X such that deg $N \geq 2g+2$ and $\Gamma(N)$ has no base points, there is a decomposition:

$$N = N_1 \otimes \cdots \otimes N_k, \qquad k \geq 2$$

where

(1) $\deg N_i = g+1, \quad 1 \leq i \leq k-1$

$$g+1 \leq \deg N_k \leq 2g+1,$$

(2) $\Gamma(N_i)$ has no base points.

(3) If J_1 (resp. J^*) is the variety of invertible sheaves of degree = degree N_1 (resp. deg N_1 - deg N_2), then for all open sets $U_1 \subset J_1$, $U^* \subset J^*$, we may assume

$$N_1 \in U_1$$

$$N_1 \otimes N_2^{-1} \in U^* .$$

__Proof__: If deg $N \leq 2g+1$, then let $k = 1$, $N_1 = N$. Now suppose deg $N = e + (g+1)$, $g+1 \leq e \leq 2g+1$. Then $k = 2$ and we must decompose $N = N_1 \otimes N_2$, deg $N_1 = g+1$, deg $N_2 = e$. Let J_2 be the variety of invertible sheaves of degree = e. Let $V_i \subset J_i$ be the set of invertible sheaves K such that $H^1(K) = (0)$ and $\Gamma(K)$ has no base points. It is well known that V_i is open and non-empty. Consider the maps:

$$f: J_1 \longrightarrow J_2, \quad \text{given by } N_1 \longmapsto N \otimes N_1^{-1}$$

$$g: J_1 \longrightarrow J^*, \quad \text{given by } N_1 \longmapsto N_1^2 \otimes N^{-1} .$$

If identity points are chosen arbitrarily on J_1, J_2, J^*, then all these varieties are canonically the same, and are nothing but the jacobian of X. Then in terms of the group law on the jacobian f becomes a map of the form $x \longmapsto a-x$, and g is of the form $x \longmapsto 2x+b$. Thus both f and g are surjective. In particular, $f^{-1}(V_2)$ and $g^{-1}(U^*)$ are non-empty. Now choose $N_1 \in U_1 \cap V_1 \cap f^{-1}(V_2) \cap g^{-1}(U^*)$, and let $N_2 = N \otimes N_1^{-1}$. Then N_1 and N_2 have all the required properties.

If $k > 2$, the proof is similar, but even simpler.

QED

To prove Theorem 8, begin by decomposing the N in the Theorem by the method of the lemma. It clearly will suffice to prove:

$$\mathcal{R}(L \otimes N_1 \otimes \cdots \otimes N_i, M) \otimes \Gamma(N_{i+1}) \longrightarrow \mathcal{R}(L \otimes N_1 \otimes \cdots \otimes N_{i+1}, M)$$

is surjective, for every i with $0 \leq i \leq k-1$. Checking degrees here, we find that we have reduced the Theorem to:

(A) If $\Gamma(N)$ has no base points, $g+1 \leq \deg N \leq 2g+1$, $\deg M \geq 2g+2$, and $\deg L - \deg N \geq 2g$, then

$$\mathcal{R}(L,M) \otimes \Gamma(N) \longrightarrow \mathcal{R}(L \otimes N, M)$$

is surjective.

We now want to apply the Proposition in §1 to interchange M and N in (A). Since $H^1(L \otimes N^{-1}) = (0)$, hence $\mathcal{S}(L,N) = (0)$, (A) is implied by:

(B) If $\Gamma(N)$ has no base points, $g+1 \leq \deg N \leq 2g+1$, $\deg M \geq 2g+2$, and $\deg L - \deg N \geq 2g$, then

$$\mathcal{R}(L,N) \otimes \Gamma(M) \longrightarrow \mathcal{R}(L \otimes M, N)$$

is surjective.

Now decompose M by the method of the lemma. To prove (B) it will suffice to prove:

(i) $\mathcal{R}(L,N) \otimes \Gamma(M_1) \longrightarrow \mathcal{R}(L \otimes M_1, N)$ surjective

(ii) $\mathcal{R}(L\otimes M_1,N) \otimes \Gamma(M_2) \longrightarrow \mathcal{R}(L \otimes M_1 \otimes M_2, N)$ surjective

---

(k) $\mathcal{R}(L \otimes M_1 \otimes \ldots \otimes M_{k-1}, N) \otimes \Gamma(M_k) \longrightarrow \mathcal{R}(L \otimes M_1 \otimes \cdots \otimes M_k, N)$ surjective.

We want to apply Theorem 4 to prove these facts. Since $\Gamma(N)$ and $\Gamma(M_i)$ are base point free, we need only check:

(i) $H^1(L \otimes N^{-1} \otimes M_1^{-1}) = (0)$

(ii) $H^1(L \otimes N^{-1} \otimes M_1 \otimes M_2^{-1}) = (0)$

(k) $H^1(L \otimes N^{-1} \otimes M_1 \otimes \cdots \otimes M_{k-1} \otimes M_k^{-1}) = (0)$.

Now $\deg(L\otimes N^{-1}\otimes M_1^{-1}) \geq 2g-(g+1) = g-1$, so if M_1 lies in a suitable open subset of the Jacobian, (i) will hold. Secondly, $\deg(L\otimes N^{-1}\otimes M_1\otimes M_2^{-1}) \geq 2g + (g+1)-(2g+1) = g$, so if $M_1 \otimes M_2^{-1}$ lies in a suitable open subset of the Jacobian, (ii) will hold. Since the lemma allows us to choose M_1 and $M_1\otimes M_2^{-1}$ in any open sets, (i) and (ii) can be achieved. As for the rest, if, for instance, $k \geq 3$,

$$\deg(L\otimes N^{-1}\otimes M_1\otimes M_2\otimes M_3^{-1}) \geq 2g + (g+1) + (g+1) - (2g+1) = 2g+1$$

so (iii) is automatic. The same holds for all the rest. Thus (B) is proven, hence (A), hence the Theorem. QED

§3. Abelian varieties: the method of theta-groups.

By definition, an abelian variety is a projective variety with a structure of a group such that $(x,y) \longmapsto x+y$ and $x \longmapsto -x$ are morphisms $X \times X \longrightarrow X$ and $X \longrightarrow X$ respectively. We first recall various basic facts about invertible sheaves on such varieties.

(I.) For every X, there is a 2nd abelian variety $\hat{X}$, called its dual, and an invertible sheaf P on $X \times \hat{X}$, called the Poincaré sheaf such that $P|_{X \times \{0\}} \cong \mathcal{O}_X$, $P|_{\{0\} \times \hat{X}} \cong \mathcal{O}_{\hat{X}}$, which is characterized by the non-degeneracy properties:

(a) If $Z \subset \hat{X}$ is a subscheme such that $P|_{X \times Z} \cong \mathcal{O}_{X \times Z}$, then $Z = \{0\}$ with reduced structure,

(b) If $Z \subset X$ is a subscheme such that $P|_{Z \times \hat{X}} \cong \mathcal{O}_{Z \times \hat{X}}$, then $Z = \{0\}$ with reduced structure.

(II.) If Pic(X) is the group of all invertible sheaves on X, there is a subgroup $\mathrm{Pic}^0(X)$ characterized by the property:

$$L \in \mathrm{Pic}^0(X) \iff T_x^* L \cong L, \quad \text{all } x \in X$$

where $T_x: X \longrightarrow X$ is the map $T_x(y) = x+y$. For all $a \in \hat{X}$, let $P_a = P|_{X \times \{a\}}$, an invertible sheaf on X. Then for all $a \in \hat{X}$, $P_a \in \mathrm{Pic}^0(X)$, and $a \longmapsto P_a$ defines an isomorphism of groups:

$$\hat{X} \cong \mathrm{Pic}^0(X).$$

(III.) For all invertible sheaves L on X, and $x,y \in X$,

$$T_{x+y}^{*}L \otimes L \cong T_x^*L \otimes T_y^*L.$$

Therefore $T_x^*L \otimes L^{-1} \in \mathrm{Pic}^o(X)$ and there is a unique homomorphism $\phi_L : X \longrightarrow \hat{X}$ characterized by:

$$P_{\phi_L(x)} \cong T_x^*L \otimes L^{-1}.$$

(IV.) The Riemann-Roch theorem for abelian varieties asserts: if $L = \mathcal{O}(D)$, D a divisor on X, then

$$\chi(L) = (D^g)/g! = \pm\sqrt{\deg \phi_L}$$

If this number is not 0, L is said to be non-degenerate. Then there is exactly one i, called the index of L, for which $H^i(L) \neq (0)$. In particular, if L is ample, then

$$\chi(L) = \dim \Gamma(L) > 0,$$
$$H^i(L) = (0),\ i \geq 1.$$

These facts are all more or less well known. Detailed proofs can be found, for example, in my book "Abelian Varieties", to be published by Oxford University Press in the series "Tata Institute Studies in Mathematics." We require, in addition, another invariant of invertible sheaves, which I call its theta-group. We treat this group first set-theoretically:

<u>Definition</u>: $\mathcal{G}(L)$ = the set of all pairs (x,ϕ), where $x \in X$ and $\phi: L \longrightarrow T_x^*L$ is an isomorphism.

The group law is given by:

$$(x,\phi) \cdot (y,\psi) = (x+y,\ T_y^*\phi \cdot \psi)$$

$$L \xrightarrow{\ \psi\ } T_y^*L \xrightarrow{\ T_y^*\phi\ } T_{x+y}^*L$$

It is easy to see that if $K(L) = \ker(\phi_L)$, then this groups fits into an exact sequence:

$$1 \longrightarrow k^* \xrightarrow{\ i\ } \mathcal{G}(L) \xrightarrow{\ \pi\ } K(L) \longrightarrow 1.$$

if $\quad i(\lambda) = (0, \text{mult. by } \lambda),$

$\quad\quad \pi(x,\phi) = x.$

Moreover, $i(k^*)$ commutes with everything in $\mathcal{G}(L)$. If, instead of using invertible sheaves, we spoke of line bundles, $\mathcal{G}(L)$ would be just the group of automorphisms of L that cover translations of X. Or if we use the language of divisors and divisor classes, then:

$\mathcal{G}(\mathcal{O}_X(D))$ = the set of pairs (x,f), $f \in k(X)$, such that

$$T_x^{-1}D = D + (f)$$

$((f)$ = divisor of poles and zeroes of $f)$.

The group law in this version is:

$$(x,f) \quad (y,g) = (x+y,\ T_y^*f \cdot g).$$

This group is well known in one case: if $L \in \mathrm{Pic}^o(X)$. In this case, $\phi_L \equiv 0$, so $K(L) = X$ and $\mathcal{G}(L)$ is an extension:

$$1 \longrightarrow k^* \longrightarrow \mathcal{G}(L) \longrightarrow X \longrightarrow 1.$$

Serre has studied this case, and has shown that $\mathcal{G}(L)$ is abelian, has a natural structure of algebraic group itself, and that

$$L \longmapsto \mathcal{G}(L)$$

defines an isomorphism:

$$\mathrm{Pic}^o(X) \longrightarrow \mathrm{Ext}^1(X, \mathbb{G}_m).$$

One can describe the non-commutativity of $\mathcal{G}(L)$ conveniently as follows: look at the commutators $xyx^{-1}y^{-1}$. Since $K(L)$ is abelian, $\pi(xyx^{-1}y^{-1}) = 1$, and $xyx^{-1}y^{-1} \in k^*$. Moreover, since $k^* \subset$ center $(\mathcal{G}(L))$, if we alter x or y by an element of k^*, $xyx^{-1}y^{-1}$ does not change. Therefore there is a map:

$$e_L: \; K(L) \times K(L) \longrightarrow k^*$$

$$\text{such that } xyx^{-1}y^{-1} = e(\pi x, \pi y), \text{ all } x,y \in \mathcal{G}(L).$$

It is easy to check that e_L is bi-multiplicative and skew-symmetric.

In treating characteristic p, we need more than a set-theoretic group $\mathcal{G}(L)$ we need a full group <u>scheme</u> $\mathcal{G}(L)$. This is defined by asking that the S-valued points of $\mathcal{G}(L)$, for every scheme S/k should be functorially isomorphic to the groups of pairs (x,ϕ), where x is an S-valued point of X, and if $T_x: X \times S \longrightarrow X \times S$ is translating by x, then

$$\phi: L \otimes \mathcal{O}_S \longrightarrow T_x^*(L \otimes \mathcal{O}_S)$$

is an isomorphism. It fits into an exact sequence of group schemes:

$$1 \longrightarrow \mathbb{G}_m \xrightarrow{\ i\ } \mathcal{G}(L) \xrightarrow{\ \pi\ } K(L) \longrightarrow 1$$

where π is smooth and surjective, and $\mathbb{G}_m$ is the kernel of ϕ_L in the category of group schemes. For details, see the last § of my book on Abelian Varieties.

The theta-group $\mathcal{G}(L)$ acts in a natural way on the cohomology groups $H^i(L)$. In fact, if $(x,\phi) \in \mathcal{G}(L)$, then define the automorphism of $H^i(L)$:

$$U_{(x,\phi)}: H^i(L) \xrightarrow[\approx]{T_x^*} H^i(T_x^*L) \xleftarrow[\approx]{H^i(\phi)} H^i(L).$$

This gives a representation of $\mathcal{G}(L)$ and it works equally well for group schemes or for ordinary groups.

I propose to divide the rest of this section in half: I shall look first in characteristic 0, where only the set-theoretic $\mathcal{G}(L)$'s are needed, and prove a theorem for these; I will then discuss the extension to characteristic p.

So let char(k) = 0 now. First we need some pure group theory: Let K be a finite abelian group, and let $\mathcal{G}$ be a central extension:

$$1 \longrightarrow k^* \longrightarrow \mathcal{G} \longrightarrow K \longrightarrow 1.$$

Call $\mathcal{G}$ <u>non-degenerate</u> if k^* is exactly the center of $\mathcal{G}$. Then if $\mathcal{G}$ is non-degenerate:

(1) explicitly, $\mathcal{G}$ has the form $\mathcal{G} \cong k^* \times A \times \hat{A}$, where A is a finite abelian group, $\hat{A} = \mathrm{Hom}(A,k^*)$, and multiplication is

$$(\lambda,x,\xi)\cdot(\mu,y,\eta) = (\lambda\mu\eta(x),\ x+y,\ \xi+\eta).$$

(2) $\mathcal{G}$ has a unique irreducible representation V in which k^* acts by its natural character. All such representations are sums of V with itself. For $k^* \times A \times \hat{A}$, this representation can be realized by:

$$V = k\text{-valued functions on } A$$

$$U_{(\lambda,x,\xi)}f\,(y) = \lambda\cdot\xi(y)\cdot f(x+y),\quad \forall\ f \in V.$$

(3) If $H \subset \mathcal{G}$ is an abelian subgroup such that $H \cap k^* = \{1\}$, then we can decompose the irreducible representation V in (2) according to the characters of H:

$$V = \bigoplus_{\lambda \in \hat{H}} V_\lambda .$$

Then each V_λ is non-empty, and if $\mathcal{G}'$ is the centralizer of H in $\mathcal{G}$, then

$$\mathcal{G}'/\{\lambda(x)^{-1}\cdot x \mid x \in H\}$$

acts on V_λ, is again a non-degenerate extension, and V_λ is its irreducible representation.

This is all elementary group theory and is easy enough to prove. (See my paper "On the equations defining abelian varieties", Inv. Math., vol. 1). The key result is:

<u>Theta-structure theorem</u>: If L is non-degenerate of index i, then $\mathcal{G}(L)$ is a non-degenerate extension and $H^i(L)$ is its unique irreducible representation, with k^* acting naturally.

We now prove in characteristic 0:

<u>Theorem 9</u>: Let L be an ample invertible sheaf on an abelian variety X. Then for all $\alpha, \beta \in \hat{X}$, all $n, m \geq 4$

$$\mathcal{S}(L^n \otimes P_\alpha, L^m \otimes P_\beta) = (0).$$

<u>Proof</u>: We require the preliminary fact:

<u>Lemma</u>: Let L and M be invertible sheaves on an abelian variety such that $\Gamma(L) \neq (0)$, $\Gamma(M) \neq (0)$, and $L \otimes M$ is ample. Then

$$\sum_{\alpha \in \hat{X}} \Gamma(L \otimes P_\alpha) \otimes \Gamma(M \otimes P_{-\alpha}) \longrightarrow \Gamma(L \otimes M)$$

is surjective.

<u>Proof of lemma</u>: If W is the image, let us show that W is invariant under the action of $\mathcal{G}(L \otimes M)$. Note that if $x \in K(L \otimes M)$, then

$$\phi_L(x) + \phi_M(x) = \phi_{L \otimes M}(x) = 0.$$

Therefore if $\beta = \alpha + \phi_L(x) = \alpha - \phi_M(x)$,

$$\begin{aligned} T_x^*(L \otimes P_\alpha) &\cong L \otimes [T_x^* L \otimes L^{-1}] \otimes P_\alpha \\ &\cong L \otimes P_{\phi_L(x)} \otimes P_\alpha \\ &\cong L \otimes P_\beta \end{aligned}$$

and

$$\begin{aligned} T_x^*(M \otimes P_{-\alpha}) &\cong M \otimes [T_x^* M \otimes M^{-1}] \otimes P_{-\alpha} \\ &\cong M \otimes P_{\phi_M(x)} \otimes P_{-\alpha} \\ &\cong M \otimes P_{-\beta} \,. \end{aligned}$$

Therefore we get a diagram:

$$\begin{array}{ccc} \Gamma(L\otimes P_\alpha) \otimes \Gamma(M\otimes P_{-\alpha}) & \longrightarrow & \Gamma(L\otimes M) \\ \downarrow T_x^* \qquad \downarrow T_x^* & & \big| T_x^* \\ \Gamma(T_x^*(L\otimes P_\alpha))\otimes\Gamma(T_x^*(M\otimes P_{-\alpha})) & \longrightarrow & \Gamma(T_x^*(L\otimes M)) \\ \wr\Vert \qquad \wr\Vert & & \wr\Vert \\ \Gamma(L\otimes P_\beta) \otimes \Gamma(M\otimes P_{-\beta}) & \longrightarrow & \Gamma(L\otimes M) \end{array}$$

In other words, under the action of an element $(x,\phi) \in \mathcal{G}(L\otimes M)$ on $\Gamma(L\otimes M)$, the image of $\Gamma(L\otimes P_\alpha) \otimes \Gamma(M\otimes P_{-\alpha})$ is taken into the image of $\Gamma(L\otimes P_\beta) \otimes \Gamma(M\otimes P_{-\beta})$. Therefore W is $\mathcal{G}(L\otimes M)$-invariant.

Now since $\Gamma(L\otimes M)$ is $\mathcal{G}(L\otimes M)$-irreducible, either $W = (0)$ or $W = \Gamma(L\otimes M)$. But if $s \in \Gamma(L)$, $s \neq 0$ and $t \in \Gamma(M)$, $t \neq 0$, then $s\otimes t \in \Gamma(L\otimes M)$ is not 0: so $W \neq (0)$.

QED

Returning to the theorem, we use the lemma to reduce the proof of the theorem to the special case $n = m = 4$. In fact, consider the diagram:

$$\begin{array}{ccc}
\sum_{\gamma\in\hat{X}} \Gamma(L^n\otimes P_\alpha)\otimes\Gamma(L^{m-1}\otimes P_{\beta+\gamma})\otimes\Gamma(L\otimes P_{-\gamma}) & \xrightarrow{a} & \sum_{\gamma\in\hat{X}} \Gamma(L^{n+m-1}\otimes P_{\alpha+\beta+\gamma})\otimes\Gamma(L\otimes P_{-\gamma}) \\
\downarrow b & & \downarrow c \\
\Gamma(L^n\otimes P_\alpha)\otimes\Gamma(L^m\otimes P_\beta) & \xrightarrow[d]{} & \Gamma(L^{n+m}\otimes P_{\alpha+\beta}).
\end{array}$$

By the lemma, c is surjective. By induction on n and m, a is surjective. Therefore d is surjective.

Now assume $n = m = 4$. We must show that the map:

$$\tau:\ \Gamma(L^4\otimes P_\alpha)\otimes\Gamma(L^4\otimes P_\beta)\longrightarrow\Gamma(L^8\otimes P_{\alpha+\beta})$$

is surjective. We need first some simple remarks. One is that if

1 L is any non-degenerate sheaf, then there is a natural isomorphism:

$$\mathcal{G}(L\otimes P_\alpha)\cong\mathcal{G}(L),\qquad \text{all } \alpha\in\hat{X}.$$

In fact, consider the diagram:

$$\begin{array}{ccccccccc}
1 & \longrightarrow & k^* & \longrightarrow & \mathcal{G}(P_\alpha) & \xrightarrow{\pi} & X & \longrightarrow & 1 \\
 & & & & & \nwarrow{\scriptstyle\rho_\alpha} & \cup & & \\
 & & & & & & K(L) & &
\end{array}$$

Since k^* is a divisible group, and $\mathcal{G}(P_\alpha)$ is abelian, it is easy to check that there is a homomorphism ρ_α such that $\pi\cdot\rho_\alpha = \mathrm{id}^*$. In other

---

*If $0\longrightarrow A\longrightarrow B\longrightarrow C\longrightarrow 0$ is an extension of abelian groups, then it splits whenever A is divisible.

words, the extension $\mathcal{G}(P_\alpha)$ <u>splits</u> over $K(L)$. Then for all $(x,\phi) \in \mathcal{G}(L)$, where $\phi\colon L \longrightarrow T_x^* L$ is an isomorphism, we get an isomorphism

$$L \otimes P_\alpha \xrightarrow{\ \phi\otimes\rho_\alpha(x)\ } T_x^*(L \otimes P_\alpha),$$

hence an element $(x, \phi\otimes\rho_\alpha(x)) \in \mathcal{G}(L \otimes P_\alpha)$.

The second remark is that τ is, in a certain sense, $\mathcal{G}(L^4)$-linear. In fact, define $\delta\colon \mathcal{G}(L^4) \longrightarrow \mathcal{G}(L^8)$ by

$$\delta(x,\phi) = (x, \phi^{\otimes 2})$$

where $\phi^{\otimes 2}\colon L^8 \longrightarrow T_x^* L^8$ is just $\phi \otimes \phi$.
Note that δ fits into a diagram:

$$\begin{array}{ccccccccc}
1 & \longrightarrow & k^* & \longrightarrow & \mathcal{G}(L^4) & \longrightarrow & K(L^4) & \longrightarrow & 1 \\
 & & \Big\downarrow{\scriptstyle \lambda\mapsto\lambda^2} & & \Big\downarrow{\scriptstyle \delta} & & \cap & & \\
1 & \longrightarrow & k^* & \longrightarrow & \mathcal{G}(L^8) & \longrightarrow & K(L^8) & \longrightarrow & 1 \\
 & & & & & & \cap & & \\
 & & & & & & X & &
\end{array}$$

Now choose splittings:

$$\rho_\alpha\colon K(L^8) \longrightarrow \mathcal{G}(P_\alpha)$$
$$\rho_\beta\colon K(L^8) \longrightarrow \mathcal{G}(P_\beta)$$

and let ρ_α, ρ_β induce a 3rd splitting:

$$\rho_{\alpha+\beta}\colon K(L^8) \longrightarrow \mathcal{G}(P_{\alpha+\beta}).$$

Use $\rho_\alpha, \rho_\beta, \rho_{\alpha+\beta}$ to define isomorphisms $\mathcal{G}(L^4) \cong \mathcal{G}(L^4 \otimes P_\alpha) \cong \mathcal{G}(L^4 \otimes P_\beta)$

and $\mathcal{G}(L^8) \cong \mathcal{G}(L^8 \otimes P_{\alpha+\beta})$. Then it is immediate that via δ, τ is $\mathcal{G}(L^4)$-linear.

The next step is to split $\mathcal{G}(L^4)$ over X_2, the group of points of X of order 2:

$$\begin{array}{ccccccccc} 1 & \longrightarrow & k^* & \longrightarrow & \mathcal{G}(L^4) & \xrightarrow{\pi} & K(L^4) & \longrightarrow & 1 \\ & & & & \nwarrow & & \cup & & \\ & & & & & \rho & X_2 & & \end{array}$$

As in the case of $\mathcal{G}(P_\alpha)$, this is possible if we check that the subgroup $\pi^{-1}(X_2)$ is abelian. But $K(L^4) = \mathrm{Ker}(\phi_{L^4}) = \mathrm{Ker}(4\cdot\phi_L)$, so $x \in K(L^4)$ if and only if $4x \in K(L)$. In particular, $X_4 \subset K(L^4)$. Therefore, if $x_1, x_2 \in X_2$ and $x_2 = 2y_2$, $y_2 \in X_4$, and

$$\begin{aligned} e_{L^4}(x_1,x_2) &= e_{L^4}(x_1,2y_2) \\ &= e_{L^4}(2x_1,y_2) \\ &= e_{L^4}(0,y_2) = 1. \end{aligned}$$

Thus $\pi^{-1}(X_2)$ is abelian and ρ exists. We may now decompose all 3 vector spaces under the action of the abelian group $\rho(X_2)$:

$$\Gamma(L^4 \otimes P_\alpha) = \bigoplus_{\ell \in \hat{X}_2} E_\ell$$

$$\Gamma(L^4 \otimes P_\beta) = \bigoplus_{\ell \in \hat{X}_2} F_\ell$$

$$\Gamma(L^8 \otimes P_{\alpha+\beta}) = \bigoplus_{\ell \in \hat{X}_2} G_\ell$$

Note that $\tau(E_\ell \otimes F_m) \subset G_{\ell+m}$, since τ is, in particular, X_2-linear.

Next, I claim that in $\mathcal{G}(L^8)$, $\delta(\mathcal{G}(L^4))$ is the centralizer of $\delta(\rho(X_2))$. Since $\delta(\mathcal{G}(L^4))$ is exactly the inverse image $\pi^{-1}(K(L^4))$ in $\mathcal{G}(L^8)$, and since e_{L^8} computes the commutators in $\mathcal{G}(L^8)$, this is equivalent to saying: $\forall\, x \in K(L^8)$

$$(*) \qquad x \in K(L^4) \Longleftrightarrow e_{L^8}(x,y) = 1 \qquad y \in X_2.$$

But if $y \in X$, then $y \in K(L^8) \Longleftrightarrow 2y \in K(L^4)$. Since X is divisible, $K(L^4) = 2 \cdot K(L^8)$. Therefore, if we abbreviate $K(L^8) = K$, (*) comes down to the assertion:

$$(**) \qquad \forall\, x \in K, \quad x \in 2K \Longleftrightarrow e(x,y) = 1, \text{ all } y \in K \text{ such that } 2y = 0.$$

Since $\mathcal{G}(L^8)$ is a non-degenerate extension, e is a non-degenerate skew-symmetric form on K, and (**) is clearly true.

We can now apply the 3rd set of statements about non-degenerate extensions that we listed above. We deduce:

1) that each E_ℓ, F_ℓ, G_ℓ is non-empty,

2) that G_ℓ is an irreducible $\delta(\mathcal{G}(L^4))$-module.

The theorem now follows. By (1), choose $s \in E_\ell$, $t \in F_m$ with $s \neq 0$, $t \neq 0$. Then $\tau(s \otimes t)$ is the section $s \otimes t$ of $L^8 \otimes P_{\alpha+\beta}$, which is not zero. So the image of τ contains at least one non-zero element of G_ℓ, for each ℓ. But the image of τ is invariant under $\delta(\mathcal{G}(L^4))$, so by (2), it contains all of G_ℓ. Thus τ is surjective.

QED

Now consider the case $\mathrm{char}(k) = p \neq 0$. To make the proof work we must use the full group scheme $\mathcal{G}(L)$. First we need some theory about group schemes $\mathcal{G}$ which are central extensions of the type:

$$1 \longrightarrow \mathbb{G}_m \longrightarrow \mathcal{G} \longrightarrow K \longrightarrow 1$$

where K is a finite commutative group scheme. As before, we call $\mathcal{G}$ <u>non-degenerate</u> if $\mathbb{G}_m$ is the full scheme-theoretic center of $\mathcal{G}$ (i.e., $\forall$ S-valued points x of $\mathcal{G}$, if x commutes with all S'-valued points y of $\mathcal{G}$ for all S'/S, then x should be a point of $\mathbb{G}_m$). There is no simple structure theorem for such $\mathcal{G}$'s. However, they do satisfy:

(2') $\mathcal{G}$ has a unique irreducible representation V in which $\mathbb{G}_m$ acts by its natural character. All such representations are sums of V with itself.

(3') If $H \subset \mathcal{G}$ is an abelian subgroupscheme such that $H \cap \mathbb{G}_m = \{1\}$ scheme-theoretically, and if $R_H = \Gamma(\mathcal{O}_H)$ regarded as a representation of H (the "regular representation"), then $V \cong R_H^m$ for some m as an H-space. In particular, for all characters $\lambda: H \longrightarrow \mathbb{G}_m$, the eigenspace $V_\lambda \subset V$ for λ is non-empty. Moreover, if $\mathcal{G}'$ is the scheme-theoretic centralizer of H in $\mathcal{G}$, then

$$\mathcal{G}'/\{\lambda(x)^{-1}\cdot x \mid x \in H\}$$

acts on V_λ, is again a non-degenerate extension, and V_λ is its irreducible representation.

Note that in (3') $V \supset \oplus V_\lambda$, but if $\mathrm{char}(k) \mid \mathrm{order}(H)$, it is possible that $V \supsetneq \oplus V_\lambda$. In compensation, we have the extra fact, $V \cong R_H^m$.

Next, we still have:

Theta-structure theorem: If L is non-degenerate of index i, then $\mathcal{G}(L)$ is a non-degenerate extension and $H^i(L)$ is its unique irreducible representation, with $\mathbb{G}_m$ acting naturally.

The proofs of these facts are, unfortunately, not yet published. Now let's generalize the proof of Theorem 9 to char(p).

(I.) The lemma remains true. However to prove it, it is necessary to show that for all rings R/k, all R-valued points α of $\mathcal{G}(L\otimes M)$, the automorphism of the R-module $\Gamma(L\otimes M) \underset{k}{\otimes} R$ induced by α takes $W\otimes R$ into itself. This follows as before provided 2
that we first prove the following:

$$(*)\quad \left\{\begin{array}{l} \text{For all } R\text{-valued points } \alpha \text{ of } \hat{X}, \text{ if } P_\alpha \text{ is the invertible} \\ \text{sheaf } (1\times\alpha)^*P \text{ on } X\times\mathrm{Spec}(R)\text{, then the image of the map:} \\ \quad \Gamma(p_1^*L\otimes P_\alpha) \otimes \Gamma(p_1^*M\otimes P_{-\alpha}) \longrightarrow \Gamma(p_1^*(L\otimes M)) \\ \qquad\qquad\qquad\qquad\qquad\qquad\qquad\qquad\qquad \| \\ \qquad\qquad\qquad\qquad\qquad\qquad\qquad\quad \Gamma(L\otimes M)\underset{k}{\otimes} R \\ \text{is contained in } W \underset{k}{\otimes} R. \end{array}\right.$$

First if R is a finitely generated integral domain over k, then the intersection of the maximal ideals in R is (0): so to prove that an element $x \in \Gamma(L\otimes M) \underset{k}{\otimes} R$ is in $W \underset{k}{\otimes} R$ for such an R, it suffices to show that for all homomorphisms $\phi: R \longrightarrow k$, the image $1 \otimes \phi(x) \in \Gamma(L\otimes M)$ is in W. And this is just a case of (*) for a k-valued point of X, i.e., it is part of the hypothesis. But since X is an integral scheme of finite type over k, for any R, and any R-valued point α of X, α is induced by an R'-valued point β of X via a homomorphism $R' \longrightarrow R$, with R' an integral domain finitely generated over k. And if (*) is true for β, it follows immediately for α. This proves (*) in general.

(II.) Once the lemma is proven, Theorem 9 is reduced to the case $n = m = 4$ exactly as before.

(III.) Next, isomorphisms $\mathcal{G}(L) \cong \mathcal{G}(L \otimes P_\alpha)$, $\alpha \in \hat{X}$, L non-degenerate, can be set up exactly as before. We need only the well-known lemma:

Lemma: If $0 \longrightarrow \mathbb{G}_m \longrightarrow \mathcal{G} \longrightarrow K \longrightarrow 0$ is an abelian extension, and K is a finite group scheme, then $\mathcal{G} \cong \mathbb{G}_m \times K$.

Moreover, we get a homomorphism of group schemes $\delta: \mathcal{G}(L^4) \longrightarrow \mathcal{G}(L^8)$ exactly as before, and τ turns out again to be $\mathcal{G}(L^4)$-linear.

(IV.) Now, if char(k) $\neq 2$, the rest of the proof works over k without alteration: $\mathcal{G}(L^4)$ splits over X_2, the vector spaces $\Gamma(L^4 \otimes P_\alpha)$, $\Gamma(L^4 \otimes P_\beta)$, $\Gamma(L^8 \otimes P_{\alpha+\beta})$ split into eigenspaces, and we apply statement (3') about the group theory of non-degenerate $\mathcal{G}$'s. <u>However</u>, if char(k) = 2, X_2, the kernel of multiplication by 2, is never a reduced group scheme. We can still split $\mathcal{G}(L^4)$ over X_2, and $\delta(\mathcal{G}(L^4))$ is still the centralizer of $\delta(\rho(X_2))$ in $\mathcal{G}(L^8)$, <u>but</u> since the representations of X_2 are not completely reducible,

$$\Gamma(L^4 \otimes P_\alpha) \underset{\neq}{\supset} \bigoplus_{\ell \in \hat{X}_2} E_\ell , \quad \text{etc.}$$

We must finish the proof in a new way. Let W = image of τ. Let $W^\perp \subset \Gamma(L^8 \otimes P_{\alpha+\beta})^*$ be the space of linear maps that kill W. Assume $W \underset{\neq}{\subset} \Gamma(L^8 \otimes P_{\alpha+\beta})$, hence $W^\perp \neq (0)$ Now W and hence $W^\perp$ is invariant under the action of $\mathcal{G}(L^4)$, hence of the action of $\rho(X_2)$. Therefore $W^\perp$ contains an eigenvector for at least one character $\ell \in \hat{X}_2$. Let $G^*_\ell \subset \Gamma(L^8 \otimes P_{\alpha+\beta})^*$ be the eigenspace for the character ℓ. Now $\Gamma(L^8 \otimes P_{\alpha+\beta})^*$ is an irreducible representation space for the opposed group to $\mathcal{G}(L^8)$, i.e., with multiplication reversed, and in this representation $\mathbb{G}_m$ acts by its natural character. Therefore applying statement (3') to this opposed group, it follows that G^*_ℓ is $\mathcal{G}(L^4)$-irreducible. Therefore $W^\perp \supset G^*_\ell$.

Now we must construct something inside W. By (3') for $\mathcal{G}(L^4)$, $\Gamma(L^4 \otimes P_\beta)$ contains a non-zero $\rho(X_2)$-invariant t. For all $s \in \Gamma(L^4 \otimes P_\alpha)$, $s \neq 0$, the element $\tau(s \otimes t) \in \Gamma(L^8 \otimes P_{\alpha+\beta})$ is not zero, so τ defines an isomorphism of $\Gamma(L^4 \otimes P_\alpha) \otimes s$ with a subspace $W_o \subset W$. As a representation space for $\rho(X_2)$, W_o is therefore isomorphic to $\Gamma(L^4 \otimes P_\alpha)$, hence to R_2^m, where R_2 denotes the regular representation of X_2. Since R_2 is an injective object in the category of representations of X_2, it follows that:

$$\Gamma(L^8 \otimes P_{\alpha+\beta}) \cong W_o \oplus \widetilde{W}$$

where $\widetilde{W}$ is also X_2-invariant. Now the dual space to R_2 contains eigenvectors for every character of X_2: so there is an element $x \in W_o^*$ which is an eigenvector for the character ℓ. Extend x to a linear map on $\Gamma(L^8 \otimes P_{\alpha+\beta})$ that is zero on $\widetilde{W}$. Then $x \in G_\ell^*$, but $x \not\equiv 0$ on W, i.e., $x \notin W^\perp$. This is a contradiction.

QED

§4. Abelian varieties: the method of the variable pencil.

First of all, we need some more results about the index of invertible sheaves. For proofs of these results, see my book on Abelian Varieties and the appendix to this paper by George Kempf.

Definition: Let L be a degenerate invertible sheaf on an abelian variety X. Let $K = K(L)^{o}$, the connected component of $K(L)$, $Y = X/K$, and $\pi: X \longrightarrow Y$ the canonical map. Then there is a non-degenerate sheaf M on Y such that $L \cong P_{\alpha} \otimes \pi^{*}M$, some $\alpha \in \hat{X}$ (cf. appendix). We define index (L) to be the interval:

$$[\text{index } (M), \text{ index } (M) + \dim K].$$

The following result is proven in the appendix:

Proposition: If $i \notin$ index (L), then $H^{i}(L) = (0)$.

Now suppose L and M are 2 invertible sheaves on X, and L is ample. Consider the collection of sheaves $L^{p} \otimes M^{q}$ and the polynomial:

$$P(p,q) = \chi(L^{p} \otimes M^{q}).$$

The following theorem is proven in §16 of my book and in the appendix to this paper:

Theorem: If $g = \dim X$, then there are $\alpha_1, \cdots, \alpha_g \in \mathbb{R}$, $\alpha_1 \geq \alpha_2 \geq \cdots \geq \alpha_g$ such that

$$P(x,y) = \prod_{i=1}^{g} (x - \alpha_i y).$$ 3

Moreover, for all $p,q \in \mathbb{Z}$, $q > 0$, if $\alpha_{i+1} = \cdots = \alpha_{i+k} = \frac{p}{q}$,

$\alpha_i > \frac{p}{q} > \alpha_{i+k+1}$, then

$$\text{index } (L^p \otimes M^q) = [i, i+k].$$

The precise result that we need is slightly stronger. I want to assume only that $\Gamma(L) \neq (0)$ (i.e., $L = \pi^* L_o$ for some $\pi: X \longrightarrow X/K$, L_o ample on X/K). In this case, I claim:

<u>Theorem</u>: Suppose L and M are 2 invertible sheaves, $\Gamma(L) \neq (0)$ and M non-degenerate. Then

4 $$P(x,y) = \prod_{i=1}^{r} (x - \alpha_i y) \cdot y^{g-r}$$

for some r and some $\alpha_i \in \mathbb{R}$ with $\alpha_1 \geq \cdots \geq \alpha_r$. For $N >> 0$, let i_o = index $(L^N \otimes M)$. Then for all p,q, $q > 0$, if $\alpha_{i+1} = \cdots = \alpha_{i+k} = p/q$, $\alpha_i > \frac{p}{q} > \alpha_{i+k+1}$, then

$$\text{index } (L^p \otimes M^q) = [i + i_o, i + k + i_o].$$

This theorem is deduced easily from the 1st one, by introducing an ample L_1 and considering all the sheaves $L^p \otimes L_1^{p_1} \otimes M^q$ and the polynomial: $P(p,p_1,q) = \chi(L^p \otimes L_1^{p_1} \otimes M^q)$. We omit this step.

The purpose of this section is to prove:

<u>Theorem 10</u>: Let X be an abelian variety, L an ample invertible sheaf and $n \geq 4$ an integer. L^n defines an immersion:

$$\phi_{L^n}: X \longrightarrow \mathbb{P}(\Gamma(L^n)) .$$

Then $\phi_{L^n}(X)$ is ideal-theoretically an intersection of quadrics of rank ≤ 4.

Proof: First, let's construct a set of quadrics containing $\phi_{L^n}(X)$. Once and for all, fix p and q, $p,q \geq 2$, such that $n = p+q$. Consider the map:

$$\Gamma(L^p \otimes P_\alpha) \otimes \Gamma(L^q \otimes P_{-\alpha}) \longrightarrow \Gamma(L^n).$$

If $s \in \Gamma(L^p \otimes P_\alpha)$, $t \in \Gamma(L^q \otimes P_{-\alpha})$, let the induced section $s\otimes t$ of $\Gamma(L^n)$ be denoted $\langle s,t\rangle$ to prevent a confusion of notation. Tnen for all $s_1,s_2 \in \Gamma(L^p \otimes P_\alpha)$, $t_1,t_2 \in \Gamma(L^q \otimes P_{-\alpha})$, we get 4 sections of L^n: $\langle s_i,t_j$, $i,j = 1$ and 2. In $\Gamma(L^{2n})$, we get an identity:

$$\langle s_1,t_1\rangle \otimes \langle s_2,t_2\rangle = \langle s_1,t_2\rangle \otimes \langle s_2,t_1\rangle.$$

Therefore

$$q_{s_1,t_1,s_2,t_2} = \langle s_1,t_1\rangle \otimes \langle s_2,t_2\rangle - \langle s_1,t_2\rangle \otimes \langle s_2,t_1\rangle \in \mathcal{R}(L^n,L^n).$$

If Q_{s_1,t_1,s_2,t_2} is the quadric in $\mathbb{P}(\Gamma(L^n))$ defining by $q_{s_1,t_1,s_2,t_2} = 0$, then we will actually prove:

$$(*)\quad \begin{cases} \phi_{L^n}(X) \text{ is the ideal-theoretic intersection of the quadrics} \\ Q_{s_1,t_1,s_2,t_2} \text{ for all } \alpha, s_i, t_i. \end{cases}$$

For most of this proof, we will deal with the fact that $\phi_{L^n}(X)$ is the set-theoretic intersection of these quadrics. At the end, we will indicate the easy extension of the method to proof that $\phi_{L^n}(X)$ is also an ideal-theoretic intersection.

The first step is to translate (*) into an assertion on X itself, not involving $\mathbb{P}(\Gamma(L^n))$. The points of $\mathbb{P}(\Gamma(L^n))$ correspond to non-zero linear maps $\ell: \Gamma(L^n) \longrightarrow k$, modulo scalars. Fix one such ℓ. Then it is easy to see that the point defined by ℓ lies on Q_{s_1,t_1,s_2,t_2} if and only if

$$\ell(\langle s_1,t_1\rangle)\cdot\ell(\langle s_2,t_2\rangle) = \ell(\langle s_1,t_2\rangle)\cdot\ell(\langle s_2,t_1\rangle).$$

Moreover, it is elementary linear algebra that this holds for all s_1,t_1,s_2,t_2 if and only if there are linear maps:

$$m_\alpha: \Gamma(L^p\otimes P_\alpha) \longrightarrow k$$
$$n_\alpha: \Gamma(L^q\otimes P_{-\alpha}) \longrightarrow k$$

such that:

$$\ell(\langle s,t\rangle) = m_\alpha(s)\cdot n_\alpha(t), \quad \text{all } s \in \Gamma(L^p\otimes P_\alpha),\ t \in \Gamma(L^q\otimes P_{-\alpha}).$$

On the other hand, what does it mean to say that the "point" ℓ is in $\phi_{L^n}(X)$? This means that there is a point $x \in X$, and an isomorphism $L^n \cong \mathcal{O}_X$ near x, such that, evaluating sections by this isomorphism:

$$\ell(s) = s(x), \quad \text{all } s \in \Gamma(L^n).$$

Thus (*) comes down to the assertion:

(**) If $\ell: \Gamma(L^n) \longrightarrow k$ is a non-zero linear map, such that for all $\alpha \in \hat{X}$, there exist linear maps $m_\alpha: \Gamma(L^p\otimes P_\alpha) \longrightarrow k$, $n_\alpha: \Gamma(L^q\otimes P_{-\alpha}) \longrightarrow k$ for which $\ell(\langle s,t\rangle) = m_\alpha(s)\cdot n_\alpha(t)$, then for some $x \in X$, $\ell(s) = s(x)$ all $s \in \Gamma(L^n)$.

In order to prove (**), the basic idea is to treat all α simultaneously, i.e., to put the m_α's and n_α's together into a single homomorphism. In fact, consider the invertible sheaves:

$$p_1^*L^p \otimes P \quad \text{and} \quad p_1^*L^q \otimes P^{-1} \quad \text{on} \quad X \times \hat{X}.$$

These have the property:

$$p_1^*L^p \otimes P\big|_{X\times(\alpha)} \cong L^p \otimes P_\alpha; \quad p_1^*L^q \otimes P^{-1}\big|_{X\times(\alpha)} \cong L^q \otimes P_{-\alpha}.$$

Define

$$E_p = p_{2,*}(p_1^*L^p \otimes P)$$

$$F_q = p_{2,*}(p_1^*L^q \otimes P^{-1}).$$

Since the higher cohomology groups of $L^p \otimes P_\alpha$, $L^q \otimes P_{-\alpha}$ are zero, E_p and F_q are locally free sheaves on S such that

$$E_p \otimes k(\alpha) \cong \Gamma(L^p \otimes P_\alpha); \quad F_q \otimes k(\alpha) \cong \Gamma(L^q \otimes P_{-\alpha}).$$

There is a natural pairing:

$$E_p \otimes_{\mathcal{O}_{\hat{X}}} F_q \longrightarrow p_{2,*}(p_1^*L^n) \cong \Gamma(L^n) \otimes_k \mathcal{O}_{\hat{X}}.$$

This is the globalized form of the individual pairings $\Gamma(L^p \otimes P_\alpha) \otimes \Gamma(L^q \otimes P_{-\alpha}) \longrightarrow \Gamma(L^n)$. In order to go further, we need:

Lemma 1: If $\ell: \Gamma(L^n) \longrightarrow k$ satisfies the condition of (**), then for all α, ℓ does not vanish identically on the image of $\Gamma(L^p \otimes P_\alpha) \otimes \Gamma(L^q \otimes P_{-\alpha})$ in $\Gamma(L^n)$.

We will prove the lemma later. Assuming this, we next globalize the m_α and n_α as follows: I claim there is an invertible sheaf K on $\hat{X}$ and surjective homomorphisms:

$$m:\ E_p \longrightarrow K$$
$$n:\ F_q \longrightarrow K^{-1}$$

such that the diagram:

$$\begin{array}{ccc} E_p \otimes F_q & \longrightarrow & \Gamma(L^n) \otimes_k \mathcal{O}_{\hat{X}} \\ \downarrow m\otimes n & & \downarrow \ell\otimes 1 \\ K\otimes K^{-1} & \xrightarrow{\approx} & \mathcal{O}_{\hat{X}} \end{array}$$

commutes. To see this, consider the composite map:

$$E_p \otimes F_q \longrightarrow \Gamma(L^n) \otimes \mathcal{O}_{\hat{X}} \xrightarrow{\ell\otimes 1} \mathcal{O}_{\hat{X}}.$$

It induces a map of locally free sheaves:

$$m':\ E_p \longrightarrow \underline{\mathrm{Hom}}\ (F_q, \mathcal{O}_{\hat{X}})\ .$$

By the hypothesis in (**), this map, after taking $\otimes k(\alpha)$, is always of rank 0 or 1; by lemma 1, it never has rank 0. Therefore, its image is an invertible subsheaf K of $\underline{\mathrm{Hom}}(F_q, \mathcal{O}_{\hat{X}})$ which is locally a direct summand. m gives a surjective homomorphism m: $E_p \longrightarrow K$. On the other hand, the inclusion of K in $\underline{\mathrm{Hom}}(F_q, \mathcal{O}_{\hat{X}})$ induces a surjection:

$$n:\ F_q = \underline{\mathrm{Hom}}(\underline{\mathrm{Hom}}(F_q, \mathcal{O}_{\hat{X}}), \mathcal{O}_{\hat{X}}) \longrightarrow \underline{\mathrm{Hom}}(K, \mathcal{O}_{\hat{X}}) = K^{-1}.$$

It is clear that the sheaf K and the homomorphisms m,n make the diagram above commute.

To motivate the next steps, let's imagine that (**) is true and see what K, m, and n ought to turn out to be. For all $x \in X$, let $Q_x = P|_{\{x\}\times\hat{X}}$. Then Q_x is an invertible sheaf on $\hat{X}$ and, if we pick an isomorphism $L^p \xrightarrow{\sim} \mathcal{O}_X$ in a neighborhood of x, then there is a natural restriction map:

$$p_1^*L^p \otimes P \longrightarrow p_1^*L^p \otimes P|_{\{x\}\times\hat{X}} \cong P|_{\{x\}\times\hat{X}}.$$

This induces a map of locally free sheaves on $\hat{X}$:

$$r_x: E_p \longrightarrow Q_x$$

which is a global form of the linear maps:

$$E_p \otimes k(\alpha) \cong \Gamma(L^p \otimes P_\alpha) \xrightarrow{\text{"evaluation at x"}} k.$$

Similarly there is a map:

$$s_x: F_q \longrightarrow Q_x^{-1}$$

which is a global form of the linear maps:

$$F_q \otimes k(\alpha) \cong \Gamma(L^q \otimes P_{-\alpha}) \xrightarrow{\text{"evaluation at x"}} k.$$

Therefore, what we want to prove is:

$$(***) \begin{cases} K \cong Q_x, \text{ for some } x \in X \\ \text{and } m \text{ is a multiple of } r_x, \quad n \text{ of } s_x. \end{cases}$$

If we prove (***), then it follows immediately that ℓ, as a point of $\mathbb{P}(\Gamma(L^n))$, equals $\phi_{L^n}(x)$. In fact, choosing an isomorphism of L^n and $\mathcal{O}_X$ near x, let $\ell': \Gamma(L^n) \longrightarrow k$ by the evaluation map $s \longmapsto s(x)$.

Then what (***) asserts is that the 2 composite homomorphisms:

$$\begin{matrix} L \\ L'' \end{matrix}: E_p \otimes F_q \longrightarrow \Gamma(L^n) \underset{k}{\otimes} \mathcal{O}_{\hat{X}} \underset{\ell'\otimes 1}{\overset{\ell\otimes 1}{\rightrightarrows}} \mathcal{O}_{\hat{X}}$$

differ by a scalar. Say $L = \lambda.L'$. Then on the image of each map

$$\Gamma(L^p \otimes P_\alpha) \otimes \Gamma(L^q \otimes P_{-\alpha}) \longrightarrow \Gamma(L^n),$$

$\ell = \lambda.\ell'$. By the lemma of §3, these images generate $\Gamma(L^n)$, so $\ell = \lambda.\ell'$ on all of $\Gamma(L^n)$ and the Theorem is proven.

To prove (***), we proceed as follows. First apply Serre duality to the morphism $p_2: X\times\hat{X} \longrightarrow \hat{X}$:

$$\begin{aligned}\Gamma(\hat{X}, \underline{\mathrm{Hom}}(E_p, K)) &= \Gamma(\hat{X}, \underline{\mathrm{Hom}}(p_{2,*}(p_1^*L^p \otimes P), K)) \\ &\cong \Gamma(\hat{X}, R^g_{p_{2,*}}(p_1^*L^{-p} \otimes P^{-1} \otimes p_2^*K))\end{aligned}$$

Since all the cohomology groups of the restriction $p_1^*L^{-p} \otimes P^{-1} \otimes p_2^*K|_{X\times\{\alpha\}} \cong L^{-p} \otimes P_{-\alpha}$ are zero, except for the gth group,

$$R^i p_{2,*}(p_1^*L^{-p} \otimes P^{-1} \otimes p_2^*K) = (0), \quad i \neq g.$$

Therefore, we conclude by the Leray spectral sequence that:

$$\Gamma(\hat{X}, \underline{\mathrm{Hom}}(E_p, K)) \cong H^g(X\times\hat{X}, p_1^*L^{-p} \otimes P^{-1} \otimes p_2^*K)$$

Similarly:

$$\Gamma(\hat{X}, \underline{\mathrm{Hom}}(F_q, K^{-1})) \cong H^g(X\times\hat{X}, p_1^*L^{-q} \otimes P \otimes p_2^*K^{-1})$$

hence by Serre duality on $X\times\hat{X}$:

$$\Gamma(\hat{X}, \underline{Hom}(F_q, K^{-1}))^* \cong H^g(X\times\hat{X}, p_1^*L^q \otimes P^{-1} \otimes p_2^*K).$$

Therefore, we have at our disposal the 2 apparently meagre bits of information:

$$H^g(X\times\hat{X}, p_1^*L^m \otimes P^{-1} \otimes p_2^*K) \neq (0), \quad \text{for } m = -p \text{ and } q.$$

But, amazingly, these facts turn out to trigger a Rube Goldberg-like set of cohomological implications that we will describe later. We summarize this part of the proof for now in:

<u>Lemma 2</u>: Let L be ample on X, K any invertible sheaf on $\hat{X}$. If there exist integers $a,b \geq 2$ such that

$$H^g(X\times\hat{X}, p_1^*L^m \otimes P^{-1} \otimes p_2^*K) \neq (0)$$

for $m = -a$ and b, then, in fact, for <u>all</u> m:

i) $p_1^*L^m \otimes P^{-1} \otimes p_2^*K$ is non-degenerate of index g,

ii) $\dim H^g(p_1^*L^m \otimes P^{-1} \otimes p_2^*K) = 1$,

iii) $K \in Pic^o(\hat{X})$.

But by the theorem of biduality, the invertible sheaf P on $X\times\hat{X}$ makes X into the dual $\hat{\hat{X}}$ of $\hat{X}$ with Poincaré sheaf still P. Therefore, all sheaves in $Pic^o(\hat{X})$ are isomorphic to Q_x, some $x \in X$, hence $K \cong Q_x$, some $x \in X$.

Finally to show that m is a multiple of r_x, and n is a multiple of s_x, it suffices to prove that

$$\dim \Gamma(\hat{X}, \underline{Hom}(E_p, K)) = 1$$
$$\dim \Gamma(\hat{X}, \underline{Hom}(F_q, K^{-1})) = 1.$$

But we saw above that these dimensions equal

$$\dim H^g(X\times\hat{X},\ p_1^*L^{-p} \otimes P^{-1} \otimes p_2^*K)$$

and
$$\dim H^g(X\times\hat{X},\ p_1^*L^{q} \otimes P^{-1} \otimes p_2^*K)$$

and these are both 1 by lemma 2. This proves (***):

We now go on to the lemmas:

<u>Proof of lemma 1</u>: Suppose $\mathfrak{s} \equiv 0$ on the image of $\Gamma(L^p\otimes P_\alpha) \otimes \Gamma(L^q\otimes P_{-\alpha})$. Since $\mathfrak{s}$ is not zero everywhere, and since $\Gamma(L^p\otimes P_\beta) \otimes \Gamma(L^q\otimes P_{-\beta})$ generate $\Gamma(L^n)$ as β varies, choose a point $\gamma \in \hat{X}$ such that

$$\mathfrak{s} \not\equiv 0 \text{ on } \Gamma(L^p\otimes P_{\alpha+\gamma}) \otimes \Gamma(L^q\otimes P_{-\alpha-\gamma})$$

By the hypothesis on $\mathfrak{s}$, $\mathfrak{s}$ on this last space is of the form $m\otimes n$, where $m \not\equiv 0$ and $n \not\equiv 0$. By the same reasoning, for <u>almost all</u> $\delta \in \hat{X}$,

$$m \not\equiv 0 \quad \text{on} \quad \Gamma(L^{p-1} \otimes P_{\alpha+\gamma+\delta}) \otimes \Gamma(L \otimes P_{-\delta}) ,$$

and again for <u>almost all</u> $\delta \in \hat{X}$

$$n \not\equiv 0 \quad \text{on} \quad \Gamma(L^{q-1} \otimes P_{-\alpha+\delta}) \otimes \Gamma(L \otimes P_{-\gamma-\delta}).$$

Choose a δ for which $m \not\equiv 0$ <u>and</u> $n \not\equiv 0$. Then it follows that $\mathfrak{s} \not\equiv 0$ on the image in $\Gamma(L^n)$ of:

$$[\Gamma(L^{p-1}\otimes P_{\alpha+\gamma+\delta}) \otimes \Gamma(L\otimes P_{-\delta})] \otimes [\Gamma(L^{q-1}\otimes P_{-\alpha+\delta}) \otimes \Gamma(L\otimes P_{-\gamma-\delta})].$$

But by interchanging the 2nd and 4th factors, this image is the same as the image in $\Gamma(L^n)$ of:

$$[\Gamma(L^{p-1}\otimes P_{\alpha+\gamma+\delta}) \otimes \Gamma(L\otimes P_{-\gamma-\delta})] \otimes [\Gamma(L^{q-1}\otimes P_{-\alpha+\delta}) \otimes \Gamma(L\otimes P_{-\delta})].$$

The map of this 4-way tensor product into $\Gamma(L^n)$ factors through $\Gamma(L^p\otimes P_\alpha) \otimes \Gamma(L^q\otimes P_{-\alpha})$, so this contradicts the assumption that $\ell \equiv 0$ on the image of this space in $\Gamma(L^n)$.

QED

Proof of lemma 2: This is where we will use the theorem quoted in the beginning of this section. First we compute $H^i(X\times\hat{X},\ P \otimes p_2^*K^{-1})$. Apply the Leray spectral sequence:

$$H^i(\hat{X},\ R^j p_{2,*}(P) \otimes K^{-1}) \Longrightarrow H^{i+j}(X\times\hat{X},\ P \otimes p_2^*K^{-1}).$$

But, as is shown in my book, §13:

$$R^j p_{2,*}(P) = (0), \quad i < g$$

$$R^g p_{2,*}(P) = k(0).$$

Therefore:

$$H^{i+g}(X\times\hat{X},\ P\otimes p_2^*K^{-1}) = \begin{cases} (0), & i < 0 \\ H^i(\hat{X},\ k(0) \otimes K^{-1}), & i \geq 0. \end{cases}$$

Hence $H^i(X\times X,\ P\otimes p_2^*K^{-1}) = (0)$ if $i \neq g$, and is 1-dimensional if $i = g$.

By Serre duality, the same is true of $P^{-1} \otimes p_2^*K$. Now consider the family of sheaves:

$$M_{p,q} = p_1^*L^p \otimes (P^{-1} \otimes p_2^*K)^q$$

and their Euler characteristics:

$$P(p,q) = \chi(M_{p,q}) .$$

We know by the above computation and by our hypothesis that:

$$\text{(I)} \qquad \begin{cases} M_{0,1} \text{ non-degenerate, index} = g \\ g \in \text{index } (M_{b,1}) \\ g \in \text{index } (M_{-a,1}) \end{cases}$$

It follows from the Theorem that $P(x,1)$ has no zeroes in the open interval $-a < x < b$. But now $P(x,1)$ is a real polynomial of x such that

i) P has only real zeroes,

ii) $P(0) = (-1)^g$,

iii) P has no zeroes with $-a < x < b$.

iv) $P(n) \in \mathbb{Z}$, for all $n \in \mathbb{Z}$.

But (i) implies that P has a unique local maximum or minimum between any 2 zeroes: let $-\alpha < 0 < \beta$ $(\alpha, \beta \in \mathbb{R}^+)$ be its zeroes of smallest absolute value. Since $-\alpha < 1 < \beta$, and $|P(1)| \geq 1 = |P(0)|$, P must have a local maximum or minimum between 0 and β; since $-\alpha < -1 < \beta$, and $|P(-1)| \geq 1 = |P(0)|$, P must also have a local maximum or minimum between $-\alpha$ and 0. This is a contradiction — <u>unless</u> P <u>is constant</u>.

Applying the theorem again, it follows that

$$\text{(II)} \qquad \begin{cases} M_{p,q} \text{ non-degenerate} \\ \text{index } (M_{p,q}) = g \\ \dim H^g(M_{p,q}) = 1 \end{cases}$$

for all $p,q \in \mathbb{Z}$, $q \neq 0$. This proves (i) and (ii) of the lemma. To prove (iii), apply the Leray spectral sequence:

If $\mathfrak{F}_i = R^i p_{1,*}(P^{-1} \otimes p_2^* K)$, then

$$E_2^{i,j} = H^i(X, L^m \otimes \mathfrak{F}_j) \Longrightarrow H^{i+j}(X \times \hat{X}, p_1^* L^m \otimes P^{-1} \otimes p_2^* K) .$$

In particular, since L is ample, $E_2^{ij} = (0)$ if $i > 0$, $m >> 0$, hence the spectral sequence reduces to :

$$H^o(X, L^m \otimes \mathfrak{F}_j) \cong H^j(X \times \hat{X}, p_1^* L^m \otimes P^{-1} \otimes p_2^* K), \text{ if } m >> 0.$$

Therefore because of (II) the whole sheaf $\mathfrak{F}_j$ must be zero if $j < g$. The spectral sequence now reduces to

$$H^i(X, L^m \otimes \mathfrak{F}_g) \cong H^{i+g}(X \times \hat{X}, p_1^* L^m \otimes P^{-1} \otimes p_2^* K), \text{ all } m ,$$

hence

$$\chi(L^m \otimes \mathfrak{F}_g) = 1, \quad \text{all } m.$$

This shows first that $\mathrm{Supp}(\mathfrak{F}_g)$ is 0-dimensional, since its Hilbert polynomial is a constant; and second, that $\dim H^o(\mathfrak{F}_g) = 1$, hence $\mathfrak{F}_g \cong k(x)$, some $x \in X$.

Now recall from EGA, Ch. 3, §7 that the cohomology of $P^{-1} \otimes p_2^* K$ along the fibre $\{x\} \times \hat{X}$ of p_1 is computed from the higher direct images by a spectral sequence:

$$\mathrm{Tor}_{-i}^{\mathcal{O}_X}(k(x), R^j_{p_{1,*}}(P^{-1} \otimes p_2^* K)) \Longrightarrow H^{i+j}(P^{-1} \otimes p_2^* K|_{\{x\} \times \hat{X}}) .$$

Since $P^{-1} \otimes p_2^*K\big|_{\{x\}\times\hat{X}} \cong Q_x^{-1} \otimes K$, and since $\mathfrak{J}_j = (0)$, $j < g$, we find

$$H^{g-i}(Q_x^{-1} \otimes K) \cong \mathrm{Tor}_i^{\mathcal{O}_X}(k(x), k(x)).$$

Thus $H^i(Q_x^{-1} \otimes K) \neq (0)$, for all i. For $i = 0$, this gives $\Gamma(Q_x^{-1} \otimes K) \neq (0)$, and for $i = g$, this gives (by Serre duality) $\Gamma(Q_x \otimes K^{-1}) \neq (0)$. Therefore $K \cong Q_x$ hence $K \in \mathrm{Pic}^0(\hat{X})$. <u>QED</u>

This completes the proof that $\phi_{L^n}(X)$ is the set-theoretic intersection of the quadrics Q_{s_1,t_1,s_2,t_2}. To prove that it is also ideal-theoretically equal to this intersection, it is enough, as we remarked in the introduction, to prove that for all $x \in X$, the tangent space to $\phi_{L^n}(X)$ at x is the intersection of the tangent spaces to the quadrics Q_{s_1,t_1,s_2,t_2} at x. Equivalently, let $R = k[\epsilon]/(\epsilon^2)$: then we must prove that for all R-valued points x of $\mathbb{P}(\Gamma(L^n))$, x is in $\phi_{L^n}(X)$ if and only if x is in all the quadrics. But such a point x is defined by a k-linear map $\lambda: \Gamma(L^n) \longrightarrow R$ such that $\mathrm{Image}(\lambda) \not\subset k\cdot\epsilon$. Translating suitably the conditions that x is in $\phi_{L^n}(X)$ and in the quadrics, we find that the assertion to be proven comes out as:

(**) If $\ell: \Gamma(L^n) \longrightarrow R$ is a k-linear map with $\mathrm{Im}(\ell) \not\subset k\cdot\epsilon$, such that for all $\alpha \in \hat{X}$, there exist linear maps $m_\alpha: \Gamma(L^p \otimes P_\alpha) \longrightarrow R$ and $n_\alpha: \Gamma(L^q \otimes P_{-\alpha}) \longrightarrow R$ for which $\ell(\langle s,t\rangle) = m_\alpha(s).n_\alpha(t)$, then for some R-valued point x of X, $\ell(s) = s(x)$, all $s \in \Gamma(L^n)$.

This is proven by a straightforward generalization of our proof for k-valued points. Lemma 1 is unchanged and one finds first an invertible sheaf K on $\hat{X}\times\mathrm{Spec}(R)$ and surjective homomorphisms:

$$m: E_p \otimes_k R \longrightarrow K$$

$$n: F_q \otimes_k R \longrightarrow K^{-1}$$

on $\hat{X} \times \mathrm{Spec}(R)$ which globalize m_α and n_α. For all R-valued points $x: \mathrm{Spec}(R) \longrightarrow X$ of X, define Q_x on $\hat{X}\times\mathrm{Spec}(R)$ to be the pull-back of P by $x \times 1_{\hat{X}}$. We get restriction maps $r_x: E_p \otimes_k R \longrightarrow Q_x$, $s_x: F_q \otimes_k R \longrightarrow Q_x^{-1}$ as before, and (**) reduces as before to:

(***) $K \cong Q_x$, for some R-valued point x of X, and $m = \mu\cdot r_x$, $n = \nu\cdot s_x$ for some units $\mu,\nu \in R$.

But by our proof for k-valued points, we know already that $K|_X \cong Q_{x_o}$ for some k-valued point x_o of X. Therefore, since Pic^o is an "open" subfunctor of Pic, and since X is the dual of $\hat{X}$, it follows immediately that $K \cong Q_x$ for some R-valued point x of X. To prove the rest of (***), it is only necessary to check that

$$\Gamma(\hat{X} \times \mathrm{Spec}(R), \underline{\mathrm{Hom}}\,(E_p, K)) \cong R$$

$$\Gamma(\hat{X} \times \mathrm{Spec}(R), \underline{\mathrm{Hom}}\,(F_q, K^{-1})) \cong R .$$

Then since the restriction m_o of m to X is a non-zero multiple of r_{x_o}, m must be a unit times r_x; and similarly for n.

As before, we compute:

$$\Gamma(\hat{X}\times\mathrm{Spec}(R), \underline{\mathrm{Hom}}(E_p,k)) \cong H^g(X\times\hat{X}\times\mathrm{Spec}(R), p_1^*L^{-p}\otimes p_{12}^*P^{-1}\otimes p_{23}^*K) .$$

We can then apply the remark:

If L is an invertible sheaf on $Z\times\mathrm{Spec}(R)$ such that $H^i(L|_Z) = (0)$, $i \neq i_o$, then $H^i(L) = (0)$ if $i \neq i_o$ and $H^{i_o}(L)$ is a free R-module.

This completes the proof of Theorem 10.

Appendix*, by George Kempf

Let X be an abelian variety, L an invertible sheaf on X, Y = connected component of K(L), and $p: X \longrightarrow X/Y$ the canonical map.

Theorem 1: (i) If $L|_Y$ is non-trivial, then $H^i(X,L) = (0)$, all i.

(ii) If $L|_Y$ is trivial, there exists a non-degenerate invertible sheaf M on X/Y with $L = p^*M$, and if $i_0 = \text{index}(M)$:

$$H^i(X,L) \cong H^{i_0}(X,M) \otimes H^{i-i_0}(Y,\mathcal{O}_Y), \quad \text{all } i.$$

Proof: The theorem follows from:

Lemma 1: $T_x^*L|_Y \cong L|_Y$ for all $x \in X$, and

Lemma 2: Let $P \xrightarrow{f} Z$ be a principal homogeneous space (in the flat topology) with structure group Y, an abelian variety. Then

$$R^if_*(\mathcal{O}_P) \cong H^i(Y,\mathcal{O}_Y) \otimes \mathcal{O}_Z.$$

By Lemma 1, we see that $L|_Y \in \text{Pic}^0(Y)$ and also that $L|_{x+Y} = L|_{p^{-1}(p(x))}$ is isomorphic to $L|_Y$. Now if $L|_Y$ is non-trivial, then

$$(0) = H^i(Y, L|_Y) = H^i(p^{-1}(p(x)), L|_{p^{-1}(p(x))})$$

for all i (see Mumford, Abelian Varieties, §13). By the theorems on cohomology and base extension, $R^ip_*(L) = (0)$ for all i. The Leray spectral sequence then implies that $H^i(X,L) = (0)$ for all i.

---

* The results in this appendix were independently discovered by C. P. Ramanujam.

If $L|_Y$ is trivial, hence $L|_{p^{-1}(p(x))}$ is trivial, the see-saw principle shows that if $M = p_*(L)$, then M is an invertible sheaf such that $L = p^*(M)$. This M is clearly non-degenerate. Note that:

$$\begin{aligned} R^i p_*(L) &= R^i p_*(p^* M) \\ &\cong R^i p_*(\mathcal{O}_X) \otimes_{\mathcal{O}_{X/Y}} M \\ &\cong H^i(Y, \mathcal{O}_Y) \otimes_k M \qquad \text{(by lemma 2).} \end{aligned}$$

Therefore

$$H^j(X/Y, R^i p_*(L)) \cong H^i(Y, \mathcal{O}_Y) \otimes_k H^j(X/Y, M),$$

and this is zero unless $j = i_o$, the index of M. Thus the Leray spectral sequence shows:

$$\begin{aligned} H^i(X,L) &\cong H^{i_o}(X/Y, R^{i-i_o} p_*(L)) \\ &\cong H^{i-i_o}(Y, \mathcal{O}_Y) \otimes_k H^i(X/Y, M). \end{aligned}$$

<u>Proof of Lemma 1</u>: If $m: X \times X \longrightarrow X$ is the addition morphism, we know that $m^*L \otimes p_1^*L^{-1} \otimes p_2^*L^{-1}$ on $X \times X$ is trivial when restricted to $Y \times X$. Define $s: Y \longrightarrow Y \times X$ by $s(y) = (y,x)$. Then

$$\begin{aligned} &s^*(m^*L \otimes p_1^*L^{-1} \otimes p_2^*L^{-1}|_{Y \times X}) \\ &\qquad \cong T_x^* L|_Y \otimes L|_Y^{-1} \otimes \mathcal{O}_Y \end{aligned}$$

is also trivial. <u>QED</u>

<u>Proof of lemma 2</u>: Since $P \times_Z P \cong Y \times P$, it will suffice to prove the stronger:

<u>Sublemma</u>: Given $f: X \longrightarrow S$ a morphism of schemes $/k$ such that there exists $\pi: S' \longrightarrow S$ where π is faithfully flat and $\pi_X(\mathcal{O}_{S'}) \cong \mathcal{O}_S$ with the property:

$\exists$ φ, Y and a diagram

$$\begin{array}{ccc} X \times_S S' & \xrightarrow[\sim]{\varphi} & Y \times_k S' \\ {\scriptstyle f'}\downarrow & & \downarrow{\scriptstyle p_2} \\ S' & = & S' \end{array}$$

where Y is proper over k. Then we have an isomorphism $R^i f_*(\mathcal{O}_X) \cong H^i(Y, \mathcal{O}_Y) \otimes_k \mathcal{O}_S$.

<u>Proof</u>: $\pi^*(R^i f_*(\mathcal{O}_X)) \cong R^i f'_*(\mathcal{O}_{X \times_S S'})$ since $S' \longrightarrow S$ is a flat base extension and $R^i f'_*(\mathcal{O}_{X \times_S S'}) \cong H^i(Y, \mathcal{O}_Y) \otimes_k \mathcal{O}_{S'}$ because of the existence of φ. Because $H^i(Y, \mathcal{O}_Y)$ is finite-dimensional, $R^i f'_*(\mathcal{O}_{X \times_S S'})$ is a vector bundle. Hence $R^i f_*(\mathcal{O}_X)$ is a vector bundle because π is faithfully flat. Now we define an isomorphism

$$\begin{aligned} R^i f_*(\mathcal{O}_X) &\xrightarrow{\sim} \pi_* \pi^* R^i f_*(\mathcal{O}_X) && (\text{since } \pi_* \mathcal{O}_{S'} = \mathcal{O}_S) \\ &\cong \pi_*[H^i(Y, \mathcal{O}_Y) \otimes_k \mathcal{O}_{S'}] \\ &\cong H^i(Y, \mathcal{O}_Y) \otimes_k \mathcal{O}_S && (\quad " \quad) \end{aligned}$$

<u>QED</u>

<u>Theorem 2</u>: Let L and M be invertible sheaves on an abelian variety X, with L ample. Let

$$P_{L,M}(n) = \chi(L^n \otimes M).$$

Then (i) all the roots of $P_{L,M}$ are real and dim K(M) is the multiplicity of 0 as a root,

(ii) Counting roots with multiplicities:

$$H^k(X,M) = (0), \quad \text{if } 0 \leq k < \text{number of positive roots}$$

$$H^{g-k}(X,M) = (0), \quad \text{if } 0 \leq k < \text{number of negative roots.}$$

<u>Proof</u>: The theorem is proven in Mumford, Abelian varieties, §16, for M non-degenerate. It is obvious when $M \in \mathrm{Pic}^0(X)$ because in this case

$$P_{L,M}(n) = \chi(L^n) = \frac{(L^g)\cdot n^g}{g!}$$

and X = K(M). Now suppose $X = X_1 \times X_2$, $L = p_1^*L_1 \otimes p_2^*L_2$ and $M = p_1^*M_1 \otimes p_2^*M_2$ where $M_1 \in \mathrm{Pic}^0(X_1)$, M_2 is non-degenerate on X_2 and L_i is ample on X_i. Then by the Künneth formula,

$$(1) \qquad P_{L,M}(n) = P_{L_1,M_1}(n)\cdot P_{L_2,M_2}(n)$$

and $K(M) = K(M_1) \times K(M_2)$. So in this case the theorem follows from the above special cases and the Künneth formula.

We shall reduce the theorem to this case. Suppose $f: Y \longrightarrow X$ is an isogeny. Then

(2) $$P_{f^*L,f^*M}(n) = \deg f \cdot P_{L,M}(n)$$

by the Riemann-Roch theorem, and $\dim K(f^*M) = \dim K(M)$. Therefore assertion (i) is invariant under an isogeny. Let Y be the identity component of K(M) and let Z be a complementary subvariety for Y in X 5
We have an isogeny $f: Y \times Z \longrightarrow X$. Now $Y \subset K(f^*M)$ and if $M_1 = f^*(M)|_Y$, then as in the proof of Theorem 1, $M_1 \in \mathrm{Pic}^0(Y)$ and M is of the form $p_1^*M_1 \otimes p_2^*M_2$ where M_2 is a non-degenerate invertible sheaf on Z. The next problem is to see that the theorem does not depend on the ample L. Then we can replace f^*L by $p_1^*L_1 \otimes p_2^*L_2$ and we have reduced the proof of (i) to a case where (i) has been proven.

Claim: $P_{L,M}$ and $P_{L',M}$ have the same number of positive, zero, and negative roots (counted with multiplicity).

Let δ (resp. δ') be the smallest positive root of $P_{L,M}$ (resp. $P_{L',M}$). Let a (resp. a') be the number of positive roots of $P_{L,M}$ (resp. $P_{L',M}$). Then

a = number of positive roots of $P_{L,M}(t+\epsilon)$, if $0 < \epsilon < \delta$

a' = " " " $P_{L',M}(t+\epsilon')$, if $0 < \epsilon' < \delta'$.

But
$$\begin{aligned} s^g P_{L,M}\left(n + \frac{r}{s}\right) &= P_{L^s,M^s}\left(n + \frac{r}{s}\right) \\ &= \chi(L^{ns+r} \otimes M^s) \\ &= \chi(L^{sn} \otimes (L^r \otimes M^s)) \\ &= P_{L,L^r\otimes M^s}(sn) \\ &= s^g P_{L,L^r\otimes M^s}(n). \end{aligned}$$

So if $0 < \frac{r}{s} < \delta$, then $L^r \otimes M^s$ is non-degenerate and

$$a = \text{number of positive roots of } P_{L, L^r \otimes M^s}$$
$$= \text{index } (L^r \otimes M^s).$$

Now let N be large enough so that $(L')^N \otimes L^{-1}$ is ample and choose r and s so that $0 < r/s < \delta$, $0 < \frac{Nr}{s} < \delta'$. Then

$$a = \text{index } (L^r \otimes M^s)$$
$$\geq \text{index } (((L')^N \otimes L^{-1})^r \otimes L^r \otimes M^s) \text{ (by Th. for non deg. M)}$$
$$= \text{index } ((L')^{Nr} \otimes M^s)$$
$$= a'.$$

By symmetry, it follows that $a = a'$. The claim is proven similarly for the multiplicity of 0 and the number of negative roots.

To prove (ii), we may assume that $M = p^*N$ for a non-degenerate N on X/Y, since otherwise M has no cohomology at all. We have the commutative diagram:

$$\begin{array}{ccc} Y \times Z & \xrightarrow{\ f\ } & X \\ \downarrow & & \downarrow p \\ Z & \xrightarrow{\ g\ } & X/Y \end{array}$$

for some isogeny g. Then g^*N is non-degenerate and index (g^*N) = index(N). So:

number of pos. rts of $P_{L,M}$ = number of pos. rts of P_{f^*L, f^*M} (by formula 2)

= number of pos. rts of $P_{L|_Z, g^*N}$ (by formula 1)

= index g^*N (Th in non-deg. case)

= index N.

Now (ii) follows from Theorem 1. <u>QED</u>

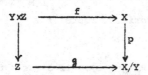

PROCEEDINGS OF THE
AMERICAN MATHEMATICAL SOCIETY
Volume 28, No. 1, April 1971

A REMARK ON MAHLER'S COMPACTNESS THEOREM

DAVID MUMFORD

ABSTRACT. We prove that if G is a semisimple Lie group without compact factors, then for all open sets $U \subset G$ containing the unipotent elements of G and for all $C>0$, the set of discrete subgroups $\Gamma \subset G$ such that

(a) $\Gamma \cap U = \{e\}$,

(b) G/Γ compact and measure $(G/\Gamma) \leq C$,

is compact. As an application, for any genus g and $\epsilon>0$, the set of compact Riemann surfaces of genus g all of whose closed geodesics in the Poincaré metric have length $\geq \epsilon$, is itself compact.

Consider the following general problem: let G be a locally compact topological group and let

$$\mathfrak{M}_G = \{\text{the set of discrete subgroups } \Gamma \subset G\}.$$

We would like to put a good topology on $\mathfrak{M}_G$ and we would like to find fairly "big" subsets of $\mathfrak{M}_G$ that turn out to be compact. Mahler studied the case $G=\boldsymbol{R}^n$, G/Γ compact, i.e., Γ is lattice (cf. Cassels [1, Chapter 5]). In this case, the group of automorphisms of G, $\mathrm{GL}(n, \boldsymbol{R})$, acts transitively on the set of lattices, so that the subset $\mathfrak{M}_G^C \subset \mathfrak{M}_G$ of lattices can be identified as a homogeneous space under $\mathrm{GL}(n, \boldsymbol{R})$; in fact:

$$\mathfrak{M}_G^C \cong \mathrm{GL}(n, \boldsymbol{R})/\mathrm{GL}(n, \boldsymbol{Z}).$$

So there is only one natural topology on $\mathfrak{M}_G^C$ and Mahler's theorem states that for all ϵ and K:

$$\left\{\Gamma \subset \boldsymbol{R}^n \,\middle|\, \begin{array}{ll} (1) & \text{if } \gamma \in \Gamma, \|\gamma\| < \epsilon \Rightarrow \gamma = 0 \\ (2) & \text{volume } (\boldsymbol{R}^n/\Gamma) \leq K \end{array}\right\} \text{ is compact.}$$

(Cassels [1, p. 137].)

Chabauty [2] has investigated generalizations of Mahler's theorem to general G and subgroups Γ such that measure $(G/\Gamma) < +\infty$.[1] We topologize $\mathfrak{M}_G$ by taking as a basis for the open sets the following:

Received by the editors April 29, 1970.

AMS 1970 *subject classifications*. Primary 22E40.

Key words and phrases. Discrete subgroups, Mahler's theorem.

[1] Although in recent years this restriction has been commonly made by people investigating automorphic functions in several variables, in the classical cases it eliminates the Fuchsian groups $\Gamma \subset \mathrm{SL}(2; \boldsymbol{R})$ of 2nd kind, and it eliminates all Kleinian groups $\Gamma \subset \mathrm{SL}(2; \boldsymbol{C})$. And $\mathfrak{M}_G$ seems very interesting in these cases.

(1) $U \subset G$ open, $S_U = \{\Gamma \in \mathfrak{M}_G \mid \Gamma \cap U \neq \emptyset\}$,

(2) $K \subset G$ compact, $T_K = \{\Gamma \in \mathfrak{M}_G \mid \Gamma \cap K = \emptyset\}$.

Then assuming that G is not too pathological,[2] Chabauty proves:

THEOREM. *Let U be an open neighborhood of e, C a positive number. Then*: $\{\Gamma \in \mathfrak{M}_G \mid \Gamma \cap U = \{e\}$ *and* measure $(G/\Gamma) \leq C\}$ *is compact.*

This is very pretty. Its main drawback, however, is that the topology on $\mathfrak{M}_G$ is so weak that it is hard to deduce things from convergence in this topology. For instance if subgroups Γ_i converge to Γ, one would like to know that suitable sets of generators of the Γ_i converge to generators of Γ. Chabauty gives some arguments about this at the end of his paper, but I believe his reasoning there is wrong. However the results of Weil [4] and Macbeath [5] show that the topology is "strong enough" on the subset

$$\mathfrak{M}_G^C = \{\Gamma \in \mathfrak{M}_G \mid G/\Gamma \text{ compact}\}.$$

THEOREM (MACBEATH [5, THEOREMS 4 AND 5]). *Assume that G is a Lie group.*[3] *Let subgroups $\Gamma_i \in \mathfrak{M}_G^C$ converge to $\Gamma \in \mathfrak{M}_G^C$. Then for i sufficiently large, there exist isomorphisms of the abstract groups*

$$\phi_i : \Gamma \xrightarrow{\sim} \Gamma_i$$

such that for all $\gamma \in \Gamma$, $\phi_i(\gamma) \in G$ converge to γ. Moreover there is a compact set $K \subset G$ and an open neighborhood $U \subset G$ of e such that $K \cdot \Gamma = G$, $K \cdot \Gamma_i = G$, $U \cap \Gamma = \{e\}$ and $U \cap \Gamma_i = \{e\}$ if i is sufficiently large.

For the application that we want, Chabauty's theorem is not the right generalization of Mahler's theorem. Instead, what we want is this:

THEOREM 1. *Let $G \subset GL(n, R)$ be a semisimple Lie group without compact factors. Let $U \subset G$ be an open set containing all unipotent elements of G and let C be a positive number. Then*

$$\{\Gamma \in \mathfrak{M}_G^C \mid \Gamma \cap U = \{e\}, \text{measure } (G/\Gamma) \leq C\}$$

is compact.

PROOF. This is an immediate consequence of Chabauty's theorem and Selberg's conjecture, proved recently by Kajdan and Margulis

[2] G satisfies the 2nd axiom of countability, and moreover $e \in G$ has a fundamental system of neighborhoods U_i such that measure $(\bar{U}_i - U_i) = 0$. In this case, $\mathfrak{M}_G$ satisfies the 2nd axiom of countability too.

[3] A Lie group is always assumed to be connected.

[3], to the effect that a discrete subgroup $\Gamma \subset G$, G as above, such that measure $(G/\Gamma) < +\infty$ but G/Γ not compact, must contain non-trivial unipotent elements of G. Q.E.D.

Instead of invoking the difficult result of Každan and Margulis, we can prove a weaker but more explicit theorem by elementary means: Let $G \subset GL(n, R)$ again be a semisimple Lie group without compact factors. Let $K \subset G$ be a maximal compact subgroup and let $X = K\backslash G$ be the associated symmetric space. Let the Killing form on G induce a metric ρ on X as usual. Define a function d on G by:

$$d(x) = \inf_{z \in X} \rho(z, z^x).$$

It is easy to see that d is continuous and $d(x) = 0$ if and only if when you decompose $x = x_s \cdot x_u$, (x_s semisimple, x_u unipotent and $x_s x_u = x_u x_s$), then x_s is in a compact subgroup of G or equivalently $x_s \in \bigcup_{y \in G} yKy^{-1}$. For all $\epsilon > 0$, define an open subset of G by:

$$U_\epsilon = \{x \in G \mid d(x) < \epsilon\}.$$

For all $C > 0$, define a compact subset of G by:

$$K_C = \{x \in G \mid \rho(K \cdot x, K \cdot e) \leqq C\}.$$

THEOREM 2. *Let $n = \dim K\backslash G$. Then there is a constant γ depending only on n such that for all $\Gamma \in \mathfrak{M}_G^C$, $\epsilon > 0$,*

$$\Gamma \cap U_\epsilon = \{e\} \Rightarrow K_C \cdot \Gamma = G$$

where $C = \gamma \cdot$ measure $(G/\Gamma)/\epsilon^{n-1}$. Hence for all positive D

1 $$\{\Gamma \in \mathfrak{M}_G \mid \overset{C}{\Gamma} \cap U_\epsilon = \{e\}, \text{ measure } (G/\Gamma) \leqq D\}$$

is compact.

PROOF. We begin by proving:

LEMMA. *Let X be a compact Riemannian manifold with all sectional curvatures $R(S) \leqq 0$. There is a constant γ depending only on $n = \dim X$ such that:*

$$\operatorname{diam}(X) \cdot (\textit{length of smallest closed geodesic on } X)^{n-1} \leqq \gamma \cdot \text{volume } (X).$$

PROOF. Let $d = \operatorname{diam}(X)$ and let $x, y \in X$ be a distance d apart. Let σ be a geodesic from x to y of length d. Let η be the length of the shortest closed geodesic on X and construct a tube T around σ of radius $\eta/4$ as the union of all geodesics perpendicular to σ of length $\eta/4$. There are 2 possibilities: either no 2 geodesics δ_1, δ_2 perpendicular to σ of length $\eta/4$ meet, or else some pair δ_1, δ_2 do meet. In the first case, we may say that the exponential map from the normal bundle N to σ in M maps an $\eta/4$-tube T_0 around the 0-section in N injectively to M. Then since all the sectional curvatures are $\leqq 0$, it follows that:

$$(*) \qquad \text{volume } X \geqq \text{volume } T \geqq \text{volume } T_0 = c_n \cdot (\eta/4)^{n-1} \cdot d$$

where c_n is the volume of the unit ball in R^{n-1}. On the other hand, suppose 2 geodesics δ_1 and δ_2 meet:

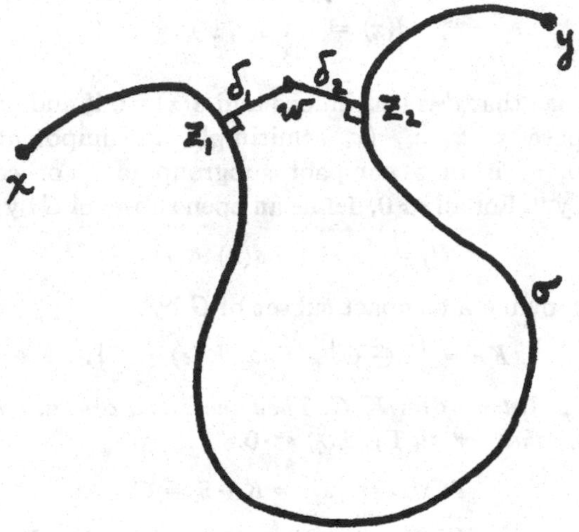

Let z_1, z_2 and w be the points indicated in the figure and let e be the distance from z_1 to z_2 along σ. Then we can go from x to y by going from x to z_1 on σ, following δ_1, then δ_2 and going from z_2 to y on σ. This has length $\leqq d-e+\eta/2$, and since σ is the shortest path from x to y, $d \leqq d-e+\eta/2$, i.e., $e \leqq \eta/2$. But then δ_1, δ_2 and the part of σ between z_1 and z_2 is a closed path τ of length at most η. τ is certainly not homotopic to 0 since on the universal covering space $\tilde{X}$ of X, the exponential from N_0 to $\tilde{X}$ is injective. Moreover, τ has corners and so is not itself a geodesic. Therefore there is a closed geodesic freely homotopic to τ of length $<\eta$. This contradicts the definition of η and so the 1st possibility must be correct. This proves (*) and the lemma. Q.E.D.

We apply the lemma to the manifold X/Γ, with the metric induced from the metric d on X. (Note that by hypothesis $\Gamma \cap U_\epsilon = \{e\}$, Γ acts freely on X, so X/Γ is a manifold.) The closed geodesics of X/Γ are all images of geodesics in X joining 2 points x, x^γ, where $x \in X$,

$z\in\Gamma$. Since $\Gamma\cap U_\epsilon=\{e\}$, these all have length at least ϵ. It follows from the lemma that:

$$\operatorname{diam}(X) \leqq \frac{\gamma \text{ volume } (X/\Gamma)}{\epsilon^{n-1}} = \frac{\gamma \text{ measure } (G/\Gamma)}{\epsilon^{n-1}} = C.$$

Therefore the projection of X onto X/Γ maps the unit ball of radius C *onto* X/Γ, hence $K_C\cdot\Gamma=G$.

Finally to prove from this that $\{\Gamma\in\mathfrak{M}_G^C \mid \Gamma\cap U_\epsilon=\{e\}$ and measure $(G/\Gamma)\leqq D\}$ is compact, it suffices by Chabauty's theorem to check that if Γ_i are in this set and $\Gamma_i\to\Gamma\in\mathfrak{M}_G$, then G/Γ is also compact. But since $K_C\cdot\Gamma_i=G$ for all i, it follows easily that $K_C\cdot\Gamma=G$ too, hence G/Γ is a quotient of K_C and is compact. Q.E.D.

I want to apply Theorem 2 to the case $G=SL(2, R)/(\pm I)$ so that Γ is a Fuchsian group. Then X is the Lobachevskian plane, and a simple calculation shows that

$U_\epsilon=$ image of A's such that $|\operatorname{tr} A|<2\cosh(\epsilon/2)$
$=$ set of elliptic and parabolic elements and those hyperbolic elements with eigenvalues t, t^{-1} for which $1<t<e^{\epsilon/2}$.

The Fuchsian groups of 1st kind which are disjoint from some U_ϵ are exactly those which act freely on X and for which X/Γ is compact. In this case X/Γ is a compact Riemann surface with its Poincaré metric, X is its universal covering space and $\Gamma\cong\pi_1(X/\Gamma)$. Moreover the map which takes an element $z\in\Gamma$ to the image mod Γ of the shortest line segment geodesic from x to x^z in X defines an isomorphism between the set of conjugacy classes in Γ and the set of closed geodesics in X/Γ. If the conjugacy class of γ corresponds to a geodesic σ, then

$$\cosh\frac{\text{length }\sigma}{2} = \left|\frac{\operatorname{Tr}\gamma}{2}\right|.$$

Moreover, by the Gauss-Bonnet theorem

$$\text{measure } (G/\Gamma) = \text{area } (X/\Gamma) = \text{cnst } (g-1)$$

where $g=$ genus of X/Γ. So in this case, the lemma in Theorem 2 says:

COROLLARY 1. *For all compact Riemann surfaces X of genus g,* $\operatorname{diam}(X)\cdot$(*length of smallest geodesic on X*) *is bounded above.*

COROLLARY 2. *For all $\epsilon>0$, $g\geqq 2$, the set of discrete subgroups* $\Gamma\subset SL(2; R)$ *such that:*

(i) *for all* $\gamma \in \Gamma$, $\gamma \neq I$, $|\mathrm{Tr}\,\gamma| \geqq 2+\epsilon$,

(ii) X/Γ *is a compact Riemann surface of genus* g,

is compact.

COROLLARY 3. *Let* $g \geqq 2$ *and let* $\mathfrak{M}_g$ *be the moduli space of compact Riemann surfaces of genus* g *(without "marking"). For all* $\epsilon > 0$, *the subset:*

$$\{X \in \mathfrak{M}_g \mid \textit{in the Poincaré metric, all geodesics on } X \textit{ have length} \geqq \epsilon\}$$

is compact.

(PROOF. Apply Theorem 1 and Corollary 1.)

This result was my motivation for looking at these questions. I originally found a completely elementary proof of this, using the method of Theorem 2, and then finding

(a) upper bounds for the number of vertices and

(b) lower bounds for the interior and exterior angles of the *Dirichlet* fundamental domain for Γ acting on X; but one reference leads to another and it turned out that $\{$elem. th.$\} \subset$ Chabauty + Weil + Každan-Margulis + Macbeath.

REFERENCES

1. J. W. S. Cassels, *An introduction to the geometry of numbers*, Springer-Verlag, Berlin, 1959. MR 28 #1175.

2. C. Chabauty, *Limite d'ensembles et géométrie des nombres*, Bull. Soc. Math. France 78 (1950), 143–151. MR 12, 479.

3. D. A. Každan and G. A. Margulis, *A proof of Selberg's conjecture*, Mat. Sb. 75 (117) (1968), 163–168 = Math. USSR Sb. 4 (1968), 147–152. MR 36 #6535.

4. A. Weil, *On discrete subgroups of Lie groups*, Ann. of Math. (2) 72 (1960), 369–384. MR 25 #1241.

5. A. M. Macbeath, *Groups of homeomorphisms of a simply connected space*, Ann. of Math. (2) 79 (1964), 473–488. MR 28 #4058.

HARVARD UNIVERSITY, CAMBRIDGE, MASSACHUSETTS 02138

INTRODUCTION TO THE THEORY OF MODULI

by

David Mumford and Kalevi Suominen [1]

5th Nordic Summer-School in Mathematics
Oslo, August 5–25, 1970

1. Endomorphisms of vector spaces

Throughout these notes, k is an algebraically closed field, varieties are reduced and irreducible k-schemes of finite type, and morphisms of varieties are k-morphisms.

A moduli problem for a class of algebraic objects consists in two parts: finding the equivalence classes of the objects under a suitable equivalence relation (usually isomorphism), and parametrizing these classes by means of a scheme (or a geometric object of more general type). In this chapter we shall be interested in the moduli of endomorphisms of vector spaces.

More precisely, let V be a vector space of dimension n over k, and let T be an endomorphism of V. The problem of classifying pairs (V, T) up to isomorphism is readily solved: there is a basis of V such that the matrix of T with respect to this basis is in the *Jordan canonical form*

$$
\left(\begin{array}{c|c|c}
\begin{matrix} \lambda_1 & \varepsilon_{11} & 0 & \cdots & 0 \\ 0 & \lambda_1 & \varepsilon_{21} & \cdots & 0 \\ 0 & 0 & \lambda_1 & \cdots & 0 \\ & & \cdots & & \\ 0 & 0 & 0 & & \varepsilon_{31} \\ 0 & 0 & 0 & \cdots & \lambda_1 \end{matrix} & 0 & 0 \\ \hline
0 & \begin{matrix} \lambda_2 & \varepsilon_{12} & \cdots & 0 \\ 0 & \lambda_2 & \cdots & 0 \\ \cdot & \cdot & \cdots & \cdot \\ 0 & 0 & \cdots & \lambda_2 \end{matrix} & 0 \\ \hline
0 & 0 &
\end{array}\right)
$$

where $\varepsilon_{ij} = 0$ or 1.

[1] Assisted by M. Hazewinkel, A. Cooper, J. H. M. Steenbrink, and F. Huikeshoven.

For the second part of the moduli problem we must introduce algebraic families of pairs (V, T). Intuitively, a family of vector spaces, parametrized by a variety S, is a vector bundle over S, and an algebraic family of endomorphisms of the fibers of this vector bundle is an endomorphism of the bundle. Now, the algebraic counterpart of a bundle is a locally free sheaf of $\mathcal{O}_S$-modules. Hence we are led to the following:

DEFINITION 1. An algebraic family of endomorphisms of n-dimensional k-vector spaces on a k-variety S is a pair $(\mathscr{E}, T)$ where $\mathscr{E}$ is a locally free $\mathcal{O}_S$-module of rank n and T is an endomorphism of $\mathscr{E}$.

For each closed point s of S, we then have a vector space $\mathscr{E} \otimes k(s)$ of dimension n over $k(s) = k$ and an endomorphism $T \otimes k(s)$ of $\mathscr{E} \otimes k(s)$.

As a first attempt in the search for a moduli space it is natural to ask if there exists a family of endomorphisms in which each isomorphism class is represented exactly once. The answer is trivially yes, e.g., the base scheme may be chosen discrete if we drop temporarily the restriction that base schemes are varieties. This is obviously no satisfactory solution to the problem. Namely, if M is a k-scheme whose closed points are in $1-1$ correspondence with the classes of endomorphisms, then for each family of endomorphisms $(\mathscr{E}, T)$ over a variety S there is a map $S(k) \to M(k)$ associating with each closed point $s \in S$ the point of M which corresponds to the pair $(\mathscr{E} \otimes k(s), T \otimes k(s))$. This map should be induced by a morphism $S \to M$!

To express this condition more exactly we introduce some functorial terminology.

For each k-variety S, we denote by $\mathscr{F}(S)$ the set of families of endomorphisms on S, modulo isomorphism. If, $f : S' \to S$ is a morphism of varieties and $(\mathscr{E}, T)$ is a family of endomorphisms on S, then $(f^*\mathscr{E}, f^*T)$ is a family of endomorphisms on S'. Thus we obtain a map $f^* : \mathscr{F}(S) \to \mathscr{F}(S')$, and $\mathscr{F}$ becomes a contravariant functor from the category of k-varieties to the category of sets.

Now, the condition stated above can be made precise:

(A) There is a morphism of contravariant functors

$$\Phi : \mathscr{F} \to h_M,$$

where $h_M(S) = \mathrm{Hom}\,(S, M)$, such that

$$\Phi(\mathrm{Spec}(k)) : \mathscr{F}(\mathrm{Spec}(k)) \to M(k)$$

is bijective.

However, this condition does not suffice to define M uniquely. In fact, it may be possible to find other solutions M' by reducing the structure sheaf: $\mathcal{O}_{M'} \subset \mathcal{O}_M$, having the underlying point set unchanged. Keeping this in mind, we write the final definition.

DEFINITION 2. A *coarse moduli space* for endomorphisms of n-dimensional k-vector spaces is a pair (M, Φ) consisting of a k-variety M and a morphism of functors $\Phi : \mathcal{F} \to h_M$ such that

$$\Phi(\operatorname{Spec}(k) : \mathcal{F}(\operatorname{Spec}(k)) \to M(k)$$

is bijective and such that for each k-variety N and each morphism of functors $\Psi : \mathcal{F} \to h_N$ there is a unique morphism $\kappa : M \to N$ which renders

$$\begin{array}{ccc} & \overset{\Phi(S)}{\nearrow} & \operatorname{Hom}(S, M) \\ \mathcal{F}(S) & & \big\downarrow {\scriptstyle \operatorname{Hom}(S, \kappa)} \\ & \underset{\Psi(S)}{\searrow} & \operatorname{Hom}(S, N) \end{array}$$

commutative for each k-variety S.

It is easy to see that a coarse moduli space is unique up to isomorphism, if it exists.

There is a priori no reason why the map

$$\Phi(S) : \mathcal{F}(S) \to \operatorname{Hom}(S, M)$$

should be bijective for k-varieties S other than $\operatorname{Spec}(k)$. That this be the case for all varieties S amounts to saying that the functor $\mathcal{F}$ is representable by M. Then the family $(\mathcal{E}, T)$ of endomorphisms on M which corresponds to the identity morphism of M is *universal* in the following sense: For each family of endomorphisms $(\mathcal{E}', T')$ on a variety S there is a unique morphism $f : S \to M$ such that $(f^*\mathcal{E}, f^*T)$ is isomorphic to $(\mathcal{E}', T')$.

DEFINITION 3. A *fine moduli space* for endomorphisms of n-dimensional k-vectorspaces is a pair (M, Φ) where M is a k-variety and $\Phi : \mathcal{F} \to h_M$ is an isomorphism of functors.

It is not difficult to see that a fine moduli space is also a coarse moduli space.

REMARK. Definitions 2 and 3 will also be applied to other functors. In any case, it is clear that a coarse moduli space is a fine moduli space if and only if the functor is representable.

Unfortunately, there is no fine moduli space for endomorphisms of vector spaces. Namely, if $(\mathcal{E}, T)$ is any family of endomorphisms on a variety S then for each invertible $\mathcal{O}_S$-Module L the family $(\mathcal{E} \otimes L, T \otimes 1)$ corresponds to the same morphism $S \to M$. Hence

$$\mathscr{F}(S) \to \operatorname{Hom}(S, M)$$

is not injective if there are non-trivial invertible sheaves on S.

But things are worse: *not even a coarse moduli space exists*! To see this, let us consider the variety $S = A^1 = \operatorname{Spec}(k[t])$ with $\mathscr{E} = \mathcal{O}_S^2$ and T defined by the matrix

$$\begin{pmatrix} 1 & t \\ 0 & 1 \end{pmatrix}$$

All the pairs $(\mathscr{E} \otimes k(s), T \otimes k(s))$ are isomorphic for closed points s of S different from 0. Hence the map $S \to M$ corresponding to the family is constant on $A^1 - 0$. By continuity, it must be constant on S, although $(\mathscr{E} \otimes k(0),\ T \otimes k(0)) = (k^2, 1)$ is *not* isomorphic to $(\mathscr{E} \otimes k(s), T \otimes k(s))$ for $s \neq 0$.

This is a typical example of the so called *jump phenomenon*, which ruins the hope of finding solutions to many moduli problems.

Similar constructions show that endomorphisms with isomorphic semi-simple parts are represented by the same point of any variety M with a morphism of functors $\mathscr{F} \to h_M$.

On the other hand, there is a variety M which separates endomorphisms with non-isomorphic semi-simple parts or, what amounts to the same, with different characteristic polynomials:

PROPOSITION 1. *There is a morphism of functors*

$$\Psi : \mathscr{F} \to h_{A^n}$$

such that $\Psi(\operatorname{Spec}(k)) : \mathscr{F}(\operatorname{Spec}(k)) \to A^n(k)$ *is given by*

$$(V, T) \to (a_1, \cdots, a_n),$$

where $X^n + a_1 X^{n-1} + \cdots + a_n$ *is the characteristic polynomial of* T.

PROOF. Let $(\mathscr{E}, T)$ be a family of endomorphisms on a variety S. There is an affine open covering (U_α) of S such that $\mathscr{E}|U_\alpha$ is free of rank n. If $T|U_\alpha$ is represented by an $n \times n$ matrix T with entries in $\Gamma(U_\alpha, \mathcal{O}_S)$, then

$$P_{T|U_\alpha}(X) = \det(X \cdot I - T_\alpha) \in \Gamma(U, \mathcal{O}_S)[X]$$

is a polynomial of degree n which is independent of the trivialization $\mathscr{E}|U_\alpha \xrightarrow{\sim} \mathcal{O}_{U_\alpha}^n$. But then $P_{T|U_\alpha}(X)$ and $P_{T|U_\beta}(X)$ coincide on $U_\alpha \cap U_\beta$, and so they may be joined together to define the *characteristic polynomial* of T:

$$P_T(X) = X^n + a_1 X^{n-1} + \cdots + a_n \in \Gamma(S, \mathcal{O}_S)[X].$$

Hence we may associate with $(\mathscr{E}, T)$ canonically a morphism

$$(a_1, \cdots, a_n) : S \to A^n.$$

The rest of the proposition follows immediately.

The partly negative results we have obtained so far are intuitively in agreement with our knowledge of the Jordan canonical form of an endomorphism. Indeed, the entries outside the diagonal are constants 0 or 1. Hence the eigenvalues, or rather their symmetric polynomials, are the only true 'moduli' of endomorphisms.

To get at least a coarse moduli space we must somehow restrict the class of endomorphisms. We shall consider two canonical possibilities: endomorphisms with all $\varepsilon_{ij} = 0$ in the Jordan canonical form, i.e., semi-simple endomorphisms, and endomorphisms with all $\varepsilon_{ij} = 1$. It is not difficult to see that the latter are exactly those endomorphisms $T : V \to V$ for which there is a cyclic vector or, more precisely, a vector $v \in V$ such that $(v, Tv \cdots, T^{n-1}v)$ is a basis of V.

Let us first consider semi-simple endomorphisms. For each variety S, let $\mathscr{F}_d(S)$ denote the set of families of endomorphisms $(\mathscr{E}, T)$ on S, modulo isomorphism, such that $T \otimes k(s)$ is semi-simple for each closed point s of S. Clearly, these sets form a subfunctor $\mathscr{F}_d$ of $\mathscr{F}$.

It follows immediately from proposition 1 that $M = A^n$ satisfies the condition (A) for the functor $\mathscr{F}_d$. But we can say more:

PROPOSITION 2. *A^n is a coarse moduli space for semi-simple endomorphisms of n-dimensional vector spaces.*

PROOF. Let $M = A^n = \text{Spec}(k[t_1, \cdots, t_n])$ and define an endomorphism T of $\mathcal{O}_M^n$ by the matrix

$$\begin{pmatrix} 0 & 0 & 0 & \cdots & 0 & -t_n \\ 1 & 0 & 0 & \cdots & 0 & -t_{n-1} \\ 0 & 1 & 0 & \cdots & 0 & -t_{n-2} \\ \cdots & \cdots & \cdots & \cdots & \cdots & \cdots \\ 0 & 0 & 0 & \cdots & 0 & -t_2 \\ 0 & 0 & 0 & \cdots & 1 & -t_1 \end{pmatrix}$$

Then the characteristic polynomial of T is

$$P_T(X) = X^n + t_1 X^{n-1} + \cdots + t_n$$

So, if $\Delta \in k[t_1, \cdots, t_n]$ is the discriminant of $P_T(X)$ and

$$U = D(\Delta) = \{x \in A^n | \Delta(x) \neq 0\},$$

the restriction $(\mathcal{O}_U^n, T|U)$ is a family of semi-simple endomorphisms.

Now, if N is any k-variety and $\Phi : \mathscr{F}_d \to h_N$ is a morphism of functors, $\Phi(U)$ of the class of $(\mathcal{O}_U^n, T|U)$ is a morphism $\varphi : U \to N$. Let $\Gamma \subset U \times N$

be its graph and denote by $\bar{\Gamma}$ the closure of Γ in $M \times N$. We claim that the projection $\bar{\Gamma} \to M$ is an isomorphism.

To prove this, put $S = \mathrm{Spec}(k[\lambda_1, \cdots, \lambda_n])$ and define by

$$T^1 = \begin{pmatrix} \lambda_1 & 0 & \cdots & 0 \\ 0 & \lambda_2 & \cdots & 0 \\ \cdots & \cdots & \cdots & \cdots \\ 0 & 0 & \cdots & \lambda_n \end{pmatrix}$$

a family $(\mathcal{O}_S^n, T^1)$ of semi-simple endomorphisms on S. There are two morphisms $f : S \to M$ and $g : S \to N$ associated with this family. Clearly, f is given by

$$X^n + t_1 X^{n-1} + \cdots + t_n = (X - \lambda_1) \cdots (X - \lambda_n).$$

Hence each λ_i is integral over $k[t_1, \cdots, t_n]$, i.e., f is finite. Then $h = (f, g) : S \to M \times N$ is finite (EGA II, 6, 1, 5 (v)) and therefore $h(S)$ is closed in $M \times N$.

On the other hand, inspecting the closed points it is not difficult to see that $\Gamma = h(V)$, where $V = f^{-1}(U) = \bigcap_{i \neq j} D(\lambda_i - \lambda_j)$. Hence $h(S) = \bar{\Gamma}$, and so $\bar{\Gamma} \to M$ is surjective with finite fibres. But then it is an isomorphism by Zariski's Main Theorem.

1 Now, $\bar{\Gamma}$ is the graph of a morphism $\psi : M \to N$ extending φ. If S is any variety, and $f : S \to M$, $g : S \to N$ are the morphisms corresponding to a family of semi-simple endomorphisms on S, then g and $\Psi \circ f$ coincide at all closed points of S, whence $\Psi \circ f = g$.

COROLLARY. *There is no fine moduli space for semi-simple endomorphisms of n-dimensional vectorspaces if $n > 1$.*

PROOF. Otherwise, there would exist a universal family $(\mathcal{E}, T)$ of semi-simple endomorphisms on A^n. Let $R = k[[t_1, \cdots, t_n]]$ be the completion of the local ring of 0 on A^n, and let $(\mathcal{E}', T')$ be the family induced by $(\mathcal{E}, T)$ on Spec (R). Since R is local, $\mathcal{E}'$ is free, and therefore $(\mathcal{E}', T')$ is isomorphic to $(\mathcal{O}^n, T'')$ where T'' is an $n \times n$ matrix with entries in the maximal ideal m of R. But then $t_n = (-1)^n \det(T'') \in m^n$, which is impossible unless $n = 1$.

REMARK. The same proof shows that in each family of endomorphisms $(\mathcal{E}, T)$ on A^n with the characteristic polynomial

$$P_T(X) = X^n + t_1 X^{n-1} + \cdots + t_n$$

the Jordan canonical form of $T \otimes k(0)$ is

$$\begin{pmatrix} 0 & 1 & 0 & \cdots & 0 \\ 0 & 0 & 1 & \cdots & 0 \\ \cdots & \cdots & \cdots & \cdots & \cdots \\ 0 & 0 & 0 & \cdots & 1 \\ 0 & 0 & 0 & \cdots & 0 \end{pmatrix}$$

We shall now examine the second possibility of restricting the functor $\mathscr{F}$: the families of endomorphisms with a cyclic vector. It is to be expected that A^n is a coarse module space for these. In fact, the remark above suggests that it might even be a fine moduli space.

However, we have seen that the existence of nontrivial invertible sheaves on a variety S prevents

$$\Psi(S) : \mathscr{F}(S) \to \mathrm{Hom}(S, A^n)$$

from being injective. To eliminate this type of redundancy we shall consider families with a 'cyclic section'. More precisely, for each variety S, let $\mathscr{F}'(S)$ denote the subset of $\mathscr{F}(S)$ represented by families of endomorphisms $(\mathscr{E}, T)$ on S such that there is a section $s \in \Gamma(S, \mathscr{E})$ for which $s, Ts, \cdots, T^{n-1}s$ span $\mathscr{E}$.

Proposition 3. *The restriction of Ψ to $\mathscr{F}$*

$$\Psi' : \mathscr{F}' \to h_{A^n}$$

is an isomorphism of functors.

Proof. Let $(\mathscr{E}, T)$ represent an element of $\mathscr{F}'(S)$ for some variety S. Then $\mathscr{E}$ is free with basis $s, Ts, \cdots, T^{n-1}s$ for some section $s \in \Gamma(S, \mathscr{E})$, and the matrix of T with respect to this basis is

$$\begin{pmatrix} 0 & 0 & 0 & \cdots & 0 & -a_n \\ 1 & 0 & 0 & \cdots & 0 & -a_{n-1} \\ 0 & 1 & 0 & \cdots & 0 & -a_{n-2} \\ \cdots & \cdots & \cdots & \cdots & \cdots & \cdots \\ 0 & 0 & 0 & \cdots & 0 & -a_2 \\ 0 & 0 & 0 & \cdots & 1 & -a_1 \end{pmatrix},$$

where the last column is determined by the characteristic polynomial of T.

$$P_T(X) = X^n + a_1 X^{n-1} + \cdots + a_n$$

as is seen by a direct computation of $\det(X, I-T)$, or by the Cayley-Hamilton theorem:

$$P_T(T) = T^n + a_1 T^{n-1} + \cdots + a_n = 0.$$

Hence the elements of $\mathscr{F}'(S)$ are in $1-1$ correspondence with the

n-tuples $(a_1, \cdots, a_n) \in \Gamma(S. \mathcal{O}_S)^n$, i.e., with the morphisms $S \to A^n$.

To explain the difference between the functors $\mathscr{F}_d$ and $\mathscr{F}'$, we note that $\mathscr{F}'$ is an open subfunctor of $\mathscr{F}$ in the following sense:

If $(\mathscr{E}, T)$ is a family of endomorphisms on a variety S, and if s is a closed point of S such that $T \otimes k(s)$ has a cyclic vector, there is an open neighborhood U of s such that $(\mathscr{E}|U, T|U)$ defines an element of $\mathscr{F}'(S)$.

In fact, if t is a section of $\mathscr{E}$ over some neighborhood of s such that $t(s) \in \mathscr{E} \otimes k(s)$ is a cyclic vector of $T \otimes k(s)$, then $t, Tt, \cdots, T^{n-1}t$ generate $\mathscr{E}$ in some neighborhood of S by Nakayama's lemma.

In the second half of this chapter we shall consider the problem of moduli of endomorphisms from another point of view, which ties up with more general theory. Namely, each family of endomorphisms is induced locally by the family $(\mathscr{E}, T)$ on $A^{n^2} = \mathrm{Spec}(k[t_{ij}])$ $(1 \leqq i, j \leqq n)$ in which $\mathscr{E} = \mathcal{O}^n$ and the matrix of T is (t_{ij}). In fact, if $(\mathscr{E}', T')$ is a family of endomorphisms on a variety S, and if (U_α) is an open covering of S such that each $\mathscr{E}'|U_\alpha$ is free, then $(\mathscr{E}'|U_\alpha, T|U_\alpha)$ is isomorphic to $(f_\alpha{}^*\mathscr{E}, f_\alpha{}^* T)$ with $f: U_\alpha \to A^{n^2}$ defined by the entries of the matrix of $T'|U_\alpha$ relative to some basis of $\mathscr{E}'|U_\alpha$. It follows that if M is a variety and $\Phi : \mathscr{F} \to h_M$ is a morphism of functors, Φ is uniquely determined by the morphism $\varphi : A^{n^2} \to M$ associated with $(\mathscr{E}, T)$. Hence the properties of Φ may be derived from a study of φ.

The set of closed points of A^{n^2} may be identified with the set $M(n)$ of $n \times n$ matrices with entries in k and the general linear group $GL(n)$ acts on $M(n)$ by $B \mapsto ABA^{-1}$, $A \in GL(n)$. Since the fibres of $(\mathscr{E}, T)$ over B and ABA^{-1} are isomorphic, $\varphi(B) = \varphi(ABA^{-1})$, i.e., φ is constant on each orbit $0(B)$.

On the other hand, if M is a variety and $\varphi : A^{n^2} \to M$ is a morphism which is constant on the orbits, then it follows from the discussion above that there is a morphism of functors $\Phi : \mathscr{F} \to h_M$ associating φ with the family $(\mathscr{E}, T)$ on A^{n^2}. Hence there is a natural $1-1$ correspondence between morphisms of functors $\Phi : \mathscr{F} \to h_M$ and morphisms $\varphi : A^{n^2} \to M$ constant on the orbits. It is then clear that the universal property of a coarse moduli space (M, Φ) (Definition 2) means that (M, φ) is a quotient of A^{n^2} by $GL(n)$ in the following sense:

Definition 4. Let G be a group operating on a variety X. A *quotient* of X by G is a pair (Y, φ) in which Y is a variety and $\varphi : X \to Y$ is a morphism satisfying:

(i) φ is constant on the orbits of the closed points of X.

(ii) given a variety Z and a morphism $\Psi : X \to Z$ constant on the orbits, there is a unique morphism $\kappa : Y \to Z$ such that $\Psi = \kappa \circ \varphi$.

The quotient of X by G is clearly unique up to isomorphism.

The moduli problem of endomorphisms of n-dimensional vector spaces has now been reduced to finding a quotient of A^{n^2} by $GL(n)$. It may be shown to be A^n (cf. proposition 1), but even without this knowledge we can easily prove the non-existence of a coarse moduli space. Indeed, a quotient (Y, φ) is a coarse moduli space if and only if Y separates non-isomorphic endomorphisms, i.e., each fibre $\varphi^{-1}(y)$ consists of a unique orbit. But the fibres of any morphism are closed whereas the orbits need not be closed in general. In fact, if B is a triangular matrix

$$B = \begin{pmatrix} \lambda_1 & \sigma_{12} & \sigma_{12} & \cdots & \sigma_{1n} \\ 0 & \lambda_2 & \sigma_{23} & \cdots & \sigma_{2n} \\ \cdots & \cdots & \cdots & \cdots & \cdots \\ 0 & 0 & 0 & \cdots & \sigma_{n-1,n} \\ 0 & 0 & 0 & \cdots & \lambda_n \end{pmatrix},$$

and $A \in GL(n)$ is

$$A = \begin{pmatrix} 1 & 0 & 0 & \cdots & 0 \\ 0 & \alpha & 0 & \cdots & 0 \\ 0 & 0 & \alpha^2 & \cdots & 0 \\ \cdots & \cdots & \cdots & \cdots & \cdots \\ 0 & 0 & 0 & \cdots & \alpha^{n-1} \end{pmatrix}$$

then

$$ABA^{-1} = \begin{pmatrix} \lambda_1 & \sigma_{12}\alpha & \sigma_{13}\alpha^2 & \cdots & \sigma_{1n}\alpha^{n-1} \\ 0 & \lambda_2 & \sigma_{23}\alpha & \cdots & \sigma_{2n}\alpha^{n-2} \\ \cdots & \cdots & \cdots & \cdots & \cdots \\ 0 & 0 & 0 & \cdots & \sigma_{n-1,n^\alpha} \\ 0 & 0 & 0 & \cdots & \lambda_n \end{pmatrix}$$

Hence, letting $\alpha \to 0$, we find that the semi-simple part B_s of B:

$$B_s = \begin{pmatrix} \lambda_1 & 0 & \cdots & 0 \\ 0 & \lambda_2 & \cdots & 0 \\ \cdots & \cdots & \cdots & \cdots \\ 0 & 0 & \cdots & \lambda_n \end{pmatrix}$$

is in the closure of the orbit $0(B)$ of B.

More precisely, we have

Proposition 4. *If B_1, B_2 are two $n \times n$ matrices, then $\overline{0(B_1)} \cap \overline{0(B_2)} \neq \phi$ if and only if $0((B_1)_s) = 0((B_2)_s)$. In any case $(B_i)_s \in \overline{0(B_i)}$ for $i = 1, 2$.*

Proof. Since each orbit contains triangular matrices, we may assume B_1, B_2 triangular, so $(B_i)_s \in \overline{0(B_i)}$ as was shown above. But then $0((B_1)_s) = 0((B_2)_s)$ implies $\overline{0(B_1)} \cap \overline{0(B_2)} \neq \phi$. The inverse implication follows e.g. from the existence of a morphism $\varphi : A^{n^2} \to A^n$ separating matrices with non-equivalent semi-simple parts (cf. proposition 1).

It is not hard to see that each fibre of the canonical morphism $\varphi : A^{n^2} \to A^n$ contains a unique closed orbit (semi-simple matrices) and a unique relatively open orbit (matrices with a cyclic vector), which coincide if all the eigenvalues are different. Furthermore, the union of the relatively open orbits is open in A^{n^2} whereas the union of the closed orbits is neither open nor closed (if $n > 1$). These facts are reflected in the moduli problem as shown by propositions 2 and 3.

We shall now consider the quotient of a variety by a group in general with applications to moduli problems in mind. We have seen that the closedness of orbits is one desirable property. For technical reasons it is convenient to impose the following conditions on a good orbit space:

DEFINITION 5. Let G be a group operating on a variety X. A *geometric quotient* of X by G is a pair (Y, φ) consisting of a variety Y and a morphism $\varphi : X \to Y$ satisfying:

(i) for each closed point $y \in Y$, $\varphi^{-1}(y)$ is an orbit, i.e., a closed invariant subset such that G acts transitively on its closed points.

(ii) for each invariant open subset $U \subset X$ there is an open subset $V \subset Y$ such that $U = \varphi^{-1}(V)$.

(iii) for each open set $V \subset Y$, $\varphi^* : \Gamma(V, \mathcal{O}_Y) \to \Gamma(\varphi^{-1}(V), \mathcal{O}_X)$ is an isomorphism of $\Gamma(V, \mathcal{O}_Y)$ onto the ring $\Gamma(\varphi^{-1}(V), \mathcal{O}_X)^G$ of invariant functions on $\varphi^{-1}(V)$.

REMARK. The condition (ii) is weaker than the corresponding condition iii) of definition 0.6 in (GIT, p.4).

The first thing to prove is

PROPOSITION 5. *A geometric quotient of a variety by a group is a quotient. In particular, it is unique up to isomorphism.*

PROOF. Let $\Psi : X \to Z$ be a morphism which is constant on the orbits of closed points. If (W_i) is an affine open covering of Z, each $\Psi^{-1}(W_i)$ is an invariant open subset of X, hence by condition (ii) of definition 5 there is an open set $V_i \subset Y$ such that $\varphi^{-1}(V_i) = \Psi^{-1}(W_i)$. Since φ is surjective by (i), (V_i) is a covering of Y.

Now, any morphism $\kappa : Y \to Z$ such that $\Psi = \kappa \circ \varphi$ must satisfy $\kappa(V_i) \subset W_i$. Hence $\kappa | V_i$ is defined by a homomorphism $h_i : \Gamma(W_i, \mathcal{O}_Z) \to \Gamma(V_i, \mathcal{O}_Y)$ such that $\varphi^* \circ h_i = \Psi^* : \Gamma(W_i, \mathcal{O}_Z) \to \Gamma(\Psi^{-1}(W_i), \mathcal{O}_X)$. Since φ^* is injective by (iii) of definition 5, h_i is uniquely determined. Hence at most one κ exists.

But Ψ^* maps $\Gamma(W_i, \mathcal{O}_Z)$ into the ring of invariant functions

$$\Gamma(\Psi^{-1}(W_i), \mathcal{O}_X)^G = \varphi^*\Gamma(V_i, \mathcal{O}_Y).$$

Therefore such an h_i exists and defines a morphism $\kappa_i : V_i \to W_i$. By uniqueness $\kappa_i = \kappa_j$ on $V_i \cap V_j$; hence $\kappa : Y \to Z$ may be constructed.

In the rest of this chapter we shall assume that G is an algebraic group (LAG, 1.1), acting algebraically on a variety X; in other words, the action is defined by a morphism of varieties $\sigma : G \times X \to X$ (LAG, 1.7). In this case, the orbits are locally closed subvarieties (LAG, 1.8).

If $G = \mathrm{Spec}(S)$ is affine, and $R = \Gamma(X, \mathcal{O}_X)$, then σ defines a k-algebra homomorphism

$$\sigma^* : R \to S \otimes R = \Gamma(G \times X, \mathcal{O}_{G\times X}).$$

More generally, an action of G on a k-vector space V is given by a linear map

$$\hat{\sigma} : V \to S \otimes V$$

such that

$$\begin{array}{ccc} V & \xrightarrow{\hat{\sigma}} & S \otimes V \\ {\scriptstyle\hat{\sigma}}\downarrow & & \downarrow{\scriptstyle 1\otimes\hat{\sigma}} \\ S \otimes V & \xrightarrow[\hat{\mu}\otimes 1]{} & S \otimes S \otimes V \end{array}$$

commutes ($\hat{\mu} : S \to S \otimes S$ is induced by the morphism $\mu : G \times G \to G$ defining the group structure) and

$$V \xrightarrow{\hat{\sigma}} S \otimes V \xrightarrow{\varepsilon\otimes 1} k \otimes V \cong V$$

is the identity ($\mathscr{E} : S \to k$ is given by $\mathscr{E}(f) = f(1)$).

Indeed, if V is finite-dimensional, with basis $e_1, \cdots, e_n$, and $\hat{\sigma}(e_i) = \sum_j a_{ij} \otimes e_j$ $(1 \leqq i \leqq n)$, then the elements $a_{ij} \in S$ define a group homomorphism $G \to GL(n)$.

In general, a closed point g of G operates on V (on the right) by

$$x \to xg = \sum_i a_i(g) \cdot x_i$$

if $\hat{\sigma}(x) = \sum a_i \otimes x_i$. It follows immediately that each vector $x \in V$ is contained in a finite-dimensional invariant subspace ($\sum kx_i$ if the x_i are linearly independent and $a_i \neq 0$). Clearly, x is invariant if and only if $\hat{\sigma}(x) = 1 \otimes x$, and a subspace $W \subset V$ is invariant if and only if $\hat{\sigma}(W) \subset S \otimes W$.

Definition 6. An affine group G is *reductive* if each action of G on a finite-dimensional vector space V is completely reductive, i.e., if $W \subset V$ is an invariant subspace, then there is an invariant subspace $W' \subset V$ such that $V = W \oplus W'$.

If the characteristic of k is 0, it may be shown that semi-simple groups are reductive (IT, 4.37).

A basic property of reductive groups is the following:

LEMMA. *If G is a reductive group acting on a vector space V, then the subspace V^G of invariant elements of V has a unique invariant complement V_G in V.*

PROOF. By Zorn's lemma there is a maximal invariant subspace $V_G \subset V$ such that $(V_G)^G = V_G \cap V^G = 0$. If $V' \subset V$ is any invariant subspace and $x \in V'$, there is a finite-dimensional invariant subspace $W \subset V'$ containing x. By complete reductivity, there is an invariant subspace $W' \subset W$ such that $W = (W \cap V_G) \oplus W'$. If $(V')^G = 0$ then $(W')^G = 0$ and therefore $(V_G \oplus W')^G = 0$. By the maximality of V_G we have $W' = 0$. Hence $V' \subset V_G$, which proves the uniqueness of V_G.

Finally, to show that $V^G \oplus V_G = V$, let $x \in V$ and let $W \subset V$ be a finite-dimensional invariant subspace containing x. Then there is an invariant subspace $W' \subset W$ such that $W = (W \cap V^G) \oplus W'$. But then $(W')^G = 0$, so $W' \subset V_G$, and therefore $x \in W \subset V^G \oplus V_G$.

The result of this lemma may be conveniently formalized by means of the *Reynolds operator* $E : V \to V$. It is the projection of V onto V^G with kernel V_G.

PROPOSITION 6. *Let G be a reductive group acting on vector spaces V and V' with Reynolds operators E and E', respectively. Then each G-linear map $u : V \to V'$ commutes with E and E':*

$$E' \circ u = u \circ E.$$

PROOF. Since $u(V^G) \subset (V')^G$, it is enough to show that $u(V_G) \subset (V')_G$. If $x \in V_G$, there is a finitedimensional invariant subspace $W \subset V_G$ containing x, and $W = (W \cap \ker(u)) \oplus W'$ for some invariant subspace $W' \subset W$. But u maps W' isomorphically onto $u(W) \subset V'$. Hence $(u(W))^G = u((W')^G) = 0$, and therefore $u(x) \in u(W) \subset V'_G$.

COROLLARY. *If a reductive group G acts on a k-algebra R by algebra automorphisms (i.e. $x \mapsto x \circ g$ is an algebra automorphism of R for each closed point $g \in G$), then the Reynolds operator E on R satisfies the Reynolds identity*

$$E(x \circ y) = x \circ E(y)$$

for $x \in R^G$, $y \in R$.

In fact, if $x \in R^G$, $y \to x \circ y$ is a G-linear map of R. Hence it commutes with E.

REMARK. If $G = \mathrm{Spec}(S)$, the assumption of the corollary means that the action $\hat{\sigma} : R \to S \otimes R$ is an algebra homomorphism.

We can now prove the main result of this chapter.

THEOREM 1. *Let G be a reductive group acting on an affine variety X with closed orbits. Then the geometric quotient (Y, φ) of X by G exists and Y is an affine variety.*

PROOF. Let $R = \Gamma(X, \mathcal{O}_X)$. Then G acts on R. Let $Y = \operatorname{Spec}(R^G)$ and define $\varphi : X \to Y$ by the inclusion $R^G \to R$. We claim that Y is an affine variety, i.e., R^G is a k-algebra of finite type.

LEMMA 1. *If S is an R^G-algebra, then S is the ring of invariants in $R \otimes_{(R^G)} S$.*

PROOF. Let E and E' be the Reynolds operators on R and $R \otimes_{(R^G)} S$ respectively. By proposition 6, $E'(a \otimes 1) = E(a) \otimes 1$ for $a \in R$. Since R is isomorphic as R^G-module to $R^G \oplus \ker E$, it follows that

$$S \cong R^G \otimes_{(R^G)} S \subset (R \otimes_{(R^G)} S)^G.$$

Conversely, if

$$f = \sum a_i \otimes b_i \in (R \otimes_{(R^G)} S)^G,$$

then

$$\begin{aligned} f &= E'(\textstyle\sum a_i \otimes b_i) = E'(\sum (a_i \otimes 1)(1 \otimes b_i)) \\ &= \textstyle\sum E'(a_i \otimes 1) \circ (1 \otimes b_i) \text{ (by Reynolds identity)} \\ &= \textstyle\sum E(a_i) \otimes b_i \in R^G \otimes_{(R^G)} S. \end{aligned}$$

If I is an ideal of R^G, then $R/IR \simeq R \otimes_{(R^G)} (R^G/I)$. Hence by lemma 1 $(R/IR)^G = R^G/I$, and therefore $IR \cap R^G = I$. This means that $I \mapsto IR$ is an order preserving injection of the set of ideals in R^G into the set of ideals in R. Since R is noetherian, R^G is also noetherian.

If $R = \sum_{n \geq 0} R_n$ is a graded k-algebra with $R_0 = k$ and the action of G preserves the gradation, $R^G = \sum_{n \geq 0} R_n^G$ is also a graded algebra. Since it is noetherian, the ideal $R_+^G = \sum_{n \geq 0} R_n^G$ is generated by a finite number of homogeneous elements $f_i \in R_{n_i}^G$ $(1 \leq i \leq r)$. By induction on n it is then easily shown that each vector space R_n^G is generated by monomials of $f_1, \cdots, f_r$. Hence R^G is finitely generated as a k-algebra.

Finally, in the general case, let $V \subset R$ be a finite-dimensional invariant subspace containing a set of generators. Then the action of G on V extends to a gradation preserving action on the symmetric algebra $R' = S(V)$, and the canonical algebra homomorphism $u : R' \to R$ is G-linear and surjective. If E and E' are the Reynolds operators on R and R', then we have

$$R^G = E(R) = E(u(R')) = u(E'(R')) = u((R')^G)$$

by proposition 6. Hence R^G is finitely generated as a quotient of a finitely generated k-algebra $(R')^G$. This proves that Y is a variety.

LEMMA 2. *If (I_i) is a family of invariant ideals in R then*

$$(\sum_i I_i) \cap R^G = \sum_i (I_i \cap R^G).$$

PROOF. If $f \in (\sum I_i) \cap R^G$, then f is a finite sum $\sum f_i$ with $f_i \in I_i$. It follows that

$$f = Ef = \sum Ef_i \in \sum_i (I_i \cap R^G)$$

since the Reynolds operator of I_i is the restriction of the Reynolds operator E on R by proposition 6.

Writing Z_i for the closed subset of X defined by I_i we obtain the following geometric statement:

If (Z_i) is a family of closed invariant subsets of X, then

$$(*) \qquad \overline{\varphi(\bigcap_i Z_i)} = \bigcap_i \overline{\varphi(Z_i)}.$$

Now, if Z is a closed invariant subset of X and $Z' = \varphi^{-1}(y)$ where y is a closed point of Y, then

$$\overline{\varphi(Z \cap Z')} = \overline{\varphi(Z)} \cap \{y\}.$$

Hence $y \in \overline{\varphi(Z)}$ implies $Z \cap Z' \neq \phi$, i.e., $y \in \varphi(Z)$. Therefore $\varphi(Z)$ is closed, and $(*)$ becomes

$$(**) \qquad \varphi(\bigcap_i Z_i) = \bigcap_i \varphi(Z_i)$$

In particular, $\varphi(X)$ is closed in Y. But φ is dominant, hence $\varphi(X) = Y$.

We now claim that the conditions (i), (ii) and (iii) of definition 5 are satisfied by (Y, φ).

(i) If y is a closed point of Y, then $\varphi^{-1}(y)$ contains at least one orbit since φ is surjective. If $Z_1, Z_2 \subset \varphi^{-1}(y)$ are two orbits, then

$$\varphi(Z_1 \cap Z_2) = \varphi(Z_1) \cap \varphi(Z_2) = \{y\},$$

since Z_1 and Z_2 are closed by assumption. Therefore $Z_1 \cap Z_2 \neq \phi$, i.e., $Z_1 = Z_2$.

(ii) If U is an invariant open subset of X, $Z = X \setminus U$ is closed and invariant. Therefore $\varphi(Z)$ is closed in Y. If V is its open complement, then $\varphi^{-1}(V) \subset U$. On the other hand, the orbit of any closed point of U is a closed invariant subset Z' of X such that $Z \cap Z' = \phi$. Therefore $\varphi(Z') \cap \varphi(Z) = \phi$ and Z' is contained in $\varphi^{-1}(V)$.

(iii) If $V = D(f)$ is an affine open subset of Y, $\Gamma(V, \mathcal{O}_Y) = R_f^G$ is the ring of invariants in $\Gamma(\varphi^{-1} V, \mathcal{O}_X) = R_f = R \otimes_{R^G} R_f^G$ by lemma 1. The same is true for any open subset V of Y by the basic properties of sheaves.

This concludes the proof of the theorem.

Remark. If the orbits of closed points of X are not assumed closed, the following is still true:

(1) If x and x' are closed points of X then $\varphi(x) = \varphi(x')$ if and only if $\overline{0(x)} \cap \overline{0(x')} \neq \phi$.

(2) For each closed point y of Y, $\varphi^{-1}(y)$ contains a unique closed orbit.

(3) There is an invariant open set $X_S \subset X$ such that a closed point x of X is in X_S if and only if the orbit $0(x)$ is closed and the stabilizer $S(x)$ is of minimal dimension. Then $Y_S = \varphi(X_S)$ is open in Y and $(Y_S, \varphi|X_S)$ is a geometric quotient of X_S by G.

In fact, (1) may be proved as (i) above. To verify (2) note that a minimal closed invariant subset of $\varphi^{-1}(y)$ is an orbit (LMG, 1.8); uniqueness follows from (1). For a proof of (3), consider the invariant open set X^{reg} which consists of the points whose stabilizers has minimal dimension (GIT, 0.9). Then $\varphi(X \setminus X^{\text{reg}})$ is closed in Y and its complement is Y_S, $X_S = \varphi^{-1}(Y_S)$. The rest of the proof is as in theorem 1.

Finally, we note that a slight generalization of the proof given above shows that (Y, φ) is a quotient of X by G even if the orbits are not closed (GIT, theorem 1.1).

2. *n* ordered points on a line

The moduli problem has led us to consider quotients of schemes by groups. The affine case was studied in chapter 1. In this chapter we shall examine the quotient of a projective variety by means of an elementary example.

The projective group $PGL(2) = GL(2)/G_m$ (over k) acts canonically on the projective line $\boldsymbol{P}^1 = P_k^1$, hence on the product $(\boldsymbol{P}^1)^n$ for each integer n. To construct a quotient of $(\boldsymbol{P}^1)^n$ under this action, we might proceed as follows: find invariant affine open sets $U_i \subset (\boldsymbol{P}^1)^n$, find quotients V_i of the U_i by $PGL(2)$ using results of chapter 1, and join the V_i together along the quotients of the $U_i \cap U_j$. In this case, however, there is a more direct method. We assume $n \geqq 3$.

Closed points of $(\boldsymbol{P}^1)^n$ are n-tuples $(x_1, \cdots, x_n)$ of closed points of $\boldsymbol{P}^1$. Let U_{123} be the invariant open subset of $(\boldsymbol{P}^1)^n$ whose closed points are those with x_1, x_2, x_3 distinct. Then the orbit of any closed point $(x_1, \cdots, x_n)$ of U_{123} contains a unique closed point of the form $(0, 1, \infty, y_1, \cdots, y_{n-3})$. It follows that the action $\sigma : PGL(2) \times (\boldsymbol{P}^1)^n \to (\boldsymbol{P}^1)^n$ induces an isomorphism

$$PGL(2) \times (\boldsymbol{P}^1)^{n-3} \xrightarrow{\sim} U_{123}$$

mapping a closed point $(g, y_1, \cdots, y_{n-3}) \in PGL(2) \times (P^1)^{n-3}$ onto $\sigma(g, 0, 1, \infty, y_1, \cdots, y_{n-3})$.

If $PGL(2)$ acts on itself by left translations and trivially on $(P^1)^{n-3}$, then the isomorphism is $PGL(2)$-linear, i.e., U_{123} is a trivial principal $PGL(2)$-bundle over $(P^1)^{n-3}$.

For each triple (i, j, k) with $1 \leqq i, j, k \leqq n$ distinct, let U_{ijk} denote the invariant open set of $(P^1)^n$ whose closed points are those with x_1, x_j, x_k distinct. Then we find as above that U_{ijk} is a trivial $PGL(2)$-bundle over a scheme P_{ijk} isomorphic to $(P^1)^{n-3}$.

Given two triples (i, j, k) and (i', j', k') the intersection

$$U' = U_{ijk} \cap U_{i'j'k'}$$

is an invariant open set. Hence its image U_1 in P_{ijk} is canonically isomorphic to its image U_2 in $P_{i'j'k'}$. Indeed, a morphism $U_1 \to U_2$ can be defined by composing the projection $U' \to U_2$ with a section $U_1 \to U'$. Joining the P_{ijk} together by these natural isomorphisms we obtain a scheme M^*.

The union $U^* = \bigcup_{1 \geqq i > j > k \geqq n} U_{ijk}$ is an invariant open subset of $(P^1)^n$, with closed points $(x_1, \cdots, x_n)$ such that at least three of the x_i are distinct. It is clear that there is a natural morphism $\tau : U^* \to M^*$ making U^* a principal fibre bundle over M^* with structure group $PGL(2)$.

Thus it seems that M^* is a reasonable (partial) solution of the quotient problem for $(P^1)^n$. In fact it is easy to see that (M^*, τ) is a geometric quotient of U^*. But the trouble is that M^* is *not separated* if $n > 3$, hence it cannot be quasi-projective.

EXAMPLE A: $n = 4$. Let us identify P_{123} with P^1 so that $\tau(0, 1, \infty, y)$ corresponds to y. If P_{134} is identified with P^1 by $\tau(0, y', \infty, 1) \leftrightarrow y'$, then the image of the diagonal map $P_{123} \cap P_{134} \to P_{123} \times P_{134} \cong (P^1)^2$ is given by $yy' = 1$, $y \neq 0$, $y' \neq 0$. Since it is not closed M^* is not separated. In fact, if y specializes to 0, then y' specializes to ∞, but the points with $x_1 = x_4$ in U_{123} are different from the points with $x_2 = x_3$ in P_{134}. Hence we get the following picture:

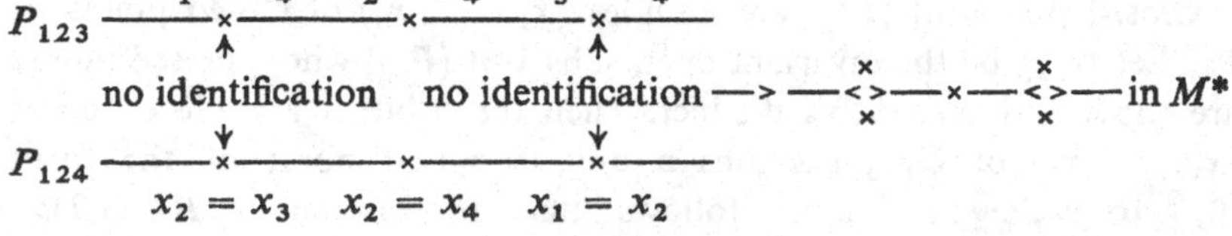

It is interesting to not e that permuting (x_1, x_2) with (x_3, x_4) leaves $P_{123} \cup P_{134}$ invariant interchanging the components of the double points.

Adding P_{124} and P_{234} brings forth another doubled point, that corresponding to points of U^* with either $x_2 = x_4$ or $x_1 = x_3$.

EXAMPLE B. $n = 5$. Let us consider P_{123} and P_{124}, both isomorphic to $P^1 \times P^1$:

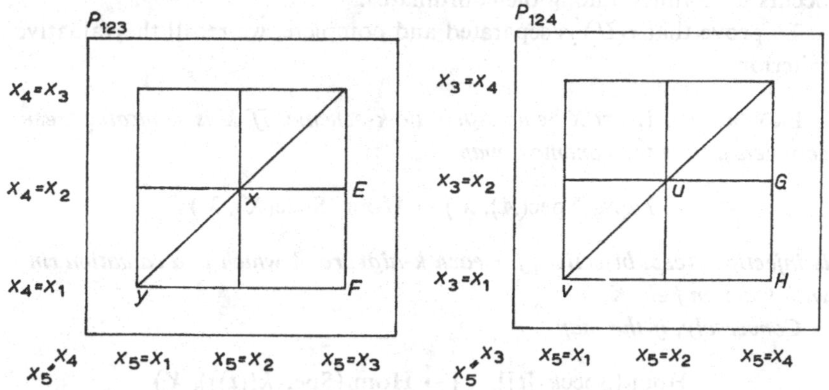

The intersection of P_{123} and P_{124} on M^* defines an isomorphism of the open subsets $P_{123} \setminus (E \cup F)$ and $P_{124} \setminus (G \cup H)$. Let $U \subset P_{123} \times P_{124}$ be the graph of this isomorphism. The complement of U in its closure P consists of two pairs of intersecting lines

$$E \times \{v\} \cup \{x\} \times H \text{ and } F \times \{u\} \cup \{y\} \times G$$

Hence $P_{123} \cup P_{124}$ is not separated as a subscheme of M^*. However: $P_{123} \setminus \{x, y\}$ and $P_{124} \setminus \{u, v\}$ are mapped isomorphically onto open subsets of P. Therefore, after omitting x, y, u, v there remains a separated subscheme which is isomorphic to $P \setminus \{(x, v), (y, u)\}$.

In the general case, to obtain a separated quotient, we must leave out part of U^*. This is quite trivial if no restriction is imposed. But if the quotient should be complete, we must be careful not to omit too many points.

PROBLEM. Does there exist an open subset $U \subset U^*$ invariant under the action of $PGL(2)$ and under permutations of the coordinates such that $\tau(U)$ is separated and complete?

Considering first the case $n = 4$, we see by example A that no such U exists. There are open subsets invariant under $PGL(2)$ such that the quotient is separated and complete, but no such set is invariant under permutations of the coordinates, since the components of the doubled points are interchanged by permutations. In fact, I conjecture that there is no solution for even integers n.

On the other hand, if n is odd, there is always a remarkable solution. Indeed, if $n = 5$, it is not hard to see that the quotient is separated and complete after the omission of the points with some three coordinates coinciding (such as x, y, u, v in example B). In general, if $n = 2e+1$, we define U as the open set whose closed points are such that no point of $\boldsymbol{P}^1$ occurs $e+1$ times among the coordinates.

To prove that $\tau(U)$ is separated and complete, we recall the valuative criterion:

PROPOSITION 1. *Let X be an algebraic k-scheme. If X is separated (resp. complete), then the canonical map*

$$\mathrm{Hom}_k(\mathrm{Spec}(A), X) \to \mathrm{Hom}_k(\mathrm{Spec}(K), X)$$

is injective (resp. bijective) for each k-algebra A which is a valuation ring with fraction field K.

Conversely, if the map

$$\mathrm{Hom}_k(\mathrm{Spec} k[[t]], X) \to \mathrm{Hom}_k(\mathrm{Spec}(k((t))), X)$$

is injective (resp. bijective), then X is separated (resp. complete).

PROPOSITION 2. $M = \tau(U)$ *is separated and complete.*

PROOF. To show that M is separated, let $R = k[[t]]$, $K = k((t))$, and consider two morphisms $\mathrm{Spec}(R) \to M$ with the same restriction to $\mathrm{Spec}(K)$. We claim that the two morphisms are equal. Using local sections of τ it is possible to lift these morphisms to R-valued points of U. By a suitable choice of a point $\infty \in \boldsymbol{P}^1$, we may assume that they factor through $(\boldsymbol{A}^1)^n \subset (\boldsymbol{P}^1)^n$, hence are of the form $x = (x_1(t), \cdots, x_n(t))$ and $y = (y_1(t), \cdots, y_n(t))$ with $x_i(t), y_i(t) \in R$ $(1 \leqq i \leqq n)$. Regarded as K-valued points, x and y have the same image in M. Hence there is a K-valued point σ of $PGL(2)$ such that $y_K = \sigma \cdot x_K$. In other words, there are elements $a(t), b(t), c(t), d(t)$ of $k((t))$ such that

$$y_i(t) = \frac{a(t)x_i(t)+b(t)}{c(t)x_i(t)+d(t)} \qquad (1 \leqq i \leqq n).$$

Clearly, we may assume that $a(t), b(t), c(t), d(t)$ are in $k[[t]]$ with constant terms $a(0), b(0), c(0), d(0)$ not all equal to 0.

Now, if $a(0) \cdot d(0) - b(0) \cdot c(0) \neq 0$, then

$$\Delta(t) = a(t)d(t) - b(t)c(t)$$

is invertible in R, and therefore σ is actually an R-valued point of $PGL(2)$. But then $\tau(x) = \tau(y)$ on $\mathrm{Spec}(R)$ as was claimed.

On the other hand, if $\Delta(0) = 0$, but $c(0) \neq 0$ or $d(0) \neq 0$, then for each i we have either

$$x_i(0) = -\frac{d(0)}{c(0)} \text{ or } y_i(0) = \frac{a(0)}{c(0)} = \frac{b(0)}{d(0)}$$

One of these conditions holds for at least $e+1 = (n+1)/2$ integers i. But this is impossible, since $(x_1(0), \cdots, x_n(0))$ and $(y_1(0), \cdots, y_n(0))$ are k-valued points of U.

Finally, if $c(0) = d(0) = 0$, then $a(0) \neq 0$, and

$$x_i(0) = -\frac{b(0)}{a(0)}$$

for each i contradicting the assumption.

To prove the completeness of M, we show that each morphism $\mathrm{Spec}(K) \to M$ may be extended to $\mathrm{Spec}(R)$, where $R = k[[t]]$ and $K = k((t))$ as above.

Any K-valued point of M may be lifted to a K-valued point of U: $(x_1(t), \cdots, x_n(t))$ with $x_i(t) \in K \cup \{\infty\}$ $(1 \leqq i \leqq n)$. Hence it will suffice to show that there is a K-valued point σ of $PGL(2)$ such that $(\sigma x_1(t), \cdots, \sigma x_n(t))$ is an R-valued point of U. In any case, each $x_i(t)$ defines a unique morphism $\mathrm{Spec}(R) \to \boldsymbol{P}^1$ since $\boldsymbol{P}^1$ is separated and complete (proposition 1). The restriction of this morphism to the closed point of $\mathrm{Spec}(R)$ is given by $x_i(0) \in k \cup \{\infty\}$. If $(x_1(0), \cdots, x_n(0))$ is a k-valued point of U, $(x_1(t), \cdots, x_n(t))$ is an R-valued point of U and there is nothing to prove.

In general, we proceed by induction on the least integer p such that no $e+1$ of the morphisms $\mathrm{Spec}(R/R \cdot t^{p+1}) \to \boldsymbol{P}^1$ defined by the $x_i(t)$ $(1 \leqq i \leqq n)$ are equal. We may assume that none of the $x_i(0)$ are infinite by a suitable choice of coordinates on $\boldsymbol{P}^1$. Writing $x_i(t) = \sum_{j=0}^{\infty} a_{ji} t^j$, we see that the condition means that no $e+1$ of the polynomials

$$\sum_{j=0}^{p} a_{ji} t^j$$

are equal. Incidentally, this shows that p is finite.

If $p = 0$, then no $e+1$ of the points $x_i(0)$ coincide, and $(x_1(t), \cdots, x_n(t))$ is an R-valued point of U as was seen above.

In case $p > 0$, there is a unique polynomial $\sum_{j=0}^{p-1} a_{ji} t^j$ occurring more than e times. We may assume for convenience that these polynomials have indices $1 \leqq i \leqq r$ with $e+1 \leqq r \leqq n$. Let σ be the K-valued point of $PGL(2)$ defined by

$$\sigma = \begin{pmatrix} 1 & -a \\ 0 & t \end{pmatrix}$$

or

$$x(t) \mapsto \sigma x(t) = \frac{x(t)-a}{t}$$

where $a \in k$ is the common value of the constants $x_i(0) = a_{0i}$ $(1 \leqq i \leqq r)$. Then it is easy to verify that $(\sigma x_1(t), \cdots, \sigma x_n(t))$ satisfies the induction assumption with p replaced by $p-1$. This completes the proof.

This result is interesting as such, but we can prove more:

THEOREM 1. *M is projective.*

PROOF. The general method of constructing ample invertible sheaves on a quotient of a scheme X is to search for ample invertible sheaves on X such that the action of the group extends to the sheaf and to apply the theory of descent. In our case, however, there is a more elementary way of producing invertible sheaves on M.

For each pair (i,j) of integers $1 \leqq i,j \leqq n$ with $i \neq j$, let D_{ij} denote the closed subset of U defined by $x_i = x_j$. Since D_{ij} is invariant under the action of $PGL(2)$, $\Delta_{ij} = \tau(D_{ij})$ is closed in M and $D_{ij} = \tau^{-1}(\Delta_{ij})$. This is easily proved by using local sections of τ. Furthermore, Δ_{ij} is irreducible of codimension one, i.e., a prime divisor on M.

Let $L_{ij} = \mathcal{O}_M(\Delta_{ij})$ be the invertible sub-$\mathcal{O}_M$-Module of the sheaf of rational functions on M whose sections are regular everwhere except for at most a simple pole at Δ_{ij}. In other words, if $f \in \Gamma(V, \mathcal{O}_M)$ is a local equation of Δ_{ij}, then the sections of L_{ij} over V are the multiples of f^{-1}. We denote by $\delta_{ij} \in \Gamma(M, L_{ij})$ the canonical section 1. To find relations among the L_{ij} we embed $\mathrm{Pic}(M)$ into $\mathrm{Pic}(U)$, which is isomorphic to $\mathrm{Pic}((P^1)^n)$, since $(P^1)^n \setminus U$ is of codimension $i = (n-1)/2 \geqq 2$ for $n \geqq 5$ (the case $n = 3$ is trivial).

LEMMA 1. $\tau^* : \mathrm{Pic}(M) \to \mathrm{Pic}(U)$ *is injective.*

PROOF. Let D be a divisor on M such that $\tau^{-1}(D)$ is linearly equivalent to 0, i.e., $\tau^{-1}(D) = \mathrm{div}(f)$ for some rational function f on U, or, what amounts to the same, on $(P^1)^n$. Then for each closed point σ of $PGL(2)$, $\mathrm{div}(\sigma(f)) = \mathrm{div}(f)$. This implies that $\sigma(f) = \chi(\sigma) \cdot f$ for some constant $\chi(\sigma) \in k^*$, since $(P^1)^n$ is complete. But then χ is a character of $PGL(2)$. [In fact, the action of $PGL(2)$ on the generic fibre $Z = \tau^{-1}(y)$ of τ induces a homomorphism

$$\begin{aligned} \hat{\sigma} : \Gamma(Z, \mathcal{O}_Z) &\to \Gamma(PGL(2) \times Z, \mathcal{O}_{PGL(2) \times Z}) \\ &= \Gamma(PGL(2), \mathcal{O}_{PGL(1)}) \otimes_k \Gamma(Z, \mathcal{O}_Z). \end{aligned}$$

If $\hat{\sigma}(f) = \sum \chi_i \otimes f_i$ where $f_i \in \Gamma(Z, \mathcal{O}_Z)$ $(1 \leqq i \leqq r)$ are linearly independent with $f_1 = f$ and $\chi_i \in \Gamma(PGL(2)\ \mathcal{O}_{PGL(2)})$ $(1 \leqq i \leqq r)$, then it is readily seen that $\chi_i = 0$ for $i > 1$ and $\chi_1 = \chi$ is an invertible section of $\mathcal{O}_{PGL(2)}$ defining a group homomorphism $PGL(2) \to G_m$].

But $PGL(2)$ has no non-trivial characters, since it coincides with its commutator subgroup (LAG, 10.8(2)). Therefore $\chi = 1$ and f is invariant under $PGL(2)$. Again, using local sections of τ, it is shown that $f = g \circ \tau$ for some rational function g on M and $D = \operatorname{div}(g)$.

There remains to study the structure of $\operatorname{Pic}((P^1)^n)$ and the image of $\operatorname{Pic}(M)$ in $\operatorname{Pic}((P^1)^n)$. It is clear that $\tau^*(L_{ij}) = \mathcal{O}_U(D_{ij})$. On the other hand, considering the rational function

$$\frac{x_0 y_1 - x_1 y_0}{x_0 y_0}$$

on $P^1 \times P^1$ with bihomogeneous coordinates $(x_0, x_1; y_0, y_1)$, we find that the diagonal D of $P^1 \times P^1$ is linearly equivalent to the divisor $P^1 \times \{0\} + \{0\} \times P^1$, or, in terms of invertible sheaves,

$$\mathcal{O}(D) \simeq p_1^*(\mathcal{O}(1)) \otimes p_2^*(\mathcal{O}(1))$$

where $p_1, p_2 : P^1 \times P^1 \to P^1$ are the projections.

Hence $L_{ij} = L_{ji}$ corresponds to $p_i^*(\mathcal{O}(1)) \otimes p_j^*(\mathcal{O}(1))$ on $(P^1)^n$. It follows that

$$L_{ij} \otimes L_{kl} \simeq L_{ik} \otimes L_{jl}$$

for each quadruple (i, j, k, l) with $i \neq j$, $k \neq l$, $i \neq k$, $j \neq l$. In fact, $\Delta_{ij} + \Delta_{kl} - \Delta_{ik} - \Delta_{jl}$ is the divisor of the rational function on M defined by the invariant crossratio of the four coordinates (x_i, x_j, x_k, x_l) of a point of U.

It may be shown that $p_1^*(\mathcal{O}(1)), \cdots, p_n^*(\mathcal{O}(1))$ are free generators of $\operatorname{Pic}((P^1)^n)$ and the image of $\operatorname{Pic}(M)$ is a subgroup of index 2. But this is not necessary for our purposes.

For any i between 1 and n, $p_i^*(\mathcal{O}(2))$ is in the image of $\operatorname{Pic}(M)$, namely, it corresponds to

$$L_{ii} = L_{ik} \otimes L_{il} \otimes L_{kl}^{-1}$$

for some k, l with i, k, l distinct. By lemma 1, L_{ii} is independent of k and l. *We claim that* $L = L_{11} \otimes \cdots \otimes L_{nn}$ *is ample.*

The proof is based on the following observation:

$$L \simeq L_{i_1 j_1} \otimes \cdots \otimes L_{i_n j_n}$$

if each integer between 1 and n occurs exactly twice in

$$(i_1, \cdots, i_n, j_1, \cdots, j_n),$$

and further, if $i_k \neq j_k$ for $1 \leq k \leq n$, then there is a section

$$\delta_{i_1 j_1} \otimes \cdots \otimes \delta_{i_n j_n} \in \Gamma(M, L_{i_1 j_1} \otimes \cdots \otimes L_{i_n j_n})$$

having $\Delta_{i_1 j_1} \cup \cdots \cup \Delta_{i_n j_n}$ as its set of zeros.

LEMMA 2. *For each closed point* $(x_1, \cdots, x_n)$ *of* U *there is a sequence* $(i_1, \cdots, i_n, j_1, \cdots, j_n)$ *where each integer between* 1 *and* n *occurs exactly twice such that* $x_{i_k} \neq x_{j_k}$ for $1 \leq k \leq n$ *and* $(i_1, i_2, i_3) = (j_3, j_1, j_2)$.

PROOF. By induction on $l = (n-1)/2$. The case $l = 1$ is trivial. If $l > 1$, we choose i_n and j_n such that x_{i_n} and x_{j_n} are two distinct points occurring with maximal multiplicity in $(x_1, \cdots, x_n)$. Since there are at most two points with multiplicity l, each of the remaining points occurs at most $l-1$ times. If one of the points x_{i_n}, x_{j_n} occurs with multiplicity one, then either all of the points x_i are distinct or exactly one of them has multiplicity greater than one (but less than $l+1$). In either case at least three distinct points remain after omitting x_{i_n} and x_{j_n}. Hence, taking $i_{n-1} = i_n, j_{n-1} = j_n$, we are reduced to the case $l-1$.

Let $(i_1, \cdots, i_n, j_1, \cdots, j_n)$ be as in lemma 2 and denote by $M_{i_1 \cdots i_n, j_1 \cdots j_n}$ the open subset of M where $\delta_{i_1 j_1} \otimes \cdots \otimes \delta_{i_n j_n}$ does not vanish, i.e.,

$$M_{i_1 \cdots i_n j_1 \cdots j_n} = M \diagdown (\Delta_{i_1 j_1} \cup \cdots \cup \Delta_{i_n j_n}).$$

These sets form an open covering of M by Lemma 2. Hence *it suffices to show that they are affine* (EGA II, 4.5.2, last statement).

Now $\tau^{-1}(M_{i_1 \cdots i_n j_1 \cdots j_n})$ is contained in $U_{i_1 i_2 i_3}$.

Therefore $M_{i_1 \cdots i_n j_1 \cdots j_n}$ is a subset of $P_{i_1 i_2 i_3}$, which is isomorphic to $(P^1)^{n-3}$. Furthermore, the complement of $M_{i_1 \cdots i_n j_1 \cdots j_n}$ in $P_{i_1 i_2 i_3}$ is the set of zeros of a section of an invertible sheaf L' such that $\tau^* L'$ is isomorphic to $p_1^*(\mathcal{O}(2)) \otimes \cdots \otimes p_n^*(\mathcal{O}(2))$ restricted to $U_{i_1 i_2 i_3}$. In fact,

$$L' \simeq \iota^*(p_1^*(\mathcal{O}(2)) \otimes \cdots \otimes p_n^*(\mathcal{O}(2)))$$

where ι is a section of τ over $P_{i_1 i_2 i_3}$. Hence L' corresponds to

$$p_1^*(\mathcal{O}(2)) \otimes \cdots \otimes p_{n-3}^*(\mathcal{O}(2))$$

under the isomorphism $P_{i_1 i_2 i_3} \to (P^1)^{n-3}$. But

$$p_1^*(\mathcal{O}(1)) \otimes \cdots \otimes p_{n-3}^*(\mathcal{O}(1))$$

is very ample (it defines the Segre morphism $(P^1)^{n-3} \to P^{2^{n-3}-1}$, cf. EGA II, 4.3). Therefore L' is ample. Thus we conclude that $M_{i_1 \cdots i_n j_1 \cdots j_n}$ is affine (EGA II, 5.5.7). This ends the proof of theorem 1.

To finish the chapter we shall discuss quotients of projective schemes more generally but without giving proofs. Let G be a reductive group (definition I.6) acting on a projective k-scheme X, and let L be an ample invertible sheaf on X to which the action of G may be lifted (in the sense of GIT, Ch. 1, § 3). For example, if E is a finite-dimensional k-vector space on which G acts, and X is a closed invariant subscheme of $P = \boldsymbol{P}(E)$ then L might be $\mathcal{O}_P(1)$ restricted to X. Conversely, if L is ample, then for some integer n, X embeds in $\boldsymbol{P}\,[\Gamma(X, L^{\otimes n})]$, on which G acts canonically.

THEOREM 2. *Let G, X, and L be as above. Then there are two canonical invariant open subsets of X: $X_S \subset X_{SS} \subset X$ such that*

(i) *a quotient (Y, π) of X_{SS} by G exists and is a projective scheme.*

(ii) *there is an open subset Y_0 of Y such that $(Y_0, \pi/X_S)$ is a geometric quotient of X_S.*

Moreover:

a) $X_S = \pi^{-1}(Y_0)$

b) *if x and y are closed points of X_{SS}, then $\pi(x) = \pi(y)$ if and only if $\overline{O(x)} \cap \overline{O(y)} \cap X_{SS} \neq \phi$.*

c) *for each closed point y of Y, $\pi^{-1}(y)$ contains a unique orbit closed in X_{SS}.*

d) *a closed point x of X_{SS} is in X_S if and only if $O(x)$ is closed in X_{SS} and the stabilizer $S(x)$ of x has minimal dimension.*

The points of X_S are called *stable* and those of X_{SS} *semistable*. For the definition X_S and X_{SS} as well as the proof of this theorem we refer to GIT, Ch. 1, § 4. The basic idea, however, is to define Y to be $\mathrm{Proj}(R^G)$, where $R = \sum \Gamma(X, L^n)$ is the homogeneous coordinate ring of X.

There is an important numerical criterion for finding X_S and X_{SS} (GIT, Ch. 2, § 1). Namely, if x is a closed point of X, then for each 1-parameter sub-group λ of G, i.e., for each homomorphism $\lambda : \boldsymbol{G}_m \to G$, the morphism $\boldsymbol{G}_m = \mathrm{Spec}(k[\alpha, \alpha^{-1}]) \to X$ defined by $\alpha \mapsto \lambda(\alpha) \cdot x$ extends uniquely to a morphism $f : \boldsymbol{A}^1 = \mathrm{Spec}(k[\alpha]) \to X$, since X is separated and complete and the local ring of 0 in $\boldsymbol{A}^1$ is a valuation ring (proposition 1). Then $z = f(0)$ is fixed under the action of $\boldsymbol{G}_m$ induced by λ, and therefore $\boldsymbol{G}_m$ acts on the 1-dimensional fiber $L \otimes k(z)$. But such an action is given by a character χ of $\boldsymbol{G}_m$; hence there is an integer r such that $\lambda(\alpha) \cdot v = \alpha^r \cdot v$ for each $v \in L \otimes k(z)$. Then the point x is stable if $r < 0$, and semi-stable if $r \leqq 0$, for each 1-parameter subgroup λ of G.

To see how this criterion works in a concrete example, let us return to the group $PGL(2)$ acting on $X = (\boldsymbol{P}^1)^n$. ($PGL(2)$ is simple, hence reductive if $\mathrm{char}(k) = 0$.)

To find an ample invertible sheaf to which the action of $PGL(2)$ extends, we first write out explicitly the action of $PGL(2)$ on $\boldsymbol{P}^1$ (cf. GIT, p. 33). If $X_0, X_1 \in \Gamma(\boldsymbol{P}^1, \mathcal{O}(1))$ are the homogeneous coordinates of $\boldsymbol{P}^1$ then

$(a_{ij}) \in GL(2)$ acts on $\Gamma(\boldsymbol{P}^1, \mathcal{O}(1))$ (on the right) by $X_i \to \sum a_{ij}X_j$. This defines an automorphism $\boldsymbol{P}^1 \to \boldsymbol{P}^1$ which depends only on the class of (a_{ij}) in $PGL(2)$. Unfortunately, there is no action of $PGL(2)$ on $\Gamma(\boldsymbol{P}^1, \mathcal{O}(1))$ compatible with the action on $\boldsymbol{P}^1$. However, the operation

$$X_i \otimes X_j \to \frac{\sum a_{ik}a_{jl}X_k \otimes X_l}{\det(a_{ij})}$$

of $GL(2)$ on $\Gamma(\boldsymbol{P}^1, \mathcal{O}(2))$ factors through $PGL(2)$. Hence the action of $PGL(2)$ lifts to $\mathcal{O}(2)$.

Let $\lambda : \boldsymbol{G}_m \to PGL(2)$ be the homomorphism such that $\lambda(\alpha)$ is the class of

$$\begin{pmatrix} 1 & 0 \\ 0 & \alpha \end{pmatrix}$$

in $PGL(2)$ for $\alpha \in k^*$. The fixed points of $\boldsymbol{P}^1$ under the action of $\boldsymbol{G}_m$ induced by λ are 0 and ∞, defined by $X_1(0) = 0$, $X_0(\infty) = 0$. If $x \in \boldsymbol{P}^1$ is different from 0 and ∞, then $\lambda(\alpha) \cdot x$ specializes to 0 as $\alpha \to 0$ and to ∞ as $\alpha \to \infty$. Moreover, the action of $\boldsymbol{G}_m$ on the fibers of $\mathcal{O}(2)$ at 0 and ∞ is given by multiplication with α^{-1} and α, respectively. To see this, it suffices to consider the behaviour of $X_0 \otimes X_0$ and $X_1 \otimes X_1$.

Now, each non-trivial 1-parameter subgroup λ' of $PGL(2)$ is conjugate to λ^m for some integer $m > 0$. Hence there are exactly two points a, b of $\boldsymbol{P}^1$ fixed under the action of $\boldsymbol{G}_m$ induced by λ'. Besides, if $x \neq a, b$ is a point of $\boldsymbol{P}^1$, $\lambda'(\alpha) \cdot x$ specializes to one of these points, say a, as $\alpha \to 0$, and so $\lambda'(\alpha) \cdot x$ specializes to b as $\alpha \to \infty$. Finally, the action of $\boldsymbol{G}_m$ on the fibers of $\mathcal{O}(2)$ at a and b is given by multiplication with α^{-m} and α^m respectively.

Returning to $(\boldsymbol{P}^1)^n$ we find that the action of $PGL(2)$ lifts to the ample invertible sheaf $L = p_1^*(\mathcal{O}(2)) \otimes \cdots \otimes p_n^*(\mathcal{O}(2))$. Let $x = (x_1, \cdots, x_n)$ be a closed point of $(\boldsymbol{P}^1)^n$ and let a, b denote the fixed points of a 1-parameter subgroup λ' of $PGL(1)$ as above. Permuting the indices if necessary, we assume that $x_i = b$ for $1 \leqq i \leqq r$ and $x_i \neq b$ for $r+1 \leqq i \leqq n$. It follows that $\lambda'(\alpha) \cdot x$ specializes to

$$z = (\overbrace{b, \cdots, b}^{r}, \overbrace{a, \cdots, a}^{n-r})$$

as $\alpha \to 0$. Considering each factor of L separately we find that the action of $\boldsymbol{G}_m$ on the fiber $L \otimes k(z)$ induced by λ' is given by multiplication with $\alpha^{m(2r-n)}$. As b may be any point of $\boldsymbol{P}^1$, we conclude:

A closed point $(x_1, \cdots, x_n)$ of $(\boldsymbol{P}^1)^n$ is stable (semistable) if and only if no point of $\boldsymbol{P}^1$ occurs with multiplicity $\geqq n/2$ $(> n/2)$ in $(x_1, \cdots, x_r)$.

If n is odd, the set of stable points $(\boldsymbol{P}^1)^n_S$ is the same as the set of semi-stable points $(\boldsymbol{P}^1)^n_{SS}$, and both coincide with the set U above. If n is even, $(\boldsymbol{P}^1)^n_S$ is a proper subset of $(\boldsymbol{P}^1)^n_{SS}$. Hence the quotient of $(\boldsymbol{P}^1)^n_S$ is not complete.

3. Elliptic curves

Let us consider an algebraic curve X over k, i.e., a reduced, irreducible, and separated algebraic k-scheme of dimension 1. We recall that X is non-singular if and only if the sheaf of differentials $\Omega_{X/k}$ is a locally free $\mathcal{O}_X$-Module of rank 1 or equivalently, if and only if the local ring $\mathcal{O}_x$ is a discrete valuation ring for each closed point x of X. In addition, if X is non-singular, x is a closed point of X, and f is an element of the maximal ideal m of $\mathcal{O}_x$, then f is a generator of m if and only if df is a local basis for $\Omega_{X/k}$ at x. It is also useful to note that a complete, non-singular algebraic curve is uniquely determined by its field of rational functions.

If X is complete and non-singular, then

$$g = \dim_k H^0(X, \Omega_{X/k})$$

is a finite integer, called the *genus* of X. An *elliptic curve* is a complete, non-singular algebraic curve of genus 1.

To examine elliptic curves in greater detail we have to know their cohomology.

Proposition 1. *Let $\mathscr{D}$ be a divisor of degree n on an elliptic curve E over k.*

(i) *For $n > 0$, $\dim_k H^0(E, \mathcal{O}_E(\mathscr{D})) = n$ and $H^1(E, \mathcal{O}_E(\mathscr{D})) = 0$.*

(ii) *For $n < 0$, $H^0(E, \mathcal{O}_E(\mathscr{D})) = 0$ and $\dim_k H^1(E, \mathcal{O}_E(\mathscr{D})) = -n$.*

(iii) *For $n = 0$, $\dim_k H^0(E, \mathcal{O}_E(\mathscr{D})) = \dim_k H^1(E, \mathcal{O}_E(\mathscr{D})) = 1$ if $\mathscr{D}$ is linearly equivalent to 0, and $H^0(E, \mathcal{O}_E(\mathscr{D})) = H^1(E, \mathcal{O}_E(\mathscr{D})) = 0$ otherwise.*

Furthermore, $H^i(E, \mathcal{O}_E(\mathscr{D})) = 0$ for $i > 2$.

The proposition follows from the Riemann-Roch formula

$$\dim_k H^0(E, \mathcal{O}_E(\mathscr{D})) - \dim_k H^1(E, \mathcal{O}_E(\mathscr{D})) = n$$

(Serre, Groupes algébriques et corps de classes, Théorème 1, p. 21) and Serre duality

$$\dim_k H^0(E, \mathcal{O}_E(\mathscr{D})) = \dim_k H^1(E, \mathcal{O}_E(K - \mathscr{D}))$$

(loc.cit., Théorème 2, p. 26) noting that the canonical class K is 0. Indeed, the degree of K is 0 (Loc. cit., p. 27) and

$$\dim_k H^0(E, \mathcal{O}_E(K)) = \dim_k H^0(E, \Omega_{E/k}) = 1$$

by assumption. Hence $K = \mathrm{div}(f)$ for some rational function $f \in H^0(E, \mathcal{O}_E(K))$.

Now let E be an elliptic curve. We fix a closed point 0 of E. The reason for this notation is the following:

PROPOSITION 2. *There is a unique abelian group structure on the set of closed points E_k of E such that the map*

$$x \mapsto \text{the class of } \mathcal{O}_E((x)-(0)),$$

where (x) is the divisor associated with x, is a group homomorphism from E_k to the Picard group $\mathrm{Pic}(E)$ *of E.*

PROOF. It is enough to show that the map

$$x \mapsto \text{the class of } (x)-(0)$$

is a bijection from E_k to the set of linear equivalence classes of divisors of degree 0. But if $\mathcal{D}$ is a divisor of degree 0 on E, then

$$\dim H^0(E, \mathcal{O}_E(\mathcal{D}+(0))) = 1$$

by proposition 1(i). Thus there is a unique principal divisor

$$\mathrm{div}(f) \geqq -\mathcal{D}-(0).$$

Then $\mathrm{div}(f)+\mathcal{D}+(0)$ is an effective divisor of degree 1, hence of the form (x) for some $x \in E_k$. Q.E.D.

Clearly, 0 is the neutral element of the group E_k.

REMARK. It may be shown that the group structure of E_k is induced by an algebraic group structure on E. It follows, in particular, that the full group of automorphisms of an elliptic curve is transitive. Hence the results are independent of the base point 0.

If A is the divisor associated with the point 0, the vector space $V = H^0(E, \mathcal{O}_E(2A))$ is 2-dimensional by proposition 1 (i). Hence there are non-constant rational functions in V, having necessarily a double pole at 0. Any such function f defines a morphism from $E-\{0\}$ to the affine line $\mathbf{A}^1$ over k, and this morphism has a unique extension π from E to the projective line $\mathbf{P}^1$ over k by proposition II.1. If f' is another non-constant function in V, and $\pi' : E \to \mathbf{P}^1$ is the corresponding morphism, there is a unique automorphism μ of $\mathbf{P}^1$ such that $\pi' = \mu \circ \pi$. In fact, since V/k is 1-dimensional, there is a unique pair $(\alpha, \beta) \in k^* \times k$ satisfying $f' = \alpha f+\beta$.

To find the fibers of π, let $\lambda \in k$ be a closed point of $\mathbf{A}^1 \subset \mathbf{P}^1$. Then $\pi^{-1}(\lambda)$ is the support of the divisor of zeros $\mathcal{D}$ of $f-\lambda$. Since the divisor of poles of $f-\lambda$ is of degree 2, $\mathcal{D} = (x)+(y)$ for some $x, y \in E_k$. But then $x+y = 0$ by proposition 2. Hence the branch points of π are the points of order 2 in the group E_k. It is shown in the theory of abelian varieties that their number is four if $\mathrm{char}(k) \neq 2$.

More directly, this may be seen by calculating the order of $\mathrm{div}(df)$. If $x \in E_k$ is not a ramification point, then df_x is a generator of $(\Omega_{E/k})_x$, hence of order 0 at x. If $x \in E_k - \{0\}$ is a ramification point, then

$$f_x = f(x)+ut^2,$$

where t is a generator of the maximal ideal m_x of $\mathcal{O}_x$ and $u \in \mathcal{O}_x^*$ is a unit, and therefore

$$df_x = 2ut\,dt+t^2du$$

is of order 1 if $2 \neq 0$ in k. Finally, if t generates $m_0 \subset \mathcal{O}_0$, we have

$$t^2 f_0 = u \in \mathcal{O}_0^*,$$
$$2tf_0 dt + t^2 df_0 = du,$$

so that df_0 is of order -3. Since the canonical divisor class on an elliptic curve is of degree 0, there are exactly 3 ramification points in addition to 0.

From now on we assume k of characteristic $\neq 2, 3$.

Let $a, b, c \in A^1$ and ∞ be the images by π of the ramification points. By a projective transformation they may be normalized to

$$0, 1, \lambda, \infty,$$

where λ is a cross-ratio of the four points a, b, c, ∞. Since the order of the points a, b, c is not specified, there are, in general, 6 different normalisations $(0, 1, \mu, \infty)$, where μ appears in the following list:

$$\lambda,\ 1-\lambda,\ 1/\lambda,\ \frac{\lambda-1}{\lambda},\ \frac{\lambda}{\lambda-1},\ \frac{1}{1-\lambda}.$$

However, any rational function $j(\lambda)$ which has the same value at each of the above points defines an invariant of the curve E. To find such a function we note that there are three series of equivalent points which are left fixed by some non-trivial substitution:

$$\tfrac{1}{2},\ 2,\ -1,$$
$$0,\ 1,\ \infty,$$
$$-\omega,\ -\omega^2$$

where $\omega^2+\omega+1 = 0$ ($\omega \neq 1$ since $\mathrm{char}(k) \neq 3$). Hence $j(\lambda)$ must be ramified at these points. If we assume that

$$j(-\omega) = j(-\omega^2) = 0,$$
$$j(0) = j(1) = j(\infty) = \infty,$$

will be proportional to

$$\frac{(\lambda^2-\lambda+1)^3}{\lambda^2\cdot(\lambda-1)^2}.$$

On the other hand, this is seen to be invariant by direct substitution (or more elegantly, noting that the function in the brackets is invariant up to a multiplicative constant determined by a character of the symmetric group S_3, hence necessarily equal to ± 1). For reasons involving the omitted characteristics 2 and 3, j is normalized by

$$j(-1) = 12^3,$$

therefore we obtain

$$j(\lambda) = 256\,\frac{(\lambda^2-\lambda+1)^3}{\lambda^2\cdot(\lambda-1)^2}.$$

Thus we have constructed a map from the set of elliptic curves over k, up to isomorphism, to the set of closed points of $A^1 = \mathrm{Spec}(k[j])$. To show that this map is bijective, we embed elliptic curves into the projective plane.

Let E be an elliptic curve over k and denote by A the divisor associated with a fixed base point 0 of E. Let $f \in H^0(E, \mathcal{O}_E(2A))$ be a rational function having a double pole at 0 as above.

By proposition 1

$$\dim_k H^0(E, \mathcal{O}_E(3A)) = 3$$

while

$$\dim_k H^0(E, \mathcal{O}_E(2A)) = 2.$$

Hence there is a rational function

$$g \in H^0(E, \mathcal{O}_E(3A))$$

with a triple pole at 0.

LEMMA 1. *For each integer $n \geqq 2$, the functions*

$$1, f, \cdots, f^k, g, fg, \cdots, f^l g,$$

where $k = [n/2]$, $l = [(n-3)/2]$, form a basis of $H^0(E, \mathcal{O}_E(nA))$.

PROOF. The orders of these functions at 0 are

$$0,\ -2, \cdots, -2k,\ -3,\ -5, \cdots, -2l-3,$$

respectively. Hence they are linearly independent. In the other hand, their number is $n = \dim H^0(E, \mathcal{O}_E(nA))$. Q.E.D.

Since $g^2 \in H^0(E, \mathcal{O}_E(6A))$, there are constants $a_i \in k$ $(1 \leqq i \leqq 6)$ such that

$$g^2 = a_1 fg + a_2 g + a_3 f^3 + a_4 f^2 + a_5 f + a_6.$$

Replacing g by $g-\frac{1}{2}a_1f-\frac{1}{2}a_2$, we may assume $a_1 = a_2 = 0$. Since g^2 has a pole of order 6 at 0, we must have $a_3 \neq 0$. Therefore, replacing g by $a_3^{-\frac{1}{2}} g$ we obtain

$$g^2 = (f-a)(f-b)(f-c)$$

for some constants $a, b, c \in k$.

It is not hard to see that a, b, and c must be distinct. Indeed, if $a = b$, then $g/(f-a)$ has a simple pole at 0 and no other poles, since its square is $f-c$. But this is impossible by proposition 1.

Let P be a projective plane over k, with homogeneous coordinates $X, Y, Z \in H^0(P, \mathcal{O}_P(1))$. Since $1, f, g$ generate the invertible sheaf $\mathcal{O}_E(3A)$, there is a unique morphism $\varphi : E \to P$ such that

$$\varphi^*(\mathcal{O}_P(1)) \simeq \mathcal{O}_E(3A)$$

with $\varphi^*(X)$, $\varphi^*(Y)$, $\varphi^*(Z)$ corresponding to f, g, and 1.

PROPOSITION 3. *The morphism φ is an embedding of E onto the cubic curve C with homogeneous equation*

$$P(X, Y, Z) = Y^2Z-(X-aZ)(X-bZ)(X-cZ) = 0.$$

PROOF. It is clear that φ factors through C. If h is a rational function on E, and $\mathcal{D}$ is its divisor of poles ($\mathcal{D} \geqq 0$), then, for each integer $n > \deg(\mathcal{D})$, there is a rational function h' with $\operatorname{div}(h') \geqq \mathcal{D}-nA$ by proposition 1. This implies that hh' is defined on $E-\{0\}$. Thus any rational function on E is a quotient of two rational functions with poles at 0 only. But such functions are polynomials of f and g by lemma 1. Hence the rational function field of E is $k(f, g)$; in other words, φ defines a birational morphism from E to C. Therefore it is enough to show that C is non-singular. This follows from the fact that each singular cubic plane curve is rational, or by direct calculation as follows:

Let (x, y, z) be a point of C. If $y \neq 0$, then $P_Y(x, y, z) = 2yz \neq 0$ unless $z = 0$, in which case $x = 0$ and $P_Z(x, y, z) = y^2 \neq 0$. On the other hand, if $y = 0$, then $x = az, bz$, or cz. Therefore $P_X(x, y, z) \neq 0$, since the constants a, b, c are distinct. Q.E.D.

It follows immediately that a, b, and c are the points of A^1 over which the morphism $\pi : E \to P^1$ defined by f is ramified.

REMARK. The assumption $\operatorname{char}(k) \neq 3$ has not been used in the proof of proposition 3.

COROLLARY. *The invariant function $j(\lambda)$ defines a bijection from the set of isomorphism classes of elliptic curves over k to the set of k-valued points of A_j^1.*

Proof. Since $\lambda \mapsto j(\lambda)$ is a surjective map from k-$\{0, 1\}$ to k, there is an elliptic curve for each $j \in k$, namely

$$P_\lambda(X, Y, Z) = Y^2Z - X(X-Z)(X-\lambda Z) = 0$$

for some $\lambda \in k - \{0, 1\}$ with $j = j(\lambda)$. Hence the map is surjective.

On the other hand, let E and E' be two elliptic curves with the same j. By proposition 3 we may assume that they are plane curves given by $P_\lambda(X, Y, Z) = 0$ and $P_{\lambda'}(X, Y, Z) = 0$. Since $j(\lambda)$ is of degree 6, $j(\lambda) = j(\lambda')$ implies that λ' is in the sequence

$$\lambda, 1-\lambda, 1/\lambda, \frac{\lambda-1}{\lambda}, \frac{\lambda}{\lambda-1}, \frac{1}{1-\lambda}.$$

Therefore E and E' are in fact projectively equivalent. Q.E.D.

Thus we have classified elliptic curves over k by the k-valued points of a scheme A_j^1. The question now arises: what is the role of the scheme structure of A_j^1? To answer this we introduce the notion of a family of elliptic curves in analogy with chapter 1.

Definition 1. *A family of elliptic curves* over a k-variety S is a morphism of k-varieties $p : E \to S$ together with a section $0 : S \to E$ such that E is proper and smooth over S, and the closed fibers of p are elliptic curves.

Remark. By the definition of an elliptic curve, the last condition implies that the closed fibers of p are non-singular. Therefore p is smooth if and only if it is flat.

To generalize the results of proposition 1 concerning the cohomology of elliptic curves for families of elliptic curves we need a base change theorem. We first introduce some notation.

If $p : E \to S$ is a morphism, we denote by E_s the fiber $p^{-1}(s)$ (considered as a subscheme of E) for each point s of S. For each $\mathcal{O}_E$-Module $\mathscr{F}$, $\mathscr{F} \otimes_{\mathcal{O}_s} k(s)$ may be regarded as an $\mathcal{O}_{E_s}$-Module denoted by $\mathscr{F}_s$. Then, for each integer i, the homomorphism

$$R^i p_*(\mathscr{F}) \to H^i(E_s, \mathscr{F}_s)$$

induced by the canonical epimorphism $\mathscr{F} \to \mathscr{F}_s$ defines a homomorphism

$$t_s^i : R^i p_*(\mathscr{F}) \underset{\mathcal{O}_s}{\otimes} k(s) \to H^i(E_s, \mathscr{F}_s).$$

Proposition 4. *Let $p : E \to S$ be a proper morphism of locally noetherian schemes, and let $\mathscr{F}$ be a coherent $\mathcal{O}_E$-Module flat over S.*

(i) *If $t_s^i : R^i p_*(\mathscr{F}) \otimes_{\mathcal{O}_s} k(s) \to H^i(E_s, \mathscr{F}_s)$ is surjective for some integer i and some point $s \in S$, then it is bijective.*

(ii) *If the condition of* (i) *is satisfied, then*

$$t_s^{i-1} : R^{i-1}p_*(\mathscr{F}) \underset{\mathcal{O}_s}{\otimes} k(s) \to H^{i-1}(E_s, \mathscr{F}_s)$$

is also surjective if and only if $R^ip_*(\mathscr{F})$ is a free $\mathcal{O}_s$-Module in a neighborhood of s.

For the proof we refer to EGA III: (i) follows from (7.7.5.3) in view of (7.7.10), and for the same reason (ii) amounts to the equivalence of (7.8.3.b) and (7.8.4.d). (or better, see Mumford, Abelian Varieties, § 5).

COROLLARY 1. *With the assumptions of proposition 4, if $H^{i+1}(E_s, \mathscr{F}_s) = 0$ for some integer i and some point $s \in S$, then t_s^i is an isomorphism.*

In fact, t_s^{i+1} is surjective, hence bijective by (i). Then $R^{i+1}p_*(\mathscr{F}) = 0$ in a neighborhood of s by Nakayawa's lemma. Therefore the conclusion from (ii), and (i) again.

COROLLARY 2. *With the assumptions of proposition 4, if $\mathscr{E}$ is a coherent $\mathcal{O}_s$-Module and $\varphi : \mathscr{E} \to p_*(\mathscr{F})$ is a homomorphism such that the induced map*

$$\mathscr{E} \underset{\mathcal{O}_s}{\otimes} k(s) \to H^0(E_s, \mathscr{T}_s)$$

is bijective for each point s of S, then φ is an isomorphism and $\mathscr{E}$ is locally free.

PROOF. The assumption implies that $t_s^0 : p_*(\mathscr{F}) \otimes k(s) \to H^0(E_s, \mathscr{F}_s)$ is surjective, hence bijective for each $s \in S$ by (i). Then

$$\varphi \otimes 1 : \mathscr{E} \otimes k(s) \to p_*(\mathscr{F}) \otimes k(s)$$

is bijective, and therefore φ is surjective by Nakayama's lemma. Finally $p_*(\mathscr{F})$ is locally free by (ii); so, if $\mathscr{G}$ is the kernel of φ, $\mathscr{G} \otimes k(s)$ is the kernel of $\varphi \otimes 1$ for each $s \in S$, whence $\mathscr{G} = 0$. Q.E.D.

Let us consider a family $p : E \to S$ of elliptic curves. Since p is proper, hence separated, the section $0 : S \to E$ is a closed immersion (EGA I, 5.4.b), i.e., it defines an isomorphism from S onto a closed subscheme A of E. Let $I \subset \mathcal{O}_E$ denote the sheaf of ideals of A.

LEMMA 2. *I is an invertible sheaf.*

PROOF. Let x be a closed point of A, $s = p(x)$, and $E_s = p^{-1}(s)$. It is clear that the maximal ideal m of $\mathcal{O}_{x,E_s}$ is generated by the image of

$$I_x \subset \mathcal{O}_{x,E} \text{ in } \mathcal{O}_{x,E_s} = \mathcal{O}_{x,E} \underset{\mathcal{O}_s}{\otimes} k(s).$$

By assumption, $\mathcal{O}_{x,E_s}$ is a discrete valuation ring, hence there is a section f of I over some open neighborhood U of x such that $f_x \otimes 1$ is a basis of

m. Then f defines a closed subscheme A' of U containing $A \cap U$ such that the fibers of A and A' over K are equal at x. As A is flat over S, it coincides with A' near x, i.e., f generates I in a neighborhood V of x.

To show that f is a free generator of I locally at x, let J denote the kernel of the epimorphism $\mathcal{O}_E|V \to I|V$ defined by f. Since I is flat over S, $J_x \otimes_{\mathcal{O}_s} k(s)$ is the kernel of

$$\mathcal{O}_{x,E_s} \xrightarrow{f_x \otimes 1} I_x \otimes_{\mathcal{O}_s} k(s) = m.$$

Therefore $J_x \otimes_{\mathcal{O}_s} k(s) = 0$, and we conclude that $J_x \otimes_{\mathcal{O}_x} k(x)$ is 0 as a quotient of $J_x \otimes_{\mathcal{O}_s} k(s)$. Hence $J = 0$ in a neighborhood of x by Nakayama's lemma. This completes the proof. Q.E.D.

This result means that A is the support of a divisor on E. The divisor will also be denoted by A, hence the sheaf of functions $\mathcal{O}_E(nA)$ with n-fold poles along A will be isomorphic to the invertible sheaf $I^{\otimes(-n)}$ for each integer n. For each n the quotient $\mathcal{O}_E(nA)/\mathcal{O}_E((n-1)A)$ may be identified with $i^*(\mathcal{O}_E(nA))$ where $i : A \to E$ is the inclusion. Hence

$$p_*(\mathcal{O}_E(nA)/\mathcal{O}_E((n-1)A))$$

is isomorphic to $i^*(\mathcal{O}_E(nA))$ and

$$R^ip_*(\mathcal{O}_E(nA)/\mathcal{O}_E((n-1)A)) = 0 \text{ for } i > 0.$$

In particular, $p_*(\mathcal{O}_E(nA)/\mathcal{O}_E((n-1)A))$ is an invertible $\mathcal{O}_S$-Module, canonically isomorphic to $\mathcal{L}^{\otimes n}$ where $\mathcal{L} = p_*(\mathcal{O}_E(A)/\mathcal{O}_E)$. Other important higher direct images are supplied by the following proposition.

Proposition 5. (i) *The canonical homomorphism*

$$\mathcal{O}_S \to p_*(\mathcal{O}_E)$$

is an isomorphism.

(ii) $p_*(\mathcal{O}_E(nA))$ *is locally free of rank* n *for* $n > 0$.

(iii) $R^1p_*(\mathcal{O}_E(nA)) = 0$ *for* $n > 0$, *and locally free of rank* 1 *for* $n = 0$.

(iv) $R^ip_*(\mathcal{O}_E(nA)) = 0$ *for* $i > 1$ *and all integers* n.

Moreover, in each case the canonical homomorphism

$$R^ip_*(\mathcal{O}_E(nA)) \otimes_{\mathcal{O}_s} k(s) \to H^i(E_s, \mathcal{O}_E(nA)_s)$$

is bijective for all $s \in S$.

Proof. (i) is an immediate consequence of corollary 2 of proposition 4. Then (ii), (iii), and (iv) follow from corollary 1 and part (ii) of proposition 4 applying proposition 1. The final assertion is established in the course of the proof. Q.E.D.

REMARK. It may be proved similarly that, for $n < 0$, $R^i p_*(\mathcal{O}_E(nA))$ is locally free of rank $-n$ if $i = 1$, and 0 otherwise.

COROLLARY 1. *The natural injection $\mathcal{O}_S \to p_*(\mathcal{O}_E(A))$ is an isomorphism.*

PROOF. Let us consider the long exact sequence

$$0 \to p_*(\mathcal{O}_E) \xrightarrow{i} p_*(\mathcal{O}_E(A)) \xrightarrow{j} p_*(\mathcal{O}_E(A)/\mathcal{O}_E) \xrightarrow{k} \\ \to R^1 p_*(\mathcal{O}_E) \to R^1 p_*(\mathcal{O}_E(A)) \to \cdots$$

By the proposition $R^1 p_*(\mathcal{O}_E(A)) = 0$ and $R^1 p_*(\mathcal{O}_E)$ is locally free of rank 1. Hence k is surjective and its kernel is locally a direct summand of $\mathcal{L} = p_*(\mathcal{O}_E(A)/\mathcal{O}_E)$. Since $\mathcal{L}$ is also invertible, $\mathrm{Ker}(k) = 0$, and therefore i is an isomorphism. Q.E.D.

COROLLARY 2. *The canonical homomorphism*

$$p_*(\mathcal{O}_E(nA)) \to p_*(\mathcal{O}_E(nA)/\mathcal{O}_E(n-1)A)) = \mathcal{L}^{\otimes n}$$

is surjective for $n > 1$.

This follows immediately from (iii) of proposition 5.

We are now ready to prove the main result of this chapter.

THEOREM 1 (*Weierstrass Normal Form*). *Let $p : E \to S$ be a family of of elliptic curves over a k-variety S. Then each point of S has an affine open neighborhood $U = \mathrm{Spec}(R)$ such that $p^{-1}(U)$ is isomorphic over U to the subscheme of $\mathbf{P}^2 \times U$ defined by*

$$Y^2Z = X^3 + aXZ^2 + bZ^3,$$

where $a, b \in R$ are unique up to the substitutions

$$a \mapsto \lambda^4 a, b \mapsto \lambda^6 b$$

for $\lambda \in R^*$. Moreover $4a^3 + 27b^2$ is invertible in R.

REMARK. It is always assumed that the subscheme A defined by the zero section $0 : S \to E$ is mapped to the set where $Z = 0$.

PROOF. Each point of S has an affine open neighborhood $U = \mathrm{Spec}(R)$ such that $\mathcal{L} = p_*(\mathcal{O}_E(A)/\mathcal{O}_E)$ is free on U. We simplify the notation by assuming that U equals S.

If t is a generator of the R-module $\Gamma(S, \mathcal{L})$, then $t^n = t^{\otimes n}$ is a basis of $\Gamma(S, \mathcal{L}^{\otimes n})$ for each integer n. By the corollaries of proposition 5 we have

$$\Gamma(E, \mathcal{O}_E(2A)) = R \oplus R \cdot f$$

where f has image t^2 in $\Gamma(E, \mathcal{O}_E(2A))/R = \Gamma(S, \mathcal{L}^{\otimes 2})$. Similarly,

$$\Gamma(E, \mathcal{O}_E(3A)) = R \oplus R \cdot f \oplus R \cdot g$$

where g projects to t^3 in $\Gamma(S, \mathscr{L}^{\otimes 3})$. Since f^2, fg, f^3 have the leading parts t^4, t^5 and t^6 respectively, it is easy to see that $1, f, g, f^2, fg, f^3$ form a basis of the R-module $\Gamma(E, \mathcal{O}_E(6A))$. In particular, we have

$$g^2 = a_1 fg + a_2 g + a_3 f^3 + a_4 f^2 + a_5 f + a_6$$

where $a_i \in R$ $(1 \leqq i \leqq b)$ are uniquely defined. Taking the leading parts in $\Gamma(S, \mathscr{L}^{\otimes 6})$ we find that $a_3 = 1$. Redefining f and g we may further simplify the equation. Replacing g by $g - \frac{1}{2}a_1 f - \frac{1}{2}a_2$ yields a new equation with $a_1 = a_2 = 0$ without affecting the principal part of g.

Similarly, if f is replaced by $f + \frac{1}{3}a_4$, we get $a_4 = 0$. Hence we may assume that

$$g^2 = f^3 + af + b$$

for some $a, b \in R$.

From the last assertion of proposition 5 we see that $1, f$, and g generate $\mathcal{O}_E(3A)$ on each geometric fiber of p. Hence they generate $\mathcal{O}_E(3A)$ and define a morphism

$$\varphi : E \to \mathbf{P}(p_*(\mathcal{O}_E(3A))) \simeq \mathbf{P}^2 \times S$$

over S. We claim that φ is an immersion, i.e., $\mathcal{O}_E(3A)$ is very ample.

Since p is proper by assumption, φ is proper (EGA II, 5.4.3). It is injective on the closed fibers (proposition 3), hence injective. So, being closed, φ is a homeomorphism of E onto a closed subspace $\varphi(E)$ of $\mathbf{P}^2 \times S$, and $\varphi_*(\mathcal{O}_E)$ is essentially the extension of $\mathcal{O}_E$ by zero. Since $\varphi_*(\mathcal{O}_E)$ is coherent on $\mathbf{P}^2 \times S$ (EGA III, 3.2.1), it is enough to show that the canonical homomorphism $u : \mathcal{O}_{\mathbf{P}^2 \times S} \to \varphi_*(\mathcal{O}_E)$ is surjective.

If s is a closed point of S, and φ_s is the restriction of φ to E_s, then

$$u_s : \mathcal{O}_{\mathbf{P}^2} \to \varphi_*(\mathcal{O}_E) \otimes k(s) = \varphi_{s,*}(\mathcal{O}_{E_s})$$

is surjective by proposition 3. Then the conclusion follows by Nakayama's lemma.

Let E' be the closed subscheme of $\mathbf{P}^2 \times S$ defined by

$$Y^2 Z = X^3 + aXZ^2 + bZ^3.$$

It is clear that $\varphi(E)$ is a subscheme of E'. But the closed fibers of $\varphi(E)$ and E' are equal by proposition 3, hence $\varphi(E) = E'$. Since the closed fibers of E are non-singular, the discriminant $4a^3 + 27b^2$ is non-zero at each closed point of S. This means that it is invertible in R.

Finally we note that the section t may be replaced by any section of the form λt where λ is a unit of R. Then f and g are replaced by $\lambda^2 f$ and $\lambda^3 g$, so $\lambda^4 a$ and $\lambda^6 b$ appear in place of a and b. Q.E.D.

Since the coefficients a and b of the Weierstrass normal form

$$Y^2 Z = X^3 + aXZ^2 + bZ^3$$

given by the theorem are defined only locally and up to multiplication by certain invertible functions, the reader may naturally suspect that they should be connected with some invertible sheaves. To find this connection we introduce global coordinates on projective bundles.

Let Y be a scheme, $\mathscr{E}$ a quasicoherent $\mathcal{O}_Y$-Module, and $S(\mathscr{E})$ the symmetric $\mathcal{O}_Y$-Algebra of $\mathscr{E}$. Then the scheme $P = \mathrm{Proj}(S(\mathscr{E}))$ is called the projective bundle over Y defined by $\mathscr{E}$ and denoted by $\boldsymbol{P}(\mathscr{E})$ (EGA II, 4.1.1). Let us assume that $\mathscr{E}$ is the direct sum of invertible $\mathcal{O}_Y$-Modules:

$$\mathscr{E} = \mathscr{L}_0 \oplus \mathscr{L}_1 \oplus \cdots \oplus \mathscr{L}_n.$$

For each i, let $\eta_i \in \Gamma(Y, \mathscr{L}_i^{-1} \otimes_{\mathcal{O}_Y} \mathscr{E})$ be the section which corresponds to the inclusion of $\mathscr{L}_i$ into $\mathscr{E}$ under the natural isomorphism

$$\mathscr{L}_i^{-1} \underset{\mathcal{O}_Y}{\otimes} \mathscr{E} \xrightarrow{\sim} \mathrm{Hom}_{\mathcal{O}_Y}(\mathscr{L}_i, \mathscr{E}).$$

If $p : P \to Y$ is the projection, the canonical epimorphism (EGA II, 4.1.5.1) $p^*(\mathscr{E}) \to \mathcal{O}_P(1)$ induces a homomorphism

$$p^*(\mathscr{L}_i^{-1} \underset{\mathcal{O}_Y}{\otimes} \mathscr{E}) \simeq p^*(\mathscr{L}_i^{-1}) \underset{\mathcal{O}_P}{\otimes} p^*(\mathscr{E}) \to p^*(\mathscr{L}_i^{-1})(1)$$

for each i. If $X_i \in \Gamma(P, p^*(\mathscr{L}_i^{-1})(1))$ is the image of η_i by this homomorphism, then $(X_0, X_1, \cdots, X_n)$ is called the gobal coordinate system of P relative to $(\mathscr{L}_0, \cdots, \mathscr{L}_n)$.

Returning to theorem 1, it is obvious that we have constructed, in effect, a canonical splitting

$$p_*(\mathcal{O}_E(3A)) = \mathcal{O}_S \oplus \mathscr{L}^{\otimes 2} \oplus \mathscr{L}^{\otimes 3}$$

where $\mathscr{L} = p_*(\mathcal{O}_E(A)/\mathcal{O}_E)$ and $\mathscr{L}^{\otimes 2}$, $\mathscr{L}^{\otimes 3}$ are locally generated by the functions f, and g, respectively. It is then easy to deduce the following variant of theorem 1.

THEOREM 1′. *Let $p : E \to S$ be a family of elliptic curves over a k-variety S. Then there is an invertible $\mathcal{O}_S$-Module $\mathscr{L}$ such that E is isomorphic over S to the subscheme of $P = \boldsymbol{P}(\mathcal{O}_S \oplus \mathscr{L}^{\otimes 2} \oplus \mathscr{L}^{\otimes 3})$ defined by*

$$Y^2Z = X^3 + aXZ^2 + bZ^3,$$

where $a \in \Gamma(S, \mathscr{L}^{\otimes(-4)})$, $b \in \Gamma(S, \mathscr{L}^{\otimes(-6)})$, and (X, Y, Z) is the global coordinate system of P relative to $(\mathscr{L}^{\otimes 2}, \mathscr{L}^{\otimes 3}, \mathcal{O}_S)$. Furthermore, $(\mathscr{L}, a, b)$ is unique up to isomorphism, and

$$4a^3 + 27b^2 \in \Gamma(S, \mathscr{L}^{\otimes(-12)})$$

is invertible.

Notice that the equation makes sense: both sides are sections of $\mathscr{L}^{\otimes(-6)}(3)$ and the divisor of their difference is associated with a subscheme of P.

For each k-variety S, let $\mathcal{M}(S)$ denote the set of isomorphism classes of families of elliptic curves over S. If E is a family of elliptic curves over S, then for each morphism $S' \to S$, $E \times_S S'$ is a family of elliptic curves over S'. Thus $\mathcal{M}$ becomes a contravariant functor from the category of k-varieties to the category of sets. Also recall that each k-variety M defines a functor h_M by $h_M(S) = \text{Hom}(S, M)$.

COROLLARY. *There is a morphism of functors*

$$\Phi : \mathcal{M} \to h_{A_j^1},$$

where $A_j^1 = \text{Spec}(k[j])$, *such that*

$$\Phi(\text{Spec}\,(k)) : \mathcal{M}(\text{Spec}\,(k)) \to A_j^1(k)$$

is the bijection given by the invariant j (cf. the corollary of proposition 3). A_j^1 is a coarse moduli space for elliptic curves over k.

PROOF. Each family of elliptic curves over S defines an invertible $\mathcal{O}_S$-Module $\mathcal{L}$ and sections $a \in \Gamma(S, \mathcal{L}^{\otimes(-4)})$, $b \in \Gamma(S, \mathcal{L}^{\otimes-(6)})$ with $4a^3+27b^2 \in \Gamma(S, \mathcal{L}^{\otimes(-12)})$ invertible. Then

$$(1) \qquad j_1 = 12^3 \frac{4a^3}{4a^3+27b^2} \in \Gamma(S, \mathcal{O}_S)$$

corresponds to a morphism $S \to A_j^1$. It is clear that this construction is functorial. The second point may be verified by direct calculation starting with the two representations of the same elliptic curve:

$$Y^2 = X'(X'-1)(X'-\lambda) \text{ and } Y^2 = X^3+aX+b$$

related by $Y = Y'$, $X = X'-(\lambda+1/3)$, and the previous definition:

$$(2) \qquad j_2(\lambda) = 256 \left[\frac{(\lambda^2-\lambda+1)^3}{\lambda^2(\lambda-1)^2}\right]^3$$

and verifying that (1) and (2) define the same j. To check the universal property of Φ, suppose $\psi : \mathcal{M} \to h_N$ is another morphism of functors. Let $S = \text{Spec}\, k[\lambda, (1/\lambda)(1/\lambda-1)]$, and let E be the subscheme of $\mathbf{P}^2 \times S$ defined by:

$$Y^2 Z = X(X-Z)(X-\lambda Z).$$

Then E is an elliptic curve over S and defines an element $(E/S) \in \mathcal{M}(S)$. Let $\psi(E/S)$ be the morphism $g : S \to N$. On the other hand, we just checked that $\Phi(E/S)$ is the morphism $j_2 : S \to A_j^1$ given by formula (2). I claim that g factors:

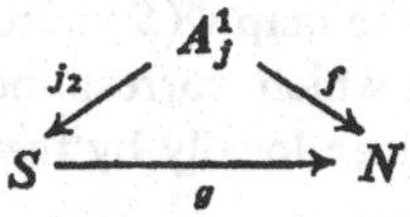

To see this, let $\Gamma \subset A_j^1 \times N$ be the image of (i_2, g). It is clear from the defining formula that j_2 is a finite morphism hence so is (j_2, g), hence Γ is closed. Since

$$j_2(\lambda) = j_2(\lambda') \Rightarrow E_\lambda \simeq E_{\lambda'} \Rightarrow g(\lambda) = g(\lambda'),$$

the projection $\Gamma \to A_j^1$ is injective. But j_2 is separable and surjective, hence $\Gamma \to A_j^1$ is separable and surjective. Therefore $\Gamma \to A_j^1$ is an isomorphism by Zariski's Main Theorem. If f is the composition

$$A_j^1 \underset{p_1}{\overset{\sim}{\longleftarrow}} \Gamma \underset{p_2}{\longrightarrow} N$$

then $f \circ j_2 = g.f$ defines a map of functors $h_{A_j^1} \to h_N$ and it follows from the definition that

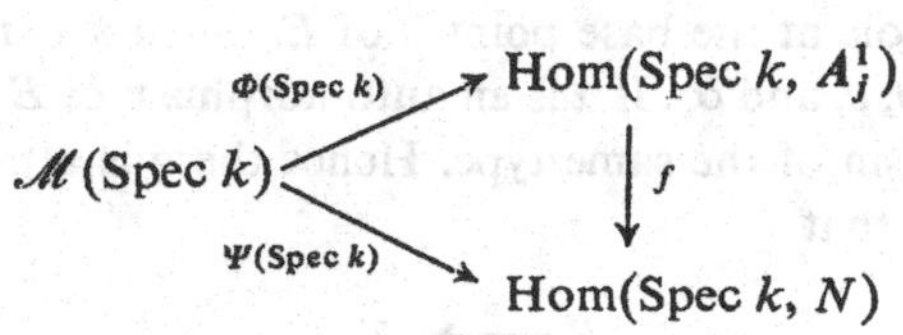

commutes. But therefore $f \circ \Phi(s) = \Psi(S)$ for any S as required. Q.E.D.

We now show by examples that the map

$$\Phi(S) : \mathcal{M}(S) \to \mathrm{Hom}\,(S, A_j^1)$$

is neither injective nor surjective in general. In particular, A_j^1 cannot be a fine moduli space for elliptic curves.

EXAMPLE A. Let A be a finitely generated integral domain containing a unit μ which has no square root in A, and put $S = \mathrm{Spec}(A)$. Then for any elliptic curve E over A with equation

$$Y^2 Z = X^3 + aXZ^2 + bZ^3$$

there is another elliptic curve E', a twisted form of E.

$$Y^2 = X^3 + \mu^2 aX + \mu^3 b$$

which is not isomorphic to E over A, but has the same j. In fact, E and E' become isomorphic over $A(\sqrt{\mu})$. Hence the map $\Phi(S)$ is not injective.

It is also easy to see that the map $\Phi(S)$ need not be surjective. In fact, each morphism $j : S \to A_j^1$ which corresponds to a family of elliptic curves over a scheme S is given locally by the formula

$$j = 12^3 \frac{4a^3}{4a^3+27b^2}$$

where a and b are sections of $\mathcal{O}_S$. If $s \in S$ is a point where $j(s) = 0$, and $\mathcal{M}$ is the maximal ideal of $\mathcal{O}_s$, then $a \in \mathcal{M}$, and therefore $j \in \mathcal{M}^3$. In other words, j must be *ramified* at s. In the same way it follows from the formula

$$j - 12^3 = -12^3 \frac{27b^2}{4a^3+27b^2}$$

that j is ramified at each point $s \in S$ with $j(s) = 12^3$.

To explain this phenomenon we study the automorphisms of elliptic curves.

Let E be an elliptic curve over an algebraically closed field k of characteristic $\neq 2,3$, and let $\pi : E \to P^1$ be the morphism defined by a function f with a double pole at the base point 0 of E. Then π is ramified over 4 distinct points a, b, c, and ∞. If α is an automorphism of E leaving 0 fixed, $\pi \circ \alpha$ is a morphism of the same type. Hence there is a projective transformation $\bar{\alpha}$ such that

$$\begin{array}{ccc} E & \xrightarrow{\alpha} & E \\ \pi \downarrow & & \downarrow \pi \\ P^1 & \xrightarrow[\bar{\alpha}]{} & P^1 \end{array}$$

commutes. Then $\bar{\alpha}$ leaves ∞ fixed and permutes the points a, b, and c. But $\bar{\alpha}$ is of order 1, 2, or 3 on the set $\{a, b, c\}$, hence on P^1:

(i) If $\bar{\alpha} = \mathrm{Id}$, $\alpha(x) = x$ or $-x$, for all x. Hence $\alpha = \mathrm{Id}$ or $\alpha = -\mathrm{Id}$. These automorphisms exist on any elliptic curve.

(ii) If $\bar{\alpha}$ is of order 2, it is a transposition on $\{a, b, c\}$. Hence we may normalize $\{a, b, c\}$ to $\{0, 1, -1\}$, $\bar{\alpha}(t) = -t$; so $\lambda = -1$ and $j = 12^3$. In fact, on $E_0 : Y^2 = X^3 - X$ there is the automorphism $X \mapsto -X$, $Y \mapsto iY$ of order 4.

(iii) If $\bar{\alpha}$ is of order 3, it is a cyclic permutation on $\{a, b, c\}$. Hence $\{a, b, c\}$ may be normalized to $\{1, \omega, \omega^2\}$ where $\omega^3 = 1$; so $j = 0$. Finally, on $E_{(12^3)} : Y^2 = X^3 - 1$, there is the automorphism $X \mapsto \omega X$, $Y \mapsto -Y$ of order 6.

Collecting the results we get:

Theorem 2. *The group of automorphisms* $\mathrm{Aut}(E_j)$ *of an elliptic curve* E_j *with invariant* j *is* $Z/2Z$ *if* $j \neq 0, 12^3$. *In addition,* $\mathrm{Aut}(E_0) = Z/6Z$ *and* $\mathrm{Aut}(E_{(12^3)}) = Z/4Z$.

It is now quite plausible that the nontrivial automorphisms of E_0 and $E_{(12^3)}$ are the cause for the non-existence of families of elliptic curves with an arbitrarily prescribed invariant $j \in \Gamma(S, \mathcal{O}_S)$. To disclose the connection more clearly we turn to complex analytic families of elliptic curves. It is sufficient for our purposes to define them in the following way.

Definition 2. An analytic family of elliptic curves over a complex analytic space S is the quotient of a line bundle L over S by a lattice $\Gamma \subset L$ (i.e., Γ is a closed discrete subgroup bundle with fibre $Z \times Z$).

Example B. Let H be the upper half plane of the complex plane C, and denote by E the quotient of the trivial line bundle $H \times C$ by the lattice generated by the sections corresponding to the constant 1 and the identity map $H \to C$. It is immediate that each analytic family of elliptic curves is induced locally by the family E.

The fiber E_z of E over a point $z \in H$ is the quotient of C by the lattice Γ_z generated by 1 and z. Any isomorphism $E_z \xrightarrow{\sim} E_{z'}$ between two fibers may be lifted to an isomorphism of C. This means that there is a constant $k \in C^*$ such that (kz, k) is a basis for $\Gamma_{z'}$, i.e.,

$$kz = az' + b$$
$$k = cz' + d,$$

for some $a, b, c, d \in Z$ with $ad - bc = 1$. Hence z is in the orbit of z' under the action of $SL(Z, 2)$ on H given by

$$z \mapsto \frac{az+b}{cz+d}$$

for

$$\begin{pmatrix} a & b \\ c & d \end{pmatrix} \in SL(Z, 2).$$

Conversely, it is clear that this is a sufficient condition for the existence of an isomorphism $E_z \xrightarrow{\sim} E_{z'}$. In fact, it may be shown that the map

$$z \mapsto (j \text{ of } E_z)$$

defines an analytic morphism $j : H \to C$ whose fibers are the orbits of the points $z \in H$ under $SL(Z, 2)$. This function j is the classical elliptic modular function (see e.g. Serre, Cours d'Arithmétique, chap. VII).

In particular, the group of automorphisms of E_z is isomorphic to the subgroup of elements

$$\begin{pmatrix} a & b \\ c & d \end{pmatrix} \in SL(\boldsymbol{Z}, 2)$$

which leave z fixed and therefore induce maps from E_z to itself.

It follows that a point $z \in H$ is left fixed by some element $\begin{pmatrix} a & b \\ c & d \end{pmatrix}$ of $SL(Z, 2)$ other than

$$\pm \begin{pmatrix} 1 & 0 \\ 0 & 1 \end{pmatrix}$$

if and only if there is a nontrivial automorphism on E_z. In each neighborhood of such a point z_0 there are always distinct points w and $(aw+b/cw+d)$ with isomorphic fibers in the family E. Hence j must be ramified at z_0. The same conclusion holds for any analytic family of elliptic curves because of the locally universal character of E.

Remark. It is not hard to see that there are two $SL(\boldsymbol{Z}, 2)$-orbits of points of H left fixed by some nontrivial element of $SL(\boldsymbol{Z}, 2)$, namely ω and i corresponding to $j = 0$ and $j = 12^3$, in agreement with theorem 2.

4. Binary quartics [2]

Thus far, we have taken an *ad hoc* approach to classifying elliptic curves, relying on explicit formulae. What happens more often in moduli problems is that in the first stage you construct either

(a) embeddings in some $\boldsymbol{P}^n$, $n > 1$, or

(b) finite morphisms to $\boldsymbol{P}^1$, with branch locus B, or

(c) some other reduction to projective data,

which is canonical except for a projective transformation. To illustrate, given an elliptic curve E, with base point 0, using an arbitrary basis of $H^0(\mathcal{O}_E(3(0)))$, we construct an isomorphism of E with a plane cubic curve C, which is then canonically determined up to a change $C' = \gamma(C)$, $\gamma \in PGL(3, k)$. Or, using an arbitrary basis of $H^0(\mathcal{O}_E(2(0)))$, we construct a finite morphism of degree 2 from E to $\boldsymbol{P}^1$, with branch locus B of degree 4; then B determines E, and conversely, E determines B up to a change $B' = \gamma(B)$, $\gamma \in PGL(2, k)$. Once the general moduli problem is reduced to the classification of some subvarieties, cycles etc. on $\boldsymbol{P}^n$ modulo $PGL(n+1, k)$, the second stage is to use the theory sketched in Chapters 1 and 2 to find moduli for these projective objects. In this chapter, I would

[2] The following chapter is completely different from Chapter 4 in the original notes and was not presented at Oslo. I have substituted it because it seems to me a much more logical continuation, tieing everything up and keeping to elementary constructions.

like to illustrate how this general procedure works when the classification of elliptic curves is carried out in 2 stages: first reduction to cycles in $\mathbf{P}^1$ of degree 4; second construction of the quotient of the space of such cycles by $PGL(2, k)$.

The space of cycles of degree n on $\mathbf{P}^1$, or what is the same, the space of *unordered* sets of n points on $\mathbf{P}^1$, is just $\mathbf{P}^n$! In fact, let

$$\mathbf{P}^n_{\langle a\rangle} = \mathbf{P}(k\cdot a_0+\cdots+k\cdot a_n)$$

be the projective space with homogeneous coordinates $a_0, \cdots, a_n$, and let

$$\mathbf{P}^1 = \mathbf{P}(k\cdot X_0+k\cdot X_1)$$

be the projective line with homogeneous coordinates X_0, X_1. Then define

$$\mathscr{D} \subset \mathbf{P}^1\times\mathbf{P}^n_{\langle a\rangle}$$

by

$$\mathscr{D} = \left\{\begin{array}{l}\text{locus of solutions of}\\ a_0X_0^n+a_1X_0^{n-1}X_1+\cdots+a_nX_1^n=0\end{array}\right\}.$$

If α is a closed point of $\mathbf{P}^1_{\langle a\rangle}$ and $D_\alpha = \mathscr{D}\cap(\mathbf{P}^1\times(\alpha))$, then D_α is the set of roots of the equation $\sum \alpha_i X_0^{n-i}X_1^i = 0$ where (α_i) are homogeneous coordinates of α. Even better though, $\mathscr{D}$ is a Cartier divisor on $\mathbf{P}^1\times\mathbf{P}^n$, i.e. it is defined locally everywhere by 1 equation, and since $\mathscr{D}\not\supset\mathbf{P}^1\times(\alpha)$ for any $(\alpha)\in\mathbf{P}^n$, we can define $D_\alpha = \mathscr{D}\cdot(\mathbf{P}^1\times(\alpha))$ as a Cartier divisor too, i.e. restrict the defining equation of $\mathscr{D}$ to $\mathbf{P}^1\times(\alpha)$. But divisors on a curve are just cycles, i.e. formal linear combinations of closed points $\sum n_i(x_i)$. It is clear that the divisor D_α is defined locally by the equation $\sum\alpha_i(X_1/X_0)^i = 0$, or by $\sum\alpha_i(X_0/X_1)^{n-i} = 0$, and as a cycle D_α is just the set of roots of the polynomial $\sum\alpha_i X_0^{n-1}X_1^i$ *counted with their multiplicities*. Now since every set of n points (λ_i, μ_i) in $\mathbf{P}^1$, with or without repetitions, is the set of solutions of a unique homogeneous polynomial (up to a scalar), viz:

$$\prod_{i=1}^n(\mu_i X_0-\lambda_i X_1),$$

it follows that every cycle of degree n on $\mathbf{P}^1$ equals D_α for one and only one $\alpha\in\mathbf{P}^n$.

From another point of view, $\mathbf{P}^n_{\langle a\rangle}$ can be viewed as the quotient of $(\mathbf{P}^1)^n$ by the group $\sum_n$ of permutations on n letters. In fact, $(\mathbf{P}^1)^n$ parametrizes the *ordered* sets of n points and $\mathbf{P}^n_{\langle a\rangle}$ the *unordered* sets. Explicitly, expand:

$$\prod_{i=1}^n(\mu_i X_0-\lambda_i X_1) = \sum_{i=0}^n a_i(\lambda,\mu)X_0^{n-i}X_1^i.$$

Then define

$$\pi : (P^1)^n \to P^n_{\langle a \rangle}$$

by

$$\pi((\lambda_1, \mu_1), \cdots, (\lambda_n, \mu_n)) = (a_0(\lambda, \mu), \cdots, a_n(\lambda, \mu)).$$

It is easy to see that $\pi(x) = \pi(y)$ if and only if $x = \sigma(n)$ for some permutation $\sigma \in \sum_n$. In fact, $P^n_{\langle a \rangle}$ is a geometric quotient of $(P^1)^n$ by $\sum_n$. On an affine level, π restricts to the map:

$$\text{res}\ \pi : (A^1) \xrightarrow{n} A^n$$

$$\pi(\lambda_1, \cdots, \lambda_n) = (\lambda_1 + \cdots + \lambda_n; \sum_{i<j} \lambda_i \lambda_j; \cdots; \lambda_1 \cdots \lambda_n),$$

the elementary symmetric functions,

and it is a classical fact that the elementary symmetric functions generate the full ring of permutation-invariant polynomials in n indeterminants $\lambda_1, \cdots, \lambda_n$.

Now consider double coverings of P^1. *Assume for the rest of this chapter* $\text{char}(k) \neq 2, 3$. We want to prove that a double covering is determined by its branch locus.

PROPOSITION 1. *Let C be a non-singular curve and let $\pi : C \to P^1$ be a finite surjective morphism of degree 2. Let $x_1, \cdots, x_n \in P^1$ be the branch points of π. Then $n = 2m$, and C can be constructed as:*

$$C = \underline{\text{Spec}}(\mathscr{A}),$$

where $\mathscr{A}$ is the sheaf of $\mathcal{O}_{P^1}$ algebras $\mathcal{O}_{P^1} \oplus \mathcal{O}_{P^1}(-m)$, where 2 functions are multiplied by the rule:

$$(*) \qquad (f_1, g_1) \cdot (f_2, g_2) = (f_1 f_2 + \phi(g_1, g_2), f_1 g_2 + f_2 g_1)$$

and ϕ is the map:

$$(**)\ \mathcal{O}_{P^1}(-m) \times \mathcal{O}_{P^1}(-m) \to \mathcal{O}_{P^1}(-2m) \cong \mathcal{O}_{P^1}(-\sum_{i=1}^{n} x_i) \subset \mathcal{O}_{P^1}.$$

PROOF. C will have an automorphism $\lambda : C \to C$ of order 2 interchanging the 2 points over each point of P^1. Now since π is a finite morphism, C is automatically equal to $\underline{\text{Spec}}(\pi_* \mathcal{O}_C)$. The automorphism λ acts as an automorphism of $\pi_*(\mathcal{O}_C)$, and since char $\neq 2$, $\pi_*(\mathcal{O}_C)$ splits into a sum $F^+ \oplus F^-$, where $\lambda(f) = f$, $f \in \Gamma(U, F^+)$; $\lambda(f) = -f$, $f \in \Gamma(U, F^-)$. Now the λ-invariant functions F^+ on C must be the functions of the form $g \circ \pi$, $g \in \Gamma(U, \mathcal{O}_{P^1})$, so $F^+ \cong \mathcal{O}_{P^1}$. Since π has degree 2, $\pi_*(\mathcal{O}_C)$ is a locally free sheaf of $\mathcal{O}_P$-modules of rank 2, hence the second factor F^- is locally free of rank 1. Therefore $F^- \cong \mathcal{O}_{P^1}(k)$, some $k \in \mathbf{Z}$. Now the product of 2 odd functions is even, hence the multiplication in

$\pi_* \mathcal{O}_C \cong \mathcal{O}_{P^1} \oplus \mathcal{O}_{P^1}(k)$ must be given by a rule of the form (*). Finally any non-zero bilinear $\phi : \mathcal{O}_P(k) \times \mathcal{O}_P(k) \to \mathcal{O}_P$ is induced by a composition:

$$\mathcal{O}_{P^1}(k) \times \mathcal{O}_{P^1}(k) \to \mathcal{O}_{P^1}(2k) \simeq \mathcal{O}_{P^1}(-\sum y_i) \subset \mathcal{O}_{P^1}$$

for some cycle $\sum y_i$ of degree $-2k$. Now reverse the construction and start from the cycle $\sum y_i$, use this to define ϕ and set $C^* = \underline{\mathrm{Spec}}(\mathcal{O}_{P^1} \oplus \mathcal{O}_{P^1}(k))$. If $y \in P^1$, t is a generator of the maximal ideal m_{y,P^1}, and s is a generator of the stalk $\mathcal{O}_{P^1}(k)_y$, then near y, C^* is given explicitly by the equation:

$$s^2 = u \cdot t^r$$

where u is a unit in $\mathcal{O}_{y,P^1}$ and

$$r = \text{mult. of } y \text{ in } \sum y_i.$$

Therefore C^* is singular if $r > 1$; C^* is non-singular and y is a branch point if $r = 1$; and C^* is non-singular and unbranched over y if $r = 0$. It follows that in the case of the proposition $\sum y_i = \sum x_i$ and $n = -2k$.

Q. E. D.

Combining this with the results of Chapter 3, we find:

COROLLARY. *The set of elliptic curves over k is canonically isomorphic to the set*:

$$\left\{ \begin{matrix} \text{cycles of degree 4 on } P^1 \\ \text{with no multiple points} \end{matrix} \right\} \Big/ PGL(2, k)$$

We could refine this result, using elliptic curves over S and the methods of Chapter 3 to show in addition:

PROPOSITION 2: *If $U \subset P^4_{\langle a \rangle}$ is the open set of cycles of degree 4 with no multiple points, then using the map of points in the previous Corollary, a geometric quotient of U by $PGL(2, k)$ becomes a coarse moduli space for elliptic curves.*

We omit the proof which follows the techniques already discussed.

The next step is to ask what Theorem 2, Ch. II says about the quotient of $P^4_{\langle a \rangle}$ by $PGL(2, k)$. We make contact at this point with some of the oldest work in invariant theory. In fact, to work out Theorem 2, first we replace $PGL(2, k)$ by its double covering $SL(2, k)$ in order that the action of the group on $P^4_{\langle a \rangle}$ should lift to the invertible sheaf $\mathcal{O}_{P^4}(1)$. Then $SL(2, k)$ acts on the whole homogeneous coordinate ring

$$k[a_0, a_1, a_2, a_3, a_4]$$

of $P^4_{\langle a \rangle}$. This action is the obvious one, i.e. let

$$\begin{pmatrix} a & b \\ c & d \end{pmatrix} \in SL(2, k)$$

It acts on $\boldsymbol{P}^1$ by the linear map

$$(X_0, X_1) \to (aX_0 + bX_1, cX_0 + dX_1)$$

in homogeneous coordinates. We must make it act on the a_i so that the form $\sum a_i X_0^{4-i} X_1^i$ has invariant meaning, since in its action on $\boldsymbol{P}^4_{\langle a \rangle}$, the subvariety $\mathscr{D} \subset \boldsymbol{P}^1 \times \boldsymbol{P}^4_{\langle a \rangle}$ should be taken into itself. In other words $\begin{pmatrix} a & b \\ c & d \end{pmatrix}$ induces the map $a_i \to a_i'$ such that

$$\sum a_i'(aX_0 + bX_1)^{4-i}(cX_0 + dX_1)^i \equiv \sum a_i X_0^{4-i} X_1^i.$$

Alternatively, $SL(2, k)$ acts on the vector space $\sum k \cdot a_i$ by the dual of the 4th symmetric power of its action on $kX_0 + kX_1$. The whole construction hinges on the subring

$$R = k[a_0, a_1, a_2, a_3, a_4,]^{SL(2,k)}$$

of invariants. In classical terminology, one wrote $f = \sum a_i X_0^{4-i} X_1$, and called f a *binary quartic*. The elements of R were called the *invariants of a binary quartics*. More generally, the elements of

$$k[X_0, X_1, a_0, a_1, a_2, a_3, a_4]^{SL(2,k)}$$

(such as f itself) were called the *covariants of a binary quartic*. Generators of both of these rings are classical. (Good references are Elliott, *Algebra of Quantics*, Grace and Young, *The algebra of invariants*, or Schur, *Vorl. über Invarianten theorie*.) They are written down quickest by an amazing technique known as the symbolic method – the only bit of linear algebra I believe that has not been thoroughly 'Bourbakized' [3]. One does this: one takes the form f and formally writes it out as though it were a power of a linear form in several different ways, i.e.

$$f = \alpha_x^4 = \beta_x^4 = \gamma_x^4 = \cdots$$

where

$$\alpha_x = \alpha_0 X_0 + \alpha_1 X_1$$
$$\beta_x = \beta_0 X_0 + \beta_1 X_1$$

etc.

One then makes a monomial in α_x, β_x, etc. and the 2×2 determinants:

$$(\alpha, \beta) = \alpha_0 \beta_1 - \alpha_1 \beta_0$$
$$(\alpha, \gamma) = \alpha_0 \gamma_1 - \alpha_1 \gamma_0$$

[3] (Added in proof) Now it has been . . .
cf. Dieudonné, Seminaire Bourbaki, June 1971.

such that the total degree in each α, β etc. used is 4. For instance:

$$\begin{aligned} f &= \alpha_x^4 \\ h &= (\alpha, \beta)^2 \alpha_x^2 \beta_x^2 \\ j &= (\alpha, \beta)^2 (\gamma, \beta) \alpha_x^2 \beta_x \gamma_x^3 \\ P &= (\alpha, \beta)^4 \\ Q &= (\alpha, \beta)^2 (\alpha, \gamma)^2 (\beta, \gamma)^2 \end{aligned}$$

Each of these can be re-interpreted as a polynomial in the coefficients a_i and the variables (X_0, X_1) by simply taking each monomial $\alpha_0^{4-i}\alpha_1^i$, $\beta_0^{4-i}\beta_1^i$, etc. and replacing it by $\binom{4}{i}^{-1} a_i$ which it equals in purely formal identity $f = \alpha_x^4, f = \beta_x^4$ etc. One must obtain in this way a covariant. Here is an example:

$$\begin{aligned} P &= (\alpha_0\beta_1 - \alpha_1\beta_0)^4 \\ &= \alpha_0^4\beta_1^4 - 4\alpha_0^3\alpha_1\beta_0\beta_1^3 + 6\alpha_0^2\alpha_1^2\beta_0^2\beta_1^2 - 4\alpha_0\alpha_1^3\beta_0^3\beta_1 + \alpha_1^4\beta_0^4 \\ &= a_0 a_4 - \tfrac{1}{4}a_1 a_3 + \tfrac{1}{6}a_2^2 - \tfrac{1}{4}a_3 a_1 + a_4 a_0 \\ &= \tfrac{1}{6}(a_2^2 - 3a_1 a_3 + 12 a_0 a_4) \end{aligned}$$

Similarly, but with a bit more sweat, Q comes out as the determinant:

$$\begin{aligned} Q &= 6 \cdot \det \begin{pmatrix} a_0 & a_1/4 & a_2/6 \\ a_1/4 & a_2/6 & a_3/4 \\ a_2/6 & a_3/4 & a_4 \end{pmatrix} \\ &= a_0 a_2 a_4 - \tfrac{3}{8} a_0 a_3^2 - \tfrac{3}{8} a_1^2 a_4 + \tfrac{1}{8} a_1 a_2 a_3 - \tfrac{1}{36} a_2^3 \end{aligned}$$

It is an old theorem that f, h, j, P and Q generate the ring of covariants. For instance another well-known invariant is the discriminant D of the form f: i.e. for the form

$$f = \prod_{i=1}^{4} (\mu_i X_0 - \lambda_i X_1)$$

let

$$D = \prod_{1 \leq i < j \leq 4} (\lambda_j \mu_i - \lambda_i \mu_j)^2$$

Then D can be rewritten as a homogeneous polynomial in the coefficients a_0, a_1, a_2, a_3, a_4 of f of degree 6. But in fact, one sees easily that:

LEMMA 1. $D = \text{const.}\,(P^3 - 6Q^2)$.

PROOF. In fact, on a k-valued form f, $D(f) = 0$ if and only if f has a double zero. And D is zero to 1st order on the irreducible variety of f's with multiple zeros. Since D, P^3 and Q^2 all have degree 6 in the a's, it suffices to show that $D | P^3 - 6Q^2$ and for this it suffices to show that $P(f)^3 = 6Q(f)^2$ if f is a form with double zeroes. But such an f is equivalent to a form $\bar{f}$ with double zero at $X_1 = 0$, i.e. with $a_0 = a_1 = 0$.

Therefore

$$P(\bar{f}) = \tfrac{1}{6}\bar{a}_2^2, Q(\bar{f}) = -\tfrac{1}{36}\bar{a}_2^3,$$

hence indeed

$$P(f)^3 = P(\bar{f})^3 = (\tfrac{1}{6})^3\bar{a}_2^6 = 6Q(\bar{f})^2 = 6Q(f)^2.$$

Q. E. D.

We will not prove or use the result on the generators of ring of covariants although it is not too hard. Instead we want to see geometrically what P and Q do. The first point is:

LEMMA 2: $P(f) = Q(f) = 0 \Leftrightarrow$ *the form f has a triple zero.*

PROOF. This is clear from lemma 1 and the fact that if f has a double zero at $X_1 = 0$, then $P(f) = \frac{1}{6}a_2^2$, so that $P(f) = 0$ if and only if the double zero is a triple zero.

Q. E. D.

Let $X_{ss} \subset \boldsymbol{P}^4_{\langle a \rangle}$ be the open set of forms f with no triple zero. Then we obtain a morphism:

$$\pi : X_{ss} \to \operatorname{Proj} k[P, Q].$$

Since $\deg P = 2$, $\deg Q = 3$, the subring of $k[P, Q]$ of elements whose degrees are multiples of 6 is just the ring $k[P^3, Q^2]$, and

$$\operatorname{Proj} k[P, Q] = \operatorname{Proj} k[P^3, Q^2] \cong \boldsymbol{P}^1.$$

Thus π is just the map from X_{ss} to $\boldsymbol{P}^1$ defined by P^3/Q^2. To examine the orbits of $PGL(2, k)$ in X_{ss}, note that any form f with at least 3 distinct zeroes is equivalent to a constant times a form

$$f_\lambda(X) = X_0 \cdot X_1 \cdot (X_0 - X_1)(X_0 - \lambda X_1)$$

by a projective transformation. But computing we find

$$a_0 = 0, a_1 = 1, a_2 = -(\lambda+1), a_3 = \lambda, a_4 = 0$$

hence

$$P(f_\lambda) = \tfrac{1}{6}(\lambda^2 - \lambda + 1)$$
$$Q(f_\lambda) = \tfrac{1}{72}(\lambda+1)(\lambda-2)(2\lambda-1)$$

and by lemma 1:

$$P(f_\lambda)^3 - 6Q(f_\lambda)^2 = \text{cnst.}\, \lambda^2(\lambda-1)^2.$$

Therefore

$$j(\lambda) = \text{cnst.}\, \frac{P(f_\lambda)^3}{P(f_\lambda)^3 - 6Q(f_\lambda)^2}.$$

But we proved in Chapter 3 that

$$j(\lambda_1) = j(\lambda_2) \Leftrightarrow \left\{\begin{array}{l}\text{the 2 sets of points } \{0, 1, \infty, \lambda_1\} \\ \text{and } \{0, 1, \infty, \lambda_2\} \text{ are projectively} \\ \text{equivalent}\end{array}\right\}$$

hence

$$P(f_{\lambda_1})^3/Q(f_{\lambda_1})^2 = P(f_{\lambda_2})^3/Q(f_{\lambda_2})^2 \Leftrightarrow j(\lambda_1) = j(\lambda_2)$$
$$\Leftrightarrow f_{\lambda_1} \text{ is equivalent to a constant times } f_{\lambda_2} \text{ by some } \sigma \in SL(2, K)$$

Finally, in X_{ss}, all orbits are represented by forms f_λ except for the forms with 2 distinct doubles zeroes, which are equivalent to a constant times the form:

$$f^*(X) = X_0^2 X_1^2.$$

It is clear that $P(f^*)^3/Q(f^*)^2 = 6$, just like the other forms with one double zero. Now define $X_s \subset X_{ss}$ to be the open set of forms f with no double zero. The we can summarize our conclusions in:

PROPOSITION 3. *Let* $\delta \in \operatorname{Proj} k[P, Q]$ *be the point defined by* $P^3/Q^2 = 6$. *Then*

(i) $X_s = \pi^{-1}(\operatorname{Proj}(k[P, Q]) - (\delta))$

(ii) *if* $x \in \operatorname{Proj} k[P, Q]$, $x \neq \delta$, *then* $\pi^{-1}(x)$ *consists of one orbit and it is closed in* X_{ss}.

(iii) $\pi^{-1}(\delta)$ *consists of* 2 *orbits, the orbit of the form* f_0 *which is* 3-*dimensional and the orbit of the form* f^* *which is* 2-*dimensional. The first is open in* $\pi^{-1}(\delta)$ *and the second is closed.*

The final step is:

PROPOSITION 4. $\operatorname{Proj} k[P, Q] - (\delta)$ *is a geometric quotient of* X_s *by* $PGL(2, k)$.

SKETCH OF PROOF. The only remaining point is that for all

$$U \subset \operatorname{Proj} k[P, Q] - (\delta),$$

invariant functions f on $\pi^{-1}(U)$ are induced by functions on U itself. This can be checked by first restricting f to the curve in X_s of forms f_λ, $(\lambda \neq 0,1)$, and noting that this curve is separable and finite over

$$\operatorname{Proj} k[P, Q] - (\delta).$$

But f is set theoretically a pull-back of a function on U, so by Zariski's Main Theorem, $f = g \circ \pi$, some $g \in \Gamma(U, \mathcal{O}_u)$. This is nothing but a rephasing of the final arguement in Ch. 3 that A_j^1 is a coarse moduli space.

This gives us a new proof of the main results of Ch. 3. It also gives us an idea of how to interpret the compactification $\boldsymbol{P}_j^1$ of $\boldsymbol{A}_j^1$ as a moduli space of a bigger moduli functor. Whenever a moduli space is not complete, the natural question is: what happens to the objects being classified when they move off to the boundary of the moduli space? As $j \to \infty$, or $P^3/Q^2 \to 6$, we see that $f_\lambda \to f_0$ or f^*, representatives of the 2 orbits in X_{ss} over δ. In the proof of Prop. 1, we saw how to construct double coverings of $\boldsymbol{P}^1$ from any cycle on $\boldsymbol{P}^1$ of even degree: if these cycles have multiplicities, the effect is to make the covering a singular curve. To be explicit, take as branch locus the cycle $f_0 = 0$, i.e. $2(0)+(1)+(\infty)$. The associated double covering is the curve C_0 which is covered by the 2 affine pieces:

(i) $\pi^{-1}(\boldsymbol{P}^1-(\infty))$ which is given by

$$Y^2 = \left(\frac{X_0}{X_1}\right)^2\left(\frac{X_0}{X_1}-1\right)$$

(ii) $\pi^{-1}(\boldsymbol{P}^1-(0))$ which is given by

$$Z^2 = \left(\frac{X_1}{X_0}\right)\left(\frac{X_1}{X_0}-1\right).$$

The two are related by $Z = i(X_1/X_0)^2 \cdot Y$. Note that C_0 is non-singular except for an ordinary double point (or node) over (0), i.e. a double point at which the tangent cone consists of 2 distinct lines. If $t = Y/(X_0/X_1)$, then $t^2 = X_0/X_1 - 1$, hence

$$\frac{X_0}{X_1} = t^2+1$$

$$Y = t(t^2+1)$$

which proves that the field of rational functions on C_0 is $k(t)$. Thus C_0 is a rational curve with 1 node. On the other hand, take as branch locus the cycle $f^* = 0$, i.e. $2(0)+2(\infty)$. The associated double covering C^* is covered by:

(i) $$Y^2 = \left(\frac{X_0}{X_1}\right)^2$$

and

(ii) $$Z^2 = \left(\frac{X_1}{X_0}\right)^2$$

related by $Z = Y \cdot (X_1/X_0)$. Thus C^* has, in fact, 2 non-singular rational components:

$$C_1^* : \text{the union of } Y = \frac{X_0}{X_1} \text{ and } Z = \frac{X_1}{X_0}$$

$$C_2^* : \text{the union of } Y = -\frac{X_0}{X_1} \text{ and } Z = -\frac{X_1}{X_0}$$

meeting transversely at 2 points

$$Y = \frac{X_0}{X_1} = 0 \text{ and } Z = \frac{X_1}{X_0} = 0.$$

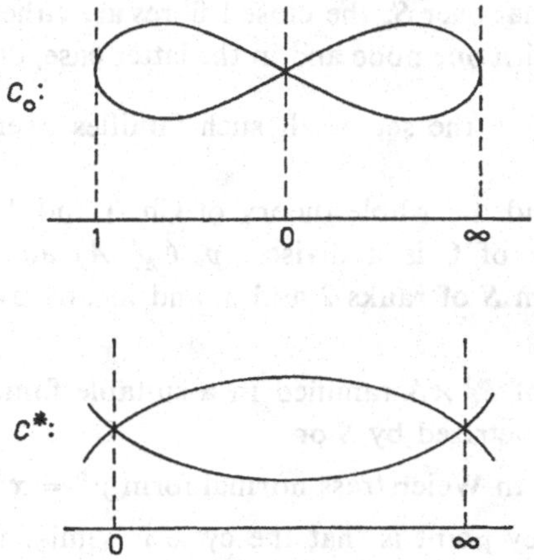

Figure 1

Moreover, consider the algebraic family of cycles $2(0)+(\mu)+(\infty)$. They are all projectively equivalent, but as $\mu \to \infty$, it 'jumps' to the cycle $2(0)+2(\infty)$. Taking double coverings, the curve C_0 can 'jump' to C^*. Topologically if $k = C$, what happens is illustrated below:

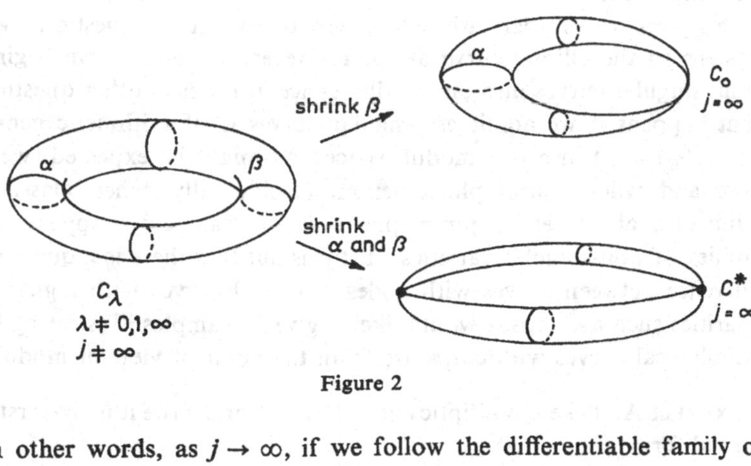

Figure 2

In other words, as $j \to \infty$, if we follow the differentiable family of surfaces C_λ, either one or two curves begin to shrink until when $j = \infty$, we wind up with a topological space which is no longer a manifold but is homeomorphic to one of the 2 spaces C_0 or C^* as illustrated.

If we return to our functor $\mathcal{M}$, there is a very natural way to enlarge $\mathcal{M}$ allowing the curve C_0 to define a new element of $\mathcal{M}(\mathrm{Spec}\ k)$.

DEFINITION. A *family of semi-elliptic curves* over a k-variety S is a morphism of k-varieties $p : E \to S$ together with a section $0 : S \to E$ such that E is proper and flat over S, the closed fibres are either elliptic curves or a rational curve with one node and in the latter case, $0(s) \neq$ the node.

DEFINITION. $\overline{\mathcal{M}}(S) =$ the set of all such families over S, up to isomorphism.

We can now extend the whole theory of Ch. 3 and 4 to semi-elliptic families: $A =$ Image of 0 is a divisor, $p_* \mathcal{O}_E(2A)$ and $p_*\mathcal{O}_E(3A)$ are locally free sheaves in S of ranks 2 and 3, and locally over S, E can be defined either

a) as a covering of $\boldsymbol{P}^1 \times S$ ramified in a suitable family of cycles of degree 4 on $\boldsymbol{P}^1$ parametrized by S or

b) by an equation in Weierstrass normal form $y^2 = x^3+ax+b$.

In case (a), the key point is that the cycle is semi-stable and has at most one double point; in case (b), $4a^3+27b^2$ need not be invertible, but a and b together should have no common zeroes. The final conclusion is that there is a canonical morphism

$$\overline{\Phi} : \overline{\mathcal{M}} \to h_{\boldsymbol{P}_j{}^1}$$

extending our previous Φ and making $\boldsymbol{P}_j^1$ into a coarse moduli space for semi-elliptic curves.

This gives us a rather satisfactory way to answer the question – what happens to the elliptic curve at ∞. However, as soon as we begin to admit singular curves into our moduli space, it raises another question – what happens if we admit *all* singular curves C of arithmetic genus 1 (i.e. $\chi(\mathcal{O}_C) = 0$) into our moduli space? As might be expected, we get more and wilder jump phenomenon. (Incidentally, when classifying 2-dimensional varieties, jump phenomenon can even appear with families of non-singular varieties). It turns out that there is a qualitative difference between curves with nodes only and curves with higher singularities such as cusps. I would like to give 2 examples illustrating how pathological curves with cusps are from the point of view of moduli.

EXAMPLE A. Take any elliptic curve E over k and write it in Weierstrass normal form:

$$y^2 = x^3+ax+b.$$

For every $\lambda \in k$, $\lambda \neq 0$,

$$y^2 = x^3 + \lambda^2 ax + \lambda^3 \cdot b$$

represents the same elliptic curve. Consider the family E of curves over $S = \operatorname{Spec} k[\lambda]$ defined by this equation.

Then

$$E \cong E_\lambda \text{ all } \lambda \neq 0.$$

But E_0 is the curve $y^2 = x^3$: an irreducible plane cubic curve with a cusp at $x = y = 0$, which is rational (if $t = y/x$, then $x = t^2$, $y = t^3$). In other words, *every elliptic curve*, without changing j, can jump to the cuspidal cubic $y^2 = x^3$. Thus j is completely indeterminant on $y^2 = x^3$ and, topologically, it is impossible to fit E_0 into the moduli space $\mathbf{P}_j^1$ even allowing non-separated schemes!

EXAMPLE B. Let $k = \mathbf{C}$.

Let C_0 be a plane curve of degree n with one cusp and no other singularities. If we choose coordinates correctly, we can normalize C_0 so that the cusp is the origin $x = y = 0$ and the affine equation of C_0 is of the form:

$$0 = x^2 + y^3 + p_4(x, y)$$

where P_4 is a polynomial whose leading terms have degree $\geqq 4$. Let C_t be the nearby curve defined by the equation

$$t = x^2 + y^3 + P_4(x, y), \ |t| < \varepsilon.$$

It is easy to see that C_t is non-singular everywhere and I would like to describe the topological situation with C_t approaching C_0. Everywhere except in a neighbourhood of (0, 0), $\bigcup_{|t|<\varepsilon} C_t$ forms a nice differentiable family of surfaces over the t-disc. However near (0, 0), since $|P_4(s, y)|$ is much less than $|x^2|$ or $|y^3|$, C_t is essentially the surface

$$C_t^0: \begin{array}{l} x = \pm\sqrt{t - y^3} \\ |y| < \eta \end{array}$$

Take t even smaller: in fact $|t| < \frac{1}{2}\eta^3$. Then x is a branched covering of the y-disc$|y| < \eta$, branched at $y = \sqrt[3]{t}, \omega \cdot \sqrt[3]{t}, \omega^2 \cdot \sqrt[3]{t}$:

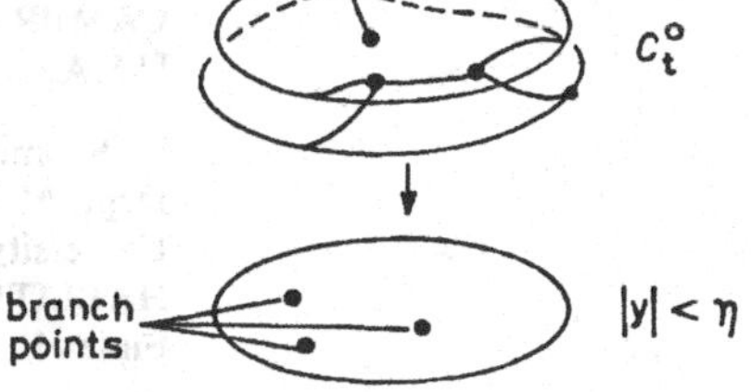

Figure 3

Topologically, the above surface with boundary is just a torus with a hole in it. Thus C_t, and its degeneration to C_0, looks like this:

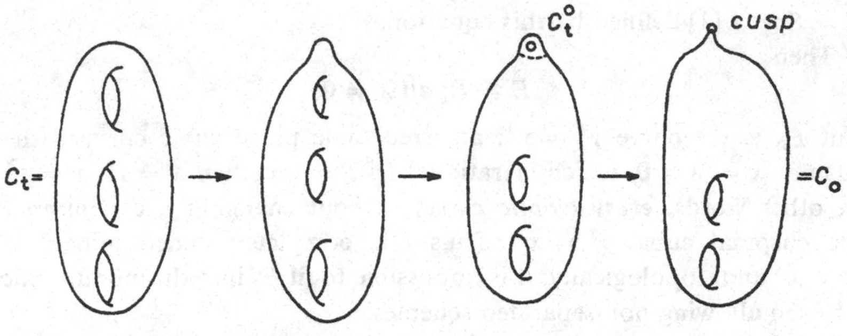

Figure 4

In other words, the cusp has swalled a whole infinitesimal elliptic curve!

REFERENCES

A. Borel
LAG – Linear algebraic groups (notes taken by H. Bass). Math. Lect. Notes series, Benjamin, 1969.

J. Fogarty
IT – Invariant theory. Math. Lect. Notes series, Benjamin, 1969.

A. Grothendieck & J. Dieudonné
EGA – Eléments de géométrie algébrique. Publ. Math. No. 4, . . ., Inst. Hat. Et. Sc., 1960, . . .

D. Mumford
GIT – Geometric invariant theory. Ergeb. Math. Bd. 34, Springer Verlag, 1965.

D. Mumford
IAG – Introduction to algebraic geometry (preliminary version of the first three chapters, lecture notes Harvard University).

(Oblatum 4–III–1971)

D. Mumford
Dept. Math.
Harvard University
2 Divinity Ave.
CAMBRIDGE, Mass., 02138
U.S.A.

K. Suominen
Dept. Math.
University of Helsinki
HALLITUSK 11–13
Finland

D. Mumford

AN EXAMPLE OF A UNIRATIONAL 3-FOLD WHICH IS NOT RATIONAL

This lecture reports on joint work with M. Artin. We have been investigating 3-folds V which admit a rational map $f: V^3 \to F^2$ to a surface whose generic fibre is a rational curve. The interesting feature of the 3-fold case is that unlike the case of a surface over a curve $f: F^2 \to C^1$ with generic fibre rational, there need not exist a unisecant (equivalently, V may not be birational to $F \times \mathbf{P}^1$).

This situation has been much studied (cf. Roth [4], Ch. 4, § 4-7), and for instance, the V's that arise correspond in a natural way to the quaternion algebras over the function field $\mathbf{C}(F)$, which ties the theory up with the Brauer Group. An important invariant of the situation is a (possibly reducible) curve Δ on F, the discriminant of the quaternion algebra, or the locus of points where the fibre of V over F is forced to become reducible. We heard this Spring that Clemens and Griffiths [2] and Iskovskih and Manin [3] had independently found rigorous proofs of the irrationality of certain unirational Fano 3-folds. C. P. Ramanujam suggested to us that our theory might yield quite elementary examples of such 3-folds if we could find non-singular V's with torsion in H^3 (hence H^4, H_2 and H_3 as well; torsion in H^2, hence in H^5, H_1 and H_4 is not possible if V is unirational as Serre [5] showed).

Such torsion does not exist on a rational variety so this is a possible method for distinguishing between rational and unirational varieties. Indeed it turns out that when the discriminant Δ is disconnected, V has 2-torsion in H^3. Furthermore, we found a unirational V of this type which was a double covering of $\mathbf{P}^3$ with quartic branch locus B, but where B has precisely 10 nodes. The details of our work will be published in [1].

Bibliograpy

[1] M. Artin and D. Mumford, *Some elementary examples of unirational varieties which are not rational*, « Proc. London Math. Society », 25, 75 (1972).

[2] H. Clemens and P. Griffiths, *The intermediate jacobian of a cubic three-fold*, « Annals of Math. », 95, 281 (1972).

[3] V. Iskovskih and Y. Manin, *Three dimensional quartics and counter-examples to the Luroth problem*, « Mat. Sbornik », 86, 140 (1971).

[4] L. Roth, *Algebraic three-folds*, Springer-Verlag, 1955.

[5] J.-P. Serre, *On the fundamental group of a unirational veriety*, « Journal London Math. Society », 34, 481 (1959).

D. Mumford, *Selected Papers*, Vol. II,

A REMARK ON THE PAPER OF M. SCHLESSINGER

by David Mumford

In the conference itself, I spoke on a theorem asserting the existence of "semi-stable" reductions for analytic families of varieties over a disc, smooth outside the origin. This talk turned out to be difficult to transcribe into a paper of moderate size and instead will be incorporated into the notes of a seminar which I am running together with G. Kempf, B. Saint-Donat, and Tai, which we will publish in the Springer Lecture Notes.

Here I would like to add a footnote to Schlessinger's calculations of versal deformations.[1] He studied the situation: $V =$ complex $n+1$-dimensional vector space; $\mathbf{P}(V) = n$-dimensional projective space of 1-dimensional subspaces of V; $Y \subset \mathbf{P}(V)$ a smooth r-dimensional variety, $r \geqq 1$; $C \subset V$ the cone over Y.

Let $L = \mathcal{O}_Y(1)$. Assume:

$$H^0(\mathbf{P}(V), \mathcal{O}_{\mathbf{P}(V)}(k)) \to H^0(Y, L^k) \text{ is surjective, } k \geqq 1$$

(We may also assume by replacing $\mathbf{P}(V)$ by a linear space that it is an isomorphism for $k = 1$). Then he proved:

a) There is a natural injection of functors:

$$\bar{H} = \left\{\begin{matrix}\text{Deformations}\\ \text{of } Y \text{ in } \mathbf{P}(V)\end{matrix}\right\} \Big/ \begin{matrix}\text{projective}\\ \text{automorphisms}\end{matrix} \to \left\{\begin{matrix}\text{Deformations}\\ \text{of } C\end{matrix}\right\}$$

b) T_C^1 has a natural graded structure

$$T_C^1 = \bigoplus_{k=-\infty}^{+\infty} (T_C^1)_k$$

such that $(T_C^1)_0 \cong$ image of Zariski tangent space to $\bar{H}$,

c) If $(T_C^1)_k = (0)$ for $k \neq 0$, then $\bar{H}$ is *isomorphic* to the functor of deformations of C, i.e., all deformations of C remain conical.

d) If $r \geqq 2$ and L is sufficiently ample on Y, then the condition in (c) is satisfied.

What I would like to show here is:

d′) If $r = 1$, L is sufficiently ample on Y and Y has genus $\geqq 2$ and is not hyperelliptic, then again the condition in (c) is satisfied.

This gives:

Corollary. There exist normal singularities of surfaces with no non-singular deformation!

To prove (d′), we let $U = C - (0)$ and use the exact sequences:

$$\begin{array}{ccccc} \Gamma(V, \theta_V) & \xrightarrow{\alpha} & \Gamma(C, N_C) & \rightarrow & T_C^1 \rightarrow 0 \\ \| & & \| & & \\ \Gamma(V-(0), \theta_V) & & \Gamma(U, N_C). & & \end{array}$$

Now $\mathbf{C}^*$ acts in a natural way on both θ_V and N_C, and if $\pi: V - (0) \rightarrow \mathbf{P}(V)$ is the projection, then both $\pi_* \theta_V$ and $\pi_* N_C$ decompose into direct sums of their eigenspaces for the various characters of $\mathbf{C}^*$. Moreover, the $\mathbf{C}^*$ invariant sections are:

$$(\pi_* \theta_V)^{\mathbf{C}^*} \cong \mathcal{O}_{\mathbf{P}(V)}(1) \otimes_{\mathbf{C}} V$$

$$(\pi_* N_C)^{\mathbf{C}^*} \cong N_Y$$

and α induces the map $\alpha' = \gamma \circ \beta$

$$\alpha' : \mathcal{O}_{\mathbf{P}(V)}(1) \otimes_{\mathbf{C}} V \xrightarrow{\beta} \theta_{\mathbf{P}(V)} \xrightarrow{\gamma} N_Y$$

(β = standard map).

Thus we get:

$$\begin{array}{ccc} \Gamma(V-(0), \theta_V) & \xrightarrow{\alpha} & \Gamma(U, N_C) \\ \| & & \| \\ \bigoplus_{\nu=-\infty}^{+\infty} \Gamma(\mathbf{P}(V), \mathcal{O}(\nu+1) \otimes_{\mathbf{C}} V) & \longrightarrow & \bigoplus_{\nu=-\infty}^{+\infty} \Gamma(Y, N_Y(\nu)). \end{array}$$

So if

$$(T_C^1)_\nu = \operatorname{coker}[\Gamma(\mathbf{P}^n, \mathcal{O}(\nu+1) \otimes_{\mathbf{C}} V) \xrightarrow{\alpha'_\nu} \Gamma(Y, N_Y(\nu))],$$

then $T_C^1 = \bigoplus_{\nu=-\infty}^{+\infty} (T_C^1)_\nu$. We must compute these groups.

The idea is to determine N_Y explicitly on Y without actually using the embedding of Y defined by L. Consider in fact $N_Y^*(1)$ via the dual of α' as a subbundle of $\mathcal{O}_Y \otimes_{\mathbf{C}} V^*$

$$N_Y^*(1) \subset \theta_{\mathbf{P}(V)}^*(1)|_Y \subset \mathcal{O}_Y \otimes_{\mathbf{C}} V^*$$

hence for every $x \in Y$:

$$[N_Y^*(1) \otimes \mathcal{O}_x/m_x] \subset [\overset{*}{\theta}_{\mathbf{P}(V)}(1) \otimes \mathcal{O}_x/m_x] \subset V^*.$$

It is easy to see that under these inclusions, if $x' \in C$ lies over x:

$$\theta^*_{\mathbf{P}(V)}(1) \otimes \mathcal{O}_x/m_x = \left\{\begin{array}{l}\text{space of linear forms } l \text{ on } V\\ \text{such that } l(x') = 0\end{array}\right\}$$

$$N_Y^*(1) \otimes \mathcal{O}_x/m_x = \left\{\begin{array}{l}\text{space of linear forms } l \text{ on } V\\ \text{such that } l(x') = 0 \text{ and}\\ l = 0 \text{ is tangent to } Y \text{ at } x\end{array}\right\}.$$

But now by assumption:

$$V^* \cong \Gamma(\mathbf{P}(V), \mathcal{O}_{\mathbf{P}(V)}(1)) \xrightarrow{\approx} \Gamma(Y, L)$$

and under this isomorphism, the linear forms l such that $l=0$ and is tangent to Y at x go over to the sections of L vanishing at x to 2nd order, i.e. $\Gamma(Y, m_x^2 \cdot L)$. Now consider

$$\begin{array}{c}\Delta \subset Y \times Y \text{ with } p_1^* L(-2\Delta)\\ \downarrow p_2\\ Y \quad \text{with } p_{2,*}[p_1^* L(-2\Delta)].\end{array}$$

Then it is easily seen that $p_{2,*}[p_1^*L(-2\Delta)]$ is a locally free sheaf on Y and that

$$\begin{aligned} p_{2,*}[p_1^*L(-2\Delta)] \otimes \mathcal{O}_x/m_x &\cong \Gamma(Y \otimes \{y\}, p_1^*L(-2\Delta) \otimes_{\mathcal{O}_Y} \mathcal{O}_x/m_x)\\ &\cong \Gamma(Y, m_x^2 \cdot L).\end{aligned}$$

Thus the two sub-bundles:

a) $p_{2,*}[p_1^*L(-2\Delta)] \subset p_{2,*}[p_1^*L] = \Gamma(Y, L) \otimes_{\mathbf{C}} \mathcal{O}_Y$

b) $N_Y^*(1) \subset V^* \otimes_{\mathbf{C}} \mathcal{O}_Y \cong \Gamma(Y, L) \otimes_{\mathbf{C}} \mathcal{O}_Y$

are equal. Now assume $r = 1$, so that Y is a curve and $\mathcal{O}(-2\Delta)$ is an invertible sheaf on $Y \times Y$. Then by Serre duality for the morphism p_2, we can identify $N_Y(-1)$ as a quotient of $V \otimes_{\mathbf{C}} \mathcal{O}_Y$ or $\Gamma(Y, L)^* \otimes_{\mathbf{C}} \mathcal{O}_Y$:

$$\begin{array}{ccccc}
V \otimes_{\mathbf{C}} \mathcal{O}_Y & \xrightarrow{\alpha'(-1)} & N_Y(-1) & \to 0\\
\wr\| & & \wr\| & \\
\Gamma(Y, L)^* \otimes_{\mathbf{C}} \mathcal{O}_Y & \longrightarrow & \mathrm{Hom}(p_{2,*}[p_1^*L(-2\Delta)], \mathcal{O}_Y) & \to 0\\
\wr\| & & \wr\| & \\
R^1 p_{2,*}(\mathrm{Hom}(p_1^*L, \Omega_{Y \times Y/Y})) & \to & R^1 p_{2,*}(\mathrm{Hom}(p_1^*L(-2\Delta), \Omega_{Y \times Y/Y})) & \to 0\\
\wr\| & & \wr\| & \\
R^1 p_{2,*}(p_1^*(\Omega_Y \otimes L^{-1})) & \longrightarrow & R^1 p_{2,*}(p_1^*(\Omega_Y \otimes L^{-1})(2\Delta)) & \to 0.
\end{array}$$

We want to show that $(T_C^1)_\nu = (0)$ if $\nu \neq 0$, i.e.,

$$\Gamma(Y, R^1 p_{2,*}[p_1^*(\Omega_Y \otimes L^{-1})] \otimes L^\nu) \to \Gamma(Y, R^1 p_{2,*}[p_1^*(\Omega_Y \otimes L^{-1})(2\Delta)] \otimes L^\nu)$$

is surjective if $\nu \neq 1$. If $\deg L > 2g$, then $p_{2,*}$ of the two sheaves in square brackets is zero, hence by the Leray spectral sequence for p_2, the above map is the same as:

$$H^1(Y \times Y, p_1^*(\Omega_Y \otimes L^{-1}) \otimes p_2^* L^\nu)$$
$$\to H^1(Y \times Y, p_1^*(\Omega_Y \otimes L^{-1}) \otimes p_2^* L^\nu \otimes \mathcal{O}(2\Delta)).$$

We treat the surjectivity in three cases:

Case I. $\nu \geqq 2$: Consider the sheaf cokernel

$$p_1^*(\Omega_Y \otimes L^{-1}) \otimes p_2^* L^\nu \to p_1^*(\Omega_Y \otimes L^{-1}) \otimes p_2^* L^\nu \otimes \mathcal{O}(2\Delta) \to K_\nu \to 0.$$

It is a sheaf of $\mathcal{O}_{2\Delta}$-modules so it lies in an exact sequence between $\mathcal{O}_\Delta \cong \mathcal{O}_Y$-modules

$$\begin{array}{ccccccccc} 0 \to & (\mathcal{O}(\Delta) \otimes \mathcal{O}_\Delta) \otimes L^{\nu-1} \otimes \Omega_Y & \to & K_\nu & \to & (\mathcal{O}(2\Delta) \otimes \mathcal{O}_\Delta) \otimes L^{\nu-1} \otimes \Omega_Y & \to 0 \\ & \wr\Vert & & & & \wr\Vert & \\ & L^{\nu-1} & & & & L^{\nu-1} \otimes (\Omega_Y)^{-1}. & \end{array}$$

So if $\deg L > 4g-4$, $H^1(K_\nu) = (0)$ when $\nu \geqq 2$.

Case II. $\nu = 0$: Consider the Leray spectral sequence for p_1. Since we have assumed Y is not hyperelliptic

a) $p_{1,*}\mathcal{O}_{Y\times Y}(2\Delta) \cong p_{1,*}\mathcal{O}_{Y\times Y}$

and

b) $R^1 p_{1,*}\mathcal{O}_{Y\times Y}(2\Delta)$ is a locally free sheaf $\mathcal{E}$ of rank $g-2$. Now we have:

$$\begin{array}{ccccccc} 0 \to H^1(Y, \Omega_Y \otimes L^{-1}) & \to & H^1(Y \times Y, p_1^*\Omega_Y \otimes L^{-1}) & \to & H^0(Y, \Omega_Y \otimes L^{-1} \otimes R^1 p_{1,*}\mathcal{O}_{Y\times Y}) & \to 0 \\ \downarrow \wr \text{ by (a)} & & \downarrow & & \downarrow & \\ 0 \to H^1(Y, \Omega_Y \otimes L^{-1} \otimes p_{1,*}\mathcal{O}(2\Delta)) & \to & H^1(Y\times Y, p_1^*\Omega_Y \otimes L^{-1}(2\Delta)) & \to & H^0(Y, \Omega_Y \otimes L^{-1} \otimes \mathcal{E}) & \to 0. \end{array}$$

Note that $\mathcal{E}$ does not depend on L. So by (b) there is an integer n_0 depending only on Y such that if $\deg L > n_0$, then $(\Omega_Y \otimes \mathcal{E}) \otimes L^{-1}$ has no sections.

Case III: $\nu \leqq -1$: Surjectivity in this case always follows from surjectivity when $\nu = 0$. In fact, if we know that

$$V \to \Gamma(Y, N_Y \otimes L^{-1}) \to 0$$

is surjective, I claim $\Gamma(Y, N_Y \otimes L^{-\nu}) = (0)$, $\nu \geqq 2$. If not, $N_Y \otimes L^{-2}$ has a non-zero section s. Then for all $t \in \Gamma(Y, L) \cong V^*$, $t \otimes s$ is a non-zero section of $N_Y \otimes L^{-1}$. Thus we must get *all* sections of $N_Y \otimes L^{-1}$ in this way. But this means that all these sections are proportional, hence do not generate $N_Y \otimes L^{-1}$. But since

$$V \otimes \mathcal{O}_Y \to N_Y \otimes L^{-1}$$

is surjective and $V \otimes \mathcal{O}_Y$ is generated by its sections, so is $N_Y \otimes L^{-1}$. This is a contradiction, so $s = 0$.

This completes the proof of (d′). Finally two remarks:

(A) If you look at the case $Y = \mathbf{P}^1$, $L = \mathcal{O}_{\mathbf{P}^1}(k)$, then C = cone over the rational curve of degree n in $\mathbf{P}^n$ and the sequences we have used enable us to compute T_C^1 easily. In fact it turns out that if $k \geqq 3$,

$$(T_C^1)_l = (0), \quad \text{if } l \neq -1$$

$$\dim (T_C^1)_{-1} = 2k - 4.$$

It seems most reasonable to conjecture that the versal deformation space of this C is a non-singular $k-1$-dimensional space but with a 0-dimensional embedded component at the origin if $k \geqq 4$.[2]

(B) What happens in the hyperelliptic case? If, for instance, $\pi: Y \to \mathbf{P}^1$ is the double covering and $L = \pi^* \mathcal{O}_{\mathbf{P}^1}(k)$, then C is itself a double covering of the rational cone considered in (A) which is known to have non-singular deformations. Do these lift to deformations of this C?

NOTES

1. M. Schlessinger, "On Rigid Singularities," in this volume, pp. 147–162.

2. H. Pinkham has recently proved that this is true if $k \geqq 5$, but if $k=4$, the versal deformation space has two components, a smooth 3-dimensional one and a smooth 1-dimensional one crossing normally at the origin! (Cf. "Deformations of cones with negative
1 grading," *J. of Algebra*, to appear.)

HARVARD UNIVERSITY

Proceedings of Symposia in Pure Mathematics
Volume 29, 1975

MATSUSAKA'S BIG THEOREM

D. Lieberman and D. Mumford

§1. INTRODUCTION

The goal of this note is to present an outline of Matsusaka's proof [4],[5] of the following Theorem:

1 THEOREM 1. Let $P(k)$ be a rational polynomial with integral values for all integers k. Then there is a k_o such that for every non-singular complex projective variety V, and every ample line bundle L on V with

$$\chi(V, L^{\otimes k}) = P(k),$$

then $L^{\otimes k}$ is very ample if $k \geq k_o$.

The proof can be divided into 2 parts. The more difficult part consists in the proof of Theorem 2 below, and a somewhat easier but still subtle part consists in checking that Theorem 2 implies Theorem 1.

THEOREM 2. Given constants $\epsilon > 0$, $\gamma, k_o, n \in \mathbb{Z}$ and $t \in \mathbb{Q}$, there is a $k_1 = k_1(\epsilon, \gamma, k_o, t, n)$ for which the following holds: Let V be any normal projective variety of dimension n over any algebraically closed field k; let C be an ample divisor on V and let D be a codimension 1 cycle on V; assume $\gamma = (C^n)$, and

$$t = \frac{(D \cdot C^{n-1})}{(C^n)} .$$

Assume

$$\dim H^o(\mathcal{O}_V(kD)) \geq \frac{(\frac{1}{2}+\epsilon)}{n!} t^n((kC)^n), \text{ for all } k \geq k_o.$$

Then for every $k \geq k_1$, one can find a subspace

$$\Lambda \subset H^o(\mathcal{O}_V(kD))$$

such that the induced rational map

$$\phi_\Lambda : V \longrightarrow \mathbb{P}^N$$

AMS (MOS) subject classifications (1970). Primary 14D20.

D. Mumford, *Selected Papers*, Vol. II,

is birational and does not blow down any codimension one subvarieties. Moreover:

$$\deg \phi_\Lambda(V) \leq \gamma k^n t^n.$$

Many special cases of Theorem 1 are well known. If V is a curve of genus g, then L is ample if and only if $\deg L \geq 1$, and it is well known that in this case $L^{\otimes k}$ is very ample if $k \geq (2g+1)/\deg L$. If V is an abelian variety, L ample, then $L^{\otimes k}$ is very ample if $k \geq 3$ (Lefschetz: cf. Mumford [8], §17). If V is a K3-surface, L ample, then again $L^{\otimes k}$ is very ample if $k \geq 3$ (Mayer [7], Saint-Donat [9]). If V is a normal surface of general type with its rational curves E with $(E^2) = -2$ blown down, and $L = \mathcal{O}_V(K)$, then $L^{\otimes k}$ is very ample if $k \geq 5$ (Enriques, Kodaira [3], Bombieri [2]). For arbitrary surfaces and ample L's in any characteristic, the Theorem was proven in Matsusaka-Mumford [6].

Once Theorem 1 is established, one may apply the theory of Hilbert schemes or of Chow varieties to conclude that the set of polarized varieties (V,L) with given Hilbert polynomial P(k) may all be parametrized by a quasi-projective scheme. (In particular this family contains all deformations of the polarized variety (V,L) because the Hilbert polynomial is invariant under deformation.)

Matsusaka's proof of Theorem 1 is non-cohomological, unlike for instance Bombieri's approach to canonically polarized surfaces. Theorem 1 would follow immediately if, for instance, one could solve directly the following:

PROBLEM: Given P(k), find k_o such that for every (V,L) with Hilbert polynomial P(k) and every $x,y \in V$,

$$H^i(V, \mathfrak{m}_x \cdot \mathfrak{m}_y \cdot L^{\otimes k}) = (0), \quad \text{all } i \geq 1,\ k \geq k_o.$$

Conversely, Matsusaka's result implies that such a k_o exists because it implies that the quadruples (V,L,x,y) form a "bounded family".

We want to add a word about the completeness of our presentation of Matsusaka's proof. We believe the careful reader can reconstruct the whole proof from what we say. However in some places we have not written out fully various details. In particular, a more complete version would include a whole section working out the elementary properties of Matsusaka's operation $\Lambda^{[j]}$ (cf. §2 below): instead we simply introduce these without proof where they are needed.

§ 2. PROOF OF THEOREM 2

(I.): The first step is to find a k_2 depending only on γ, k_o, t such that for all $k \geq k_2$, the rational map

$$\phi_{kD}: V \longrightarrow \mathbb{P}(H^o(\mathcal{O}_V(kD)))$$

satisfies $\dim \phi_{kD}(V) = n$. We shall in fact prove:

LEMMA 2.1. If $\Lambda \subset H^o(\mathcal{O}_V(kD))$ and

$$\dim \Lambda > \max_{1 \leq i \leq n-1} (i+\gamma t^i k^i)$$

then $\dim \phi_\Lambda(V) = n$.

Because of our assumed lower bound on $h^o(kD)$ one gets immediately a $k_2(\gamma, k_o, t)$ such that $\Lambda = H^o(\mathcal{O}_V(kD))$ satisfies this for $k \geq k_2$.

To prove the lemma, let $W = \phi_\Lambda(V)$. We show in fact that if $\dim W = i$, then

$$\dim \Lambda \leq i + \gamma t^i k^i.$$

Firstly, recall the well known fact (cf. [6], Th. 3) that for any projective variety $X \subset \mathbb{P}^n$, X not in any hyperplane:

$$n+1 \leq \deg X + \dim X.$$

In particular, if $X = W$, $\mathbb{P}^n = \mathbb{P}(\Lambda)$, then

$$\dim \Lambda \leq \deg W + i,$$

so it suffices to prove

$$\deg W \leq \gamma t^i k^i.$$

To transform the inequality on deg W into an estimate on V itself, Matsusaka introduces an interesting new concept of the <u>variable</u> j-<u>fold intersection cycle</u> $\Lambda^{[j]}$ of a linear system $\Lambda \subset \Gamma(V, M)$. This is the codimension j cycle, defined only up to rational equivalence on V, obtained in either of the following ways: let $B \subset V$ be the base points of the linear system Λ so that Λ defines a morphism

$$\phi_\Lambda: V\text{-}B \longrightarrow \mathbb{P}(\Lambda).$$

Take the closure in V of $\phi_\Lambda^{-1}(L)$, $L \subset \mathbb{P}(\Lambda)$ a general codimension j linear space; or take the closure in V of the intersection cycle $V(s_1)\cdot\cdots\cdot V(s_j)$ in V-B, where the s_i are j general element of

Λ. If $\Lambda = \Gamma(V, \mathcal{O}_V(E))$ for some divisor E, write $(E)^{[j]}$ for $\Lambda^{[j]}$. Note that if $\Lambda_1 \subset \Lambda_2$ are 2 linear systems, then

$$\Lambda_2^{[i]} \underset{\text{rat.eq.}}{\sim} \Lambda_1^{[i]} + \text{eff. cycle.}$$

With this concept, we find:

$$\begin{aligned} \deg W &\leq \#\text{ of components of } \Lambda^{[i]} \\ &\leq (C^{n-i}.\Lambda^{[i]}) \\ &\leq (C^{n-i}.(kD)^{[i]}) , \end{aligned}$$

hence Lemma 2.1 follows by taking E = kD in the following key result:

PROPOSITION 2.2: Let V be a normal projective variety of dimension n, C an ample divisor and E a codimension one cycle on V such that $\dim \phi_{kE}(V) \geq i$ for $k \gg 0$. Then

$$(E^{[i]}.C^{n-i}) \leq (C^n)\cdot\left(\frac{(C^{n-1}\cdot E)}{(C^n)}\right)^i$$

<u>Proof</u>: Replacing C by ℓC multiplies both sides by ℓ^{n-i} so we may assume C very ample. Let V' be a general intersection of n-i divisors $C_1,\cdots,C_{k-i} \in |C|$, let $C' = V'.C$ and let $E' = V'.E$. Then one sees easily that

$$(E^{[i]}.C^{n-i})_V = (E')^{[i]}_{V'}$$

$$(C^n)_V = (C'^i)_{V'}$$

$$(C^{n-1}.E)_V = (C'^{i-1}.E')_{V'} ,$$

hence replacing V by V', we may assume i = n. Now $\dim \phi_{kE}(V) = n$ for $k \gg 0$, hence in fact ϕ_{kE} is birational for $k \gg 0$ [i.e., if $W_k = \phi_{kE}(V) \subset \mathbb{P}_{N_k}$ and if $\pi: W_k' \longrightarrow W_k$ is the normalization of W_k in the field k(V), then $\pi^*(\mathcal{O}_{W_k}(1))$ is ample on W_k' and

$$\Gamma(W_k', \pi^*(\mathcal{O}_{W_k}(\ell)) \subset \Gamma(V, \mathcal{O}(k\ell E)).]$$

We may also replace E by kE to prove the Proposition because:

$$(kE)^{[n]} \geq k^n(E)^{[n]}$$

$$(C^{n-1}.kE)^n = k^n(C^{n-1}.E)^n$$

(because the base locus of $|kE|$ is contained in the base locus of $|E|$, and the variable intersection of n general divisors in $|kE|$ specializes to k^n times the variable intersection of n general divisors in $|E|$ plus some components in the base locus of $|E|$.) So we may assume ϕ_E is birational.

Now let $W = \phi_E(V)$. Since ϕ_E is birational, $\deg W = (E^{[n]})$. Moreover, if $k \gg 0$:

$$\begin{aligned} h^0(V,\mathcal{O}_V(kE)) &\geq h^0(W,\mathcal{O}_W(k)) \\ &= \deg W\cdot\frac{k^n}{n!} + \text{lower terms} \\ &= E^{[n]}\cdot\frac{k^n}{n!} + \text{lower terms.} \end{aligned}$$

The Proposition now follows from considering the upper bound on $h^0(\mathcal{O}_V(kE))$ as $k \longrightarrow \infty$, which is given by

PROPOSITION 2.3 ("Q-estimate"): Let V be a projective variety of dimension n, let C be a hyperplane section of V and let $\mathfrak{I}$ be a torsion-free rank 1 sheaf on V. Then

$$h^0(\mathfrak{I}) \leq \binom{[t]+n}{n}\gamma + \binom{[t]+n-1}{n-1}$$

where

$$t = \deg \mathfrak{I}/\deg V$$
$$\gamma = \deg V\ .$$
(degree measured via C as in Kleiman, Annals, 1966)

<u>Proof</u>: For $n = 1$, the inequality reads

$$h^0(\mathfrak{I}) \leq ([t]+1)\gamma + 1$$

which follows from the Riemann-Roch estimate:

$$h^0(\mathfrak{I}) \leq \deg \mathfrak{I} + 1 = t\gamma + 1.$$

We proceed by induction, assuming the result true on a general hyperplane section C. First we need to find a hyperplane C such that C is again a variety and $\mathfrak{I} \otimes \mathcal{O}_C$ is still torsion-free. Indeed almost all C's are varieties (Seidenberg's Theorem) and for $\mathfrak{I}\otimes\mathcal{O}_C$ to be torsion-free, it suffices to make sure $\operatorname{depth}_{\mathcal{O}_x}(\mathfrak{I}_x) \geq 2$ for all $x \in C$ except the generic point of C. Since there are only finitely many $x \in V$ with $\operatorname{codim}_V\overline{\{x\}} \geq 2$ and $\operatorname{depth}_{\mathcal{O}_x}(\mathfrak{I}_x) = 1$ (cf. for instance EGA, Ch. 4, §10.8), this is possible. Then one has the exact sequences:

$$0 \longrightarrow \mathcal{J}(-k-1) \longrightarrow \mathcal{J}(-k) \longrightarrow \mathcal{J}\otimes\mathcal{O}_C(-k) \longrightarrow 0,$$

hence

$$h^0(\mathcal{J}(-k)) \leq h^0(\mathcal{J}(-k-1)) + h^0(\mathcal{J}\otimes\mathcal{O}_C(-k)).$$

But $h^0(\mathcal{J}(-k)) \neq 0$ implies there is a homomorphism

$$0 \longrightarrow \mathcal{O}_V(kC) \longrightarrow \mathcal{J}$$

hence $\deg \mathcal{J} \geq k \deg V$, hence $[t] \geq k$. Thus

$$h^0(\mathcal{J}) \leq \sum_{k=0}^{[t]} h^0(\mathcal{J}\otimes\mathcal{O}_C(-k)).$$

Using the estimate on C, we get:

$$h^0(\mathcal{J}) \leq \sum_{k=0}^{[t]} \binom{[t]-k+n-1}{n-1}\gamma + \binom{[t]-k+n-2}{n-2}$$

$$= \binom{[t]+n}{n}\gamma + \binom{[t]+n-1}{n-1} \qquad \text{QED}$$

(II.): This completes the first step: if $W_k = \phi_{kD}(V)$, we have a k_2 such that if $k \geq k_2$, $\dim W_k = n$. The second step is to find a k_3 also depending only on ϵ, γ, k_o, t such that if $k \geq k_3$, then ϕ_{kD} is <u>birational</u>. We will in fact produce an ℓ_o such that $\phi_{\ell_o k_2 D}$ is birational. Note that since $k \geq k_o$ implies kD is effective, then for $k \geq k_3 = \ell_o k_2 + k_o$, $\Gamma(\mathcal{O}(kD)) \supseteq \Gamma(\mathcal{O}(\ell_o k_2 D))$, hence ϕ_{kD} is also birational.

To produce ℓ_o, consider for each $k \geq k_2$, $\ell > 1$, the diagram of rational maps:

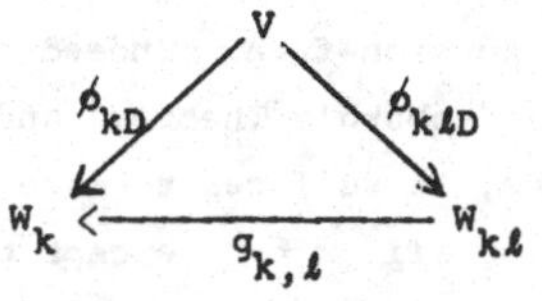

Note that $\deg(\phi_{k\ell D}) = \deg(\phi_{kD}).\deg(g_{k,\ell})$.

LEMMA 2.4: There is an integer ℓ (depending on ϵ, n and t) such that if $\deg(\phi_{kD}) > 1$ and $\deg(g_{k,\ell}) = 1$, then one must have

$$\frac{(k\ell D)^{[n]}}{k^n \ell^n} > (1+\epsilon)^{1/n} \cdot \frac{(kD)^{[n]}}{k^n} .$$

Proof: Choose ℓ such that

$$\binom{\ell(1+\epsilon)^{1/n}+n+1}{n} + \frac{2}{t^n}\binom{\ell(1+\epsilon)^{1/n}+n}{n-1} < \frac{1+2\epsilon}{n!} \ell^n.$$

This is possible because for $\ell >> 0$, the left hand side grows like $(1+\epsilon)\ell^n/n!$. Now W_k and $W_{k\ell}$ both have explicit projective embeddings:

$$W_k \subset \mathbb{P}(H^o(\mathcal{O}_V(kD)))$$

$$W_{k\ell} \subset \mathbb{P}(H^o(\mathcal{O}_V(k\ell D))).$$

Since $g_{k,\ell}$ is birational by assumption, let $U \subset W_k$ be the domain of definition of $g^{-1}_{k,\ell}$. Then the morphism

$$U \longrightarrow W_{k\ell} \subset \mathbb{P}(H^o(\mathcal{O}_V(k\ell\ D)))$$

is defined by an invertible sheaf $\mathfrak{I}'$ on U and $h^o(k\ell D)$ sections s_i' of $\mathfrak{I}'$ generating $\mathfrak{I}'$. There is a unique torsion free sheaf $\mathfrak{I}$ on W_k plus sections s_i generating $\mathfrak{I}$, which restrict on U to $\{\mathfrak{I}', s_i'\}$. Thus $h^o(\mathfrak{I}) \geq h^o(k\ell D)$. On the other hand, for the given projective embedding of W_k, we calculate:

deg $\mathfrak{I}$ = # of intersections on V outside base locus of $|kD|$ of (n-1) general sections of $\mathcal{O}(kD)$, one section of $\mathcal{O}(k\ell D)$,

hence:

$$\ell^{n-1}.\deg \mathfrak{I} \leq (k\ell D)^{[n]} .$$

Now combine the assumed lower bound on $h^o(k\ell D)$, and the upper bound on $h^o(\mathfrak{I})$ given by the Q-estimate to get:

$$\frac{\frac{1}{2}+\epsilon}{n!}\, t^n (k\ell)^n \gamma \;\le\; h^o(k\ell D)$$
$$\le\; h^o(\mathfrak{J})$$
$$\le \binom{[\frac{\deg \mathfrak{J}}{\deg W_k}]+n}{n} \deg W_k + \binom{[\frac{\deg \mathfrak{J}}{\deg W_k}]+n-1}{n-1} .$$

Moreover:

$$\deg W_k = (kF)^{[n]}/\deg(\phi_{kD})$$
$$\le \frac{t^n k^n \gamma}{2} \quad \text{by Prop. 2.2}$$

and

$$[\frac{\deg \mathfrak{J}}{\deg W_k}] \le \frac{(k\ell D)^{[n]}}{\ell^{n-1}(kD)^{[n]}} + 1$$
$$= \ell R + 1$$

if

$$R = \frac{(k\ell D)^{[n]}}{\ell^n (kD)^{[n]}} .$$

Hence

$$\frac{\frac{1}{2}+\epsilon}{n!}\, t^n k^n \ell^n \gamma \;\le\; \binom{\ell R+n+1}{n}\frac{t^n k^n \gamma}{2} + \binom{\ell R+n}{n-1}$$

hence

$$\frac{1+2\epsilon}{n!}\,\ell^n \;\le\; \binom{\ell R+n+1}{n} + \frac{2}{t^n}\binom{\ell R+n}{n-1} .$$

If $R \le (1+\epsilon)^{1/n}$, this contradicts the inequality that ℓ was chosen to satisfy. Thus $R > (1+\epsilon)^{1/n}$. QED

However, for all k,

$$\frac{(kD)^{[n]}}{k^n} \le \gamma t^n$$

by Prop. (2.2). Hence starting at any ϕ_{kD}, we see that:

$$\deg \phi_{(k\ell^e D)} < \deg \phi_{kD} \quad \text{if} \quad e > \frac{n \log(\gamma t^n k^n)}{\log(1+\epsilon)} .$$

Since we know that ϕ_{k_2D} is finite-to-one and

$$\deg \phi_{k_2D} \leq (k_2D)^{[n]} \leq \gamma t^n k_2^n,$$

it follows that $\phi_{(k_2+e)D}$ is birational if

$$e > n\gamma t^n k_2^n \frac{\log(\gamma t^n k^n)}{\log(1+\epsilon)} .$$

(III): This completes the second step: we have a k_3 such that if $k \geq k_3$, ϕ_{kD} is birational. The third step is to find a k_4 also depending only on ϵ,γ,k_o,t such that if $k \geq k_4$ then there is a $\Lambda \subset |kD|$ such that ϕ_Λ is birational and does not blow down any codimension 1 subvarieties of V. We will in fact only produce a k_5 such that if $k \geq k_5$, then there is a $\Lambda' \subset |kD|$ such that $\dim \phi_{\Lambda'}(V) = n$ and $\phi_{\Lambda'}$ does not blow down any codimension 1 subvarieties. Setting $k_4 = k_3+k_5$ and Λ = minimal sum of $|k_3D|$ and Λ', we get the Λ with all the properties. The proof is very similar to the beautifully simple Method of Albanese by which for any projective n-dimensional variety X one constructs a $\Lambda \subset H^o(\mathcal{O}_X(k))$ such that $\phi_\Lambda(X)$ is birational to X and has no points of multiplicity $> n!$ (cf. [1],(12.4.4)) We show in fact:

LEMMA (2.5). Choose k such that

$$tk \geq 2n\cdot n!(1+\tfrac{1}{\gamma}) \quad \text{and} \quad k \geq k_o.$$

Then there is a $\Lambda \subset H^o(\mathcal{O}_V(kD))$ such that $\dim \phi_\Lambda(V) = n$ and ϕ_Λ does not blow down any codimension 1 subvarieties.

Proof: Note by the assumption on k,

$$n\gamma(tk)^{n-1}+n \leq \gamma(tk)^{n-1}(n+\tfrac{n}{\gamma}) \leq \frac{\gamma(tk)^n}{2n!} < h^o(kD).$$

Also $tk \geq 1$, so by Lemma (2.1) any $\Lambda \subset H^o(kD)$ for which

$$\dim \Lambda \geq n + \gamma(tk)^{n-1}$$

has the property $\dim \phi_\Lambda(V) = n$. So any Λ such that

$$\operatorname{codim} \Lambda \leq (n-1)\gamma(tk)^{n-1}$$

has this property too. What we will do is this: starting

* See also Lecture 1, §5 of Lipman's article, " Introduction to resolution of singularities", in this volume.

with $\Lambda_o = H^o(\mathcal{O}_V(kD))$, choose a sequence of subspaces

$$\Lambda_o \supset \Lambda_1 \supset \Lambda_2 \supset \cdots$$

with $\dim \Lambda_i/\Lambda_{i+1} = 1$, until we reach the desired Λ. In fact, say Λ_r is chosen but there is still an $E \subset V$, $\dim E = n-1$, such that

$$\dim \phi_{\Lambda_r}(E) = i-1, \qquad n-1 \geq i \geq 1.$$

If s is the multiplicity to which E occurs as a fixed component of Λ_r, let $\Lambda_r' \subset H^o(\mathcal{O}_V(kD-sE))$ be the linear system such that $\Lambda_r = (e^{\otimes s}) \otimes \Lambda_r'$, e = canonical section of $\mathcal{O}_V(E)$. Let x be a general point of E. Define

$$\Lambda_{r+1}' = \{s \in \Lambda_r' \mid s(x) = 0\},$$

$$\Lambda_{r+1} = (e^{\otimes s}) \otimes \Lambda_{r+1}' .$$

Note that ϕ_{Λ_r} is defined at x, so define

$$Z = \text{closure of } \phi_{\Lambda_r}^{-1}(\phi_{\Lambda_r}(x)).$$

Then $\dim Z = n-i$ if x is sufficiently general, and if $s \in \Lambda_r'$ vanishes at x, it vanishes on all of Z. Thus

$$(\Lambda_r')^{[i]} \underset{\text{rat.eq.}}{\sim} (\Lambda_{r+1}')^{[i]} + Z + \text{eff. cycle}.$$

But $\Lambda_r^{[i]} = (\Lambda_r')^{[i]}$ and $\Lambda_{r+1}^{[i]} = (\Lambda_{r+1}')^{[i]}$, so it follows

$$(\Lambda_r^{[i]}.C^{n-i}) > (\Lambda_{r+1}^{[i]}.C^{n-i}).$$

Since for each j,

$$(\Lambda_r^{[j]}.C^{n-j}) \geq (\Lambda_{r+1}^{[j]}.C^{n-j}),$$

it follows that the invariant

$$\delta(\Lambda) = \sum_{j=1}^{n-1} (\Lambda^{[j]}.C^{n-j})$$

decreases when you pass from Λ_r to Λ_{r+1}. But by Prop. (2.2)

$$\delta(\Lambda) \leq \delta(kD) \leq \sum_{j=1}^{n-1} \gamma(kt)^j \leq (n-1)\gamma(kt)^{n-1}.$$

Since we have "this much room" in $H^o(\mathfrak{O}(kD))$, we can find a Λ for which no E is blown down. QED

§3. TH. 2 ⟹ TH. 1

This is the part of the proof that involves char. 0 because we want to apply Kodaira's Vanishing Theorem. The idea is to apply Theorem 2 to V with C,D chosen so that

$$L = \mathfrak{O}_V(C)$$

$$L^{m_o} \otimes \Omega_V^n = \mathfrak{O}_V(D).$$

Here m_o will be chosen below depending only on P so as to make Theorem 2 apply. Note that for $m > 0$ by Kodaira Vanishing and Serre duality:

$$(*) \qquad \begin{aligned} \dim H^o(L^m \otimes \Omega_V^n) &= \chi(L^m \otimes \Omega_V^n) \\ &= (-1)^n \chi(L^{-m}) \\ &= (-1)^n P(-m): \text{ call this } P'(m). \end{aligned}$$

Moreover, by Riemann-Roch,

$$P(k) = (C^n)\frac{k^n}{n!} - \frac{(K_V.C^{n-1})}{2(n-1)!} k^{n-1} + \text{lower degree terms}$$

hence P determines the integer $\gamma = (C^n)$ and, once m_o is chosen, P determines $(D.C^{n-1})$ and hence $(D.C^{n-1})/(C^n)$ too. Finally, we need a lower bound for $\dim H^o(\mathfrak{O}_V(kD))$ of the type used in Th. 2. This is obtained as follows -

a) Say $P'(m_1) > 0$, so that by (*) in divisor notation m_1C+K is an effective divisor. Then:

$$\begin{aligned} kD &= k(m_oC + K) \\ &= (k-1)(m_1C+K) + ((k(m_o-m_1)+m_1)C + K). \end{aligned}$$

The first term is an effective divisor, so

$$\begin{aligned}\dim H^o(\mathcal{O}_V(kD)) &\geq \dim H^o(\mathcal{O}_V((k(m_o-m_1)+m_1)C + K)) \\ &= P'(k(m_o-m_1)+m_1) \\ &= \frac{(C^n)}{n!}(k(m_o-m_1)+m_1)^n + \text{lower degree terms in } k \\ &= \frac{((kC)^n)}{n!}(m_o-m_1)^n + \text{lower degree terms in } k.\end{aligned}$$

b) But

$$\begin{aligned} t \underset{\text{def}}{=} \frac{(D.C^{n-1})}{(C^n)} &= \frac{((m_oC+K).C^{n-1})}{(C^n)} \\ &= m_o + \frac{(K.C^{n-1})}{(C^n)},\end{aligned}$$

i.e.,

$$\dim H^o(\mathcal{O}_V(kD)) \geq \frac{((kC)^n)}{n!}\left[t-m_1-\frac{(K.C^{n-1})}{(C^n)}\right]^n + \text{lower degree terms in } k.$$

If m_o and hence t is large enough, the term $[\ \ ]^n$ is at least $\frac{3}{4}t^n$ and then for k_o large enough, we certainly obtain:

$$\dim \Gamma(\mathcal{O}_V(kD)) \geq \frac{((kC)^n)}{n!}\cdot\frac{5}{8}t^n, \quad \text{if } k \geq k_o.$$

Thus Theorem 2 applies for some m_o and k_o readily computed in terms of the polynomial P alone. Thus we can find k, so that for every (V,L)

$$\begin{aligned}\exists\ \Lambda \subset \Gamma(V,\mathcal{O}_V(k_1D)) &= \Gamma(V,\mathcal{O}_V(k_1(m_oC+K))) \\ &= \Gamma(V,L^{k_1m_o}\otimes(\Omega_V^n)^{k_1})\end{aligned}$$

for which ϕ_Λ is birational and does not blow down any divisors - we abbreviate this to "Λ is quasi-ample".

Now let's analyze the projective variety $U = \phi_\Lambda(V)$. By Prop. (2.2) we know:

$$(**) \qquad \deg U \leq \gamma k_1^n t^n.$$

Automatically then, the ambient space $\mathbb{P}(\Lambda)$ has its dimension bounded as follows:

$$\dim \mathbb{P}(\Lambda) \leq \deg U + n - 1 \underset{\text{def}}{=} N.$$

It follows that the set of varieties U lies in a bounded family when

(V,L) varies over all pairs with Hilbert polynomial P! This is the key point, from which we want to argue backwards, obtaining eventually the boundedness of the set of pairs (V,L). From this point on, we leave the area in which we can make explicit estimates, and rely on general results asserting that various numbers are bounded when calculated for some set of varieties and divisors in a bounded family. The first point is that if U_{nor} is the normalization of U, then there is a k_2 such that for all U with degree bounded by (**), the pullback of $\mathcal{O}_U(k_2)$ to U_{nor} is very ample. It follows that if we choose a suitable $\Lambda' \subset \Gamma(V,\mathcal{O}_V(k_1k_2D))$, then $\phi_{\Lambda'}(V) \cong U_{nor}$. Replacing k_1 by k_1k_2 and Λ by Λ', this means we may assume that U is always normal. Call these Λ "normally quasi-ample". In that case, working with "Weil"-divisors on U, i.e., cycles of codimension 1, we may define the total transform $\phi_\Lambda(E)$ for every divisor E on V; and because ϕ_Λ does not contract any divisors, this defines an injection of the groups of Weil-divisors

$$\mathrm{Div}(V) \xrightarrow[\phi_\Lambda]{} \mathrm{Div}(U)$$

such that

a) $\phi_\Lambda(E)$ eff. $\Longleftrightarrow$ E eff.

b) $\phi_\Lambda((f)_V) = (f)_U$.

Thus ϕ_Λ^* sets up an isomorphism between

$$\Gamma(V,\mathcal{O}_V(E)) \xrightarrow{\approx} \Gamma(U,\mathcal{O}_U(\phi_\Lambda E)).$$

Moreover, if $U_o \subset U$ is the maximal open set such that

$$\phi_\Lambda^{-1}: U_o \longrightarrow V$$

is a morphism, then codim $U-U_o \geq 2$. ϕ_Λ^{-1} then defines an injection:

$$(\phi_\Lambda^{-1})^*: \Omega_V^n \longrightarrow \Omega_U^n|_{U_o}.$$

This implies that

$$\phi_\Lambda(K_V) = K_U + (\text{eff. divisor}).$$

It follows that if $E \in |\ell C+K_V|$, then

$$\begin{aligned}\deg \phi_\Lambda(E) &= \ell \deg \phi_\Lambda(C) + \deg \phi_\Lambda(K_V)\\ &\le \ell(C.\Lambda^{[n-1]}) + \deg K_U\\ &\le \gamma k_1^{n-1} t^{n-1} \ell + \deg K_U.\end{aligned}$$

Of course, $\deg K_U$ is bounded when U varies over all U's with degree bounded by (**): call this bound κ.

We can now reveal the diagram on which the rest of the proof is based. We consider 3 sets, related by 2 maps, as follows:

$$B = \left\{\begin{array}{l}\text{Set of all } (V,L,s_0,\cdots,s_N,E_0,E_1)\\ V,L \text{ as in } A;\ E_0 \in |m_1C+K_V|;\\ E_1 \in |(m_1+1)C+K_V|; \text{ and } s_0,\cdots,s_N \in \Gamma(V,\mathcal{O}_V(k_1D))\\ \text{spanning a normally quasi-ample } \Lambda\end{array}\right\}$$

α: ignore(s_i),(E_i) (from B to A); β (from B to C)

$$A = \left\{\begin{array}{l}\text{Set of all } (V,L),\\ V \text{ non-sing., proj.,}\\ L \text{ ample with}\\ \text{Hilbert Polyn. } P\end{array}\right\}$$

$$C = \left\{\begin{array}{l}\text{Set of } (U,F_0,F_1), \text{ where}\\ U \subset \mathbb{P}^N \text{ is an } n\text{-dimensional}\\ \text{normal subvariety, } F_i \text{ divisors on } U\\ \deg F_i \le \gamma k_1^{n-1} t^{n-1}(m_1+i)+\kappa\\ \deg U \le \gamma k_1^n t^n\\ h^0(F_0+\ell(F_1-F_0)) \ge P'(\ell+m_1)\\ \text{all } \ell \ge 0\end{array}\right\}$$

Here m_1,k_1,N,γ, and t are chosen as above, in particular so that α is surjective. β is defined by

$$U = \phi_\Lambda(V)$$

$$F_i = \phi_\Lambda(E_i).$$

Note that this is OK because

$$\begin{aligned}H^0(U,\mathcal{O}_U(F_0+\ell(F_1-F_0)) &\cong H^0(U,\mathcal{O}_U(\phi_\Lambda(E_0+\ell(E_1-E_0))))\\ &\cong H^0(V,\mathcal{O}_V((m_1+\ell)C+K_V))\end{aligned}$$

which has dimension exactly $P'(m_1+\ell)$. Also, set C is isomorphic to the set of points on a <u>locally closed subset of a union of 3-way products of Chow Varieties</u>, i.e., each U,F_1,F_2 has a Chow form, normal is an open condition and $h^0(\cdots) \ge c$ are all closed conditions. Thus C has a natural structure of a (reducible) variety.

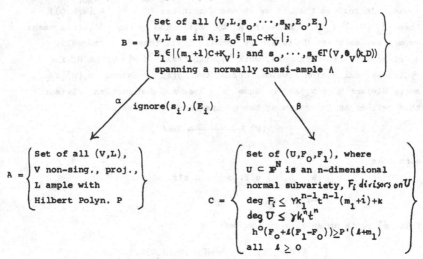

LEMMA 3.1: β is injective.

Proof: In fact, to recover V from (U,F_1,F_2), let $\psi_\ell: U \longrightarrow \mathbb{P}^M$ be the rational map defined by $H^0(U,\mathcal{O}_U(\ell(F_1-F_0)))$. Then if $\ell >> 0$, $V = \psi_\ell(U)$ (using the fact that

$$H^0(U,\mathcal{O}_U(\ell(F_1-F_0))) \cong H^0(V,L^{\otimes \ell})$$

and that L is ample on V). Moreover $\phi_\Lambda = \psi_\ell^{-1}$, L is the line bundle associated to $\psi_\ell(F_1-F_0)$, $E_i = \psi_\ell(F_i)$, and $(s_0,\cdots,s_N)$ are the sections of $\mathcal{O}_V(k_1D)$ corresponding to the canonical sections $(X_0,\cdots,X_N)$ of $\mathcal{O}_U(1)$ via

$$\phi_\Lambda^*: H^0(\mathcal{O}_U(1)) \longrightarrow H^0(\mathcal{O}_V(k_1D)). \qquad \underline{\text{QED}}$$

LEMMA 3.2: The image of β is Zariski-open.

Proof: It is elementary to see that the image of β is a countable union of locally closed subsets of the set C. Therefore it is enough to show that for any valuation ring R and morphism ϕ: Spec R $\longrightarrow$ C, if the closed point is in the image of β, then so is the generic point. Then over R, we get a flat family of normal varieties (by Hironaka's lemma) $\mathcal{U} \longrightarrow$ Spec R, plus divisors $\mathfrak{F}_1,\mathfrak{F}_2$ on $\mathcal{U}$. For every ℓ,m, let

$$M_{\ell,m} = H^0(\mathcal{U},\mathcal{O}(m\mathfrak{F}_0+\ell m(\mathfrak{F}_1-\mathfrak{F}_0))).$$

$M_{\ell,m}$ is a finitely generated torsion-free and hence free R-module, and $\sum_m M_{\ell,m} = R_\ell$ is an R-algebra. If $k = R/M$ is the residue field, $\bar{U},\bar{F}_0,\bar{F}_1$ is the induced triple over k, we get an injection:

$$\sigma_{\ell,m}: M_{\ell,m} \otimes_R k \hookrightarrow H^0(\bar{U},\mathcal{O}_{\bar{U}}(m\bar{F}_0+\ell m(\bar{F}_1-\bar{F}_0))).$$

Let $(\bar{U},\bar{F}_0,\bar{F}_1) = \beta(\bar{V},\bar{L},\bar{s}_0,\cdots,\bar{s}_N;\bar{E}_1,\bar{E}_2)$. If K is the fraction field of R, U^*,F_0^*,F_1^* is the induced triple over K, we get an isomorphism:

$$M_{\ell,m}\otimes_R K \xrightarrow{\approx} H^0(U^*,\mathcal{O}_{U^*}(mF_0^*+\ell m(F_1^*-F_0^*))).$$

But then it follows that

$$\begin{aligned}\dim_k M_{\ell,1}\otimes_R k &\le \dim_k H^0(\bar{U},\mathcal{O}_{\bar{U}}(\bar{F}_0+\ell(\bar{F}_1-\bar{F}_0)))\\ &= \dim_k(H^0(\bar{V},\mathcal{O}_{\bar{V}}((m_1+\ell)\bar{C}+K_{\bar{V}}))\\ &= P'(m_1+\ell)\\ &\le \dim_K H^0(U^*,\mathcal{O}_{U^*}(F_0^*+\ell(F_1^*-F_0^*)))\\ &= \dim_K M_{\ell,1}\otimes_R K.\end{aligned}$$

Since $M_{\ell,1}$ is free, the 2 extremes are equal, so equality holds everywhere. In particular, $\sigma_{\ell,1}$ is an isomorphism

$$\sigma_{\ell,1}\colon\ M_{\ell,1}\otimes_R k \xrightarrow{\approx} H^0(\bar{U},\mathcal{O}_{\bar{U}}(\bar{F}_0+\ell(\bar{F}_1-\bar{F}_0))).$$

Now on $\bar{V}$, since $\bar{C}$ is ample, for $\ell \gg 0$ it follows that the ring

$$\sum_{m=0}^{\infty} H^0(\bar{V},\mathcal{O}_{\bar{V}}((m_1+\ell)\bar{C}+K_{\bar{V}})^{\otimes m})$$

is generated by its elements of degree 1 and that $\bar{V}$ is its Proj. This implies that the ring

$$\sum_{m=0}^{\infty} H^0(\bar{U},\mathcal{O}_{\bar{U}}(m\bar{F}_0+m\ell(\bar{F}_1-\bar{F}_0)))$$

is generated by its elements of degree 1. But since $\sigma_{\ell,1}$ is surjective, this implies that $\sigma_{\ell,m}$ is surjective too: i.e., if $\ell \gg 0$, there is an isomorphism of rings:

$$\sigma_\ell\colon\ R_\ell\otimes_R k \xrightarrow{\approx} \sum_{m=0}^{\infty} H^0(\bar{U},\mathcal{O}_{\bar{U}}(m\bar{F}_0+m\ell(\bar{F}_1-\bar{F}_0))).$$

Therefore $\mathrm{Proj}(R_\ell\otimes_R k)\cong\bar{V}$. So $\mathcal{V}=\mathrm{Proj}(R_\ell)$ itself is a flat family of schemes of Spec R with special fibre $\bar{V}$. Moreover since $R_\ell\otimes_R k$ is generated by its elements of degree 1, R_ℓ is also generated by its elements of degree 1. Therefore $\mathrm{Proj}(R_\ell)$ comes equipped with a line bundle $\mathcal{O}_{\mathcal{V}}(1)$, which on the closed fibre $\bar{V}$ is just $\mathcal{O}_{\bar{V}}((m_1+\ell)\bar{C}+K_{\bar{V}})$, i.e., $\bar{L}^{m_1+\ell}\otimes\Omega^n_{\bar{V}}$. Since $\bar{V}$ is non-singular, $\mathcal{V}$ is smooth over R. Moreover by deformation theory $\bar{L}$ lifts to a unique invertible sheaf $\mathcal{L}$ on $\mathcal{V}$ such that

$$\mathcal{O}_{\mathcal{V}}(1)\cong\mathcal{L}^{m_1+\ell}\otimes\Omega^n_{\mathcal{V}/R}.$$

Let (V^*,L^*) be the generic fibre of $(\mathcal{V},\mathcal{L})$. It is now easy to see that the rational map

$$\mathcal{V} \longrightarrow \mathcal{U}$$

defines $s_0,\dots,s_N,\mathcal{E}_1,\mathcal{E}_2$ on $\mathcal{V}$, hence $s_0^*,\dots,s_N^*,E_0^*,E_1^*$ on V^* such that $(U^*,F_0^*,F_1^*) = \beta(V^*,L^*,s_0^*,\dots,s_N^*,E_0^*,E_1^*)$. <u>QED</u>

Heuristically, this shows that $\beta(B)$ is a "limited family", hence so is B, hence so is A. To be precise, note that <u>all</u> elements of B can be
2 parametrized a suitable countably infinite set of families each defined over a base space B_α which is an algebraic variety. Then $\beta(B_\alpha)$ is at least a constructible subset of $C_0 = \operatorname{Im}\beta$. But assuming the ground field k is uncountable*, then a (reducible) variety C_0 which is a countable union of constructible subsets $\beta(B_\alpha)$ is also a finite union of them: hence B is a finite union of B_α's.

---

*The other way of arguing is to look at 2 countable algebraically closed ground field $\bar{\mathbb{Q}} \subset k$, where $\bar{\mathbb{Q}}$ = field of algebraic numbers and k has infinite transcendence degree over $\mathbb{Q}$. Considering k-rational points, we get a bijection

$$\beta:\ B(k) \longrightarrow C_0(k)$$

but each B_α may be assumed to be defined over $\bar{\mathbb{Q}}$. Apply the elementary compactness assertion: if any set of $\bar{\mathbb{Q}}$-rational constructible sets covers $C_0(k)$, a finite subset already covers $C_0(k)$.

[1] S. Abhyankar, Resolution of singularities of embedded algebraic surfaces, Academic Press, 1966.

[2] E. Bombieri, Canonical models of surfaces of general type, Publ. IHES, 42 (1973), p. 171.

[3] K. Kodaira, Pluricanonical systems on algebraic surfaces of general type, J. Math. Soc. Japan, 20 (1968), p. 170.

[4] T. Matsusaka, On canonically polarized varieties II, Am. J. Math., 92 (1970), p. 283.

[5] T. Matsusaka, Polarized varieties with a given Hilbert polynomial, Am. J. Math., 94 (1972), p. 1027.

[6] T. Matsusaka and D. Mumford, 2 fundamental theorems on deformations of polarized varieties, Am. J. Math., 86 (1964), p. 668.

[7] A. Mayer, Families of K3 surfaces, Nagoya Math. J., 48 (1972), p. 1.

[8] D. Mumford, Abelian Varieties, Tata Studies in Math., Oxford Univ. Press, 1970.

[9] B. Saint-Donat, Projective models of K-3 surfaces, Amer. J. Math. 96 (1974), pp. 602-639.

BRANDEIS UNIVERSITY

HARVARD UNIVERSITY

Math. Proc. Camb. Phil. Soc. (1975), **78**, 117
MPCPS 78–10
Printed in Great Britain

The self-intersection formula and the 'formule-clef'

BY A. T. LASCU, D. MUMFORD AND D. B. SCOTT
Université de Montréal, Harvard University, University of Sussex

(*Received* 25 *September* 1974)

Introduction. We shall consider exclusively algebraic non-singular quasi-projective irreducible varieties over an algebraically closed field. If V is such a variety $\mathscr{A}(V)$ will be the Chow ring of rational equivalence classes of cycles of V and

$$\phi_*: \mathscr{A}(V_1) \to \mathscr{A}(V_2)$$

the group homomorphism defined by any proper morphism $\phi: V_1 \to V_2$. Also

$$\phi^*: \mathscr{A}(V_2) \to \mathscr{A}(V_1)$$

denotes the ring homomorphism defined by ϕ.

Let X, Y be two varieties. Assume that Y is a subvariety of X (by subvariety, we will always mean *closed* subvariety), let $i: Y \to X$ be the inclusion and E the normal bundle of Y in X. Then Grothendieck conjectured the *self-intersection formula*

$$i^*i_*(y) = yc_r(E), \tag{1}$$

for any $y \in \mathscr{A}(Y)$, where $r = \operatorname{codim}_X(Y)$ and $c_r(E)$ is the rth Chern class of E.

Consider the blowing-up diagram

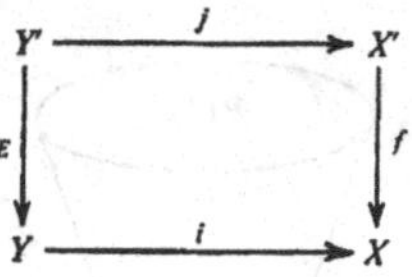

where $Y' = P(E)$ is the projective bundle associated to E. The normal bundle of Y' in X' is the tautological bundle $\check{L}_E$ of E and one has the exact sequence

$$0 \to \check{L}_E \to \rho_E^* E \to E^{(1)} \to 0$$

which defines $E^{(1)}$ (4). Then Grothendieck also conjectured:

$$f^*i_*(y) = j_*(\rho^*(y)\, c_{r-1}(E^{(1)})) \tag{2}$$

for any $y \in \mathscr{A}(Y)$, which he called the 'formule-clef' for calculating $\mathscr{A}(X')$.

Our aim is to prove both formulae. Especially (1) looks so innocuous that it is hard to believe it is not false or trivial; for instance they are both well known in singular cohomology when the ground field is $\mathbb{C}$. They were first conjectured by Grothendieck in 1957 ((2), exposé 0), and were proven *modulo torsion* by a very roundabout method in (2), cf. esp. exposé XIV (4·4). They were subsequently used in (6). Their analogues in étale cohomology are apparently to be published in SGA5. Our proof is completely elementary, but requires a good deal of manipulation.

1. *Preliminaries*. The notation will be that of (7) as far as possible. If E is a vector bundle

$$C(t,E) = \sum_{i=0}^{r} c_i(E)t^i,$$

where $r = \text{rank}(E)$, will be its Chern polynomial and $[\lambda C](t,E)$ the polynomial in which an operator λ has been applied to its coefficients. The point of this notation is that if a suitable element x is substituted for t then in $[\lambda C](x,E)$ the operator λ does not apply to the powers of x. The reversed Chern polynomial is defined as

$$\bar{C}(t,E) = t^r C(t^{-1},E).$$

One has

$$[\rho_E^* C](t,E) = C(t,\rho_E^* E) = C(t,E^{(1)}).C(t,\check{L}_E)$$

by applying the additivity formula to the exact sequence of the introduction. Let $\xi_E = c_1(L_E)$ so that $C(t,\check{L}_E) = 1 - t\xi_E$. Multiplying by $(1-t\xi_E)^{-1}$ and comparing constant terms, we get:

$$c_{r-1}(E^{(1)}) = \sum_{i=0}^{r-1} \rho_E^* c_i(E).\xi_E^{r-1-i}. \tag{1·1}$$

Let $G = E \oplus 1_Y$ where 1_Y is the trivial line bundle over Y. Then $\bar{E} = P(G)$ is the projective closure of E. The canonical inclusions of E and 1_Y in G give two subvarieties of $\bar{E}$ isomorphic to $P(E)$ and Y respectively, which can be called the 'roof' and the 'floor' of $\bar{E}$. We shall identify the roof with $P(E)$ and the floor with Y.

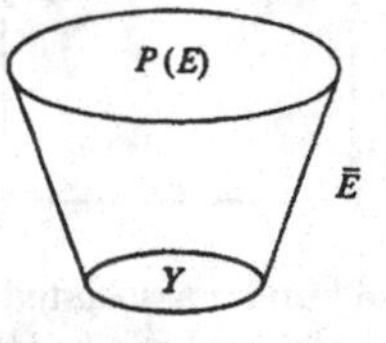

If $\bar{\imath}: Y \to \bar{E}$ and $\bar{\jmath}: P(E) \to \bar{E}$ are the inclusion maps then

$$\rho_G \bar{\imath} = Id_Y, \quad \rho_G \bar{\jmath} = \rho_E, \tag{1·2}$$

where ρ_G is the projection of $P(G)$. One has

$$\bar{\jmath}_*(1) = \xi_G. \tag{1·3}$$

and

$$\bar{\imath}_*(1) = [\rho_G^* \bar{C}](\xi_G, E) \tag{1·4}$$

according to Scott's formula ((5) and (7)). The first equality implies

$$\bar{\imath}^*(\xi_G) = 0 \tag{1·5}$$

because $P(E) \cap Y = \varnothing$ in $\bar{E}$. One can now prove the self intersection formula for the special case where Y is regarded as a subvariety of the variety $\bar{E}$ as follows.† For any $y \in \mathscr{A}(Y)$, $\bar{\imath}_*(y) = \bar{\imath}_*\bar{\imath}^*\rho_G^*(y)$ by (1·2). By projection formula $\bar{\imath}_*(y) = \rho_G^*(y).\bar{\imath}_*(1)$ hence

$$\bar{\imath}_*(y) = \rho_G^*(y).[\rho_G^*\bar{C}](\xi_G, E), \tag{1·6}$$

using also (1·4). Applying $\bar{\imath}^*$ on both sides and taking into account (1·5),

$$\bar{\imath}^*\bar{\imath}_*(y) = \bar{\imath}^*\rho_G^*(y).\bar{\imath}^*\rho_G^* c_r(E) = \bar{\imath}^*\rho_G^*(yc_r(E)) = yc_r(E),$$

since $\bar{\imath}^*\rho_G^* = Id$. Hence

$$\bar{\imath}^*\bar{\imath}_*(y) = y.c_r(E) \tag{1·7}$$

for any $y \in \mathscr{A}(Y)$.

A useful remark in this situation is that $\bar{\jmath}_*$ is injective. In fact, if α is any element of $\mathscr{A}(P(E))$ then $\alpha = \sum_0^{r-1} \rho_E^* a_i.\xi_E^i$, so $\alpha = \bar{\jmath}^* \sum_0^{r-1} \rho_G^* a_i.\xi_G^i = \bar{\jmath}^*\beta$, say. Then

$$\bar{\jmath}_*\alpha = \bar{\jmath}_*\bar{\jmath}^*\beta = \beta.\bar{\jmath}_*(1) = \beta\xi_G$$

by (1·3). So $\bar{\jmath}_* \sum_0^{r-1} \rho_E^* a_i.\xi_E^i = \sum_0^{r-1} \rho_G^* a_i.\xi_G^{i+1}$. As the minimal equations of ξ_E and ξ_G are respectively of degrees $r-1$ and r it follows that $\bar{\jmath}_*$ is injective.

LEMMA 1·1. ('Excision lemma' cf. (3), 4–30, Lemma 4). *Let U, V be two varieties, V a subvariety of U, $W = U \setminus V$ and Z a cycle of U such that its restriction Z_W to W is rationally equivalent to zero. Then there exists a cycle Γ of V which is rationally equivalent to Z on U.*

COROLLARY 1·2. *Consider the blowing-up diagram*

$$\begin{array}{ccc} V' & \longrightarrow & U' \\ \downarrow & & \downarrow \phi \\ V & \longrightarrow & U \end{array}$$

and let Z be a cycle of U such that no irreducible component of Z is contained in V. Let Z' be the proper transform of Z by ϕ. Assume that the restriction Z_W of Z to $W = U \setminus V$ is rationally equivalent to zero. Then there is a cycle Γ of V' such that Z' is rationally equivalent to Γ.

Proof. $W' = U' \setminus V'$ is isomorphic to W by ϕ and the restriction $Z'_{W'}$ of Z' to W' corresponds to Z_W. Hence $Z'_{W'}$ is rationally equivalent to zero. Apply (1·1) to Z', U' and V'.

2. *The construction.* Let $\tilde{X} = X \times \mathbb{P}^1$ and blow it up along $Y \times 0$ to get a morphism $f: \tilde{X}' \to \tilde{X}$. One can identify X with $X \times 0$ and its proper transform by f^{-1} with X'.

† A no more difficult argument enable us to establish the self intersection formula in the case where X and Y are both projective bundles over the same base space.

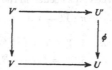

Then $\tilde{f}|X' = f$ and the total transform of X is $\tilde{f}^{-1}(X) = X' + \bar{E}$, because the normal bundle of $Y \times 0$ in $\tilde{X}$ is $E \oplus 1_Y$, $Y \times 0$ is a simple subvariety of $X \times 0$ and

$$\operatorname{codim}_{\tilde{X}}(X \times 0) = 1.$$

Also $\bar{E}$ is attached to X' along Y' its roof, $\tilde{Y}'$ is the proper transform of $Y\,(= Y \times \mathbb{P}^1)$ and $\tilde{f}$ induces an isomorphism $\tilde{Y}' \xrightarrow{\sim} \tilde{Y}$.

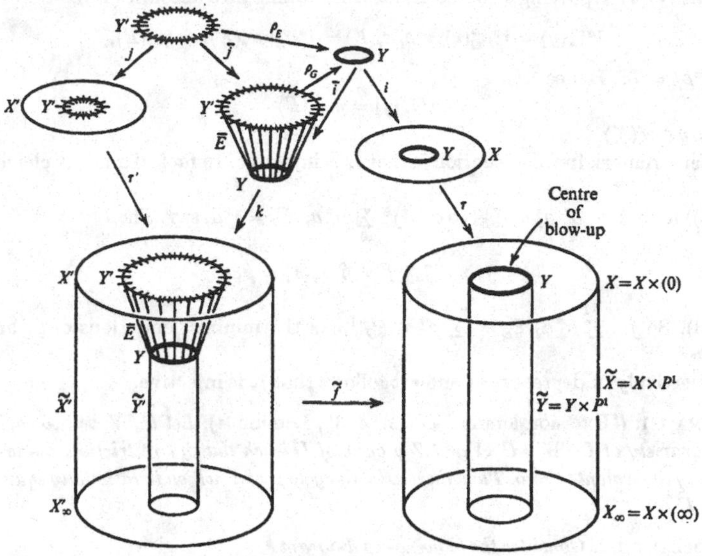

For convenience we shall list the maps needed in the sequel: The projection

$$\pi' : \tilde{Y}' \to Y.$$

The inclusions†

$$i: Y \to X, \quad \bar{\imath}: Y \to \bar{E}, \quad \tilde{\imath}': \tilde{Y}' \to \tilde{X}',$$
$$j: Y' \to X', \quad \bar{\jmath}: Y' = P(E) \to \bar{E},$$
$$k: \bar{E} \to \tilde{X},$$
$$\beta: X_\infty \to \tilde{X}, \quad \beta': X'_\infty \to \tilde{X}',$$
$$\tau: X \to \tilde{X}, \quad \tau': X' \to \tilde{X}'.$$

Lemma 2·1. $k_*(1) = \beta'_*(1) - \tau'_*(1)$.

Proof. $\tau_*(1) = \beta_*(1)$ and $\tilde{f}^*\tau_*(1) = k_*(1) + \tau'_*(1)$, $\tilde{f}^*\beta_*(1) = \beta'_*(1)$.

Lemma 2·2. (i) $\bar{\imath}^* = \pi'_* \tilde{\imath}'^* k_*$, (ii) $\tau'^* k_* = j_* \bar{\jmath}^*$.

Proof. (i). Any subvariety Z of $\bar{E}$ properly intersecting Y on $\bar{E}$, also intersects $\tilde{Y}'$ properly on $\tilde{X}'$ because $\operatorname{codim}_{\bar{E}}(Y) = \operatorname{codim}_{\tilde{X}'}(\tilde{Y}')$ and $Z \cap Y = Z \cap \tilde{Y}'$. The proof of (ii) is similar taking into account that $\operatorname{codim}_{\bar{E}}(Y') = \operatorname{codim}_{\tilde{X}'}(X') = 1$.

† β and τ are just abbreviations for 'bottom' and 'top'.

LEMMA 2·3. $\bar{\imath}^*k^*k_* = 0$.

Proof. By Lemma 2·2, (i) for any $u \in \mathscr{A}(\bar{E})$, $\bar{\imath}^*k^*k_*u = \pi'_*\bar{\imath}'^*k_*k^*k_*u$. Then

$$k_*k^*k_*u = k_*(u)\,k_*(1)$$

by projection formula, $k_*(u).k_*(1) = k_*(u)(\beta'_*(1) - \tau'_*(1))$ by Lemma 2·1 and $k_*(u).\beta'_*(1) = 0$, because $\bar{E} \cap X'_\infty = \varnothing$. It follows that $k_*k^*k_*u = -k_*(u)\tau'_*(1)$. Finally $\pi'_*\bar{\imath}'^*k_*k^*k_*u = -\pi'_*\bar{\imath}'^*(k_*(u).\tau'_*(1)) = 0$ because $k_*(u).\tau'_*(1)$ is represented by a cycle on X' which will be disjoint from $\tilde{Y}' = \operatorname{Im}(\bar{\imath}')$.

3. *The self intersection formula.* For any cycle A of X denote $A \times \mathbb{P}^1$ by $\tilde{A}$ and let $\tilde{A}'$ be the proper transform of $\tilde{A}$ by $\tilde{f}$. The following lemma is straightforward to verify:

LEMMA 3·1. (i) *If A is a subvariety of Y, then $\tilde{A}'$ is isomorphic with $\tilde{A}$ and it intersects the floor of $\bar{E}$ along A; moreover $\tilde{A}'.\bar{E}$ is defined on $\tilde{X}'$, $\tilde{A}'.\bar{E} = A$ and $\tilde{A}'.X' = 0$.*

(ii) *If A intersects Y properly on X then the cycle $\tilde{f}^{-1}(\tilde{A})$ is defined and $\tilde{f}^{-1}(\tilde{A}) = \tilde{A}'$; the cycle $\tilde{A}'.\bar{E}$ is defined on $\tilde{X}'$ and $\tilde{A}'.\bar{E} = \rho_G^{-1}(A.Y)$.*

(iii) *Under the assumption of (ii) the cycle $X'.\tilde{A}'$ is defined on $\tilde{X}'$ and $X'.\tilde{A}' = f^{-1}(A)$.*

THEOREM 1. *For any $y \in \mathscr{A}(Y)$*

$$i^*i_*(y) = yc_r(E),$$

where E is the normal bundle of Y in X.

Proof. One can assume that y is the class of a subvariety B of Y. Let B_1 be a cycle of X properly intersecting Y and rationally equivalent to B. According to Lemma 3·1, (i)

$$\tilde{B}'.\bar{E} = B \tag{3·1}$$

and

$$\tilde{B}'.X' = 0. \tag{3·2}$$

By (ii) and (iii) of the same lemma

$$\tilde{f}^{-1}(\tilde{B}_1) = \tilde{B}'_1, \tag{3·3}$$

$$\tilde{B}'_1.\bar{E} = \rho_G^{-1}(B_1.Y), \tag{3·4}$$

$$\tilde{B}'_1.X' = f^{-1}(B_1). \tag{3·5}$$

Also the class of $B_1.Y$ in $\mathscr{A}(Y)$ is

$$cl_Y(B_1.Y) = i^*i_*(y). \tag{3·6}$$

The cycle $\tilde{B} - \tilde{B}_1$ is rationally equivalent to zero on $\tilde{X}$. By Corollary 1·2, with $\phi = \tilde{f}$ and $Z = \tilde{B} - \tilde{B}_1$, there exists a cycle Γ of $\bar{E}$ rationally equivalent to $\tilde{B}' - \tilde{B}'_1$ on $\tilde{X}'$, i.e. in rational equivalence classes

$$cl_{\tilde{X}'}(\tilde{B}') - cl_{\tilde{X}'}(\tilde{B}'_1) = k_*(\gamma), \tag{3·7}$$

where $\gamma \in \mathscr{A}(\bar{E})$ is the class of Γ. According to (3·1) $k^*cl_{\tilde{X}'}(\tilde{B}') = \bar{\imath}_*(y)$. Also from (3·4) and (3·6) one deduces $k^*cl_{\tilde{X}'}(\tilde{B}'_1) = \rho_G^*i^*i_*(y)$. These give, by applying k^* to (3·7),

$$\bar{\imath}_*(y) - \rho_G^*i^*i_*(y) = k^*k_*(\gamma). \tag{3·8}$$

Applying $\bar{\imath}^*$ on both sides and taking into account that $\bar{\imath}^* k^*k_* = 0$ by Lemma 2·3, $\bar{\imath}^*\bar{\imath}_*(y) - \bar{\imath}^*\rho_G^* i^* i_*(y) = 0$, hence $\bar{\imath}^*\bar{\imath}_*(y) = i^*i_*(y)$ because $\rho_G \bar{\imath} = Id_Y$. To conclude apply (1·7).

Note. Since $i^*i_*(y) = yc_r(E)$ one can write (3·8) as follows

$$\bar{\imath}_*(y) = \rho_G^*(yc_r(E)) + k^*k_*(\gamma). \tag{3·9}$$

4. *The 'formule-clef' for $\mathscr{A}(X')$.*

Lemma 4·1. *With notations of section 3*

$$-\bar{\jmath}^*(\gamma) = \rho_E^*(y) \, . \, c_{r-1}(E^{(1)}).$$

Proof. By (3·9) $k^*k_*\gamma = \bar{\imath}_*(y) - \rho_G^*(y \, . \, c_r(E))$. Also

$$k^*k_*(\gamma) = -\gamma \, . \, \xi_G = -\gamma \, . \, \bar{\jmath}_*(1) = -\bar{\jmath}_*\bar{\jmath}^*(\gamma)$$

using the self-intersection formula, the fact that $\check{L}_G$ is the normal bundle of $\bar{E}$ in $\tilde{X}'$ and the formula $c_1(\check{L}_G) = -\xi_G$ for the first equality; and using (1·3) for the second formula. By (1·6) and (1·3) again:

$$\begin{aligned}\bar{\imath}_*(y) - \rho_G^*(y \, . \, c_r(E)) &= \rho_G^*(y) \, . \, ([\rho_G^* \bar{C}](\xi_G, E) - \rho_G^* c_r(E)) \\ &= \rho_G^*(y) \, . \, \xi_G \, . \left(\sum_{i=0}^{r-1} \xi_G^{r-1-i} \, . \, \rho_G^* c_i(E)\right) \\ &= \rho_G^*(y) \, . \, \bar{\jmath}_* \bar{\jmath}^* \left(\sum_{i=0}^{r-1} \xi_G^{r-1-i} \, . \, \rho_G^* c_i(E)\right).\end{aligned}$$

But $\bar{\jmath}^*\xi_G = \xi_E$ and $\rho_E = \rho_G j$, so putting everything together and using (1·1), we get:

$$\begin{aligned}-\bar{\jmath}_*\bar{\jmath}^*(\gamma) &= \bar{\imath}_*(y) - \rho_G^*(y \, . \, c_r(E)) \\ &= \bar{\jmath}_* \left(\rho_E^*(y) \, . \sum_{i=0}^{r-1} \xi_E^{r-1-i} \, . \, \rho_E^* c_i(E)\right) \qquad (4·1) \\ &= \bar{\jmath}_*(\rho_E^*(y) \, . \, c_{r-1}(E^{(1)}).\end{aligned}$$

But $\bar{\jmath}_*$ is injective as remarked in section 1, so we may cancel it in (4·1).

Theorem 2. *For any* $y \in \mathscr{A}(Y)$

$$j_*(\rho^*(y) \, . \, c_{r-1}(E^{(1)})) = f^*i_*(y).$$

Proof. By applying τ'^* to (3·7), $-\tau'^* cl_{\tilde{X}'}(\tilde{B}_1') = \tau'^*k_*(\gamma)$ since $\tilde{B}'$ is a cycle of $\tilde{Y}'$ and $\tilde{Y}' \cap X' = \varnothing$. Also $\tau'^* cl_{\tilde{X}'}(\tilde{B}_1') = f^*i_*(y)$ according to (3·5). Hence

$$-f^*i_*(y) = \tau'^*k_*(\gamma).$$

By (ii) of Lemma 2·2, $\tau'^*k_*(\gamma) = j_*\bar{\jmath}^*(\gamma)$. But in Lemma (4·1) we have shown that $-\bar{\jmath}^*(\gamma) = \rho_E^*(y) \, c_{r-1}(E^{(1)})$. Putting this together, we get the 'formule-clef'.

REFERENCES

(1) Borel, A. and Serre, J.-P. Le théorème de Riemann-Roch. *Bull. Soc. Math. France* **86** (1958), 97–136.

(2) Berthelot, A., Grothendieck, A. and Illusie, L. Théorie des intersections et théorème de Riemann-Roch. SGA6, Springer Lecture Notes no. 225.

(3) Grothendieck, A. Sur quelques propriétés fondamentales en théorie des intersections. *Anneaux de Chow et applications. Séminaire C. Chevalley*, 2e année (1958).

(4) Grothendieck, A. La théorie des classes de Chern. *Bull. Soc. Math. France* **86** (1958), 137–159.

(5) Ilori, S., Ingleton, A. W. and Lascu, A. T. On a formula of D. B. Scott. *J. London Math. Soc.* (2), **8** (1974), 539–544.

(6) Jouanolou, J. P. Riemann–Roch sans dénominateurs. *Inventiones Math.* **11** (1970), 15–26.

(7) Lascu, A. T. and Scott, D. B. An algebraic correspondence with applications to projective bundles and blowing-up Chern classes. *Annali di Matematica pura ed applicata* (to appear).

Proceedings of Symposia in Pure Mathematics
Volume 28, 1976

HILBERT'S FOURTEENTH PROBLEM - THE FINITE GENERATION OF SUBRINGS SUCH AS RINGS OF INVARIANTS

David Mumford[1]

1. INTRODUCTION

The precise statement of the problem is this:

Let k be a field

Let K be a subfield of the rational functions in n-variables over k:

$$k \subset K \subset k(x_1,\cdots,x_n).$$

(n.b. all such K are automatically finitely generated over k as fields)

Is the ring:

$$K \cap k[x_1,\cdots,x_n]$$

finitely generated over k?

The motivation for this question came from its affirmative answer by Hilbert and others in certain very interesting cases: e.g., say $\mathrm{char}(k) = 0$, suppose $G = SL(m)$ is acting linearly on k^n, and suppose K is defined as the field of G-invariant rational functions. Then $K \cap k[x_1,\cdots,x_n]$ is just the ring of G-invariant polynomials and Hilbert had proven that this was finitely generated. Unfortunately, it turns out that the answer is, in general, NO: $K \cap k[x_1,\cdots,x_n]$ may require an infinite number of generators. A beautiful counter-example was discovered by M. Nagata [13] in 1959. It would appear that after Hilbert's discovery of the extremely general finiteness principle on which his proof in the $SL(m)$-invariant case was based, namely "Hilbert's basis theorem" on the finite generation of all ideals in $k[x_1,\cdots,x_n]$, Hilbert was overly optimistic about finiteness results in other algebraic contexts. However my belief is that it was not at all a blind alley: that on the one hand its failure reveals some very significant and far-reaching subtleties in the category of varieties; and that the search for cases where it and

AMS (MOS) subject classifications (1970) 13B99, 14C20, 14E05.

[1]Research supported by the National Science Foundation under grant GP-36269X2.

related geometric questions are correct is a very important area of research in algebraic geometry. In fact, my guess is that it was Hilbert's idea to take a question that heretofore had been considered only in the narrow context of invariant theory and thrust it out into a much broader context where it invited geometric analysis and where its success or failure had to have far-reaching algebro-geometric significance. We will discuss the problem in 3 sections — first in the case of invariant theory where K is the field of G-invariant functions for some G, second in its geometric form involving linear systems formulated and analyzed first by Zariski [23], and thirdly as a special case of the general problem of forming quotient spaces of varieties by algebraic equivalence relations.

2. INVARIANT THEORY

Hilbert's proof of the finiteness when K is the field of G-invariant functions, $G = SL(m)$, $char(k) = 0$ is so very elegant and simple that it should really be part of every mathematician's bag of tricks. So I would like to begin by running through this marvelous proof: to begin with, it is known that if V is any finite-dimensional polynomial representation of $SL(m)$ in char. 0, then V is completely reducible. In particular, there is a unique decomposition:

$$V = V^G \oplus V_1$$

where V^G is the subspace of invariant vectors and V_1 is a G-stable subspace containing no invariants. Let ρ_V be the projection of V onto V^G with kernel V_1. Next, let $R = k[X_1,\cdots,X_n]$, and let $R^G \subset R$ be the ring of invariants. R and R^G are graded rings, i.e.,

$$R = \oplus R_k, \quad R_k = \text{vector space of homogeneous degree } k \text{ polynomials}$$

$$\text{and } R^G = \oplus R_k^G, \quad R_k^G = G\text{-invariants in } R_k.$$

Thus the operators

$$\rho_{R_k}: R_k \longrightarrow R_k^G$$

patch together into a projection

$$\rho_R: R \longrightarrow R^G.$$

A simple argument using the uniqueness of ρ shows that ρ_R satisfies the identity:

$$\rho_R(fg) = f\rho_R(g), \quad f \in R^G, \ g \in R.$$

Now we let

$$R^G_+ = \bigoplus_{k>0} R^G_k$$

and let $I = R^G_+ .R$ be the ideal in R generated by all invariants of positive degree. Hilbert's Basis Theorem asserts that

$$I = \sum_1^N f_i .R$$

for some $f_1, \cdots, f_N \in I$; we can assume if we like that each f_i is in fact in R^G and homogeneous of some degree d_i. Then Hilbert asserts that these f_i generate R^G as ring! He proves this by induction on degree: choose $g \in R^G_n$ and assume all $h \in R^G_{n'}$ for $n' < n$ are polynomials in the f_i's. Then $g \in I$, hence there is an expression:

$$g = \sum_1^N a_i f_i, \quad a_i \in R_{n-d_i}.$$

Apply ρ_R:

$$g = \rho_R g = \sum \rho_R(a_i f_i) = \sum (\rho_R a_i) f_i.$$

Then $\rho_R a_i \in R^G_{n-d_i}$ which is a polynomial in the f_i's by induction, hence so is g!

What was the history of invariant theory after Hilbert? First of all, Hilbert did not give the above abstract description of ρ, but rather an explicit construction of ρ, called "Cayley's Ω-process" in which ρ appears in the Universal enveloping algebra of $s\ell(m)$. As mentioned in Hilbert's problem itself, A. Hurwitz [7] had already observed and H. Weyl was later to use effectively the fact that if $k = \mathbb{C}$, (and we can reduce easily any char. 0 case to the case $k = \mathbb{C}$), then

$$\rho x = \int_{g \in SU(m)} g^*(x) \cdot dg$$

$SU(m)$ = special unitary group

dg = Haar measure

Via the fact that any reductive algebraic group over $\mathbb{C}$ has a Zariski-dense compact subgroup, this gives us an explicit construction for the projection ρ for any such groups, hence a proof of finiteness. The final step - to observe that no explicit formula for ρ is needed but one merely must know the complete reducibility of all finite-dimensional representations to construct ρ abstractly - was taken by M. Schiffer in 1933 (unpublished; it appeared in H. Weyl's "Classical Groups" [22], Supplement C).

In char. p, no semi-simple group has the property that all its representations are completely reducible. For instance, think of SL(2) acting on the 3-dimensional space of quadratic forms

$$V = k\cdot x^2 + k\cdot xy + k\cdot y^2.$$

In char. 2, $k\cdot x^2 + k\cdot y^2$ is an invariant subspace with no complement. Therefore the Schiffer-Hilbert method breaks down. However, very recently, W. Haboush [25] has succeeded in proving the following Theorem which I conjectured in [9]:

<u>THEOREM</u>: If a semi-simple (or even reductive) algebraic group G acts on a vector space V and leaves fixed a vector $v \in V$, there is a polynomial function f on V such that:

i) $f(v) \neq 0$
ii) f is G-invariant.

In char. 0, f exists and may be taken linear by complete reducibility. Seshadri [17] had previously proven that such f's exist when G = SL(2). Nagata [14] has proven that if G has the property of the Theorem (this is sometimes stated as "G is semi-reductive"), then the ring of G-invariants* is finitely generated, i.e., whenever G acts linearly on $kx_1 + \cdots + kx_n$, then $k[x_1,\cdots,x_n]^G$ is finitely generated. Therefore, it follows that the ring of invariants is finitely generated for G reductive.

*We have not made precise before whether by G-invariants we meant polynomials $f(x_1,\cdots,x_n)$ which were <u>identically</u> invariant, i.e.,

$$f(g(x)) - f(x) \equiv 0 \quad \text{as function of } g,$$

or f's which were invariant separately under every $g \in G(k)$ (the k-rational points of G). If k is infinite, G(k) is Zariski-dense in G and there is no difference between these 2 concepts. But if k is finite there is a difference: in this case G(k) is finite and I wish G-invariant to mean <u>identically</u> invariant.

For groups G which are not semi-simple or reductive (i.e., which have a "unipotent radical"), very little is known even in char. 0 about finiteness of the ring of invariants. I know of only 2 results --
a) Weitzenbock [21] (cf. Also [16]) proved $k[x_1,\cdots,x_n]^G$ finitely generated if $G = \mathbb{G}_a$ (i.e., $\mathbb{G}_a$ = the additive group of the ground field),
b) Nagata's counter-example [13] is a non-finitely generated ring $k[x_1,\cdots,x_n]^G$ where G is commutative, but G is a product of many groups $\mathbb{G}_a$ and many groups $\mathbb{G}_m$ (here $\mathbb{G}_m$ = the multiplicative group of the ground field*).

3. ZARISKI'S FORMULATION WITH LINEAR SYSTEMS

We recall that if X is a non-singular projective variety (or more generally if X is normal) and D is a positive divisor on X (i.e., $D = \Sigma n_i E_i$, $E_i \subset X$ a subvariety of codimension 1 and $n_i \geq 0$), then we define:

$$\mathcal{L}(D) = \left\{\begin{array}{l}\text{vector space of rational functions f on X} \\ \text{with poles bounded by D, i.e., } \forall\, E \subset X \text{ of} \\ \text{codimension 1,} \\ \quad \operatorname{ord}_E f \geq -(\text{mult. of E in D}).\end{array}\right\}$$

(Either $\mathcal{L}(D)$ or the family of divisors that occurs as the zeroes of the functions $f \in \mathcal{L}(D)$ is called a linear system on X.) Zariski introduced the 2 rings:

$$R(D) = \bigcup_{n=0}^{\infty} \mathcal{L}(nD) = \left\{\begin{array}{l}\text{ring of rational functions f with} \\ \text{poles of any order but only on D}\end{array}\right\}$$

$$R^*(D) = \bigoplus_{n=0}^{\infty} \mathcal{L}(nD).$$

---

* In concrete terms, a representation of $\mathbb{G}_a^n$ is a commutative group of matrices all of the form

$$\begin{pmatrix} 1 & & * \\ & \ddots & \\ 0 & & 1 \end{pmatrix}$$

in a suitable basis of $kx_1 + \cdots + kx_n$. A representation of $\mathbb{G}_m^n$ is a commutative group of diagonal matrices

$$\begin{pmatrix} * & & 0 \\ & \ddots & \\ 0 & & * \end{pmatrix}$$

in a suitable basis of $kx_1 + \cdots + kx_n$.

The ring $R^*(D)$, though apparently much bigger than $R(D)$, is easily shown to be isomorphic to $R(D_1)$ for a suitable divisor D_1 on a variety X_1 which is a $\mathbb{P}^1$-bundle over the variety X you start with. So the class of rings $R^*(D)$ is really a subset of the class of rings $R(D)$. More generally, for any divisors* $D_1,\cdots,D_k$, we can define a k-times graded ring:

$$R^*(D_1,\cdots,D_k) = \bigoplus_{n_1=0}^{\infty} \cdots \bigoplus_{n_k=0}^{\infty} \mathcal{L}(\Sigma n_i D_i)$$

and this is also isomorphic to $R(D_1)$ for a suitable D_1 on an X_1 (which is now a $\mathbb{P}^k$-bundle over X.) In his penetrating article [23], Zariski showed that Hilbert's rings $K \cap k[x_1,\cdots,x_n]$ were isomorphic to rings of the form $R(D)$ for a suitable X and D; asked more generally whether all the rings $R(D)$ might not be finitely generated; and proved $R(D)$ finitely generated if $\dim X = 1$ or 2. I want to outline the procedure for finding X and D such that:

$$K \cap k[x_1,\cdots,x_n] \cong R(D).$$

First of all, X is to be a suitable projective variety with function field K^{**}. For any such X, the inclusion of fields

$$K \subset k(X_1,\cdots,X_n)$$

defines a "rational map"

$$\pi:\ \mathbb{P}^n \longrightarrow X$$

i.e., π is a many-valued map whose graph in $\mathbb{P}^n \times X$ is a subvariety and which is single-valued on a Zariski -open subset $U \subset \mathbb{P}^n$. Let $r = \dim X = \mathrm{tr.d.}_k K$. Then if X is chosen "sufficiently blown up", one can make π^{-1} nice in the sense:

$\forall\, x \in X$, the full inverse image $W_x = \pi^{-1}[x]$ has dimension $n-r$.

Roughly speaking, we have a fibration of $\mathbb{P}^n$ by $(n-r)$-dimensional algebraic sets W_x such that K is the field of rational functions constant on each

---

* If D has some negative coefficients, an $f \in \mathcal{L}(D)$ should have corresponding <u>zeroes</u> of order at least that coefficient.

** If resolution is known for this dimension and characteristic one would take X non-singular; if not, one takes X to be normal and $R(D)$ is defined as before.

W_x, i.e., invariant generically under the equivalence relation defined by belonging to the same W_x. Of course, these W_x's may become singular and in general will meet at certain "bad" points of $\mathbb{P}^n$, namely where the map π is not single-valued. Now let $D_1,\cdots,D_k$ be the subvarieties of X of codimension 1 such that $\pi^{-1}[D_i] \subset$ (the hyperplane at ∞, $\mathbb{P}^n - \mathbb{A}^n$). Then for all rational functions f on X, f has poles only on $\bigcup D_i$ if and only if $f\cdot\pi$ has poles only at ∞, hence

$$K \cap k[x_1,\cdots,x_n] = R(\sum D_i).$$

Unfortunately, it was precisely by focusing so clearly the divisor-theoretic content of Hilbert's 14th problem that Zariski cleared the path to counter-examples. The history is this -

i) Rees [15] in 1958 found a 3-dimensional X and a D with R(D) infinitely generated. His X was birational to $\mathbb{P}^2 \times E$ (E an elliptic curve).

ii) Nagata [13] in 1959 found that for suitable points $P_1,\cdots,P_r \in \mathbb{P}^2$, if X is the surface obtained by blowing up each P_i into a rational curve E_i, then

$$R^*(\ell, -\sum_1^r E_i) \quad (\ell \text{ a line not through any } P_i)$$

is infinitely generated; and that this ring was a ring of invariants $k[x_1,\cdots,x_{2r}]^G$ as mentioned in §1.

iii) Zariski [24] in 1962 returned to the problem and pursuing some constructions which had been considered in different contexts by Grauert [3] and Nagata [12], found that it was not at all uncommon for $R^*(D)$ to be infinitely generated when dim X = 2 (hence for R(D) to be infinitely generated when dim X = 3).

I would like to describe the situation Zariski looked at because it is a very useful source of counter-examples to several problems and illustrates some basic facts about the category of algebraic varieties. Suppose you have

a) a non-singular surface X,

b) a curve $E \subset X$ of genus $g > 0$ such that

i) $(E^2) < 0$ (i.e., the normal bundle to E in X has negative curvature)

ii) Pic X $\longrightarrow$ Pic E is injective (i.e., if a line bundle L on X is trivial on E, then it is trivial on X).

Such a situation is not hard to obtain: start with any sufficiently general hypersurface section H_o on X_o and blow up enough generic points on H_o to make its normal bundle negative. First of all, here is what Grauert observed about this situation: analytically, E can be blown down, i.e., there is a normal analytic surface X_1 and $\pi: X \longrightarrow X_1$ mapping E to a point x but bijective elsewhere. But X_1 is not a variety: if it were, x would have an affine neighborhood $U \subset X$, hence $C = X_1 - U$ would be a curve not containing x, hence $\pi^{-1}(C)$ would be a curve on X disjoint from E, hence "twisting by $\pi^{-1}(C)$" we get a line bundle $\mathcal{O}_X(\pi^{-1}C)$ trivial on E but not trivial on X: contradiction.

Zariski did this: let $H \subset X$ be a hyperplane section, let $a = (H.E)$, (the intersection number of H and E), let $(E^2) = -b$. Then he showed

$$R^*(bH + aE)$$

is not finitely generated. The reason is this — look for functions f on X with poles kbH + kaE, some $k \geq 1$. If at some $P \in E$, $x = 0$ is the local equation of E, expand f:

$$f = \frac{g_o}{x^{ak}} + \frac{g_1}{x^{ak-1}} + \cdots\cdots$$

and consider the function g_o on E. Suitably interpreting what g_o means, g_o comes out as a section of a line bundle on E; in fact the line bundle $\mathcal{O}_X(kbH+kaE)$ on X restricted to E. This has degree 0 but by assumption (b ii) is not trivial. So it has no sections and $g_o \equiv 0$, i.e., f can have at most poles of type $kbH + (ka-1)E$. On the other hand, Zariski showed that there is a fixed k_o such that for all k, there are functions f with poles of type $kbH + \max(0, ka-k_o)E$. To see the implications of this, say for instance that $k_o = 1$: then for all k, let

$$f_k \in \mathcal{L}(kbH + kaE)$$

have a pole $kbH + (ka-1)E$. Then for all k,

1 $$f_k \notin \left(\begin{array}{c}\text{subring of } R^*(bH + aE) \text{ generated by} \\ 1, f_1, \cdots, f_{k-1}\end{array}\right)$$

since every function in the degree k piece of the subring has a pole of at most $kbH + (ka-2)E$. Taking into account that $R^*(bH+aE)$ is <u>graded</u>, it requires at least one generator in each degree, hence is not finitely generated!

Are there any positive results asserting that R(D) and $R^*(D)$ are finitely generated in some cases? When $\dim X = 2$, Zariski's paper [24] gives a thorough analysis of when $R^*(D)$ is finitely generated. In higher dimensions, at the moment, the best results are numerical criteria on D implying that D is ample, which in turn implies very quickly that both R(D) and $R^*(D)$ are finitely generated. Here "D ample" means that for some $n \geq 1$, nD is a hyperplane section of X in a suitable projective embedding. These criteria use intersection numbers and are as follows:

1.) Nakai's Criterion: if for every subvariety $Y \subset X$, $(Y.D^r) > 0$ where $r = \dim Y$, then D is ample.
2.) Seshadri's Criterion: if there is an $\epsilon > 0$ such that for every curve $C \subset X$, $(C.D) > \epsilon \cdot \left[\max_{P \in C}(\text{mult. of } P \text{ on } C)\right]$, then D is ample.

For proofs, see Hartshorne's book [4], Chapter I.

4. QUOTIENT SPACES BY ALGEBRAIC EQUIVALENCE RELATIONS

Another way of generalizing Hilbert's problem is to ask: given a variety X, and

$$R \subset X \times X, \quad R = \begin{cases} 1) \text{ a finite union of subvarieties of } X \times X \\ 2) \text{ set-theoretically, an equivalence relation on } X \end{cases}$$

when is there another variety Y and a surjective morphism $f: X \longrightarrow Y$ such that

$$R = \left\{(x_1,x_2) \,\middle|\, f(x_1)=f(x_2)\right\}?$$

For short, we speak* of Y as X/R. Two cases of particular interest are i) a group G acts on X and $R = \left\{(x,gx) \,\middle|\, x \in X, g \in G\right\}$, and ii) E is a subvariety of X to be "blown down" and R = (diagonal) $\cup$ $(E \times E)$. In Hilbert's case, $X = \mathbb{A}^n$ (affine space) but one is given R only generically by specifying the subfield K (i.e., $R = \left\{(x_1,x_2) \,\middle|\, x_1,x_2 \text{ belong to some } W_x\right\}$ in the notation of §2); Hilbert's problem can be broken up into 2 steps – first extend this equivalence relation nicely to one on all of $\mathbb{A}^n$, second prove $\mathbb{A}^n/R$ exists and is an affine variety, in which case Hilbert's ring $k[X_1,\cdots,X_n] \cap K$ is just the affine coordinate ring of $\mathbb{A}^n/R$.

Returning to the general case, it is always possible to find a Zariski open subset $U \subset X$ stable under R such that U/R exists (this may be proven for instance using Chow coordinates of the equivalence classes).

---

*The requirements do not determine Y uniquely, but in all cases that arise, there are natural extra conditions one imposes that make Y unique if it exists at all.

Equivalently the field of rational functions K on X/R is easy to construct and then any model Y of K realizes X/R on some sufficiently small Zariski-open U in X. The real problem is a birational one of finding a Y which works everywhere. However, as in Zariski's divisor formulation of the problem, one is confronted straightway by a raft of counter-examples:

1.) Grauert's example [3] described in §3 of an $E \subset X$, where $\dim X = 2$, $(E^2) < 0$* and E can be blown down analytically but not algebraically,
2.) Hironaka [6] found a beautiful example of a complete (though non-projective) variety X on which $\mathbb{Z}/2\mathbb{Z}$ acts freely, but $X/(\mathbb{Z}/2\mathbb{Z})$ is not a variety at all,
3.) Nagata and I found ([9], p. 83) examples of $PGL(n)$ acting freely on quasi-projective varieties X such that the orbit space $X/PGL(n)$ is not a variety.

In rough outline, here is the idea of Hironaka: take a 3-dimensional projective variety X_o with 2 curves C_1, C_2 in it crossing transversely at 2 points P_1, P_2 and with $\mathbb{Z}/2\mathbb{Z}$ acting on X_o interchanging the C's and the P's:

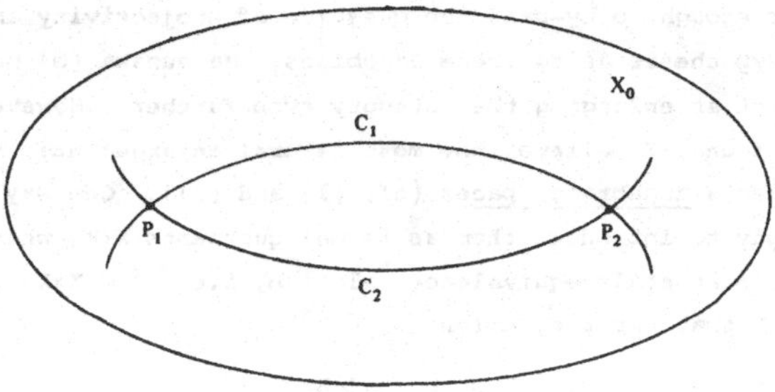

We then blow up C_1 and C_2 in X_o to obtain X. However, where the C's cross, we must specify the order in which the C's are blown up — so at P_1, we blow up C_1 first, then in the resulting variety we blow up C_2; at P_2,

---

*If $(E^2) \geq 0$, then E cannot be blown down even analytically so of course one cannot construct X/R, $R = (\text{diag}) \cup (E \times E)$, algebraically. For general equivalence relations R one asks first that R have some reasonable properties ensuring that X/R exists in the analytic context.

we blow up C_2 first, then in the result, we blow up C_1. Then $\mathbb{Z}/2\mathbb{Z}$ still acts on X. However, say Y = X mod ($\mathbb{Z}/2\mathbb{Z}$) were a variety. Since X_o is projective, it can be shown (cf. e.g. [10], p. 111) that X_o mod($\mathbb{Z}/2\mathbb{Z}$) is a variety Y_o. In Y_o, C_1 and C_2 have the same image D and P_1 and P_2 have the same image Q. Then Y would be obtained from Y_o by blowing up D; but at Q, the 2 branches of D must be blown up in a definite order. As D is an irreducible curve, these 2 branches cannot be distinguished by rational functions! This turns out to mean that Y in fact does not exist in the category of algebraic varieties.

Confronted with these counter-examples, people have had 2 reactions: a) find criteria for X/R to exist as a variety, or b) instead enlarge the category you are working in. The ploy (b) was most notably successful in Weil's hands in his 2nd proof of the Riemann hypothesis for curves over finite fields [20]. His idea here required the construction of the Jacobian variety of such a curve. At that time, only affine and projective varieties had been considered. Weil invented the category of what he called <u>abstract varieties</u> — now called simply varieties — and constructed the Jacobian as one of these. Subsequently he and Chow independently showed that the Jacobian was actually a projective variety; however, at the time, Weil instead developed the theory of "abstract" varieties far enough to by-pass the question of projectivity and prove the Riemann Hypothesis using these Jacobians. Matsusaka [8] made an initial attempt at enlarging the category even further. However it was M. Artin who found, I believe, the most natural enlargement: he calls these new objects <u>algebraic spaces</u> (cf. [1] and [2]). One way to define these is simply to introduce them as formal quotients X/R, where X is a scheme and R is an etale equivalence relation, i.e., $R \subset X \times X$ is a subscheme such that the projection

$$p: R \longrightarrow X$$

is etale — essentially makes R into an unramified covering over X. Artin then went on to show that the category of algebraic spaces is closed under apparently all "reasonable" further quotient operations $X \longrightarrow X/R$. For details we refer the reader to his papers, which make algebraic spaces into a very effective and powerful tool.

Still you may have a sentimental attachment to familiar old varieties. It would appear especially that <u>projective</u> varieties play such a central technical role in algebraic geometry that it may be virtually impossible to eliminate their use even if you wanted to. In any case, it is very interesting to prove, when possible, that X/R is actually a projective variety. I would like to state one such result concerning orbit spaces:

Suppose:

X = a projective variety over k

G = a semi-simple (or more generally reductive) algebraic group over k, acting on X

$X \subset \mathbb{P}^n$: an embedding such that the action of G on X extends to an action on $\mathbb{P}^n$.

Then there are canonical open subsets

$X_s \subset X$: the set of "stable" points

$X_{ss} \subset X$: the set of "semi-stable" points

such that X_s, X_{ss} are G-invariants, and there is a diagram:

$$\begin{array}{ccccc} X_s & \subset & X_{ss} & \subset & X \\ \downarrow & & \downarrow & & \\ X_s/G & \subset & \overline{X_s/G} & & \end{array}$$

where $\overline{X_s/G}$ is a projective variety, X_s/G is an open subset of $\overline{X_s/G}$, $X_s \longrightarrow X_s/G$ makes X_s/G into an orbit space by G, and $X_{ss} \longrightarrow \overline{X_s/G}$ makes $\overline{X_s/G}$ into the quotient of X_{ss} by a cruder equivalence relation $\sim$ defined by:

$$x \sim y \text{ if } \overline{O^G(x)} \cap \overline{O^G(y)} \cap X_{ss} \neq \emptyset$$

(here $O^G(x)$ = G-orbit of x). This theorem is proven in my book [9] when char. = 0, and the part about X_s is proven by Seshadri [18] when char. = p. This X_{ss}-part in char. p follows from the recent results of Haboush[25] discussed in §1. See [11] for examples and a discussion of this result. This result has proven very useful for proving that various moduli spaces are quasi-projective varieties (and not "just" algebraic spaces).

The above theorem is in fact a natural extension of Hilbert's own ideas about the ring of invariants, especially as developed in his last big paper on the subject, "Über die vollen invariantensystemen" [5]. To indicate this, let me define X_{ss}. Assume for simplicity that there is actually a representation of G on k^{n+1} which induces the action of G on the $\mathbb{P}^n$ ambient to X. We then make the definition:

If $x \in X$, then

$$x \in X_{ss} \Longleftrightarrow \begin{cases} \forall \text{ homomorphisms } \lambda\colon \mathbb{G}_m \longrightarrow G, \text{ let } x(\lambda) = \lim\limits_{t \to 0} \lambda(t)(x). \\ \text{Let } x(\lambda)^* \in k^{n+1} \text{ be homogeneous coordinates for } x(\lambda), \\ \text{so that } \lambda(t)[x(\lambda)^*] = t^r \cdot [x(\lambda)^*] \text{ for some } r \in \mathbb{Z}. \\ \text{We ask } r < 0 \text{ for all } \lambda. \end{cases}$$

Now let R be the homogeneous coordinate ring of X, and let R_+^G be the invariants with no constant term. Then we have Hilbert's result:

$$X - X_{ss} = V(R_+^G . R).$$

Contrary to the usual credo that Hilbert eliminated the interest in studying special cases in invariant theory, my belief is that some of the most challenging problems still open in invariant theory concern special cases. I would like to raise two rather broad questions:

PROBLEM Let S be the parameter space for a family $\{X_s \mid s \in S\}$ of non-singular projectively normal subvarieties $X_s \subset \mathbb{P}^n$. Assume PGL(n+1) acts on S so that for all $g \in PGL(n+1)$, $X_{g(s)} = g(X_s)$. Assume this action is proper. Then is the quotient S/PGL(n+1) always a variety?

PROBLEM Now that we have computers, is there a practical way to actually find generators of such classical rings of invariants as those of a binary or ternary n-ic (i.e., SL(2) or SL(3) acting on the space of homogeneous degree n polynomials in 2 or 3 variables)? After an extraordinary effort, Shioda [19] only recently found these for binary octics.

2 Added in proof: Independent of W. Haboush's work, E. Formanek and C. Procesi have recently in a preprint entitled "Mumford's Conjecture for the general linear group" given another very beautiful proof of the semi-reductivity of GL(n) and SL(n).

REFERENCES

[1] M. Artin, Algebraic Spaces, Yale Math. Monographs, Yale Univ. Press

[2] M. Artin, Algebraization of formal moduli, I, in Global Analysis, Princeton Univ. Press, 1969; II in Annals of Math., 91 (1970), p. 88.

[3] H. Grauert, Über Modifikationen und exzeptionelle analytische Mengen, Math. Annalen, 146 (1962), p. 331.

[4] R. Hartshorne, Ample Subvarieties of Algebraic Varieties, Springer Lecture Notes 156, Springer, 1970.

[5] D. Hilbert, Über die vollen Invariantensystemen, Math. Annalen, 42 (1893), p. 313.

[6] H. Hironaka, An example of a non-kählerian deformation of kählerian complex structures, Annals of Math., 75 (1962), p. 190.

[7] A. Hurwitz, Über die Erzeugung der Invarianten durch Integration, Nachr. Gött. Ges. Wissensch. 1897, p. 71.

[8] T. Matsusaka, Theory of Q-varieties, Publ. Math. Soc. of Japan, 1965.

[9] D. Mumford, Geometric Invariant Theory, Springer-Verlag, 1965.

[10] D. Mumford, Abelian Varieties, Tata Inst. Studies in Math., Oxford Univ. Press, 1970.

[11] D. Mumford and K. Suominen, Introduction to the theory of moduli, in Algebraic Geometry, Oslo 1970, ed. by F. Oort, Noordhoff, 1972.

[12] M. Nagata, Existence theorems for nonprojective complete algebraic varieties, Ill. J. Math., 2 (1958), p. 490.

[13] M. Nagata, On the 14th problem of Hilbert, Am. J. Math., 81 (1959), p. 766.

[14] M. Nagata, Invariants of a group in an affine ring, J. Math. Kyoto Univ. 3 (1964), p. 369.

[15] D. Rees, On a problem of Zariski, Ill. J. Math., 2 (1958), p. 145.

[16] C.S. Seshadri, On a theorem of Weitzenböck in invariant theory, J. Math. Kyoto Univ., 1 (1962), p. 403.

[17] C.S. Seshadri, Mumford's conjecture for GL(2) and applications, in Algebraic Geometry, Tata Inst. Studies in Math., Oxford Univ. Press, 1969.

[18] C.S. Seshadri, Quotient spaces modulo reductive algebraic groups, Annals of Math., 95 (1972), p. 511.

[19] T. Shioda, On the graded ring of invariants of binary octavics, Am. J. Math., 89 (1967), p. 1022.

[20] A. Weil, Variétés Abéliennes et Combes Algébriques, Hermann, 1948.

[21] R. Weitzenbock, Über die Invarianten von Linearen Gruppen, Acta Math., 58 (1932), p. 230.

[22] H. Weyl, The Classical Groups, Princeton Univ. Press, 2nd Ed. 1946.

[23] O. Zariski, Interprétations algébrico-géometriques du 14ième problème de Hilbert, Bull. Sci. Math., 78 (1954), p. 155; collected papers vol. II.

[24] O. Zariski, The theorem of Riemann-Roch for high multiples of an effective divisor on an algebraic surface, Annals of Math., 76 (1962), p. 560; collected papers vol. II.

3 [25] W.J. Haboush, Reductive groups are semi-reductive, preprint.

MATH. SCAND. 39 (1976), 19–55

THE PROJECTIVITY OF THE MODULI SPACE OF STABLE CURVES
I: PRELIMINARIES ON "det" AND "Div"

FINN KNUDSEN and DAVID MUMFORD

Introduction.

This paper is the first in a sequence of three. In the last paper Mumford will prove that the coarse moduli space of "stable" curves is a projective variety. The proof is a direct application of the very powerful Grothendieck relative Riemann-Roch Theorem.

The notion of a stable curve was introduced by Deligne and Mumford [1]. A stable curve is a reduced, connected curve with at most ordinary double points such that every non-singular rational component meets the other components in at least 3 points.

In this first paper we deal with some essential preliminary constructions which may also have other applications.

In the first paragraph we give the details of a construction whose existence was asserted by Grothendieck and described in the unpublished expose of Ferrand in SGA "Theorie des Intersections —". The construction is to assign to every perfect complex $\mathscr{F}^{\cdot}$ an invertible sheaf det $\mathscr{F}^{\cdot}$ in such a way that det becomes a functor from the category of perfect complexes and isomorphisms (in the derived categorical sense) to the category of invertible sheaves and isomorphisms. Roughly det $\mathscr{F}^{\cdot}$ is the alternating tensor product of the top exterior products of a locally free resolution of $\mathscr{F}^{\cdot}$. However in making this precise a certain very nasty problem of sign arises. The authors' first solution to these sign problems was described by Grothendieck in a letter as very alambicated* and he suggested to use the "Koszul rule of signs" which we follow in this paper.

The second paragraph deals with a generalization of Chow's construction assigning a "chow form" to every subvariety of $\mathbf{P}^n$. We functorialize this and analyse the invertible sheaves involved, following some ideas in an unpublished letter of Grothendieck to Mumford (1962) and in

*) This apparently means similar to an alchemical apparatus.

Received April 10, 1975.

[3, p. 109]. Finally we must mention that we have several overlaps with J. Fogarty "Truncated Hilbert functors" [4]. He analyses the relation between Div and Chow in the case $\mathscr{F}^{\cdot}$ is an 0-dimensional perfect complex, i.e. a coherent sheaf of finite Tor. dimension. In his notation Div and Chow correspond to ∇ and ω respectively.

Chapter I: det.

Let X be a scheme. We denote by $\mathscr{P}_X$ the category of graded invertible $\mathcal{O}_X$-modules. An object of $\mathscr{P}_X$ is a pair (L,α) where L is an invertible $\mathcal{O}_X$-module and α is a continouus function:

$$\alpha:\ X \to \mathbf{Z}\,.$$

A homomorphism $h:(L,\alpha) \to (M,\beta)$ is a homomorphism of $\mathcal{O}_X$-modules such that for each $x \in X$ we have:

$$\alpha(x) \neq \beta(x) \Rightarrow h_x = 0\,.$$

We denote by $\mathscr{P}is_X$ the subcategory of $\mathscr{P}_X$ whose morphisms are isomorphisms only.

The tensor product of two objects in $\mathscr{P}_X$ is given by:

$$(L,\alpha) \otimes (M,\beta) = (L \otimes M, \alpha+\beta)\,.$$

For each pair of objects $(L,\alpha),(M,\beta)$ in $\mathscr{P}_X$ we have an isomorphism:

$$\psi_{(L,\alpha),(M,\beta)}:\ (L,\alpha) \otimes (M,\beta) \xrightarrow{\ \sim\ } (M,\beta) \otimes (L,\alpha)$$

defined as follows: If $l \in L_x$ and $m \in M_x$ then

I $$\psi(l \otimes m) = (-1)^{\alpha(x)+\beta(x)} \cdot m \otimes l\,.$$

Clearly:

$$\psi_{(M,\beta),(L,\alpha)} \cdot \psi_{(L,\alpha),(M,\beta)} = 1_{(L,\alpha)\otimes(M,\beta)}\,.$$

We denote by 1 the object $(\mathcal{O}_X, 0)$. A right inverse of an object (L,α) in $\mathscr{P}_X$ will be an object (L',α') together with an isomorphism

$$\delta:\ (L,\alpha) \otimes (L',\alpha') \xrightarrow{\ \sim\ } 1\,.$$

Of course $\alpha' = -\alpha$.

A right inverse will be considered as a left inverse via:

$$(L',\alpha') \otimes (L,\alpha) \xrightarrow[\sim]{\ \psi\ } (L,\alpha) \otimes (L',\alpha') \xrightarrow[\sim]{\ \delta\ } 1\,.$$

We denote by $\mathscr{C}_X$ the category of finite locally free $\mathcal{O}_X$-modules, and by $\mathscr{C}is_X$ the subcategory whose morphisms are isomorphisms only.

If $F \in \mathrm{ob}(\mathscr{C}_X)$ we define:

$$\det{}^*(F) = (\Lambda^{\max} F, \operatorname{rank} F)$$

$$(\text{where}(\Lambda^{\max} F)_x = \Lambda^{\operatorname{rank} F_x} F_x)\,.$$

It is well known that det* is a functor from $\mathscr{C}is_X$ to $\mathscr{P}is_X$.

For every short-exact sequence of objects in $\mathscr{C}_X$

$$0 \to F' \xrightarrow{\alpha} F \xrightarrow{\beta} F'' \to 0$$

we have an isomorphism:

$$i^*(\alpha,\beta): \det^* F' \otimes \det^* F'' \xrightarrow{\sim} \det^* F$$

such that locally,

$$i^*(\alpha,\beta)((e_1 \wedge \ldots \wedge e_t) \otimes (\beta f_1 \wedge \ldots \wedge \beta f_s)) = \alpha e_1 \wedge \ldots \wedge \alpha e_t \wedge f_1 \wedge \ldots \wedge f_s$$

for $e_i \in \Gamma(U,F')$ and $f_j \in \Gamma(U,F)$.

The following proposition is well known:

Proposition 1. i) i* *is functorial,* i.e., *given a diagram:*

$$\begin{array}{ccccccccc} 0 & \to & F' & \xrightarrow{\alpha} & F & \xrightarrow{\beta} & F'' & \to & 0 \\ & & \wr\downarrow \lambda' & & \wr\downarrow \lambda & & \wr\downarrow \lambda'' & & \\ 0 & \to & G' & \xrightarrow{\gamma} & G & \xrightarrow{\delta} & G'' & \to & 0 \end{array}$$

where the rows are short-exact sequences of objects in $\mathscr{C}_X$, *and the columns are isomorphisms, the diagram:*

$$\begin{array}{ccc} \det^* F' \otimes \det^* F'' & \xrightarrow[\sim]{i^*(\alpha,\beta)} & \det^* F \\ \wr\downarrow \det^*\lambda' \otimes \det^*\lambda'' & & \wr\downarrow \det^*\lambda \\ \det^* G' \otimes \det^* G'' & \xrightarrow[i^*(\gamma,\delta)]{\sim} & \det^* G \end{array}$$

commutes.

ii) *Given a commutative diagram of objects in* $\mathscr{C}_X$

$$\begin{array}{ccccccccc} & & 0 & & 0 & & 0 & & \\ & & \downarrow & & \downarrow & & \downarrow & & \\ 0 & \to & F' & \xrightarrow{\alpha'} & G' & \xrightarrow{\beta'} & H' & \to & 0 \\ & & \downarrow \gamma' & & \downarrow \gamma & & \downarrow \gamma'' & & \\ 0 & \to & F & \xrightarrow{\alpha} & G & \xrightarrow{\beta} & H & \to & 0 \\ & & \downarrow \delta' & & \downarrow \delta & & \downarrow \delta'' & & \\ 0 & \to & F'' & \xrightarrow{\alpha''} & G'' & \xrightarrow{\beta''} & H'' & \to & 0 \\ & & \downarrow & & \downarrow & & \downarrow & & \\ & & 0 & & 0 & & 0 & & \end{array}$$

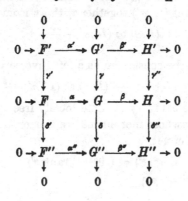

where each row and each column is a short-exact sequence, the diagram

$$\begin{array}{ccc}
\det^*F' \otimes \det^*F'' \otimes \det^*H' \otimes \det^*H'' & \xrightarrow[\sim]{i^*(\gamma',\delta')\otimes i^*(\gamma'',\delta'')} & \det^*F \otimes \det^*H \\
\Big\downarrow \wr\, {\scriptstyle i^*(\alpha',\beta')\otimes i^*(\alpha'',\beta'')\cdot(1\otimes\psi_{\det^*F'',\det^*H'}\otimes 1)} & & \Big\downarrow \wr\, {\scriptstyle i^*(\alpha,\beta)} \\
\det^*G' \otimes \det^*G'' & \xrightarrow[i^*(\gamma,\delta)]{\sim} & \det^*G
\end{array}$$

commutes.

iii) $\det^*$ *and* i^* *commute with base change.*

The isomorphism i^* is a special case of a more general canonical isomorphism: suppose E is a locally free $\mathcal{O}_X$-module and:

$$(0) = F^0E \subset F^1E \subset \ldots \subset F^rE = E$$

is a filtration such that $F^iE/F^{i-1}E$ are all locally free. Then there is a canonical isomorphism:

$$i^*(\{FE\})\colon \otimes_{i=1}^{r} \det^*(F^iE/F^{i-1}E) \xrightarrow{\approx} \det^*(E)\,.$$

Moreover these isomorphisms satisfy the following basic compatibility generalizing (ii) above: suppose $\{F^0E\}$ and $\{G^0E\}$ are 2 filtrations on E such that for all i,j

$$G^{i,j} = F^iE \cap G^jE/(F^{i-1}E \cap G^jE) + (F^iE \cap G^{j-1}E)$$

is locally free. For each fixed i, the $G^{i,j}$ are the graded objects associated to a filtration on $F^iE/F^{i-1}E$, and for each fixed j, they are the graded objects associated to a filtration on $G^jE/G^{j-1}E$. Thus the i's give us a diagram:

$$\begin{array}{ccc}
\otimes_{i,j}\det^*(G^{i,j}) & \xrightarrow{\sim} & \otimes_i \det^*(F^iE/F^{i-1}E) \\
\Big\downarrow \wr & & \Big\downarrow \wr \\
\otimes_j \det^*(G^jE/G^{j-1}E) & \xrightarrow{\sim} & \det^*E
\end{array}$$

This then commutes. We will not enter into the details here however, because the general isomorphism i can be defined inductively as a composition of the special isomorphisms i associated to short filtrations:

$$(0) = F^0E \subset F^1E \subset F^2E = E\,,$$

which is then just the i associated to the exact sequence:

$$0 \to F^1E \to E \to E/F^1E \to 0\,.$$

Moreover, the general compatibility property is just a formal consequence of the special one – (ii) above.

Next we consider the category $\mathscr{C}^{\cdot}_X$ of bounded complexes of objects in $\mathscr{C}_X$, morphisms being all maps of complexes. A map of complexes which induces an isomorphism in cohomology will be called a *quasi-isomorphism*. The subcategory of $\mathscr{C}^{\cdot}_X$ whose maps are quasi-isomorphisms will be called $\mathscr{C}^{\cdot}is_X$.

DEFINITION 1. A determinant functor from $\mathscr{C}^{\cdot}is$ to $\mathscr{P}is$ consists of the following data:

I) For each scheme X a functor f_X from $\mathscr{C}^{\cdot}is_X$ to $\mathscr{P}is_X$.

II) For each scheme X and for each short-exact sequence:

$$0 \to F^{\cdot\prime} \xrightarrow{\alpha} F^{\cdot} \xrightarrow{\beta} F^{\cdot\prime\prime} \to 0$$

in $\mathscr{C}^{\cdot}_X$ an isomorphism:

$$\mathrm{i}_X(\alpha,\beta): \mathrm{f}(F^{\cdot\prime}) \otimes \mathrm{f}(F^{\cdot\prime\prime}) \xrightarrow{\sim} \mathrm{f}(F^{\cdot})\,.$$

This data is to satisfy the following requirements:
i) Given a commutative diagram:

$$\begin{array}{ccccccccc} 0 & \to & F^{\cdot\prime} & \xrightarrow{\alpha} & F^{\cdot} & \xrightarrow{\beta} & F^{\cdot\prime\prime} & \to & 0 \\ & & \downarrow{\scriptstyle \lambda'} & & \downarrow{\scriptstyle \lambda} & & \downarrow{\scriptstyle \lambda''} & & \\ 0 & \to & G^{\cdot\prime} & \xrightarrow{\gamma} & G^{\cdot} & \xrightarrow{\delta} & G^{\cdot\prime\prime} & \to & 0 \end{array}$$

where the rows are short-exact sequences of objects in $\mathscr{C}^{\cdot}_X$ and λ', λ and λ'' are quasi-isomorphisms, the diagram:

$$\begin{array}{ccc} \mathrm{f}(F^{\cdot\prime}) \otimes \mathrm{f}(F^{\cdot\prime\prime}) & \xrightarrow[\sim]{\mathrm{i}_X(\alpha,\beta)} & \mathrm{f}(F^{\cdot}) \\ \wr\downarrow{\scriptstyle \mathrm{f}(\lambda')\otimes \mathrm{f}(\lambda'')} & & \wr\downarrow{\scriptstyle \mathrm{f}(\lambda)} \\ \mathrm{f}(G^{\cdot\prime}) \otimes \mathrm{f}(G^{\cdot\prime\prime}) & \xrightarrow[\sim]{\mathrm{i}_X(\gamma,\delta)} & \mathrm{f}(G^{\cdot}) \end{array}$$

commutes.
ii) Given a commutative diagram:

$$
\begin{array}{ccccccccc}
 & & 0 & & 0 & & 0 & & \\
 & & \downarrow & & \downarrow & & \downarrow & & \\
0 & \to & F^{\cdot\prime} & \xrightarrow{\alpha'} & G^{\cdot\prime} & \xrightarrow{\beta'} & H^{\cdot\prime} & \to & 0 \\
 & & \downarrow{\scriptstyle\gamma'} & & \downarrow{\scriptstyle\gamma} & & \downarrow{\scriptstyle\gamma''} & & \\
0 & \to & F^{\cdot} & \xrightarrow{\alpha} & G^{\cdot} & \xrightarrow{\beta} & H^{\cdot} & \to & 0 \\
 & & \downarrow{\scriptstyle\delta'} & & \downarrow{\scriptstyle\delta} & & \downarrow{\scriptstyle\delta''} & & \\
0 & \to & F^{\cdot\prime\prime} & \xrightarrow{\alpha''} & G^{\cdot\prime\prime} & \xrightarrow{\beta''} & H^{\cdot\prime\prime} & \to & 0 \\
 & & \downarrow & & \downarrow & & \downarrow & & \\
 & & 0 & & 0 & & 0 & &
\end{array}
$$

where each row and each column is a short-exact sequence, the diagram:

$$
\begin{array}{ccc}
f(F^{\cdot\prime}) \otimes f(F^{\cdot\prime\prime}) \otimes f(H^{\cdot\prime}) \otimes f(H^{\cdot\prime\prime}) & \xrightarrow[\sim]{i_X(\gamma',\delta')\, i_X(\gamma'',\delta'')} & f(F^{\cdot}) \otimes f(H^{\cdot}) \\
\wr\Big\downarrow {\scriptstyle i_X(\alpha',\beta') \otimes i_X(\alpha'',\beta'') \otimes 1 \otimes \psi_{f(F^{\cdot\prime\prime}),\, f(H^{\cdot\prime})} \otimes 1} & & \wr\Big\downarrow {\scriptstyle i_X(\alpha,\beta)} \\
f(G^{\cdot\prime}) \otimes f(G^{\cdot\prime\prime}) & \xrightarrow[i_X(\gamma,\delta)]{\sim} & f(G^{\cdot})
\end{array}
$$

commutes.

iii) f and i both commute with base change.

iv) f and i are normalized as follows:

a) $f(0^{\cdot}) = 1$

b) For the exact sequence:

$$0 \to F^{\cdot} \xrightarrow{1_{F^{\cdot}}} F^{\cdot} \xrightarrow{0} 0^{\cdot} \to 0$$

the map

$$f(F^{\cdot}) \otimes 1 \xrightarrow[i_X(1,0)]{\sim} f(F^{\cdot})$$

is the canonical one,

b′) For the exact sequence:

$$0 \to 0^{\cdot} \xrightarrow{0} F^{\cdot} \xrightarrow{1_{F^{\cdot}}} F^{\cdot} \to 0$$

the map

$$f(F^{\cdot}) \otimes 1 \xrightarrow[i_X(1,0)]{\sim} f(F^{\cdot})$$

is the canonical one.

v) We consider $\mathscr{C}is$ as a full subcategory of $\mathscr{C}^{\cdot}is$ by viewing objects of $\mathscr{C}is$ as complexes with only one nonvanishing term, this term being placed in degree zero. Then for such objects:

$$
\begin{aligned}
f(F) &= \det{}^{*} F \\
i_X(\alpha,\beta) &= i^{*}(\alpha,\beta)\,.
\end{aligned}
$$

The main theorem of this chapter is

THEOREM 1. *There is one and, up to canonical isomorphism, only one determinant functor* (f,i), *which we will write* (det,i).

Let X be a scheme, $H^\cdot$ an acyclic object in $\mathscr{C}^\cdot_X$. If (f,i) is a determinant functor, we have an isomorphism:

$$\mathrm{f}(0)\colon \mathrm{f}(H^\cdot) \to 1 .$$

If

$$0 \to H'^\cdot \xrightarrow{\alpha} H^\cdot \xrightarrow{\beta} H''^\cdot \to 0$$

is an exact sequence of acyclic objects it follows from Definition 1, i) and iv a) that the diagram

$$\begin{array}{ccc}
\mathrm{f}(H'^\cdot) \otimes \mathrm{f}(H''^\cdot) & \xrightarrow[\sim]{\mathrm{i}_X(\alpha,\beta)} & \mathrm{f}(H^\cdot) \\
\wr\downarrow & & \wr\downarrow \\
1 \otimes 1 & \xrightarrow[\sim]{\text{mult.}} & 1
\end{array}$$

commutes.

Let $\alpha\colon F^\cdot \to G^\cdot$ be an injective quasi-isomorphism such that the cokernel is again an object of $\mathscr{C}^\cdot_X$, i.e., we have a short-exact sequence:

$$0 \to F^\cdot \xrightarrow{\alpha} G^\cdot \xrightarrow{\beta} H^\cdot \to 0$$

such that $H^\cdot$ is acyclic.

From the diagram:

$$\begin{array}{ccccccc}
0 \to & F^\cdot & \longrightarrow & F^\cdot & \longrightarrow & 0^\cdot & \to 0 \\
 & \downarrow{\scriptstyle 1_{F^\cdot}} & & \downarrow{\scriptstyle \alpha} & & \downarrow{\scriptstyle 0} & \\
0 \to & F^\cdot & \xrightarrow{\alpha} & G^\cdot & \xrightarrow{\beta} & H^\cdot & \to 0
\end{array}$$

we get a commutative diagram:

$$\begin{array}{ccc}
\mathrm{f}(F^\cdot) \otimes 1 & \xrightarrow[\sim]{\text{mult.}} & \mathrm{f}(F^\cdot) \\
\wr\downarrow & & \wr\downarrow \\
\mathrm{f}(F^\cdot) \otimes \mathrm{f}(H^\cdot) & \xrightarrow[\sim]{\mathrm{i}_X(\alpha,\beta)} & \mathrm{f}(G^\cdot)
\end{array}$$

hence we see that $\mathrm{f}(\alpha)$ is determined by the maps $\mathrm{i}_X(\alpha,\beta)$ and $\mathrm{f}(0)\colon \mathrm{f}(H^\cdot) \to 1$.

Let $\lambda\colon F^\cdot \to G^\cdot$ be an arbitrary quasi-isomorphism. We denote by $Z_\lambda^\cdot$ the following complex:

$$Z_\lambda^i = F^i \oplus G^i \oplus F^{i+1}$$

$$d^i_{Z_\lambda} = \begin{pmatrix} d^i & 0 & -1 \\ 0 & d^i & \lambda^{i+1} \\ 0 & 0 & -d^{i+1} \end{pmatrix}$$

Consider the diagram:

$$F^\cdot \xrightarrow{\alpha} Z_\lambda^\cdot \underset{\beta}{\overset{\beta'}{\rightleftarrows}} G^\cdot$$

where

$$\alpha = \begin{pmatrix} 1 \\ 0 \\ 0 \end{pmatrix}, \quad \beta = \begin{pmatrix} 0 \\ 1 \\ 0 \end{pmatrix}, \quad \beta' = (\lambda, 1, 0)\,.$$

We leave to the reader to check that these are all quasi-isomorphisms and furthermore,

$$\beta' \circ \alpha = \lambda, \quad \beta' \circ \beta = 1_G\,.$$

Hence we have:

$$\begin{aligned} \mathrm{f}(\lambda) &= \mathrm{f}(\beta')\circ\mathrm{f}(\alpha) = \mathrm{f}(\beta')\circ\mathrm{f}(\beta)\circ\mathrm{f}(\beta)^{-1}\circ\mathrm{f}(\alpha) \\ &= \mathrm{f}(\beta'\circ\beta)\circ\mathrm{f}(\beta)^{-1}\circ\mathrm{f}(\alpha) = \mathrm{f}(\beta)^{-1}\circ\mathrm{f}(\alpha)\,. \end{aligned}$$

Hence, since both α and β are injective quasi-isomorphisms, the map $\mathrm{f}(\lambda)$ is determined by the maps i and $\mathrm{f}(0)$ from $\mathrm{f}(H^\cdot) \to 1$ for acyclic $H^\cdot$. We summarize this in the following:

Lemma 1. *Let* (f,i) *and* (g,j) *be two determinant functors from* $\mathscr{C}$*'is to* $\mathscr{P}$*is. Suppose we are given* θ *as follows:*

i) *For each scheme X and each object* $F^\cdot$ *in* $\mathscr{C}_X$ *we have an isomorphism:*

$$\theta_{X,F^\cdot}: \mathrm{f}(F^\cdot) \xrightarrow{\sim} \mathrm{g}(F^\cdot)\,.$$

ii) *For all acyclic* $H^\cdot$ *the diagram:*

$$\begin{array}{ccc} \mathrm{f}(H^\cdot) & \xrightarrow{\theta_{X,H^\cdot}} & \mathrm{g}(H^\cdot) \\ \wr\downarrow{\scriptstyle \mathrm{f}(0)} & & \wr\downarrow{\scriptstyle \mathrm{g}(0)} \\ 1 & = & 1 \end{array}$$

commutes.

iii) *For all short-exact sequences:*

$$0 \to F^\cdot \xrightarrow{\alpha} G^\cdot \xrightarrow{\beta} H^\cdot \to 0$$

with $H^\cdot$ *acyclic, the diagram:*

$$\begin{array}{ccc} \mathrm{f}(F^{\cdot}) \otimes \mathrm{f}(H^{\cdot}) & \xrightarrow[\sim]{1_X(\alpha,\beta)} & \mathrm{f}(G^{\cdot}) \\ \wr\big\downarrow \theta_{X,F^{\cdot}} \otimes \theta_{X,H^{\cdot}} & & \wr\big\downarrow \theta_{X,G^{\cdot}} \\ \mathrm{g}(F^{\cdot}) \otimes \mathrm{g}(H^{\cdot}) & \xrightarrow[\sim]{1_X(\alpha,\beta)} & \mathrm{g}(G^{\cdot}) \end{array}$$

commutes.

iv) θ *commutes with base change.*

Then for all quasi-isomorphisms $\lambda\colon F^{\cdot} \to G^{\cdot}$ *the diagram:*

$$\begin{array}{ccc} \mathrm{f}(F^{\cdot}) & \xrightarrow[\sim]{\mathrm{f}(\lambda)} & \mathrm{f}(G^{\cdot}) \\ \wr\big\downarrow \theta_{X,F^{\cdot}} & & \wr\big\downarrow \theta_{X,G^{\cdot}} \\ \mathrm{g}(F^{\cdot}) & \xrightarrow[\sim]{\mathrm{g}(\lambda)} & \mathrm{g}(G^{\cdot}) \end{array}$$

commutes.

As a side remark, notice that these methods prove:

Proposition 2. *Let* (f, i) *be a determinant functor from* $\mathscr{C}$*is to* $\mathscr{P}$*is, and let*

$$\lambda, \mu\colon F^{\cdot} \rightrightarrows G^{\cdot}$$

be two quasi-isomorphism such that locally on X, λ *is homotopic to* μ, *then*

$$f(\lambda) = \mathrm{f}(\mu)\,.$$

Proof. Two maps being equal is a local property, and since f commutes with base change we may assume that X is affine. However in the affine case locally homotopic maps are homotopic so let H be such a homotopy, i.e.,

$$\lambda - \mu = dH + Hd\,.$$

We leave to the reader to check that we have an isomorphism of complexes:

$$Z_{\lambda}^{\cdot} \xrightarrow{\sim} Z_{\mu}^{\cdot}$$

given by the matrix:

$$\begin{pmatrix} 1 & 0 & 0 \\ 0 & 1 & H \\ 0 & 0 & 1 \end{pmatrix}$$

such that the diagram

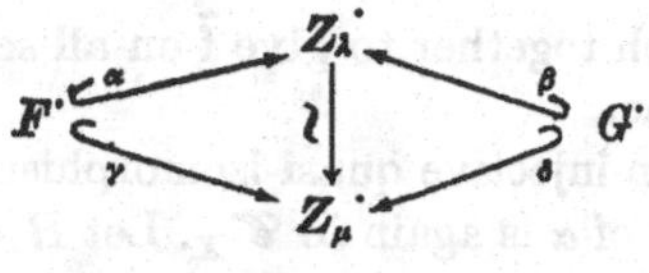

commutes. But we have already seen that

$$f(\lambda) = f(\beta)^{-1} \circ f(\alpha) \quad \text{and} \quad f(\mu) = f(\delta)^{-1} \circ f(\gamma)$$

hence the proposition.

Lemma 2. *Suppose we are given a pair* (f,i) *satisfying all the axioms of definition* 1 *except*:

I) *is replaced by*:

I′) *For each scheme X we have a map*

$$f_X : \mathrm{ob}(\mathscr{C}_X^{\cdot}) \to \mathrm{ob}(\mathscr{P}_X)$$

such that $f_X(0^{\cdot}) = 1$ *and for each acyclic complex* $H^{\cdot}$ *on X an isomorphism*:

$$f_X(0) : f_X(H^{\cdot}) \xrightarrow{\sim} 1 .$$

i) *is replaced by*

i′) *For each scheme X and for each short-exact sequence of acyclic objects*:

$$0 \to H''^{\cdot} \xrightarrow{\alpha} H^{\cdot} \xrightarrow{\beta} H'''^{\cdot} \to 0$$

the diagram

$$\begin{array}{ccc} f(H''^{\cdot}) \otimes f(H'''^{\cdot}) & \xrightarrow[\sim]{i_X(\alpha,\beta)} & f(H^{\cdot}) \\ \wr \downarrow {\scriptstyle f(0)\otimes f(0)} & & \wr \downarrow {\scriptstyle f(0)} \\ 1 \otimes 1 & \xrightarrow[\sim]{\text{mult.}} & 1 \end{array}$$

commutes.

(The rest is left unaltered.)

Then there exists up to caninical isomorphism a unique determinant functor $(\tilde{f}, i)$ *such that for all* $F^{\cdot}$ *we have*

$$\tilde{f}(F^{\cdot}) = f(F^{\cdot})$$

and for each quasi-isomorphism

$$H^{\cdot} \xrightarrow{0} 0 .$$

we have:

$$\tilde{f}(0) = f(0) .$$

Proof. Uniqueness follows immediately from Lemma 1. Suppose we have defined $\tilde{f}$ for all affine schemes, then since $\tilde{f}$ commutes with base

change, the maps patch together to give $\tilde{\mathrm{f}}$ on all schemes, hence we may assume that X is affine.

Let $F^\cdot \xrightarrow{\alpha} G^\cdot$ be an injective quasi-isomorphism. We will say that α is good if the cokernel of α is again in $\mathscr{C}^\cdot_X$. Let $H^\cdot =$ cokernel of α. Then we get a short-exact sequence of complexes

$$0 \to F^\cdot \xrightarrow{\alpha} G^\cdot \xrightarrow{\beta} H^\cdot \to 0$$

such that $H^\cdot$ is acyclic. We define $\mathrm{f}'(\alpha)$ via:

$$\begin{array}{ccccccc}
\mathrm{f}(F^\cdot) & \xleftarrow[\sim]{\text{mult.}} & \mathrm{f}(F) \otimes 1 & \xleftarrow[\sim]{1 \otimes \mathrm{f}(0)} & \mathrm{f}(F^\cdot) \otimes \mathrm{f}(H^\cdot) & \xleftarrow[\sim]{i_X(\alpha, \beta)} & \mathrm{f}(G^\cdot) \\
\big| & & & & & & \uparrow \\
\multicolumn{7}{c}{\xrightarrow[\sim]{\qquad\qquad\qquad \mathrm{f}'(\alpha) \qquad\qquad\qquad}}
\end{array}$$

Let $\alpha: E^\cdot \to F^\cdot$ and $\beta: F^\cdot \to G^\cdot$. be two good injective quasi-isomorphisms. We have a commutative diagram:

$$\begin{array}{ccccccccc}
 & & 0 & & 0 & & 0 & & \\
 & & \downarrow & & \downarrow & & \downarrow & & \\
0 & \to & E & \xrightarrow{\alpha} & F & \xrightarrow{\gamma} & H & \to & 0 \\
 & & \| & & \downarrow{\scriptstyle\beta} & & \downarrow{\scriptstyle\xi} & & \\
0 & \to & E & \xrightarrow{\beta\circ\alpha} & G & \xrightarrow{\delta} & K & \to & 0 \\
 & & \downarrow & & \downarrow{\scriptstyle\varepsilon} & & \downarrow{\scriptstyle\zeta} & & \\
0 & \to & 0 & \longrightarrow & L & = & L & \to & 0 \\
 & & \downarrow & & \downarrow & & \downarrow & & \\
 & & 0 & & 0 & & 0 & &
\end{array}$$

so by axiom ii), iv) and i′) we have:

(**) $$\mathrm{f}'(\beta)\circ\mathrm{f}'(\alpha) = \mathrm{f}'(\beta\circ\alpha)\,.$$

If $\lambda: F^\cdot \to G^\cdot$ is an arbitrary quasi-isomorphism, we have a diagram:

$$F^\cdot \xrightarrow{\alpha} Z_\lambda^\cdot \xrightarrow{\beta} G^\cdot$$

where

$$\alpha = \begin{pmatrix} 1 \\ 0 \\ 0 \end{pmatrix} \quad \text{and} \quad \beta = \begin{pmatrix} 0 \\ 1 \\ 0 \end{pmatrix}.$$

Clearly α and β are both good injective quasi-isomorphisms, and we define

$$\tilde{\mathrm{f}}(\lambda) = \mathrm{f}'(\beta)^{-1}\cdot\mathrm{f}'(\alpha)\,.$$

To see that $\tilde{\mathrm{f}}$ is functorial, let

$$\lambda : E^{\cdot} \to F^{\cdot} \quad \text{and} \quad \mu : F^{\cdot} \to G^{\cdot}$$

be quasi-isomorphisms: we define a complex $W^{\cdot}$ as follows:

$$W^i = E^i \oplus F^i \oplus G^i \oplus E^{i+1} \oplus F^{i+1}$$

$$d_W = \begin{bmatrix} d & 0 & 0 & -1 & 0 \\ 0 & d & 0 & \lambda & -1 \\ 0 & 0 & d & 0 & \mu \\ 0 & 0 & 0 & -d & 0 \\ 0 & 0 & 0 & 0 & -d \end{bmatrix}$$

We then have a commutative diagram:

$$\begin{array}{ccccc} & Z_\lambda^{\cdot} & \longleftarrow & F^{\cdot} & \\ \nearrow & & \searrow^{p} & & \searrow \\ E^{\cdot} & & W^{\cdot} & \xleftarrow{\;q\;} & Z_\mu^{\cdot} \\ \searrow & & \nearrow_{r} & & \nearrow \\ & Z^{\cdot}_{\mu\circ\lambda} & \longleftarrow & G^{\cdot} & \end{array}$$

where

$$p = \begin{bmatrix} 1 & 0 & 0 \\ 0 & 1 & 0 \\ 0 & 0 & 0 \\ 0 & 0 & 1 \\ 0 & 0 & 0 \end{bmatrix}, \quad q = \begin{bmatrix} 0 & 0 & 0 \\ 1 & 0 & 0 \\ 0 & 1 & 0 \\ 0 & 0 & 0 \\ 0 & 0 & 1 \end{bmatrix}, \quad r = \begin{bmatrix} 1 & 0 & 0 \\ 0 & 0 & 0 \\ 0 & 1 & 0 \\ 0 & 0 & 1 \\ 0 & 0 & \lambda \end{bmatrix}.$$

The fact that $\tilde{\mathrm{f}}(\mu \circ \lambda) = \tilde{\mathrm{f}}(\mu) \circ \tilde{\mathrm{f}}(\lambda)$ now follows from this diagram and the functoriality of f'. We leave to the reader to check axiom i): this is not hard. It is also easy to check that $\tilde{\mathrm{f}} = \mathrm{f}'$ where f' is defined, but this is not needed.

For each scheme X and each object L in $\mathscr{P}_X$ we fix a right inverse L^{-1} of L, i.e., an isomorphism

$$\delta_L : L \otimes L^{-1} \xrightarrow{\sim} 1 .$$

If $\alpha : L \xrightarrow{\sim} M$ is an isomorphism in $\mathscr{P}_X$ we denote by α^{-1} the unique isomorphism making the diagram:

$$\begin{array}{ccc} L \otimes L^{-1} & \xrightarrow{\sim} & 1 \\ \wr \big\downarrow {\scriptstyle \alpha \otimes \alpha^{-1}} & & \big\| \\ M \otimes M^{-1} & \xrightarrow{\sim} & 1 \end{array}$$

commutative.

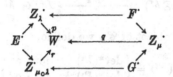

For every pair of objects L, M we denote by $\theta_{L,M}$ the unique isomorphism making the diagram

$$\begin{array}{ccc} (M\otimes L)\otimes(M\otimes L)^{-1} & \xrightarrow[\sim]{\delta} & 1 \\ \wr\downarrow 1\otimes\theta_{M,L} & & \wr\downarrow(\delta\cdot\delta)^{-1} \\ (M\otimes L)\otimes(M^{-1}\otimes L^{-1}) & \xrightarrow[\sim]{1\otimes\psi_{L,M}\otimes 1} & M\otimes M^{-1}\otimes L\otimes L^{-1} \end{array}$$

Then $^{-1}$ is a functor from $\mathscr{P}is$ to $\mathscr{P}is$ which commutes with base change, and for each pair L, M the diagram:

$$\begin{array}{ccc} (M\otimes L)^{-1} & \xrightarrow[\sim]{\theta_{M,L}} & M^{-1}\otimes L^{-1} \\ \wr\downarrow(\psi_{M,L})^{-1} & & \wr\downarrow\psi_{M^{-1},L^{-1}} \\ (L\otimes M)^{-1} & \xrightarrow[\sim]{\theta_{L,M}} & L^{-1}\otimes M^{-1} \end{array}$$

commutes.

If F^i is an indexed object of $\mathscr{C}_X$ we define:

$$\det(F^i) = \begin{Bmatrix} \det^*(F^i) & \text{for } i \text{ even} \\ \det^*(F^i)^{-1} & \text{for } i \text{ odd} \end{Bmatrix}$$

If

$$0 \to F^{i'} \xrightarrow{\alpha^i} F^i \xrightarrow{\beta^i} F^{i''} \to 0$$

is an indexed short-exact sequence of objects in $\mathscr{C}_X$, we define

$$\mathrm{i}(\alpha^i,\beta^i) = \begin{Bmatrix} \mathrm{i}^*(\alpha^i,\beta^i) & \text{for } i \text{ even} \\ \mathrm{i}^*(\alpha^i,\beta^i)^{-1} & \text{for } i \text{ odd} \end{Bmatrix}$$

If $F^\cdot$ is an object of $\mathscr{C}^\cdot_X$ we define

$$\det(F^\cdot) = \ldots\otimes\det(F^{i+1})\otimes\det(F^i)\otimes\det(F^{i-1})\otimes\ldots$$

Finally if

$$0 \to F^{\cdot\prime} \xrightarrow{\alpha} F^\cdot \xrightarrow{\beta} F^{\cdot\prime\prime} \to 0$$

is a short exact sequence of objects in $\mathscr{C}_X^\cdot$ we define

$$\mathrm{i}(\alpha,\beta) : \det(F^{\cdot\prime})\otimes\det(F^{\cdot\prime\prime}) \xrightarrow{\sim} \det(F^\cdot)$$

to be the composite:

$$\begin{aligned} \det(F^{\cdot\prime})\otimes\det(F^{\cdot\prime\prime}) &= \ldots\otimes\det(F^{i\prime})\otimes\det(F^{i-1\prime})\otimes\ldots \\ \otimes\det(F^{i\prime\prime})\otimes\det(F^{i-1\prime\prime})\otimes\ldots &\xrightarrow{\sim} \ldots\otimes\det(F^{i\prime})\otimes\det(F^{i\prime\prime}) \\ \otimes\det(F^{i-1\prime})\otimes\det(F^{i-1\prime\prime})\otimes\ldots &\xrightarrow[\sim]{\otimes_i \mathrm{i}(\alpha^i,\beta^i)} \ldots\otimes\det(F^i) \\ &\otimes\det(F^{i-1})\otimes\ldots = \det(F^\cdot)\,. \end{aligned}$$

the most amazing thing is that we can construct for each acyclic object $H^{\cdot}$ in $\mathscr{C}^{\cdot}_X$ an isomorphism:

$$\det_X(0) : \det_X(H^{\cdot}) \xrightarrow{\sim} 1$$

such that all the axioms of lemma 2 holds.

These axioms are all trivially verified execpt for I′) and i′). We will verify these simultaneously and we use induction with respect to the length of the complexes.

Step 1. Complexes of length 2.

Consider first an acyclic complex

$$H^{\cdot} = \ldots \to 0 \to H^i \xrightarrow{d} H^{i+1} \to 0 \to \ldots$$

with i an odd integer. Since d is an isomorphism we get an isomorphism:

$$\begin{aligned}\det(H^{\cdot}) = \det{}^*H^{i+1})\otimes\det{}^*(H^i)^{-1} &\xrightarrow[\sim]{1\otimes\det^*(d)^{-1}} \\ \det{}^*(H^{i+1})\otimes\det{}^*(H^{i+1})^{-1} &\xrightarrow{\sim} 1.\end{aligned}$$

We define this isomorphism to be det(0).

Given a short exact sequence of acyclic length 2-complexes:

$$\begin{array}{ccccccccc}
 & & \vdots & & \vdots & & \vdots & & \\
 & & \downarrow & & \downarrow & & \downarrow & & \\
0 & \to & 0 & \longrightarrow & 0 & \longrightarrow & 0 & \longrightarrow & 0 \\
 & & \downarrow & & \downarrow & & \downarrow & & \\
0 & \to & F^i & \xrightarrow{\alpha^i} & G^i & \xrightarrow{\beta^i} & H^i & \longrightarrow & 0 \\
 & & \downarrow{\scriptstyle d} & & \downarrow{\scriptstyle d} & & \downarrow{\scriptstyle d} & & \\
0 & \to & F^{i+1} & \longrightarrow & G^{i+1} & \xrightarrow{\beta^i} & H^{i+1} & \to & 0 \\
 & & \downarrow & & \downarrow & & \downarrow & & \\
0 & \to & 0 & \longrightarrow & 0 & \longrightarrow & 0 & \longrightarrow & 0 \\
 & & \downarrow & & \downarrow & & \downarrow & & \\
 & & \vdots & & \vdots & & \vdots & &
\end{array}$$

we get a diagram:

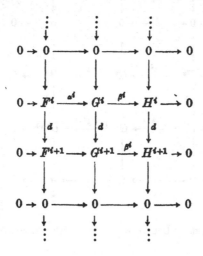

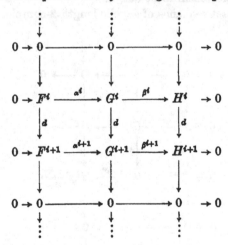

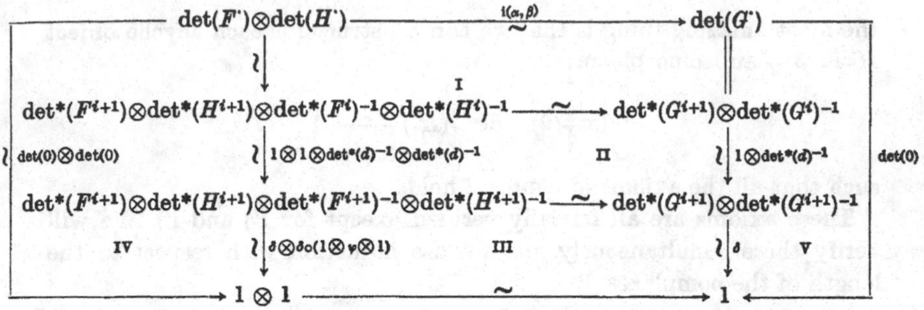

The square I is commutative by the definition of i, and the squares IV and V are commutative by the definition of det(0). The square III is commutative by the definition of i^{-1} and finally II is commutative by axioms iii) of definition 1. Hence the whole diagram commutes. If

$$H^{\cdot} = \ldots \to 0 \to H^i \xrightarrow{d} H^{i+1} \to 0 \to \ldots$$

is an acyclic complex with i even we define det(0) to be the composite

$$\det(H^{\cdot}) = \det{}^*(H^{i+1})^{-1} \otimes \det{}^*(H^i) \xrightarrow[\sim]{1 \otimes \det^*(d)} \det{}^*(H^{i+1})^{-1} \otimes \det{}^*(H^{i+1})$$
$$\xrightarrow[\sim]{\psi} \det{}^*(H^{i+1}) \otimes \det{}^*(H^{i+1})^{-1} \xrightarrow[\sim]{\delta} 1 \ldots .$$

Given a short-exact sequence of acyclic length 2-complexes

$$\begin{array}{ccccccc}
 & & \vdots & & \vdots & & \vdots & \\
 & & \downarrow & & \downarrow & & \downarrow & \\
0 \to & & 0 & \longrightarrow & 0 & \longrightarrow & 0 & \to 0 \\
 & & \downarrow & & \downarrow & & \downarrow & \\
0 \to & & F^i & \xrightarrow{\alpha^i} & G^i & \xrightarrow{\beta^i} & H^i & \to 0 \\
 & & \downarrow d & & \downarrow d & & \downarrow d & \\
0 \to & & F^{i+1} & \xrightarrow{\alpha^{i+1}} & G^{i+1} & \xrightarrow{\beta^{i+1}} & H^{i+1} & \to 0 \\
 & & \downarrow & & \downarrow & & \downarrow & \\
0 \to & & 0 & \longrightarrow & 0 & \longrightarrow & 0 & \to 0 \\
 & & \downarrow & & \downarrow & & \downarrow & \\
 & & \vdots & & \vdots & & \vdots &
\end{array}$$

with i even we get just as before a commutative diagram.

$$\begin{array}{ccc}
\det(F^{\cdot})\otimes\det(H^{\cdot}) & \xrightarrow{i(\alpha,\beta)} & \det(G^{\cdot}) \\
\Big\downarrow{\scriptstyle \det(0)\otimes\det(0)} & & \Big\downarrow{\scriptstyle \det(0)} \\
1\otimes 1 & \xrightarrow{\sim} & 1
\end{array}$$

Hence I′) and i′) holds for all acyclic complexes of length 2.

STEP 2. Suppose I′) and i′) hold for all acyclic complexes of length $\leqq n$, and let

$$H^{\cdot} = \ldots \to 0 \to H^i \to H^{i+1} \to H^{i+2} \to \ldots \to H^{i+n} \to 0 \to \ldots$$

be an acyclic complex of length $n+1$. We then get a short-exact sequence of complexes:

$$\begin{array}{ccccccccc}
 & & \vdots & & \vdots & & \vdots & & \\
 & & \downarrow & & \downarrow & & \downarrow & & \\
0 & \to & 0 & \to & 0 & \to & 0 & \to & 0 \\
 & & \downarrow & & \downarrow & & \downarrow & & \\
0 & \to & H^i & = & H^i & \to & 0 & \to & 0 \\
 & & \downarrow & & \downarrow & & \downarrow & & \\
0 & \to & H^i & \to & H^{i+1} & \to & H^{i+1}/H^i & \to & 0 \\
 & & \downarrow & & \downarrow & & \downarrow & & \\
0 & \to & 0 & \to & H^{i+2} & \to & H^{i+2} & \to & 0 \\
 & & \downarrow & & \downarrow & & \downarrow & & \\
 & & \vdots & & \vdots & & \vdots & & \\
 & & \downarrow & & \downarrow & & \downarrow & & \\
0 & \to & 0 & \to & H^{i+n} & \to & H^{i+n} & \to & 0 \\
 & & \downarrow & & \downarrow & & \downarrow & & \\
0 & \to & 0 & \to & 0 & \to & 0 & \to & 0 \\
 & & \downarrow & & \downarrow & & \downarrow & & \\
 & & \vdots & & \vdots & & \vdots & & \\
 & & \| & & \| & & \| & & \\
0 & \to & H^{\cdot}_{\mathrm{I}} & \xrightarrow{\alpha} & H^{\cdot} & \xrightarrow{\beta} & H^{\cdot}_{\mathrm{II}} & \to & 0
\end{array}$$

Since $H^{\cdot}_{\mathrm{I}}$ and $H^{\cdot}_{\mathrm{II}}$ are of length $\leqq n$ we define det(0) so as to make the diagram

$$\begin{array}{ccc}
\det(H^{\cdot}_{\mathrm{I}})\otimes\det(H^{\cdot}_{\mathrm{II}}) & \xrightarrow[\sim]{i(\alpha,\beta)} & \det(H^{\cdot}) \\
\Big|{\scriptstyle \det(0)\otimes\det(0)} & & \Big|{\scriptstyle \det(0)} \\
1\otimes 1 & \xrightarrow[\sim]{\text{mult.}} & 1
\end{array}$$

commutative. It is then easy to check that i′) follows from axiom ii) of definition 2. Now by Lemma 2, the pair (det,i) is a determinant

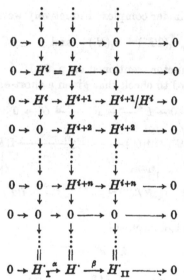

functor $\mathscr{C}is$ to $\mathscr{P}is$. Now say (f,j) is any determinant functor. If $E^\cdot$ is any complex we define an isomorphism

$$\theta_1 : \mathrm{f}(E^\cdot) \xrightarrow{\sim} \mathrm{f}(TE^\cdot)^{-1}$$

in such a way that the diagram:

$$\begin{array}{ccccc}
\mathrm{f}(E^\cdot)^{-1}\otimes\mathrm{f}(E^\cdot) & \xrightarrow[\sim]{\varphi} & \mathrm{f}(E^\cdot)\otimes\mathrm{f}(E^\cdot)^{-1} & \xrightarrow[\sim]{\delta} & 1 \\
\Big\downarrow{\scriptstyle 1\otimes\theta_1} & & & & \Big\| \\
\mathrm{f}(E^\cdot)^{-1}\otimes\mathrm{f}(TE^\cdot)^{-1} & \xrightarrow{\sim} & \mathrm{f}(C^\cdot_{1_{E^\cdot}})^{-1} & \xrightarrow{\mathrm{f}'(0)} & 1
\end{array}$$

commutes. Here T stands for the shift operator defined by

$$(TE^\cdot)^n = E^{n+1} \quad \text{and} \quad Td = -d\,.$$

$C^\cdot$ is the mapping cylinder complex. Inductively we define

$$\theta_n(E^\cdot) = \theta_{n-1}(TE^\cdot)^{-1}\cdot\theta_1(E^\cdot) \quad \text{and} \quad \theta_{-n}(E^\cdot) = \theta_n(T^{-n}E^\cdot)^{-1}$$

(note this $^{-1}$ is the functor mentioned on p. 31.)

It is straightforward to check that given a short-exact sequence

$$0 \to E^\cdot \xrightarrow{\alpha} F^\cdot \xrightarrow{\beta} G^\cdot \to 0$$

the diagram

$$\begin{array}{ccc}
\mathrm{f}(E^\cdot)\otimes\mathrm{f}(G^\cdot) & \xrightarrow[\sim]{\mathrm{i}(\alpha,\beta)} & \mathrm{f}(F^\cdot) \\
\wr\Big\downarrow{\scriptstyle \theta_1\otimes\theta_1} & & \wr\Big\downarrow{\scriptstyle \theta_1} \\
\mathrm{f}(TE^\cdot)^{-1}\otimes\mathrm{f}(TG^\cdot)^{-1} & \xrightarrow[\sim]{\mathrm{i}(T\alpha,T\beta)^{-1}} & \mathrm{f}(TF^\cdot)^{-1}
\end{array}$$

commutes.

And for any quasi-isomorphism

$$\lambda : E^\cdot \to F^\cdot$$

the diagram

$$\begin{array}{ccc}
\mathrm{f}(E^\cdot) & \xrightarrow[\sim]{\mathrm{f}(\lambda)} & \mathrm{f}(F^\cdot) \\
\wr\Big\downarrow{\scriptstyle \theta_1} & & \wr\Big\downarrow{\scriptstyle \theta_1} \\
\mathrm{f}(TE^\cdot)^{-1} & \xrightarrow[\sim]{\mathrm{f}(T\lambda)^{-1}} & \mathrm{f}(TF^\cdot)^{-1}
\end{array}$$

commutes.

We proceed to define an isomorphism of functors:

$$\eta : (\mathrm{f},\mathrm{j}) \xrightarrow{\sim} (\det,\mathrm{i})\,.$$

First consider a complex $E^\cdot$ concentrated in degree i.

We define η as follows:

$$\begin{array}{ccccc}
f(E^\cdot) & \xrightarrow[\sim]{\theta_i} & f(T^iE^\cdot)^{(-1)^i} & = & \det(E^\cdot) \\
\big| & & & & \uparrow \\
\big\lfloor & & \xrightarrow[\sim]{\eta} & & \big\rfloor
\end{array}$$

It is then obvious that restricted to all complexes which are concentrated in a single degree, η is an isomorphism of functors. If

$$E^\cdot = \ldots \to 0 \to E^i \to E^{i+1} \to \ldots \to E^{i+n} \to 0$$

is a complex of length $n+1$, we get a short-exact sequence of complexes

$$\begin{array}{ccccccccc}
 & & \vdots & & \vdots & & \vdots & & \\
 & & \downarrow & & \downarrow & & \downarrow & & \\
0 & \to & 0 & \longrightarrow & 0 & \longrightarrow & 0 & \longrightarrow & 0 \\
 & & \downarrow & & \downarrow & & \downarrow & & \\
0 & \to & 0 & \longrightarrow & E^i & \longrightarrow & E^i & \longrightarrow & 0 \\
 & & \downarrow & & \downarrow & & \downarrow & & \\
0 & \to & E^{i+1} & \to & E^{i+1} & \to & 0 & \longrightarrow & 0 \\
 & & \downarrow & & \downarrow & & \downarrow & & \\
0 & \to & E^{i+2} & \to & E^{i+2} & \to & 0 & \longrightarrow & 0 \\
 & & \downarrow & & \downarrow & & \downarrow & & \\
 & & \vdots & & \vdots & & \vdots & & \\
 & & \downarrow & & \downarrow & & \downarrow & & \\
0 & \to & E^{i+n} & \to & E^{i+n} & \to & 0 & \longrightarrow & 0 \\
 & & \downarrow & & \downarrow & & \downarrow & & \\
0 & \to & 0 & \longrightarrow & 0 & \longrightarrow & 0 & \longrightarrow & 0 \\
 & & \downarrow & & \downarrow & & \downarrow & & \\
 & & \vdots & & \vdots & & \vdots & & \\
 & & \| & & \| & & \| & & \\
0 & \to & E^\cdot_{\mathrm{I}} & \xrightarrow{\alpha} & E^\cdot & \xrightarrow{\beta} & E^\cdot_{\mathrm{II}} & \to & 0
\end{array}$$

Inductively we can define η such that the diagram

$$\begin{array}{ccc}
f(E^\cdot_{\mathrm{I}}) \otimes f(E^\cdot_{\mathrm{II}}) & \xrightarrow[\sim]{i(\alpha,\beta)} & f(E^\cdot) \\
\wr\Big\downarrow \eta\otimes\eta & & \wr\Big\downarrow \eta \\
\det(E^\cdot_{\mathrm{I}})\otimes\det(E^\cdot_{\mathrm{II}}) & \xrightarrow[\sim]{i(\alpha,\beta)} & \det(E^\cdot)
\end{array}$$

commutes.

Using axiom ii) it is easy to check that for all short-exact sequences of complexes

$$0 \to E^\cdot \xrightarrow{\alpha} F^\cdot \xrightarrow{\beta} G^\cdot \to 0$$

the diagram

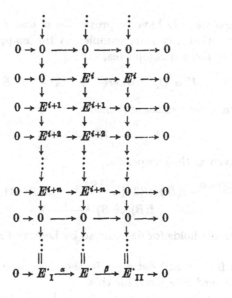

$$\begin{array}{ccc} \mathrm{f}(E^{\cdot})\otimes\mathrm{f}(G^{\cdot}) & \xrightarrow[\sim]{\mathfrak{i}(\alpha,\beta)} & \mathrm{f}(F^{\cdot}) \\ \wr\downarrow{\scriptstyle \eta\otimes\eta} & & \downarrow{\scriptstyle \eta} \\ \det(E^{\cdot})\otimes\det(G^{\cdot}) & \xrightarrow[\sim]{i(\alpha,\beta)} & \det(F^{\cdot}) \end{array}$$

commutes.

Finally we want to show that for each acyclic complex $H^{\cdot}$ the diagram

$$\begin{array}{ccc} \mathrm{f}(H^{\cdot}) & \xrightarrow[\sim]{\eta} & \det(H^{\cdot}) \\ \wr\Big|{\scriptstyle \mathrm{f}(0)} & & {\scriptstyle \det(0)}\Big|\wr \\ & 1 & \end{array}$$

commutes.

By induction we only have to prove this in case of a length 2 acyclic complex. Note that any such complex is the mapping cylinder of an isomorphism of pointed complexes, say:

$$H^{\cdot} = C_{\lambda}^{\cdot} \quad \text{where} \quad \lambda : A^{\cdot} \xrightarrow{\sim} B^{\cdot}.$$

We have then a short-exact sequence

$$0 \to B^{\cdot} \xrightarrow{\alpha} H^{\cdot} \xrightarrow{\beta} TA^{\cdot} \to 0$$

and f(0) is given as the composite:

$$\mathrm{f}(H^{\cdot}) \xrightarrow[\sim]{\mathfrak{i}(\alpha,\beta)} \mathrm{f}(B^{\cdot})\otimes\mathrm{f}(TA^{\cdot}) \xrightarrow{1\otimes\theta_{-1}} \mathrm{f}(B)\otimes\mathrm{f}(A)^{-1} \xrightarrow{1\otimes\mathrm{f}(\lambda)^{-1}}$$
$$\mathrm{f}(B)\otimes\mathrm{f}(B)^{-1} \xrightarrow{\delta} 1$$

The same formula holds for det, and so by Lemma 1 η is an isomorphism of functors.

We can in fact extend det even further. We need some preliminaries concerning derived categories for this.

Let $\mathscr{A}$ be an abelian category; we denote by $\mathscr{A}_3$ the following category.

i) The objects of $\mathscr{A}_3$ are sequences of the form

$$E'' \xrightarrow{\alpha} E \xrightarrow{\beta} E'$$

such that $\beta\cdot\alpha = 0$.

ii) The morphisms in $\mathscr{A}_3$ are triples of maps in $\mathscr{A}$ making the resulting diagram commute.

Definition 2. The subcategory of $D(\mathscr{A}_3)$ whose objects are short-exact sequences of complexes will be denoted by $VT(\mathscr{A})$ and, we will call it the category of true triangles of $D(\mathscr{A})$.

Remark. Let

$$X = 0 \to E''' \xrightarrow{\alpha} E^{\cdot} \xrightarrow{\beta} E'' \to 0$$

be a true triangle. Taking the mapping cylinder of the first map we get an ordinary triangle

$$\ldots \to E''' \to E^{\cdot} \to C_\alpha^{\cdot} \to TE''' \to TE^{\cdot} \to TC_\alpha^{\cdot} \to \ldots.$$

If $1_{E'''}$ is the identity map on E''' we have a short-exact sequence

$$0 \to C_{1_{E'''}} \to C_\alpha^{\cdot} \xrightarrow{u} E'' \to 0\,.$$

But $C_{1_{E'''}}$ is acyclic so u is a quasi-isomorphism, and hence the composition

$$E'' \xrightarrow{u^{-1}} C_\alpha^{\cdot} \to TE'''$$

gives us a triangle which we call

$$\delta(X) = \to E''' \xrightarrow{\alpha} E^{\cdot} \xrightarrow{\beta} E'' \to TE''' \to \ldots.$$

In fact δ is a functor from true triangles to the category $\mathrm{TD}(\mathscr{A})$ of triangles in $\mathrm{D}(\mathscr{A})$. Note that the homomorphism

$$\delta : \mathrm{Hom}_{\mathrm{VT}(\mathscr{A})}(XY) \to \mathrm{Hom}_{\mathrm{TD}(\mathscr{A})}(\delta(X),\delta(Y))$$

is in general neither injective nor surjective.

Proposition 3. *Let $f: X \to Y$ be a morphism of schemes and let* $\mathrm{Mod}(X)$, $\mathrm{Mod}(Y)$ *be the category of $\mathcal{O}_X$- and $\mathcal{O}_Y$-modules. Then left and right derived functors*

$$\mathrm{L}f^* : \mathrm{VT}(\mathrm{Mod}(Y))^- \to \mathrm{VT}(\mathrm{Mod}(X))^-$$
$$\mathrm{R}f_* : \mathrm{VT}(\mathrm{Mod}(X))^+ \to \mathrm{VT}(\mathrm{Mod}(Y))^+$$

exists.

Proof. According to Hartshorne: Residues and Duality, Chapter I, Theorem 5.1, the proposition follows if each true triangle bounded below allows a quasi-isomorphism into a true triangle consisting of injective $\mathcal{O}_Y$-modules, respectively each true triangle bounded above is quasi-isomorphic to a true triangle consisting of flat $\mathcal{O}_X$-modules. The fact that such quasi-isomorphisms exist follows from the following:

i) A short-exact sequence of injective $\mathcal{O}_X$-modules is an injective object in the category $\mathrm{Mod}(X)_3$, and every object of $\mathrm{Mod}(X)_3$ with the first map injective admits an embedding into a short-exact sequence of injectives.

ii) Every object of $\mathrm{Mod}(X)_3$ with the last map surjective is the quotient of a short-exact sequence of flat $\mathcal{O}_X$-modules.

This proves the proposition.

Recall the definition of a perfect complex $\mathscr{F}^{\cdot}$ on a scheme X [2]. This means that $\mathscr{F}^{\cdot}$ is a complex of $\mathcal{O}_X$-modules (not necessary quasi coherent) such that locally on X there exists a bounded complex $\mathscr{G}^{\cdot}$ of finite free $\mathcal{O}_X$-modules and a quasi isomorphism:

$$\mathscr{G}^{\cdot} \to \mathscr{F}^{\cdot}|_U$$

We denote by Parf_X the full subcategory of D(Mod X) whose objects are perfect complexes. We leave the proof of the following result to the reader.

Proposition 4. a) *Let X be any affine scheme and $\mathscr{F}^{\cdot}$ a perfect complex on X. Then there exists a bounded complex of locally free, finitely generated $\mathcal{O}_X$-modules $\mathscr{G}^{\cdot}$ and a quasi-isomorphism:*

$$\mathscr{G}^{\cdot} \to \mathscr{F}^{\cdot}$$

(i.e., globally on X:)

Let $\alpha : \mathscr{F}^{\cdot} \to \mathscr{F}^{\cdot\prime}$ be a map in the category Parf_X, and suppose we are given quasi-isomorphisms:

$$\mathrm{p}: \mathscr{G}^{\cdot} \to \mathscr{F}^{\cdot} \quad \textit{and} \quad \mathrm{p}': \mathscr{G}^{\cdot\prime} \to \mathscr{F}^{\cdot\prime}$$

where $\mathscr{G}^{\cdot}$ and $\mathscr{F}^{\cdot\prime}$ are bounded complexes of locally free $\mathcal{O}_X$-modules, then there exists up to homotopy a unique map

$$\beta : \mathscr{G}^{\cdot} \to \mathscr{G}^{\cdot\prime}$$

such that $\mathrm{p}'\beta = \alpha\mathrm{p}$ in Parf_X.

b) *If*

$$0 \to \mathscr{F}^{\cdot\prime} \to \mathscr{F}^{\cdot} \to \mathscr{F}^{\cdot\prime\prime} \to 0$$

is a true triangle of perfect complexes there exists a true triangle of bounded complexes of finite locally free $\mathcal{O}_X$-modules.

$$0 \to \mathscr{G}^{\cdot\prime} \to \mathscr{G}^{\cdot} \to \mathscr{G}^{\cdot\prime\prime} \to 0$$

and an isomorphism in the category $\mathrm{VT}(\mathrm{Parf}_X)$

$$\begin{array}{ccccccccc} 0 & \to & \mathscr{G}^{\cdot\prime} & \to & \mathscr{G}^{\cdot} & \to & \mathscr{G}^{\cdot\prime\prime} & \to & 0 \\ & & \downarrow{\scriptstyle \mathrm{p}'} & & \downarrow{\scriptstyle \mathrm{p}} & & \downarrow{\scriptstyle \mathrm{p}''} & & \\ 0 & \to & \mathscr{F}^{\cdot\prime} & \to & \mathscr{F}^{\cdot} & \to & \mathscr{F}^{\cdot\prime\prime} & \to & 0 \end{array}$$

Moreover if

$$\begin{array}{ccccccccc} 0 & \to & \mathcal{F}^{\cdot\prime} & \to & \mathcal{F}^{\cdot} & \to & \mathcal{F}^{\cdot\prime\prime} & \to & 0 \\ & & \downarrow{\scriptstyle \alpha'} & & \downarrow{\scriptstyle \alpha} & & \downarrow{\scriptstyle \alpha''} & & \\ 0 & \to & \mathcal{H}^{\cdot\prime} & \to & \mathcal{H}^{\cdot} & \to & \mathcal{H}^{\cdot\prime\prime} & \to & 0 \end{array}$$

is any morphism in $\mathrm{VT}(\mathrm{Parf}_X)$ *and*

$$\begin{array}{ccccccccc} 0 & \to & \mathcal{K}^{\cdot\prime} & \to & \mathcal{K}^{\cdot} & \to & \mathcal{K}^{\cdot\prime\prime} & \to & 0 \\ & & \downarrow{\scriptstyle q'} & & \downarrow{\scriptstyle q} & & \downarrow{\scriptstyle q''} & & \\ 0 & \to & \mathcal{H}^{\cdot\prime} & \to & \mathcal{H}^{\cdot} & \to & \mathcal{H}^{\cdot\prime\prime} & \to & 0 \end{array}$$

is an isomorphism with $\mathcal{K}^{\cdot\prime}, \mathcal{K}^{\cdot}$ *and* $\mathcal{K}^{\cdot\prime\prime}$ *bounded complexes of locally free* $\mathcal{O}_X$*-modules. Then there exists up to homotopy a unique map:*

$$\begin{array}{ccccccccc} 0 & \to & \mathcal{G}^{\cdot\prime} & \to & \mathcal{G}^{\cdot} & \to & \mathcal{G}^{\cdot\prime\prime} & \to & 0 \\ & & \downarrow{\scriptstyle \beta'} & & \downarrow{\scriptstyle \beta} & & \downarrow{\scriptstyle \beta''} & & \\ 0 & \to & \mathcal{K}^{\cdot\prime} & \to & \mathcal{K}^{\cdot} & \to & \mathcal{K}^{\cdot\prime\prime} & \to & 0 \end{array}$$

such that $\alpha' p' = q'\beta', \alpha p = q\beta$ *and* $\alpha'' p'' = q''\beta''$ *in* Parf_X.

c) *Same for diagrams of the form*

$$\begin{array}{ccccccccc} & & 0 & & 0 & & 0 & & \\ & & \downarrow & & \downarrow & & \downarrow & & \\ 0 & \to & \mathcal{F}^{\cdot\prime} & \to & \mathcal{F}^{\cdot} & \to & \mathcal{F}^{\cdot\prime\prime} & \to & 0 \\ & & \downarrow & & \downarrow & & \downarrow & & \\ 0 & \to & \mathcal{G}^{\cdot\prime} & \to & \mathcal{G}^{\cdot} & \to & \mathcal{G}^{\cdot\prime\prime} & \to & 0 \\ & & \downarrow & & \downarrow & & \downarrow & & \\ 0 & \to & \mathcal{H}^{\cdot\prime} & \to & \mathcal{H}^{\cdot} & \to & \mathcal{H}^{\cdot\prime\prime} & \to & 0 \\ & & \downarrow & & \downarrow & & \downarrow & & \\ & & 0 & & 0 & & 0 & & \end{array}$$

DEFINITION 4. An extended determinant functor (f,i) from Parf-is to $\mathcal{P}is$ consists of the following data:

I) For every scheme X a functor

$$f_X : \text{Parf-is}_X \to \mathcal{P}is_X$$

such that $f_X(0) = 1$.

II) For every true triangle in Parf-is_X

$$0 \to F \xrightarrow{\alpha} G \xrightarrow{\beta} H \to 0$$

we have an isomorphism:

$$i_X(\alpha,\beta) : f_X(F) \otimes f_X(H) \xrightarrow{\sim} f_X(G)$$

such that for the particular true triangles

$$0 \to H = H \to 0 \to 0$$

and

$$0 \to 0 \to H = H \to 0$$

we have:

$$\mathrm{i}_X(1,0) = \mathrm{i}_X(0,1) = 1_{\mathrm{f}_X(H)}\,.$$

We require that:

i) Given an isomorphism of true triangles*:

$$\begin{array}{ccccccccc} 0 \to & F & \xrightarrow{\alpha} & G & \xrightarrow{\beta} & H & \to 0 \\ & \downarrow{\scriptstyle u} & & \downarrow{\scriptstyle v} & & \downarrow{\scriptstyle w} & \\ 0 \to & F' & \xrightarrow{\alpha'} & G' & \xrightarrow{\beta'} & H' & \to 0 \end{array}$$

the diagram

$$\begin{array}{ccc} \mathrm{f}_X(F)\otimes\mathrm{f}_X(H) & \xrightarrow[\sim]{\mathrm{i}_X(\alpha,\beta)} & \mathrm{f}_X(G) \\ \wr\downarrow{\scriptstyle \mathrm{f}_X(u)\otimes\mathrm{f}_X(w)} & & \wr\downarrow{\scriptstyle \mathrm{f}_X(v)} \\ \mathrm{f}_X(F')\otimes\mathrm{f}_X(H') & \xrightarrow[\mathrm{i}_X(\alpha',\beta')]{\sim} & \mathrm{f}_X(G') \end{array}$$

commutes.

ii) Given a true triangle of true triangles, i.e. a commutative diagram

$$\begin{array}{ccccccccc} & 0 & & 0 & & 0 & \\ & \downarrow & & \downarrow & & \downarrow & \\ 0 \to & F & \xrightarrow{\alpha} & G & \xrightarrow{\beta} & H & \to 0 \\ & \downarrow{\scriptstyle u} & & \downarrow{\scriptstyle u'} & & \downarrow{\scriptstyle u''} & \\ 0 \to & F' & \xrightarrow{\alpha'} & G' & \xrightarrow{\beta'} & H' & \to 0 \\ & \downarrow{\scriptstyle v} & & \downarrow{\scriptstyle v'} & & \downarrow{\scriptstyle v''} & \\ 0 \to & F'' & \xrightarrow{\alpha''} & G'' & \xrightarrow{\beta''} & H'' & \to 0 \\ & \downarrow & & \downarrow & & \downarrow & \\ & 0 & & 0 & & 0 & \end{array}$$

the diagram:

$$\begin{array}{ccc} \mathrm{f}_X(F)\otimes\mathrm{f}_X(H)\otimes\mathrm{f}_X(F'')\otimes\mathrm{f}_X(H'') & \xrightarrow{\mathrm{i}_X(\alpha,\beta)\otimes\mathrm{i}_X(\alpha'',\beta'')} & \mathrm{f}_X(G)\otimes\mathrm{f}_X(G'') \\ \wr\downarrow{\scriptstyle \mathrm{i}_X(u,v)\otimes\mathrm{i}_X(u'',v'')\otimes(1\otimes\psi\otimes 1)} & & \wr\downarrow{\scriptstyle \mathrm{i}_X(u',v')} \\ \mathrm{f}_X(F')\otimes\mathrm{f}_X(H') & \xrightarrow[\mathrm{i}_X(\alpha'\beta')]{\sim} & \mathrm{f}_X(G') \end{array}$$

* This means this diagram commutes as $\mathcal{O}_X$-modules and not just $v\cdot\alpha=\alpha'\cdot u$ in $\mathrm{D}(\mathrm{Mod}\,X)$: in fact, even assuming $v\cdot\alpha$ and $w\cdot\beta$ homotopic to $\alpha'\cdot u$ and $\beta'\cdot v$ respectively and all sheaves locally free this property will *not* hold for det!

commutes.

iii) f and i commute with base change. Written out this means: For every morphism of schemes

$$g: X \to Y$$

we have an isomorphism

$$\eta(g): \mathrm{f}_X \cdot \mathrm{L}g^* \xrightarrow{\sim} g^*\mathrm{f}_X$$

such that for every true triangle

$$0 \to F^\cdot \xrightarrow{u} G^\cdot \xrightarrow{v} H^\cdot \to 0$$

the diagram:

$$\begin{array}{ccc} \mathrm{f}_X(\mathrm{L}g^*F^\cdot)\otimes \mathrm{f}_X(\mathrm{L}g^*H^\cdot) & \xrightarrow[\sim]{\mathrm{i}_X(\mathrm{L}g^*(u,v))} & \mathrm{f}_X(\mathrm{L}g^*G^\cdot) \\ \wr\downarrow \eta\cdot\eta & & \wr\downarrow \eta \\ g^*\mathrm{f}_Y(F^\cdot)\otimes g^*\mathrm{f}_Y(H^\cdot) & \xrightarrow[\sim]{\mathrm{i}_Y(u,v)} & g^*\mathrm{f}_Y(G^\cdot) \end{array}$$

commutes. Moreover if

$$X \xrightarrow{g} Y \xrightarrow{h} Z$$

are two consecutive morphisms, the diagram:

$$\begin{array}{ccccc} \mathrm{f}_X(\mathrm{L}g^*\mathrm{L}h^*) & \xrightarrow[\sim]{\eta(g)} & g^*\mathrm{f}_Y\mathrm{L}h^* & \xrightarrow[\sim]{g^*\eta(h)} & g^*h^*\mathrm{f}_Z \\ \wr\downarrow \mathrm{f}_X(\theta) & & & & \wr\downarrow \\ \mathrm{f}_X(\mathrm{L}(g\cdot h)^*) & & \xrightarrow{\sim} & & (g\cdot h)^*\mathrm{f}_Z \end{array}$$

commutes where θ is the canonical isomorphism

$$\theta: \mathrm{L}g^*\cdot \mathrm{L}h^* \xrightarrow{\sim} \mathrm{L}(g\cdot h)^*,$$

iv) On finite complexes of locally free $\mathcal{O}_X$-modules,

$$\mathrm{f} = \det \quad \text{and} \quad \mathrm{i} = \mathrm{i}.$$

Then using Proposition 4, one proves easily:

THEOREM 2. There is one, and, up to canonical isomorphism, only one extended determinant functor (f,i), which we will write (det,i) again.

REMARK. If $\mathscr{F}$ is a perfect complex and you filter it with subcomplexes such that the successive quotients $\mathrm{gr}^n(\mathscr{F}^\cdot)$ are all perfect, then there is a canonical isomorphism:

$$\det(\mathscr{F}^\cdot) \xrightarrow{\approx} \otimes \det(\mathrm{gr}^n\mathscr{F}^\cdot).$$

This is constructed easily by induction on the number of steps in the filtration, using the isomorphisms $i(\alpha,\beta)$ at each stage and it has the compatibility property described after Proposition 1 above for ordinary det*. In particular:

a) if each $\mathscr{F}^n$ is itself perfect, i.e., has locally a finite free resolution, then

$$\det(\mathscr{F}^{\cdot}) \cong \otimes_n \det{}^*(\mathscr{F}^n)^{(-1)^n}$$

b) if the cohomology sheaves $H^n(\mathscr{F}^{\cdot})$ of the complex are perfect — we call these complexes the objects of the subcategory $\text{Parf}^0 \subset \text{Parf}$ — then

$$\det(\mathscr{F}^{\cdot}) \cong \otimes_n \det{}^*(H^n(\mathscr{F}^{\cdot}))^{(-1)^n}.$$

This has various easy consequences:

Corollary 1. *Let* $\mathscr{F}^{\cdot}$ *and* $\mathscr{G}^{\cdot}$ *be two objects of* $\text{Parf}^0{}_X$ *and suppose* α *and* β

$$\alpha,\beta : \mathscr{F}^{\cdot} \rightrightarrows \mathscr{G}^{\cdot}$$

are two quasi-isomorphisms such that $H^i(\alpha) = H^i(\beta)$ *for each* i. *Then* $\det(\alpha) = \det(\beta)$.

Corollary 2. *Let*

$$\to \mathscr{F}_1^{\cdot} \xrightarrow{u} \mathscr{F}_2^{\cdot} \xrightarrow{v} \mathscr{F}_3^{\cdot} \xrightarrow{w} T\mathscr{F}_1^{\cdot} \to$$

be an ordinary triangle in Parf_X *such that the* $\mathscr{F}_j$ *are in* $\text{Parf}^{\circ}{}_X$. *We then have an isomorphism*

$$\det(\mathscr{F}_1^{\cdot}) \otimes \det(\mathscr{F}_3^{\cdot}) \xrightarrow[\sim]{i_X(u,v,w)} \det(\mathscr{F}_2^{\cdot})$$

which is functorial with respect to such triangles.

Proof.

$$\det(\mathscr{F}_1^{\cdot}) \otimes \det(\mathscr{F}_3^{\cdot}) \cong [\otimes \det{}^*(H^n(\mathscr{F}_1^{\cdot}))^{(-1)^n}] \otimes [\otimes \det{}^*(H^n(\mathscr{F}_3^{\cdot}))^{(-1)^n}]$$

and

$$\det(\mathscr{F}_2^{\cdot}) \cong \otimes \det{}^*(H^n(\mathscr{F}_2^{\cdot}))^{(-1)^n}.$$

But the long exact cohomology sequence $H^{\cdot}(u,v,w)$ is an acyclic complex with perfect sheaves at each stage, so

$$\begin{aligned} 1_X &\xrightarrow{\sim} \det(H^{\cdot}(u,v,w)) \\ &\cong \otimes \det{}^*(n^{\text{th}} \text{ sheaf of } H^{\cdot}(u,v,w))^{(-1)^n} \\ &\cong [\otimes \det{}^*(H^n(\mathscr{F}_1^{\cdot}))^{(-1)^n}] \otimes [\otimes \det{}^*(H^n(\mathscr{F}_2^{\cdot}))^{(-1)^{n+1}}] \\ &\quad \otimes [\otimes \det{}^*(H^n(\mathscr{F}_3^{\cdot}))^{(-1)^n}]. \end{aligned}$$

We tried for some time to extend i to ordinary triangles, but in general this is not possible. It is true that for each ordinary triangle we can find an isomorphism but it is by no means functorial or unique (cf. footnote to Definition 4 above). We have seen that i extends when the complexes are good, we will now see that it also extends when the schemes are good (i.e., reduced).

PROPOSITION 6. *Let X be a reduced scheme $\mathscr{F}^\cdot$ and $\mathscr{G}^\cdot$ perfect complexes, α and β two quasi-isomorphism*

$$\alpha, \beta : \mathscr{F}^\cdot \rightrightarrows \mathscr{G}^\cdot$$

such that

a) *For each integer i there are finite filtrations*

$$F^\cdot(H^i(\mathscr{F}^\cdot)) \quad \text{and} \quad F^\cdot(H^i(\mathscr{G}^\cdot)) .$$

b) *For each generic point $x \in X$, the maps*

$$H^i(\alpha) \otimes 1_{k(x)} \quad \text{and} \quad H^i(\beta) \otimes 1_{k(x)}$$

are compatible with the induced filtrations on $H^i(\mathscr{F}^\cdot) \otimes k(x)$ and $H^i(\mathscr{G}^\cdot) \otimes k(x)$. (Note that $k(x) = \mathcal{O}_{X,x}$) and we have

$$\mathrm{gr}(H^i(\alpha) \otimes 1_{k(x)}) = \mathrm{gr}(H^i(\beta) \otimes 1_{k(x)})$$

for each i.
Then

$$\det(\alpha) = \det(\beta) .$$

PROOF. Since X is reduced and det commutes with base change, we may as well assume $X = \mathrm{Spec}(k)$ where k is a field. However in this case we have

$$\mathrm{Parf}_X = \mathrm{Parf}^0{}_X$$

and so the proposition follows from the last one.

PROPOSITION 7. *Let X be a reduced scheme, then for each triangle of perfect complexes:*

$$\mathscr{F}^\cdot \xrightarrow{u} \mathscr{G}^\cdot \xrightarrow{v} \mathscr{H}^\cdot \xrightarrow{w} T\mathscr{F}^\cdot \rightarrow$$

we have a unique isomorphism.

$$\mathrm{i}_X(u,v,w) : \det(\mathscr{F}^\cdot) \otimes \det(\mathscr{H}^\cdot) \xrightarrow{\sim} \det(\mathscr{G}^\cdot)$$

which is functorial with respect to isomorphisms of triangles.

PROOF. First we represent the mapping w by a diagram of real maps

$$T^{-1}(\mathscr{H}^\cdot) \xrightarrow{\ \mu\ } I^\cdot \longleftarrow \mathscr{F}^\cdot$$

where $I^\cdot$ is injective.

The mapping cylinder of μ gives us a true triangle:

$$\begin{array}{ccccccccccc}
\ldots T^{-1}(\mathscr{H}^\cdot) & \to & I^\cdot & \to & C_\mu^\cdot & \to & \mathscr{H}^\cdot & \to & TI^\cdot & \to & \ldots \\
\| & & \downarrow & & & & \| & & \downarrow & & \\
\ldots T^{-1}(\mathscr{H}^\cdot) & \to & \mathscr{F}^\cdot & \to & \mathscr{G}^\cdot & \to & \mathscr{H}^\cdot & \to & T\mathscr{F}^\cdot & \to & \ldots
\end{array}$$

By the second axiom for triangles there exists a map (necessarily an isomorphism) $\lambda\colon C^\cdot \to \mathscr{G}^\cdot$ making the diagram above into an isomorphism of triangles. By Proposition 6 the map $\det(\lambda)$ does not depend on the choice of λ. If we represent w by a different diagram say:

$$T^{-1}(\mathscr{H}^\cdot) \xrightarrow{\ \mu'\ } I'^\cdot \longleftarrow \mathscr{F}^\cdot$$

we get a homotopy commutative diagram

$$\begin{array}{ccc}
 & \xrightarrow{\ \mu\ } & I^\cdot \\
T^{-1}(\mathscr{H}^\cdot) & & \downarrow \alpha \\
 & \xrightarrow{\ \mu'\ } & I'^\cdot
\end{array}$$

If H is a homotopy we get a commutative diagram

$$\begin{array}{ccccccccc}
0 & \to & I^\cdot & \longrightarrow & C_\mu & \longrightarrow & \mathscr{H}^\cdot & \to & 0 \\
 & & \downarrow \alpha & & \downarrow \begin{pmatrix} \alpha & H \\ 0 & 1 \end{pmatrix} & & \| \, 1 & & \\
0 & \to & I'^\cdot & \longrightarrow & C_{\mu'} & \longrightarrow & \mathscr{H}^\cdot & \to & 0
\end{array}$$

i.e., a map of true triangles. It follows that if λ' is a map from $C_{\mu'}$ to $\mathscr{G}^\cdot$ making

$$\begin{array}{ccccc}
I'^\cdot & \longrightarrow & C_{\mu'} & \longrightarrow & \mathscr{H}^\cdot \\
\downarrow & & \downarrow \lambda' & & \| \\
\mathscr{F}^\cdot & \longrightarrow & \mathscr{G}^\cdot & \longrightarrow & \mathscr{H}^\cdot
\end{array}$$

into an isomorphism of triangles, then the diagram:

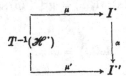

$$\begin{array}{ccccccc} \det(\mathscr{F}^\cdot)\otimes\det(\mathscr{H}^\cdot) & \longrightarrow & \det(I^\cdot)\otimes\det(\mathscr{H}^\cdot) & \longrightarrow & \det(C^\cdot_\mu) & \xrightarrow{\det(\lambda)} & \det(\mathscr{G}^\cdot) \\ \| & & \downarrow\wr & & \downarrow\wr & & \| \\ \det(\mathscr{F}^\cdot)\otimes\det(\mathscr{H}^\cdot) & \longrightarrow & \det(I'^\cdot)\otimes\det(\mathscr{H}^\cdot) & \longrightarrow & \det(C^\cdot_{\mu'}) & \xrightarrow{\det(\lambda')} & \det(\mathscr{G}^\cdot) \end{array}$$

commutes.

The composite map above we define to be $i(u,v,w)$. It is clearly functorial.

Let $p: X \to Y$ be a proper morphism of finite Tor-dimension with Y noetherian. Recall that if $\mathscr{F}^\cdot$ is a perfect complex on X then $R^\cdot p_*\mathscr{F}^\cdot$ is again perfect (cf. Proposition 4.8, SGA 6, expose 3 (Lecture Notes in Mathematics 225, p. 257, Springer-Verlag, Berlin-Heidelberg-New York). Hence to every perfect complex on X we can associate a graded invertible sheaf on Y

$$\det(R^\cdot p_*(\mathscr{F}^\cdot)) .$$

True triangles on X have injective resolutions so $R^\cdot p^*$ maps true triangles to true triangles. Hence for every true triangle

$$0 \to \mathscr{F}^\cdot \xrightarrow{\alpha} \mathscr{G}^\cdot \xrightarrow{\beta} \mathscr{H}^\cdot \to 0$$

on X we have an isomorphism:

$$i_Y(\alpha,\beta) : \det(R^\cdot p_*\mathscr{F}^\cdot)\otimes\det(Rp_*\mathscr{H}^\cdot) \xrightarrow{\sim} \det(Rp_*\mathscr{G}^\cdot)$$

which is functorial with respect to isomorphisms in $\mathrm{VT}(\mathrm{Parf}_X)$.

This operation commutes with base change too, i.e., given a morphism of noetherian schemes, $g: Y' \to Y$, let

$$\begin{aligned} X' &= X \times_Y Y' , \\ g' &= p_1 : X \times_Y Y' \to X , \\ p' &= p_2 : X \times_Y Y' \to Y' . \end{aligned}$$

Then there are canonical isomorphisms:

$$g^*\big(\det_Y(Rp_*(\mathscr{F}^\cdot))\big) \cong \det_{Y'}\big(Lg^*(Rp_*(\mathscr{F}^\cdot))\big) \cong \det_{Y'}(Rp'_*(Lg'^*\mathscr{F}^\cdot)) .$$

The last result of this chapter we state in the

Proposition 8. *Let $p: X \to Y$ be a proper morphism of noetherian schemes and suppose that Y is a regular scheme. We then have a functorial isomorphism:*

$$\det(R^\cdot p_*\mathscr{F}^\cdot) \xrightarrow{\sim} \otimes_{p,q}\det(R^q p_* H^p(\mathscr{F}^\cdot))^{(-1)^{p+q}} .$$

PROOF. The proof is easy by observing that on a noetherian regular scheme we have:

$$\mathrm{Parf}_X = \mathrm{Parf}^0{}_X ,$$

and using the spectral sequence

$$R^q p^*(H^p(\mathscr{F}^\cdot)) \Rightarrow R^{p+q} p_*(\mathscr{F}^\cdot) .$$

Chapter II: Div and Chow.

Let X be a notherian scheme, and

$$\lambda : \mathscr{F}^\cdot \to \mathscr{G}^\cdot$$

a map of perfect complexes in the derived categorical sense. We define the open set $U(\lambda)$ as follows:

$$U(\lambda) = \{x \in X \mid \text{there exists a neighbourhood } V \text{ of } x \text{ in } X \text{ such that } \lambda \text{ restricted to } V \text{ is an isomorphism in } D(\mathrm{Mod}(V))\} .$$

We define the support of λ to be the closed set:

$$\mathrm{Supp}(\lambda) = X - U(\lambda) .$$

Finally we say that λ is a *good* map if $\mathrm{Supp}(\lambda)$ contains no points of depth 0 or equivalently $U(\lambda)$ contains all points of depth 0.

Let again $\lambda : \mathscr{F}^\cdot \to \mathscr{G}^\cdot$ be a *good* map of perfect complexes, and let x be a point in X. By the very definition of a perfect complex, we can find a neighbourhood V containing x and two bounded complexes of coherent free $\mathcal{O}_X$-modules, say $\mathscr{E}_1{}^\cdot$ and $\mathscr{E}_2{}^\cdot$ plus, restricted to V, quasi-isomorphisms

$$\mathscr{E}_1{}^\cdot|_V \xrightarrow{\alpha} \mathscr{F}^\cdot|_V \quad \text{and} \quad \mathscr{E}_2{}^\cdot|_V \xrightarrow{\beta} \mathscr{G}^\cdot|_V .$$

By choosing basis for the various $\mathscr{E}_j{}^i$'s we get an isomorphism:

$$\mathcal{O}_X|_{V\cap U(\lambda)} \xrightarrow{\sim} \det(\mathscr{E}_1{}^\cdot)|_{V\cap U(\lambda)} \xrightarrow[\sim]{\det(\alpha)} \det(\mathscr{F}^\cdot)|_{V\cap U(\lambda)} \xrightarrow[\sim]{\det(\lambda)}$$
$$\det(\mathscr{G}^\cdot)|_{V\cap U(\lambda)} \xrightarrow[\sim]{\det(\beta)^{-1}} \det(\mathscr{E}_2{}^\cdot)|_{V\cap U(\lambda)} \xrightarrow{\sim} \mathcal{O}_X|_{V\cap U(\lambda)}$$

and this isomorphism determines a section $s \in \Gamma(V \cap U(\lambda), \mathcal{O}_X{}^*)$.

Since $V \cap U(\lambda)$ contains all points of depth 0 in V, $s = 0$ defines a Cartier divisor $\delta(s)$ in V. Clearly $\delta(s)$ does not depend on the choice of $\mathscr{E}_1{}^\cdot$ and $\mathscr{E}_2{}^\cdot$, so we have defined a global divisor via the formula:

$$\mathrm{Div}(\lambda)|_V = \delta(s) .$$

It follows immediately from the definition that the canonical map on $U(\lambda)$

$$\det(\lambda) : \det(\mathscr{F}^{\cdot})|_{U(\lambda)} \xrightarrow{\approx} \det(\mathscr{G}^{\cdot})|_{U(\lambda)}$$

extends to an isomorphism on the whole of X:

$$\det(\lambda) : \det(\mathscr{F}^{\cdot})(\mathrm{Div}(\lambda)) \xrightarrow{\approx} \det(\mathscr{G}^{\cdot}) .$$

In particular:

(i) $\mathrm{Supp}(\mathrm{Div}(\lambda)) \subset \mathrm{Supp}(\lambda)$

(ii) $\mathscr{O}(\mathrm{Div}(\lambda)) \approx \det(\mathscr{G}^{\cdot}) \otimes (\det(\mathscr{F}^{\cdot}))^{-1}$.

If $\mathscr{F}^{\cdot}$ is a perfect complex on X such that the zero map:

$$0^{\cdot} \to \mathscr{F}^{\cdot}$$

is a good map of complexes, we simply write

$$\mathrm{Div}(\mathscr{F}^{\cdot}) = \mathrm{Div}(0^{\cdot} \to \mathscr{F}^{\cdot})$$

and we have a canonical map:

$$\det(0) : \mathscr{O}(\mathrm{Div}(\mathscr{F}^{\cdot})) \xrightarrow{\approx} \det(\mathscr{F}^{\cdot}) .$$

This association of a divisor to every good map of perfect complexes satisfies some properties which we will summarize in the following:

THEOREM 3. (i) *Let $\lambda : \mathscr{F}^{\cdot} \to \mathscr{G}^{\cdot}$ and $\mu : \mathscr{G}^{\cdot} \to \mathscr{H}^{\cdot}$ be two good maps of perfect complexes, then the composition is good too and we have:*

$$\mathrm{Div}(\mu \cdot \lambda) = \mathrm{Div}(\mu) + \mathrm{Div}(\lambda) .$$

(ii) *Consider a strictly commutative diagram of short-exact sequences of perfect complexes:*

$$\begin{array}{ccccccc} 0 \to \mathscr{F}^{\cdot} & \longrightarrow & \mathscr{G}^{\cdot} & \longrightarrow & \mathscr{H}^{\cdot} & \to 0 \\ \downarrow{\scriptstyle\alpha} & & \downarrow{\scriptstyle\beta} & & \downarrow{\scriptstyle\gamma} & \\ 0 \to \mathscr{F}^{\cdot\prime} & \longrightarrow & \mathscr{G}^{\cdot\prime} & \longrightarrow & \mathscr{H}^{\cdot\prime} & \to 0 . \end{array}$$

Then of any two if the vertical maps are good, so is the third and we have:

$$\mathrm{Div}(\alpha) - \mathrm{Div}(\beta) + \mathrm{Div}(\gamma) = 0 .$$

(iii) *Let*

$$0 \to \mathscr{F}^{\cdot} \xrightarrow{\lambda} \mathscr{G}^{\cdot} \xrightarrow{\mu} \mathscr{H}^{\cdot} \to 0$$

be a short exact sequence of perfect complexes such that λ is good, then $0^{\cdot} \to \mathscr{H}^{\cdot}$ is good and we have:

$$\mathrm{Div}(\lambda) = \mathrm{Div}(\mathscr{H}^{\cdot}) .$$

(v) *Let $f: X \to Y$ be a morphism of noetherian schemes, $\lambda: \mathscr{F}^{\cdot} \to \mathscr{G}^{\cdot}$ a good map of perfect complexes on Y. Suppose that for each $x \in X$ of depth 0, $f(x) \in U(\lambda)$, then the map:*

$$Lf^*(\lambda) : Lf^*(\mathscr{F}^{\cdot}) \to Lf^{\cdot}(\mathscr{G}^{\cdot})$$

is good too, and we have:

$$\mathrm{Div}(Lf^*(\lambda)) = f^*(\mathrm{Div}(\lambda)) .$$

(vi) *Let X be a normal noetherian scheme and $\mathscr{F}^{\cdot}$ a good perfect complex on X. For every point x in X of depth 1 recall that $\mathcal{O}_X$ is a discrete rank 1 valuation ring, and since $\mathscr{F}^{\cdot}$ is good $H^i(\mathscr{F}^{\cdot})_X$ is a torsion $\mathcal{O}_X$-module of finite length, say:*

$$\mathrm{length}(H^i(\mathscr{F}^{\cdot})_x) = r_x^i(\mathscr{F}^{\cdot}) .$$

We define the number:

$$\mathrm{r}_x(\mathscr{F}^{\cdot}) = \textstyle\sum_{i=-\infty}^{\infty}(-1)^i \mathrm{r}_x^i(\mathscr{F}^{\cdot}) .$$

Since X is a normal noetherian scheme the group of Cartier-divisors injects into the group of Weil-divisors and we have

$$(*) \qquad \mathrm{Div}(\mathscr{F}^{\cdot}) = \sum_{\substack{x \in X \\ \mathrm{depth}(x)=1}} \mathrm{r}_x(\mathscr{F}^{\cdot}) \cdot \overline{\{x\}} .$$

Proof. Everything is obvious except for v. Since a divisor is determined by its values at points of depth 1 we may assume that $X = \mathrm{Spec}(\mathcal{O})$ where $\mathcal{O}$ is a regular local ring of dimension 1.

For every good perfect complex $\mathscr{F}^{\cdot}$ on X we define:

$$\mathrm{Div}(\mathscr{F}^{\cdot}) = \mathrm{r}_x(\mathscr{F}^{\cdot}) \cdot x$$

where x is the unique closed point of X. Clearly Div satisfies (i), (ii), and (iii). Since every coherent sheaf $\mathscr{F}$ on X with $\mathrm{Supp}(\mathscr{F}) \subset \{x\}$ can be considered as a perfect complex, it follows by induction that we can reduce the proof of the equality (*) to the case where $\mathscr{F}^{\cdot}$ is a complex of length 1, that is $\mathscr{F}^{\cdot} = \tilde{M}$ in degree 0 and 0 otherwise where M is a torsion $\mathcal{O}$-module. By the structure theorem for such modules we can find integers n_i, $1 \leq i \leq s$ such that

$$M \approx \textstyle\sum_{i=1}^{s} \mathcal{O}/\pi^{n_i}\mathcal{O} .$$

We then have a free resolution of M

$$0 \to \mathcal{O}_s \xrightarrow{d} \mathcal{O}_s \longrightarrow M \to 0$$

where d is given by the matrix

$$\begin{pmatrix} \pi^{n_1} & 0 & 0 & 0 \\ 0 & \pi^{n_2} & 0 & 0 \\ 0 & 0 & \pi^{n_3} & 0 \\ \vdots & \vdots & \ddots & \vdots \\ 0 & 0 & 0 \dots & \pi^{n_k} \end{pmatrix}.$$

It follows that the local equation of $\mathrm{Div}(\tilde{M})$ is $\det(d) = \pi^{\Sigma n_i}$. Since length $M = \sum n_i$, the equality (*) follows.

Let $f: X \to Y$ be a morphism of noetherian schemes, and $\mathscr{F}^{\cdot}$ a perfect complex on X. We put:

$$\mathrm{Supp}(\mathscr{F}^{\cdot}) = \bigcup_i \mathrm{Supp}(H^i(\mathscr{F}^{\cdot})) .$$

For any point $y \in Y$ consider the fibre product

$$\begin{array}{ccc} \mathrm{Supp}(\mathscr{F}^{\cdot})_Y & \longrightarrow & \mathrm{Supp}(\mathscr{F}^{\cdot}) \\ \downarrow & & \downarrow \\ \mathrm{Spec}(\,k(y)) & \longrightarrow & Y \end{array}$$

DEFINITION. Let f, X, Y and $\mathscr{F}^{\cdot}$ be as above. We will say that $\mathscr{F}^{\cdot}$ satisfies condition $Q_{(r)}$ if the following holds:

1) For each point $y \in Y$ of depth 0

$$\dim(\mathrm{Supp}(\mathscr{F}^{\cdot})_y) \leq r .$$

2) For each point $y \in Y$ of depth 1

$$\dim(\mathrm{Supp}(\mathscr{F}^{\cdot})_y) \leq r+1 .$$

PROPOSITION 9. *Let $f: X \to Y$ be a proper morphism of finite* Tor-*dimension. If $\mathscr{F}^{\cdot}$ is a perfect complex on X satisfying condition $Q_{(-1)}$ for the morphism f, then*

a) $\mathrm{Div}(\mathrm{R}f_*(\mathscr{F}^{\cdot}))$ *is defined,*
b) *for all line bundles $\mathscr{H}$ on X,*

$$\mathrm{Div}(\mathrm{R}f_*(\mathscr{F}^{\cdot})) = \mathrm{Div}(\mathrm{R}f_*(\mathscr{F}^{\cdot} \otimes \mathscr{H})) .$$

PROOF. a) is clear and to prove (b), we may make a base change and replace Y by $\mathrm{Spec}\, \mathcal{O}_{y,Y}$, where $y \in Y$ has depth 0 or 1. Then $\mathrm{Supp}(\mathscr{F}^{\cdot})$ is finite over Y, hence there is an open neighborhood U

$$\mathrm{Supp}(\mathscr{F}^{\cdot}) \subset U \subset X$$

and an isomorphism of $\mathscr{H}|_U$ with $\mathcal{O}_U$. Therefore there is a sheaf of ideals $\mathscr{I} \subset \mathcal{O}_X$ such that $\operatorname{Supp} \mathcal{O}_X/\mathscr{I} \subset X - U$ and a homomorphism φ as follows:

$$0 \to \mathscr{I} \to \mathscr{H} \to \mathscr{K} \to 0$$

$$\operatorname{Supp}(\mathscr{K}) \subset X - U .$$

Then $\mathscr{F}^{\cdot} \otimes^L \mathcal{O}_X/\mathscr{I}$ and $\mathscr{F}^{\cdot} \otimes^L \mathscr{K}$ are acyclic, hence $\mathscr{F}^{\cdot} \otimes \mathscr{H}$ is quasi-isomorphic first to $\mathscr{F} \otimes^L \mathscr{I}$, and second to $\mathscr{F}^{\cdot}$. This proves (b).

Let $f: X \to Y$ be a morphism of noetherian schemes, $\mathscr{F}^{\cdot}$ a perfect complex on X and consider the function $Y \to \mathbf{Z}$ given by

$$y \to \dim(\operatorname{Supp}(\mathscr{F}^{\cdot})_y) .$$

It will be convenient to compute this function in a slightly different manner. Consider the fibre product:

$$\begin{array}{ccc} X_y & \xrightarrow{\;i\;} & X \\ \downarrow & & \downarrow \\ \operatorname{Spec}(k(y)) & \longrightarrow & Y \end{array}$$

Lemma 1. *With the notations as above we have:*

$$\dim(\operatorname{Supp}(\mathscr{F}^{\cdot})_y) = \dim \operatorname{Supp}(\mathrm{L}i^*\mathscr{F}^{\cdot}) .$$

Proof. We may assume X and Y affine, so let $X = \operatorname{Spec}(S)$, $Y = \operatorname{Spec}(R)$, and let $Y = [p]$. Let k be the field $k(y)$

$$k(y) = R_p/p \cdot R_p$$

we then have:

$$X_y = \operatorname{Spec}(S \otimes_R k) .$$

Also we may assume that $\mathscr{F}^{\cdot} = \tilde{M}^{\cdot}$ where $M^{\cdot}$ is a bounded complex of finite free S-modules, hence $\mathrm{L}i^*\mathscr{F}^{\cdot}$ is represented by $M^{\cdot} \otimes_S (S \otimes_R k) = M^{\cdot} \otimes_R k$. But there is a spectral sequence:

$$E_2^{-p,q} = \operatorname{Tor}^S{}_p(H^q(M^{\cdot}), S \otimes_R k) \Rightarrow H^n(M^{\cdot} \otimes_R k) .$$

If $x \in \bigcup \operatorname{Supp}(H^i(M^{\cdot}) \otimes_R k)$, let i_0 be the maximum of the indexes i such that:

$$x \in \operatorname{Supp}(H^i(M^{\cdot}) \otimes_R k) .$$

Then

$$x \in \operatorname{Supp}\big(\operatorname{Tor}^S{}_p(H^i(M), S \otimes_R k)\big)$$

for $i > i_0$. Consequently we have

$$x \in \bigcup_{p+q=i_0} \mathrm{Supp}(E_r^{p,q})$$

for all r, and hence

$$x \in \mathrm{Supp}(H^{i_0}(M^{\cdot} \otimes_R k)).$$

Conversely, if $x \in \bigcup \mathrm{Supp}(H^i(M^{\cdot} \otimes_R k))$ we have $x \in \bigcup_{p,q} \mathrm{Supp}(E_r^{p,q})$ for all r, Since

$$\mathrm{Supp}\big(\mathrm{Tor}^S{}_r(H^q(M^{\cdot}), S \otimes_R k)\big) \subset \mathrm{Supp} H^q(M^{\cdot}) \otimes_R k$$

we are done.

Now we come to the main application of our techniques, namely to "*Chow points*". Let Y be a noetherian scheme, and E a locally free rank $n+1$ sheaf of $\mathcal{O}_Y$-modules. These define:

$P = \mathbf{P}(E)$, a $\mathbf{P}^n$-bundle over Y,
$\pi: P \to Y$ the projection,
$\mathcal{O}_p(1)$, the "tautological" line bundle (s.t. $\pi_*\mathcal{O}_p(1) = E$),
$\check{P} = \mathbf{P}(\check{E})$ the dual, $\mathcal{O}_{\check{P}}(1)$ its tautological line bundle,
$H \subset P \times_Y \check{P}$ the universal hyperplane, i.e.,

$$E \otimes \check{E} \simeq \mathcal{O}_Y \oplus [\text{trace zero subsp. of } E \otimes \check{E}] \quad \text{canonically,}$$

and if $1 \in \Gamma(\mathcal{O}_Y)$ corresponds to

$$\delta \in \Gamma(Y, E \otimes \check{E}) = \Gamma(P \times_Y \check{P}, p_1^*\mathcal{O}(1) \otimes p_2^*\mathcal{O}(1))$$

then $H = V(\delta)$.
$\mathcal{K}^{\cdot}_{(1)}$: the complex on $P \times_Y \check{P}$:

$$\begin{array}{ccccc} 0 \to & p_1^*\mathcal{O}_P(-1) \otimes p_2^*\mathcal{O}_{\check{P}}(-1) & \xrightarrow{\ \otimes\delta\ } & \mathcal{O}_{P\times\check{P}} & \to 0 \\ & \| & & \| & \\ & \mathcal{K}^{-1}_{(1)} & & \mathcal{K}^{0}_{(1)} & \end{array}$$

which resolves $\mathcal{O}_H$.
$P \times_Y (\check{P})^k =$ the fibre product over Y,
$\mathcal{K}^{\cdot}_{(k)}$: the complex $\otimes_{i=2}^{k+1} p_{1,i}^*(\mathcal{K}^{\cdot}_{(1)})$ on $P \times_Y \check{P}^k$.
This complex is a resolution of $\mathcal{O}_{H_k}$, where

$$H_k = \bigcap_{i=2}^{k+1} p_{1,i}^{-1}(H).$$

So much for the "universal" elements of our construction. Now say $\mathcal{F}^{\cdot}$ is a perfect complex on P and define:

$$\mathcal{F}^{\cdot}_{(k)}(n) = \mathrm{L}p_1^*(\mathcal{F}^{\cdot}(n)) \otimes^L \mathcal{K}^{\cdot}_{(k)}, \quad \text{on } P \times_Y \check{P}^k,$$

$$\mathscr{L}_{(k)}(n) = \det(\mathrm{R}p_{2*}\mathscr{F}^{\cdot}_{(k)}(n)), \quad \text{on } \check{P}^k ,$$
$$\mathscr{L}(n) = \det(\mathrm{R}\pi_*\mathscr{F}^{\cdot}(n)), \quad \text{on } Y .$$

Lemma 2. *If $\mathscr{F}^{\cdot}$ satisfies condition $Q_{(r)}$ for the morphism $\pi: P \to Y$, and $r \geq k-1$, then $\mathscr{F}^{\cdot}_{(k)}$ satisfies condition $Q_{(r-k)}$ for the morphism $p_2: P \times_Y \check{P}^k \to \check{P}^k$.*

Proof. By induction it is sufficient to prove the Lemma in case $k=1$ (with $\check{P}^{k-1}$ as the new Y and $\mathscr{F}^{\cdot}_{(k-1)}$ as the new $\mathscr{F}^{\cdot}$). If x is a point of $\hat{P}$, let $y=\pi(x) \in Y$ and let $k = k(y)$.
Identifying the fibre of $\check{P}$ over y with $\check{\mathsf{P}}_k{}^n$, we get the diagram:

$$\begin{array}{ccccc}
\mathrm{Spec}(k(x)) & \longrightarrow & \check{\mathsf{P}}_k{}^n & \longrightarrow & \check{P} \\
 & & \downarrow & & \downarrow{\scriptstyle p_2} \\
 & & \mathrm{Spec}(k) & \longrightarrow & Y
\end{array}$$

Since p_2 is flat, it follows from E.G.A., Chapitre IV, Proposition 6.3.1 that:

$$\mathrm{depth}(\mathscr{O}_{Y,y}) + \mathrm{depth}(\mathscr{O}_{\check{\mathsf{P}}_k{}^n,x}) = \mathrm{depth}(\mathscr{O}_{\check{P},x}) .$$

From this and the previous lemma it follows that we may assume $Y = \mathrm{Spec}(k), P = \mathsf{P}_k{}^n, k$ a field, in which case the Lemma is straight-forward.

Corollary-Definition. *If $\mathscr{F}^{\cdot}$ satisfies condition $Q_{(r)}$, then $\mathscr{F}^{\cdot}_{(r+1)}$ satisfies $Q_{(-1)}$, hence we can define the Chow divisor*

$$\mathrm{Chow}(\mathscr{F}^{\cdot}) = \mathrm{Div}(\mathrm{R}p_{2*}\mathscr{F}^{\cdot}_{(r+1)})$$

on $\check{P}^{r+1}$. Then $\mathrm{Chow}(\mathscr{F}^{\cdot}(n)) = \mathrm{Chow}(\mathscr{F}^{\cdot})$ and there is a canonical isomorphism:

$$\mathscr{O}_{\check{P}^{r+1}}(\mathrm{Chow}(\mathscr{F}^{\cdot})) \cong \mathscr{L}_{(r+1)}(n), \quad \textit{for every } n .$$

Next, we would like to compute $\mathscr{L}_{(k)}(n)$ in another way: since $\mathscr{K}^{\cdot}_{(k)}$ is locally free, each term $\mathscr{F}^{\cdot} \otimes \mathscr{K}_{(k)}{}^l$ is perfect, hence there is a canonical isomorphism

$$\mathscr{L}_{(k)}(n) = \det(\mathrm{R}p_{2*}\mathscr{F}^{\cdot}_{(k)}(n)) \cong \otimes_{l=0}^{k} \det(\mathrm{R}p_{2*}\mathrm{L}p_1{}^*\mathscr{F}^{\cdot}(n) \otimes^L \mathscr{K}^{-l}_{(k)})^{(-1)^l} .$$

On $\check{P}^k$, let $\mathscr{H}_i$ be the invertible sheaf $\mathscr{O}_{\check{P}}(1)$ pulled up from the ith factor. Then by definition:

$$\mathscr{K}^{-l}_{(k)} = p_1{}^*(\mathscr{O}_P(-l)) \otimes p_2{}^* \textstyle\sum_{1<i_1<\dots<i_l\leq k} \mathscr{H}_{i_1}^{-1} \otimes \dots \otimes \mathscr{H}_{i_l}^{-1}$$

hence if $\check{\pi}: \check{P}^k \to Y$ denotes the projection:

$$\begin{aligned}\mathscr{L}_{(k)}(n) &\cong \otimes_{l=0}^{k} \otimes_{1<i_1<\ldots<i_l\le k} \det\big(\mathrm{R}p_{2*}(\mathrm{L}p_1{}^*\mathscr{F}^{\cdot}(n-l)\\ &\quad \otimes p_2{}^*(\mathscr{H}_{i_1}^{-1}\otimes\ldots\otimes\mathscr{H}_{i_l}^{-1}))\big)^{(-1)^l}\\ &\cong \otimes_{l=0}^{k} \otimes_{1<i_1<\ldots<i_l\le k} \det\big(\mathrm{L}\check{\pi}^*(\mathrm{R}\pi_*\mathscr{F}^{\cdot}(n-l))\\ &\quad \otimes\mathscr{H}_{i_1}^{-1}\otimes\ldots\otimes\mathscr{H}_{i_l}^{-1}\big)^{(-1)^l}\end{aligned}$$

On the other hand, it is easy to check that for any perfect complex and invertible sheaf:

$$\det(\mathscr{G}^{\cdot}\otimes\mathscr{L}) \cong \det(\mathscr{G}^{\cdot})\otimes\mathscr{L}^{\mathrm{rk}(\mathscr{G}^{\cdot})} .$$

Note that

$$\mathrm{rk}(\mathscr{L}(n)) = \chi(\mathscr{F}^{\cdot}(n))$$

i.e. = the continuous function $Y \to \mathbf{Z}$ given by

$$y \to \Sigma(-1)^i \dim_{k(y)} H^i(\mathscr{F}^{\cdot}\otimes^L P_{k(y)}) .$$

We abbreviate this to $\chi(n)$. Therefore we have canonical isomorphisms:

$$\mathscr{L}_{(k)}(n) \cong \otimes_{l=0}^{k} \otimes_{1\le i_1<\ldots<i_l\le k} \check{\pi}^*\mathscr{L}(n-l)^{(-1)^l}\otimes(\mathscr{H}_{i_1}\otimes\ldots\otimes\mathscr{H}_{i_l})^{(-1)^{l+1}\chi(n-l)} .$$

Now defined by induction:

a) "difference" sheaves:

$$\begin{aligned}\Delta\mathscr{L}(n) &= \mathscr{L}(n)\otimes\mathscr{L}(n-1)^{-1}\\ \Delta^k\mathscr{L}(n) &= \Delta^{k-1}\mathscr{L}(n)\otimes\Delta^{k-1}\mathscr{L}(n-1)^{-1}\\ &\cong \Delta^{k-2}\mathscr{L}(n)\otimes\Delta^{k-2}\mathscr{L}(n-1)^{-2}\otimes\Delta^{k-2}\mathscr{L}(n-2)\\ &\quad \ldots\\ &\cong \otimes_{l=0}^{k}\mathscr{L}(n-l)^{(-1)^l\binom{k}{l}}\end{aligned}$$

b) difference functions:

$$\begin{aligned}\chi_1(n) &= \chi(n)-\chi(n-1)\\ \chi_k(n) &= \chi_{k-1}(n)-\chi_{k-1}(n-1)\\ &\quad \ldots\\ &= \Sigma_{l=0}^{k}(-1)^l\tbinom{k}{l}\chi(n-l)\end{aligned}$$

Then it follows easily that:

$$\mathscr{L}_{(k)}(n) \cong \check{\pi}^*(\Delta^k\mathscr{L}(n))\otimes(\mathscr{H}_1\otimes\ldots\otimes\mathscr{H}_k)^{\chi_{k-1}(n-1)} .$$

Combining this with above Corollary, if $\mathscr{F}^{\cdot}$ satisfies $Q_{(r)}$, then we find

$\chi_r(n)$ is independent of n
Up to canonical isomorphisms, $\pi^*(\Delta^{r+1}\mathscr{L}(n))$ is independent of n.

Since $\check{\pi}_*(\mathcal{O}_{\check{P}_k}) = \mathcal{O}_Y$, this implies that:

Up to canonical isomorphisms, $\Delta^{r+1}\mathcal{L}(n)$ independent of n. Going backwards, this implies that χ is a polynomial of degree at most r and that $\mathcal{L}(n)$ can be expanded as in the following final Theorem:

THEOREM 4. *Let Y be a noetherian scheme, E a locally free sheaf of rank $n+1$ on $Y, P = \mathbf{P}(E)$ and $\mathcal{F}^\cdot$ a perfect complex on P satisfying condition $Q_{(r)}$ for $\pi: P \to Y$. Then there are sheaves $\mathcal{M}_0, \ldots, \mathcal{M}_{r+1}$ on Y and canonical and functorial isomorphisms:*

$$\det(R\pi_*\mathcal{F}^\cdot(n)) \cong \otimes_{k=0}^{r+1} \mathcal{M}_k^{\binom{n}{k}}.$$

Moreover the leading term $\mathcal{M}_{r+1}$ is related to the Chow divisor by a canonical isomorphism:

$$\check{\pi}^*\mathcal{M}_{r+1} \otimes (\mathcal{H}_1 \otimes \ldots \otimes \mathcal{H}_{r+1})^d \cong \mathcal{O}_{\check{P}^{r+1}}(\mathrm{Chow}(\mathcal{F}^\cdot))$$

where $\check{\pi}: \check{P}^{r+1} \to Y$ is the projection,

$$\mathcal{H}_i = i^{\text{th}} \textit{ sheaf } \mathcal{O}_{\check{P}}(1) \textit{ on } \check{P}^{r+1},$$

$d\binom{n}{r}$ = leading term of the Hilbert polynomial $\chi(\mathcal{F}^\cdot(n))$.

REFERENCES

1. P. Deligne and D. Mumford, *The irreducibility of the space of curves of a given genus*, Inst. Hautes Études Sci. Publ. Math. 36 (1969), 75–109.
2. P. Berthelot, A. Grothendieck and L. Illusie, SGA 6, *Théorie des intersectiones et Théoreme de Riemann-Roch*, Lecture Notes in Mathematics 225, Springer-Verlag, Berlin, Heidelberg New York, 1969.
3. D. Mumford, *Geometric Invariant Theory*, Ergebnisse der Math. 34, Springer-Verlag, Berlin-Göttingen-Heidelberg, 1965.
4. J. Fogarty, *Truncated Hilbert functors*, J. Reine und Angew. Math. 234 (1969), 65–88.

UNIVERSITY OF PENNSYLVANIA, PHILADELPHIA, PENNSYLVANIA, U.S.A.
AND
HARVARD UNIVERSITY, CAMBRIDGE, MASSACHUSETTS, U.S.A.

Intl. Symp. on Algebraic Geometry
Kyoto, 1977, pp. 115–153

An algebro-geometric construction of commuting operators and of solutions to the Toda lattice equation, Korteweg deVries equation and related non-linear equations

By D. MUMFORD

(Received June 6, 1977)

A remarkable "dictionary" was discovered by I.M. Krichever [7], following suggestions in the work of Zaharov-Shabat [11], where they attempted to find a common formalism for the inverse scattering method of integrating certain non-linear partial differential equations. Subsequently, a characteristic p analog of this dictionary was discovered by V.G. Drinfeld, and a matrix analog was worked out by P. van Moerbeke and myself. Preceding this stage, a legion of authors have worked previously in the hyperelliptic-degree 2 operator case: much of this can be traced through the recent articles [3], [5] and [10]. It is not entirely inaccurate to say that initial insight behind this and related discoveries was the work of the first electronic computer! This lecture is a report on these 3 dictionaries with only a brief discussion of their applications. (see Added in proof)

This dictionary is a one-one correspondence between 2 types of superficially totally unrelated sorts of data: on one side of the dictionary, one has an algebraic curve, one or more points on it, and a vector bundle over it; on the other side, one has commutative subring of some big non-commutative ring of operators. This correspondence seems to me remarkable for many reasons. Firstly, it appears, as mentioned above, in at least 3 quite distinct cases. Secondly, it enables one, generally in terms of theta functions, to construct solutions both to equations formed from operators in these commutative subrings and to equations formed from flows in the space of all operators in the big non-commutative ring. Thirdly, it gives a new parametrization of the moduli space of the curves involved and/or their jacobians, vector bundle moduli spaces, etc.. We will discuss this in more detail below.

To make the idea precise, we state here the results in the simple case where the bundle is a line bundle, for all 3 types of operators:

(I) *Difference Operator case.* Let k be any field. Let $M^{\Delta}_{\infty}(k)$ be the ring of finite difference operators over k, i.e., maps $A: \prod_{-\infty}^{+\infty} k \to \prod_{-\infty}^{+\infty} k$ given by

$$A(x)_n = \sum_{m=n+N_1}^{n+N_2} A_{nm} x_m, \qquad \text{all } n \in \mathbf{Z}.$$

If $[N_1, N_2]$ is the smallest interval such that $A_{nm}=0$ if $m-n \notin [N_1, N_2]$, we say $[N_1, N_2]$ is the support of A. If moreover $A_{n,n+N_1}\neq 0$ and $A_{n,n+N_2}\neq 0$ for all $n \in \boldsymbol{Z}$, we say that A is *properly bordered*. Then there is a natural bijection between sets of data as follows:

Data A. a) *X a complete curve over k (i.e., X reduced and irreducible, one-dimensional, proper over k).*

b) *$P, Q \in X$, smooth k-rational points,*

c) *$\mathscr{F}$ torsion-free rank 1 sheaf on X such that*

$$\chi(\mathscr{F})=0$$
$$h^1(\mathscr{F}(nP-nQ))=0, \qquad \text{all } n \in \boldsymbol{Z}.$$

Data B. *Commutative subrings $R\subset M^{\Delta}_{\infty}(k)$, with $k\subset R$ and such that $\exists\, A, B \in R$*
1 *which are properly bordered, with supports $[a_1, a_2]$, $[b_1, b_2]$ such that $(a_1, b_1)=1$, $(a_2, b_2)=1$ and $a_2b_1<a_1b_2$; two subrings $R_1, R_2\subset M^{\Delta}_{\infty}(k)$ being identified, however, if for some invertible element:*

$$\Lambda=(\lambda_n\delta_{nm}), \qquad \lambda_n \in k^*,$$

we have

$$R_1=\Lambda\circ R_2\circ\Lambda^{-1}.$$

(II) *Differential Operator case* (*Krichever*). Let k be any field of characteristic zero. Let $k[[t]][d/dt]$ be the ring of formal linear ordinary differential operators over k. Then there is a natural bijection between sets of data as follows:

Data A. a) *X a complete curve over k,*

b) *$P \in X$, smooth k-rational point, and an isomorphism*

2 $$T_{x,p}\cong k,$$

c) *$\mathscr{F}$ torsion-free rank 1 sheaf on X such that*

$$h^0(\mathscr{F})=h^1(\mathscr{F})=0.$$

Data B. *Commutative subrings $R\subset k[[t]][d/dt]$, with $k\subset R$ and such that $\exists A, B \in R$, operators of form:*

$$A=\left(\frac{d}{dt}\right)^{\alpha}+a_1(t)\left(\frac{d}{dt}\right)^{\alpha-1}+\cdots+a_{\alpha}(t)$$
$$B=\left(\frac{d}{dt}\right)^{\beta}+b_1(t)\left(\frac{d}{dt}\right)^{\beta-1}+\cdots+b_{\beta}(t)$$

with $(\alpha, \beta)=1$; two subrings $R_1, R_2\subset k[[t]][d/dt]$ being identified, however, if for some $u(t) \in k[[t]]$, $u(0)\neq 0$, we have

$$R_1 = u(t) \circ R_2 \circ u(t)^{-1}.$$

(III) *Field Operator case (Drinfeld).* Let k be any field, $\sigma \in \operatorname{Aut}(k)$ an automorphism of infinite order and let k_0 be the fixed field. Let $k\{\sigma\}$ be the ring of maps $A : k \to k$ of the form

$$A(x) = \sum_{i=0}^{N} a_i \sigma^i(x).$$

Then there is a natural bijection between sets of data as follows:

Data A. a) *X_0 a complete curve over k_0,*
b) *$P_0 \in X_0$ a smooth k_0-rational point,*
c) *$\mathscr{F}$ torsion-free rank 1 sheaf on $X \overset{\text{def}}{=} X_0 \times_{k_0} k$ such that $h^0(\mathscr{F}) = h^1(\mathscr{F}) = 0$,*
d) *an isomorphism:*

$$r : (1_{X_0} \times \sigma)^* \mathscr{F} \xrightarrow{\approx} \mathscr{F}(P_0 - P_1)$$

for some smooth point $P_1 \in X$, $P_1 \neq P_0$. (Here $1_{X_0} \times \sigma : X \to X$ is the map given by $1_{X_0} : X_0 \to X_0$ and $\sigma : \operatorname{Spec} k \to \operatorname{Spec} k$.)

Data B. *Commutative subrings $R \subset k\{\sigma\}$, with $k_0 \subset R$ and such that $\exists\, A, B \in R$, operators of form*

$$A = a_n \sigma^n + \cdots + a_0, \qquad a_n \neq 0$$
$$B = b_m \sigma^m + \cdots + b_0, \qquad b_m \neq 0$$

with $(n, m) = 1$; two subrings $R_1, R_2 \subset k\{\sigma\}$ being identified, however, if for some $a \in k^*$, we have:

$$R_1 = a \cdot R_2 \cdot a^{-1}.$$

§ 1. Difference operator case

Let me first explain, in the rank 1 case, how one goes from Data A to Data B. This construction will give the essence of everything that follows and we can sketch the generalizations fairly rapidly. We consider the infinite sequence of sheaves:

$$\begin{array}{ccccccccccc} & \mathscr{F}(2Q-P) & & \mathscr{F}(Q) & & \mathscr{F}(P) & & \mathscr{F}(2P-Q) & & \mathscr{F}(3P-2Q) & \\ \cup & & \supset \;\; \cup & & \cup \;\; \cup & & \supset \;\; \cup & & \supset \;\; \cup & & \supset \\ \cdots & & \mathscr{F}(Q-P) & & \mathscr{F} & & \mathscr{F}(P-Q) & & \mathscr{F}(2P-2Q) & & \cdots \end{array}.$$

Those on the bottom row have no H^0 or H^1. Therefore those on the top row have a 1-dimensional H^0, and if

$$s_n \in H^0(X, \mathscr{F}((n+1)P - nQ))$$

is a non-zero section, s_n generates 1-dimensional vector spaces:

$$\mathcal{F}((n+1)P-nQ)/\mathcal{F}(nP-nQ)\cong\mathcal{F}((n+1)P-nQ)\otimes_{\mathcal{O}_X}K(P)$$

and

$$\mathcal{F}((n+1)P-nQ)/\mathcal{F}((n+1)P-(n+1)Q)\cong\mathcal{F}((n+1)P-nQ)\otimes_{\mathcal{O}_X}K(Q).$$

The first follows from the sequence:

$$\begin{aligned}0\to H^0(\mathcal{F}(nP-nQ))\to H^0(\mathcal{F}((n+1)P-nQ))\\ \to H^0(\mathcal{F}((n+1)P-nQ))\otimes K(P)\to H^1(\mathcal{F}(nP-nQ))\end{aligned}$$

and the second from the similar sequence with $K(Q)$. As a result, it follows that $\{s_n\}_{n\in Z}$ is a k-basis of the infinite-dimensional vector space $M=\Gamma(X-P-Q,\mathcal{F})$. In fact, starting with any $s\in\Gamma(X-P-Q,\mathcal{F})$, let k,l be least so that s extends to

$$s\in\Gamma(X,\mathcal{F}(kP+lQ)).$$

Then for a suitable $a\in k$, as_{k-1} and s will have the same pole at P, i.e.,

$$s-as_{k-1}\in\Gamma(X,\mathcal{F}((k-1)P+l(Q))).$$

Similarly, for suitable $b\in k$, bs_{-l} and s will have the same pole at Q, i.e.,

$$s-as_{k-1}-bs_{-l}\in\Gamma(X,\mathcal{F}((k-1)P+(l-1)Q)).$$

Continuing in this way, we eventually find a section of $\mathcal{F}$. Since $H^0(\mathcal{F})=(0)$, this is zero and s is written as a combination of the $\{s_n\}$. Now let

$$R=\Gamma(X-P-Q,\mathcal{O}_X).$$

Clearly M is an R-module, so for all $a\in R$, $n\in Z$, we can write:

$$a\cdot s_n=\sum_{m=n-N_1}^{n+N_2}A_{nm}s_m.$$

In fact, it is easy to see that N_1,N_2 may be taken to be the order of poles of a at P and Q:

$$(a)=N_1Q+N_2P-D,\qquad\text{some } D\geq 0 \text{ supported on } X-P-Q,$$

and that in this case $A_{n,n-N_1}\neq 0$ and $A_{n,n+N_2}\neq 0$, all n. (If a had a zero at either P or Q, N_1 or N_2 can be taken to be negative and the matrix A is upper or lower triangular.) Now consider the map:

$$\begin{aligned}R&\longrightarrow M_\infty^\Delta(k)\\ a&\longmapsto A.\end{aligned}$$

We may check that it is a homomorphism as follows:

$$(ab)s_n = b\Big(\sum_k A_{nk}s_k\Big)$$
$$= \sum_k A_{nk}(bs_k)$$
$$= \sum_k A_{nk}\Big(\sum B_{km}s_m\Big)$$
$$= \sum_m \Big(\sum_k A_{nk}B_{km}\Big)s_m.$$

Note that the only choice we made in defining this map was that of the $\{s_n\}$. If s_n is replaced by $\lambda_n s_n$, $\lambda_n \in k^*$, then the matrix A is replaced by $(A')_{nm} = \lambda_n \lambda_m^{-1} A_{nm}$, i.e., $A' = \Lambda A \Lambda^{-1}$. Finally, for all N_1, N_2 sufficiently large, there are functions $a \in R$ with poles at P and Q of order exactly N_2, N_1: so the image of R in $M_\infty^A(k)$ has the properties required in Data B. Incidentally, if X is smooth, it is an arduous task, but not deep, to give *explicit* formulae for the entries A_{nm} in terms of theta functions associated to X. This can be done following the methods of Fay [4].

The spectral properties of the rings R which we get in this way are very simple and help to understand how to reconstruct $(X, P, Q, \mathscr{F})$ from R. Since all the operators $A \in R$ commute, you can expect to find simultaneous eigenvectors $\vec{x}$ for all $A \in R$, at least over suitable extension fields $K \supset k$. We put no convergence restriction on $\vec{x}$, but seek vectors $\vec{x} \in \prod_{-\infty}^{+\infty} K$ for some field $K \supset k$, such that

$$A\vec{x} = \lambda_A \cdot \vec{x}, \qquad \text{all } A \in R.$$

In this case, the eigenvalues λ_A together give a homomorphism $\lambda: R \to K$, hence define a K-valued point of $X - P - Q$. The following holds:

3 **Proposition.** *Let* Data $\{X, P, Q, R\}$ *define* $R \subset M_\infty^A(k)$ *as above. Let* $K \supset k$ *be a field,* $\lambda: R \to K$ *be a* K*-valued point of* $X - P - Q$*, lying over* $x \in X - P - Q$ (x *defined by the prime ideal* Ker λ). *Then there is an isomorphism between*

a) *the eigenspace* $\{\vec{x} \in \prod_{-\infty}^{+\infty} K \mid A\vec{x} = \lambda_A \cdot \vec{x}, \text{ all } A \in R\}$

and

b) $\mathrm{Hom}_R(\mathscr{F}_x / m_x \mathscr{F}_x, K)$. (*Here* K *is an* R*-module via* λ).

Proof. In fact

$$\mathrm{Hom}_R(\mathscr{F}_x / m_x \mathscr{F}_x, K) \cong \mathrm{Hom}_R(\Gamma(X - P - Q, \mathscr{F}), K).$$

Using the basis $\{s_n\}$ of $\Gamma(X - P - Q, \mathscr{F})$, this comes out as:

$$\cong \left\{\begin{array}{l} \text{maps } s_n \mapsto x_n \in K \text{ such that for all } a \in R \\ \text{if } as_n = \sum A_{nm}s_m, \text{ then} \\ \qquad \sum A_{nm}x_m = \lambda(a) \cdot x_n \end{array}\right\}$$

$$= \text{Eigenspace for eigenvalue } \lambda.$$

This suggests how to go back from Data B to Data A. The idea is simply to take $X-P-Q$ to be Spec R and to complete it to X. For each point of $X-P-Q$, we consider the corresponding eigenspace, and "glue" these together into a bundle over $X-P-Q$. Then $\mathscr{F}$ is just the sheaf of functions on this bundle, linear on each fibre (e.g., generated by the functions $x \mapsto x_n$). If $k=C$, the 2 points at infinity on X have a spectral meaning in that their neighborhoods are given by the set of
4 all eigenfunctions growing exponentially as $n \to +\infty$ or $\to -\infty$: $|x_n| \geq C|x_{n-1}|$ or $|x_{n+1}| \geq C|x_n|$, for C getting larger and larger. Over any k, the 2 points at infinity correspond to the 2 valuations on R given by assigning to each matrix A the least integers N_1, N_2 such that A is supported on $[-N_1, N_2]$.

It seems difficult to make the above rigorous by a direct attack. We take a much more algebraic approach as follows: define 2 filtrations on R:

$$R_n = \{A \in R \mid A \text{ supported on } (-\infty, n]\}$$
$$R^m = \{A \in R \mid A \text{ supported on } [-m, +\infty)\}.$$

Using the 2 given elements $A, B \in R$ where

$$A \in R_{a_2} \cap R^{a_1}$$
$$B \in R_{b_2} \cap R^{b_1}$$

and $(a_2, b_2)=1$, $(a_1, b_1)=1$, one proves:

Lemma. i) *Every $C \in R$ is properly bordered,*
ii) *for all n,* $\dim R_{n+1}/R_n \leq 1$, *equality if $n \gg 0$,*
iii) *for all n,* $\dim R^{n+1}/R^n \leq 1$, *equality if $n \gg 0$.*

To prove this, simply note that if C has support $[-c_1, c_2]$ then

$$AC=CA \Rightarrow a_{n,n+a_2} c_{n+a_2,n+a_2+c_2} = c_{n,n+c_2} \cdot a_{n+c_2,n+c_2+a_2}$$
$$\Rightarrow (c_{n,n+c_2}=0 \text{ iff } c_{n+a_2,n+a_2+c_2}=0).$$

So

5 $$\left.\begin{matrix} AC=CA \\ BC=CA \end{matrix}\right\} \Rightarrow (c_{n,n+c_2}=0 \text{ for one } n \text{ iff } c_{n,n+c_2}=0 \text{ for all } n)$$

hence C is properly bordered. Now if C, C' have support $[*, c_2]$, then some combination $\alpha C+\beta C'$ has one zero along $c_{n,n+c_2}$, hence it has support $[*, c_2-1]$. A similar argument appllies to the bottom border. Finally, monomials $A^i B^j$ give us all supports $[*, n]$, $n \geq n_0$ and $[-m, *]$, $m \geq m_0$.

Corollary. *R is an integral domain, the subring of R generated by A and B is isomorphic to $k[X, Y]/(f)$, f irreducible, and R mod this ring is finite-dimensional over k. In particular, R is a finitely generated k-algebra.*

Proof. The fact that every $C \in R$ is properly bordered shows R is a domain. The lemma shows $\dim R_n \cap R^m \leq n+m+1$, and a simple count of the set of monomials A^iB^j in $R_n \cap R^n$ $n \gg 0$, shows their number grows like n^2: so A and B satisfy some identity. Finally, using the inequality $a_2b_1 > a_1b_2$, choose positive integers λ, μ such that

$$\frac{a_2}{a_1} > \frac{\lambda}{\mu} > \frac{b_2}{b_1}.$$

Using $(a_2, b_2) = (a_1, b_1) = 1$, one finds monomials A^iB^j with $i \gg 0$, $0 \leq j < a_2$ in

$$(R_{\lambda n+k+1} \cap R^{\mu n}) - (R_{\lambda n+k} \cap R^{\mu n})$$

for all k, $0 \leq k \leq \lambda - 1$, $n \gg 0$, and likewise, taking $0 \leq i < b_1$, $j \gg 0$, in

$$(R_{\lambda(n+1)} \cap R^{\mu n+k+1}) - (R_{\lambda(n+1)} \cap R^{\mu n+k})$$

for all k, $0 \leq k \leq \mu - 1$, $n \gg 0$. Thus these monomials plus the subspace $R_{\lambda n_0} \cap R^{\mu n_0}$ span R as a k-vector space. Q.E.D.

We can now define $X - P - Q$ to be Spec R. To define the whole of X, the most convenient way seems to be as Proj of a *graded* ring. As in the proof of the Corollary, fix $\lambda, \mu \geq 1$ such that

6 $$\frac{a_2 - 1}{a_1} \geq \frac{\lambda}{\mu} \geq \frac{b_2}{b_1 - 1},$$

and define

$$\mathcal{R}_n = R_{\lambda n} \cap R^{\mu n}$$

$$\mathcal{R} = \bigoplus_{n=0}^{\infty} \mathcal{R}_n.$$

In particular $\mathcal{R}$ contains

a) the element 1 in $\mathcal{R}_1 = R_\lambda \cap R^\mu$: we call this e,

7 b) $A^\lambda \in \mathcal{R}_{a_1}$,

c) $B^\mu \in \mathcal{R}_{b_1}$.

An argument like that above shows that $\mathcal{R}$ is generated, as a module over $k[e, A^\lambda, B^\mu]$, by a subspace $\mathcal{R}_0 \oplus \cdots \oplus \mathcal{R}_{n_0}$, $n_0 \gg 0$. Thus $\mathcal{R}$ is also a finitely generated domain over k. Define

$$X = \text{Proj}\,(\mathcal{R}).$$

X contains the affine piece $e \neq 0$, which is, by definition:

$$\operatorname{Spec}\left(\mathcal{R}\left[\frac{1}{e}\right]\right)_0$$

($_0$ signifies the degree 0 component), and

$$\mathcal{R}\left[\frac{1}{e}\right]_0 = \varinjlim_{\text{mult by } e} (R_{\lambda n} \cap R^{\mu n}) = R.$$

To see what we have put at infinity, note that X is covered by the 3 affine pieces $e \neq 0$, $A^\lambda \neq 0$ and $B^\mu \neq 0$. Since $A^\lambda B^\mu \in R_{\lambda(a_2+b_1-1)} \cap R^{\mu(a_2+b_1-1)}$, we get *in* $\mathcal{R}$:

$$A^\lambda \cdot B^\mu = e \cdot C, \qquad C \in \mathcal{R}_{a_2+b_1-1}.$$

Thus outside the affine $e \neq 0$, X has points $e = A^\lambda = 0$, $B^\mu \neq 0$ and points $e = B^\mu = 0$, $A^\lambda \neq 0$. I claim there is exactly one of each, and that it is a smooth k-rational point. To see this, check first that the direct systems

$$\begin{array}{ccccccc}
\mathcal{R}_0 & \longrightarrow & \mathcal{R}_{a_2} & \xrightarrow{A^\lambda} \cdots \longrightarrow & R_{ka_2} & \xrightarrow{A^\lambda} & \cdots \\
\| & & \| & & \| & & \\
R_0 \cap R^0 & & R_{\lambda a_2} \cap R^{\mu a_2} & & R_{k\lambda a_2} \cap R^{k\mu a_2} & & \\
\cap & & \cap & & \cap & & \\
R_0 & \longrightarrow & R_{\lambda a_2} & \xrightarrow{A^\lambda} \cdots \longrightarrow & R_{k\lambda a_2} & \xrightarrow{A^\lambda} & \cdots
\end{array}$$

have the same direct limit, so the affine ring of $A^\lambda \neq 0$ is:

$$\mathcal{R}\left[\frac{1}{A^\lambda}\right]_0 = \varinjlim_n (R_{\lambda n a_2}, \text{ mult. by } A^\lambda) = \begin{pmatrix}\text{ring of fractions} \\ C/A^k,\ C \in R_{ka_2}\end{pmatrix}.$$

In this ring, the homogeneous ideal (e) defines the ideal of elements C/A^k, $C \in R_{ka_2-\lambda}$. Choose positive integers σ, τ such that

$$\sigma a_2 + \tau b_2 = k a_2 - 1$$

and set $C = A^\sigma B^\tau$. It follows that for all n,

$$\mathcal{R}\left[\frac{1}{A^\lambda}\right]_0 \cong k \cdot 1 \oplus k\left(\frac{C}{A^k}\right) \oplus \cdots \oplus k\left(\frac{C}{A^k}\right)^{n-1} \oplus \left\{\begin{matrix}\text{ideal } C/A^k \\ C \in B_{ka_2-n}\end{matrix}\right\},$$

thus

$$\left\{\text{Completion of } \mathcal{R}\left[\frac{1}{A^\lambda}\right]_0 \text{ in the } e\text{-adic topology}\right\} \cong k\left[\left[\frac{C}{A^k}\right]\right]$$

which proves our claim for the points $e = B^\mu = 0$. The other case is similar. Let P be the point $e = B^\mu = 0$ and let Q be the point $e = A^\lambda = 0$. Note that e vanishes to order λ at P and μ at Q. Thus

$$\mathcal{O}_X(1) \cong \mathcal{O}_X((e)) = \mathcal{O}_X(\lambda P + \mu Q).$$

Incidentally, describing $X - P - Q = \operatorname{Spec} R$, then the valuations $f \mapsto \operatorname{ord}_P f$ and $f \mapsto \operatorname{ord}_Q f$, for $f \in R$, are easily seen to be just the upper and lower limits of support of f. Note that the ideal of P is

$$\bigoplus_{n=0}^{\infty} R_{\lambda n - 1} \cap R^{\mu n}.$$

To get the the sheaf $\mathscr{F}$ on X, let M be the vector space of column vectors and consider it as a module over R. Filter it like R:

$$M_n = \{(a_i) \mid a_i = 0, i > n\}$$
$$M^n = \{(a_i) \mid a_i = 0, i < -n\}.$$

Introduce the graded $\mathscr{R}$-module:

$$\mathfrak{M}_n = M_{\lambda n} \cap M^{\mu n - 1}$$
$$\mathfrak{M} = \bigoplus_{n=0}^{\infty} \mathfrak{M}_n.$$

One checks immediately that if $n \geq \max(a_2 + 1, b_1 + 1)$, then:

$$\mathfrak{M}_n = e \cdot \mathfrak{M}_{n-1} + A^{\lambda} \cdot \mathfrak{M}_{n-a_2} + B^{\mu} \cdot \mathfrak{M}_{n-b_1}.$$

It works like this:

$$[-\mu(n-1)+1, \cdots\cdots\cdots\cdots, \lambda(n-1)]$$
$$\downarrow \times e$$
$$[-\mu n + 1, \cdots, \mu(n-1)+1, \qquad , \lambda(n-1), \cdots, \lambda n]$$
$$\uparrow \times B^{\mu} \qquad\qquad \uparrow \times A^{\lambda}$$
$$[-\mu(n-b_1)+1, \cdots, 0] \qquad [0, \cdots, \lambda(n-a_2)]$$

Define $\mathscr{F}$ to be $\tilde{\mathfrak{M}}$. Since $\dim \mathfrak{M}_n = (\lambda + \mu)n$, $n \geq 0$, it follows that the Hilbert polynomial $\chi(\mathscr{F}(n))$ is $(\lambda + \mu)n$ for all n. In particular, $rk \mathscr{F} = 1$ and $\chi(\mathscr{F}) = 0$. Finally, we may define related sheaves by:

$$\mathfrak{M}_n^{(a,b)} = M_{\lambda n + a} \cap M^{\mu n + b - 1}$$
$$\mathfrak{M}^{(a,b)} = \bigoplus \mathfrak{M}_n^{(a,b)}$$
$$\mathscr{F}^{(a,b)} = \tilde{\mathfrak{M}}^{(a,b)}.$$

Since $\mathfrak{M}^{(a,b)} \subset \mathfrak{M}^{(a+1,b)}$, $\mathfrak{M}^{(a,b)} \subset \mathfrak{M}^{(a,b+1)}$, we get $\mathscr{F}^{(a,b)} \subset \mathscr{F}^{(a+1,b)}$, $\mathscr{F}^{(a,b)} \subset \mathscr{F}^{(a,b+1)}$, and it is easy to check that

$$\begin{array}{ccc} (\mathfrak{M}^{(a+1,b)}/\mathfrak{M}^{(a,b)})_n & \underset{n \text{ large}}{\cong} & M_{\lambda n+a+1}/M_{\lambda n+a} \\ & & \approx \uparrow \text{mult. by } e_{a+1} \\ (\mathscr{R}/\text{ideal of } P)_n & \cong & R_{\lambda n}/R_{\lambda n-1} \end{array}$$

(where $e_k \in M$ is the kth unit column vector). Therefore

$$\mathscr{F}^{(a+1,b)} \cong \mathscr{F}^{(a,b)}(P)$$

and by induction:

$$\mathscr{F}^{(a,b)} \cong \mathscr{F}(aP+bQ).$$

Moreover, if $a \geq -b$, then $e_{a+1} \in M_{a+1} \cap M^{b-1}$, hence

$$e_{a+1} \in \Gamma(X, \mathscr{F}^{(a+1,b)})$$

is a section that doesn't vanish at P. Then using the exact sequence

$$0 \to \mathscr{F}^{(a,b)} \to \mathscr{F}^{(a+1,b)} \to K(P) \to 0$$

and the existence of the section e_{a+1} of $\mathscr{F}^{(a+1,b)}$, we find

$$H^1(\mathscr{F}^{(a,b)}) \xrightarrow{\sim} H^1(\mathscr{F}^{(a+1,b)}).$$

But if $a+b$ is large enough, H^1 is zero. So H^1 is zero whenever $a+b \geq 0$, i.e.,

$$H^1(X, \mathscr{F}(aP-aQ)) = (0), \qquad \text{all } a \in \mathbf{Z}.$$

This completes the construction of Data A. We leave it to the reader to verify that our maps between Data A and Data B are inverse to each other.

The dictionary can be greatly extended. Here is one much more general correspondence:

Data A′. a) *X a one-dimensional scheme, proper over k, without embedded components. Let*

$$R = \bigoplus_{\substack{\eta \in X \\ \text{generic}}} \mathcal{O}_{\eta,X}$$

be its total ring of fractions,

b) *$S, T \subset X$ disjoint finite closed subsets meeting every component of X. Let*

$$\mathcal{O}_S = \{f \in R \mid f \in \mathcal{O}_{x,X}, \quad \text{all } x \in S\}$$
$$\mathcal{O}_T = \{f \in R \mid f \in \mathcal{O}_{x,X}, \quad \text{all } x \in T\},$$

c) *$\mathscr{F}$ a coherent sheaf on X such that $\chi(\mathscr{F})=0$ and $\mathscr{F}$ has no zero-dimensional associated points,*

d) *a flag of $\mathcal{O}_S$-modules*

$$\mathcal{F}_S = K_0 \supset K_1 \supset K_2 \supset \cdots \supset K_\alpha = f \cdot K_0,$$

$f \in \mathcal{O}_S$ a non-zero divisor zero at every $x \in S$, $\dim_k (K_l / K_{l+1}) = 1$, and a flag of $\mathcal{O}_T$-modules:

$$\mathcal{F}_T = L_0 \supset L_1 \supset L_2 \supset \cdots \supset L_\beta = g \cdot L_0$$

$g \in \mathcal{O}_T$ a non-zero divisor zero at every $x \in T$, $\dim_k (L_l / L_{l+1}) = 1$.

We put 2 requirements on this: first

$$\mathcal{O}_S = \{a \in R \mid aK_l \subset K_l, 0 \leq l \leq \alpha\}$$
$$\mathcal{O}_T = \{a \in R \mid aL_l \subset L_l, 0 \leq l \leq \beta\}.$$

Secondly, if we define K_l, L_l for all $l \in \boldsymbol{Z}$ by

$$K_{k+\alpha} = f \cdot K_k, \quad L_{l+\beta} = g \cdot L_l$$

and sheaves $\mathcal{F}^{(k,l)}$ by

$$\mathcal{F}^{(k,l)} = \begin{cases} K_k \text{ at } S \\ L_l \text{ at } T \\ \mathcal{F} \text{ elsewhere} \end{cases}$$

then

$$h^0(\mathcal{F}^{(k,-k)}) = 0.$$

Data B′. *Commutative subrings $R \subset M_\infty^\Delta(k)$ with $k \subset R$ such that $\exists\, A, B \in R$ where A is properly bordered above with support $[a_1, a_2]$ and B is properly bordered below with support $[b_1, b_2]$ and $a_2 b_1 < a_1 b_2$; two subrings R_1, R_2 being identified if*

$$R_1 = \Lambda \circ R_2 \circ \Lambda^{-1}$$

(Λ diagonal) as before.

To go from A′ to B′, as before we just choose

$$s_n \in H^0(X, \mathcal{F}^{(n+1,-n)})$$

and verify that $\{s_n\}$ is a k-basis of $H^0(X-S-T, \mathcal{F})$. Hence $R = H^0(X-S-T, \mathcal{O}_X)$ acts as a ring of matrices on the $\{s_n\}$ and this is data B'.

To go from B' to A', define $R_n, R^n, \mathcal{R}$ and $M_n, M^n, \mathfrak{M}^{(a,b)}$ as before. Instead of the lemma above, the argument only shows:

$$\dim R_{n+1}/R_n \leq a_2, \quad \text{all } n$$
$$\dim R^{n+1}/R^n \leq b_1, \quad \text{all } n.$$

One then proves as before that R is a finite module over $k[A, B]$ and that A, B satisfy some non-zero identity. Similarly $\mathscr{R}$ and $\mathfrak{M}^{(a,b)}$ are finite modules over $k[e, A^\lambda, B^\mu]$. Set $X=\operatorname{Proj}\mathscr{R}$, $\mathscr{F}^{(a,b)}=\tilde{\mathfrak{M}}^{(a,b)}$. As before, the open set $e\neq 0$ in X is just $\operatorname{Spec} R$. Moreover we have

$$A^\lambda\cdot B^\mu=e\cdot C, \qquad C\in\mathscr{R}_{a_2+b_1-1}$$

so the divisor $e=0$ breaks up into S defined by $e=B^\mu=0$ and T defined by $e=A^\lambda=0$.

The open set $A^\lambda\neq 0$ is Spec of the ring

$$R^{(A)}=\varinjlim_n (R_{\lambda n a_2}, \text{ mult by } A^\lambda)=\left\{\begin{matrix}\text{ring of fractions}\\ C/A^k,\ C\in R_{ka_2}\end{matrix}\right\},$$

and $\mathscr{F}^{(a,b)}$ on this open set comes from the module:

$$M_a^{(A)}=\varinjlim_n (M_{\lambda n a_2+a}, \text{ mult by } A^\lambda)=\left\{\begin{matrix}\text{module of fractions}\\ m/A^k,\ m\in M_{ka_2+a}\end{matrix}\right\}.$$

These are independent of b, have isomorphic localizations on the complement of S (the set S is defined in this open piece by $e=0$, which becomes $1/A\neq 0$ in this affine ring). Therefore these define a flag $\{K_l\}$ as required. Moreover

$$\frac{1}{A}\cdot\left\{\begin{matrix}\text{module of}\\ m/A^k,\ m\in M_{ka_2+a}\end{matrix}\right\}=\left\{\begin{matrix}\text{module of}\\ m/A^k,\ m\in M_{ka_2+a-a_2}\end{matrix}\right\}$$

so the α in Data A' is a_2 and the f is the function defined by $1/A$ in the above affine ring. $\{K_l\}$, β and g are defined similarly. The calculation of χ and vanishing of h^0's goes through as in the special case. Note that $X-S-T$ is the affine $e=0$, so each component X_i of X meets either S or T. But if, for instance, only S met X_i, then $\chi(\mathscr{F}^{(k,-k)}|_{X_i})$ would go to $+\infty$ as $k\to\infty$, so $h^0(\mathscr{F}^{(k,-k)})\to\infty$ which is not so. Thus both S and T meet all X_i. Finally, note that if $C\in R$, then $C\in R_l$ iff $C\cdot(M)_a\subset M_{a+l}$, all a; hence:

$$R^{(A)}=\left\{f\in R\left[\frac{1}{A}\right]\middle| f\cdot M_a^{(A)}\subset M_a^{(A)},\text{ all } a\right\};$$

from which the requirement

$$\mathcal{O}_S=\{f\in R\,|\,f\cdot K_l\subset K_l,\text{ all } l\}$$

follows directly.

There is one particularly nice case of the dictionary. This relates to arbitrary *periodic* difference operators A: we say A has *period* n if

$$A_{k+n,l+n}=A_{k,l}, \qquad \text{all } k, l\in \mathbf{Z}.$$

If S is the shift operator:

$$S_{k,l}=\delta_{k,l+1},$$

then to say A has period n is equivalent to saying

$$A\cdot S^n=S^n\cdot A.$$

Therefore, if A is any properly bordered periodic difference operator, with support
8 $[-a_1, a_2]$, such that $(n, a_1)=(n, a_2)=0$, the ring $R=k[A, S^n, S^{-n}]$ is an example of Data B. It is easy to see that the corresponding curves X are those such that

a) $nP\equiv nQ$

b) there is a function f on X with poles a_1Q+a_2P.

Another interesting case is when $k=C$ and we strengthen the hypothesis $h^1(\mathscr{F}(nP-nQ))=0$. Suppose:

a) *X is a smooth curve of genus g.*

b) *$P, Q\in X$. Let $\phi: X\to \mathrm{Pic}^1(X)$ be the canonical map and let $\alpha=\phi(P)-\phi(Q)\in \mathrm{Pic}^0(X)$.*

c) *$\mathscr{F}$ is an invertible sheaf on X of degree $g-1$ defining a point $[\mathscr{F}]\in \mathrm{Pic}^{g-1}(X)$.*

Let

$$\Sigma=\overline{\{[\mathscr{F}]+n\alpha\}_{n\in Z}}\subset \mathrm{Pic}^{g-1}(X)$$

where $^{-}$ denotes closure in the complex topology. Let $\Theta\subset\mathrm{Pic}^{g-1}(X)$ be the theta divisor, i.e., the set of divisor classes with $h^1>0$. Then assume

$$\Sigma\cap\Theta=\phi.$$

As we have remarked, periodic matrices arise when Σ is finite. However, whenever $\Sigma\cap\Theta=\phi$, I claim that the matrices $A\in M_\infty^d(C)$ that arise will be almost periodic. By definition this means that:

$$\begin{aligned}&\forall\varepsilon>0, \exists N_1<N_2<N_3<\cdots \quad \text{such that}\\ &|A_{k,l}-A_{k+N_i,l+N_i}|<\varepsilon, \quad \text{all } k, l\\ &\text{and } N_{i+1}-N_i \text{ are bounded.}\end{aligned}$$

The most interesting aspect of the dictionary, however, is to analyze what it does to the *Jacobian flows*. Again take $k=C$ and consider the general Data A′, B′ correspondence. The Jacobian variety $\mathrm{Pic}^0(X)$ acts on Data A′ by

$$\mathscr{F}\longmapsto\mathscr{F}\otimes L$$

L an invertible sheaf on X, at least "generically", i.e., for most L, $\mathscr{F}\otimes L$ will still satisfy the vanishing hypothesis. This means that the tangent space to $\mathrm{Pic}^0(X)$, which is canonically $H^1(\mathcal{O}_X)$, defines a vector space of commuting vector fields on the manifold of Data A′, although, when you integrate these into flows, they may be incomplete. It is very interesting to express these vector fields in terms of Data B′. The result is this: let $R\subset M_\infty^\Delta(C)$ be an instance of Data B′. Regard R as a fixed abstract ring, but consider deformations of its embedding in $M_\infty^\Delta(C)$:

$$\phi_t : R\mapsto M_\infty^\Delta(C).$$

Fix one element $C\in R$. For any $X\in M_\infty^\Delta(C)$, let

$$(X_+)_{ij}=\begin{cases}X_{ij}, & i<j\\ 0, & i\geq j\end{cases};\quad (X_0)_{ij}=\begin{cases}X_{ii}, & i=j\\ 0, & i\neq j\end{cases};\quad (X_-)_{ij}=\begin{cases}0, & i\leq j\\ X_{ij}, & i>j\end{cases}.$$

Then the flows in Data B′ are defined by the differential equations:

$$\frac{d}{dt}\phi_t(A)=[\phi_t(C)_+,\phi_t(A)].$$

This is not hard to prove:

a) describe $H^1(\mathcal{O}_X)$ by Cech co-cycles via the covering $\mathcal{U}=\{X-S, X-T\}$, giving:

$$\begin{aligned}R&=\Gamma((X-S)\cap(X-T),\mathcal{O}_X)\\ &=Z^1(\mathcal{U};\mathcal{O}_X)\longrightarrow H^1(\mathcal{O}_X).\end{aligned}$$

Corresponding to $c\in R$, we get the tangent vector to $\mathrm{Pic}^0(X)$ described by the invertible sheaf L on $X\times_k k[\varepsilon]$:

$$L\cong\begin{cases}\mathcal{O}_X\otimes_k k[\varepsilon] \text{ on } X-S,\ X-T\\ \text{glued by mult. by } 1+\varepsilon c.\end{cases}$$

b) Let $\mathcal{F}^{(k,l)}$, $s_k\in\Gamma(\mathcal{F}^{(k+1,-k)})$ define $\phi_0 : R\rightarrow M_\infty^\Delta(k)$. Deform these to

$$\begin{aligned}&\mathcal{F}^{(k,l)}\otimes_{\mathcal{O}_X}L \text{ on } X\times_k k[\varepsilon]\\ &s_k^*\in\Gamma(\mathcal{F}^{(k+1,-k)}\otimes_{\mathcal{O}_X}L).\end{aligned}$$

Via $L\xrightarrow{\sim}\mathcal{O}_X\otimes k[\varepsilon]$ on $X-S$ (resp. $X-T$), write

$$\begin{aligned}s_k^*&=s_k+\varepsilon s_k',\ s_k'\in\Gamma(X-S,\mathcal{F}^{(k+1,-k)})\\ &=s_k+\varepsilon s_k'',\ s_k''\in\Gamma(X-T,\mathcal{F}^{(k+1,-k)})\end{aligned}$$

where

$$s_k+\varepsilon s_k'=(1+\varepsilon c)(s_k+\varepsilon s_k''),$$

or

$$s'_k - s''_k = cs_k.$$

c) If $\phi_0(c) = C_{k,l}$, then by definition

$$cs_k = \sum C_{k,l} s_l.$$

Write $C = C_+ + C_0 + C_-$. Then

$$cs_k = \underbrace{\sum_{k<l} C_{k,l} s_l}_{(C_+s)_k} + \underbrace{\sum_{k \geq l} C_{k,l} s_l}_{((C_0+C_-)s)_k}.$$

Since

$$(C_+s)_k \in \Gamma\Big(\sum_{l>k} \mathscr{F}^{(l+1,-l)}\Big) \subset \Gamma(X-S, \mathscr{F}^{(k+1,-k)})$$

$$((C_0+C_-)s)_k \in \Gamma\Big(\sum_{l\leq k} \mathscr{F}^{(l+1,-l)}\Big) \subset \Gamma(X-T, \mathscr{F}^{(k+1,-k)})$$

we may define

$$s'_k = (C_+s)_k, \qquad s''_k = -((C_0+C_-)s)_k$$

and get s^*_k with the required property.

d) To determine the change in the matrices $\phi(f)$ associated to $f \in R$, we must write:

$$f \cdot s^*_k = \sum (\phi_0(f)_{kl} + \varepsilon \phi_1(f)_{kl})(s^*_l).$$

This works out to say

$$f \cdot s'_k = \sum \phi_1(f)_{kl} s_l + \sum \phi_0(f)_{kl} s'_l$$

or

$$([C_+, \phi_0(f)]s)_k = \sum (C_+)_{kl} f \cdot s_l - \sum \phi_0(f)_{kl} (C_+)_{ln} s_n = \sum \phi_1(f)_{kn} s_n$$

as required.

Note that as $[\phi_t(C), \phi_t(A)] = 0$ and $[\phi_t(C)_0, \phi_t(A)]$ generates a flow in the direction of equivalent subrings, the flow in Data B′ may also be written:

$$\frac{d}{dt}\phi_t(A) = \frac{1}{2}[\phi_t(C)_+ - \phi_t(C)_-, \phi_t(A)].$$

To see the connection with the Toda lattice equations, as promised in the title, take $A = C$ to be an n-periodic symmetric matrix with support $[-1, +1]$ ("tridiagonal"). Let A be

$$\begin{pmatrix} & & \cdots\cdots\cdots & & & \\ e^{a_{n-1}-a_n} & b_n & e^{a_n-a_1} & 0 & 0 & 0 \\ 0 & e^{a_n-a_1} & b_1 & e^{a_1-a_2} & 0 & 0 \\ 0 & 0 & e^{a_1-a_2} & b_2 & e^{a_2-a_3} & 0 \\ 0 & 0 & 0 & e^{a_2-a_3} & b_3 & e^{a_3-a_4} \\ & & \cdots\cdots\cdots & & & \end{pmatrix},$$

$$\text{where} \quad \sum_{i=1}^{n} a_i=0 \quad \sum_{i=1}^{n} b_i=0.$$

Then one readily calculates that the flow is given by

$$\dot{a}_k=b_k$$
$$\dot{b}_k=e^{a_{k-1}-a_k}-e^{a_k-a_{k+1}}$$

which are the Toda lattice equations describing a set of n particles on a circle, each pair being connected by a spring-like force that tends to keep the $(k+1)^{\text{st}}$ ahead of the kth, rising exponentially if they get closer or even get in the wrong order, but relaxing exponentially as they get farther apart in the right order.

It appears that particularly nice solutions of these equations arise by taking X to be a singular curve whose smooth model is $\boldsymbol{P}^1$. Apparently, if X has p ordinary double points we get the so-called p-soliton solutions of these non-linear equations. And if X is "unicursal", i.e., the map $\boldsymbol{P}^1 \to X$ is bijective, hence $\mathrm{Pic}^0(X)$ is an additive group, then we apparently get solutions in which the entries A_{ij} of the matrix are rational functions of i, j. This has not been fully worked out as yet.

§ 2. Differential operator case (Krichever)

Again let us start with Data A. Our first goal is to construct a deformation of the sheaf $\mathscr{F}$ to a sheaf $\mathscr{F}^*$ over $X\times_k k[[t]]$, plus a differential operator

$$\nabla : \mathscr{F}^* \longrightarrow \mathscr{F}^*(P)$$

such that:

1) $\nabla(as)=a\cdot\nabla s+\dfrac{\partial a}{\partial t}\cdot s, \qquad \forall a\in\mathcal{O}_X\otimes_k k[[t]],\ s\in\mathscr{F}^*.$

Moreover, if $z\in\mathfrak{M}_{P,X}-\mathfrak{M}^2_{P,X}$ is a local coordinate so that $\partial/\partial z\in T_{P,X}$ is the basis given by $T_{P,X}\cong k$, then we require:

2) $\nabla(s)=\dfrac{s}{z}+(\text{section of } \mathscr{F}^*).$

If $k=\boldsymbol{C}$, we can describe $\mathscr{F}^*$ analytically in a very simple way: let $U\subset X$ be a small complex neighborhood of P in which z is still a local coordinate. Define $\mathscr{F}^*$ on $X\times C$ to be $\mathscr{F}\otimes\mathcal{O}_C$ on $U\times C$ and on $(X-P)\times C$, but glue $\mathscr{F}^*$ to itself on $(U-P)\times C$ by the transition function $e^{t/z}$. Define ∇ on sections of $\mathscr{F}^*$ on $(X-P)\times C$ to be $\partial/\partial t$. Since

$$e^{-t/z}\frac{\partial}{\partial t}(e^{t/z}f(z,t))=\frac{1}{z}f(z,t)+\frac{\partial}{\partial t}f(z,t),$$

∇ extends to an operator from $\mathscr{F}^*$ to $\mathscr{F}^*(P)$ as required. To do this algebraically, we do the same thing regarding $e^{t/z}$ as a formal power series in t. This gives us a formal sheaf on the formal completion of $X\times_k \mathbf{A}_k^1$ along $X\times(0)$. By Grothendieck's formal existence theorem, it defines a sheaf $\mathscr{F}^*$ on $X\times_k k[[t]]$.

Then $H^i(X\times_k k[[t]],\mathscr{F}^*)=(0)$, $i=0,1$, so the map:

$$\begin{array}{c} H^0(X\times_k k[[t]],\mathscr{F}^*(P))\rightarrow H^0(\mathscr{F}^*(P)/\mathscr{F}^*) \\ \phantom{H^0(X\times_k k[[t]],\mathscr{F}^*(P))\rightarrow{}} \wr\Vert \\ \phantom{H^0(X\times_k k[[t]],\mathscr{F}^*(P))\rightarrow{}} k[[t]] \end{array}$$

is an isomorphism. Let s_0 be a generator of this $k[[t]]$-module. Define

$$s_n\in H^0(X\times_k k[[t]],\mathscr{F}^*((n+1)P))$$

by

$$s_n=\nabla^n(s_0).$$

Note that $s_n=s_0/z^n+$(lower terms), hence $s_0,\cdots,s_n$ are a $k[[t]]$-basis of $H^0(X\times_k k[[t]],\mathscr{F}^*((n+1)P))$. Now let $R=\Gamma(X-P,\mathcal{O}_X)$. Then for every $a\in R$, if $a=\alpha/z^n+$(lower terms) at P, then $as_0\in H^0(\mathscr{F}^*((n+1)P))$ so

$$\begin{aligned} as_0&=\alpha s_n+\sum_{i=0}^{n-1}a_i(t)s_i \\ &=\left(\alpha\nabla^n+\sum_{i=0}^{n-1}a_i(t)\nabla^i\right)s_0. \end{aligned}$$

Define an embedding of R in $k[[t]][d/dt]$ by taking a to:

$$D(a)=\alpha\left(\frac{d}{dt}\right)^n+\sum_{i=0}^{n-1}a_i(t)\left(\frac{d}{dt}\right)^i.$$

It is easy to verify that this is a homomorphism and that if s_0 is changed to $u(t)\cdot s_0$, $u(0)\neq 0$, then $D(a)$ is replaced by

$$u(t)\circ D(a)\circ u(t)^{-1}$$

so we get an equivalent ring.

To see intuitively how to go backwards from R to $(X,P,\mathscr{F})$, consider as in § 1 the spectral properties of the differential operators in R. Let $K\supset k$ be an extension field and look for formal power series $f(t)=\sum a_i t^i$, $a_i \in K$, such that

$$Df=\lambda_D f, \qquad \text{all } D \in R.$$

Then $D \mapsto \lambda_D$ is a homomorphism $R \to K$, hence a K-valued point of $X-P$ and the following holds:

Proposition. *Let* Data $(X, P, \mathscr{F})$ *define* $R \subset k[[t]][d/dt]$ *as above. Let* $K \supset k$ *be a field,* $\lambda: R \to K$ *be a K-valued point of* $X-P$ *lying over* $x \in X-P$ *(x defined by the prime ideal* Ker λ*). Then there is an isomorphism between:*

a) *the eigenspace* $\{f \in K[[t]] \mid Df=\lambda(D)\cdot f \text{ all } D \in R\}$,

and

b) $\mathrm{Hom}_R(\mathscr{F}_x/m_x\mathscr{F}_x, K)$ *(K an R-module via* λ*).*

Proof. Start with $\phi: \mathscr{F}_x/m_x\mathscr{F}_x \to K$. This is the same as an R-linear map:

$$\phi: \bigoplus_{n=0}^{\infty} s_n \cdot k = \Gamma(X-P, \mathscr{F}) \to K.$$

Any such map extends uniquely to an $R[[t]]$-linear map

$$\phi^*: \bigoplus_{n=0}^{\infty} s_n \cdot k[[t]] = \Gamma((X-P)\times k[[t]], \mathscr{F}^*) \to K[[t]]$$

such that

$$\phi^*(\nabla a) = \frac{d}{dt}\phi^*(a).$$

Such a ϕ^* is determined by the value $f(t)=\phi^*(s_0)$ and conversely, given $f(t)$, ϕ^* must map

$$\sum s_n \cdot a_n(t) \mapsto \sum a_n(t)\left(\frac{d}{dt}\right)^n f(t).$$

For this to be R-linear, however, means:

$$\phi^*(a\cdot s_0) = \lambda(a)\phi^*(s_0), \qquad \text{all } a \in R,$$

i.e., if $a\cdot s_0 = (\sum a_i(t)\nabla^i)s_0$, then

$$\sum a_i(t)\left(\frac{d}{dt}\right)^i f(t) = \lambda(a)\cdot f(t)$$

which means that f is in the λ-eigenspace of R. Q.E.D.

Now to go backwards from R to $(X, P, \mathscr{F})$, the intuitive picture is this: $X-P$

is just Spec R. $D \mapsto \deg D$ is a valuation on R and thus X is just Spec R plus one point P such that

$$\mathcal{O}_{P,X}=\left\{\frac{D_1}{D_2}\,\middle|\, D_1, D_2 \in R,\ \deg D_1 \leq \deg D_2\right\}.$$

To get $\mathscr{F}$, associate to each point of $X-P$ the corresponding eigenspace of R and "glue" these into a bundle. $\mathscr{F}$ is to be the sheaf of functions on this bundle, linear in each fibre, generated by the functions $f \mapsto f^{(n)}(0)$.

Alternatively, if M is the vector space of singular distributions on the t-line supported at $t=0$, M is an R-module and $\mathscr{F}$ on $X-P$ is just $\tilde{M}$.

However, rather than following this approach, it seems easier to use a more abstract approach better suited to generalizations. Starting with $R \subset k[[t]][d/dt]$, and $A, B \in R$ as in Data B, let:

$$R_n=\{D \in R \mid \deg D \leq n\}.$$

Then we have:

Lemma. i) *For all* $D \in R$, $D=\alpha(d/dt)^n+$(lower terms), $\alpha \in k$.
ii) *For all* n, $\dim R_{n+1}/R_n \leq 1$, *equality holding if* n *is large.*

Proof. Let $A, B \in R$ be the given operators of relatively prime degree. If
9 $\deg A=\alpha$, calculate the term of degree $a+n+1$ in $DA-AD$ and we find that if $D=a_n(t)(d/dt)^n+$(lower terms), then $\alpha \cdot a_n'(t)=0$, hence $a_n(t)$ is a constant. Thus $\dim R_{n+1}/R_n \leq 1$. The monomials A^iB^j give us operators of arbitrary sufficiently large degree.

Introduce the graded ring:

$$\mathscr{R}=\sum_{n=0}^{\infty} R_n.$$

Then as in § 1, we have:

Corollary. *R is a finite $k[A]$-module. $\mathscr{R}$ is a finite $k[e, A]$-module where $e \in \mathscr{R}_1$ represents the operator 1, $A \in \mathscr{R}_\alpha$ represents A. Hence R and $\mathscr{R}$ are finitely generated integral domains over k of transcendence degree 1 and 2 respectively.*

We now define

$$X=\operatorname{Proj}(\mathscr{R}).$$

The affine open $e \neq 0$ is given by:

$$\begin{pmatrix}\text{open subset}\\ e\neq 0\end{pmatrix}=\operatorname{Spec}\left(\mathscr{R}\left[\frac{1}{e}\right]\right)_0$$
$$\cong \operatorname{Spec} R,$$

and the affine open $A\neq 0$ is given by:

$$\begin{pmatrix}\text{open subset}\\ A\neq 0\end{pmatrix}=\operatorname{Spec}\left(\mathscr{R}\left[\frac{1}{A}\right]_0\right)$$
$$\cong\operatorname{Spec}\{\text{ring of fractions } C/A^k,\ \deg C\leq k\alpha\}.$$

As in § 1, if $C=A^iB^j$ has degree $k\alpha-1$, then the completion of this last ring in the e-adic topology is just $k[[z]]$, z being the local coordinate corresponding to C/A^k. So $e=0$ consists in one smooth k-rational point P, and the sheaf $\mathcal{O}_X(1)$ is just $\mathcal{O}_X(P)$.

Next, let M denote the big ring $k[[t]][d/dt]$, but considered now as a module over various rings by left and right multiplication. Define:

$$M_n=\{D\in M\mid \deg D\leq n\}$$
$$\mathfrak{M}=\bigoplus_{n=0}^{\infty} M_n.$$

We consider M as a $k[[t]]\otimes_k R$-module by letting $k[[t]]$ act by left-multiplication and R by right multiplication. Similarly, we consider $\mathfrak{M}$ as a $k[[t]]\otimes_k\mathscr{R}$-module. It is immediate that these modules are finitely generated: let $\mathscr{F}^*=\tilde{\mathfrak{M}}$ be corresponding sheaf over $X\times_k k[[t]]$. We have canonical maps:

$$\phi_n: M_n\to\Gamma(X\times_k k[[t]],\mathscr{F}^*(n))=\Gamma(X\times_k k[[t]],\mathscr{F}^*(nP)).$$

Let s_n be the image of $(d/dt)^n$. For $n\gg 0$, this is an isomorphism and $H^1(\mathscr{F}^*(nP))=(0)$. For each n, s_n generates $\mathscr{F}^*(n)/\mathscr{F}^*(n-1)$. Hence by descending induction on n, ϕ_n is an isomorphism for all $n\geq 0$ and $H^1(\mathscr{F}^*(nP))=(0)$. Also, by the Hilbert polynomial, $\mathscr{F}^*$ is a rank 1 sheaf.

Next, consider $(d/dt)_{\text{left-mult.}}:\mathfrak{M}\to\mathfrak{M}[1]$. It induces a map

$$\nabla:\mathscr{F}^*\to\mathscr{F}^*(P)$$

and from the identities

$$\left(\frac{d}{dt}\right)_{\text{left}}\cdot D_{\text{right}}=D_{\text{right}}\cdot\left(\frac{d}{dt}\right)_{\text{left}}$$
$$\left(\frac{d}{dt}\right)_{\text{left}}\cdot a(t)_{\text{left}}=a'(t)_{\text{left}}+a(t)_{\text{left}}\cdot\left(\frac{d}{dt}\right)_{\text{left}}$$
$$\left(\frac{d}{dt}\right)_{\text{left}}=\left(\frac{d}{dt}\right)_{\text{right}}+(\text{operator from }\mathfrak{M}\text{ to }\mathfrak{M}),$$
$$A^k_{\text{right}}=C_{\text{right}}\cdot\left(\frac{d}{dt}\right)_{\text{right}}+(\text{lower order operator})$$

we deduce that ∇ satisfies:

$$\nabla(as)=\frac{\partial a}{\partial t}\cdot s+a\cdot\nabla(s)$$

$$\nabla(s)=\frac{s}{z}+(\text{section of } \mathscr{F}^*).$$

We are now essentially back where we started: I claim that any pair $(\mathscr{F}^*, \nabla)$ with these properties is constructed as a deformation of $\mathscr{F}$ on X as in the beginning of this section. We omit this verification.

Extensions of this Dictionary to rank d sheaves $\mathscr{F}$ and commutative rings R in which all operators have degrees divisible by d can be made. However an additional complication arises from the possibility that the sheaf $\mathscr{F}$ may be unstable. We have only worked out the "generic case" where the bundles involved are all stable, and, moreover, have not characterized the rings of this generic type. However we can give a procedure for constructing certain rings of commuting operators from vector bundles. We need some definitions: let X be a curve over k (char $k=0$), and assume for simplicity that X is smooth, and irreducible of genus $g>0$. Consider the set of all stable rank r bundles with parabolic structure at P: i.e.,

$$E_0\subset E_1\subset\cdots\subset E_r=E_0(P)$$

$$\chi(E_i)=i$$

E_i stable locally free of rank r for all i, meaning for all non-zero subsheaves $F\subsetneq E_i$,

$$\chi(F)<\frac{\mathrm{rk}F}{r}\cdot\chi(E_i).$$

The set of all these forms a smooth quasi-projective moduli space V^r of dimension $r^2(g-1)+1+r(r-1)/2$. To each point $\{E_*\}\in V^r$, we may associate an infinite flag of bundles by requiring

$$E_{i+r}=E_i(P), \qquad \text{all } i\in Z.$$

Let

$$\mathrm{End}^k(E_*)$$

be the sheaf which is just $\mathrm{End}(E_0)$ on $X-P$ where, near P, the endomorphism λ is required to satisfy

$$\lambda(E_i)\subset E_{i+k}.$$

Then it is well known that the tangent bundle T_V can be identified canonically via:

$$T_{V,\{E_*\}}\cong H^1(X,\mathrm{End}^0(E_*)).$$

Look at the exact sequence:

$$0 \to \mathrm{End}^0(E_*) \to \mathrm{End}^1(E_*) \to \bigoplus_{k=0}^{r-1} \mathrm{Hom}\left(\frac{E_k}{E_{k-1}}, \frac{E_{k+1}}{E_k}\right) \to 0.$$

Lemma. $H^0(\mathrm{End}^1(E_*)) = k$.

Proof. In fact, take any

$$\lambda : E_0|_{X-P} \to E_0|_{X-P}$$

which extends to P so that $\lambda(E_k) \subset E_{k+1}$, all k. If $\lambda(E_k) \subset E_k$ for some k, then since E_k is stable, $\lambda = \alpha \cdot \mathrm{id.}$, some $\alpha \in k$. If not, then in a neighborhood $U \subset X$ of P, choose

$$e_1 \in \Gamma(U, E_{-r+1}) - \Gamma(U, E_{-r}).$$

Then $e_1, \lambda e_1, \cdots, \lambda^{r-1} e_1$ will have non-zero image in the quotients E_{-r+1}/E_{-r}, $E_{-r+2}/E_{-r+1}, \cdots, E_0/E_{-1}$, hence will give a basis of $E_0/E_0(-P)$. Thus they are a basis of E_0 in some smaller $U_1 \subset U$. Likewise, if z is a local coordinate at P, then $(1/z)e_1, \lambda e_1, \cdots, \lambda^{r-1} e_1$ are a basis of E_1 near P; so in terms of this basis, λ is given by a matrix

$$\lambda = \begin{bmatrix} 0 & 1 & 0 \cdots 0 \\ 0 & 0 & 1 \cdots 0 \\ 0 & 0 & 0 \cdots 1 \\ a_1/z & a_2 & a_3 \cdots a_r \end{bmatrix}, \qquad a_i \in \Gamma(U_1, \mathcal{O}_X).$$

Since $\lambda(E_0) \not\subset E_0$, $a_1(P) \neq 0$.

Now $\det(\lambda)$ is a rational function on X with poles only at P. Then the above shows that it has a simple pole at P, and this is impossible since $g > 0$. Q.E.D.

Now taking cohomology, we find:

$$0 \to \bigoplus_{k=0}^{r-1} \mathrm{Hom}\left(\frac{E_k}{E_{k-1}}, \frac{E_{k+1}}{E_k}\right) \to T_{V,\{E_*\}} \to H^1(\mathrm{End}^1(E_*)) \to 0.$$

Globally, this defines r sub-line bundles $L_i \subset T_V$, hence a rank r distribution:

10 $$\bigoplus_{i=1}^{r-1} L_i \subset T_V.$$

Now define:

Data A′ (*smooth stable case*):

a) *X a complete smooth curve over k,*

b) *$P \in X$ a k-rational point,*

c) *ϕ: Spec $k[[t]] \to V^r$ a morphism such that*

$$\dot{\phi}(\partial/\partial t) \in \bigoplus_{i=1}^{r-1} L_i$$

$$\dot{\phi}(\partial/\partial t)|_{t=0} \notin \Big(\bigoplus_{k \text{ omitted}} L_i\Big) \qquad \text{for any } 0 \le k \le r-1$$

$$\phi(0) \in (\text{open set where } h^0(E_0) = h^1(E_0) = 0).$$

Data B′. *Commutative subrings* $R \subset k[[t]][d/dt]$ *such that* $r \mid \deg C$, *all* $C \in R$ *and* $\exists\, A, B \in R$ *of form*

$$A = a_0(t)\Big(\frac{d}{dt}\Big)^n + \cdots + a_n(t)$$

$$B = b_0(t)\Big(\frac{d}{dt}\Big)^m + \cdots + b_m(t)$$

where $(n, m) = r$, $a_0(0) \neq 0$, $b_0(0) \neq 0$, *modulo*

$$R \sim u(t) \circ R \circ u(t)^{-1}, \qquad u(0) \neq 0.$$

We claim merely that every piece of Data A′ defines a piece of Data B′. To see this, let ϕ define the family of vector bundles with parabolic structure $\{E_k\}$ over $X \times \operatorname{Spec} k[[t]]$. Let $\vec{e}_\alpha$ be a basis of $E_0|_{U_\alpha}$ where $\{U_\alpha\}$ is a sufficiently fine covering of $X \times \operatorname{Spec} k[[t]]$ and assume that if $P \in U_\alpha$, then

$$\frac{e_{\alpha,1}}{z}, \ldots, \frac{e_{\alpha,i}}{z}, e_{\alpha,i+1}, \ldots, e_{\alpha,r}$$

is a basis of $E_i|_{U_\alpha}$, $0 \le i \le r$. Let

$$\vec{e}_\alpha = A_{\alpha\beta} \cdot \vec{e}_\beta$$

on $U_\alpha \cap U_\beta$. Then

$$\dot{\phi}(\partial/\partial t) \in H^1(X \times \operatorname{Spec} k[[t]], \operatorname{End}^0(E_*))$$

is given by the 1-co-cycle:

$$A_{\alpha\beta}^{-1} \cdot \frac{\partial A_{\alpha\beta}}{\partial t}$$

and the assumption that this lies in $\oplus L_i$ means that it dies in $H^1(\operatorname{End}^1)$, i.e.,

$$(*) \qquad A_{\alpha\beta}^{-1} \frac{\partial A_{\alpha\beta}}{\partial t} = A_{\alpha\beta}^{-1} D_\alpha A_{\alpha\beta} - D_\beta$$

$$D_\alpha \in \Gamma(U_\alpha, \operatorname{End}^1(E_*)).$$

Note that if $P \notin U_\alpha$, D_α is a matrix of regular functions and if $P \in U_\alpha$, D_α has the form

$$(**)\qquad D_\alpha=\begin{bmatrix} a_{11} & a_{12} & \cdots & a_{1,r-1} & a_{1r}/z \\ a_{21} & a_{22} & \cdots & a_{2,r-1} & a_{2,r} \\ za_{31} & a_{32} & \cdots & a_{3,r-1} & a_{3,r} \\ \vdots & & & \vdots & \vdots \\ za_{r1} & za_{r2} & \cdots & a_{r,r-1} & a_{rr} \end{bmatrix}.$$

Then define ∇ by $\nabla\bar{e}_\alpha=D_\alpha\bar{e}_\alpha$: (*) shows that ∇ is a global differential operator and (**) shows that $\nabla(E_k)\subset E_{k+1}$. Moreover, note that $a_{21}(P)\neq 0, \cdots, a_{r,r-1}(P)\neq 0$, $a_{1,r}(P)\neq 0$. For if one of these were zero, then $\dot{\phi}(\partial/\partial t)|_{t=0}$ would die already in H^1 of $\mathrm{End}^1(E_*)\cap\mathrm{End}(E_k)$ and hence $\dot{\phi}(\partial/\partial t)|_{t=0}$ would lie in $\oplus L_i$ (k omitted) contrary to assumption. Therefore, the *polar part* of ∇:

$$\bar{\nabla}:\bigoplus_{k=0}^{q-1}E_k/E_{k-1}\rightarrow\bigoplus_{k=0}^{r-1}E_{k+1}/E_k$$

is an isomorphism.

Now choose a generator s_0 of $\Gamma(X\times\mathrm{Spec}\,k[[t]], E_1)$. By the 3rd assumption on ϕ, s_0 generates E_1/E_0. Define $s_n\in\Gamma(E_n)$ by

$$s_n=\nabla^n s_0.$$

Then $\{s_n\}$ are a $k[[t]]$-basis of $\Gamma((X-P)\times\mathrm{Spec}\,k[[t]], E_0)$, hence the ring R is defined as before via:

$$\forall a\in\Gamma(X-P,\mathcal{O}_X),\quad\text{if}$$
$$a\cdot s_0=\sum a_i(t)\cdot s_i=(\sum a_i(t)\nabla^i)s_0,$$

then let

$$D(a)=\sum a_i(t)\left(\frac{d}{dt}\right)^i$$

and let

$$R=\mathrm{Image}\,(D).$$

In the case where $r=1$, V^r reduces to the Jacobian and ϕ reduces to a 1-parameter group on the Jacobian. Interestingly enough, if $r>1$, the distribution $\oplus L_i$ is contained in the tangent space to the fibres of the map

$$\pi: V\rightarrow(\text{Jacobian})$$
$$E_*\mapsto\Lambda^r E_0,$$

hence the curve ϕ is a curve on one of the *rational* varieties $\pi^{-1}(pt.)$. Thus whereas an explicit description of the operators in the rings R may be expected to involve the theta function when $r=1$, the operators must be very different (perhaps more elementary?) when $r>1$. The differential geometry of the highly non-integrable distribution $\oplus L_i$ on V has not yet been studied as far as I know.

Suppose $r=1$, X is smooth and $k=\mathbf{C}$. As in the difference operator case, we may strengthen the hypothesis $h^i(\mathscr{F})=0$ as follows:

Let $\phi: \mathbf{R}\to\mathrm{Pic}^0(X)$ be the 1-parameter group given by the line bundles $e^{t/z}$ defined above.

Let $\Sigma\subset\mathrm{Pic}^{g-1}(X)$ be the closure in the complex topology of the locus of points $[\mathscr{F}]+\phi(t)$, $t\in\mathbf{R}$.

Assume

$$\Sigma\cap\Theta=\phi.$$

In this case, it is to be expected that the coefficients of the differential operators that arise are almost periodic functions of t. On the other hand, if $r=1$, $k=\mathbf{C}$ and X is rational with double points, these coefficients should be rational functions of exponential functions $e^{\lambda_i t}$; and if X is unicursal, these coefficients should be rational functions of t.

The most interesting aspect of the dictionary, however, is its effect on the *Jacobian flows*. Let $k=\mathbf{C}$ and consider the general Data A′→Data B′ mapping (or for singular curves X, we may consider the Data A→Data B mapping). For every invertible sheaf L, L acts on V^r by $E_*\mapsto E_*\otimes L$, hence it acts on the set of all possible ϕ. As in § 1, this means that the tangent space $H^1(\mathcal{O}_X)$ to $\mathrm{Pic}^0(X)$ acts as a space of commuting vector fields on the manifold of all possible Data A′'s. These vector fields are very beautiful when expressed in terms of Data B′. The result is this: let $R\subset\mathbf{C}[[t]][d/dt]$ be an instance of Data B′. Regard R as a fixed abstract ring, but consider deformations of its embedding in $\mathbf{C}[[t]][d/dt]$:

$$D_s: R\to\mathbf{C}[[t]]\left[\frac{d}{dt}\right].$$

Fix one element $b\in R$ and some $l\geq 1$; let $k=\text{order } D_s(b)$. We shall define, in a minute, for every ordinary differential operator D of order k whose leading coefficient $a(t)(d/dt)^k$ satisfies $a(0)\neq 0$, an approximate $(l/k)^{\text{th}}$ power $(D^{l/k})_+$ of D. Then the flows in Data B′ are defined by the differential equation of Lax type*:

$$(\natural)\qquad \frac{d}{ds}D_s(a)=[(D_s(b)^{l/k})_+, D_s(a)],\qquad \text{all } a\in R.$$

What is $(D^{l/k})_+$? We may introduce a formal symbol $(d/dt)^{-1}$ subject to the commutation relations:

* I am told that this description of the Jacobian flows has been discovered also by Gel'fand and collaborators.

$$(d/dt)^{-1}\cdot a = a\cdot(d/dt)^{-1} - (d/dt)^{-1}\cdot\frac{da}{dt}\cdot(d/dt)^{-1}.$$

or, solving inductively:

$$\left(\frac{d}{dt}\right)^{-1}\cdot a = \sum_{k=0}^{\infty}(-1)^k\frac{d^k a}{dt^k}\cdot\left(\frac{d}{dt}\right)^{-k-1}.$$

We get this way a very large non-commutative ring of formal operators whose elements we write:

$$D = \sum_{i=-\infty}^{+k} a_i(t)\left(\frac{d}{dt}\right)^i, \qquad a_i(t)\in \mathbf{C}[[t]].$$

If the a_i are replaced by C^∞ functions of t, this may be interpreted as the ring of pseudo-differential operators in t mod the ideal of C^∞-integral operators. Let PsD $\{t\}$ denote our formal ring. Then we have the following lemma:

Lemma. *Let $D\in \mathbf{C}[[t]][d/dt]$, $D=a_0(t)(d/dt)^n+\cdots+a_n(t)$, $a_0(0)\neq 0$. Then, up to an nth root of 1, D has a unique n^{th} root $D^{1/n}\in$ PsD $\{t\}$. Moreover the commutator subring Z_D of D in PsD $\{t\}$ is the commutative ring of operators:*

$$\sum_{i=-\infty}^{+k} a_i D^{i/n}, \qquad a_i\in \mathbf{C}.$$

Proof. The main point is the calculation:

$$\left[D, c(t)\left(\frac{d}{dt}\right)^m\right] = (na_0(t)c'(t) - mc(t)a_0'(t))\left(\frac{d}{dt}\right)^{n+m-1} + \text{lower terms}.$$

From this it follows by easy induction that Z_D has, mod scalars and lower order terms, a unique element of each degree $m\in \mathbf{Z}$, and that it has the form $a_0(t)^{m/n}(d/dt)^m$ + (lower order terms). If $E\in Z_F$ has degree 1 and $E'\in Z_D$ has degree -1, it follows that $E\cdot E' = c+N$, $\deg N<0$, $c\in \mathbf{C}$, $c\neq 0$. Therefore

$$E^{-1} = E'\cdot\frac{1}{c}\cdot\left(\sum_{i=0}^{\infty}(-1)^i N^i/c^i\right)\in Z_D,$$

hence,

$$Z_D \supset \left\{\text{ring of Laurent series } \sum_{i=-\infty}^{+k} c_i E^i \text{ in } E\right\},$$

hence "=" holds here because each side has one new element in each degree. Thus Z_D is commutative. Finally, D itself is in Z_D so

$$D = \sum_{i=-\infty}^{+n} c_i E^i, \qquad c_i\in \mathbf{C},\ c_n\neq 0,$$

and, in a ring of Laurent series, such an element has a unique n^{th} root (up to a root of unity):

$$D = \sqrt[n]{c_n} \cdot E \cdot \left(1 + \frac{c_{n-1}}{c_n} E^{-1} + \frac{c_{n-2}}{c_n} E^{-2} + \cdots\right)^{1/n}$$

where the last term can be expanded by the binomial theorem. Q.E.D.

Definition. For all D as above, set

$$D^{k/n} = (D^{k/n})_+ + (D^{k/n})_-$$

where $(D^{k/n})_+ \in \mathbf{C}[[t]][d/dt]$, and $(D^{k/n})_- \in \text{PsD}\{t\}$ has negative degree.

To prove equation (♮), we first extend the isomorphism

$$\Gamma(X - P, \mathcal{O}_X) \xrightarrow{\approx} R \subset \mathbf{C}[[t]]\left[\frac{d}{dt}\right]$$
$$a \longmapsto D(a)$$

to an isomorphism:

$$\begin{array}{c} \hat{K}_{P,X} \xrightarrow{\approx} Z_R \subset \text{PsD}\{t\} \\ \Big\| \text{def} \\ \text{fraction field} \\ \text{of } \hat{\mathcal{O}}_{P,X} \end{array}.$$

To do this, for all k, let

$$\hat{E}_k = E_k \otimes (\hat{\mathcal{O}}_{P,X} \otimes \mathbf{C}[[t]])$$

and note that ∇ extends to an isomorphism:

$$\nabla : \hat{E}_k \to \hat{E}_{k+1}.$$

Then define

$$s_{-k} \in E_{-k+1}$$

by

$$s_{-k} = \nabla^{-k} s_0$$

and note that for all k:

$$\hat{E}_k \cong \left\{\text{module of elements } \sum_{i=-\infty}^{k-1} a_i(t) \cdot s_i\right\}.$$

In particular, for all $a \in \hat{K}_{P,X}$, if a has a k-fold pole at P, then $as_0 \in \hat{E}_{k+1}$, so

$$as_0 = \left(\sum_{i=-\infty}^{k} a_i(t)\nabla^i\right)s_0.$$

Set

$$D(a) = \sum_{i=-\infty}^{k} a_i(t)\left(\frac{d}{dt}\right)^i \in \mathrm{PsD}\,\{t\}.$$

With this preparation, we can easily check (♮):

a) describe $H^1(\mathcal{O}_X)$ by the acyclic resolutions:

$$0 \to \mathcal{O}_X \to \mathcal{O}_X(nP) \to \mathcal{O}_X(nP)/\mathcal{O}_X \to 0, \qquad n \gg 0,$$

giving

$$\begin{aligned} H^1(\mathcal{O}_X) &\cong \varinjlim_n \operatorname{coker}\left\{H^0(\mathcal{O}_X(nP)) \to \frac{\mathcal{O}_X(nP)}{\mathcal{O}_X}\right\} \\ &\cong \frac{\hat{K}_{P,X}}{\hat{\mathcal{O}}_{P,X} + \Gamma(X-P, \mathcal{O}_X)}. \end{aligned}$$

Then $c \in \hat{K}_{P,X}$ defines a tangent vector to $\mathrm{Pic}^0(X)$ described by an invertible sheaf L on $X \times_C C[\varepsilon]/(\varepsilon^2)$ by:

$$L \cong \begin{cases} \mathcal{O}_X \otimes_C C[\varepsilon] \text{ on } X-P, \text{ over } \operatorname{Spec} \hat{\mathcal{O}}_{P,X} \\ \text{glued by mult. by } 1+\varepsilon c. \end{cases}$$

b) Via L, we may deform E_* to $E_* \otimes_{\mathcal{O}_X} L$ over $X \times \operatorname{Spec} C[[t, \varepsilon]]/(\varepsilon^2)$, and we may extend ∇ to

$$\nabla : E_k \otimes_{\mathcal{O}_X} L \to E_{k+1} \otimes_{\mathcal{O}_X} L$$

by $(a \otimes b) = (\nabla a) \otimes b$, for any section b of L over $X \times \operatorname{Spec} C[\varepsilon]$, i.e., b not depending on t. Moreover, we may lift $s_0 \in \Gamma(X \otimes C[[t]], E_1)$ to a section s_0^* of $E_1 \otimes_{\mathcal{O}_X} L$ given by

$$\begin{aligned} &s_0^*|_{X-P} = (s_0 + \varepsilon s_0') \in \Gamma((X-P) \times C[[t, \varepsilon]]/(\varepsilon^2), E_1) &&\text{via } L|_{X-P} \cong \mathcal{O}_X|_{X-P} \\ &s_0^*|_{\mathcal{O}_{X,P}} = (s_0 + \varepsilon s_0'') \in \hat{E}_1 \otimes C[\varepsilon]/(\varepsilon^2) &&\text{via } L|_{\mathcal{O}_{X,P}} \cong \hat{\mathcal{O}}_{X,P} \end{aligned}$$

where

$$(s_0 + \varepsilon s_0'') = (1 + \varepsilon c)(s_0 + \varepsilon s_0')$$

or

$$s_0'' = cs_0 + s_0'.$$

c) But we may write

$$cs_0 = \sum_{i=-\infty}^{k} c_i(t)s_i.$$

If we *define*

$$s_0'' = \sum_{i=-\infty}^{-1} c_i(t)s_i \qquad s_0' = -\sum_{i=0}^{k} c_i(t)s_i$$

then $s_0^* = (s_0 + \varepsilon s_0', s_0 + \varepsilon s_0'')$ gives the required lifting of the section s_0. Define $s_i^* = \nabla^i s_0^*$.

d) To determine the change in the differential operators $D(a)$, $a \in \Gamma(X-P, \mathcal{O}_X)$, we must solve

$$as_0^* = \sum_{i=0}^{k} (a_i(t) + \varepsilon a_i'(t))s_i^*.$$

Expanding this over $X-P$, it says:

$$as_0' = \sum_{i=0}^{l} a_i'(t)\nabla^i s_0 + \sum_{i=0}^{l} a_i(t)\nabla^i s_0'.$$

But

$$\begin{aligned} as_0' &= -\Big(\sum_{j=0}^{k} c_j(t)\nabla^j\Big)(as_0) \\ &= -\Big(\sum_{j=0}^{k} c_j(t)\nabla^j\Big)\Big(\sum_{i=0}^{l} a_i(t)\nabla^i\Big)s_0 \end{aligned}$$

while

$$\sum_{i=0}^{l} a_i(t)\nabla^i s_0' = -\Big(\sum_{i=0}^{l} a_i(t)\nabla^i\Big)\Big(\sum_{j=0}^{k} c_j(t)\nabla^j\Big)s_0.$$

Thus, if we set

$$(*) \qquad \sum_{i=0}^{l} a_i'(t)\nabla^i = \Big[\sum_{i=0}^{l} a_i(t)\nabla^i, \sum_{j=0}^{k} c_j(t)\nabla^j\Big]$$

we have a solution (the higher degree terms in the commutator are zero because $\sum_{j=-\infty}^{k} c_j(t)\nabla^j$ commutes with $\sum_{i=0}^{l} a_i(t)\nabla^i$).

e) But

$$\sum_{j=0}^{k} c_j(t)\Big(\frac{d}{dt}\Big)^j = D(c)_+,$$

the differential operator part of the pseudo-differential operator assigned to c. If we choose $b \in \Gamma(X-P, \mathcal{O}_X)$ with an h-fold pole at P and let $c = b^{k/h}$, then

$$\sum_{j=0}^{k} c_j(t)\left(\frac{d}{dt}\right)^j = (D(b)^{k/h})_+,$$

so (*) reads:

$$\frac{d}{d\varepsilon}D(a) = [D(a), (D(b)^{k/h})_+]$$

as required.

To see the connection of the general theory with the Korteweg-deVries equation, as promised in the title, we take $D(a)=D(b)$ to be the second order operator

$$D=\left(\frac{d}{dt}\right)^2 + a(t),$$

and take $k=3$, $h=2$. Then one can solve mechanically for $D^{1/2}$, finding:

$$D^{1/2}=\left(\frac{d}{dt}\right)+\frac{a(t)}{2}\left(\frac{d}{dt}\right)^{-1}-\frac{a'(t)}{4}\left(\frac{d}{dt}\right)^{-2}+\frac{a''(t)-a(t)^2}{8}\left(\frac{d}{dt}\right)^{-3}$$
$$+\frac{6a(t)\cdot a'(t)-a'''(t)}{16}\left(\frac{d}{dt}\right)^{-4}+\cdots$$

whence

$$D^{3/2}=\left(\frac{d}{dt}\right)^3+\frac{3a(t)}{2}\left(\frac{d}{dt}\right)+\frac{3a'(t)}{4}+\frac{a''(t)+3a(t)^2}{8}\left(\frac{d}{dt}\right)^{-1}+\cdots$$

and

$$[D, (D^{3/2})_+] = -\tfrac{1}{4}(a'''(t)+6a(t)\cdot a'(t)).$$

Therefore, if

$$D_s=\left(\frac{d}{dt}\right)^2 + a(s, t),$$

the Jacobian flow is given by:

$$\frac{\partial a}{\partial s} = -\frac{1}{4}\left(\frac{\partial^3 a}{\partial t^3}+6a\cdot\frac{\partial a}{\partial t}\right)$$

which, (up to coefficients which can be normalized away) is the Kortweg-de Vries equation.

Now whenever X is a hyperelliptic curve and $P \in X$ is a Weierstrass point, then there is a function a on X which a double pole at P only. Then with suitable normalization:

$$D(a)=\left(\frac{d}{dt}\right)^2+a_0(t).$$

Thus if we follow a Jacobian flow on Pic (X), we get a 1-parameter family of operators:

$$D_s(a)=\left(\frac{d}{dt}\right)^2+a_0(s,t)$$

where a_0 satisfies the Korteweg-de Vries equation. For smooth X, these solutions were discovered by McKean and van Moerbecke [9] and others; for singular X of type

$$y^2=xf(x)^2$$

these appear to be the n-soliton solutions of Kay and Moses [6]; and for a unicursal X of type

$$y^2=x^{2n+1},$$

these appear to be the rational solutions of Airault, McKean and Moser [1]. These connections have not yet been investigated in detail. (see Added in proof)

§ 3. **Field operator case** (Drinfeld)

As in the introduction, let k be a field, $\sigma \in \mathrm{Aut}(k)$ of infinite order, k_0 the fixed field. Generalizing the dictionary in the introduction, consider:

Data A′. a) *X_0 a reduced and irreducible complete curve over k_0. Let $X=X_0\times_{k_0}k$: we assume this is reduced* and irreducible.*

b) *$P_0 \in X_0$ a regular closed point. Let $P=P_0\times_{k_0}k\subset X$.*

c) *A torsion-free sheaf $\mathcal{F}$ on X such that*

$$h^0(\mathcal{F})=h^1(\mathcal{F})=0.$$

d) *A maximal flag of subsheaves:*

$$\mathcal{F}=\mathcal{F}_0\supsetneqq\mathcal{F}_{-1}\supsetneqq\cdots\supsetneqq\mathcal{F}_{-d}=\mathcal{F}(-P),$$

where length $(\mathcal{F}_{k+1}/\mathcal{F}_k)=1$.

e) *A homomorphism of sheaves*

* As pointed out to me by J. Tate, "X reduced" is automatic because whenever k_0 is the fixed field of some σ, k is separable over k_0. (To see this, suppose on the contrary there were $x_1,\cdots,x_n \in k_0^{1/p}$ which are linearly independent over k_0 but dependent over k: $x_1+a_2x_2+\cdots+a_nx_n=0$, $a_i \in k$. Assume n is minimal too. Then $x_1+\sigma a_2\cdot x_2+\cdots+\sigma a_n\cdot x_n=0$ so $(\sigma a_2-a_2)x_2+\cdots+(\sigma a_n-a_n)x_n=0$, so either $x_2,\cdots,x_n$ are also dependent or $\sigma a_k=a_k$, all k, hence $a_k \in k_0$. Both cases are impossible.)

$$\alpha: (1_{X_0}\times\sigma)^*\mathscr{F}|_{X-P}\to\mathscr{F}|_{X-P}$$

on $X-P$, which is not surjective, such that, on X, α carries $\mathscr{F}_k$ to $\mathscr{F}_{k+1}$. Here σ: Spec $k\leftarrow$Spec k is induced by $\sigma: k\to k$, so that $(1_{X_0}\times\sigma)^*\mathscr{F}$ is a sheaf on X conjugate to $\mathscr{F}$ via σ.

Data B′. *A commutative subring $R\subset k\{\sigma\}$, with $R\supsetneq k_0$ and $R\cap k=k_0$, modulo the identification:*

$$R\sim aRa^{-1}, \qquad a\in k^*.$$

We claim these 2 sorts of Data are equivalent as before. Before proving this, however, we want to prove a remarkable observation of Drinfeld—that in Data A′, the assumption $h^0(\mathscr{F})=h^1(\mathscr{F})=0$ follows from the a priori weaker assumption that $\chi(\mathscr{F})=0$. To see this, first define $\mathscr{F}_n$ for all $n\in Z$ by requiring

$$\mathscr{F}_{n+d}=\mathscr{F}_n(P).$$

Note that $\chi(\mathscr{F}_n)=n$. Since $\mathscr{F}_n\subset\mathscr{F}_{n+1}$, $h^0(\mathscr{F}_n)\le h^0(\mathscr{F}_{n+1})$ and $h^1(\mathscr{F}_n)\ge h^1(\mathscr{F}_{n+1})$. Let n_0 be the smallest n such that $h^0(\mathscr{F}_n)\neq 0$ and let $s_0\in\Gamma(\mathscr{F}_{n_0})$. Certainly $n_0\le 1$ because $h^0(\mathscr{F}_1)\ge\chi(\mathscr{F}_1)=1$. Consider the maps

$$\alpha: (1_{X_0}\times\sigma)^*\mathscr{F}_n\to\mathscr{F}_{n+1}.$$

Define inductively sections

$$s_n\in\Gamma(\mathscr{F}_{n+n_0})$$

by

$$s_n=\alpha((1_{X_0}\times\sigma)^*s_{n-1}).$$

By assumptions e, since $\chi(\mathscr{F}_{n+1})=\chi((1_{X_0}\times\sigma)^*\mathscr{F}_n)+1$, $l(\text{coker }\alpha)=1$ and $\alpha|_{X-P}$ is *not* surjective, α must be surjective at P. Thus

$$\mathscr{F}_{n+1}=\mathscr{F}_n+\alpha((1_{X_0}\times\sigma)^*\mathscr{F}_n).$$

From this it follows also that intersecting in $\mathscr{F}_{n+1}$:

$$\alpha((1_{X_0}\times\sigma)^*\mathscr{F}_{n-1})=\mathscr{F}_n\cap\alpha((1_{X_0}\times\sigma)^*\mathscr{F}_n).$$

Therefore the sequence of sections s_n satisfies the implications:

$$\begin{aligned} s_n\in\Gamma(\mathscr{F}_{n+n_0-1}) &\Rightarrow \alpha(1_{X_0}\times\sigma)^*s_{n-1}\in\Gamma(\mathscr{F}_{n+n_0-1}\cap\alpha(1_{X_0}\times\sigma)^*\mathscr{F}_{n+n_0-1})\\ &\Rightarrow s_{n-1}\in\Gamma(\mathscr{F}_{n+n_0-2}).\end{aligned}$$

Since $s_0\notin\Gamma(\mathscr{F}_{n_0-1})$, it follows that for all n, $s_n\notin\Gamma(\mathscr{F}_{n+n_0-1})$. But therefore the sections $\{s_0, s_1, \cdots, s_n\}$ are linearly independent. Thus

$$
\begin{aligned}
n+1 &\leq h^0(\mathscr{F}_{n+n_0}) \\
&= \chi(\mathscr{F}_{n+n_0}) \qquad \text{if } n \gg 0 \\
&= n+n_0.
\end{aligned}
$$

Thus $n_0 \geq 1$. Putting this together, $n_0=1$, i.e., $h^0(\mathscr{F}_0)=0$ as asserted.

Now to go from Data A′ to Data B′, construct $s_n \in \Gamma(\mathscr{F}_{n+1})$ as above. Notice that $h^1(\mathscr{F}_n) \leq h^1(\mathscr{F}_0)=0$ if $n \geq 0$ so by the argument just given, $\{s_0, \cdots, s_n\}$ is a basis of $\Gamma(\mathscr{F}_{n+1})$, all $n \geq 0$. Therefore $\{s_n\}_{n\geq 0}$ is a basis of $\Gamma(X-P, \mathscr{F})$. Now let $R=\Gamma(X_0-P_0, \mathcal{O}_{X_0})$ and consider the action of R on $\Gamma(X-P, \mathscr{F})$. This is given by:

$$
f \cdot s_0 = \sum_{n=0}^{N} a_n(f) \cdot s_n,
$$

for all $f \in R$, and suitable $a_n(f) \in k$. Define a map from R to $k\{\sigma\}$ by

$$
f \mapsto \sum_{n=0}^{N} a_n(f) \sigma^n.
$$

As in the previous cases, it is easy to see that this is an injective homomorphism and the image is an example of Data B′.

Concerning "eigenvalues" of the operators $R \subset k\{\sigma\}$, the corresponding problem is to look for solutions in some extension field $K \supset k$ of equations of the form

$$
\sum_{i=0}^{n} a_i \cdot \sigma^i \xi = 0, \qquad a_i \in k.
$$

We have the following result:

Proposition. *Let* Data $(X_0, P_0, \{\mathscr{F}_i\}, \alpha)$ *define* $R \subset k\{\sigma\}$ *as above. Let* $K \supset k$ *be a field and* $\sigma: K \to K$ *an extension of* σ *to* K. *Let* $x_0 \in X_0 - P_0$ *be a closed point with ideal* m_{x_0}. *Then there is an isomorphism between*

a) *the* k_0*-vector space of* $\xi \in K$ *such that* $\sum a_i \sigma^i \xi = 0$, *all* $\sum a_i \sigma^i \in R$ *corresponding to functions* $a \in m_{x_0}$

and

b) $\operatorname{Hom}_{(k,\sigma)}(\mathscr{F}/m_{x_0} \cdot \mathscr{F}, K)$.

Proof. Since $\Gamma(X-P, \mathscr{F}) = \oplus_{n=0}^{\infty} k s_n$ is a free $k\{\sigma\}$-module with basis s_0,

$$
\begin{aligned}
\operatorname{Hom}_{(k,\sigma)}(\mathscr{F}/m_{x_0} \cdot \mathscr{F}, K) &\cong (k,\sigma)\text{-maps } \lambda: \Gamma(X-P, \mathscr{F}) \to K \\
&\qquad \text{with } m_{x_0} \cdot \Gamma(X-P, \mathscr{F}) \subset \operatorname{Ker} \lambda \\
&\cong \text{elements } \xi \in K \text{ killed by } m_{x_0}.
\end{aligned}
$$

Q.E.D.

To go backwards, start from R. Define

$$R_n = \{x \in R \mid \deg x \leq n\}$$
$$\mathscr{R} = \bigoplus_{n=0}^{\infty} R_n$$
$$X_0 = \operatorname{Proj} \mathscr{R}.$$

Note that R is an integral domain and $x \mapsto \deg x$ is a valuation on R. Let s be the g.c.d. of the values $\{\deg x\}$, $x \in R$. For any elements $A_1, A_2 \in R_{ks}$, write them:

$$A_i = a_i \sigma^{ks} + (\text{lower terms}).$$

Using the commutativity of R, it follows that

$$\sigma^s(a_1/a_2) = a_1/a_2.$$

Let $k_2 \subset k$ be the fixed field of σ^s: then k_2 is a Galois extension of k_0 with group Z/sZ. Let $k_1 \subset k_2$ be the subfield generated by the ratios a_1/a_2. Then $s = d \cdot r$, and k_1 will be Galois of degree r, for some factorization of s. In particular, for all n:

$$\dim_{k_0} (R_{(n+1)s}/R_{ns}) \leq r$$

with equality for n large. Now let $e \in \mathscr{R}_1$ represent 1 and take some non-constant operator $A \in \mathscr{R}_{\alpha s}$ with $\deg A = \alpha s$. Then as before we see that $\mathscr{R}$ is a finite $k_0[e, A]$-module, hence it is a finitely generated k_0-algebra as well as an integral domain. Thus X_0 is a reduced and irreducible curve proper over Spec k_0. The affine open set $e \neq 0$ is just:

$$\begin{pmatrix}\text{open subset} \\ e \neq 0\end{pmatrix} = \operatorname{Spec} \mathscr{R}\left[\frac{1}{e}\right]_0$$
$$\cong \operatorname{Spec} R$$

and the affine open set $A \neq 0$ is just:

$$\begin{pmatrix}\text{open subset} \\ A \neq 0\end{pmatrix} = \operatorname{Spec} \mathscr{R}\left[\frac{1}{A}\right]_0$$
$$\cong \operatorname{Spec} \left\{\begin{matrix}\text{ring of fractions} \\ C/A^k,\ \deg C \leq k\alpha s\end{matrix}\right\}.$$

Since k_0 is algebraically closed in R, it follows that $X \underset{\text{def}}{=} X_0 \times_{k_0} k$ is also irreducible, and, as remarked in the footnote earlier, k is separable over k_0 so X is also reduced. On the other hand the Cartier divisor $e = 0$ on X_0 is given by:

$$\begin{aligned} V(e) &= \operatorname{Proj} (\mathscr{R}/e\mathscr{R}) \\ &= \operatorname{Proj} \left(\bigoplus_{n=0}^{\infty} R_n/R_{n-1}\right) \\ &= \operatorname{Proj} (\text{subring of } k_1[t] \text{ of finite codim., } \deg t = s) \\ &= \operatorname{Spec} k_1. \end{aligned}$$

Since this is reduced and irreducible, $V(e)$ consists in one regular point P_0, with residue field k_1.

For the next step, we rename the ring $k\{\sigma\}$ as M and regard it as a module over $k \otimes_{k_0} R$: namely, let k act by left multiplication and let R act by right multiplication. Moreover, let

$$M_n = \{x \in M \mid \deg x \leq n\}$$

$$\mathfrak{M} = \bigoplus_{n=0}^{\infty} M_n$$

$$\mathfrak{M}[n] = \mathfrak{M} \text{ with grading shifted by } n \ (\mathfrak{M}[n]_k = \mathfrak{M}_{n+k})$$

$$\mathscr{F}_{n+1} = \widetilde{\mathfrak{M}[n]} \text{ on } X.$$

It is easy to check that $\mathfrak{M}$ is a finitely generated $\mathscr{R}$-module (in fact, it is finitely generated over $k[e, A]$), so all the sheaves $\mathscr{F}_n$ are coherent. Multiplication by $e \in \mathscr{R}$ defines a degree-preserving injection:

$$e : \mathfrak{M}[n] \to \mathfrak{M}[n+1]$$

hence an injection

$$\mathscr{F}_{n+1} \hookrightarrow \mathscr{F}_{n+2},$$

which reduces to an isomorphism on the open set $e \neq 0$, i.e., on $X - P$. Moreover, coker (e) is a graded module, all of whose graded pieces are isomorphic to k, so $\mathscr{F}_{n+2}/\mathscr{F}_{n+1}$ is a sheaf isomorphic to k, hence has length 1. To check $\mathscr{F}_{n-s} \cong \mathscr{F}_n(-P)$, you have to be careful because $\mathscr{F}_n(-P)$ does not correspond to:

$$(\text{Graded ideal of } P) \cdot (\text{Graded module of } \mathscr{F}_n).$$

This is because $\mathscr{R}$ is not generated by elements of degree 1 so X does not carry an invertible sheaf $\mathcal{O}(1)$! You have to take a sufficiently large, sufficiently divisible l. Then working with degrees divisible by l:

$$\mathscr{F}_n(-P) = (\text{graded ideal of } P) \cdot (\text{graded module of } \mathscr{F}_n)^{\sim}$$

$$= \left[\bigoplus_{m=1}^{\infty} \operatorname{Im}(R_{ml-1} \to R_{ml})\right] \cdot \left[\bigoplus_{n=0}^{\infty} M_{n+ml}\right]^{\sim}.$$

But if $s \mid l$, $l \gg 0$, then $R_{l-1} = R_{l-s}$ and $R_{l-s} \cdot M_k = M_{l-s+k}$ for all $k \gg 0$. Thus

$$\mathscr{F}_n(-P) = \left[\bigoplus_{m=1}^{\infty} \operatorname{Im}(R_{ml-s} \to R_{ml})\right] \cdot \left[\bigoplus_{m=0}^{\infty} M_{n+ml}\right]^{\sim}$$

$$= \left[\bigoplus_{m=0}^{\infty} R_{l-s} \cdot M_{n+(m-1)l}\right]^{\sim}$$

$$= \left[\bigoplus_{m=0}^{\infty} M_{n+ml-s}\right]^{\sim}$$

$$= \mathscr{F}_{n-s}.$$

Now using the dictionary of *FAC*, we have:

$$(*)\qquad \left.\begin{array}{l} M_n \cong \Gamma(X, \mathscr{F}_{n+1}) \\ H^1(X, \mathscr{F}_{n+1}) = (0) \end{array}\right\} n \gg 0.$$

Moreover, comparing n and $(n-1)$, if $(*)$ holds for n and $n \geq 0$, then we have:

$$\begin{array}{ccccccccc} 0 \longrightarrow & H^0(X, \mathscr{F}_n) & \longrightarrow & H^0(X, \mathscr{F}_{n+1}) & \longrightarrow & H^0(X, \mathscr{F}_{n+1}/\mathscr{F}_n) & \longrightarrow & H^1(X, \mathscr{F}_n) & \longrightarrow 0 \\ & \uparrow \alpha & & \uparrow \wr & & \uparrow \beta & & & \\ 0 \longrightarrow & M_{n-1} & \longrightarrow & M_n & \longrightarrow & k(\sigma^n) & \longrightarrow 0. & & \end{array}$$

Then since $\mathscr{F}_{n+1}/\mathscr{F}_n = (\mathfrak{M}[n]/\mathfrak{M}[n-1])^{\sim}$ and since $\sigma^n \in \mathfrak{M}[n]_0$ generates this module, it follows that β is an isomorphism. Thus the diagram shows that $(*)$ holds for $n-1$ also. Continuing down, it follows eventually that $h^0(\mathscr{F}_0) = h^1(\mathscr{F}_0) = 0$. A Corollary of this is that $\mathscr{F}_0$ is torsion free. Note incidentally that P consists of r distinct regular points, so the sheaves $\mathscr{F}_n$ must be locally free of rank $d = s/r$ in a neighborhood of P. Finally, left multiplication by σ gives a degree preserving map

$$\mathfrak{M}[n] \rightarrow \mathfrak{M}[n+1]$$

which is linear with respect to R, σ-semi-linear with respect to k. Thus it defines a homomorphism α as required. Over the affine piece $X-P$, all the $\mathscr{F}_n$ reduce to $\tilde{M}$ and α is again left multiplication by σ. Thus its cokernel is $\widetilde{M/\sigma M}$, which is just the sheaf k sitting in fact at the point of Spec $R \otimes_{k_0} k$ defined by the ideal

$$\begin{array}{l} \operatorname{Ker}\,[R \otimes_{k_0} k \xrightarrow{\phi} k] \\ \phi(a_0 + a_1\sigma + \cdots + a_n\sigma^n) \otimes b = a_0 b. \end{array}$$

The most interesting case of this dictionary is when k_0 is the finite field $\mathbf{F}_q$, $k \supset k_0$ is any extension, and $\sigma(x) = x^q$. In this case, Data A′ is essentially what Drinfeld calls a "Shtuka" and Data B′ is exactly what he calls an "Elliptic module" [2]. The point is that if $(X_0, P_0, \{\mathscr{F}_n\}, \alpha)$ is an example of Data A′, then the whole tower $\{\mathscr{F}_n\}$ is derived simply from the diagram:

$$(*)\qquad \begin{array}{ccc} (1_{X_0} \times \sigma)^* \mathscr{F}_0 & & \\ & \searrow^{\alpha} & \\ & & \mathscr{F}_1 \\ & \nearrow_{\beta} & \\ \mathscr{F}_0 & & \end{array}$$

(β = given inclusion). In fact, we saw that

$$\alpha(1_{X_0} \times \sigma)^* \mathscr{F}_{-1} = \alpha(1_{X_0} \times \sigma)^* \mathscr{F}_0 \cap \mathscr{F}_0 \qquad \text{(intersection in } \mathscr{F}_1)$$

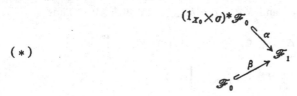

$$\alpha(1_{X_0}\times\sigma)^*\mathscr{F}_{-2}=\alpha(1_{X_0}\times\sigma)^*\mathscr{F}_{-1}\cap\mathscr{F}_{-1} \qquad \text{(intersection in } \mathscr{F}_1).$$
etc.

Now Drinfeld defines quite generally a *Shtuka* to be a pair of vector bundles $\mathscr{F}_0, \mathscr{F}_1$ on X, plus a diagram like (∗) such that $l(\operatorname{coker}\alpha)=l(\operatorname{coker}\beta)=1$. The support of coker ($\alpha$) is called the zero of the shtuka and the support of coker (β) is called the pole. The shtuka arising from towers $\{\mathscr{F}_n\}$ are easily seen to be characterized by 2 properties:

1) Let $P=\{$the pole and all its conjugates over $F_q\}$. Then the zero is disjoint from P.
2) Restricted to P, $\alpha^{-1}\cdot\beta$ defines a q^{-1}-semi-linear map of the F_q-vector space $\mathscr{F}_0/m_P\mathscr{F}_0$ into itself. This map should be nilpotent.

The purpose of the twin tools of elliptic modules and shtuka in Drinfeld's papers is to set up a non-abelian reciprocity law, i.e., prove Langland's conjecture for the field $F_q(X_0)$. I don't want to say anything about this except to indicate why, in the rank one case-Data A and B, the dictionary gives a new method of constructing explicitly the *abelian* extensions of the field $F_q(X_0)$. Let us rephrase the idea of Data A once again, assuming now that $\mathscr{F}$ is an invertible sheaf on X. Let $\operatorname{Pic}^0(X_0)$ be the jacobian of X_0, considered as parametrizing invertible sheaves of degree 0. As usual, map the regular points $(X_0)_{\text{reg}}$ of X_0 to $\operatorname{Pic}^0(X)$ by taking y to the point representing the sheaf $\mathcal{O}_X(y-P_0)$: call this ψ. Let $\sigma:\operatorname{Pic}^0(X_0)\to\operatorname{Pic}^0(X_0)$ be the F_q-morphism induced by pull-back by σ on sheaves: $\mathscr{F}\mapsto(1_{X_0}\times\sigma)^*\mathscr{F}$. Following Lang [8], we consider the diagram

$$\begin{array}{ccc} (Y_0)_{\text{reg}} & \longrightarrow & \operatorname{Pic}^0(X) \\ \downarrow & & \downarrow{\scriptstyle 1-\sigma} \\ (X_0)_{\text{reg}} & \xrightarrow{\ \psi\ } & \operatorname{Pic}^0(X) \end{array}$$

where $(Y_0)_{\text{reg}}$ is the fibre product. Now note

$$\begin{pmatrix}\text{A } k\text{-valued point}\\ y \text{ of } Y_{\text{reg}}\end{pmatrix}=\begin{pmatrix}\text{A pair } (\mathscr{F},x),\ \mathscr{F} \text{ an invertible sheaf}\\ \text{on } X \text{ of degree } 0,\ x\in X_{\text{reg}} \text{ such that}\\ (1-\sigma)([\mathscr{F}])=\psi(x)\end{pmatrix}$$
$$=\begin{pmatrix}\text{A pair } (\mathscr{F},x) \text{ where}\\ \mathscr{F}\otimes(1_{X_0}\times\sigma)^*\mathscr{F}^{-1}\cong\mathcal{O}_X(x-P_0)\end{pmatrix}.$$

Associating the sheaf $\mathscr{G}=\mathscr{F}((g-1)P_0)$ to $\mathscr{F}$, we carry the identification further:

$$=\begin{pmatrix}\text{A pair } (\mathscr{G},x),\ \mathscr{G} \text{ of degree } g-1,\ x\in X_{\text{reg}} \text{ such that}\\ (1_{X_0}\times\sigma)^*\mathscr{G}\cong\mathscr{G}(P_0-x)\end{pmatrix}.$$

Thus $(Y_0)_{\text{reg}}$ is the scheme classifying all possible examples of Data A′. But Lang's geometric class field theory states that the curve Y_0 is the maximal abelian covering of X_0, such that (1) it is unramified over $(X_0)_{reg}$, (2) with certain bounds on the ramification over the singular points (these bounds getting as weak as you wish as the points get more singular) and (3) with no residue field extension over P_0. The dictionary now states that $(Y_0)_{reg}$ is equally the scheme classifying all possible examples of Data B. But these are readily described by equations: write

$$\Gamma(X_0-P_0, \mathcal{O}_{X_0}) \cong F_q[Z_1, \cdots, Z_n]/(f_1, \cdots, f_k).$$

Let $\bar{Z}_i$ have a pole of order n_i at P_0. Then Data B is given by assigning:

$$Z_i \mapsto A_i = \sum_{j=0}^{n_i} a_{ij}\sigma^j$$

where $A_i \in k\{\sigma\}$ satisfy

$$(*)\qquad \begin{aligned} &A_iA_j = A_jA_i \\ &f_k(A_1, \cdots, A_n) = 0 \\ &a_{i,n_i} \neq 0. \end{aligned}$$

We may normalize this mod $R \sim aRa^{-1}$, e.g., by picking $\bar{Z}_1, \bar{Z}_2$ so that $(n_1, n_2)=1$ and $\bar{Z}_1^{n_2}/\bar{Z}_2^{n_1}(P_0)=1$, and then requiring

$$a_{1,n_1} = a_{2,n_2} = 1.$$

Then the equations $(*)$, written out as equations in the a_{ij} define a scheme over k_0, which is precisely the affine piece of the abelian covering Y_0 over $(X_0)_{reg} - P_0$. This is one of the simpler results in Drinfeld's extraordinary paper [2].

Appendix (added on Oct. 15, 1977)

Professor E. Coddington has kindly given me references to 3 very beautiful papers of J. L. Burchnall and T. W. Chaundy, all entitled "Commutative Ordinary Differential Operators", which appear in

Proceedings London Mathematical Society 21 (1922), p. 420
Proceedings Royal Society London (A), 118 (1928), p. 557
Proceedings Royal Society London (A), 134 (1931), p. 471.

It appears that virtually all the results described in § 2 are in fact due to them: in particular the correspondence given in the introduction between Data A and B for differential operators was established by them. Even more remarkably, they even recognized the fact that when the curve X has singularities, there are several classes of commutative rings of operators in Data B corresponding to which strata of the compactified Pic (X) $\mathcal{F}$ lies in: see their 3rd paper where the case of the curve X given by $x^m = y^n$, $(m, n)=1$, is analyzed at length. In their papers, and in a note by H. F. Baker following their 2nd paper, the explicit construction of these operators via theta-functions and related abelian functions is given in detail. The one point they do not explore is the infinitesimal deformation of a pair of commuting operators hence they were not led to a Lax equation or to the link with the Korteweg-deVries equation. Instead they discuss at length a procedure for relating 2 rings $R_1, R_2 \subset C[[t]][d/dt]$ namely via an auxiliary operator T such that: $R_1 \circ T = T \circ R_2$. It would seem that once this link is made, their work anticipates a large amount of the recent work on degenerate-spectrum Sturm-Liouville operators and exact solutions of the Korteweg-deVries equation.

Added in proof. 1) Krichever's work had been anticipated in some nearly forgotten papers of Burchnall and Chaundy in the 20's—cf. Appendix. Also at this point I would like to thank D. Kajdan for introducing me to these ideas and
11 sharing his many insights. 2) cf. H. McKean, *Theta functions, Solitons, and Singular Curves*, to appear.

References

12 [1] H. Airault, H. McKean, and J. Moser, Rational and elliptic solutions of the Korteweg-de Vries equation, to appear.

[2] V. G. Drinfeld, Elliptic modules, Mat. Sb., 94 (1974); transl., 23 (1974), p. 561.

[3] B. Dubrovin, V. Matveev and S. Novikov, Non-linear equations of the Korteweg-de Vries type, finite zone operators and abelian varieties, Russian Math. Surveys, 31 (1976), p. 59.

[4] J. Fay, Theta functions on Riemann surfaces, Springer Lecture Notes, 352, 1973.

[5] C. S. Gardner, J. Greene, M. Kruskal and R. Miura, Korteweg-de Vries Equation and Generalizations VI: Methods for exact solution, Comm. Pure Appl. Math., 27 (1974), p.97.

[6] I. Kay and H. Moses, Reflectionless transmission through dielectrics, J. Appl. Physics, 27 (1956), p. 1503.

[7] I. M. Krichever, Algebro-geometric construction of the Zaharov-Shabat equations and their periodic solutions, Doklady Akad. Nauk SSSR, 1976.

[8] S. Lang, Unramified class field theory over function fields in several variables, Ann. of Math., 64 (1956).

[9] H. McKean and P. Van Moerbeke, The spectrum of Hill's equations, Invent. Math., 1975, p. 1.

[10] H. McKean and P. Van Moerbeke, Sur le spectre de quelques opérateurs et les variétées de Jacobi, Sem. Bourbaki 1975/ No. 474.

[11] V. E. Zaharov and A. B. Shabat, A scheme for integrating the non-linear equations of math. physics by the method of the inverse scattering problem I, Functional Anal. Appl., 8 (1974) (transl., 1975, p. 226).

HARVARD UNIVERSITY

THE WORK OF C. P. RAMANUJAM IN ALGEBRAIC GEOMETRY

By D. MUMFORD

IT WAS a stimulating experience to know and collaborate with C. P. Ramanujam. He loved mathematics and he was always ready to take up a new thread or to pursue an old one with infectious enthusiasm. He was equally ready to discuss a problem with a first year student or a colleague, to work through an elementary point or to puzzle over a deep problem. On the other hand, he had very high standards. He felt the spirit of mathematics demanded of him not merely routine developments but the *right* theorem on any given topic. He wanted mathematics to be beautiful and to be clear and simple. He was sometimes tormented by the difficulty of these high standards, but, in retrospect, it is clear to us how often he succeeded in adding to our knowledge, results both new, beautiful and with a genuinely original stamp.

Our lives and researches intertwined considerably. I first met him in Bombay in 1967-68, when he took notes on my course in Abelian Varieties and we worked jointly on refining and understanding better many points related to this theory. Later, in 1970-71, we were together in Warwick where he ran seminars on étale cohomology and on classification of surfaces. His excitement and enthusiasm was one of the main factors that made that "Algebraic Geometry year" a success. We discussed many topics involving topology and algebraic geometry at that time, and especially Kodaira's Vanishing Theorem. My wife and I spent many evenings together with him, talking about life, religion and customs both in India and the West and we looked forward to a warm and continuing friendship. His premature death was a great shock to all who knew him. I will always miss his companionship and collaboration in the enterprise of mathematics.

I will give a short survey of his contributions to algebraic geometry. Perhaps his most perfect piece of work is his proof that a smooth

D. Mumford, *Selected Papers*, Vol. II,

affine complex surface X, which is contractable *and* simply connected at ∞, is isomorphic to the plane $\mathbf{C}^2$. The proof of this is not simple and uses many techniques; in particular, it shows how well he knew his way about in the classical geometry of surfaces! What is equally astonishing is his very striking counter-example showing that the hypothesis "simply connected at ∞" cannot be dropped. The position of this striking example in a general theory of 4-manifolds and particularly in a general theory of the topology of algebraic surfaces is yet to be understood. As mentioned above, the Kodaira Vanishing Theorem was an enduring interest of his. Both of us were particularly fascinated by this "deus ex machina", an intrusion of analytic tools (i.e., harmonic forms) to prove a purely algebraic theorem. His two notes on this subject went a long way to clarifying this theorem: (a) he proves it by merely topological, not analytic, techniques and (b) he finds a really satisfactory definitive extension of the theorem to a large class of non-ample divisors on surfaces. This second point is absolutely essential for many applications and was used immediately and effectively by Bombieri in his work on the pluricanonical system $| nK |$ for surfaces of general type. His result is that if D is a divisor on X, such that $(D^2) > 0$ and $(D. C) \geqslant 0$ for all effective curves C, then $H^1(X, \mathcal{O}(-D)) = (0)$.

His earliest paper, on automorphisms group of varieties, is a definitive analysis of the way this group inherits an algebraic structure from the variety itself. This work employs the techniques of functors, e.g., families of automorphisms developed by Gröthendieck at about the same time. His paper "On a certain purity theorem" addresses itself to a question of Lang that puzzled almost all algebraic geometers at that time: *given a proper surjective morphism $f: X \to Y$ between smooth varieties, is the set*

$$\{y \in Y \mid f^{-1}(Y) \text{ singular}\}$$

of codimension 1 *in* Y? Here he provides a topologico-algebraic analysis of one good case where it is true, and describes a counter-example to the general case worked out jointly with me. We again see his fascination with the interactions between purely topological techniques and algebro-geometric ones.

This interest comes out again in his joint paper with Le Dung-Trang, whose Main Result is described in the title: "The invariance of Milnor's number implies the invariance of the topological type." Here they are concerned with a family of hypersurfaces in $\mathbf{C}^{n+1}$: $F_t(z_0, \dots, z_n) = 0$, with isolated singularities at the origin, whose coefficients are C^∞ functions of $t \in [0, 1]$. They show that when Milnor's number μ_t, giving the number of vanishing cycles at the origin, is independent of t, then if $n \neq 2$, the germs of the maps F_t: $\mathbf{C}^{n+1} \to \mathbf{C}$ near 0 are independent of t, up to homeomorphism (if $n = 2$, they get a slightly weaker result). A beautiful and intriguing corollary is that the Artin local ring

$$\mathbf{C}\,[[z_0, \dots, z_n]] \Big/ \left(F, \frac{\partial F}{\partial z_0}, \dots, \frac{\partial F}{\partial z_n}\right)$$

already determines the topology of the map F near 0.

Finally, his paper "On a geometric interpretation of multiplicity" proves essentially the following elegant theorem: *If $Y \subset X$ is a closed subscheme defined by $I_Y \subset \mathcal{O}_X$, which blows up to a divisor $E \subset X'$, then*

$$\frac{(-1)^{n-1}(E^n)}{n!} = \left[\begin{array}{c}\textit{leading coefficient of the polynomial} \\ P(k) = X(\mathcal{O}_X/I_Y^k),\ k \gg 0\end{array}\right].$$

In addition to these published papers, Ramanujam made many contributions to my book "Abelian Varieties", while writing up notes from my lectures. Reprinted here is the Appendix by him on Tate's Theorem on abelian varieties over finite fields; and the following extraordinary theorem: It had been proven by Weil that if X is a projective variety and m: $X \times X \to X$ is a morphism, then if m makes X into a group, m must satisfy the commutative law too. Ramanujam proved that if m merely possessed a 2-sided identity ($m(x, e) = m(e, x) = x$), then m must also have an inverse and satisfy the associative law, hence make X into a group!

SOME FOOTNOTES TO THE WORK OF C. P. RAMANUJAM

By D. MUMFORD

THIS PAPER consists of a series of remarks, each of which is connected in some way with the work of Ramanujam. Quite often, in the last few years, I have been thinking on some topic, and suddenly I realize—Yes, Ramanujam thought about this too—or—This really links up with his point of view. It is uncanny to see how his ideas continue to work after his death. It is with the thought of embellishing some of his favourite topics that I write down these rather disconnected series of results.

I

The first remark is a very simple example relevant to the purity conjecture (sometimes called Lang's conjecture) discussed in Ramanujam's paper [10]. The conjecture was—let

$$f\colon X^n \to Y^m$$

be a proper map of an n-dimensional smooth variety onto an m-dimensional smooth variety with all fibres of dimension $n - m$. Assume the characteristic is zero. Then show

$$\{y \in Y \mid f^{-1}(y) \text{ is singular}\}$$

has codimension one in Y. When $n = m$, this result is true and is known as "purity of the branch locus"; when $n = m + 1$, it is also true and was proven by Dolgačev, Simha and Ramanujam. When $n = m + 2$, Ramanujam describes in [10] a counter-example due to us jointly. Here is another counter-example for certain large values of $n - m$.

We consider the following very special case for f. Start with $Z^r \subset \mathbf{P}^m$ an arbitrary subvariety. Let $\check{\mathbf{P}}^m$ be the dual projective space—the space of hyperplanes in $\mathbf{P}^m$. The dual variety $\check{Z} \subset \check{\mathbf{P}}^m$ is, by definition the Zariski-closure of the locus of hyperplanes H such that, at some smooth point $x \in Z^r$, $T_{x,H} \supset T_{x,Z}$. It is apparently well known, although I don't know a reference, that in characteristic 0,

$$\check{\check{Z}} = Z.$$

Consider the special case where Z is smooth and spans $\mathbf{P}^m$. Then we don't need to take the Zariski-closure in the above definition and, in fact, the definition of $\check{Z}$ can be reformulated like this:

Let

$$I \subset \mathbf{P}^m \times \check{\mathbf{P}}^m$$

be the universal family of hyperplanes, i.e., if $(X_0, \ldots, X_m)$, resp. $(\xi_0, \ldots, \xi_m)$ are coördinates in $\mathbf{P}^m$, resp. $\check{\mathbf{P}}^m$, then I is given by

$$\Sigma \xi_i X_i = 0.$$

Let

$$X = I \cap (Z^r \times \check{\mathbf{P}}^m).$$

Note that I and $Z^r \times \check{\mathbf{P}}^m$ are smooth subvarieties of $\mathbf{P}^m \times \check{\mathbf{P}}^m$ of codimension 1 and $m - r$ respectively. One sees immediately that they meet transversely, so X is smooth of dimension $m + r - 1$. Consider

$$p_2 \colon X \to \check{\mathbf{P}}^m.$$

Its fibres are the hyperplane sections of Z, all of which have dimension $r - 1$. Thus p_2 is a morphism of the type considered in the conjecture. In this case

$$\{\xi \in \check{\mathbf{P}}^m \mid p_2^{-1}(\xi) \text{ singular}\} = \{\xi \in \check{\mathbf{P}}^m \mid \text{if } \xi \text{ corresponds to } H \subset \mathbf{P}^m, \text{ then } Z.H \text{ is singular}\}$$

$$= \check{Z}.$$

Thus the conjecture would say that the dual $\check{Z}$ of a smooth variety Z spanning $\mathbf{P}^m$ is a hypersurface.

I claim this is false, although I feel sure it can only be false in very rare circumstances. In fact, I don't know any cases other than the following example where it is false.† Simply take

Z^r = [Grassmannian of lines in $\mathbf{P}^{2k}$, $k > 1$].

Here $r = 2(2k-1)$, $m = k(2k+1)-1$ and the embedding $i: Z^r \subset \mathbf{P}^m$ is the usual Plücker embedding. In vector space form, let

V = a complex vector space of dimension $2k+1$

Z = set of 2-dimensional subspaces $W_2 \subset V$

$\mathbf{P}^m$ = set of 1-dimensional subspaces $W_1 \subset \Lambda^2 V$

i = map taking W_2 to $W_1 = \Lambda^2 W_2$.

Note that we may identify

$\check{\mathbf{P}}^m$ = set of 1-dimensional subspaces $W_1' \subset \Lambda^2 V^*$, where

$\Lambda^2 V^*$ = space of skew-symmetric 2-forms $A: V \times V \to \mathbf{C}$.

Write $[W_2] \in Z$ for the point defined by W_2, and $H_A \subset \mathbf{P}^m$ for the hyperplane defined by a 2-form A. Then it is immediate from the definitions that

$$i([W_2]) \in H_A \Leftrightarrow \mathrm{res}_{W_2} A \text{ is zero.}$$

To determine when moreover,

$$i_*(T_{W_2, Z}) \subset T_{i(W_2), H_A},$$

let $v_1, v_2 \in W_2$ be a basis, and make a small deformation of W_2 by taking $v_1 + \epsilon v_1'$, $v_2 + \epsilon v_2'$ to be a basis of $\widetilde{W}_2 \subset V \otimes \mathbf{C}[\epsilon]$. Then $\widetilde{W}_2$ represents a tangent vector t to Z at $[W_2]$ and

$$i_*(t) \subset T_{i(W_2), H_A} \Leftrightarrow A(v_1 + \epsilon v_1', v_2 + \epsilon v_2') \equiv 0 \pmod{\epsilon^2}$$

$$\Leftrightarrow A(v_1', v_2) + A(v_1, v_2') = 0,$$

Thus:

$$i_*(T_{W_2, Z}) \subset T_{i(W_2), H_A} \Leftrightarrow \text{for all } v_1', v_2', \in V,$$

$$A(v_1, v_2') + A(v_1', v_2) = 0$$

$$\Leftrightarrow W_2 \subset (\text{nullspace of } A).$$

†M. Reid has indicated to me another set of examples: Suppose

$$Z = \mathbf{P}(E)$$

where E is a vector bundle of rank s on y^t, so $r = s + t - 1$, and the fibres of $\mathbf{P}(E)$ 1
are embedded linearly. Then if $s > t + 2$, $\check{\dot{Z}}$ is not a hypersurface.

Therefore

$$H_A \text{ is tangent to } i(Z) \Leftrightarrow \dim(\text{nullspace}) \geqslant 2.$$

Now the nullspace of A has odd dimension, and if it is 3, one counts the dimension of the space of such A as follows:

$$\dim\begin{pmatrix}\text{space of } A\text{'s with}\\ \dim(\text{nullspace}) = 3\end{pmatrix} = \dim\begin{pmatrix}\text{space of}\\ W_3 \subset V\end{pmatrix} + \dim \Lambda^2 (V/W_3)$$

$$= 3(2k-2) + \frac{(2k-2)(2k-3)}{2}$$

$$= 2k^2 + k - 3.$$

Thus $\dim\ \check{Z} = m - 3$, and $\operatorname{codim}\ \check{Z} = 3$! (Compare this with Buchsbaum-Eisenbud [3], where it is shown that $\check{Z} \subset \check{\mathbf{P}}^m$ is a "universal codimension 3 Gorenstein scheme".)

II

The second remark concerns the Kodaira Vanishing Theorem. We want to show that Ramanujam's strong form of Kodaira Vanishing for surfaces of Char. 0 is a consequence of a recent result of F. Bogomolov. In particular, this is interesting because it gives a new completely algebraic proof of this result, and one which uses the Char. 0 hypothesis in a new way (it is used deep in Bogomolov's proof, where one notes that if $V^3 \to F^2$ is a ruled 3-fold and $D \subset V^3$ is an irreducible divisor meeting the generic fibre set-theoretically in one point, then D is birational to F). Ramanujam's result [11] is this: let F be a smooth surface of Char. 0, D a divisor on F.

Then

2 $$\left.\begin{array}{l}(D^2) > 0\\ (D.C) \geqslant 0, \text{ all curves } C \subset F\end{array}\right\} \Rightarrow H^1(F, \mathcal{O}(-D)) = (0). \qquad (1)$$

Bogomolov's theorem is that if F is a smooth surface of char. 0, E a rank 2 vector bundle on F, then

$C_1(E)^2 > 4C_2(E) \Rightarrow E$ is unstable, meaning $\exists$ an extension

$$0 \to L(D) \to E \to I_Z L \to 0,$$

I_Z = ideal sheaf of a 0-dim. subscheme $Z \subset F$,

L invertible sheaf, D a divisor

$D \in$[num. pos. cone, $(D^2) > 0$, $(D.H) > 0$].

(2)

(See Bogomolov [2], Reid [13]; another proof using reduction mod p instead of invariant theory has been found by D. Gieseker.) 3

To prove (2) $\Rightarrow$ (1), suppose D_1 is given satisfying the conditions 4
of (1). Take any element $\alpha \in H^1(F, \mathcal{O}(-D_1))$ and via α, form an extension

$$0 \longrightarrow \mathcal{O}_F \xrightarrow{\mu} E \xrightarrow{\nu} \mathcal{O}_F(D_1) \longrightarrow 0.$$

Note that $C_1(E) = D_1$, $C_2(E) = 0$, $(D_1^2) > 0$, so E satisfies the conditions of (2). Therefore, by Bogomolov's theorem, E is unstable: this gives an exact sequence

$$0 \longrightarrow L(D_2) \xrightarrow{\sigma} E \xrightarrow{\tau} I_Z L \longrightarrow 0$$

$$D_2 \in \text{(num. pos. cone)}.$$

Note that the subsheaf $\sigma(L(D_2))$ of E cannot equal the subsheaf $\mu(\mathcal{O}_F)$ in the definition of E, because this would imply, comparing the 2 sequences, that $D_2 \equiv -D_1$ whereas both D_1, D_2 are in the numerically positive cone. Therefore, the composition

$$L(D_2) \xrightarrow{\sigma} E \xrightarrow{\nu} \mathcal{O}_F(D_1)$$

is not zero, hence

$$L \simeq \mathcal{O}_F(D_1 - D_2 - D_3),\ D_3 \text{ an effective divisor.}$$

Next, comparing Chern classes of E in its 2 presentations, we find

$$2C_1(L) + D_2 \equiv C_1(E) \equiv D_1 \qquad (3a)$$

$$(C_1(L) + D_2).\ C_1(L) + \deg Z = C_2(E) = 0. \qquad (3b)$$

By (3a), we find $D_1 - D_2 - 2D_3 \equiv 0$, hence $L \simeq \mathcal{O}_F(+D_3)$; by (3b), we find $(D_1 - D_3).D_3 < 0$. But

$$\det\begin{vmatrix}(D_1^2) & (D_1.D_3)\\ (D_1.D_3) & (D_3^2)\end{vmatrix} = (D_1^2)[(D_3^2)-(D_1.D_3)]+(D_1.D_3)[(D_1^2)-2(D_1.D_3)] + (D_1.D_3)^2$$

while $(D_3^2) - (D_1.D_3) \geqslant 0$ (by 3b)

$(D_1^2) - 2(D_1.D_3) = (D_1.D_2) > 0$ (since D_1, D_2 num. pos.)

$(D_1.D_3) \geqslant 0$ (by the assumptions on D_1).

On the other hand, this det is $\leqslant 0$ by Hodge's Index Theorem. Therefore $(D_1.D_3) = 0$ and $\det = 0$. From the latter, D_3 is numerically equivalent to λD_1, $\lambda \in \mathbf{Q}$, hence $(D_1.D_3) = \lambda(D_1^2)$. Thus $\lambda = 0$ and since D_3 is effective, $D_3 = 0$. Therefore the subsheaf $\sigma(L(D_2))$ is isomorphic to $\mathcal{O}_F(D_1)$ and defines a splitting of the original exact sequence. Therefore the extension class $\alpha \in H^1(\mathcal{O}_F(-D_1))$ is 0, so $H^1(\mathcal{O}_F(-D_1)) = (0)$.

III

The last two remarks are applications of Kodaira's Vanishing Theorem. To me it is quite amazing how this cohomological assertion has such strong consequences, both for geometry and for local algebra. Here is a geometric application. This application is a link between the recent paper of Arakelov [1] (proving Shafarevich's finiteness conjecture on the existence of families of curves over a fixed base curve, with prescribed degenerations), and Raynaud's counter-example [12] to Kodaira Vanishing for smooth surfaces in char. p. What I claim is this (this remark has been observed by L. Szpiro also):

Proposition. *Let $p: F \to C$ be a proper morphism of a smooth surface F onto a smooth curve C over a field k of arbitrary characteristic. Let $E \subset F$ be a section of p and assume the fibres of p have positive arithmetic genus. Let F_0 be the normal surface obtained by blowing down all components of fibres of p not meeting E. Then*:

Kodaira's Vanishing Theorem for ample divisors on F_0 $\Longrightarrow (E^2) < 0$.

If Char $(k) = 0$, then Kodaira's Vanishing Theorem holds for F_0 (cf. [9]), so $(E^2) < 0$ follows. This result, and its refinement — $(E^2) < 0$ unless all the smooth fibres of p are isomorphic — are due to Arakelov [1], who proved them by a very ingenious use of the Weierstrass points of the fibres $p^{-1}(x)$. On the other hand, if char(k) $=p$, Raynaud has shown how to construct examples of morphisms p: $F \to C$ and sections $E \subset F$, where all the fibres of p are irreducible but singular (thus $F = F_0$), and $(E^2) > 0$. Thus Kodaira Vanishing is false for this F. If char$(k) = 2$ or 3, he finds in fact quasi-elliptic surfaces F of this type. This Proposition is, in fact, merely an elaboration of the last part of Raynaud's example.

Proof of Proposition: Suppose $(E^2) > 0$. Let p_0: $F_0 \to C$ be the projection and let E stand for the image of E in F_0 too. Consider divisors on F_0 of the form

$$H = E + p_0^{-1}(\mathfrak{A}), \deg \mathfrak{A} > 0.$$

Then $(H^2) > 0$ and $(H.C) > 0$ for all curves C on F_0, so H is ample by the Nakai-Moisezon criterion. On the other hand, let's calculate $H^1(F_0, \mathcal{O}(-H))$. We have

$$0 \to H^1(C, p_{0,*}\mathcal{O}(-H)) \to H^1(F_0, \mathcal{O}(-H)) \to H^0(C, R^1 p_{0,*}\mathcal{O}(-H)) \to 0.$$

Clearly

$$p_{0,*}\mathcal{O}(-H) = (0) \text{ and } R^1 p_{0,*}\mathcal{O}(-H) \simeq (R^1 p_{0,*}\mathcal{O}(-E)) \otimes \mathcal{O}_C(-\mathfrak{A}).$$

Now using the sequences:

$$0 \to \mathcal{O}_{F_0}(-E) \to \mathcal{O}_{F_0} \to \mathcal{O}_E \to 0$$
$$0 \to \mathcal{O}_{F_0} \to \mathcal{O}_{F_0}(E) \to \mathcal{O}_E((E^2)) \to 0$$

we find

$$\begin{array}{ccccccccc} p_{0,*}\,\mathcal{O}_{F_0} & \xrightarrow{\alpha} & p_{0,*}\,\mathcal{O}_E & \longrightarrow & R^1 p_{0,*}\,\mathcal{O}(-E) & \xrightarrow{\beta} & R^1 p_{0,*}\,\mathcal{O}_{F_0} & \longrightarrow & 0 \\ \| & & \| & & & & & & \\ \mathcal{O}_C & & \mathcal{O}_C & & & & & & \end{array}$$

so α and β are isomorphisms, and (using the fact that the genus of the fibres is positive):

$$\begin{array}{ccccccccc} 0 & \longrightarrow & p_{0,*}\,\mathcal{O}_{F_0} & \xrightarrow{\gamma} & p_{0,*}\,\mathcal{O}(E) & \longrightarrow & p_{0,*}\,\mathcal{O}_E((E^2)) & \xrightarrow{\delta} & R^1 p_{0,*}\,\mathcal{O}_{F_0} \\ & & \| & & \| & & & & \\ & & \mathcal{O}_C & & \mathcal{O}_C & & & & \end{array}$$

so γ is an isomorphism, and δ is injective. Now via the isomorphism
5 resp: $E \to C$, let the divisor class (E^2) on E correspond to the divisor class $\mathfrak{A}$ on C. Then $p_{0,*}\,\mathcal{O}_E((E^2)) \simeq \mathcal{O}_C(\mathfrak{A})$, and we see that

$$\mathcal{O}_C(\mathfrak{A}) \subset R^1 p_{0,*}\,\mathcal{O}_{F_0} \simeq R^1\, p_{0,*}\,\mathcal{O}_{F_0}(-E)$$

hence

$$\mathcal{O}_C \subset R^1\, p_{0,*}\,\mathcal{O}_{F_0}(-H)$$

hence

6 $$H^1(C, \mathcal{O}(-H)) \neq (0).$$

Q.E.D.

IV

The last remark is an application of Kodaira's Vanishing Theorem to local algebra. It seems to me remarkable that such a global result should be useful to prove local statements about the non-existence of local rings, but this is the case. The question I want to study is that of the *smoothability of non-Cohen-Macauley surface singularities*. In other words, given a surface F, $P \in F$ a non-CM-singular point, when does there exist a flat family of surfaces F_t parametrized by $k\,[[t]]$ such that $F_0 = F$ while the generic F_t is smooth. More locally, the problem is:

Given a complete non-CM-purely†-2-dimensional local ring $\mathcal{O}$ without nilpotents, when does there exist a complete 3-dimensional local ring $\mathcal{O}'$ and a non-zero divisor $t \in \mathcal{O}'$ such that

a) $\mathcal{O} \simeq \mathcal{O}'/t\mathcal{O}'$

b) $\mathcal{O}_{\mathfrak{p}}$ regular for all prime ideals $\mathfrak{p} \subset \mathcal{O}$ with $t \notin \mathfrak{p}$.

† i.e. $\mathcal{O}/I$ is 2-dimensional for all minimal prime ideals of $I \subset \mathcal{O}$

First of all, let

$$\mathcal{O}^* = \bigoplus_{\substack{I \subset \mathcal{O} \\ \text{minimal} \\ \text{prime ideals}}} \begin{pmatrix} \text{integral closure of } \mathcal{O}/I \\ \text{in its fraction field} \end{pmatrix}$$

and let

$$\widetilde{\mathcal{O}} = \left\{ a \in \mathcal{O}^* \,\middle|\, \begin{array}{l} m^n a \subset \mathcal{O} \text{ for some } n \geq 1 \\ m = \text{maximal ideal in } \mathcal{O} \end{array} \right\}$$

$$= \Gamma\,(\text{Spec } \mathcal{O}\text{-closed pt.}, \mathcal{O})$$

Note that $\widetilde{\mathcal{O}}$ is a finite $\mathcal{O}$-module and $m^n . \widetilde{\mathcal{O}} \subset \mathcal{O}$ for some large n, so that $\widetilde{\mathcal{O}}/\mathcal{O}$ is an $\mathcal{O}$-module of finite length. Moreover, it is easy to see that $\widetilde{\mathcal{O}}$ is a semi-local Cohen-Macauley ring. It has been proven by Rim [14] (cf. also Hartshorne [6], Theorem 2.1, for another proof) that:

$$\mathcal{O} \text{ smoothable} \Rightarrow \widetilde{\mathcal{O}} \text{ local.}$$

The result we want to prove is:

THEOREM. ***Assume*** **char $(\mathcal{O}/m) = 0$, Spec $\mathcal{O}$** ***has an isolated singularity at its closed point and that $\mathcal{O}$ is smoothable, so that, by the remarks above, $\widetilde{\mathcal{O}}$ is a normal local ring. Let π: $X^* \to$ Spec $\mathcal{O}$ be a resolution and let*** 7

$$p_a\,(\widetilde{\mathcal{O}}) = l\,(R^1\pi_*\,\mathcal{O}_X)$$

be the genus of the singularity $\widetilde{\mathcal{O}}$. Then

$$l\,(\widetilde{\mathcal{O}}/\mathcal{O}) < p_a\,(\widetilde{\mathcal{O}}).$$

Actually, for our applications, we want to know this result for rings $\mathcal{O}$ where Spec $\mathcal{O}$ has ordinary double curves too, with a suitable definition of p_a. We will treat this rather technical generalization in an appendix.

For example, the theorem shows:

COROLLARY. ***Let $\widetilde{\mathcal{O}} = k[[x, y]]$, char $k = 0$. Let $I \subsetneq (x, y)$ be an ideal of finite codimension. Then if $\mathcal{O} = k + I$, $\mathcal{O}$ is not smoothablé.***

On the other hand, if $F \subset \mathbf{P}^n$ is an elliptic ruled surface and $\mathcal{O}'$ is the completion of the local ring of the cone over F at its apex, then $\mathcal{O}'$ is a normal 3-dimensional ring which is not Cohen-Macauley. If $C = V(t) = (F.\mathbf{P}^{n-1})$ is a generic hyperplane section of F, then $t \in \mathcal{O}'$ and $\mathcal{O} = \mathcal{O}'/t\mathcal{O}'$ is the completion of the local ring of the cone over C at its apex. Now C is an elliptic curve, but embedded by an incomplete linear system — in fact, C is a projection of an elliptic curve $\widetilde{C}$ in $\mathbf{P}^n$ from a point not on $\widetilde{C}$ — this follows from the exact sequence:

$$0 \to H^0(\mathcal{O}_F) \to H^0(\mathcal{O}_F(1)) \to H^0(\mathcal{O}_C(1)) \to \underset{\substack{\| \\ \mathbf{C}}}{H^1(\mathcal{O}_F)} \to 0.$$

Let $\widetilde{\mathcal{O}}$ be the completion of the local ring of the cone over $\widetilde{C}$ at its apex. Then $\widetilde{\mathcal{O}}$ is a normal 2-dimensional ring, in fact an "elliptic singularity", i.e., $p_a(\widetilde{\mathcal{O}}) = 1$; moreover $\widetilde{\mathcal{O}} \supset \mathcal{O}$ and $\dim \widetilde{\mathcal{O}}/\mathcal{O} = 1$. This shows that there are smoothable singularities $\mathcal{O}$ with

$$l(\widetilde{\mathcal{O}}/\mathcal{O}) = p_a(\widetilde{\mathcal{O}}) = 1.$$

Proof of Theorem. Let $\mathcal{O} \simeq \mathcal{O}'/t\mathcal{O}'$ give the smoothing of $\mathcal{O}$. The proof is based on an examination of the exact sequence of local cohomology groups:

$$(*) \ldots \longrightarrow H^1_{\{x\}}(\mathcal{O}') \longrightarrow H^1_{\{x\}}(\mathcal{O}) \longrightarrow H^2_{\{x\}}(\mathcal{O}') \xrightarrow{\ t\ } H^2_{\{x\}}(\mathcal{O}')$$
$$\xrightarrow{\ \alpha\ } H^2_{\{x\}}(\mathcal{O}) \longrightarrow \ldots$$

where $x \in \operatorname{Spec} \mathcal{O} \subset \operatorname{Spec} \mathcal{O}'$ represents the closed point.

What can we say about each of these groups?

(a) $H^1_{\{x\}}(\mathcal{O}')$ is zero since $\mathcal{O}'$ is an integrally closed ring of dimension 3, hence has depth at least 2.

(b) To compute $H^1_{\{x\}}(\mathcal{O})$, use

$$H^0(\operatorname{Spec} \mathcal{O}, \mathcal{O}) \to H^0(\operatorname{Spec} \mathcal{O} - \{x\}, \mathcal{O}) \to H^1_{\{x\}}(\mathcal{O}) \to 0$$

which gives us:

$$H^1_{\{x\}}(\mathcal{O}) \simeq \widetilde{\mathcal{O}}/\mathcal{O}.$$

(c) As for $H^2_{\{x\}}(\mathcal{O}')$, it measures the degree to which $\mathcal{O}'$ is not Cohen-Macauley. A fundamental fact is that it is of finite length—cf. Theoreme de finitude, p. 89, in Grothendieck's seminar [15].

(d) As for $H^2_{\{x\}}(\mathcal{O})$, we can say at least:

$$\begin{aligned} H^2_{\{x\}}(\mathcal{O}) &\simeq H^1(\operatorname{Spec}\mathcal{O} - \{x\}, \mathcal{O}) \\ &\simeq H^1(\operatorname{Spec}\widetilde{\mathcal{O}} - \{x\}, \widetilde{\mathcal{O}}) \\ &\simeq H^2_{\{x\}}(\widetilde{\mathcal{O}}) \end{aligned}$$

but unfortunately this group is huge: it is not even an $\widetilde{\mathcal{O}}$-module of finite type.

However, for any local ring $\mathcal{O}$ with residue characteristic O and with isolated singularity, we can define, by using a resolution of Spec $\mathcal{O}$, important subgroups:

$$H^i_{\{x\},\,\mathrm{int}}(\mathcal{O}).$$

Namely, let $\pi: X \to \operatorname{Spec}\mathcal{O}$ be a resolution and set

$$H^i_{\{x\},\,\mathrm{int}}(\mathcal{O}) = \operatorname{Ker}[\pi^*: H^i_{\{x\}}(\mathcal{O}) \to H^i_{\pi^{-1}x}(\mathcal{O}_X)].$$

This is independent of the resolution, as one sees by comparing any 2 resolutions $\pi_i: X_i \to \operatorname{Spec}\mathcal{O}$, $i = 1,2$, via a 3rd:

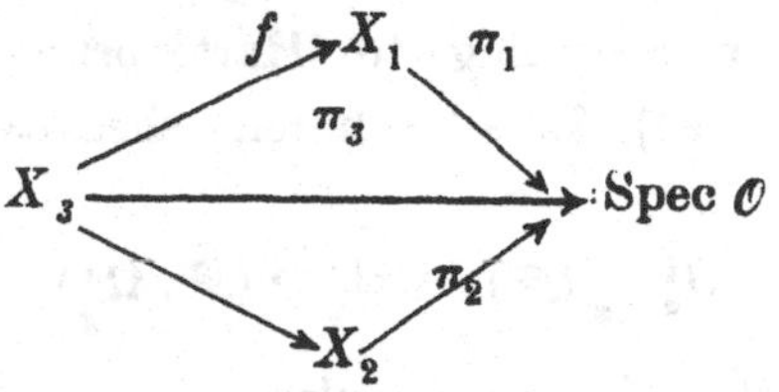

and using the Leray spectral sequence

$$H^p_{\pi_1^{-1}(x)}(X_1, R^q f_*(\mathcal{O}_{X_3})) \Rightarrow H^*_{\pi_3^{-1}(x)}(X_3, \mathcal{O}_{X_3})$$

plus Matsumura's result $R^q f_* \mathcal{O}_{X_3} = (0)$, $q > 0$ when X_1 and X_3 are 8
smooth and characteristic zero. Moreover, when $\mathcal{O} \simeq \mathcal{O}'/I$, then the restriction map

$$H^i_{\{x\}}(\mathcal{O}') \longrightarrow H^i_{\{x\}}(\mathcal{O})$$

gives

$$H^i_{\{x\},\mathrm{int}}(\mathcal{O}') \longrightarrow H^i_{\{x\},\mathrm{int}}(\mathcal{O})$$

because we can find resolutions fitting into a diagram:

$$\begin{array}{ccc} X & \hookrightarrow & X' \\ \downarrow & & \downarrow \\ \mathrm{Spec}\,\mathcal{O} & \hookrightarrow & \mathrm{Spec}\,\mathcal{O}'. \end{array}$$

Next, we prove using the Kodaira Vanishing Theorem and following Hartshorne and Ogus ([16], p. 424):

Lemma. *Assume x is the only singularity of $\mathcal{O}$, $\dim \mathcal{O} = n$ and $\pi: X \to \mathrm{Spec}\,\mathcal{O}$ is a resolution. Then*

$$H^i_{\{x\},\mathrm{int}}(\mathcal{O}) \simeq H^i_{\{x\}}(\mathcal{O}),\ 0 < i < n-1$$

9 *and*

$$H^i_{\{x\},\mathrm{int}}(\mathcal{O}) \simeq R^{i-1}\pi_*(\mathcal{O}_X)_x,\ 2 < i < n.$$

Proof. Because $\mathcal{O}$ has an isolated singularity, we may assume $\mathcal{O} \simeq \hat{\mathcal{O}}_{x,X_0}$, where X_0 is an n-dimensional projective variety with x its only singular point. We may assume our resolution is global:

$$\pi : X \to X_0.$$

Let $\hat{X} = X \times_{X_0} \mathrm{Spec}\,\mathcal{O}$ and let I be the injective hull of $\mathcal{O}_{x,X_0}/m_{x,X_0}$ as $\mathcal{O}_{x,X_0}$-module. Then according to Hartshorne's formal duality theorem (cf. [7], p. 94), for all coherent sheaves $\mathcal{F}$ on X, the 2 $\mathcal{O}$-modules

10 $$H^i_{\pi^{-1}x}(\mathcal{F}),\ \mathrm{Ext}^{n-i}_{\mathcal{O}_{\hat{X}}}(\hat{\mathcal{F}}, \Omega^n_{\hat{X}})$$

are dual via Hom $(-, I)$. In particular,

$$H^i_{\pi^{-1}x}(\mathcal{O}_X),\ H^{n-i}(\Omega^n_{\hat{X}})$$

are dual. But

$$H^{n-i}(\Omega^n_{\hat{X}}) \simeq R^{n-i}\pi_*(\Omega^n_X) \otimes_{\mathcal{O}_{X_0}} \mathcal{O}$$

and it has been shown by Grauert and Riemenschneider [5] that $R_i\pi_*(\Omega^n_X) = (0)$, $i > 0$. (This is a simple consequence of Kodaira's Vanishing Theorem because if L_0 is an ample invertible sheaf on X_0

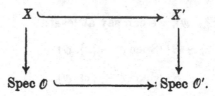

with $H^i(X_0, L_0\otimes\pi_*\Omega^n_X)=(0)$, $i>0$, then by the Leray Spectral Sequence:

$$H^i(X, \pi^* L_0\otimes\Omega^n_X) \simeq H^0(X_0, L_0\otimes R^i\pi_*\Omega^n_X)$$

$$\Big\updownarrow \text{dual}$$

$$H^{n-i}(X, (\pi^* L_0)^{-1})$$

and Kodaira's Vanishing Theorem applies to all invertible sheaves M such that $\Gamma(X, M^n)$ is base point free and defines a birational morphism, $n>0$ (cf. [9]).) Recapitulating, this shows $H^{n-i}(\Omega^n_{\hat{X}})=(0)$, $i<n$, hence $H^i_{\pi^{-1}x}(\mathcal{O}_X)=(0)$, $i<n$, hence $H^i_{\{x\}\text{int}}(\mathcal{O}) \xrightarrow{\approx} H^i_{\{x\}}(\mathcal{O})$ is an isomorphism.

To get the second set of isomorphisms, we use the Leray Spectral Sequence:

$$H^p_{\{x\}}(X_0, R^q\pi_*\mathcal{O}_X) \Rightarrow H^{\bullet}_{\pi^{-1}x}(X, \mathcal{O}_X).$$

The only non-zero terms occur for $p=0$ or $q=0$, so we get a long exact sequence

$$\dots\to H^0_{\{x\}}(R^{i-1}\pi_*\mathcal{O}_x)\to H^i_{\{x\}}(\pi_*\mathcal{O}_x)\to H^i_{\pi^{-1}x}(\mathcal{O}_X)\to H^0_{\{x\}}(R^i\pi_*\mathcal{O}_x)$$
$$\to H^{i+1}_{\{x\}}(\pi_*\mathcal{O}_X)\to\dots$$

Using the first part, plus the isomorphism:

$$\begin{aligned} H^i_{\{x\}}(\pi_*\mathcal{O}_X) &\simeq H^{i-1}(\operatorname{Spec}\mathcal{O}-\{x\}, \pi_*\mathcal{O}_X)\\ &\simeq H^{i-1}(\operatorname{Spec}\mathcal{O}-\{x\}, \mathcal{O})\\ &\simeq H^i_{\{x\}}(\mathcal{O}),\ i>2,\end{aligned}$$

we get the results. *Q.E.D.*

We now go back to the sequence (*). It gives us:

$$0\longrightarrow\tilde{\mathcal{O}}/\mathcal{O}\longrightarrow H^2_{\{x\},\text{int}}(\mathcal{O}')\xrightarrow{t} H^2_{\{x\},\text{int}}(\mathcal{O}')\longrightarrow R^1\pi_*(\mathcal{O}_X)_x\longrightarrow\dots$$

where $\pi: X\to\operatorname{Spec}\mathcal{O}$ is a resolution. Therefore

$$\begin{aligned} l(\tilde{\mathcal{O}}/\mathcal{O}) &= l(\ker \text{ of } t \text{ in } H^2_{\{x\},\text{int}}(\mathcal{O}'))\\ &= l(\text{Coker of } t \text{ in } H^2_{\{x\},\text{int}}(\mathcal{O}'))\\ &\leq l(R^1\pi_*(\mathcal{O}_X)_x) = p_a(\tilde{\mathcal{O}}).\end{aligned}$$

Q.E.D.

APPENDIX

The purpose of this appendix is to make a rather technical extension of the result in §IV, which seems to be better for use in applications. Let X be an affine surface, reduced, with at most ordinary double curves, plus one point $P \in X$ about which we know nothing. Let

$$\tilde{X} = \operatorname{Spec} \Gamma(X-P, \mathcal{O}_X)$$

so that we get

$$\pi_1 : \tilde{X} \to X,$$

an isomorphism outside P, everywhere a finite morphism, with $\tilde{X}$ Cohen-Macauley. Our goal is to show that in certain cases X is not smoothable near P, i.e. $\nexists$ an analytic family

$$f: X' \to \Delta = \text{disc in the } t\text{-plane},$$

where $f^{-1}(0) \approx$ (neighborhood of P in X), and $f^{-1}(t)$ is smooth, $t \neq 0$. (Here we work in the analytic setting rather than the formal one to be able below to take an exponential.) We know that a necessary condition for X' to exist is that $\pi_1^{-1}(P)$ is one point $\tilde{P}$, so henceforth we assume this too. Next blow up $\tilde{X}$, but only at $\tilde{P}$ and at centers lying over $\tilde{P}$: it is not hard to see that we arrive in this way at a birational proper morphism

$$\pi_2 : X^* \to \tilde{X}$$

such that X^* has at most ordinary double curves and pinch points (points like $z^2 = x^2y$), these pinch points moreover lying over $\tilde{P}$. Define

$$p_a(\mathcal{O}_{\tilde{P}}) = \dim_{\mathbf{C}}[R^1\pi_{2,*}(\mathcal{O}_{X^*})_{\tilde{P}}].$$

It is easy to verify that this number is independent of the choice of X^*. (However, this would *not* be true if $\tilde{X}$ had cuspidal lines—in this case, there is no bound on dim $R^1\pi_*$ as you blow up $\tilde{X}$ more and more!) We claim the following

THEOREM. *If X is smoothable near P, then $l(\mathcal{O}_{\tilde{P}}/\mathcal{O}_P) \leq p_a(\mathcal{O}_{\tilde{P}})$.*

PROOF. We follow the same plan as in the case where $\mathcal{O}$ has an isolated singularity, except that, for an arbitrary local ring $\mathcal{O}$, we set

$$H^i_{\{x\},\,\mathrm{int}}(\mathcal{O}) = \bigcup_{\left(\substack{\text{modifications}\\ \pi: X\to \operatorname{Spec}\mathcal{O},\\ \text{where}\\ X-\pi^{-1}(x) \xrightarrow{\approx} \operatorname{Spec}\mathcal{O}-\{x\}}\right)} [\ker: H^i_{\{x\}}(\mathcal{O}) \to H^i_{\pi^{-1}(x)}(\mathcal{O}_X)]$$

The proof is then the same as before except that we cite only the following case of the lemma:

THEOREM (Boutot [17]): *Let $\mathcal{O}$ be a normal excellent local k-algebra, with residue field k, and* char $(k) = 0$. *Then*:

$$H^2_{\{x\},\mathrm{int}}(\mathcal{O}) = H^2_{\{x\}}(\mathcal{O}).$$

This result is a Corollary of Proposition 2.6, Chapter V [17]. Since 11
char$(k) = 0$, we may disregard "red" in that Proposition and apply it to the values of the functor on the dual numbers. It tells us that there is a blow-up $\pi: X \to \operatorname{Spec}(\mathcal{O})$ concentrated at the origin such that

$$\operatorname{Pic}_{X/k}(k[\epsilon]/(\epsilon^2)) \longrightarrow \operatorname{Pic}_{\operatorname{Spec}(\mathcal{O})-\{x\}}(k[\epsilon]/(\epsilon^2))$$

is an isomorphism. In other words, in the sequence:

$$\longrightarrow H^1(X, \mathcal{O}_X) \xrightarrow{\alpha} H^1(X-\pi^{-1}(x), \mathcal{O}_X) \xrightarrow{\beta} H^2_{\pi^{-1}(x)}(\mathcal{O}_X) \longrightarrow$$

α is surjective, hence β is zero.

BIBLIOGRAPHY

1. S. JU. ARAKELOV: Families of algebraic curves with fixed degeneracies, *Izvest. Akad. Nauk*, 35 (1971).

2. F. BOGOMOLOV, to appear. 12

13 3. D. Buchsbaum and D. Eisenbud: Algebra structures for finite free resolutions and some structure theorems for ideals of codimension 3, *Amer. J. Math.* (to appear).

4. R. Fossum: The divisor class group of a Krull domain, *Springer Verlag*, (1973).

5. H. Grauert and O. Riemenschneider: Verschwindungssätze für analytische Kohomologiegruppen auf komplexen Räumen, *Inv. Math.*, 11 (1970).

6. R. Hartshorne: Topological conditions for smoothing algebraic singularities, *Topology*, 13 (1974).

7. R. Hartshorne: Ample subvarieties of algebraic varieties, *Springer Lecture Notes 156* (1970).

8. H. Hironaka: Resolution of singularities of an algebraic variety over a field of char. zero, *Annals of Math.*, 79 (1964).

9. D. Mumford: Pathologies III, *Amer. J. Math.*, 89 (1967).

10. C. P. Ramanujam: On a certain purity theorem, *J. Indian Math. Soc.*, 34 (1970).

11. C. P. Ramanujam: Remarks on the Kodaira Vanishing Theorem, *J. Indian Math. Soc.*, 36 (1972) and 38 (1974).

12. M. Raynaud: Contre-example au "Vanishing Theorem" en caractérisque $p > 0$, this volume.

14 13. M. Reid: Bogomolov's theorem $C^2_1 \leqslant 4C_2$, to appear in *Proc. Int. Colloq. in Alg. Geom.*, Kyoto, (1977).

14. D. Rim: Torsion differentials and deformation, *Trans. Amer. Math. Soc.*, 169 (1972).

15. SGA 2, Cohomologie locale des faisceaux coherents., by A. Grothendieck and others, North-Holland Publishing Co., (1968).

16. R. Hartshorne and A. Ogus: On factoriality of local rings of small embedding codimension, *Comm. in Algebra*, 1 (1974).

15 17. J. F. Boutot: Schéma de Picard Local, thesis, Orsay, (1977).

Fields Medals (IV): An Instinct for the Key Idea

Pierre Deligne was born in Brussels, Belgium, in 1944. When he was 14 an enthusiastic high school teacher, M. J. Nijs, lent him several volumes of the *Elements of Mathematics* by N. Bourbaki. This work develops a solid foundation for all of modern mathematics, in a most logically efficient manner, proceeding from the general to the particular; for example, the real number system is discussed only in the fourth chapter of the third long book, after general topology and abstract algebra have been extensively treated. In the whole treatment there is (except perhaps for the excellent historical notes) no motivation given at all, other than the internal logic of the development itself. That Deligne not only survived but even thrived on his exposure to such a work at such a tender age was perhaps already an indication of his genius, as well as of Nijs' good judgment.

Thus when Deligne went to the University of Brussels he already knew the fundamentals of most of modern mathematics. There he learned much from group theorist Jaques Tits now at the College de France, and Tits gave him excellent advice on his general mathematical development. In 1965, at Tits' suggestion, Deligne went to Paris to pursue further his interests in algebraic geometry and number theory. It would be hard to imagine a better place for this at the time. Among other activities there were

0036-8075/78/1117-0737$00.50/0

D. Mumford, *Selected Papers*, Vol. II,

the seminars in algebraic geometry of Alexander Grothendieck (Fields Medal, 1966) and the lectures of Jean-Pierre Serre (Fields Medal, 1954), which had a more number-theoretical flavor. Deligne was strongly influenced by both these men.

Deligne's association with Grothendieck during the late 1960's at the IHES (European Institute for Advanced Study, in Bures-sur-Yvette just south of Paris) was especially close. We personally first heard of Deligne in 1966 from Grothendieck, who was more impressed than we had ever seen him be by a young mathematician. At that time Deligne was 21 and Grothendieck immediately recognized him as his equal. The significance of this and of their collaboration will be clearer if we explain the situation in algebraic geometry at this time. In the 1930's algebraic geometry had an antiquated air, with many appealing charming results but an embarrassingly handmade and dusty look. During the period 1940 to 1960 several of the greatest mathematicians of this century contributed to building suitable foundations for algebraic geometry and fitting it into the abstract conceptual framework that had by then been built for most of the rest of mathematics. After the great contributions of Oscar Zariski now at Harvard University, André Weil, and Serre of the Institute for Advanced Study in Princeton, it was Grothendieck who pushed this program through to its ultimate logical conclusion. Grothendieck was an untiring, implacably logical, almost fanatical force. He was guided in his thinking perhaps more than any other mathematician has ever been by the desire to view each concept in the greatest possible degree of generality with no artificial restrictions—that is, no restrictions not absolutely forced by the logic of the situation. The result, as Grothendieck wrote his monumental works on the foundations of algebraic geometry, was an utter transformation of the subject. As he pursued the ultimate in generality the volume of the work increased exponentially, and algebraic geometry became a vast structure, gleaming, hard to grasp, overpowering. The key ideas seemed hidden, let alone the appealing artifacts of the previous century.

Deligne mastered this structure of Grothendieck's seemingly without effort, but his style was not to add a whole new layer of systematic development to the theory unless it was absolutely necessary. He preferred to find an elegant fundamental new idea suddenly clarifying a whole area or an old problem. Deligne was able to use the extensive developments of Grothendieck as well as any one, but his own ideas were often more concise, more particular. To contrast their styles metaphorically, one could say that Grothendieck liked to cross a valley by filling it in, Deligne by building a suspension bridge.

Pierre Deligne

During the next few years Deligne touched on virtually all areas of algebraic geometry, making extraordinary contributions. In 1970, at the age of 26, he was promoted to a permanent professorship at the IHES, the position he now holds. We will not try to describe his early work but will focus instead on his most exciting and deepest result, his proof in 1973 of the last and hardest of Weil's conjectures. Fortunately this result is relatively easy to state in simple language, and it may convey an idea of the almost mystical flavor of the direction in which this frontier of mathematics is growing.

One starts with a set of one or more simultaneous polynomial equations in several unknowns. This could be something as simple as one equation in two unknowns, such as $y^2 - x^3 + 1 = 0$, but in general would be $f_1(x, y, z, \ldots) = 0$, $f_2(x, y, z, \ldots) = 0, \ldots$. The f_i's, as stated, are to be polynomials, and we assume that their coefficients are whole numbers. The oldest question in arithmetic is to find, or give procedures for finding, all solutions in which the unknowns $x, y, z, \ldots$ are whole numbers. But this has turned out to be intractable in all but some elementary cases. Another question is to consider the set of solutions in which $x, y, z, \ldots$ are complex numbers. These solutions form a continuum, or manifold, X, of a certain dimensionality, called an algebraic variety because it is described by algebraic equations (sometimes one adds points at infinity to X to "complete" it). Such manifolds have been extensively studied, and in particular certain properties of X are described by its so-called Betti numbers $B_0, B_1, B_2, \ldots$. Thus B_0 is the number of connected pieces of X, and B_1 describes how many essentially different loops X contains. For example, in the case of the single equation $y^2 - x^3 + 1 = 0$, X turns out to be two-dimensional (remember that we are allowing complex values for x and y, not only real values) and to be like the surface of a doughnut (a space called a torus). In this case $B_0 = 1$, because X is connected, and $B_1 = 2$, because there are really two different ways around a torus (Fig. 1).

There is a third type of solution to our equations $f_1 = f_2 = \ldots = 0$ that is very important: one tries to put the unknowns $x, y, z, \ldots$ equal to whole numbers, but requires only that the values $f_i(x, y, z, \ldots)$ of the polynomials be divisible by a fixed prime number p (that is, be congruent to zero modulo p) instead of being 0. If $(x, y, z, \ldots)$ is one such set of values for the unknowns, then adding multiples of p to them, for example $(x + 2p, y - 3p, z + p, \ldots)$, gives another such set of values. So one can restrict $x, y, z, \ldots$ to be one of the p whole numbers $0, 1, 2, \ldots, p - 1$ and not miss anything. We then have in all only a finite set of values for the $x, y, z, \ldots$ to try, and there will be a finite number Np of solutions in the sense just described. For example, try the possible values 0, 1, and 2 for x and for y, and out of the nine possibilities you will find three of them such that 3 divides $y^2 - x^3 + 1$. Thus in this case $N_3 = 3$. With a bit more patience you can check $N_5 = 5$, $N_2 = 2$, and $N_7 = 3$ for the same equation.

We can now state a famous result of Weil, which is the leitmotiv of this whole development. Take the case of one irreducible polynomial equation in two variables. Also modify the number Np slightly to take into account infinite solutions and singularities; we omit describing this. Then

$$|Np - (p + 1)| \leq B_1 \sqrt{p} \qquad (1)$$

where B_1 is the first Betti number of the complex variety associated to the same equation. This variety will be like the surface of a doughnut with a certain number of holes, and B_1 is twice the number of holes. The point that is so startling here is that this sets up a connection between the solutions modulo p with whole numbers and the geometry of the continuum of complex solutions. What other cases can one find of such a miraculous connection between arithmetic and geometry? This question tantalizes many mathematicians today.

What should one expect for a general set of equations of the type we are considering? Weil guessed the answer in 1949, and Deligne proved that his guess was correct 24 years later. To explain this guess we must view the number Np described above in a more sophisticated way, as the number of solutions to our equations in the finite field with p elements. For each positive integer r there is an essentially unique finite field with p^r elements, and if Np^r denotes the number of solutions with $x, y, z, \ldots$ in that field, Weil conjectured that for each prime p there should exist complex number α_{ij} such that for each r

$$Np^r = \sum_{j=1}^{n} (-1)^j \sum_{i=1}^{B_j} \alpha_{ij}^r \qquad (2)$$

where n is the dimension of the space X of complex solutions and the B_j are the Betti numbers of X. Moreover the absolute values of the numbers x_{ij} should be given by

$$|\alpha_{ij}| = p^{j/2} \qquad (3)$$

(In this brief statement of Weil's conjectures we have exaggerated a bit: one must desingularize X and add some points at infinity, and make the corresponding modifications in counting the solutions in finite fields; also one must exclude a finite set of primes p, those for which X does not have "good reduction modulo p.") In the case of one equation in two unknowns, $n = 2$, $B_0 = B_2 = 1$, $x_{10} = 1$, and $x_{12} = p$, so that Eq. 1 is a consequence of Eqs. 2 and 3. A formula of the same type as Eq. 2 was proved by Bernard Dwork of Princeton University in 1959, and Eq. 2 was proved by Grothendieck in 1965. However, Eq. 3 is much harder, and it is this result for which Deligne is justly famous. Clearly, Eqs. 2 and 3 strengthen and confirm the link between the arithmetical problem of solving polynomial equations modulo p and the geometry of their complex solutions.

To see how Deligne proved Eq. 3, we must go back again to Grothendieck. It was in order to prove a formula like Eq. 2, and with the hope of using it to prove Eq. 3, that Grothendieck began doing algebraic geometry. Weil had pointed out that Eq. 3 could be obtained as a "Lefschetz fixed point formula," if one had a "cohomology theory of varieties in characteristic p" (indeed Np^r is just the number of fixed points of the transformation F^r, where F is the Frobenius map of the set of solutions in characteristic p into itself). At the start of his work Grothendieck had guessed that such a cohomology theory could be obtained by systematically confusing the two mathematical senses in which the word covering is used (Fig. 2). This was the kind of abstract idea at which Grothendieck excelled, and in this case he was absolutely right. With the aid of Michael Artin of the Massachusetts Institute of Technology and Jean Louis Verdier of the University of Paris he constructed a new cohomology theory, known as "étale cohomology," yielding the numbers x_{ij} in a natural way. This theory was one of the building blocks of Deligne's proof.

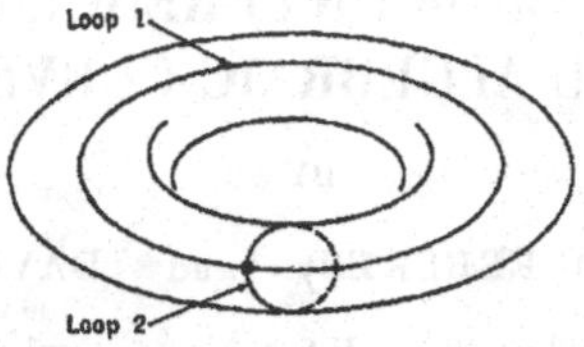

Fig. 1. The two ways to go around a torus.

The other main ingredient came from a little-known prewar (1939) paper of Robert Rankin in the *Proceedings of the Cambridge Philosophical Society*, in which Rankin made some progress on an analogous conjecture of the Indian mathematician Srinivara Ramanujan, by a squaring trick. It is hard to imagine two mathematical schools more different in spirit and outlook than were those of the British analytic number theorists in the 1930's and of the French algebraic geometers in the 1960's. That Deligne's proof is a blend of ideas from both is an indication of the universality of his mathematical taste and understanding. He had a clue to the connection because already in 1968 he had shown that Weil's conjectures implied Ramanujan's. The ideas behind this were due to the Japanese mathematicians Kuga, Sato, Shimura, and Ihara, but it was Deligne who had the technical power to carry them out, and it was Serre who realized this and urged him to do it. At any rate, Deligne saw that Rankin's method could be understood geometrically and could be greatly extended. Combining this with a very delicate analysis of the cohomology via so-called Lefschetz pencils, using also a theorem of David Kazhdan now at Harvard University and Margoulis (one of this year's Fields Medalists), Deligne put together his sensational proof of Weil's conjecture: Besides its own intrinsic interest, this result has also already yielded several important consequences in number theory and algebraic geometry.

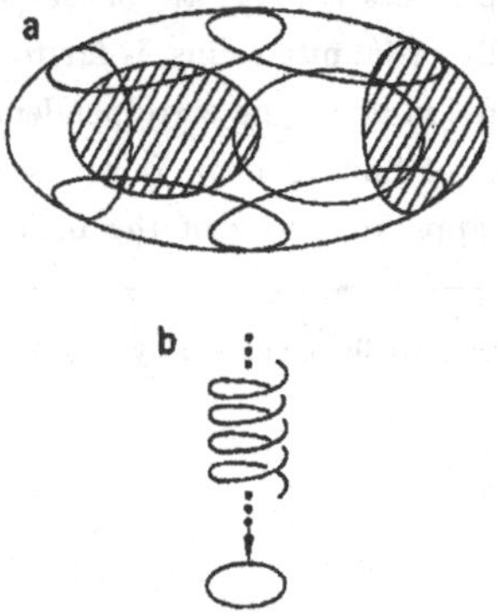

Fig. 2. (a) Covering of type I. A set of pieces that fill the whole. In this case, an oval region covered by nine smaller oval regions, two of which are shaded. (b) Covering of type II. One space lying smoothly over another. In this case, an infinite spring covering a closed loop.

Since 1973 Deligne's center of interest has shifted slightly from geometry toward number theory. He has made several key contributions to problems connected with the vast program of Robert P. Langlands of the Institute for Advanced Study to relate the way in which the numbers x_{ij} mentioned above vary with p to the theory of automorphic forms.

Deligne's economy and clarity of thought are amazing. His writings contain few unnecessary words, little or no redundancy. The ideas are there, simply and clearly stated, but so densely that almost every phrase is relevant.

Deligne's nonmathematical interests and activities exhibit the same simplicity. For years he has cultivated a large vegetable garden in the rich soil of the housing project of the IHES. He enjoys organizing Easter egg hunts for the children living there. For transportation he prefers a bicycle to a car, and his vacations are usually spent hiking. There is nothing artificial about him. He is self-assured but modest and able and willing to discuss almost any mathematical subject with anyone. There are few subjects that his questions and comments do not clarify, for he combines powerful technique, broad knowledge, daring imagination, and unfailing instinct for the key idea.

DAVID MUMFORD
JOHN TATE

Department of Mathematics, Harvard University, Cambridge, Massachusetts 02138

THE SPECTRUM OF DIFFERENCE OPERATORS AND ALGEBRAIC CURVES

BY

PIERRE VAN MOERBEKE(1) and DAVID MUMFORD

Brandeis University, Waltham, Mass., U.S.A. *Cambridge, Mass., U.S.A.*

The explicit linearization of the Korteweg–de Vries equation [10, 18] and the Toda lattice equations [10, 12, 22] led to a theory relating periodic second order (differential and difference) operators to hyperelliptic curves with branch points given by the periodic and antiperiodic spectrum of the original operator. As a result the periodic second order operators with a given spectrum form a torus (except for a lower dimensional submanifold) which is the Jacobi variety of the defining curve. Krichever [15, 16, 17], motivated by further examples in the work of Zaharov-Shabat [30], showed how curves with certain properties lead to commuting differential operators reconfirming forgotten work by Burchnell and Chaundy [6]. Inspired by Krichever's ideas, Mumford [24] establishes then a dictionary between commutative rings of (differential and difference) operators and algebraic curves using purely algebraic methods. As an example, the Hill's operator whose spectrum consists of a finite number of non-degenerate bands leads to a finite number of independent differential operators commuting with the original Hill's operator and this commutative ring defines a curve of finite genus. However, the generic Hill's operator has an infinite number of bands and must be analyzed in terms of a hyperelliptic curve of infinite genus; see McKean and Trubowitz [21]. These analytical techniques have not yet been extended to higher order differential operators so that the correspondence between differential operators and curves, generically of infinite genus, is far from being understood. In view of this, it is important to discuss in detail the correspondence between periodic *difference* operators and algebraic curves (of finite genus). In the second order case, the periodic difference operators are good approximations of the periodic differential operators and

(1) Research for this paper was partially supported by NSF Grant No. MCS-75-05576 A01.

D. Mumford, *Selected Papers*, Vol. II,

the corresponding curves are also hyperelliptic (see McKean and van Moerbeke [20]). Hopefully periodic differential operators will lead to infinite genus versions of the curves suggested by the difference operators.

In this work, we show that every so-called regular periodic difference operator of any order and not necessarily symmetric leads to a spectral curve $\mathcal{R}$ of a given type and a "regular" point on its Jacobi variety Jac ($\mathcal{R}$) and vice-versa. The regularity is a condition on the "symbol" of the difference operator, which in turn provides information about the infinite points of the spectral curve. Except for a finite number of translates of the theta-divisor, every point of Jac ($\mathcal{R}$) is regular. As a consequence, the isospectral class of regular difference operators C of a given order with a given h-spectrum for all Floquet multipliers h parametrizes the regular points of Jac ($\mathcal{R}$). This is the content of § 2.

How does a linear flow on Jac ($\mathcal{R}$) translate in terms of an isospectral deformation of the difference operators C? It translates into a system of ordinary differential equations, given by Lax-type commutation relations on the original difference operator: $\dot{C}=[C, A^+]$, where A^+ is the upper-triangular part of some operator A, constructed as follows: the linear flow above picks out a specific meromorphic function on $\mathcal{R}$, which is holomorphic on the affine part; this function then maps into the difference operator A. However there is more to it: these flows all derive from Hamiltonians and a symplectic structure reminiscent of the Kostant-Kirillov method of orbits for the group of upper-triangular matrices; for this method, see Kostant [14] and Abraham-Marsden [1]. It can be summarized as follows: the usual Bruhat decomposition of $SL(n, \mathbf{R})$ leads to a natural symplectic structure on the orbits in $\mathcal{N}^*$ in $sl(n, \mathbf{R})^*$ under the action of the triangular subgroup N; this fact does not apply as such because of the periodic and not necessarily symmetric nature of the difference operators, but it is nevertheless suggestive. The result is that the coefficients of the algebraic expressions for $\mathcal{R}$ can be regarded as Hamiltonians (depending on the difference operator C) in involution for the symplectic structure above; they lead to Hamiltonian flows, each of which is linearizable on Jac ($\mathcal{R}$); moreover all linear flows on Jac ($\mathcal{R}$) derive from such Hamiltonians. Its proper group-theoretiinterpretcal ation relates to the Kac-Moody extension of $sl(n, \mathbf{R})$, which will be developed, also for other classical groups, in a forthcoming paper by Adler and van Moerbeke [3]. For relations of this symplectic structure with the Gelfand–Dikii [8] symplectic structure and its group theoretical content, consult Adler [2]. All these considerations specialized to hyperelliptic curves leads then to the explicit linearization of the periodic Toda lattice equations.

In § 3, we deal with a number of interesting special cases. Whenever the curve $\mathcal{R}$ comes from a symmetric difference operator, it carries a natural involution, which, in turn, defines a linear subvariety of Jac ($\mathcal{R}$), called the Prym variety of $\mathcal{R}$. Then the manifold of isospectral *symmetric* operators coincides with Prym ($\mathcal{R}$) and all linear isospectral deformations are generated by meromorphic functions on $\mathcal{R}$, holomorphic on the affine part, as above, and moreover invariant under the involution. A similar statement holds for curves $\mathcal{R}$ defined by self-adjoint operators.

The entries of the difference operators can be regarded as Abelian functions on the Jacobi variety of the corresponding curve; then, using classical formulas (see Fay [11]), the entries can be expressed as quotients of theta functions (§ 5). Most of the results sketched above relate to *periodic* difference operators; its periodic nature is responsible for the division properties of the curve, as will be explained in § 2; in the Toda hyperelliptic case the latter amounts to the existence of two points P and Q on the curve, such that some integer multiple of $Q-P$ vanishes on Jac ($\mathcal{R}$). When the division properties do not hold and for a somewhat more restricted class of divisors, the associated difference operators are merely almost periodic. It remains an interesting open question to characterize those almost periodic difference operators which lead to finite genus curves; this is unknown even for the second order difference (and differential) operator case. Results close to those in paragraphs 2, 3 and 5 have been obtained by Krichever [16].

The relation between difference operators and curves (special curves) have been extended by Mumford [26] to a connection between two-dimensional difference operators and algebraic surfaces (spectral surface). As pointed out for one-dimensional operators, its "symbol" is a zero-dimensional difference operator and it defines the non-affine part of the curve; in the same way, in two dimensions, the "symbol" will be one-dimensional and the "symbol of the symbol" zero-dimensional; they lead to the non-affine behaviour of the spectral surface, which is crucial in the study of the Picard variety for the spectral surface (analogous to the Jacobi variety for curves). In fact, unlike for spectral curves, the spectral surface has trivial Picard variety, so that generic periodic two-dimensional difference operators do not admit isospectral deformations; for a fairly elementary exposition of Mumford's result, see P. van Moerbeke [23].

The first author thanks Professor P. Deligne for many helpful conversations, especially with regard to § 4.

Table of contents:

§ 1. Introduction

Let f be an infinite column vector $f=(\dots f_{-1}, f_0, f_1, \dots)^T$. Let D operate on f as the shift $Df_k=f_{k+1}$. Consider the difference operator C defined by

$$(Cf)_n=\sum_{k=-M'}^{M} c_{n,n+k}f_{n+k}=\left(\sum_{k=-M'}^{M} c_{n,n+k}D^k\right)f_n, \quad c_{i,j}\in \mathbf{C};$$

C acts on f as an infinite band matrix (c_{ij}) acting on f, zero outside the band $-M'\leqslant i-j\leqslant M$; C is said to have support $[-M', M]$. Assume C to be periodic of period N, i.e. $c_{i+N,j+N}=c_{ij}$; this amounts to the commutation relation $CS=SC$, where $S=D^N$. Let $(M,N)=n$ and $(M',N)=n'$; let $M_1n=M$, $M_1'n'=M'$ and $N_1n=N$, $N_1'n'=N$.

A difference operator C will be called *regular*, if the n quantities

$$\sigma_i=c_{i,i+M}c_{i+M,i+2M}\dots c_{i+(N_1-1)M,i+N_1M}, \quad 1\leqslant i\leqslant n$$

are all different from zero and different from each other and the same for the n' quantities

$$\sigma_i'=c_{i,i-M'}c_{i-M',i-2M'}\dots c_{i-(N_1'-1)M',i-N_1'M'}, \quad 1\leqslant i\leqslant n'.$$

They involve only boundary elements, i.e., elements on the outer diagonals. Note that $\sigma_{i+n}=\sigma_i$ and $\sigma'_{i+n'}=\sigma'_i$.

A square matrix C_h of order N will be used throughout this paper. It is constructed as follows: if $N>M+M'$ consider the square matrix of order N taken from C, having c_{11} for upper left corner and c_{NN} for lower right corner, put the upper-left and lower-right triangular corners (see Figure 1) respectively in the upper-right and lower-left corner of the square block after multiplication by h^{-1} and h. In general, we write:

$$(C_h)_{i,j}=\sum_{k=-\infty}^{+\infty} h^k\cdot c_{i,j+kN}$$

In fact C_h contains all the information contained in C. Also observe that $C_hD_h=(CD)_h$ for any two difference operators C and D. The determinant of C_h-zI is readily seen to be a polynomial expression in z, h and h^{-1}, which has the form

$$\begin{aligned}F(h,h^{-1},z)&\equiv\det(C_h-zI)\\&=A_0h^M+A_1(z)h^{M-1}+\dots+A_M(z)+A_{M+1}(z)h^{-1}+\dots+A_{M+M'}h^{-M'}=0,\end{aligned}$$

where

$$A_0=(-1)^{M(N-M)}\prod_{k=1}^{N}c_{k,k+M}=(-1)^{M(N-M)}\prod_{i=1}^{n}\sigma_i\neq 0,$$

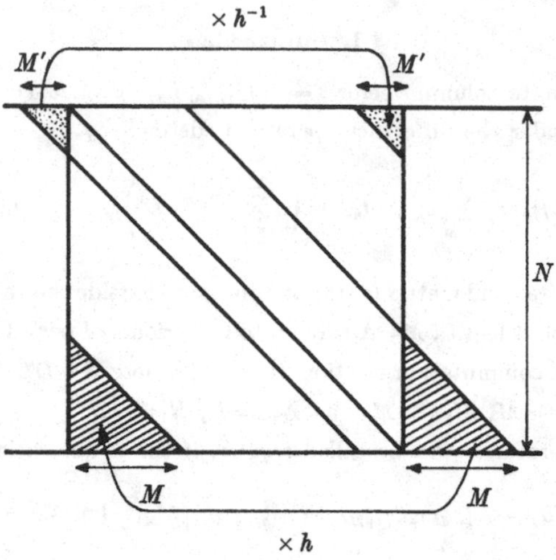

Figure 1

$$A_{M+M'} = (-1)^{M'(N-M')} \prod_{k=1}^{N'} c_{k,k-M'} = (-1)^{M'(N-M')} \prod_{i=1}^{n'} \sigma'_i \neq 0$$

and

$$A_M(z) = (-1)^N z^N + \ldots .$$

Further information about the polynomials $A_j(z)$ is contained in Lemma 1, § 2.

For later use, we introduce some geometrical notations. Let $\mathcal{R}$ be an algebraic curve of genus g. We will allow $\mathcal{R}$ to be singular and even reducible, but we will always require $\mathcal{R}$ to be connected and reduced, i.e., no nilpotents in its structure sheaf. We also require that its singular points will be locally isomorphic to singular points of plane curves. In the singular case, the genus g will be the "arithmetic genus" of $\mathcal{R}$, i.e., dim $H^1(O_{\mathcal{R}})$ or dim $H^0(\omega_{\mathcal{R}})$ where $\omega_{\mathcal{R}}$ are the 1-forms η on $\mathcal{R}$ with poles only at singular points P of $\mathcal{R}$ and at those points

$$\sum_{\substack{\text{Branches } \gamma \\ \text{of } \mathcal{R} \text{ at } P}} \operatorname{res}_\gamma (f\eta) = 0 \quad \text{all} \quad f \in O_{P,\mathcal{R}}. \tag{0}$$

At each singular point P, there is a 1-form η with "highest poles at P", i.e., every other 1-form η' satisfying (0) equals $f\eta$, for some $f \in O_{p,\mathcal{R}}$ (cf. Serre [28]). Let Jac ($\mathcal{R}$) be the Jacobian variety of $\mathcal{R}$ (the generalized Jacobian [27, 28], if $\mathcal{R}$ is singular). We will be interested in positive divisors $\mathcal{D}$ on $\mathcal{R}$ of degree g which are sufficiently generic. If $\mathcal{R}$ is smooth, a positive divisor $\mathcal{D}$ is just $\sum_{i=1}^{d} \nu_i$, $\nu_i \in \mathcal{R}$. In the singular case, $\mathcal{D}$ is given by such

7 – 792907 *Acta mathematica* 143. Imprimé le 28 Septembre 1979

an expression *and* if k of the ν_i's equal a singular point P, then in addition to P occurring in $\mathcal{D}$ with multiplicities k, we must also give a k-dimensional space of "allowable" poles at P, i.e., a module $M_P(\mathcal{D})$ over $O_{P,R}$ such that

$$O_{P,R} \subset M_P(\mathcal{D}) \subset \mathbb{C}(\mathcal{R})$$

with $\mathbb{C}(\mathcal{R})$ being the field of meromorphic functions on $\mathcal{R}$ and such that

$$\dim M_P(\mathcal{D})/O_{P,R} = k.$$

A general divisor $\mathcal{D}$ is an expression $\Sigma \pm \nu_i$, plus for all singular points, a finitely generated $O_{P,R}$-module $M_P(\mathcal{D}) \subset \mathbb{C}(\mathcal{R})$ such that if P occurs with multiplicity k in $\mathcal{D}$, then

$$k = \dim (M_P + O_P/O_P) - \dim (M_P + O_P/M_P).$$

For every such $\mathcal{D}$, we define the space of functions with poles at $\mathcal{D}$ as:

$$\mathcal{L}(\mathcal{D}) = \{f \in \mathbb{C}(\mathcal{R}) \,|\, (f) + \mathcal{D} \geqslant 0\}.$$

Here if $\mathcal{R}$ is singular, then at every singular point P, $(f) + \mathcal{D} \geqslant 0$ at P means $f \in M_P(\mathcal{D})$. We define the space of differentials with zeroes at $\mathcal{D}$ as

$$\Omega(-\mathcal{D}) = \{\text{meromorphic differentials } \eta \text{ on } \mathcal{R} \,|\, (\eta) \geqslant \mathcal{D}\}.$$

Here at singular points P, $(\eta) \geqslant \mathcal{D}$ means that for all $f \in M_P(\mathcal{D})$

$$\sum_{\substack{\text{Branches } \gamma \\ \text{of } R \text{ at } P}} \text{res}_\gamma (f\eta) = 0;$$

the Riemann–Roch theorem tells us as usual that

$$\dim \mathcal{L}(\mathcal{D}) - \dim \Omega(-\mathcal{D}) = \deg \mathcal{D} - g + 1.$$

Now let $\mathcal{D}$ be a positive divisor of degree g. $\mathcal{D}$ is *general* if $\dim \mathcal{L}(\mathcal{D}) = 1$, i.e., $\dim \Omega(-\mathcal{D}) = 0$. $\mathcal{D}$ will be called *regular* with regard to two infinite sequences of smooth points $\{P_i\}_{i\in\mathbb{Z}}$ and $\{Q_i\}_{i\in\mathbb{Z}}$ if [1]

$$\dim \mathcal{L}\left(\mathcal{D} + \sum_{i=1}^{k} P_i - \sum_{i=0}^{k} Q_i\right) = 0$$

---

[1]
$$\begin{aligned} \sum_{i=1}^{k} P_i &= \sum_{1}^{k} P_i, \quad k \geqslant 1 \\ &= 0, \quad k = 0 \\ &= -P_0 - P_{-1} - \dots - P_{k+1} \quad \text{for} \quad k \leqslant -1 \end{aligned}$$

moreover
$$\begin{aligned} \sum_{i=0}^{k} Q_i &= \sum_{0}^{k} Q_i, \quad k \geqslant 0 \\ &= 0 \quad k = -1 \\ &= -Q_{-1} - Q_{-2} - \dots - Q_{k+1} \quad \text{for} \quad k \leqslant -2. \end{aligned}$$

§ 2. The correspondence between difference operators and curves

THEOREM 1. *There is a one-to-one correspondence between the two sets of data:*

(a) *a regular difference operator C of support $[-M', M]$ and period N, modulo conjugation by diagonal periodic operators.*

(b) *a curve $\mathcal{R}$, $(n+n')$ points on $\mathcal{R}$, a divisor $\mathcal{D}$ on $\mathcal{R}$ and two functions h, z on $\mathcal{R}$ subject to several conditions. $\mathcal{R}$ may be singular, but always has genus:*

$$g=\frac{(N-1)(M+M')-(n+n')+2}{2}.$$

The $(n+n')$ points $P_1, \ldots, P_n$ and $Q_1, \ldots, Q_{n'}$ are smooth and have a definite ordering. We define P_i(resp Q_i) for all $i \in \mathbf{Z}$ by $P_{i+n}=P_i$ (resp $Q_{i+n'}=Q_i$). $\mathcal{D}$ has degree g and is regular for these sequences. The functions h and z have zeroes and poles as follows:

$$(h)=-N_1\sum_{i=1}^{n} P_i+N_1'\sum_{i=1}^{n'} Q_i$$

and

$$(z)=-M_1\sum_{i=1}^{n} P_i-M_1'\sum_{i=1}^{n'} Q_i+\text{a positive divisor not containing the } P_i\text{'s and } Q_i\text{'s}.$$

Finally, $z^{N_1}h^{-M_1}$ (resp. $z^{N_1'}h^{M_1'}$) should take on distinct values at the P_i's (resp. the Q_i's).

Remark. The condition that $\mathcal{D}$ be regular reduces in this instance to the vanishing of a finite number of determinants involving differentials.

Proof. We first give the entire proof of this theorem, assuming for (a) ⇒ (b) that $F(h, h^{-1}, z)$ defines a non-singular curve in $\mathbf{C}^* \times \mathbf{C}$ and for (b) ⇒ (a) that $\mathcal{R}$ is non-singular. After we add a few words on the modifications necessary to deal with the singular case. First we show that (a) implies (b). The eigenvalues z and h such that

$$Cf=zf \quad \text{and} \quad Sf=hf \tag{1}$$

satisfy

$$\sum_{k=-M'}^{M} c_{n,n+k} f_{n+k}=zf_n, \quad 1 \leqslant n \leqslant N \tag{2}$$

with

$$f_{n+k}=h^{\alpha} f_{n+k-\alpha N} \tag{3}$$

where α is the integer such that $\alpha N<n+k<(\alpha+1)N$. So, (2) can be rewritten

$$\sum_{k=-M'}^{M} c_{n,n+k} h^{\alpha} f_{n+k-\alpha N}=zf_n, \quad 1 \leqslant n \leqslant N$$

or, what is the same

$$C_h f = z f$$

where $f = (f_1, \ldots, f_N)^T$. Therefore the eigenvalues (z, h) of (1) satisfy

$$\det (C_h - zI) = F(h, h^{-1}, z) = 0, \tag{4}$$

which determines an algebraic curve over $\mathbf{C}$; vice versa, any couple (z, h) satisfying this algebraic relation provides a couple of eigenvalues for (1). Since A_0 and $A_{M+M'}$ in (4) are nonzero, the function h has its poles or zeros only at $z = \infty$. Therefore $F(h, h^{-1}, z)$ with z, h and $h^{-1} \in \mathbf{C}$, defines the affine part $\mathcal{R}_0$ of an algebraic curve $\mathcal{R}$. The equation

$$G(h, z) \equiv h^{M'} f(h, h^{-1}, z)$$

shows that $\mathcal{R}$ is an $M + M'$-sheeted covering of $\mathbf{C}$.

We now turn to the behaviour of the curve at the boundary. For this, we need to analyze the coefficients $A_j(z)$ of h^{M-j} in F very closely:

LEMMA 1. *The functions $A_j(z)$ are polynomials in z of degree k_j satisfying*

$$k_j \leqslant \frac{Nj}{M}, \quad 0 \leqslant j \leqslant M$$

and

$$k_{M+M'-j} \leqslant \frac{Nj}{M'}, \quad 0 \leqslant j \leqslant M'$$

with equality if and only if the right hand side is an integer, i.e., when $j = 0, M_1, 2M_1, \ldots, nM_1 = M$ in the first case and when $j = 0, M_1', 2M_1', \ldots, n'M_1' = M'$ in the second case. For $j = \alpha M_1$, $0 \leqslant \alpha \leqslant n$, the coefficient of

$$z^{k_j} h^{M-j} = z^{\alpha N_1} h^{(n-\alpha)M_1}$$

in $F(h, h^{-1}, z) = 0$ is the symmetric polynomial of degree $n - \alpha$ in σ_i defined as

$$\tau_\alpha = \sum_{1 \leqslant i_1, \ldots, i_\alpha \leqslant n} \prod_{i \neq i_1, \ldots, i_\alpha} \sigma_i, \quad 0 \leqslant \alpha \leqslant n.$$

Likewise, for $j = \alpha M_1'$, $0 \leqslant \alpha \leqslant n'$, the coefficient of

$$z^{k_{M+M'-j}} h^{-M'+j} = z^{\alpha N_1'} h^{-(n'-\alpha)M_1'}$$

is the symmetric polynomial of degree $n' - \alpha$ in σ_i', defined in a similar way as above.

Proof. It proceeds by induction. Let T_i be a typical term of det $(C_h - zI) = 0$, containing h^i and consider how many times it appears in the determinant of $C_h - zI$. Clearly $T_0 = (-z)^N$. A term T_1 of maximal degree in g is formed by picking one of the entries $(N-k_1+1, k_2)$, i.e., $C_{N-k_1+1, N+k_2}h$ with $1 \leqslant k_i \leqslant M$ and $k_1 + k_2 \leqslant M+1$ and keeping the largest possible number of entries on the diagonal of $C_h - zI$. This choice excludes the entries (k_1, k_2) and $(N-k_1+1, N-k_1+1)$ of the diagonal and forces one to take the elements of the upper outer-diagonal. Since every column must have a representation in T_1, take the entry $(N-k_1+1-M, N-k_1+1)$ of the upper outer-diagonal, which excludes the entry $(N-k_1+1-M, N-k_1+1-M)$ of the diagonal. More generally, if T_1 contains the entry $(N-k_1+1-iM, N-k_1+1-(i-1)M)$, it does not contain the entry $(N-k_1+1-iM, N-k_1+1-iM)$ of the diagonal, as long as $1 \leqslant i \leqslant i_0$, where i_0 is the largest i such that $N-k_1+1-iM \geqslant k_2$. Two cases must now be distinguished: (a) if $N-k_1+1-i_0M = k_2$ the process is terminated and at least i_0+1 number of elements of the diagonal have been excluded and the degree of z in T_1 is bounded above by

$$N - i_0 - 1 = N - 1 - \left[\frac{N-k_1-k_2+1}{M}\right];$$

(b) if $N-k_1+1-i_0M > k_2$, the elements (k_2, k_2) and $(N-k_1+1-i_0M, N-k_1+1-i_0M)$ of the diagonal must be excluded, so that the degree of z in T_1 is bounded above by

$$N - i_0 - 2 = N - 2 - \left[\frac{N-k_1-k_2+1}{M}\right].$$

In either case, these estimates will be maximal provided $k_1 + k_2$ assumes its largest possible value $M+1$, so that the degree k_{M-1} of z in T_1 is bounded by $N - N/M$.

The rest of the argument goes by induction: if $k_{M-j} \leqslant N(M-j)/M$, then $k_{M-j-i} \leqslant N(M-j-1)/M$; this is done using the same method as above; pick the entry $(N-k_1+1, k_2)$ in det $(C_h - zI)$ containing h which does not appear yet in T_j; this excludes a number of diagonal entries, bounded below by N/M, so that the degree k_{M-j-1} of z in T_{j+1} is bounded above by

$$\frac{Nj}{M} - \frac{N}{M}.$$

It remains to establish the second part of Lemma 1. Whenever Nj/M is an integer, this estimate is exact: $k_j = Nj/M$. It is done by exhibiting the term of exact degree Nj/M in z and $M-j$ in h. Consider the expression denoted by σ_i in C; let $\tilde{\sigma}_i$ be the expression obtained in the same fashion in C_h. A factor $C_{i+\alpha M, i+(\alpha+1)M}$ yields an h in $\tilde{\sigma}_i$ as soon as $i + \alpha M \leqslant kN < i + (\alpha+1)M$ for some integer k. Therefore $\tilde{\sigma}_i$ will be of degree

$\#\{\alpha \in \mathbf{Z};\ 0 \leqslant \alpha \leqslant N_1 - 1$ such that $i + \alpha M \leqslant kN < 1 + (\alpha+1)M$ for some integer $k\} = M_1$

in h. But $\tilde{\sigma}_i$ can be completed to a term in $\det(C_h - zI)$ by multiplying σ_i by the maximal possible elements of the diagonal. Every factor $C_{i+\alpha M,\, i+(\alpha+1)M}$ in $\tilde{\sigma}_i$ excludes exactly one diagonal element, because the integers $\{i + \alpha M \mid 0 \leqslant \alpha \leqslant N_1 - 1\}$ are all different modulo N. Therefore $\det(C_h - zI)$ contains a term of degree $N - N_1 = (n-1)N_1$ in z and M_1 in h. All possible such terms are obtained by making a sum over the index i from 1 to n.

Moreover every term in $\det(C_h - zI)$ of degree βM_1 $(1 \leqslant \beta \leqslant n)$ in h and $(n-\beta)N_1$ in z is obtained in a similar way from considering $\sigma_{i_1} \dots \sigma_{i_\beta}$ and the corresponding combination $\tilde{\sigma}_{i_1} \dots \tilde{\sigma}_{i_\beta}$ in C_h and to complete it with diagonal elements of $C_h - zI$ to form a term in $\det(C_h - zI)$. This finishes the proof of Lemma 1.

We now turn to the behaviour of the curve at the boundary: the lemma implies that there are n distinct points $P_1, \dots, P_n$ covering $z = \infty$, where $h = \infty$ and n' other points $Q_1, \dots, Q_{n'}$ covering $z = \infty$, where $h = 0$. To check this fact, define a local parameter t near each point P_i as follows

$$z = t^{-M_1} \quad \text{and} \quad h = C_i t^{-N_1} + \dots, \quad \text{where} \quad C_i^{M_1} = \frac{1}{\sigma_i}. \tag{5}$$

Near each point Q_i, define another local parameter t such that

$$z = t^{-M_1'} \quad \text{and} \quad h = C_i' t^{N_1'} + \dots, \quad \text{where} \quad C_i'^{\,M_1'} = \sigma_i'.$$

A typical term of $F = 0$ containing h^{M-j} $(0 \leqslant j \leqslant M)$ looks like

$$z^i h^{M-j} \quad \text{with} \quad 0 \leqslant i \leqslant k_j \leqslant \frac{Nj}{M};$$

expressed in the local parameter t it appears as $t^{-M_1 i - N_1(M-j)}$; the exponent satisfies $M_1 i + N_1(n-j) \leqslant N_1 M$ with equality if and only if $i = k_j$ with $j = \alpha M_1$, $\alpha = 0, 1, \dots, n$. Therefore F can be expressed near $(z, h) = (\infty, \infty)$ as

$$\sum_{\alpha=0}^{n} (-1)^\alpha \tau_\alpha z^{\alpha N_1} h^{(n-\alpha)M_1} + \text{lower order terms}.$$

Then

$$\begin{aligned} z^{-N} F(h, h^{-1}, z) &= \sum_{\alpha=0}^{n} (-1)^\alpha \tau_\alpha \left(\frac{h^{M_1}}{z^{N_1}}\right)^{n-\alpha} + \text{lower order terms} \\ &= \prod_{i=1}^{n} \left(\sigma_i \frac{h^{M_1}}{z^{N_1}} - 1\right) + \text{lower order terms}, \end{aligned}$$

which implies that $h^{M_1} z^{-N_1}$ assumes n distinct values $1/\sigma_i \neq 0$. Therefore the point at $(z, h) = (\infty, \infty)$ separates into n distinct points $P_1, \dots, P_n$.

The same analysis can now be applied to the point $(z, h) = (\infty, 0)$; there the upshot is that $h^{M_1'} z^{N_1'}$ assumes n' distinct values $\sigma_i' \neq 0$ and therefore the point at $(z, h) = (\infty, 0)$ separates into n' distinct points $Q_1, \ldots, Q_{n'}$. Let Q_i correspond to σ_i'. It follows that

$$(h) = -N_1 \sum_{i=1}^{n} P_i + N_1' \sum_{i=1}^{n'} Q_i$$

and

$$(z) = -M_1 \sum_{i=1}^{n} P_i - M_1' \sum_{i=1}^{n'} P_i + (\text{a positive divisor on } \mathcal{R}_0). \tag{6}$$

We are now in a position to compute the genus of $\mathcal{R}$ from Hurwicz's formula. Relation (6) implies at once that the ramification index of $\mathcal{R}_\infty = \mathcal{R} \backslash \mathcal{R}_0$ equals

$$V_\infty = n(M_1 - 1) + n'(M_1' - 1);$$

whereas the ramification index V_0 of $\mathcal{R}_0$ is given by the number of zeros of the different $\Delta = G_h'(h, z)$ or what is the same, by the number of poles of Δ. Near the point P_i, Δ behaves as

$$\begin{aligned}
\Delta &= \text{constant } (\neq 0) \times h^{M'} \frac{d}{dh} \prod (\sigma_i h^{M_1} - z^{N_1}) + \text{lower order terms} \\
&= \text{constant } (\neq 0) \times h^{M'} \sum_{k=1}^{n} \sigma_k M_1 \prod_{i \neq k} \left(\frac{\sigma_i}{\sigma_k} z^{N_1} + z^{N_1} \right) h^{M_1 - 1} + \text{lower order terms} \\
&= \text{constant } (\neq 0) \times h^{M' + M_1 - 1} z^{(n-1)N_1} + \text{lower order terms} \\
&= \text{constant } (\neq 0) \times t^{-N_1(M_1 + M' - 1) - M_1 N_1 (n-1)} + \text{lower order terms,}
\end{aligned}$$

and, using a similar argument, near Q_i

$$\Delta = \text{constant} (\neq 0) \times t^{-N_1'} + \text{lower order terms.}$$

Therefore

$$V_0 = N(M + M')$$

and

$$g = \frac{V_0 + V_\infty}{2} - (M + M') + 1 = \frac{(N-1)(M + M') - (n + n') + 2}{2}$$

The eigenvectors f common to C and S can be regarded as column vectors of meromorphic functions; using the normalization $f_0 = 1$, $f = (f_1, f_2, f_3, \ldots, f_{N-1}, h)^{\mathrm{T}}$. Since f satisfies

$$(C_h - zI) f = 0,$$

f_k can be expressed as follows

$$f_k = \frac{\Delta_{1,k}}{\Delta_{1,i}} f_i = \frac{\Delta_{2,k}}{\Delta_{2,i}} f_i = \ldots = \frac{\Delta_{N,k}}{\Delta_{N,i}} f_i, \quad 1 \leqslant i, k \leqslant N$$

where $\Delta_{i,j}=(-1)^{i+j}\times(i,j)$th minor of C_h-zI. In particular

$$f_k=\frac{\Delta_{N,k}}{\Delta_{N,N}}h=\frac{\Delta_{k,k}}{\Delta_{k,N}}h$$

which expresses each f_k as a rational function in z and h. In order to find the divisor $\mathcal{D}$ on $\mathcal{R}$, it is important to investigate the nature of the poles and zeros of these functions on $\mathcal{R}_\infty$.

LEMMA 2. *The meromorphic functions f_k satisfy the following conditions at infinity, If* $0\leqslant i\leqslant n-1$, $\alpha\in\mathbf{Z}$

(i) *at* P_k, order $\left(\frac{f_{k+n\alpha+i}}{f_k}\right)\geqslant-\alpha$, *with equality if* $i=0$

(ii) *at* Q_k, order $\left(\frac{f_{k+n\alpha+i}}{f_k}\right)\geqslant\alpha$, *with equality if* $i=0$.

Proof. In order to investigate the poles at $P=\bigcup P_i$, consider the new set of coordinates

$$\alpha=z^{N_1}h^{-M_1}\quad\text{and}\quad\beta=z^{\delta}h^{\gamma}$$

with $\gamma N_1+\delta M_1=-1$ with $N_1>\delta\geqslant 0$. (It is always possible to find two such integers γ and δ, since N_1 and M_1 are relatively prime.) These coordinates are most convenient, because near a point P_i, α and β behave as follows.

$$\begin{aligned}\alpha&=\sigma_i+O(t)\\ \beta&=\text{constant}(\neq 0)t+O(t^2).\end{aligned}\tag{7}$$

The meromorphic functions α and β can also be regarded as the eigenvalues of the commuting operators $A=C^{N_1}S^{-M_1}$ and $B=C^{\delta}S^{\gamma}$ with entries a_{ij} and b_{ij} respectively. In fact A is a lower triangular difference operator (i.e., $a_{ij}=0$, for $i<j$), whereas B is a strictly lower triangular difference operator, whose first non-zero subdiagonal is at $-n$ (i.e., $b_{ij}=0$ for $i-n<j$). Moreover $a_{kk}=\sigma_k$, because a_{kk} is obtained from C and S as

$$c_{k,k+M}D^M c_{k,k+M}D^M\ldots c_{k,k+M}D^M S^{-M_1}=\sigma_k D^0;$$

therefore A has an n-periodic diagonal with entries $\sigma_k\neq 0$ different from one another. The fact that A and B commute induces relations between the a_{ij}'s and b_{ij}'s. The first one expresses that $a_{kk}=a_{k+n,k+n}$ which is the same as $\sigma_k=\sigma_{k+n}$. From (7), β itself can be used as a local parameter near P_m; let α admit the following Taylor expansion in β near P_m

$$\alpha=\sum_{i=0}^{\infty}\alpha_i\beta^i\quad\text{with}\quad\alpha_0=\sigma_m.$$

Express the fact that $Bf=\beta f$ and $Af=\alpha f$ starting with f_0:

$$\begin{aligned}
&\dots+b_{0,-n-2}f_{-n-2}+b_{0,-n-1}f_{-n-1}+b_{0,-n}f_{-n} &&=\beta f_0\\
&\dots+b_{-1,-n-2}f_{-n-2}+b_{-1,-n-1}f_{-n-1} &&=\beta f_{-1}\\
&\dots+b_{-2,-n-2}f_{-n-2} &&=\beta f_{-2}\\
&\text{etc.}
\end{aligned}$$

and

$$\begin{aligned}
&\dots+a_{0,-n}f_{-n}\ +a_{0,-n+1}f_{-n+1}+\dots+a_{0,-2}f_{-2}+a_{0,-1}f_{-1}+\sigma_n f_0 &&=(\alpha_0+\alpha_1\beta+\alpha_2\beta^2+\dots)f_0\\
&\dots a_{-1,-n}f_{-n}\ +a_{-1,-n+1}f_{-n+1}+\dots+a_{-1,-2}f_{-2}+\sigma_{n-1}f_{-1} &&=(\alpha_0+\alpha_1\beta+\alpha_2\beta^2+\dots)f_{-1}\\
&\dots a_{-2,-n}f_{-n}\ +a_{-2,-n+1}f_{-n+1}+\dots+\sigma_{n-2}f_{-2} &&=(\alpha_0+\alpha_1\beta+\alpha_2\beta^2+\dots)f_{-2}\\
&\dots &&\dots\qquad\dots\\
&\dots+a_{-n+1,-n}f_{-n}+\sigma_1 f_{-n+1} &&=(\alpha_0+\alpha_1\beta+\alpha_2\beta^2+\dots)f_{-n+1}\\
&\dots+\sigma_n f_{-n} &&=(\alpha_0+\alpha_1\beta+\alpha_2\beta^2+\dots)f_{-n}
\end{aligned}$$

This is in fact a finite system of equations, because multiplication by h^{-1} shifts all the indices by $-N$. First consider the point P_n, where $\alpha_0=\sigma_n$. The result of this lemma will be established in this case; the extension to the other points P_i will then be straightforward.

Step 1

In the proof of this lemma, the following statement will be used at several occasions: fix $k\in\mathbf{Z}$, $k\leqslant 1$; if at P_n

$$\operatorname{order}\left(\frac{f_{-(\alpha n+i)}}{f_{-\alpha n}}\right)\geqslant k\quad\forall i\geqslant n$$

then the same inequality holds for $i\geqslant 1$.

Suppose the contrary; then the second system of equations leads to an homogeneous triangular system of $n-1$ equations in $n-1$ unknowns $f_{-i}^{(k_1)}$, $1\leqslant i\leqslant n-1$, where

$$\frac{f_{-(\alpha n+i)}}{f_{-\alpha n}}=f_{-i}^{(k_1)}\beta^{k_1}+\text{lower order terms}\quad\text{with}\quad k_1<k,$$

to wit,

$$\sum_{j=n-1}^{i-1}a_{-i,-j}f_{-j}^{(k_1)}+(\sigma_{n-i}-\sigma_n)f_{-i}^{(k_1)}=0,\quad 1\leqslant i\leqslant n-1.$$

Its only solution is given by $f_{-j}^{(k_1)}=0$ $(1\leqslant j\leqslant n-1)$, since its determinant equals

$$\prod_{i=1}^{n-1}(\sigma_{n-i}-\sigma_n)\neq 0.$$

Step 2

Next, we show that at P_n

$$\text{order}\left(\frac{f_{-(\alpha n+i)}}{f_{-\alpha n}}\right)\geqslant 1 \quad \text{for } i\geqslant 1,$$

with equality if $i=n$.

In view of Step 1, it suffices to show the statement for $i\geqslant n$. Suppose the contrary; then for some $\gamma\geqslant n$

$$\frac{f_{-(\alpha n+\gamma)}}{f_{-\alpha n}}=f_\gamma^{(-k)}\beta^{-k}+\text{lower order terms},$$

with $k\geqslant 0$ and $\gamma\geqslant n$ and with $f_\gamma^{(-k)}\neq 0$. Then $\beta f_{-(\alpha n+\gamma)}f_{-\alpha n}^{-1}$ is of order $-k+1$ and the γth equation of the first system tells you that for some $\gamma_1\geqslant 2n$,

$$\text{order}\left(\frac{f_{-(\alpha n+\gamma_1)}}{f_{-\alpha n}}\right)=-k+1,\quad \gamma_1\geqslant 2n$$

and by induction

$$\text{order}\left(\frac{f_{-(\alpha n+\gamma_i)}}{f_{-\alpha n}}\right)=-k+i \quad \text{for some} \quad \gamma_i\geqslant(i+1)n \quad \text{with} \quad i\geqslant 1.$$

In particular for $i=N_1-1$, we find some $\gamma_i\geqslant N$ so that

$$\text{order}\left(\frac{f_{-(\alpha n+\gamma_i)}}{f_{-\alpha n}}\right)=-k+N_1-1.$$

This is a contradiction, because for $k\geqslant N$,

$$\text{order}\left(\frac{f_{-(\alpha n+k)}}{f_{-\alpha n}}\right)=\text{order}\left(\frac{f_{-(\alpha n+k-N)}}{f_{-\alpha n}}h^{-1}\right)\geqslant -k+N_1.$$

To show the equality in Step 2, when $i=n$, notice that for $k\geqslant 1$

$$\text{order}\left(\frac{f_{-(\alpha+1)n-k}}{f_{-\alpha n}}\right)=\text{order}\left(\frac{f_{-(\alpha+1)n-k}}{f_{-(\alpha+1)n}}\right)+\text{order}\left(\frac{f_{-(\alpha+1)n}}{f_{-\alpha n}}\right)\geqslant 2.$$

But in order to satisfy the αnth equation of the first system (which is analogous to the first one of the same system), you must have

$$\text{order}\left(\frac{f_{-(\alpha+1)n}}{f_{-\alpha n}}\right)=1.$$

Step 3

It is now straightforward to extend Steps 1 and 2 to the other points P_j. More specifically, at the point P_{n-i} (where $\sigma_0=\sigma_{n-1}$), divide both systems by the function f_{-i}. Then the systems so obtained are the same as the ones above, in which f_{-k} is replaced by

$$g_{-k}=\frac{f_{-k-i}}{f_{-i}}$$

so that at P_{n-i}

$$\text{order}\left(\frac{f_{-(\alpha n+i+j)}}{f_{-(\alpha n+i)}}\right)\geqslant 1 \quad \text{for} \quad j\geqslant 1 \quad \text{with equality if} \quad j=n.$$

By multiplication of the numerator and denominator with the same power of h, one observes that the result holds for any $\alpha\in\mathbf{Z}$. Therefore at P_k, letting $k=n-i$, $0\leqslant j\leqslant n-1$

$$\begin{aligned}\text{order}\,\frac{f_{k+\beta n+j}}{f_k}&=\text{order}\left(\frac{f_{-[-(\beta+2)n+i+(n-j)]}}{f_{-(-n+i)}}\right)\\&=\text{order}\left(\frac{f_{-[-(\beta+2)n+i+(n-j)]}}{f_{-[-(\beta+2)+i]}}\right)+\sum_{\alpha=1}^{\beta+1}\text{order}\left(\frac{f_{-[-(\alpha+1)n+i]}}{f_{-[-\alpha n+i]}}\right)\\&\geqslant 1-(\beta+1)=-\beta\end{aligned}$$

with equality if $j=0$.

A similar analysis can be made near the points $Q_1, \ldots, Q_{n'}$, by considering the new coordinates

$$\alpha'=z^{N'_1}h^{M'_1} \quad \text{and} \quad \beta=z^{\delta'}h^{\gamma'}$$

where $\gamma' N'_1-\delta' M'_1=1$. Near a point Q_i, α and β behave as follows

$$\begin{aligned}\alpha'&=\sigma'_i+O(t)\\ \beta&=\text{constant }(\neq 0)\,t+O(t^2);\end{aligned}$$

the operators $C^{N_1}S^{M_1}$ and $C^{\delta'}S^{\gamma'}$ are both upper triangular with eigenvalues α' and β' and with a nonzero diagonal in the first case and with a nonzero n'th subdiagonal (above the main diagonal) in the second case. This establishes the result of Lemma 2.

Define now $\mathcal{D}$ to be the *minimal positive divisor* on $\mathcal{R}$ such that [1]

$$(f_k)+\mathcal{D}\geqslant -\sum_{i=1}^{k}P_i+\sum_{i=0}^{k-1}Q_i \quad \text{for all} \quad k\in\mathbf{Z}.$$

---

[1] Recall the convention in the footnote of § 1.

It is finite, since it suffices to consider the functions $f_1, \ldots, f_{N-1}$ only. Note that, by Lemma 2, in the kth inequality above, the 2 divisors have *equal orders* at P_k and at Q_k.

LEMMA 3. *$\mathcal{D}$ is a divisor of order g.*

Proof. In a first step, one shows that every function f in

$$L = \{f \text{ meromorphic with } (f)+\mathcal{D} \geqslant \text{any linear combination of } P_i \text{ and } Q_i \text{ with coefficients in } \mathbf{Z}\}$$

can be expressed as a linear combination of functions f_k. Let L_1 be the linear span of the functions f_k and define the ring

$$R = \mathbf{C}[h, h^{-1}, z]/F(h, h^{-1}, z);$$

R contains z and h. The space L_1 is an R-module, because $h^{\pm 1} f_k = f_{k\pm N}$ and $z f_k = (Cf)_k$. Moreover $L_1 \supset R$, because $f_{aN} = h^a$ and $z = z f_0 = (Cf)_0$. If $L_1 \subsetneqq L$, then there is a maximal ideal m in R such that

$$L_1 \subset mL;$$

a maximal ideal m in R is naturally associated with a point $p \in \mathcal{R}_0$ (i.e., $\mathcal{R} \backslash \cup P_i \backslash \cup Q_i$) such that

$$m = \{g \in R \mid g(p) = 0\}.$$

In fact $\mathcal{D} - p \geqslant 0$; because if not, all functions in L_1 would vanish at p; but this is a contradiction, since the functions h and h^{-1} have no common zeros on $\mathcal{R}$ and they both belong to L_1. Therefore every function f_k would be such that

$$(f_k) + (\mathcal{D} - p) \geqslant -\sum_{i=1}^{k} P_i + \sum_{i=0}^{k-1} Q_i \quad \forall k \in \mathbf{Z}.$$

This contradicts the fact that $\mathcal{D}$ is minimal. Therefore $L = L_1$. Choose integers k_1 and k_2 such that $k_1 + k_2 + \text{order}(\mathcal{D}) > 2g-2$. In this next step, the dimension of

$$\mathcal{L}(\mathcal{E}) \equiv \mathcal{L}\left(\sum_{i=1}^{k_1} P_i + \sum_{i=0}^{k_2-1} Q_i + \mathcal{D}\right)$$

will be counted in two different ways: on the one hand using the Rieman-Roch theorem [1]

$$\dim \mathcal{L}(\mathcal{E}) = k_1 + k_2 + \text{order}(\mathcal{D}) - g + 1 \tag{11}$$

---

[1] Since order $(\mathcal{E}) > 2g-2$, $\dim \Omega(-\mathcal{E}) = 0$.

and on the other hand in a direct way to be explained below. Any function φ in $\mathcal{L}(\mathcal{E})$ can be expressed as a linear combination $\varphi=\sum_{i=-k_2}^{k_1} a_i f_i$. To prove this, let φ have a pole of maximal order([1]), say β, among the points P_i and let P_j be the one with maximal index. Then subtract from φ an appropriate multiple of $f_{n(\beta-1)+j}$; the latter belongs to the space $\mathcal{L}(\mathcal{E})$, since $n(\beta-1)+j\leqslant k_1$. The new function obtained in this way belongs to $\mathcal{L}(\mathcal{E}-P_{k_1})$. The same procedure can now be repeated over and over again until you get a function

$$\psi=[\varphi-\text{some linear combination of } f_k \ (-k_2\leqslant k\leqslant k_1)]\in \mathcal{L}(\mathcal{D})\subset L.$$

Since $L=L_1$, ψ is a linear combination of the functions $f_k\in L_1$; but no $k\neq 0$ can occur; indeed considering the form f_k with $|k|$ maximal which occurs and considering the pole at P_k (if $k>0$) or at Q_k (if $k<0$), we find $\psi\notin\mathcal{L}(\mathcal{D})$. Therefore $\psi=\text{constant}=\text{constant}\times f_0$. This shows that $\varphi=\sum a_i f_i$ where the summation ranges from $-k_2$ to k_1. But every f_j $(-k_2\leqslant j\leqslant k_1)$ is in $\mathcal{L}(\mathcal{E})$ and they are independent. Therefore, this second count yields

$$\dim \mathcal{L}(\mathcal{E})=k_1+k_2+1. \tag{12}$$

Comparing (11) and (12) leads to the conclusion that

$$\text{order }(\mathcal{D})=g.$$

LEMMA 4. *$\mathcal{D}$ is a regular divisor.*

Proof. Firstly, one shows that $\mathcal{D}$ is general. Consider an integer k_1 such that $k_1>g-2$; then

$$\dim \mathcal{L}\left(\mathcal{D}+\sum_{i=1}^{k_1} P_i\right)=k_1+1$$

because

$$\text{order }\left(\mathcal{D}+\sum_{i=1}^{k_1} P_i\right)=k_1+g>2g-2.$$

Then

$$\dim \mathcal{L}\left(\mathcal{D}+\sum_{i=1}^{j} P_i\right)\leqslant j+1 \quad \text{for} \quad 1\leqslant j\leqslant k_1$$

because $\mathcal{L}(\mathcal{D}+\sum_{i=1}^{j+1} P_i)$ is strictly larger than $\mathcal{L}(\mathcal{D}+\sum_{i=1}^{j} P_i)$, because f_{j+1} belongs to the first space and not to the second. Therefore letting the index j go down by one lowers the dimension by at least one unit. It follows that

$$1\leqslant \dim \mathcal{L}(\mathcal{D})\leqslant 1.$$

This is to say that $\mathcal{D}$ is general.

([1]) Here this statement must be understood as follows: φ has a pole of order β at P_k if the actual order of pole at P_k is $\beta+\gamma$ where γ is the number of times P_k occurs in $\mathcal{D}$.

It remains to be shown that $\mathcal{D}$ is regular. To do this, it suffices to show that

$$\dim \mathcal{L}\left(\mathcal{D}+\sum_{i=1}^{k} P_i-\sum_{i=0}^{k} Q_i\right)=0 \quad \text{for} \quad k\geqslant 0.$$

For $k=0$, we have that

$$\dim \mathcal{L}(\mathcal{D})=1$$

and

$$\mathcal{L}(\mathcal{D}-Q_1)\subsetneqq \mathcal{L}(\mathcal{D})$$

since the function $f_0=1$ belongs to the second space, but not the first. For the induction step note that

$$\dim \mathcal{L}\left(\mathcal{D}+\sum_{i=1}^{k+1} P_i-\sum_{i=0}^{k} Q_i\right)\leqslant \dim \mathcal{L}\left(\mathcal{D}+\sum_{i=1}^{k} P_i-\sum_{i=0}^{k} Q_i\right)+1=1$$

(because $\dim \mathcal{L}(\mathcal{E})$ increases by at most one when you allow one further pole). Since f_{k+1} belongs to $\mathcal{L}(\mathcal{D}+\sum_{i=1}^{k+1} P_i-\sum_{i=0}^{k} Q_i)$ but not to $\mathcal{L}(\mathcal{D}+\sum_{i=1}^{k+1} P_i-\sum_{i=0}^{k+1} Q_i)$, we also have that

$$\dim \mathcal{L}\left(\mathcal{D}+\sum_{i=1}^{k+1} P_i-\sum_{i=0}^{k+1} Q_i\right)=0.$$

This ends the proof that (a) implies (b). The converse statement (that (b) implies (a)) derives from the following observation.

From the Riemann–Roch theorem, from the fact that allowing one extra pole increases $\dim \mathcal{L}$ by at most one, and from the regularity of $\mathcal{D}$ (in that order) one has

$$1\leqslant \dim \mathcal{L}\left(\mathcal{D}+\sum_{i=1}^{k} P_i-\sum_{i=0}^{k-1} Q_i\right)\leqslant \dim \mathcal{L}\left(\mathcal{D}+\sum_{i=1}^{k-1} P_i-\sum_{i=0}^{k-1} Q_i\right)+1=1.$$

Let f_k be the unique element of $\mathcal{L}(\mathcal{D}+\sum_{i=1}^{k} P_i-\sum_{i=0}^{k-1} Q_i)$ up to scalars. It is clear that h is the unique function in $\mathcal{L}(\sum_{i=1}^{N} P_i-\sum_{i=0}^{N-1} Q_i)$. Normalize f_k such that $hf_k=f_{k+N}$ for every $k\in\mathbf{Z}$.

Lemma 5. *If*

$$f\in\mathcal{L}\left(\mathcal{D}+\sum_{i=1}^{r} P_i-\sum_{i=0}^{s-1} Q_i\right) \quad \textit{with} \quad r\geqslant s, \quad r,s\in\mathbf{Z}$$

then f is a linear combination of f_k ($s\leqslant k\leqslant r$).

Proof. If $r=s$, the result is trivial. Suppose $r>s$; then, since

$$(f_r)\geqslant -\mathcal{D}-\sum_{1}^{r} P_i+\sum_{0}^{s-1} Q_i=\mathcal{E}$$

and since no meromorphic function has a divisor $\geqslant \mathcal{E}+P_r$, the function f_r has a pole at P_r of exact order equal to

$$\left(\# \text{ of } P_r \text{ appearing in } \sum_1^r P_i\right) + (\# \text{ of } P_r \text{ appearing in } \mathcal{D}).$$

The function f has at P_r a pole of order, at worst, the integer given above. Therefore for some constant c_r

$$f - c_r f_r \in \mathcal{L}\left(\mathcal{D} + \sum_{i=1}^{r-1} P_i - \sum_{i=0}^{s-1} Q_i\right).$$

The same argument can now be applied over and over again so as to find constants c_k such that

$$f - \sum_{k=s}^{r} c_k f_k \in \mathcal{L}\left(\mathcal{D} + \sum_{i=1}^{r-1} P_i - \sum_{i=0}^{s-1} Q_i\right).$$

This implies that

$$f - \sum_{k=s}^{r} c_k f_k = 0$$

which establishes Lemma 5.

The rest of the proof is now straightforward. Consider any meromorphic function u, holomorphic on $\mathcal{R}_0$. Then, for K and K' large enough

$$u f_k \in \mathcal{L}\left(\mathcal{D} + \sum_{i=1}^{k+K} P_i - \sum_{i=0}^{k-K'} Q_i\right)$$

admits an expansion as above. In particular,

$$(z)_\infty = -\sum_{i=1}^{M} P_i - \sum_{i=0}^{M'-1} Q_i$$

so that

$$z f_k = \sum_{i=-M'}^{M} c_{k,k+i} f_i$$

with $c_{k,k+M}$ and $c_{k,k-M'} \neq 0$. Moreover, the difference operator $C = (c_{ij})$ is regular by the last hypothesis of (b). C is periodic as a result of the normalization $h f_k = f_{k+N}$. The functions f_k $(0 \leqslant k \leqslant n-1)$ are defined up to some nonzero multiplicative constant. Such a change of basis, due to multiplying f_k with some nonzero constant, amounts to conjugating C with a diagonal operator of period N.

As promised at the start of the proof, we want to add a few words on the modifications necessary to deal with the singular case. In general, we can always define $\mathcal{R}_0$ to be the affine scheme given by $F(h, h^{-1}, z)=0$. Since, near $h=\infty$, F can be re-written:

$$z^{-N}F=\prod\left(\sigma_i\frac{h^{M_i}}{z^{N_i}}-1\right)+\text{lower order terms},$$

it follows that even if F is reducible, F can have no multiple factors as long as the σ_i remain distinct. Then as above we get a reduced algebraic curve $\mathcal{R}$ with n smooth points over $h=\infty$, n' smooth points over $h=0$. Next, this $\mathcal{R}$, although possibly reducible, is at least connected. To see this, note by the constructions already given that any polynomial F subject to the restrictions of Lemma 1 arises from an operator C. Thus for generic choices of C, $\mathcal{R}$ is certainly irreducible. If you approximate an arbitrary C by a sequence C_i whose curves $\mathcal{R}_i$ are irreducible, $\mathcal{R}$ appears as the limit of the $\mathcal{R}_i$'s. Thus $\mathcal{R}$ must be connected. In the singular case, the genus g of $\mathcal{R}$ is to be interpreted as the arithmetic genus, i.e., $g=\dim H^1(O_{\mathcal{R}})=\dim\Omega$, and $2g-2$ is the degree of the divisor of any differential (η). Then the calculation of g by Hurwicz's formula can be interpreted as calculating the degree of the divisor of (dz), and as such works in all cases. Next, in the definition of the divisor $\mathcal{D}$, we must be careful what we mean at singular points P: at each such P, the "multiplicity" of $\mathcal{D}$ is given more precisely by a module $M_P(\mathcal{D})$. In our case, we define $M_P(\mathcal{D})$ to be the O_p-module generated by the functions f_k. Thus the space L in the proof of Lemma 3 is by definition

$$L=\bigcup_{P\in\mathcal{R}_0} M_P(\mathcal{D}),$$

and the rest of the argument goes through without change. For instance, the Riemann–Roch theorem is valid over any such singular $\mathcal{R}$ so the dimension counts all work as before. Q.E.D.

Note that in the correspondence of the theorem, the spectrum of the operators C_h determines the curve $\mathcal{R}$, together with the points $P_1, \ldots, P_n, Q_1, \ldots, Q_{n'}$, and the functions z and h on it. The divisor $\mathcal{D}$ plays the role of the auxiliary parameters which must be given in addition to the spectrum in order to fully recover the operator C. If $\mathcal{R}$ is non-singular, the set of all regular $\mathcal{D}$'s is given by a *Zariski*-open subset of the Jacobian Jac $(\mathcal{R})$ of $\mathcal{R}$. When $\mathcal{R}$ is singular, however, one must distinguish between those operators C which correspond to divisiors $\mathcal{D}$ which are *principal* (i.e., for all singular points P, the module $M_P(\mathcal{D})$ has one generator) and those C corresponding to non-principal $\mathcal{D}$. The first set is again parametrized by a *Zariski*-open subset of the so-called generalized Jacobian, Jac $(\mathcal{R})$, an algebraic group which is an extension of an abelian variety part, and a part isomorphic to $\mathbf{C}^n \times (\mathbf{C}^*)^m$ (cf. Serre [28]). The second set is harder to parametrize: it corresponds to a *Zariski*-open set in the boundary of the *compactified* Jacobian $\overline{\mathrm{Jac}\,(\mathcal{R})}$ (cf. DeSouza [7], Altman–Kleiman [4, 5]). It can be shown that the operator-theoretic meaning of this distinction is that $\mathcal{D}$ is principal if and only if for those h for which C_h has a multiple eigenvalue, the minimal polynomial of C_h is still its characteristic polynomial (see Mumford [25]). For use later, we will say that C is *of principal type* if the divisor $\mathcal{D}$ is principal. (Note that this always holds if $\mathcal{R}$ is non-singular.)

Besides the divisor $\mathcal{D}$ and the points P_i, Q_i, the curve $\mathcal{R}$ has various other elements of structure on it which are important for later analysis. One of these is the holomorphic differential form

$$\zeta = \frac{dz}{h\dfrac{\partial F}{\partial h}} = -\frac{dh}{h\dfrac{\partial F}{\partial z}}.$$

Clearly, when $\mathcal{R}$ is non-singular, $\partial F/\partial h$ and $\partial F/\partial z$ have no common zeroes on $\mathcal{R}_0$, so ζ *has neither zeroes nor poles in* $\mathcal{R}_0$. When $\mathcal{R}$ is singular, the same thing holds if we interpret zeroes and poles in the sense described in the introduction. Otherwise put, ζ is a generator on $\mathcal{R}_0$ of the sheaf of 1-forms η satisfying, for all $P \in \mathcal{R}_0$:

$$\sum_{\substack{\text{Branches}\\ \text{of } \mathcal{R} \text{ at } P}} \operatorname{res}_\gamma (f \cdot \eta) = 0, \quad \text{all} \quad f \in O_p.$$

LEMMA 7. *If* $\alpha = NM_1 - M_1 - 1$, $\alpha' = NM_1' - M_1' - 1$, *then*

$$(\zeta) = \alpha \sum_{i=1}^{n} P_i + \alpha' \sum_{i=1}^{n'} Q_i.$$

Proof. To study the order of zero or pole of ζ at P_i, we use the expansion

$$z^{-N} F(h, h^{-1}, z) = \prod_{i=1}^{n} \left(\sigma_i \frac{h^{M_1}}{z^{N_1}} - 1 \right) + \text{lower order terms}$$

described above. Since at P_i, in terms of a local parameter t,

$$z = t^{-M_1} + \ldots, \quad h = C_i t^{-N_1} + \ldots,$$

we calculate from this formula that:

$$\begin{aligned} \frac{\partial F}{\partial h} &= (\text{constant}) \cdot z^{N-N_1} \cdot h^{M_1 - 1} + \text{lower order poles} \\ &= (\text{constant}) \cdot t^{-M_1 N + N_1} + \text{lower order poles.} \end{aligned}$$

8 – 792907 *Acta mathematica* 143. Imprimé le 28 Septembre 1979

Substituting into the formula for ζ, one checks that ζ has an α-fold zero at P_i as required. The proof for the Q_i's is similar.

LEMMA 8. *The following inequality holds:*

$$(\Delta_{11}\zeta)_\infty \geqslant -P_1-Q_1.$$

Proof. Consider the minor Δ_{11} obtained by removing the first row and the first column from $C_h - zI$. The computation of Δ_{11} follows the same argument as in Lemma 1, § 1; the only difference is that all the terms involving σ_1 in det $(C_h - zI)$ are absent. Therefore, the leading terms in Δ_{11} will be

$$\sum_{\alpha=0}^{n-1} (-1)^\alpha \bar{\tau}_\alpha z^{\alpha N_1 - 1} h^{(n-1-\alpha)M_1} = z^{-1} \prod_{i=2}^{n} (\sigma_i h^{M_1} - z^{N_1})$$

where $\bar{\tau}_\alpha$ denotes the symmetric polynomial of degree α in $(\sigma_2, ..., \sigma_n)$ (instead of $\sigma_1, ..., \sigma_n$ as in Lemma 1). From the expression for the leading term one reads off that Δ_{11} will have a pole of order $M_1(N-1)$ at worst if $h^{-M_1}z^{N_1} = \sigma_1$ (i.e., at P_1) and $M_1(N-1)-1$ at worst if $h^{-M_1}z^{N_1} = \sigma_i$, $i \neq 1$ (i.e., at P_i with $i \neq 1$). An analogous statement can be made about the points Q_i.

PROPOSITION 1. *Every regular difference operator leads to* $2N$ *regular divisors* $\mathcal{D}_1^{(i)}$ and $\mathcal{D}_2^{(i)}$ $(1 \leqslant i \leqslant N)$ *of degree* g *(where* $\mathcal{D} = \mathcal{D}_1^{(N)}$*) having the property that for* $1 \leqslant i, j \leqslant N$:

$$\mathcal{D}_1^{(i)} + \mathcal{D}_2^{(j)} - \sum_{k=j}^{i} P_k + \sum_{k=j+1}^{i-1} Q_k \quad \textit{if} \quad i > j$$

$$\mathcal{D}_1^{(i)} + \mathcal{D}_2^{(j)} + \sum_{k=i+1}^{j-1} P_k - \sum_{k=i}^{j} Q_k \quad \textit{if} \quad i < j$$

$$\mathcal{D}_1^{(i)} + \mathcal{D}_2^{(i)} - P_i - Q_i \quad \textit{if} \quad i = j$$

is the divisor of some meromorphic differential ω_{ij}. *Then*

$$\omega_{ij} = \Delta_{ji}\zeta.$$

Moreover,

$$(f_k) = \mathcal{D}_1^{(k)} - \mathcal{D} - \sum_{i=1}^{k} P_i + \sum_{i=0}^{k-1} Q_i.$$

Proof. Since $\mathcal{D}_1^{(N)} = \mathcal{D}$ is general

$$0 \leqslant \dim \mathcal{L}(\mathcal{D} - P_N - Q_N) \leqslant \dim \mathcal{L}(\mathcal{D}) - 1 \leqslant 0.$$

Therefore by the Riemann–Roch theorem

$$\dim \Omega(-\mathcal{D}+P_N+Q_N)=1.$$

Let ω_{NN} be the unique differential, up to some multiplicative constant such that

$$(\omega_{NN})\geqslant \mathcal{D}-P_N-Q_N.$$

This differential has poles at P_N and Q_N, otherwise $(\omega_{NN})\geqslant\mathcal{D}-P_N$ or $\geqslant\mathcal{D}-Q_N$ which would contradict the fact that $\mathcal{D}$ is general. Define the positive divisor $\mathcal{D}_2^{(N)}$ such that

$$(\omega_{NN})=\mathcal{D}_1^{(N)}-P_N-Q_N+\mathcal{D}_2^{(N)} \tag{13}$$

and the meromorphic differential

$$\omega_{kN}=f_k\omega_{NN}.$$

These differentials enjoy the property that

$$(\omega_{kN})\geqslant\mathcal{D}_2^{(N)}-\sum_{i=0}^{k}P_i+\sum_{i=1}^{k-1}Q_i \tag{14}$$

and they are the only ones with this property. Define the positive divisor $\mathcal{D}_1^{(k)}$ such that

$$(\omega_{kN})=\mathcal{D}_1^{(k)}-\sum_{i=0}^{k}P_i+\sum_{i=1}^{k-1}Q_i+\mathcal{D}_2^{(N)}.$$

From (13), it follows also $\mathcal{D}_2^{(N)}$ is general and since ω_{kN} satisfying (14) is unique, $\mathcal{D}_1^{(k)}$ is also general. Therefore we may define ω_{kk} to be the only differential such that

$$(\omega_{kk})\geqslant\mathcal{D}_1^{(k)}-P_k-Q_k$$

and $\mathcal{D}_2^{(k)}$ such that

$$(\omega_{kk})=\mathcal{D}_1^{(k)}-P_k-Q_k+\mathcal{D}_2^{(k)}.$$

So far we have defined the last column and the diagonal of $\Omega=(\omega_{ij})$. The remaining differentials ω_{ij} are defined such that

$$\frac{\omega_{ij}}{\omega_{Nj}}=\frac{\omega_{iN}}{\omega_{NN}}.$$

The next step is to show that $\omega_{ii}=\Delta_{ii}\zeta$ up to some multiplicative constant, to begin with for $i=N$. By the uniqueness it suffices to prove that

$$(\Delta_{NN}\zeta)\geqslant\mathcal{D}_1^{(N)}-P_N-Q_N.$$

Any regular divisor $\mathcal{D}_1^{(N)}$ can be approximated by regular divisors in $\mathcal{R}_0$. The corresponding operators also approach the original one. Therefore *it is legitimate to assume* $\mathcal{D}_1^{(N)}$ *in* $\mathcal{R}_0$. By Lemma 8,

$$(\Delta_{NN}\zeta)_\infty \geqslant -P_N - Q_N. \tag{15}$$

the zeros of $\Delta_{NN}\zeta$ on $\mathcal{R}_0$ come from zeros of Δ_{NN}, since ζ never vanishes on $\mathcal{R}_0$ (by Lemma 7). Since

$$f_i = \frac{\Delta_{Ni}}{\Delta_{NN}} h = \frac{\Delta_{ii}}{\Delta_{iN}} h, \quad 1 \leqslant i \leqslant N,$$

since $\Delta_{Ni}h$ never has any poles on $\mathcal{R}_0$ and since every point of $\mathcal{D}_1^{(N)}$ appears as a pole of some f_i, the minor Δ_{NN} vanishes at each of the points of $\mathcal{D}_1^{(N)}$;

$$(\Delta_{NN})_0 \geqslant \mathcal{D}_1^{(N)},$$

combined with (15), leads to

$$(\Delta_{NN}\zeta) \geqslant \mathcal{D}_1^{(N)} - P_N - Q_N.$$

To show this statement for $1 \leqslant i \leqslant N-1$, shift the matrix $C_h - zI$ up i levels and to the left i steps (1) and call $(f_1^{(i)}, \ldots, f_{N-1}^{(i)}, h)$ its eigenvector. Then

$$f_k^{(i)} = f_{k+i} f_i^{-1} = \frac{\Delta_{i,k+i}}{\Delta_{i,i}} \quad \text{and} \quad (f_k^{(i)}) = \mathcal{D}_1^{(k+i)} - \mathcal{D}_1^{(i)} - \sum_{j=i+1}^{k+i} P_j + \sum_{j=i}^{k+i-1} Q_j.$$

Clearly the shifted operator is also regular, leads to the same curve and defines the points P_i and Q_i in the shifted order $P_{i+1}, \ldots, P_n, P_1, \ldots, P_i$ and $Q_{i+1}, \ldots, Q_n, Q_1, \ldots, Q_i$. The functions $f_k^{(i)}$ define a regular divisor of order g, which from the relation above must be $\mathcal{D}_1^{(i)}$. Therefore assuming again that by some small deformation $\mathcal{D}_1^{(i)}$ is in $\mathcal{R}_0$,

$$(\Delta_{ii})_0 \geqslant \mathcal{D}_1^{(i)}.$$

But, since

$$(\Delta_{ii}\zeta)_\infty \geqslant -P_i - Q_i$$

it follows that

$$(\Delta_{ii}\zeta) \geqslant \mathcal{D}_1^{(i)} - P_i - Q_i.$$

Therefore $\Delta_{ii}\zeta = \omega_{ii}$. This establishes the fact that $\Delta^T\zeta = \Omega$.

To show that every divisor $\mathcal{D}_2^{(N)}$ is regular, consider the transposed difference operator C^T. It leads to the same curve and the eigenvector is given by

$$f_k' = \frac{\Delta_{kN}}{\Delta_{NN}} h^{-1} = \frac{\Delta_{kk}}{\Delta_{Nk}} h^{-1} = \frac{\omega_{Nk}}{\omega_{NN}} h^{-1}$$

---

(1) The diagona l entries of the new matrix are then $(c_{i+1,i+1} - z, \ldots, c_{NN} - z, c_{11} - z, \ldots, c_{ii} - z)$.

with divisor

$$(f'_k) = \mathcal{D}_2^{(k)} - \mathcal{D}_2^{(N)} - \sum_1^k Q_j + \sum_0^{k-1} P_j;$$

hence $\mathcal{D}_2^{(N)}$ and therefore every $\mathcal{D}_2^{(i)}$ $(1 \leqslant i \leqslant N)$ is regular.

This last part of the above proof may be rephrased as asserting that if in the correspondence of Theorem 1, the 2 sets of data:

$$\{\mathcal{R}, \{P_i\}, \{Q_i\}, z, h, \mathcal{D}_1^{(N)}\} \quad \text{and} \quad C$$

correspond to each other, then the modified data

$$\{\mathcal{R}, \{Q_i\}, \{P_i\}, z, h^{-1}, \mathcal{D}_2^{(N)}\} \quad \text{and} \quad C^T$$

also correspond to each other.

Also note that the definition of $\mathcal{D}$ as the least positive divisor such that

$$(f_k) + \mathcal{D} \geqslant -\sum_{i=1}^{k} P_i + \sum_{i=0}^{k-1} Q_i$$

shows immediately that the set of divisors $\mathcal{D}_1^{(k)}$ $(k \in \mathbf{Z})$ have no common points; the same holds for the divisors $\mathcal{D}_2^{(k)}$ $(k \in \mathbf{Z})$.

Finally, in case $\mathcal{R}$ is singular, we must make the definition of $\mathcal{D}_2^{(N)}$ more precise: (1)

$$\mathcal{D}_2^{(N)} = \text{set of points where } (\omega_{NN}) \gneqq \mathcal{D}_1^{(N)} - P_N - Q_N,$$

and if P is a singular point in this set, then

$M_P(\mathcal{D}_2^{(N)})$ = module of meromorphic functions f such that for all

$$g \in M_P(\mathcal{D}_1^{(N)}), \ (f \cdot g \cdot \omega_{NN}) \geqslant 0 \text{ at } P.$$

It follows that:

$$M_P(\mathcal{D}_1^{(N)}) = \text{module of } g \text{ such that for all } f \in M_P(\mathcal{D}_2^{(N)}), \ (f \cdot g \cdot \omega_{NN}) \geqslant 0 \text{ at } P.$$

So the relation between $\mathcal{D}_1^{(N)}$, $\mathcal{D}_2^{(N)}$ is symmetric and we still write this

$$(\omega_{NN}) = \mathcal{D}_1^{(N)} + \mathcal{D}_2^{(N)} - P_N - Q_N.$$

---

(1) In the language of coherent sheaves,

$$O(\mathcal{D}_2^{(N)}) = \underline{\text{Hom}}\,(O(\mathcal{D}_1^{(N)}), \Omega(P_N + Q_N)).$$

In checking the above proof for the singular case, one must use Serre duality for $\mathcal{R}$. For torsion free sheaves $\mathfrak{F}$, it says

$$\underline{\text{Ext}}^i\,(\mathfrak{F}, \Omega) = 0, \quad i > 0$$

$$H^i(\mathfrak{F}) \text{ dual to } H^{1-i}\,(\text{Hom}\,(\mathfrak{F}, \Omega)).$$

§ 3. Symmetric and self-adjoint difference operators with examples

There are two interesting special cases of difference operators: the symmetric and self-adjoint difference operators. They both lead to curves with involutions and to divisors with special properties. This is the topic of this chapter. At the end of it some examples will be discussed with applications to inverse spectral problems.

THEOREM 2. *There is a one-to-one correspondence between the following sets of data:*

(1) *a regular, symmetric difference operator C of support $[-M, M]$ of period N, modulo conjugation by periodic diagonal operators, with entries ± 1.*

(2) *a curve $\mathcal{R}$ possibly singular of genus*

$$g = M(N-1)-n+1$$

with ordered smooth points $P_1, \ldots, P_n$ and $Q_1, \ldots, Q_n$, two meromorphic functions h and z with the properties given in theorem 1 with $N_1 = N_1'$ and $M_1 = M_1'$ and a regular divisor $\mathcal{D}$. Moreover, $\mathcal{R}$ has an involution τ such that $h^\tau = h^{-1}$, $z^\tau = z$, $\tau(P_k) = Q_k$ and $\mathcal{D}$ has the property that

$$\mathcal{D} + \mathcal{D}^\tau - P_n - Q_n$$

is the divisor of some differential ω on $\mathcal{R}$.

Proof. This theorem results from combining Theorem 1 with the last remark of the previous section. We are simply dealing in (1) with a regular difference operator C mod conjugation by periodic diagonal Λ, such that $C^T = \Lambda \cdot C \cdot \Lambda^{-1}$. In fact, any such C is conjugate to a symmetric C' and this C' is unique up to conjugation by a Λ with entries ± 1. In (2), we are dealing with data $\{\mathcal{R}, \{P_i\}, \{Q_i\}, z, h, \mathcal{D}\}$ such that $\mathcal{R}$ has an automorphism τ carrying this data to $\{\mathcal{R}, \{Q_i\}, \{P_i\}, z, h^{-1}, \mathcal{D}'\}$.

It is useful to see more explicitly how τ arises. In fact, since C is symmetric, the algebraic equation

$$F(h, h^{-1}, z) = \det(C_h - zI)$$

is symmetric with regard to h and h^{-1}, i.e., F is now a function of $h + h^{-1}$ and z. Hence, the map

$$(z, h)^\tau \equiv (z, h^{-1})$$

maps $\mathcal{R}$ into $\mathcal{R}$ and is an involution since $\tau^2 = \text{identity}$. The formula for the genus with $M = M'$ and $n = n'$ simplifies to the one above. Note that $\sigma_i = \sigma'$ and that, because $z^{N_1} h^{-M_1}$ has value σ_i at P_i (resp. $z^{N_1} h^{M_1}$ has value σ_i' at Q_i), therefore $\tau(P_i) = Q_i$.

Finally, because C is symmetric it follows that

$$\Delta_{ij}(z, h) = \Delta_{ji}(z, h^{-1})$$

i.e., as a function on $\mathcal{R}$, $\Delta_{ij} \circ \tau = \Delta_{ji}$. Therefore by Proposition 1, § 1, for all i and j, τ acting on the divisor of ω_{ij} is the divisor of ω_{ji}, i.e.

$$(\mathcal{D}_1^{(i)} + \mathcal{D}_2^{(j)})^\tau = \mathcal{D}_2^{(i)} + \mathcal{D}_1^{(j)}.$$

Since the divisors $\mathcal{D}_1^{(j)}$ have no common points, this implies that $(\mathcal{D}_1^{(i)})^\tau = \mathcal{D}_2^{(i)}$ and in particular

$$\mathcal{D} + \mathcal{D}^\tau - P_n - Q_n = (\omega_{NN}).$$

Remark. Let $\mathcal{R}$ have an involution τ and let O be some origin in Jac ($\mathcal{R}$). Then the *Prym variety* Prym ($\mathcal{R}$) of $\mathcal{R}$ over S (quotient of $\mathcal{R}$ by τ) may be defined as the set of principal $\mathcal{D} \in \text{Jac}(\mathcal{R})$, $\mathcal{D}$ a divisor $\sum_{i=1}^{g} \nu_i$ considered modulo linear equivalence, such that

$$\sum_{i=1}^{g} \int_{0_i}^{\nu_i} \omega + \sum_{i=1}^{g} \int_{0_i^\tau}^{\nu_i^\tau} \omega = 0, \quad \text{(mod periods)}.$$

It is a linear subvariety of Jac ($\mathcal{R}$). Moreover every holomorphic differential on S can be lifted to a holomorphic differential on $\mathcal{R}$ which is invariant under τ. Therefore it is possible to find a basis $\omega_1, \ldots, \omega_g$ such that $\omega_i^\tau = \omega_i$ for $1 \leqslant i \leqslant g_0$ and $\omega_i^\tau = -\omega_i$ for $g_0 < i \leqslant g$. Since the relation above is trivially satisfied for ω's such that $\omega^\tau = -\omega$, it, in fact, reduces to g_0 conditions

$$\sum_{i=1}^{g} \int_{0_i}^{\nu_i} \omega_k = 0, \quad \text{(mod periods)}, \quad 1 \leqslant k \leqslant g_0.$$

We show that the regular part of Prym ($\mathcal{R}$) can be parametrized by symmetric regular difference operators of principal type (p. 123).

The symmetric regular difference operators of principal type lead to curves with an involution and to a principal divisor $\mathcal{D} \in \text{Jac}(\mathcal{R})$ such that

$$\mathcal{D} + \mathcal{D}^\tau - P_n - Q_n$$

is the divisor of some differential $\omega_0 = \Delta_{NN}\zeta$; Δ_{NN} is a meromorphic function having for divisor

$$(\Delta_{NN}) = (\omega_0) - (\zeta) = \mathcal{D} + \mathcal{D}^\tau - \alpha \sum_1^n P_i - P_n - \alpha \sum_1^n Q_i - Q_n.$$

Choose the origin to be $\alpha \sum_1^n P_i + P_n$; then Abel's theorem tells you that $\mathcal{D} \in \text{Prym}(\mathcal{R})$; also $\mathcal{D}$ is regular. Conversely, consider a regular principal divisor $\mathcal{D}$ in Prym $(\mathcal{R})$, defined with regard to the origin $\alpha \sum_1^n P_i + P_n$. This implies the existence of a meromorphic function Δ having for divisor

$$\mathcal{D} + \mathcal{D}^\tau - \alpha \sum_1^n P_i - P_n - \alpha \sum_1^n Q_i - Q_n.$$

Then, $\omega_0 = \Delta\zeta$ satisfies

$$(\omega_0) = \mathcal{D} + \mathcal{D}^\tau - P_n - Q_n$$

and therefore $\mathcal{D}$ leads to a symmetric principal difference operator.

The next theorem deals with self-adjoint difference operators:

THEOREM 3. *There is a one-to-one correspondence between the following set of data*
(a) *a regular self-adjoint difference operator C of support $[-M, M]$ of period N*
(b) *a curve, possible singular, of genus*

$$g = M(N-1) - n + 1$$

with ordered smooth points $P_1, \ldots, P_n$ and $Q_1, \ldots, Q_{n'}$, two meromorphic functions h and z with the properties given in Theorem 1 with $N_1' = N_1$ and $M_1 = M_1'$ and a divisor $\mathcal{D}$ of degree g. Moreover $\mathcal{R}$ is endowed with an antiholomorphic involution $\tilde{}$ for which $\mathcal{R} \backslash \mathcal{R}_{\mathbf{R}} = \mathcal{R}_+ \cup \mathcal{R}_-$ (disconnected) (define $\mathcal{R}_{\mathbf{R}} = \{p \in \mathcal{R} \mid \tilde{p} = p\}$), such that $\tilde{P}_i = Q_i$ with $P_i \in \mathcal{R}_+$ and $Q_i \in \mathcal{R}_-$, and such that if $\varphi^*(P) = \overline{\varphi(\tilde{P})}$, then $hh^* = 1$ and $z = z^*$. The divisor $\mathcal{D}$ has the property that

$$\mathcal{D} + \tilde{\mathcal{D}} - P_n - Q_n$$

is the divisor of some differential on $\mathcal{R}$, which is real positive(¹) *on $\mathcal{R}_{\mathbf{R}}$.*

Remark. Note that the *regularity* of $\mathcal{D}$ is not assumed. In this case, we will prove it using the relation

$$\mathcal{D} + \tilde{\mathcal{D}} - P_n - Q_n = (\omega)$$

where $\omega \geq 0$ on $\mathcal{R}_{\mathbf{R}}$.

The proof of this statement goes in two steps: first $\mathcal{D}$ is shown to be general. Indeed, since $\omega \geq 0$ on $\mathcal{R}_{\mathbf{R}}$,

$$2\pi i \operatorname{Res}_{P_n} \omega = -2\pi i \operatorname{Res}_{Q_n} \omega = \int_{\mathcal{R}_{\mathbf{R}}} \omega > 0;$$

(¹) This makes sense because $\mathcal{R}_{\mathbf{R}}$ inherits a natural orientation as the boundary of the oriented surface $\mathcal{R}_+$.

P_n and Q_n are poles of ω and therefore neither P_n nor Q_n appears in $\mathcal{D}$ or $\tilde{\mathcal{D}}$. To show that $\mathcal{D}$ is general, it suffices to show that $\mathcal{L}(\mathcal{D}-Q_n)=\{0\}$. Suppose $\mathcal{L}(\mathcal{D}-Q_n)\neq\{0\}$ and let φ satisfy $(\varphi)\geqslant -\mathcal{D}+Q_n$; then $(\varphi)=-\mathcal{D}+Q_n+\mathcal{E}$ for some positive divisor $\mathcal{E}$ and

$$\begin{aligned}(\varphi\varphi^*\omega) &= -\mathcal{D}+Q_n+\mathcal{E}-\mathcal{D}+P_n+\tilde{\mathcal{E}}+\mathcal{D}+\mathcal{D}-P_n-Q_n\\ &= \mathcal{E}+\tilde{\mathcal{E}},\end{aligned}$$

contradicting

$$\int_{R_{\mathbf{R}}} \varphi\varphi^*\omega = \int_{R_{\mathbf{R}}} |\varphi|^2\omega > 0.$$

In the second step we show that

$$\mathcal{L}\left(\mathcal{D}+\sum_1^{k-1} P_i-\sum_0^{k-1} Q_i\right)=\{0\}$$

by induction. According to Step 1, this holds for $k=1$ and N. So, assume that

$$\mathcal{L}\left(\mathcal{D}+\sum_1^{k} P_i-\sum_0^{k} Q_i\right)=\{0\}.$$

This fact and the Riemann–Roch theorem imply that

$$\dim \mathcal{L}\left(\mathcal{D}+\sum_1^{k} P_i-\sum_0^{k-1} Q_i\right)=1.$$

Let f_k be the unique function in this space. Define

$$\mathcal{D}^{(k)}=(f_k)+\mathcal{D}+\sum_1^{k} P_i-\sum_0^{k-1} Q_i.$$

Then in Jac $(\mathcal{R})$,

$$\mathcal{D}^{(k)}-P_k\equiv\mathcal{D}+\sum_1^{k-1} P_i-\sum_0^{k-1} Q_i.$$

This implies that

$$\dim \mathcal{L}(\mathcal{D}^{(k)})=1$$

and, hence

$$\dim \mathcal{L}(\mathcal{D}^{(k)}-P_k)=0,$$

since P_k does not appear in $\mathcal{D}^{(k)}$; indeed $f_k f_k^*\omega\geqslant 0$ on $\mathcal{R}_{\mathbf{R}}$ and

$$(f_k f_k^*\omega)=\tilde{\mathcal{D}}^{(k)}+\mathcal{D}^{(k)}-P_k-Q_k,$$

so that the integral of $f_k f_k^*\omega$ over $\mathcal{R}_{\mathbf{R}}$ is strictly positive and therefore $f_k f_k^*\omega$ must have at least one pole, the only possible ones being P_k and Q_k.

This also implies that neither P_k, nor Q_k ever appear in $\mathcal{D}^{(k)}$ and, in particular, $\mathcal{D}$ does not contain P_n or Q_n.

Proof of Theorem 3. Given a self-adjoint operator C, we construct a curve $\mathcal{R}$ with the given properties. Since C is self-adjoint,

$$\det (C_h - zI) = F(h, h^{-1}, z) = 0$$

has the following form:

$$A_0 h^M + \bar{A}_0 h^{-M} + \sum_{i=1}^{M} (A_i(z) h^{M-i} + (\overline{A_i(\bar{z})}) h^{-(M-i)}) = 0. \tag{16}$$

Therefore, the map

$$\tilde{\ }: (z, h) \to (\bar{z}, \bar{h}^{-1})$$

defines an anti-holomorphic involution from $\mathcal{R}$ into $\mathcal{R}$. But if $|h| = 1$ the finite matrix C_h is self-adjoint and therefore has a real spectrum. Therefore the fixed points $\mathcal{R}_{\mathbf{R}}$ for this map are given by

$$\mathcal{R}_{\mathbf{R}} = \{(z, h) \mid h = \bar{h}^{-1}, z = \bar{z}\}$$
$$= \{(z, h) \mid |h| = 1\}$$

and

$$\mathcal{R} \backslash \mathcal{R}_{\mathbf{R}} = \{|h| > 1\} \cup \{|h| < 1\} = \mathcal{R}_+ \cup \mathcal{R}_-$$

with $\mathcal{R}_+ \ni P_i$ and $\mathcal{R}_- \ni Q_i$; this defines two distinct regions $\mathcal{R}_+$ and $\mathcal{R}_-$, whose boundary is given by $\mathcal{R}_{\mathbf{R}}$ and therefore $\mathcal{R}_{\mathbf{R}}$ is homologous to zero. $\mathcal{R}_{\mathbf{R}}$ will consist of possibly several circles σ_i with a definite orientation as noted in the footnote above. Since $\sigma_i = \bar{\sigma}_i$, we have $\tilde{P}_i = Q_i$; moreover this involution extends to the field of meromorphic functions as follows

$$\varphi^*(p) = (\overline{\varphi(\tilde{p})})$$

and to meromorphic differentials as follows:

$$(\varphi d\psi)^* = \varphi^* d\psi^*.$$

With this definition

$$h^* = h^{-1} \quad \text{and} \quad z^* = z.$$

By Theorem 1, the difference operator C maps into a regular divisor $\mathcal{D}$ such that

$$(\omega) = (\omega_{NN}) = (\zeta \Delta_{NN}) = \mathcal{D} + \mathcal{D}' - P_n - Q_n. \tag{17}$$

Next, we show that $\mathcal{D}' = \tilde{\mathcal{D}}$. But, because C is self-adjoint, it follows that

$$\Delta_{ij}^* = \Delta_{ji}.$$

Therefore, by Proposition 1, for all i and j, $\tilde{}$ acting on the divisor of ω_{ij} is the divisor of ω_{ji}, i.e.,

$$\tilde{\mathcal{D}}_1^{(i)}+\tilde{\mathcal{D}}_2^{(j)}=\tilde{\mathcal{D}}_2^{(i)}+\tilde{\mathcal{D}}_1^{(j)}.$$

Since the divisors $\mathcal{D}_1^{(j)}$ have no common points this implies that $\mathcal{D}_1^{(i)}=\mathcal{D}_2^{(i)}$ and in particular $\mathcal{D}=\tilde{\mathcal{D}}$.

Finally, we show that $\omega\geqslant 0$ on $\mathcal{R}_{\mathbf{R}}$ for some appropriate normalization of ζ. To do this take the differential of F: since z only appears on the diagonal C_h-zI,

$$-\sum_{i=1}^{N}\Delta_{ii}\,dz+h\frac{\partial F}{\partial h}\frac{dh}{h}=0.$$

Using this relation, and choosing $\zeta=-i\,dz(h\,\partial F/\partial h)^{-1}$ and using Proposition 1,

$$\omega=\zeta\Delta_{NN}=-i\Delta_{NN}\frac{dz}{h\dfrac{\partial F}{\partial h}}=\frac{-i\,dh/h}{\sum\limits_{i=1}^{N}\dfrac{\Delta_{ii}}{\Delta_{NN}}}=\frac{-i\,dh/h}{\sum\limits_{i=1}^{N}\dfrac{\Delta_{ii}}{\Delta_{iN}}\dfrac{\Delta_{iN}}{\Delta_{NN}}}=\frac{-i\,dh/h}{\sum\limits_{i=1}^{N}\dfrac{\Delta_{Ni}}{\Delta_{NN}}\cdot\dfrac{\Delta_{iN}}{\Delta_{NN}}}$$

$$=\frac{-i\,dh/h}{\sum\limits_{i=1}^{N}\dfrac{\Delta_{Ni}}{\Delta_{NN}}\cdot\left(\dfrac{\Delta_{Ni}}{\Delta_{NN}}\right)^{*}}\quad(\text{since } \Delta_{Ni}^{*}=\Delta_{iN},\quad 1\leqslant i\leqslant N)$$

$$=\frac{-i\,dh/h}{\sum\limits_{i=1}^{N}f_if_i^{*}}$$

Note that this formula shows that $\omega^{*}=\omega$. We now show that $\omega\geqslant 0$ on $\mathcal{R}_{\mathbf{R}}$. Indeed on $\mathcal{R}_{\mathbf{R}}$

$$\sum_{i=1}^{N}f_if_i^{*}=\sum|f_i|^2\geqslant 0.$$

To show that $-i\,dh/h\geqslant 0$ on $\mathcal{R}_{\mathbf{R}}$, let $h=\varrho e^{i\theta}$; at all but a finite number of points, h is a local parameter on $\mathcal{R}$, and θ is a local coordinate on $\mathcal{R}_{\mathbf{R}}$. Since $-i\,dh/h=d\theta$, $\omega\geqslant 0$ at these points, hence by continuity, at all points.

Consider now the converse. The curve $\mathcal{R}$ has the properties listed in (b), in particular it has an antiholomorphic involution $p\to\tilde{p}$ such that $\tilde{P}_i=Q_i$. The curve of fixed points $\mathcal{R}_{\mathbf{R}}=\{p\,|\,p=\tilde{p}\}$ divides $\mathcal{R}$ into two distinct regions $\mathcal{R}_+$ and $\mathcal{R}_-$, the first containing the points P_i and the second the points Q_i. The curve $\mathcal{R}_{\mathbf{R}}$ can thus be regarded as the boundary of $\mathcal{R}_+$ or $\mathcal{R}_-$, (thus $\mathcal{R}_{\mathbf{R}}$ is homologous to zero).

Choose any regular $\mathcal{D}$ such that

$$\mathcal{D}+\tilde{\mathcal{D}}-P_n-Q_n=(\omega)$$

where $\omega = \omega^*$ and it is real positive on $\mathcal{R}_{\mathbf{R}}$. This means that if $z_0 \in \mathcal{R}_{\mathbf{R}}$ and t is a local parameter at z_0 where t is real on $\mathcal{R}_{\mathbf{R}}$, $\operatorname{Im} t > 0$ on $\mathcal{R}_+$, $\operatorname{Im} t < 0$ on $\mathcal{R}_-$, then

$$\omega = a(t)\,dt, \quad a(\bar{t}) = \overline{a(t)}, \quad a(t) \geq 0 \quad \text{if } t \in \mathbf{R}.$$

Let f_k be the usual meromorphic functions associated to $\mathcal{D}$, i.e., such that

$$(f_k) \geq -\sum_1^k P_i + \sum_0^{k-1} Q_i - \mathcal{D}.$$

Let

$$\mathcal{D}_k \equiv (f_k) + \mathcal{D} + \sum_1^k P_i - \sum_0^{k-1} Q_{k-1}.$$

Normalize f_k as before such that $f_{k+N} = h f_k$. We now define a scalar product between the f_k's, i.e.,

$$(f_k, f_l) = \int_{\mathcal{R}_{\mathbf{R}}} f_k f_l^* \,\omega = \int_{\mathcal{R}_{\mathbf{R}}} f_k \bar{f}_l \,\omega.$$

When $k \neq l$, $(f_k, f_l) = 0$; indeed, for $k > l$

$$\begin{aligned}(f_k f_l^* \omega) &= \mathcal{D}_k - \mathcal{D} - \sum_1^k P_i + \sum_0^{k-1} Q_i + \tilde{\mathcal{D}}_l - \tilde{\mathcal{D}} - \sum_1^l \tilde{P}_i + \sum_0^{l-1} \tilde{Q}_i + \mathcal{D} + \tilde{\mathcal{D}} - P_n - Q_n \\ &= \mathcal{D}_k + \tilde{\mathcal{D}}_l - \sum_l^k P_i + \sum_{l+1}^{k-1} Q_i;\end{aligned}$$

it tells you that $f_k f_l^* \omega$ has no other poles but at some of the points P_i, i.e., in the region $\mathcal{R}_+$ only. Since $\mathcal{R}_{\mathbf{R}}$ is the boundary of that region and homologous to zero and since

$$\sum_{P_i} \operatorname{Res} f_k f_l^* \,\omega = 0,$$

the conclusion above follows. A similar argument proves the assertion when $k < l$. For $k = l$,

$$(f_k, f_k) = \int_{\mathcal{R}_{\mathbf{R}}} f_k f_k^* \,\omega = \int_{\mathcal{R}_{\mathbf{R}}} |f_k|^2 \omega > 0$$

again because on $\mathcal{R}_{\mathbf{R}}$, ω is non-negative.

Since

$$2\pi i \operatorname{Res}_{P_k} (f_k f_k^* \omega) = \int_{\mathcal{R}_{\mathbf{R}}} |f_k|^2 \omega > 0$$

the f_k's can be multiplied with positive real constants such that

$$\operatorname{Res}_{P_k} (f_k f_k^* \omega) = -i. \tag{20}$$

This is compatible with the normalization $hh^* = 1$; it suffices to multiply ω with a real multiplicative constant (which can still be done) such that

$$\mathrm{Res}_{P_n}(\omega) = -i.$$

Consider now φ such that $\varphi^* = \varphi$. The matrix of the operator $C[\varphi]$ associated to φ is defined by:

$$\varphi f_k = \sum_l a_{kl} f_l. \tag{21}$$

Now

$$\sum_{i=1}^{n} \mathrm{Res}_{P_i}(\varphi f_k f_j^* \omega) = \sum_l a_{kl} \sum_{i=1}^{n} \mathrm{Res}_{P_i}(f_l f_j^* \omega) = -i a_{kj}. \tag{22}$$

Also

$$\begin{aligned}\overline{\sum_{i=1}^{n} \mathrm{Res}_{P_i}(\varphi f_k f_j^* \omega)} &= \sum_{i=1}^{n} \mathrm{Res}_{Q_i}(\varphi f_k f_j^* \omega)^* \\ &= \sum_{i=1}^{n} \mathrm{Res}_{Q_i}(\varphi f_k^* f_j \omega) \\ &= -\sum \mathrm{Res}_{P_i}(\varphi f_j f_k^* \omega) \\ &= i a_{jk}.\end{aligned}$$

Therefore $\bar{a}_{kj} = a_{jk}$, i.e., the operator $C[\varphi]$ associated to φ is self-adjoint. In particular, $z^* = z$; therefore the operator C is self-adjoint. The rest of this chapter will be devoted to the application of these theorems to a few examples.

1. Consider a second order symmetric difference operator[1] of period N, i.e.,

$$(Cf)_k = b_{k-1} f_{k-1} + a_k f_k + b_k f_{k+1} \quad \text{with} \quad a_{k+N} = a_k,\ b_{k+N} = b_k \in \mathbf{C}.$$

Here $M = M' = 1$ and $n = 1$. It is regular as soon as $b_k \neq 0$ $(1 \leq k \leq N)$. Then

$$C_h - zI = \begin{bmatrix} a_1 - z & b_1 & 0 & & \cdots & & b_N h^{-1} \\ b_1 & a_2 - z & b_2 & & & & \vdots \\ 0 & b_2 & a_3 - z & & & & \\ \vdots & & & \ddots & & & \\ & & & & a_{N-2} - z & b_{N-2} & 0 \\ & & & & b_{N-2} & a_{N-1} - z & b_{N-1} \\ b_N h & \cdots & & & 0 & b_{N-1} & a_N - z \end{bmatrix}$$

(1) Observe that any second order difference operator can be symmetrized by conjugation with a diagonal matrix.

is a tridiagonal period matrix, with determinant

$$F(h, h^{-1}, z) = (-1)^{N+1}\left(\prod_1^N b_i(h+h^{-1}) - P(z)\right) = 0,$$

where $P(z)$ is a polynomial of degree N with leading coefficient $=1$. Setting $\prod_1^N b_i = A \neq 0$

$$h(z) = \frac{1}{2A}(P(z) \pm \sqrt{P(z)^2 - 4A^2}) = \frac{1}{\frac{1}{2A}(P(z) \mp \sqrt{P(z)^2 - 4A^2})}$$

Therefore the curve is hyperelliptic of genus $g = N-1$ with two points P and Q at infinity; besides $(h) = -NP + NQ$. Moreover, switching the sign of the radical in the formula above amounts to changing h into h^{-1}; therefore the involution τ coincides with the hyperelliptic involution. The fixed points for this involution are given by the 2N points where $h = \pm 1$, i.e., the branch points. Let α_i and β_i be homology cycles $(1 \leqslant i \leqslant g)$; then the fact that $(h) = -NP + NQ$ implies the existence of a closed loop $\sum n_i\alpha_i + m_i\beta_i$ such that

$$\sum_1^g n_i \int_{\alpha_i} \omega + \sum_1^g m_i \int_{\beta_i} \omega = N \int_P^Q \omega$$

for every holomorphic differential. This amounts to g relations between the branch points. So, any hyperelliptic curve coming from such a tridiagonal matrix satisfies these relations and vice-versa.

The inverse problem, as discussed in [12, 22], is an immediate consequence of Theorem 2. Let the spectrum of C_1 (i.e., C_h for $h=1$) and $\prod_1^N b_i$ be known. Let also the spectrum of the matrix C^0 be known; C^0 is formed from C_1 after removal of the last row and column. Then, the matrix C is completely known up to at most 2^{N-1} ambiguities.

Indeed, the knowledge of $A = \prod_1^N b_i$ and the spectrum of C_1 determines the equation

$$A(h + h^{-1}) - P(z) = 0$$

and therefore the hyperelliptic curve $\mathcal{R}$. The matrix C^0 has $N-1 = g$ spectral points in $\mathbf{C}$. They can be lifted up to $\mathcal{R}$ in 2^g different ways, if no one of them coincides with the branch points and if no two of them coincide. Each of these ways leads to a regular divisor $\mathcal{D}$ of order g such that

$$\mathcal{D} + \mathcal{D}^\tau - P - Q = \left(\frac{\Delta_{NN}(z)}{\sqrt{P(z)^2 - 4A^2}} dz\right).$$

Each one of the regular divisions $\mathcal{D}$ determines in a unique way a periodic tridiagonal matrix C modulo conjugation by diagonal matrices with entries ± 1, such that the spectrum of C^0 is the one given above. Observe also that Prym $(\mathcal{R})$ = Jac $(\mathcal{R})$.

2. Let C be a self-adjoint periodic difference operator [1]

$$(Cf)_k = \bar{b}_{k-1}f_{k-1} + a_k f_k + b_k f_{k+1} \quad \text{with} \quad a_{k+N} = a_k \in \mathbf{R},\ b_{k+N} = b_k \neq 0.$$

Then

$$F(h, h^{-1}, z) = (-1)^{N+1}(Ah + \bar{A}h^{-1} - P(z)) = 0$$

where $P(z) = z^N + \ldots$ is a real polynomial of degree N. This defines a hyperelliptic curve whose branch points are located at the values of z where $h(z) = \pm|A|/A$; for each of these values of h, the matrix $C_h - zI$ is self-adjoint. Therefore the branch points are real. The involution $\tilde{}$ transforms a point of $\mathcal{R}$ as follows: take the complex conjugate in $\mathbf{C}$ and flip sheets.

3. Next consider a symmetric fourth-order difference operator

$$(Cf)_k = c_{k-2}f_{k-2} + b_{k-1}f_{k-1} + a_k f_k + b_k f_{k+1} + c_k f_{k+2}$$

with $a_{k+N} = a_k$, $b_{k+N} = b_k$ and $c_{k+N} = c_k$. Here $M = M' = 2$, so that $(N, M) = n = 1$ or 2; so, a distinction must be made between N odd and N even. When N is odd, the regularity reduces to the condition that $c_k \neq 0$ $(1 \leqslant k \leqslant N)$ and when N is even, it reduces to $c_k \neq 0$ $(1 \leqslant k \leqslant N)$ and

$$c_1 c_3 c_5 \ldots c_{N-1} \neq c_2 c_4 c_6 \ldots c_N.$$

In either case

$$F(h, h^{-1}, z) = \det(C_h - zI) = A(h + h^{-1})^2 + P_1(z)(h + h^{-1}) + P_2(z)$$

with $P_2(z) = (-z)^N + \ldots$ and $\deg P_1(z) \leqslant [N/2]$ with equality if N is even. This implies that $\mathcal{R}$ is a double covering of the hyperelliptic curve $\mathcal{S}$, defined by

$$Ag^2 + P_1(z)g + P_2(z) = 0$$

ramified at the $2N$ points on $\mathcal{S}$ where $g = h + h^{-1} = \pm 2$, i.e., where $h = \pm 1$. When N is odd, $\mathcal{S}$ is ramified at infinity, so that $\mathcal{R}$ has two points P and Q covering $z = \infty$. If N is even, $\mathcal{S}$ is not ramified at infinity, so that $\mathcal{R}$ has 4 points P_1, P_2, Q_1 and Q_2 covering infinity. Then

$$\begin{aligned}(h) &= -NP + NQ & N \text{ odd}\\ &= -\frac{N}{2}(P_1 + P_2) + \frac{N}{2}(Q_1 + Q_2) & N \text{ even.}\end{aligned}$$

$\mathcal{R}$ has genus $g = 2N - 2$ or $2N - 3$ according to whether N is odd or even.

---

(1) Unlike in the symmetric case, a second order difference operator cannot necessarily be made self-adjoint by conjugation with a diagonal matrix.

The inverse problem can now be formulated as follows. From the knowledge of the spectrum, the antiperiodic spectrum of C and $A = \prod_1^N c_i$, you can reconstruct the curve $\mathcal{R}$. A generic set of g points in $\mathbf{C}$ leads to 4^g regular fourth order difference operators.

Indeed from A, from the periodic and antiperiodic spectrum of C you know the polynomials

$$4A + 2P_1(z) + P_2(z) = (-z)^N + \dots$$

and

$$4A - 2P_1(z) + P_2(z) = (-z)^N + \dots$$

Therefore P_1 and P_2 are known; this defines $\mathcal{R}$ completely. The generic set of g points in $\mathbf{C}$ can be lifted up to $\mathcal{R}$ in 4^g ways, defining 4^g different divisors $\mathcal{D}$. Each one of those leads to a fourth order periodic difference operator which is not necessarily symmetric. Only, when $\mathcal{D}$ is in Prym $(\mathcal{R})$, C can be made symmetric by conjugation with a diagonal operator. This imposes $g_0 = [(N-1)/2]$ conditions on the choice of the g points in $\mathbf{C}$, expressing the fact that they must be roots of the function $\Delta_{NN}(z)$; the latter is symmetric in h and h^{-1} if it is to come from a symmetric operator.

§ 4. Flows on the Jacobi variety and symplectic structures

As a result of § 2, the Jacobi variety (except for a lower dimensional manifold) can be parametrized by difference operators of a given order with the same h-spectrum. Therefore the linear flows on Jac $(\mathcal{R})$ (with regard to the group structure) can be regarded as isospectral flows on the space of difference operators. This section shows that these flows can be expressed in terms of Lax-type commutation relations. Let $\mathcal{A}$ denote the ring of meromorphic functions on $\mathcal{R}$, holomorphic on $\mathcal{R}_0$. Since z, h, h^{-1} are affine coordinates on $\mathcal{R}_0$, $\mathcal{A}$ is the ring of polynomials in z, h, h^{-1}. Let $\{\omega_k\}$ be a basis for the space of holomorphic differentials.

THEOREM 4. *Every linear flow on* Jac $(\mathcal{R})$

$$\sum_{i=1}^{g} \int_{\nu_i(0)}^{\nu_i(t)} \omega_k = a_k t, \quad 1 \leqslant k \leqslant g,$$

is associated with a function $u = P(z, h, h^{-1})$ *in* $\mathcal{A}$ *such that*

$$a_k = \sum_{i=1}^{n} \operatorname{Res}_{P_i} (\omega_k u).$$

This flow is equivalent to the system of differential equations, given by (1)

$$\dot{C} = [C[u]^+, C] \quad \text{or} \quad \dot{C} = [C[u]^{[+]}, C]$$

where $C[u] = P(C, S, S^{-1})$.

These two equations give flows in the space of periodic difference operators that differ by conjugation by a periodic diagonal $\Lambda(t)$.

This theorem is equally valid whether $\mathcal{R}$ is singular or not. It is important to realize that by considering integrals of the differentials $\eta \in \Omega$, we get for singular $\mathcal{R}$ also an Abel mapping from principal divisors mod linear equivalence to points of the generalized Jacobian variety. A good reference for this is Serre [28] or Rosenlicht [27].

In the proof of the theorem, to avoid questions of convergence, it is best to approximate the given $\mathcal{D}$ by divisors $\mathcal{D}(s)$ made up of smooth points of $\mathcal{R}$, and correspondingly approximate C by $C(s)$. If the flow through $C(s)$ is given by our formula, then by passing to the limit, so is the flow through C. We first prove 2 simple lemmas:

LEMMA 1. *Let* $\mathcal{D} = \sum \nu_i$ *be a regular point of* Jac ($\mathcal{R}$). *Let* $\mathcal{D}(t) = \sum \nu_i(t)$ *and* $\mathcal{D}'(t) = \sum \nu_i'(t)$ *be in a small enough neighborhood of* $\mathcal{D}$, *such that*

$$\sum_{i=1}^{g} \int_{\nu_i}^{\nu_i(t)} \omega_j = a_j t + O(t^2), \quad 1 \leqslant j \leqslant g$$

and

$$\sum_{i=1}^{g} \int_{\nu_i(t)}^{\nu_i'(t)} \omega_j = O(t^2), \quad 1 \leqslant j \leqslant g$$

then

$$f_k(t) - f_k = O(t)$$

and

$$f_k(t) - f_k'(t) = O(t^2)$$

uniformly over any open V *such that* $\bar{V} \subset \mathcal{R}_0 \backslash \mathcal{D}$, *where* f_k, $f_k(t)$ *and* $f_k'(t)$ *correspond to* $\mathcal{D}$, $\mathcal{D}(t)$ *and* $\mathcal{D}'(t)$ *respectively.*

The proof follows at once from the fact that the functions f_k depend analytically on their poles. This dependence will be made explicit in § 4, where f_k will be expressed as quotients of theta functions.

---

(1) For any difference operator C, define

$$C^+_{ij} = C_{ij} \ \text{if } i<j, \quad = 0 \ \text{if } i \geqslant j \qquad \text{and} \qquad (C^{[+]})_{ij} = C_{ij} \ \text{if } i<j, \quad = \tfrac{1}{2}C_{ii} \ \text{if } i=j, \quad = 0 \ \text{if } i>j.$$

Define C^- in the same way and $C^{[-]} = C - C^{[+]}$

9 – 792907 *Acta mathematica* 143. Imprimé le 28 Septembre 1979

Lemma 2. *Consider a point $P\in C$, a holomorphic differential $\omega=\varphi dz$ in the neighborhood V of P and an analytic function u in V with a pole of order n at P. Consider $t\in\mathbb{C}$ small enough, so that the n points $P_i(t)$, where $u(P_i(t))+t^{-1}=0$, belong to V. Then*

$$\lim_{t\to 0}\frac{d}{dt}\sum_{i=1}^{n}\int_P^{P_i(t)}\omega=-\mathrm{Res}_P\,\omega u.$$

Proof. Let $\omega=d\psi$ with $\psi(P)=0$. Then for any path π enclosing the zeros $P_i(t)$ of $u+t^{-1}$,

$$\begin{aligned}\frac{1}{t}\sum_{i=1}^{n}\int_P^{P_i(t)}\omega&=\frac{1}{t}\sum\psi(P_i(t))\\&=\frac{1}{t}\sum\mathrm{Res}_{P_i(t)}\frac{u'}{u+\dfrac{1}{t}}\psi\\&=\sum\mathrm{Res}_{P_i(t)}\frac{u'}{1+tu}\psi\\&=\frac{1}{2\pi i}\int_\pi\frac{u'}{1+tu}\psi\,dz.\end{aligned}$$

When t tends to zero, the right hand side tends to

$$\frac{1}{2\pi i}\int_\pi u'\psi\,dz=-\frac{1}{2\pi i}\int_\pi u\omega=-\mathrm{Res}_P\,(u\omega)$$

Proof of Theorem 4. Consider $u\in\mathcal{A}$. Assume $\mathcal{D}$ in $\mathcal{R}_0$; the proof extends easily to the case where $\mathcal{D}$ is not in $\mathcal{R}_0$. Then u splits into two functions g_+ and g_- according to

$$u=uf_0=\sum_{i=-K'}^{K}c_{N,i}f_i=\sum_{i=-K'}^{0}c_{N,i}f_i+\sum_{i=1}^{K}c_{N,i}f_i=g_-+g_+.$$

Then, since

$$(g_+)\geqslant-\sum_1^K P_i+Q_1-\mathcal{D}\quad\text{and}\quad(g_-)\geqslant-\sum_{-1}^{-K'}Q_i-\mathcal{D}\tag{25}$$

we have that

$$\left(g_++\frac{1}{t}\right)=-\sum_1^K P_i-\mathcal{D}+\sum_1^K P_i(t)+\mathcal{D}(t)\tag{26}$$

and

$$\left(g_--\frac{1}{t}\right)=-\sum_{-1}^{-K'}Q_i-\mathcal{D}+\sum_{-1}^{-K'}Q_i(t)+\mathcal{D}'(t)$$

where $\mathcal{D}(t)$ (resp. $\mathcal{D}'(t)$) is a divisor of order g, near $\mathcal{D}$ and each $P_i(t)$ (resp. $Q_i(t)$) is near P_i (resp. Q_i), as near as you wish by choosing t small enough. Let $\mathcal{D}=\sum_1^g\nu_i(0)=\sum_1^g\nu_i$,

$\mathcal{D}(t) = \sum_1^g \nu_i(t)$ and $\mathcal{D}'(t) = \sum_1^g \nu_i'(t)$. Then $\nu_i(t)$ as a function of t, is holomorphic in t, because near ν_i, the function g_+ behaves as

$$g_+ = \frac{b_{-1}}{s} + b_0 + b_1 s + \dots$$

in the local parameter s. Putting $g_+ + t^{-1} = 0$ leads to an expansion of s as a power series in t. Using Abel's theorem

$$\sum \int_{\nu_i}^{\nu_i(t)} \omega = -\sum_{i=1}^{K} \int_{P}^{P_i(t)} \omega \quad \text{and} \quad \sum \int_{\nu_i}^{\nu_i'(t)} \omega = -\sum_{i=1}^{K'} \int_{Q_i}^{Q_i(t)} \omega \tag{27}$$

for every holomorphic differential ω. From Lemma 4 it follows that

$$\lim_{t \downarrow 0} \frac{1}{t} \sum_{i=1}^{K} \int_{P_i}^{P_i(t)} \omega = -\sum_{i=1}^{n} \operatorname{Res}_{P_i} (\omega g^+). \tag{28}$$

The same argument applies to the function $g_- - t^{-1}$, yielding

$$\lim_{t \to 0} \frac{1}{t} \sum_{i=1}^{K'} \int_{Q_i}^{Q_i(t)} \omega = \sum_{i=1}^{n} \operatorname{Res}_{Q_i} (\omega g^-).$$

Observe the change in sign with regard to (28), as a result of considering the function $g_- - t^{-1}$. In view of (27), (28) and the fact that $\mathcal{D}$ belongs to $\mathcal{R}_0$, we conclude that

$$\sum_{i=1}^{g} \int_{\nu_i}^{\nu_i(t)} \omega = -t \sum_{i=1}^{n} \operatorname{Res}_{P_i} (\omega u) + O(t^2) \tag{29}$$

and

$$\sum_{i=1}^{g} \int_{\nu_i}^{\nu_i'(t)} \omega = t \sum_{i=1}^{n} \operatorname{Res}_{Q_i} (\omega u) + O(t^2).$$

Therefore also

$$\sum \int_{\nu_i(t)}^{\nu_i'(t)} \omega = O(t^2).$$

Since the regular divisors form an open subset in Jac $(\mathcal{R})$, take t small enough, such that $\mathcal{D}(t)$ and $\mathcal{D}'(t)$ are still regular. Denote by $f_k(t)$ (resp. $f_k'(t)$) the unique meromorphic function corresponding to $\mathcal{D}(t)$ (resp. $\mathcal{D}'(t)$). Then

$$((1 + tg_+) f_k(t)) \geqslant -\sum_1^k P_i - \sum_1^K P_i - \mathcal{D} + \sum_0^{k-1} Q_i + \sum_1^K P_i(t)$$

and

$$((1-tg_-)f'_k(t)) \geqslant \sum_0^{k-1} Q_i - \sum_{-1}^{-K'} Q_i - \mathcal{D} - \sum_1^k P_i + \sum_{-1}^{-K'} Q_i(t).$$

From Lemma 5 (§ 1), it follows that $(1+tg_+)f_k(t)$ and $(1-tg_-)f_k(t)$ have an expansion in terms of the $f_i(0)$'s

$$(1+tg_+)f_k(t) = \sum_{i \geqslant k} a^+_{ki}(t) f_i(0)$$

and

$$(1-tg_-)f'_k(t) = \sum_{i \leqslant k} a^-_{ki}(t) f_i(0).$$

The difference of these two equations reads

$$\begin{aligned}\sum (a^+_{ki}(t) - a^-_{ki}(t)) f_i(0) &= f_k(t) - f'_k(t) + t(g_+ f_k(t) + g_- f'_k(t)) \\ &= tuf_k(0) + tu(f_k(t) - f_k(0)) + (f_k(t) - f'_k(t))(1-tg_-) \\ &= t \sum C[u]_{ki} f_i(0) + O(t^2)\end{aligned}$$

uniformly over any open set such that $\bar{V} \subset \mathcal{R}_0 \backslash \mathcal{D}$. Since the functions $f_i(0)$ are independent, we conclude that

$$a^+_{ki}(t) - a^-_{ki}(t) = tC[u]_{ki} + O(t^2).$$

Since $f_k(t)$ is defined up to multiplication with some function of t, it can be determined such that $a^+_{kk}(t) = 1 + O(t^2)$ or $a^+_{kk}(t) = 1 + \frac{1}{2}tC[u]_{kk} + O(t^2)$. Therefore $a^+_{ki}(t) = \delta_{ki} + tC[u]_{ki} + O(t^2)$ or $a^+_{ki}(t) = \delta_{ki} + tC[u]^+_{ki} + O(t^2)$.

Finally write the column vector $(1+tg_+)zf(t)$ in two different ways, using the results above. On the one hand

$$\begin{aligned}(1+tg_+)zf(t) = (1+tg_+)C(t)f(t) &= C(t)(1+tg_+)f(t) \\ &= C(t)(I + tC[u]^+ + O(t^2))f(0)\end{aligned}$$

and on the other hand

$$\begin{aligned}z(1+tg_+)f(t) &= z(I + tC[u]^+ + O(t^2))f(0) \\ &= (I + tC[u]^+ + O(t^2))C(0)f(0).\end{aligned}$$

Both relations are valid for all $(z, h) \in V$. In each of the cases $C[u]^+$ may be replaced by $C[u]^{[+]}$. Then also

$$(I + tC[u]^+ + O(t^2))^{-1} C(t)(I + tC[u]^+ + O(t^2))f(0) = C(0)f(0). \tag{30}$$

So, the dependence of $\mathcal{D}(t)$ on t given by (26) (at least for small enough t) can be expressed equivalently as (29) or (30). Differentiating both (25) and (30)with regard to t and letting $t\downarrow 0$, we conclude that the flow

$$\dot{C}=[C[u]^{+},C] \quad \text{or} \quad \dot{C}=[C[u]^{[+]},C]$$

is equivalent to the motion

$$\sum_i \dot{\nu}_i(0)\,\omega(\nu_i(0)) = -\sum_j \operatorname{Res}_{P_j}(\omega u) \quad \text{for every holomorphic } \omega.$$

The proof of Theorem 4 is finished, if every flow is shown to occur in this fashion. It suffices to show that the mapping

$$u \rightarrow \sum_{i=1}^{n} \operatorname{Res}_{P_i}(\omega_k u)$$

maps $\mathcal{A}$ *onto* $\mathbf{C}^g$. Observe that it is possible to find a function in $\mathcal{A}$ with arbitrarily prescribed polar parts at all of the points P_i, provided arbitrarily large poles are allowed at the points Q_i. But the power series expansions of $\omega_1, \ldots, \omega_g$ at P_1, say, are linearly independent, so their Nth order truncations are independent for $N \gg 0$. Thus a suitable u with Nth order poles at P_1 and regular at the other P_i will give any sequence of g constants

$$\sum_{i=1}^{n} \operatorname{Res}_{P_i}(\omega_k u).$$

It is useful to have an explicit basis for the space of holomorphic differential forms on $\mathcal{R}$. The result is this: the forms

$$\begin{gathered} h^k z^i \eta, \quad i \geqslant 0 \\ kN_1 + iM_1 \leqslant \alpha \equiv NM_1 - M_1 - 1 \\ -kN_1' + iM_1' \leqslant \alpha' \equiv NM_1' - M_1' - 1 \end{gathered}$$

are such a basis. One way to check this is to note that these forms are linearly independent and to prove there are g of them, using a counting argument. Another method comes from the toroidal embedding. We explain both methods. The first one is based on a combinatorial lemma:

LEMMA 3.

$$\#\{i,k \geqslant 1 \ \text{ such that } \ N_1 k + M_1 i \leqslant N_1 M_1 n - 1\} = \frac{n}{2}(nN_1M_1 - N_1 - M_1 - 1) + 1$$

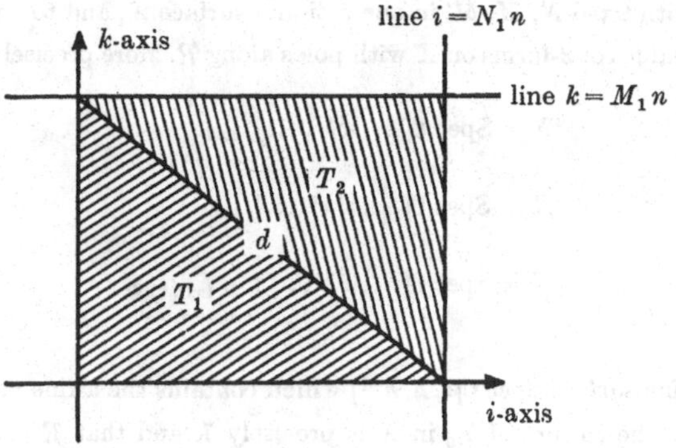

Figure 2

Proof. In Figure 2, we observe that

$\#$ lattice points on $T_1 \cup T_2 \cup d = (N_1 n - 1)(M_1 n - 1)$

$\#$ lattice points on $T_1 = \#$ lattice points on T_2, because $(i, k) \to (N_1 n - i, M_1 n - k)$ interchanges them.

$\#$ lattice points on $d = n - 1$.

The result follows at once from this count.

The rest of the argument goes as follows: the set of (i, k) such that

$$i \geqslant 0, \quad kN_1 + iM_1 \leqslant nN_1 M_1 - M_1 - 1$$
$$-kN_1' + iM_1' \leqslant n'N_1' M_1' - M_1 - 1$$

can be decomposed into

$$\{i \geqslant 0,\ k \geqslant 1 \text{ such that } kN_1 + (i+1)M_1 \leqslant nN_1 M_1 - 1\},$$
$$\{i \geqslant 0,\ k \leqslant -1 \text{ such that } -kN_1' + (i+1)M_1' \leqslant n'N_1' M_1' - 1\}$$

and

$$\{i \geqslant 0, \text{ such that } (i+1)M_1 \leqslant nN_1 M_1 - 1\}$$

whose total cardinal—according to Lemma 3—equals

$$\frac{n}{2}(nN_1 M_1 - N_1 - M_1 - 1) + 1 + \frac{n'}{2}(n'N_1' M_1' - N_1' - M_1' - 1) + 1 + N - 1 = g$$

As announced, the second method comes from the theory of toroidal embeddings (Kempf et al. [13]). We merely sketch this. The idea is to embed all the Riemann Surfaces

$\mathcal{R}$ asociated to C with fixed N, M, M' in one rational surface X, and to write holomorphic 1-forms on $\mathcal{R}$ as residues of 2-forms on X with poles along $\mathcal{R}$. More precisely, X is the union of 3-affine pieces:

$$X_1 = \operatorname{Spec} \mathbf{C}[\ldots, z^i h^k, \ldots]_{\substack{i \geqslant 0 \\ -M_1' i + N_1' k \geqslant 0}}$$

$$X_2 = \operatorname{Spec} \mathbf{C}[\ldots, z^i h^k, \ldots]_{\substack{i \geqslant 0 \\ -M_1 i - N_1 k \geqslant 0}}$$

$$X_3 = \operatorname{Spec} \mathbf{C}[\ldots, z^i h^k, \ldots]_{\substack{-M_1' i + N_1' k \geqslant 0 \\ -M_1 i - N_1 k \geqslant 0}}.$$

X contains the affine surface Spec $\mathbf{C}[z, h, h^{-1}]$ which contains the affine curve $\mathcal{R}_0$. It is not hard to check that the closure of $\mathcal{R}_0$ in X is precisely $\mathcal{R}$ and that $\mathcal{R}$ misses the singular points of X. The maps $z \mapsto \lambda z$, $h \mapsto \mu h$ extend to automorphisms of X, so X is a "torus embedding" in the sense of [13]; in fact in the notation of the book, it is the one associated to the simplicial subdivision of the plane into the 3 sectors:

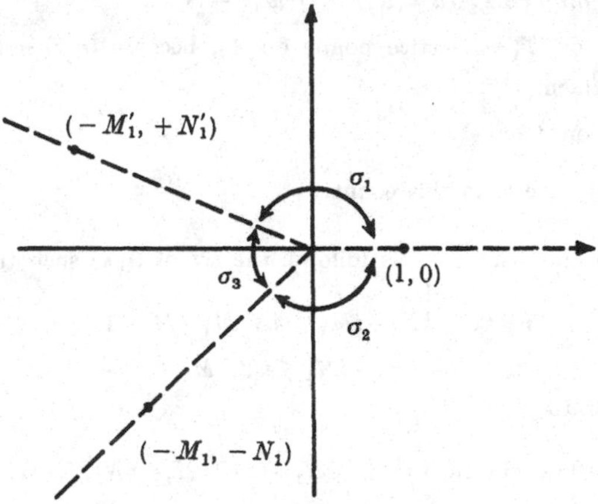

Figure 3

If $\omega(\mathcal{R})$ is the sheaf of meromorphic 2-forms on X, holomorphic outside the singular points of X except for simple poles at $\mathcal{R}$, then residue sets up an exact sequence:

$$H^0(\omega) \to \Gamma(\omega(\mathcal{R})) \to \Gamma(\Omega^1_{\mathcal{R}}) \to H^1(\omega)$$

and as $H^i(\omega)$, $H^{2-i}(O_x)$ are dual and $H^1(O_x) = H^2(O_x) = 0$ (cf. [13], p. 44), it follows that

$$\text{res:} \quad \Gamma(\omega(\mathcal{R})) \xrightarrow{\approx} \Gamma(\Omega_{\mathcal{R}}).$$

Explicitly this means that every holomorphic 1-form on $\bar{R}$ is uniquely expressible as

$$\mathrm{Res}_R\left(\frac{g\cdot dz\wedge dh}{F(h,h^{-1},z)\cdot z\cdot h}\right)$$

where $g=g(z,h,h^{-1})$ is chosen so that the 2-form in parenthesis has no poles other than $F=0$. There are 3 possible curves: (1) $h\in\mathbf{C}^*$, $z=0$, (2) $h^{M_1'}\cdot z^{N_1'}\in\mathbf{C}^*$, $h=0$, (3) $h^{M_1}z^{-N_1}\in\mathbf{C}^*$, $h^{-1}=0$. Checking the order of pole at each of these, we find that if

$$g=\sum a_{ik}h^k z^i$$

then $a_{ik}\neq 0$ only if

$$i\geqslant 1$$
$$-iM_1'+kN_1'\geqslant 1-NM_1'$$
$$-iM_1-kN_1\geqslant 1-NM_1.$$

Theorem 4 shows that the isospectral flows for difference operators (written in the Lax form) can actually be linearized: they are linear flows on the Jacobi variety of the corresponding curve. In the second part of this section, we show that these flows derive from Hamiltonians according to a co-symplectic structure on the space of all periodic difference operators suggested by group-theoretical considerations. Consider the group G of lower triangular invertible matrices (including the diagonal) of order N. Let g be the Lie algebra of lower triangular matrices and g' be its dual, namely the space of upper triangular matrices. g and g' are paired by the trace of the product. Let $\sigma\in G$, $X\in g$ and $f\in g'$; the adjoint action amounts to conjugation $\sigma\cdot X=\sigma^{-1}X\sigma$ and its coadjoint action amounts to conjugation and projection on g', i.e. $\sigma\cdot f=(\sigma^{-1}f\sigma)-(\sigma^{-1}f\sigma)^{-}$. Fix an element $f\in g'$ and consider the orbit $G\cdot f\subset g'$ of f under the action of the group G. According to a theorem by Kirillov and Kostant [14] the orbit $G\cdot f$ is endowed with a natural symplectic form, i.e., an alternating two-form in the tangent plane T_h to the orbit $G\cdot f$ at a given point h. Since G acts on g' by conjugation,

$$T_h=\{\text{locus of points } \xi_h A,\ A\in g, \text{ where } \xi_h A=[h,A]-[h,A]^{-}\};$$

let $\xi_h A$ and $\xi_h B\in T_h$; then

$$\omega(\xi_h A,\xi_h B)\equiv \mathrm{Tr}\,(h[A,B])$$

is a non-singular alternating 2-form, so that the orbit $G\cdot f$ is even dimensional. Instead of introducing a skew-symmetric form on the tangent planes to each orbit $G\cdot f$, it is completely equivalent to introduce a skew-symmetric form on the co-tangent planes. Such a form is just a skew-symmetric Poisson-bracket on pairs of functions on $G\cdot f$, whose value at each point is bilinear in the differentials of these functions. This is called a co-symplectic structure, and this structure fits together into one co-symplectic structure on g', inducing all the separate ones on the orbits. It is given by

$$\{f, g\}(x) = \mathrm{Tr}\,(x\cdot[df(x), dg(x)])$$

(where $df(x)$ and $dg(x)$, being linear functions on g', can be identified with elements of g, hence can be bracketed).

The definition of this co-symplectic form can be adapted after some changes to the case of periodic symmetric difference operators and to the case of periodic non-symmetric difference operators. This leads to the Poisson bracket defined below. Before proceeding we need the following definitions.

Let $\mathfrak{M}$ be the vector space of N-periodic infinite matrices C such that

$$C_{ij} = 0 \quad \text{if } |i-j| > K \text{ for some } K.$$

Define Tr (C), for $C \in \mathfrak{M}$ to be $\sum_{i=1}^{N} C_{ii}$.

Put on $\mathfrak{M}$ an inner product

$$\langle C, D\rangle = \mathrm{Tr}\,(CD^{\mathrm{T}}) = \sum_{\substack{(i,j)\in \mathbf{Z}^2 \\ \text{in cosets of} \\ (N,N)\mathbf{Z}}} C_{ij} D_{ij}.$$

A functional F on $\mathfrak{M}$ is called differentiable, if there is a matrix $\partial F/\partial C$ in $\mathfrak{M}$ such that for all D

$$\lim_{\varepsilon\downarrow 0} \frac{F(C+\varepsilon D) - F(C)}{\varepsilon} = \left\langle \frac{\partial F}{\partial C}, D\right\rangle.$$

Taking D given by

$$D_{ij} = \begin{cases} 1 & \text{if} \quad (i,j) = (i_0 + kN, j_0 + kN), \quad \text{some } k \\ 0 & \text{otherwise} \end{cases}$$

it follows that

$$\left(\frac{\partial F}{\partial C}\right)_{i_0, j_0} = \frac{\partial F}{\partial C_{i_0 j_0}}.$$

A simple identity which will be useful is: $\langle [A, B], C\rangle = \langle [A^{\mathrm{T}}, C], B\rangle$. Define the following bracket between two differentiable functionals F and G on $\mathfrak{M}$

$$\{F, G\} = \left\langle \left[\left(\frac{\partial F}{\partial C}\right)^{[+]}, \left(\frac{\partial G}{\partial C}\right)^{[+]}\right] - \left[\left(\frac{\partial F}{\partial C}\right)^{[-]}, \left(\frac{\partial G}{\partial C}\right)^{[-]}\right], C\right\rangle.$$

9† – 792907 *Acta mathematica* 143. Imprimé le 28 Septembre 1979

Lemma 5. $\{\ ,\ \}$ satisfies the Jacobi identity.

Proof. In general, when we are dealing with a Poisson bracket structure on a vector space, we can make a preliminary reduction in the proof of Jacobi's identity as follows: write

$$\{f,g\}(x)=\sum A_{ij}(x)\cdot\frac{\partial f}{\partial x_i}\cdot\frac{\partial g}{\partial x_j}.$$

Then

$$\frac{\partial}{\partial x_k}\{f,g\}=\sum\frac{\partial A_{ij}}{\partial x_k}\cdot\frac{\partial f}{\partial x_i}\cdot\frac{\partial g}{\partial x_j}+A_{ij}\cdot\frac{\partial^2 f}{\partial x_i\partial x_k}\cdot\frac{\partial g}{\partial x_j}+A_{ij}\cdot\frac{\partial f}{\partial x_i}\cdot\frac{\partial^2 g}{\partial x_j\partial x_k}.$$

We claim that when you evaluate $\{f,\{g,h\}\}+\{g,\{h,f\}\}+\{h,\{f,g\}\}$ the terms involving the 2nd derivatives of f, g and h *always* cancel out. This is easy to check directly, and it also follows because Jacobi's identity is equivalent to $d\omega=0$, ω the dual 2-form; and $d\omega=0$ is automatic when the coefficients A_{ij} are constant.

In our case

$$\frac{\partial}{\partial C}\{F,G\}=(\text{2nd derivative terms})+\left[\left(\frac{\partial F}{\partial C}\right)^{[+]},\left(\frac{\partial G}{\partial C}\right)^{[+]}\right]-\left[\left(\frac{\partial F}{\partial C}\right)^{[-]},\left(\frac{\partial G}{\partial C}\right)^{[-]}\right].$$

and

$$\left(\frac{\partial}{\partial C}\{F,G\}\right)^{[+]}=(\text{2nd derivative terms})+\left[\left(\frac{\partial F}{\partial C}\right)^{[+]},\left(\frac{\partial G}{\partial C}\right)^{[+]}\right]$$

(since neither of the Lie bracket terms have any diagonal entries). Thus

$$\begin{aligned}\{H,\{F,G\}\}&=(\text{2nd derivative terms})\\&\quad+\left\langle\left[\left(\frac{\partial H}{\partial C}\right)^{[+]},\left[\left(\frac{\partial F}{\partial C}\right)^{[+]},\left(\frac{\partial G}{\partial C}\right)^{[+]}\right]\right]+\left[\left(\frac{\partial H}{\partial C}\right)^{[-]},\left[\left(\frac{\partial F}{\partial C}\right)^{[-]},\left(\frac{\partial G}{\partial C}\right)^{[-]}\right]\right],C\right\rangle.\end{aligned}$$

Writing this out for $\{H,\{F,G\}\}$, $\{F,\{G,H\}\}$, $\{G,\{H,F\}\}$ and summing, the right hand side is zero by the usual Jacobi identity for Lie brackets.

Theorem 5. *The linear flows on* Jac $(\mathcal{R})$ *are Hamiltonian flows with regard to the Poisson bracket* $\{\ ,\ \}$. *In particular, in Poisson bracket notation, a typical flow*

$$\dot{C}=[C,(S^{-k}C^l)^{[+]}]$$

can be written as

$$\dot{c}_{ij}=\{F,c_{ij}\}$$

where

$$F(C)=\frac{1}{l+1}\operatorname{Tr}(S^{-k}C^{l+1}).$$

Proof. To begin with,

$$\frac{\partial}{\partial C}\operatorname{Tr}(S^{-k}C^{l+1}) = (l+1)(S^{-k}C^{l})^{\mathrm{T}}.$$

Indeed, by direct calculation,

$$\frac{\partial}{\partial c_{\alpha\beta}}\operatorname{Tr}(S^{-k}C^{l+1}) = \frac{\partial}{\partial c_{\alpha\beta}}\sum c_{i,i_1}c_{i_1,i_2}\dots c_{i_{l-1},i_l}c_{i_l,i+kN}$$

where the sum extends over $1\leqslant i\leqslant N$, $|i-i_1|\leqslant K$, $|i_1-i_2|\leqslant K$, ..., and $|i_l-(i+kN)|\leqslant K$,

$$= \sum_{m=0}^{l}\sum c_{i,i_1}c_{i_1,i_2}\dots c_{i_{m-1},\alpha}c_{\beta,i_{m+2}}\dots c_{i,i+kN}$$

the latter sum extending over $1\leqslant i\leqslant N$, $|i-i_1|\leqslant K$, $|i_1-i_2|\leqslant K$, ..., $|i_{m-1}-\alpha|\leqslant K$, $|\beta-i_{m+2}|\leqslant K$, ..., $|i_l-(i+kN)|\leqslant K$

$$= \sum_{m=0}^{l}\sum c_{\beta,i_{m+2}}\dots c_{i_l,i+kN}c_{i+kN,i_1+kN}\dots c_{i_{m-1}+kN,\alpha+kN}$$

$$= (l+1)(S^{-k}\cdot C^{l})_{\beta,\alpha}.$$

Let E_{ij} be the "elementary" matrix

$$(E_{ij})_{k,l} = \begin{cases}1 & \text{if } (k,l) = (i+lN,\, j+lN), \quad \text{some } l\in\mathbf{Z}\\ 0 & \text{otherwise.}\end{cases}$$

We check that the derivative of the functional $C\mapsto c_{ij}$ is given by:

$$\frac{\partial}{\partial C}(c_{ij}) = E_{ij}.$$

Thus we can calculate $\{F, c_{ij}\}$:

$$\{F, c_{ij}\} = \langle[((S^{-k}\cdot C^{l})^{\mathrm{T}})^{[+]}, E_{ij}^{[+]}] - [((S^{-k}\cdot C^{l})^{\mathrm{T}})^{[-]}, E_{ij}^{[-]}], C\rangle.$$

If $i<j$, $E_{ij}^{[-]}=0$, $E_{ij}^{[+]}=E_{ij}$, so

$$\begin{aligned}\{F, c_{ij}\} &= \langle[((S^{-k}\cdot C^{l})^{[-]})^{\mathrm{T}}, E_{ij}], C\rangle\\ &= \langle[(S^{-k}\cdot C^{l})^{[-]}, C], E_{ij}\rangle\\ &= [(S^{-k}\cdot C^{l})^{[-]}, C]_{ij}\end{aligned}$$

If $i>j$, $E_{ij}^{[+]}=0$, $E_{ij}^{[-]}=E_{ij}$, so

$$\begin{aligned}\{F, c_{ij}\} &= -\langle[((S^{-k}\cdot C^{l})^{[+]})^{\mathrm{T}}, E_{ij}], C\rangle\\ &= -\langle[(S^{-k}\cdot C^{l})^{[+]}, C], E_{ij}\rangle\\ &= -[(S^{-k}\cdot C^{l})^{[+]}, C]_{ij}.\end{aligned}$$

If $i=j$, we get the sum of half of each. Using the fact that C commutes with $S^{-k}C^l$, we find

$$[(S^{-k}\cdot C^l)^{[-]}, C] = -[(S^{-k}\cdot C^l)^{[+]}, C],$$

so in all cases, we get

$$\{F, c_{ij}\} = -[(S^k\cdot C^l)^{[+]}, C]_{ij} = [C, (S^k\cdot C^l)^{[+]}]_{ij}$$

as asserted.

THEOREM 6. *Any two functionals* $\mathrm{Tr}\,(S^{-k}C^{l+1})$ *have Poisson bracket zero: i.e., we have a set of Hamiltonians in involution. If to every* $C\in\mathfrak{M}$ *we associate the coefficients of* h^kz^l *in* $\det(C_h-zI)$, *these functionals are also in involution and through each* C, *generate the same set of flows.*

Proof. The first step is to show that any two expressions of the form

$$H_i = \mathrm{Tr}\,(S^{-k_i}C^{l_i+1}) \tag{32}$$

commute with one another for the symplectic structure. Consider the Hamiltonian vector field $\mathbf{X}_i$ derived from H_i, acting on differentiable functionals:

$$\mathbf{X}_i(F) = \{H_i, F\}.$$

Theorem 5 tells you that this vector field acts on C as follows.

$$\dot{C} = [C, (S^{-k}C^l)^{[+]}]. \tag{33}$$

This flow preserves the h-periodic spectrum of C (i.e., the spectrum of C_h) for every $h\in\mathbf{C}$. Therefore it preserves $\mathrm{Tr}\,(C_h)^\alpha = \mathrm{Tr}\,(C^\alpha)_h$ for every non-negative integer α, and in particular the coefficient of h^β in $\mathrm{Tr}\,(C_h)^\alpha$, namely $\mathrm{Tr}\,(S^{-\beta}C^\alpha)$. Therefore this flow leaves invariant every H_i, i.e., $\{H_i, H_j\}=0$; also, the Lie bracket $\{\mathbf{X}_i, \mathbf{X}_j\}$ vanishes, because

$$\begin{aligned}\{\mathbf{X}_i, \mathbf{X}_j\}F &= (\mathbf{X}_i\mathbf{X}_j - \mathbf{X}_j\mathbf{X}_i)F\\ &= \{H_i\{H_j, F\}\} - \{H_j, \{H_i, F\}\}\\ &= \{\{H_i, H_j\}, F\}\\ &= 0\end{aligned}$$

using Jacobi's identity. Finally the coefficient of z^{N-l} in $\det(C_h-zI)$ is a polynomial in $\mathrm{Tr}\,(C_h)^i$ for $1\leqslant i\leqslant l$ and $\mathrm{Tr}\,(C_h)^l$ is a polynomial in the coefficients of z^{N-i}, $1\leqslant i\leqslant l$. Therefore the coefficient of h^kz^l in $\det(C_h-zI)$ will be a polynomial in the coefficients of h^k in $\mathrm{Tr}\,C_h^i$ for $1\leqslant i\leqslant l$; i.e. in the quantities $\mathrm{Tr}\,S^{-k}C^i$. This proves the second assertion of the theorem. Observe that since C_h is $N \times N$, the expressions $\operatorname{Tr}(C_h)^i$ for $i > N$ are linear combinations of $\operatorname{Tr}(C_h)^i$ with $i \leq N$. So, no new functionals arise by considering $\operatorname{Tr} S^{-k}C^i$ for $i > N$.

Remark 1. Some of the coefficients appearing in $\det(C_h - zI) = 0$ lead to identically zero vector fields. For instance the coefficients of z^{N-1} equal $\operatorname{Tr} C$; but since $\partial(\operatorname{Tr} C)/\partial C = I$, its Hamiltonian vector field vanishes.

Remark 2. Consider the special case of symmetric difference operators. Let $\mathcal{A}_0 \subset \mathcal{A}$ be the subring of functions u such that $u^\tau = u$. The functions of $\mathcal{A}_0$ lead to linear flows in Prym $(\mathcal{R})$, because for $1 \leq k \leq g_0$ (for which $\omega_k^\tau = \omega_k$)

$$\sum_{i=1}^{n} \operatorname{Res}_{P_i}(\omega_k u) = \sum_{i=1}^{n} \operatorname{Res}_{Q_i}(\omega_k^\tau u^\tau) = \sum_{i=1}^{n} \operatorname{Res}_{Q_i}(\omega_k u)$$

and, moreover

$$\sum_{i=1}^{n} \operatorname{Res}_{P_i}(\omega_k u) + \sum_{i=1}^{n} \operatorname{Res}_{Q_i}(\omega_k u) = 0.$$

Therefore

$$\sum_{i=1}^{g} \int_{O_i}^{v_i} \omega_k = 0 \quad \text{for} \quad 1 \leq k \leq g_0.$$

Since $\mathcal{A}$ is the polynomial ring in h, h^{-1} and z, the ring $\mathcal{A}_0$ is the polynomial ring in $h + h^{-1}$ and z. Therefore all the flows in Prym $(\mathcal{R})$ translate into flows of the type

$$\begin{aligned}\dot{C} &= [C, ((S^k + S^{-k})C^i)^+] \\ &= \tfrac{1}{2}[C, ((S^k + S^{-k})C^i)^+ - ((S^k + S^{-k})C^i)^-].\end{aligned}$$

Notice that if C is symmetric, $(S^k + S^{-k})C^i$ is also symmetric and the flow above is generated by an antisymmetric operator, which indeed preserves the symmetry of C.

Examples.

1. Let C be an infinite (generic) tridiagonal matrix of period N. Let $\mathcal{R}$ be the hyperelliptic curve associated with it; let P and Q be the two points at infinity. Moreover, with the notation used in Example 1 (§ 2), a basis of holomorphic differentials is given by

$$\omega_k = \frac{z^{k-1}\,dz}{\sqrt{R(z)}}, \quad \text{where } R(z) = P(z)^2 - 4A^2.$$

Moreover, since the order of zero of ω_k at P or Q equals $g - k$,

$$\begin{aligned} a_k = \operatorname{Res}_P(\omega_k z^j) &= 0 \quad \text{for } k < g - j + 1 \quad 1 \leq j \leq g \\ &\neq 0 \quad \text{for } k = g - j + 1 \quad 1 \leq j \leq g. \end{aligned}$$

Therefore a complete set of flows is given by the functions $z, z^2, \ldots, z^g$, so that the most general isospectral flow for C (i.e., leaving the spectrum of C and A unchanged) is given by a polynomial $P(z)$ of degree at most g:

$$\dot{C} = \tfrac{1}{2}[C, P(C)^{+} - P(C)^{-}].$$

The Poisson Bracket between two functionals F and G has the simplified form

$$\{F, G\} = \left\langle \begin{pmatrix} \dfrac{\partial F}{\partial a} \\ \dfrac{\partial F}{\partial b} \end{pmatrix}^{\mathrm{T}}, J \begin{pmatrix} \dfrac{\partial G}{\partial a} \\ \dfrac{\partial G}{\partial b} \end{pmatrix} \right\rangle$$

where $\partial F/\partial a$ and $\partial F/\partial b$ are the column vectors whose elements are given by $\partial F/\partial a_i$ and $\partial F/\partial b_i$ respectively and J is defined as the $2n \times 2n$ antisymmetric matrix

$$\begin{pmatrix} 0 & B \\ -B^{\mathrm{T}} & 0 \end{pmatrix}$$

where

$$B = 2 \begin{pmatrix} b_1 & 0 & 0 & \cdots & -b_N \\ -b_1 & b_2 & 0 & & \vdots \\ 0 & -b_2 & b_3 & & \vdots \\ \vdots & & & & \vdots \\ 0 & & \cdots & -b_{N-1} & b_N \end{pmatrix}$$

The symplectic structure is given by

$$\omega = \sum_{i=2}^{N} da_i \wedge \sum_{j \leqslant i \leqslant N} \frac{db_i}{b_i}.$$

The g independent quantities in involution, leading to g independent flows, are given by $\beta_2, \ldots, \beta_N$, where

$$\det(C_h - zI)|_{h=i} = A_1(z) = (-1)^N z^N + \sum_{i=1}^{N} \beta_i z^{N-i}.$$

An equivalent set is given by $N-1$ points chosen from the spectrum of C_1 or C_{-1} (i.e., $N-1$ branch points of the hyperelliptic curve) or, alternatively, by the quantities

$$\operatorname{Tr} C^k, \quad 2 \leqslant k \leqslant N.$$

2. Consider the symmetric fourth order difference operator

$$(Cf)_k = c_{k-2}f_{k-2} + b_{k-1}f_{k-1} + a_kf_k + b_kf_{k+1} + c_kf_{k+2}$$

and assume N odd. Then $\mathcal{R}$ is a double covering of the hyperelliptic curve

$$F(g, z) \equiv Ag^2 + P_1(z)g + P_2(z) = 0,$$

where $\deg P_1 \leqslant (N-1)/2$ (generically, $=$) and $\deg P_2 = N$. The differentials on $\mathcal{R}_0$ or the symmetric (sheet invariant) differentials on $\mathcal{R}$ are given by linear combinations of

$$\omega_k = z^{k-1}(F'_g)^{-1}dz, \quad 1 \leqslant k \leqslant \frac{N-1}{2} = g_0;$$

this basis can be completed with the antisymmetric (for the involution τ) differentials

$$\omega = \frac{g^i}{h-h^{-1}} z^{k-1}(F'_g)^{-1}dz \begin{cases} i=1, & 1 \leqslant k \leqslant \dfrac{N-1}{2} \quad \text{or} \\ i=0, & 1 \leqslant k \leqslant N-1. \end{cases}$$

Since the dimension of the Prym variety equals $\frac{3}{2}(N-1)$, one expects to find $\frac{3}{2}(N-1)$ functions in $\mathcal{A}_0$ leading to independent flows in Prym ($\mathcal{R}$), namely

$$z, z^2, \ldots, z^{N-1}$$

and

$$(h+h^{-1})z^{(N+1)/2}, \ldots, (h+h^{-1})z^{N-1}.$$

The second sequence starts with the power $(N+1)/2$ in z, because this is the smallest possible power for which $h^{-1}z^{(N+1)/2}$ has an upper triangular part. This set can be completed to a set of flows spanning the whole of Jac ($\mathcal{R}$), by adding $(N-1)/2$ independent flows transversal to Prym ($\mathcal{R}$); they are generated by the functions

$$h^{-1}z^{(N+1)/2}, \ldots, h^{-1}z^{N-1}.$$

Finally, a set of integrals in involution spanning out all the linear flows in Jac ($\mathcal{R}$) is given by the coefficients of h^kz^i in the algebraic expression $F(h, h^{-1}, z) = 0$: the $(N-1)/2$ coefficients of $P_1(z)$ (except for the highest order coefficient, which leads to a zero vector field) counted twice (once as coefficient of hz^k and once of $h^{-1}z^k$) and the $N-1$ coefficients of $P_2(z)$ (except for the coefficient in z^N and z^{N-1}, which again lead to zero vector fields).

3. A symmetric sixth order difference operator leads to a double covering of the curve

$$F(g,z)=Ag^3+P_1(z)g^2+P_2(z)g+P_3(z)=0.$$

Assume that N is not a multiple of 3. Then all the flows in Prym ($\mathcal{R}$) are given by linear combinations of the functions

$$z, z^2, \ldots, z^{N-1}, \quad (h+h^{-1})z^{i_0}, \ldots, (h+h^{-1})z^{N-1},$$

where i_0 is the smallest integer $>N/3$, and

$$(h+h^{-1})^2z^{i_1}, \ldots, (h+h^{-1})^2z^{N-1},$$

where i_1 is the smallest integer $>2N/3$, and the flows in Jac ($\mathcal{R}$) by these and the transversal flows to Prym ($\mathcal{R}$) generated by

$$h^{-1}z^{i_0}, \ldots, h^{-1}z^{N-1}$$

and

$$h^{-2}z^{i_1}, \ldots, h^{-2}z^{N-1}.$$

§ 5. Theta functions and difference operators

Certain theta-identities allow us to provide explicit formulas for the operator C in terms of the curve $\mathcal{R}$. They are very similar to Cor. 2.19, p. 33 in Fay [11]. To fix notations, we assume a basis $\{\omega_i\}$ of holomorphic 1-forms chosen; we write Abel's mapping from the curve $\mathcal{R}$ to its Jacobian Jac ($\mathcal{R}$) by

$$P \mapsto \int_{P_0}^{P} \omega.$$

We fix an *odd* theta characteristic $\begin{bmatrix}\alpha\\ \beta\end{bmatrix}$ such that the theta function $\theta\begin{bmatrix}\alpha\\ \beta\end{bmatrix}$ does not vanish identically at all points

$$\int_{P_0}^{P} \omega$$

(this exists; cf. Fay [11], p. 16). We write θ for $\theta\begin{bmatrix}\alpha\\ \beta\end{bmatrix}$ for short. The theta-identity we want is this:

PROPOSITION. *There is a constant* c_1 *depending only on* $\mathcal{R}$ *and* $\begin{bmatrix}\alpha\\ \beta\end{bmatrix}$ *such that for all*

$$x_1, \ldots, x_M, \; P_1, \ldots, P_M, \; Q_1, \ldots, Q_{M-1} \in \mathcal{R}, \; \mathfrak{e} \in C^g$$

we have

$$\det_{1\leq i,j\leq M}\left\{\theta\left(\mathfrak{e}+\sum_{\alpha=1}^{j-1}\int_{P_\alpha}^{Q_\alpha}\boldsymbol{\omega}+\int_{P_j}^{x_i}\boldsymbol{\omega}\right)\cdot\prod_{\alpha=1}^{j-1}\theta\left(\int_{Q_\alpha}^{x_i}\boldsymbol{\omega}\right)\prod_{\alpha=j+1}^{M}\theta\left(\int_{P_\alpha}^{x_i}\boldsymbol{\omega}\right)\right\}$$

$$=c_1\cdot\prod_{j=1}^{M-1}\theta\left(\mathfrak{e}+\sum_{\alpha=1}^{j}\int_{P_\alpha}^{Q_\alpha}\boldsymbol{\omega}\right)\cdot\prod_{1\leq i<j\leq M}\theta\left(\int_{x_i}^{x_j}\boldsymbol{\omega}\right)\cdot\prod_{1\leq\beta<\alpha\leq M}\theta\left(\int_{Q_\alpha}^{P_\alpha}\boldsymbol{\omega}\right)\cdot\theta\left(\mathfrak{e}+\int_{\Sigma_1^M P_\alpha}^{\Sigma_1^M x_i}\boldsymbol{\omega}\right)$$

The proof follows the standard classical procedure (cf. Fay [11], Prop. 2.16 and the references given there): we check that the right hand side of the equation and all terms in the determinant on the left hand side are in the same line bundle over $\mathcal{R}^{3M-1}\times\mathrm{Jac}\,(\mathcal{R})$ and that they have the same zeroes. To see the first, we assume more generally that we have 3M variable points of Jac $(\mathcal{R})$ and consider both sides as sections of a line bundle on Jac $(\mathcal{R})^{3M}$: these bundles are products of pull-backs by linear maps Jac $(\mathcal{R})^{3M}\to$ Jac $(\mathcal{R})$ of the standard line bundle with section θ, and to check they are equal it suffices to check the corresponding assertion for the Hermitian forms representing the 1st Chern class of these bundles. If B is the Hermitian form of the standard ample bundle on Jac $(\mathcal{R})$, this comes down to checking that all the bilinear forms

$$\sum_{i=1}^{M}\left[B\left(e+x_{\sigma_i}-\sum_{\alpha=1}^{i}P_\alpha+\sum_{\alpha=1}^{i-1}Q_\alpha\right)+\sum_{\alpha=1}^{i-1}B(x_{\sigma_i}-Q_\alpha)+\sum_{\alpha=i+1}^{M}B(x_{\sigma_i}-P_\alpha)\right],$$

σ any permutation of $\{1,\ldots,M\}$

and

$$\sum_{i=1}^{M-1}B\left(e-\sum_{\alpha=1}^{i}P_\alpha+\sum_{\alpha=1}^{i}Q_\alpha\right)+\sum_{1\leq i<j\leq N}B(x_i-x_j)+\sum_{1\leq\beta<\alpha\leq M}B(P_\alpha-Q_\beta)+B\left(e-\sum_{\alpha=1}^{M}P_\alpha+\sum_{i=1}^{M}x_i\right)$$

are all equal. This is elementary. As for the 2nd step, fix P_i, Q_j and $\mathfrak{e}$ and consider as functions of $x\in\mathcal{R}$:

$$\psi_k(x)=\theta\left(\mathfrak{e}+\sum_{\alpha=1}^{k-1}\int_{P_\alpha}^{Q_\alpha}\boldsymbol{\omega}+\int_{P_k}^{x}\boldsymbol{\omega}\right)\cdot\prod_{\alpha=1}^{k-1}\theta\left(\int_{Q_\alpha}^{x}\boldsymbol{\omega}\right)\cdot\prod_{\alpha=k+1}^{M}\theta\left(\int_{P_\alpha}^{x}\boldsymbol{\omega}\right).$$

Let 0 be some origin on $\mathcal{R}$. Then recall (for instance from Siegel [29]) that for some constant $\mathfrak{c}$ (Riemann's constant):

$$\theta(\mathfrak{t}-\mathfrak{c})=0$$

if and only if

$$\mathfrak{t}=\sum_{2}^{g}\int_{\eta_i}^{0}\boldsymbol{\omega}\quad\text{(mod. periods)}$$

for some positive divisor $\sum_2^g \eta_i$ of order $g-1$. Moreover the g roots $q_1, \ldots, q_g$ of

$$\theta\left(\int_0^q \omega - \mathfrak{s} - \mathfrak{c}\right) = 0$$

satisfy the relation

$$\sum_1^g \int_0^{q_i} \omega = \mathfrak{s} \quad \text{(mod. periods)}.$$

The vector $\mathfrak{e} \in C^g$ defines a divisor $\sum_1^g q_i - 0$ of degree $g-1$ and hence a line bundle $L_\mathfrak{e}$ if we write

$$\mathfrak{e} + \mathfrak{c} = \sum_1^g \int_{q_i}^0 \omega.$$

Then the zeros of

$$x \to \theta\left(\mathfrak{e} + \sum_1^{k-1} \int_{P_\alpha}^{Q_\alpha} \omega + \int_{P_k}^x \omega\right)$$

satisfy the relation

$$\sum_1^g \int_0^{q_i} \omega = -\mathfrak{e} - \sum_1^{k-1} \int_{P_\alpha}^{Q_\alpha} \omega + \int_0^{P_k} \omega - \mathfrak{c} \quad \text{(mod. periods)},$$

i.e., this function is a section of $L_\mathfrak{e}(\sum_1^k P_\alpha - \sum_1^{k-1} Q_\alpha)$. Moreover x will be a zero if and only if

$$\mathfrak{e} + \sum_1^{k-1} \int_{P_\alpha}^{Q_\alpha} \omega + \int_{P_k}^x \omega = \sum_2^g \int_{\eta_i}^0 \omega,$$

i.e., if and only if (by Abel's theorem),

$$\Gamma\left(L_\mathfrak{e}\left(\sum_1^k P_\alpha - \sum_1^{k-1} Q_\alpha - x\right)\right) \neq (0).$$

Similarly for any point $R \in \mathcal{R}$,

$$x \to \theta\left(\int_R^x \omega\right)$$

is a section of $L_0(R)$. But as θ is an odd theta function, $\theta(0) = 0$; and this section of $L_0(R)$ is zero at R, i.e., L_0 itself has a section φ with $g-1$ zeros $\mathcal{D}_0$ and the function $\theta(\int_R^x \omega)$ has its zeros at $\mathcal{D}_0 + R$. Thus ψ_k is a section of

$$L_\mathfrak{e}\left(\sum_{\alpha=1}^k P_\alpha - \sum_{\alpha=1}^{k-1} Q_\alpha\right) \otimes \mathcal{O}\left(\sum_{\alpha=1}^{k-1} (\mathcal{D}_0 + Q_\alpha)\right) \otimes \mathcal{O}\left(\sum_{\alpha=k+1}^{M} (\mathcal{D}_0 + P_\alpha)\right)$$

$$\cong L_\mathfrak{e}\left(\sum_{\alpha=1}^M P_\alpha\right) \otimes \mathcal{O}((M-1)\mathcal{D}_0).$$

In fact, the last $(M-1)$ factors all vanish on $\mathcal{D}_0$, so ψ_k comes just from a section of

$$L_e\left(\sum_{\alpha=1}^{M} P_\alpha\right).$$

This bundle has degree $g-1+M$, so by Riemann–Roch, we may expect $\psi_1, \ldots, \psi_M$ to be a basis of its sections. Now $\det(\psi_i(x_j)) \neq 0$ if and only if no linear combination of the sections ψ_i is zero at all the points $x_1, \ldots, x_M$. Consider the various ways the left-hand side can be zero:

1. If $\theta(e+\sum_{\alpha=1}^{j}\int_{P_\alpha}^{Q_\alpha}\omega)=0$, then by formula 45, Fay [11], one sees that

$$\theta\left(e+\sum_{\alpha=1}^{j}\int_{P_\alpha}^{Q_\alpha}\omega+\int_{P_{j+1}}^{x}\omega\right)\cdot\theta\left(\int_{Q_j}^{x}\omega\right)$$

and

$$\theta\left(e+\sum_{\alpha=1}^{j-1}\int_{P_\alpha}^{Q_\alpha}\omega+\int_{P_j}^{x}\omega\right)\cdot\theta\left(\int_{P_{j+1}}^{x}\omega\right)$$

are linearly dependent. (Take Fay's e to be our $(e+\sum_1^j\int_{P_\alpha}^{Q}\omega)$, and take his y to be suitably general.) Thus ψ_j and ψ_{j+1} are linearly dependent and the determinant is zero.

2. If $\theta(\int_{x_i}^{x_j}\omega)=0$, then either $x_i=x_j$ and the determinant is zero or $x_i \in \mathcal{D}_0$. The left-hand side vanishes to order $M-1$ along the divisor $x_i \in \mathcal{D}_0$, but so does the whole ith row of the determinant.

3. If $\theta(\int_{Q_{\beta_1}}^{P_{\alpha_1}}\omega)=0$, then either $P_{\alpha_1}=Q_{\beta_1}$ or $P_{\alpha_1}\in\mathcal{D}_0$ or $Q_{\beta_1}\in\mathcal{D}_0$. If $P_{\alpha_1}=Q_{\beta_1}$, $\beta_1<\alpha_1$, then every section ψ_k vanishes at P_{α_1}. If e is sufficiently general, this means that all ψ_k are sections of

$$L_e\Big(\sum_{\alpha\neq\alpha_1} P_\alpha\Big)$$

of degree $g-2+M$, hence with only $M-1$ sections, hence the ψ_k's are linearly dependent and the determinant is zero. For special values of e, but $P_{\alpha_1}=Q_{\beta_1}$, the determinant is still zero by continuity. The left-hand side vanishes to order α_1-1 along the divisor $P_{\alpha_1}\in\mathcal{D}_0$, but (α_1-1)-columns (given by $j+1\leqslant\alpha_1$) of the determinant also vanish here.

4. Finally, if $\theta(e+\sum_1^M\int_{P_i}^{x_i}\omega)=0$, then $L_e(\sum_1^M P_\alpha-\sum_1^M x_i)$ has a section. To show the determinant is zero, we may assume $\theta(e+\int_{P_1}^{Q_1}\omega)\neq 0$, i.e. $\Gamma(L_e(P_1-Q_1))=(0)$. But then

$$\dim\Gamma\left(L_e\left(\sum_1^M P_\alpha\right)\right)\leqslant\dim\Gamma\left(\frac{L_e\left(\sum_1^M P_\alpha\right)}{L_e(P_1-Q_1)}\right)=M.$$

So either the ψ_k are linearly dependent and we are done, or they span $\Gamma(L_e(\sum_1^M P_\alpha))$. In this case, some combination is zero at all the points $x_1, \ldots, x_M$, so again $\det \psi_k(x_i) = 0$.

This proves the proposition because the divisor of the left-hand side is greater than or equal to that of the right, but both divisors come from zeroes of sections of the same line bundle. Renumbering the P_α's, Q_α's from $-M'$ to M instead of 1 to M, and shifting $\mathfrak{e}$ by

$$\sum_{\alpha=-M'}^{0} \int_{P_\alpha}^{Q_\alpha},$$

the formula reads:

Corollary.

$$\det_{\substack{-M'\leqslant j\leqslant M\\ 1\leqslant i\leqslant M+M'+1}} \left\{\theta\left(\mathfrak{e}+\sum_{\alpha=1}^{j-1}\int_{P_\alpha}^{Q_\alpha}\omega+\int_P^{x_i}\omega\right)\cdot\prod_{\alpha=-M'}^{j-1}\theta\left(\int_{Q_\alpha}^{x_i}\omega\right)\cdot\prod_{\alpha=j+1}^{M}\theta\left(\int_{P_\alpha}^{x_i}\omega\right)\right\}$$

$$= c_1\prod_{j=-M'}^{M-1}\theta\left(\mathfrak{e}+\sum_{\alpha=1}^{j}\int_{P_\alpha}^{Q_\alpha}\omega\right)\cdot\prod_{1\leqslant i<j\leqslant M+M'+1}\theta\left(\int_{x_i}^{x_j}\omega\right)\cdot\prod_{-M'\leqslant\beta<\alpha\leqslant M}\theta\left(\int_{Q_\beta}^{P_\alpha}\omega\right)\theta\left(\mathfrak{e}+\int_{\sum_{-M}^{0}Q_\alpha+\sum_1^M P_\alpha}^{\sum_1^{M+M'+1}x_i}\omega\right)$$

Now we apply this to our curve $\mathcal{R}$ with given points $P_1, \ldots, P_n, Q_1, \ldots, Q_{n'}$ and function z. By assumption

$$(z) = -M_1\sum_{i=1}^{n}P_i - M_1\sum_{i=1}^{n'}Q_i + \sum_{i=1}^{M+M'}R_i$$

for some set of points $R_1, \ldots, R_{M+M'}$. In this case, the function z can be expanded:

$$z(x) = c_2\cdot\frac{\prod_{i=1}^{M+M'}\theta\left(\int_{R_i}^{x}\omega\right)}{\left(\prod_{\alpha=1}^{n}\theta\left(\int_{P_\alpha}^{x}\omega\right)\right)^{M_1}\cdot\left(\prod_{\alpha=1}^{n'}\theta\left(\int_{Q_\alpha}^{x}\omega\right)\right)^{M_1'}}$$

where c_2 is a suitable constant. Moreover, if $\mathcal{D}$ is a regular divisor of degree g and we define

$$\mathfrak{e} = \int_{\mathcal{D}}^{Q_0+D_0}\omega;$$

then we claim that for suitable constants λ_k:

$$f_k(x) = \lambda_k\cdot\frac{\theta\left(\mathfrak{e}+\sum_{\alpha=1}^{k-1}\int_{P_\alpha}^{Q_\alpha}\omega+\int_{P_k}^{x}\omega\right)\prod_{\alpha=0}^{k-1}\theta\left(\int_{Q_\alpha}^{x}\omega\right)}{\theta\left(\mathfrak{e}+\int_{Q_0}^{x}\omega\right)\prod_{\alpha=1}^{k}\theta\left(\int_{P_\alpha}^{x}\omega\right)}.$$

(Here $k \geq 1$; an obvious modification holds if $k \leq 0$.) In fact, it is immediate by the functional equation of θ that this is a meromorphic function on $\mathcal{R}$. The factors on the right give it zeroes at $Q_0, \ldots, Q_{k-1}$ and poles at $P_1, \ldots, P_k$. The other factor in the denominator satisfies

$$\theta\left(\mathfrak{e}+\int_{Q_0}^{x} \omega\right)=\theta\left(\int_{\mathcal{D}}^{x+\mathcal{D}_0} \omega\right)$$

which is zero if $x \in \mathcal{D}$ because then

$$x+\mathcal{D}_0-\mathcal{D} \equiv \mathcal{D}_0 - \text{positive divisor of degree } g-1.$$

These properties characterize f_k up to scalars. Now apply the Corollary replacing $\mathfrak{e}$ by $\mathfrak{e}+\sum_{\alpha=1}^{k-1} \int_{P_\alpha}^{Q_\alpha} \omega$, renumbering P_α, Q_α by $P_{\alpha+k}$, $Q_{\alpha+k}$, letting $x_i = R_i$, $1 \leq i \leq M+M'$ and $x_{M+M'+1}=x$, and expanding the determinant along the 1st row:

$$\sum_{j=-M'+k}^{M+k} (-1)^j \theta\left(\mathfrak{e}+\sum_{\alpha=1}^{j-1} \int_{P_\alpha}^{Q_\alpha} \omega+\int_{P_j}^{x} \omega\right) \cdot \prod_{\alpha=-M'+k}^{j-1} \theta\left(\int_{Q_\alpha}^{x} \omega\right) \cdot \prod_{\alpha=j+1}^{M+k} \theta\left(\int_{P_\alpha}^{x} \omega\right)$$

$$\times \det_{\substack{-M'+k \leq l \leq M+k \\ 1 \leq i \leq M+M' \\ l \neq j}}\left\{\theta\left(\mathfrak{e}+\sum_{\alpha=1}^{l-1} \int_{P_\alpha}^{Q_\alpha} \omega+\int_{P_l}^{R_i} \omega\right) \cdot \prod_{\alpha=-M'+k}^{l-1} \theta\left(\int_{Q_\alpha}^{R_i} \omega\right) \cdot \prod_{\alpha=l+1}^{M+k} \theta\left(\int_{P_\alpha}^{R_i} \omega\right)\right\}$$

$$= \pm c_1 \prod_{j=-M'+k}^{M+k-1} \theta\left(\mathfrak{e}+\sum_{\alpha=1}^{j} \int_{P_\alpha}^{Q_\alpha} \omega\right) \cdot \prod_{1 \leq i<j \leq M+M'} \theta\left(\int_{R_i}^{R_j} \omega\right) \cdot \prod_{i=1}^{M+M'} \theta\left(\int_{R_i}^{x} \omega\right)$$

$$\times \prod_{-M'+k \leq \beta<\alpha \leq M+k} \theta\left(\int_{Q_\beta}^{P_\alpha} \omega\right) \cdot \theta\left(\mathfrak{e}+\sum_{\alpha=1}^{k-1} \int_{P_\alpha}^{Q_\alpha} \omega+\int_{P_k}^{x} \omega\right)$$

or

$$z(x) \cdot f_k(x)=\sum_{j=-M'+k}^{M+k}(-1)^j \frac{c_2}{c_1} \frac{\lambda_k}{\lambda_j}$$

$$\times \frac{\det\left\{\theta\left(\mathfrak{e}+\sum_{\alpha=1}^{j-1} \int_{P_\alpha}^{Q_\alpha} \omega+\int_{P_j}^{R_i} \omega\right) \cdot \prod_{\alpha=-M'+k}^{j-1} \theta\left(\int_{Q_\alpha}^{R_i} \omega\right) \cdot \prod_{\alpha=j+1}^{M+k} \theta\left(\int_{P_\alpha}^{R_i} \omega\right)\right\}}{\prod_{l=-M'+k}^{M+k-1} \theta\left(\mathfrak{e}+\sum_{\alpha=1}^{l} \int_{P_\alpha}^{Q_\alpha} \omega\right) \cdot \prod_{1 \leq i<j \leq M+M'} \theta\left(\int_{R_i}^{R_j} \omega\right) \cdot \prod_{-M'+k \leq \beta<\alpha \leq M+k} \theta\left(\int_{Q_\beta}^{P_\alpha} \omega\right)} \cdot f_j(x).$$

Thus the operator C, up to a constant and suitably conjugated is given by

THEOREM 7.

$$c_{kj}=\frac{(-1)^j \det_{\substack{-M'+k \leq l \leq M+k \\ 1 \leq i \leq M+M' \\ l \neq j}} \theta\left(\mathfrak{e}+\sum_{\alpha=1}^{l-1} \int_{P_\alpha}^{Q_\alpha} \omega+\int_{P_l}^{R_i} \omega\right) \cdot \prod_{\alpha=-M'+k}^{l-1} \theta\left(\int_{Q_\alpha}^{R_i} \omega\right) \cdot \prod_{\alpha=l+1}^{M+k} \theta\left(\int_{P_\alpha}^{R_i} \omega\right)}{\prod_{l=-M'+k}^{M+k-1} \theta\left(\mathfrak{e}+\sum_{\alpha=1}^{l} \int_{P_\alpha}^{Q_\alpha} \omega\right) \cdot \prod_{-M'+k \leq \beta<\alpha \leq M+k} \theta\left(\int_{Q_\beta}^{P_\alpha} \omega\right)}.$$

§ 6. Almost periodic difference operators

Non-singular curves $\mathcal{R}$ with the properties listed in Theorem 1, but without the existence of a meromorphic function h, lead to *almost periodic difference operators*, in the following sense: for every $\varepsilon>0$, there is an integer $T>0$ such that for every interval $I(T)\subseteq \mathbf{Z}$ of length T you can find $\alpha\in I(T)$ with the property

$$|c_{k,k+i}-c_{k+\alpha,k+i+\alpha}|<\varepsilon \quad \forall k,\, k+i\in \mathbf{Z}.$$

Considering the Jacobian Jac $(\mathcal{R})$ as a moduli space for divisor classes of degree $g-1$, the Theta-divisor $\Theta\subset$ Jac $(\mathcal{R})$ is the subvariety of positive divisors in Jac $(\mathcal{R})$ of order $g-1$. Whenever one considers a regular divisor $\mathcal{D}$, the corresponding sequence of meromorphic functions f_k and the associated sequence of regular divisors

$$\mathcal{D}_k=(f_k)+\mathcal{D}+\sum_1^k P_i-\sum_0^{k-1} Q_i, \quad k\in \mathbf{Z},$$

then

$$\mathcal{L}(\mathcal{D}_k-Q_k)=\{0\}$$

or what is the same

$$\{\mathcal{D}_k\}\notin Q_k+\Theta.$$

where $\{\mathcal{D}_k\}\in$ Jac $(\mathcal{R})$ is the point corresponding to $\mathcal{D}_k$.

Now, we define a *uniformly* regular divisor $\mathcal{D}$ with regard to the same sequence to be a regular divisor, with the property that

$$\mathcal{L}(\mathcal{D}'-Q_k)=\{0\}$$

for every k $(1\leqslant k\leqslant n)$ and for every:

$$\{\mathcal{D}'\}\in \overline{\bigcup_{p\in \mathbf{Z}}\{\mathcal{D}_{k+pn}\}}$$

or, equivalently, for every k $(1\leqslant k\leqslant n')$,

$$(\Theta+Q_k)\cap \overline{\bigcup_{p\in \mathbf{Z}}\{\mathcal{D}_{k+pn}\}}=\emptyset.$$

As we shall see later, there are many such *uniformly* regular divisors.

THEOREM 8. *Let $\mathcal{R}$ be a non-singular curve with points $P_1, \ldots, P_n$ and $Q_1, \ldots, Q_{n'}$; let z be a function on $\mathcal{R}$ subject to*

$$(z)=-M_1\sum_1^n P_i-M_1'\sum_1^{n'} Q_i+d,$$

where d is a positive divisor not containing the P_i's and Q_i's. Then every uniformly regular divisor $\mathcal{D}$ determines an almost periodic difference operator C.

Proof. Let ν be the l.c.m. of n and n'. Observe that in Jac $(\mathcal{R})$,

$$\mathcal{D}_{k+\nu}-\mathcal{D}_k \equiv \frac{\nu}{n}\sum_1^n P_i - \frac{\nu}{n}\sum_1^{n'} Q_i.$$

Let $\mathfrak{D}$ be the divisor on the right hand side. The divisors $\ldots \mathcal{D}_{k-\nu}, \mathcal{D}_k, \mathcal{D}_{k+\nu}, \ldots$ form a linear sequence of points in Jac $(\mathcal{R})$. The transformation obtained by adding $\mathfrak{D}$ to a given point on the torus is periodic or almost periodic. If this transformation would be periodic, it would imply the existence of a meromorphic function h having for divisors some multiple of $\mathfrak{D}$. If we fix some measure of distance on Jac $(\mathcal{R})$, then at least we can say that for any $\varepsilon>0$, there is a positive integer T such that for every interval $I(T)$ of length T there is an integer $P\in I(T)\cap \mathbf{Z}$ with the property that in Jac $(\mathcal{R})$

$$\begin{aligned}\left|\left(\mathcal{D}+\sum_1^k P_\alpha-\sum_0^k Q_\alpha\right)-\left(\mathcal{D}+\sum_1^{k+p\nu} P_\alpha-\sum_0^{k+p\nu} Q_\alpha\right)\right| &= |(\mathcal{D}_k-Q_k)-(\mathcal{D}_{k+p\nu}-Q_k)|\\ &=|\mathcal{D}_k-\mathcal{D}_{k+p\nu}|\\ &=|p\mathfrak{D}|<\varepsilon \quad \forall k\in\mathbf{Z}.\end{aligned}$$

Consider the closure $\overline{\{p\mathfrak{D}\}}$ of the sequence of points $p\mathfrak{D}$ in Jac $(\mathcal{R})$: this will be the union of a finite number of cosets of the real subtorus $P\subset$ Jac $(\mathcal{R})$. We wish to prove first that, with a suitable choice of A- and B-periods on $\mathcal{R}$, P is contained in the real sub-torus given by the A-periods alone. In fact, our hypothesis that $\mathcal{D}$ is uniformly regular means that certain cosets of P are disjoint from Θ. This means that the cohomology class of Θ restricted to P is zero. But we have

$$H_1(P,\mathbf{Z})\subset H_1(\text{Jac}(\mathcal{R}),\mathbf{Z})\cong H_1(\mathcal{R},\mathbf{Z})$$

and the cohomology class of Θ on Jac $(\mathcal{R})$ is just given by the 2-form: intersection product $(a,b)\to(\mathrm{a}\cdot\mathrm{b})$ on $H_1(\mathcal{R},\mathbf{Z})$. Thus this triviality means $H_1(P,\mathbf{Z})$ is an isotropic subspace. Any maximal isotropic subspace of $H_1(\mathcal{R},\mathbf{Z})$ can be taken as the set of A-periods, so this proves our assertion. This choice of A-periods means that θ is a *periodic* function in the P-coordinates, hence for any $\mathfrak{e}$, the values

$$\theta\left(\mathfrak{e}+\sum_{\alpha=1}^{l}\int_{P_\alpha}^{Q_\alpha}\omega\right)$$

for *all* l are equal to their values in some compact fundamental domain mod periods. This plus the explicit formula for c_{ij} in the last section proves Theorem 8.

THEOREM 9. *Consider a curve $\mathcal{R}$, $2n$ points $P_1, \ldots, P_n, Q_1, \ldots, Q_n$ on $\mathcal{R}$, a meromorphic function z having the properties above with $M_1 = M_1'$. Let $\tilde{}$ be an antiholomorphic involution for which $\mathcal{R}\backslash\mathcal{R}_{\mathbf{R}} = \mathcal{R}_+ \cup \mathcal{R}_-$ where $\mathcal{R}_{\mathbf{R}} = \{p \in \mathcal{R} \mid \tilde{p} = p\}$ such that $\tilde{P}_i = Q_i$ with $P_i \in \mathcal{R}_+$ and $Q_i \in \mathcal{R}_-$ and such that $\overline{z(p)} = z(\tilde{p})$. Consider a divisor $\mathcal{D}$ having the property that*

$$\mathcal{D} + \tilde{\mathcal{D}} - P_n - Q_n$$

is the divisor of some differential on $\mathcal{R}$, which is real positive on $\mathcal{R}_{\mathbf{R}}$. Then $\mathcal{D}$ determines a self-adjoint, almost periodic difference operator C, whose L^2-spectrum is the set of values of z on $\mathcal{R}_{\mathbf{R}}$.

LEMMA. *Any divisor $\mathcal{D}$ on the curve $\mathcal{R}$ satisfying the conditions of Theorem 9 is uniformly regular.*

Proof. Recall from the proof of Theorem 3 that

$$\mathcal{D}^{(k)} + \tilde{\mathcal{D}}^{(k)} = P_k + Q_k + (\omega_k),$$

for some 1-form ω_k, with $\omega_k \geqslant 0$ on $\mathcal{R}_{\mathbf{R}}$. Therefore all the divisors $\mathcal{D} = \mathcal{D}^{(k+pn)}$, $p \in \mathbf{Z}$, satisfy:

$$\mathcal{D}' + \tilde{\mathcal{D}}' = P_k + Q_k + (\omega), \quad \omega \geqslant 0 \quad \text{on} \quad \mathcal{R}_{\mathbf{R}}.$$

Passing to the limit of any sequence, it follows that this condition still holds. But by the argument in Theorem 3, any such ω must have a non-zero residue at Q_k, hence Q_k doesn't occur in the divisor $\mathcal{D}'$, hence $\mathcal{D}' - Q_k \notin \Theta$.

Proof of Theorem 9. From the Lemma it follows that $\mathcal{D}$ is uniformly regular; by Theorem 3 (§ 2) and 8 it maps into a self-adjoint almost periodic difference operator C. Consider now the space of meromorphic functions

$$\mathcal{L} = \{f \mid (f) > -\mathcal{D} - \sum k_i P_i - \sum l_i Q_i \quad \text{with } k_i, l_i \in \mathbf{Z} \text{ arbitrary}\}.$$

Any $f \in \mathcal{L}$ maps into a sequence λ_n, zero for almost all n (Lemma 5, § 1) such that

$$f = \sum_{i=-\infty}^{\infty} \lambda_i f_i.$$

Let $\mathcal{D}$ not contain any point of $\mathcal{R}_{\mathbf{R}}$. Then using the inner product defined in Theorem 3,

$$\frac{1}{2\pi}\int_{\mathcal{R}_{\mathbf{R}}} |f|^2 \omega = \frac{1}{2\pi}\sum_n \sum_m \lambda_n \bar{\lambda}_m \int_{\mathcal{R}_{\mathbf{R}}} f_n f_m^* \omega = \sum |\lambda_n|^2 < \infty.$$

$\mathcal{L}$ is a space of complex-valued functions on $\mathcal{R}_{\mathbf{R}}$, separating points and closed under conjugation, so by the Stone–Weierstrass theorem the space $\mathcal{L}$ naturally completes to the space $L^2(\mathcal{R}_{\mathbf{R}})$ of L^2-complex valued functions on $\mathcal{R}_{\mathbf{R}}$; the space of almost everywhere vanishing sequences completes to $l^2(\mathbf{Z})=\{\{\lambda_n\} \mid \sum|\lambda_n|^2<\infty\}$. A basis for the space $L^2(\mathcal{R}_{\mathbf{R}})$ is given by the functions f_k. This defines now a unitary transformation from $l^2(\mathbf{Z})$ to $L^2(\mathcal{R}_{\mathbf{R}})$. The difference operator C acts on $l^2(\mathbf{Z})$ as follows

$$(C\lambda)_n=\sum_{k=n-M}^{n+M}\lambda_k c_{kn}=\sum_{k=n-M}^{n+M}\overline{c_{nk}}\lambda_k$$

and C acts on $L^2(\mathcal{R}_{\mathbf{R}})$ as a multiplication operator. Indeed for $f=\sum\lambda_n f_n$

$$\begin{aligned} Cf &= \sum \lambda_n(Cf_n) \\ &= \sum \lambda_n z f_n \\ &= zf. \end{aligned}$$

This operator is bounded and self-adjoint, since $\mathcal{R}_{\mathbf{R}}$ does not contain P_i or Q_i. The spectrum of this operator is the range of z, defined on the cycles $\mathcal{R}_{\mathbf{R}}$.

If $\mathcal{D}$ contains a point of $\mathcal{R}_{\mathbf{R}}$, we may argue by a limiting process that the theorem still holds, or, noting that ω has zeroes where f has poles, we replace $L^2(\mathcal{R}_{\mathbf{R}})$ by the space of functions f on $\mathcal{R}_{\mathbf{R}}-\mathcal{D}\cap\mathcal{R}_{\mathbf{R}}$ such that

$$\int_{\mathcal{R}_{\mathbf{R}}}|f|^2\omega<\infty.$$

Then the proof goes through as before.

References

[1]. Abraham, R. & Marsden, J., *Foundations of Mechanics*. Benjamin, San Francisco, 1978.

[2]. Adler, M., On a trace functional for pseudo-differential operators and the symplectic structure of the Korteweg-De Vries type equations. *Inventiones Math.* (1979).

[3]. Adler, M. & van Moerbeke, P., The Kac-Moody extension for the classical groups and algebraic curves. To appear.

[4]. Altman, A. & Kleiman, S., Compactifying the Jacobian. *Bull. A.M.S.*, 82 (1976), 947–949.

[5]. —— Compactifying the Picard scheme. *Advances in Math.* To appear.

[6]. Burchnall, J. L. & Chaundy, T. W., Commutative ordinary differential operators. *Proc. London Math. Soc.*, 81 (1922), 420–440; *Proc. Royal Soc. London* (*A*), 118 (1928), 557–593 (with a note by H. F. Baker); *Proc. Royal Soc. London* (*A*), 134 (1931), 471–485.

[7]. De Souza, M., Compactifying the Jacobian. Thesis at the Tata Institute (1973).
[8]. Dikii, L. A. & Gel'fand, I. M., Fractional Powers of Operators and Hamiltonian systems. *Funk. Anal. Priloz.*, 10 (1976).
[9]. Drinfeld, V. G., Elliptic modules. *Mat. Sbornik*, 94 (1974); translation 23 (1976), 561.
[10]. Dubrovin, B. A., Matveev, V. B. & Novikov, S. P., Non-linear equations of the Korteweg-de Vries type, finite zone linear operators, and Abelian manifolds. *Uspehi Mat. Nauk*, 31 (1976); *Russian Math. Surveys*, 31 (1976), 55–136.
[11]. Fay, J., *Theta functions on Riemann surfaces*. Springer Lecture Notes 352 (1973).
[12]. Kac, M. & van Moerbeke, P., On periodic Toda lattices. *Proc. Nat. Acad. Sci. U.S.A.*, 72 (1975), 1627–1629, and A complete solution of the periodic Toda problem. *Proc. Nat. Acad. Sci. U.S.A.*, 72 (1975), 2875–2880.
[13]. Kempf, G., Knudsen, P., Mumford, D. & Saint-Donat, B. *Toroidal embeddings I*. Berlin–Heidelberg–New York: Springer vol. 339 (1973).
[14]. Kostant, B., Quantization and unitary representation, *Lectures on Modern Analysis and Applications III*, Berlin–Heidelberg–New York, Springer vol. 170 (1970).
[15]. Krichever, I. M., Algebro-geometric construction of the Zaharov-Shabat equations and their periodic solutions. *Sov. Math. Dokl.*, 17 (1976), 394–397.
[16]. —— *Uspekhi Mat. Nauk*, (1978).
[17]. —— Methods of Algebraic geometry in the theory of non-linear equations. *Uspekhi Mat. Nauk*, 32: 6 (1977), 183–208. Translation: *Russian Math Surveys*, 32: 6 (1977), 185–213.
[18]. McKean, H. P. & van Moerbeke, P., The spectrum of Hill's equation. *Inventiones Math.*, 30 (1973), 217–274.
[19]. —— Sur le spectre de quelques operateurs et les varietes de Jacobi. *Sem Bourbaki*, 1975–76, No, 474, 1–15.
[20]. —— About Toda and Hill curves. *Comm. Pure Appl. Math.*, (1979) to appear.
[21]. McKean, H. P. & Trubowitz, E., The spectrum of Hill's equation, in the presence of infinitely many bands. *Comm. Pure. Appl. Math.*, 29 (1976), 143–226.
[22]. van Moerbeke, P., The spectrum of Jacobi matrices. *Inventiones Math.*, 37 (1976), 45–81.
[23]. —— About isospectral deformations of discrete Laplacians. *Proc. of a conference on non-linear analysis*, Calgary, June 1978. Springer Lecture Notes, 1979.
[24]. Mumford, D., An algebro-geometrical construction of commuting operators and of solutions to the Toda lattice equation, Korteweg-de Vries equation and related non-linear equations. *Proc. of a Conference in Algebraic Geometry*, Kyoto, 1977, publ. by Japan Math. Soc.
[25]. —— *Abelian varieties*, Tata Institute; Oxford University Press, (1970).
[26]. —— The Spectra of Laplace-like periodic partial difference operators and algebraic surfaces, to appear.
[27]. Rosenlicht, M., Generalized Jacobian varieties. *Ann. of Math.*, 59 (1954), 505–530.
[28]. Serre, J. P., *Groupes algebriques et Corps de Classes*, Paris. Hermann (1959).
[29]. Siegel, C. L., *Topics in complex function theory, vol. 2*. New York, Wiley (1971).
[30]. Zaharov, V. E. & Shabat, A. B., A scheme for integrating the non-linear equations of math. physics by the method of the inverse scattering problem I. *Funct. Anal. and its Appl.* 8 (1974) (translation 1975, p. 226).

Received August 28, 1978

APPENDIX: PROOF OF THE CONVEXITY THEOREM

By David Mumford

In this Appendix, we prove the convexity theorem of Section 5 by using the techniques of geometric invariant theory. Half of the proof in fact can be viewed as giving a purely algebraic definition of the convex set Image ($\check{m}$). First fix the following notation:

G a reductive algebraic group, Lie algebra $\mathfrak{g}$
K a maximal compact subgroup, Lie algebra $\mathfrak{k}$
T a maximal torus such that T = complexification of $K \cap T$
B a Borel subgroup containing T
$\mathfrak{t}$ the Lie algebra of T
$\mathfrak{t}^+$ the positive Weyl chamber determined by B
$\mathbf{V}$ a representation space for G
$\langle , \rangle$ an inner product on $\mathfrak{g}$ satisfying (12).

As in Section 5, we get moment maps

$$m^*: \mathbf{P}(V) \to i\mathfrak{k}$$

$$\check{m}: \mathbf{P}(V) \to i\mathfrak{t}^+.$$

Let $X \subset \mathbf{P}(V)$ be a G-invariant subvariety. We wish to prove that $\check{m}(X)$ is a convex rational polytope in $i\mathfrak{t}^+$. The proof consists in 2 steps.

Step I: Let $w \in$ Weyl group carry $\mathfrak{t}^+$ to $-\mathfrak{t}^+$ and let an integral point (in the "weight" lattice) $\alpha \in \mathfrak{t}$ (define a character of T, hence a line bundle L^α on G/B. Then I claim:

(A1) $\forall \alpha \in i\mathfrak{t}^+$, $\alpha = \beta/n$, β *integral*
$(-w)\alpha \in \check{m}(x) \Leftrightarrow$ *the generic point of* $X \times G/B$ *is* $\mathcal{O}_x(n) \otimes L^\beta$*-semi-stable.*

Step II: Let G act on $\mathbf{V}^{(1)}, \ldots, \mathbf{V}^{(m)}$ and let $X \subset \mathbf{P}(\mathbf{V}^{(1)}) \times \cdots \times \mathbf{P}(\mathbf{V}^{(m)})$ be a G-invariant subvariety. Then there is a rational convex polyhedral cone $C \subset \mathbf{R}^m_+$ such that for all $k_i \in \mathbf{Z}$, $k_i \geq 0$,

D. Mumford, *Selected Papers*, Vol. II,

(A2) $(k_1, \ldots, k_m) \in C \Leftrightarrow$ *the generic point of x is* $\mathcal{O}_{\mathbf{P}}(k_1) \otimes \cdots \otimes \mathcal{O}_{\mathbf{P}}(k_m)$*-semi-stable.*

(A1) and (A2) together prove that the *rational* points in $\check{m}(X)$ are the rational points in a rational convex polytope. In view of lemma 2.1, m and hence $\check{m}$ is generically regular and hence rational points will be dense in $\check{m}(X)$, unless the intersection of K and the generic stabilizer G_x is positive dimensional. In the rather special case that $\dim K \cap G_x > 0$, all x, we can still argue that rational points are dense in $\check{m}(x)$ as follows: Replacing x by kx, some $k \in x$, we may assume $m(x) \in it^*$ and apply lemmas 1.2 and 2.1 for the moment map just for the torus T. By lemma 1.2, the components of m in directions in $T \cap G_x$ are rational and constant on $T \cdot x$; by lemma 2., the other components vary independently hence can be made rational by replacing x by $t \cdot x$, $t \in T$ arbitrarily close to e.

Proof of Step I. This follows from 2 easy lemmas:

Lemma A3. *Embed* $\mathbf{P}(\mathbf{V}_1) \times \mathbf{P}(\mathbf{V}_2)$ *in* $\mathbf{P}(\mathbf{V}_1 \otimes \mathbf{V}_2)$ *as usual. Then the moment map for* $\mathbf{P}(\mathbf{V}_1) \times \mathbf{P}(\mathbf{V}_2)$ *is the sum of the moment maps for* $\mathbf{P}(\mathbf{V}_1)$, $\mathbf{P}(\mathbf{V}_2)$:

$$m_{\mathbf{P}(\mathbf{V}_1 \otimes \mathbf{V}_2)}(x \otimes y) = m_{\mathbf{P}(\mathbf{V}_1)}(x) + m_{\mathbf{P}(\mathbf{V}_2)}(y).$$

This follows from the formula of Section 1.

Lemma A4. *Let* $\alpha \in \mathfrak{t}^+$ *be integral and let it define* L^α *on* G/B. *Map* G/B *to* $\mathbf{P}(\mathbf{V})$, $\mathbf{V} = \Gamma(G/B, L^\alpha)$: *call this* $\phi: G/B \to \mathbf{P}(\mathbf{V})$. *Then*

$$m^*_{\mathbf{P}(\mathbf{V})(\phi(G/B))} = K\text{-orbit of } i\alpha \text{ in } i \cdot \mathfrak{k}$$

This follows from the calculations of Section 8. Putting these together, we see that

gen. pt. of $X \times G/B$ *is* $\mathcal{O}_x(n) \otimes L^\beta$-semi-stable $\Leftrightarrow$ Image $(m_{X \times G/B})$ where $x \times G/B$ *is mapped to* $\mathbf{P}(\mathbf{V})$; $\mathbf{V} = \Gamma(\mathcal{O}_x(n)) \otimes \Gamma(L^\beta)$

$\Leftrightarrow \exists x \in X, y \in G/B$ *such that* $n \cdot m_x(x) + m_{G/B}(y) = 0$

$\Leftrightarrow \exists x \in X$ *such that* $n \cdot m_x(x) + \beta = 0$

$\Leftrightarrow \exists x \in X$ *such that* $n \cdot m_x(x) + \omega(\beta) = 0$ *or* $m_x(x) = (-\omega)(\alpha)$. Q.E.D.

Proof of Step II. Decompose $\mathbf{V}^{(\alpha)}$, into weight spaces with respect to T:

$$\mathbf{V}^{(\alpha)} = \bigoplus_{i \in W_\alpha} \mathbf{V}^{(\alpha)}, \qquad W_\alpha \subset \mathfrak{t}^*.$$

For each $x \in X$, write the coordinates of x in the Segre embedding as

$$\Big(\sum_{i \in W_1} v_i^{(1)}\Big) \otimes \cdots \otimes \Big(\sum_{i \in W_m} v_i^{(m)}\Big), \qquad v_i^{(\alpha)} \in \mathbf{V}_i^{(\alpha)}.$$

Define

$$\mathcal{S}_\alpha(x) = \{ i \in W_\alpha \,|\, v_i^{(\alpha)} \neq 0 \}$$

and call the m-tuples of subsets $(\mathcal{S}_1(x), \ldots, \mathcal{S}_n(x))$ the "T-state" of the point x. As there are only finitely many possible T-states, X decomposes into locally closed pieces

$$X = \Pi X_\ell$$

where all points on X_ℓ have the same T-state. For each ℓ, either $G \cdot X_\ell$ is dense in X or is part of a finite set of proper subvarieties. Call the T-states of the X_ℓ such that $G \cdot X_\ell$ is dense the generic *T-states*, and write then $(\mathcal{S})_{1\beta}, \ldots, \mathcal{S}_{n\beta})$, $\beta \in I$. We now apply the numerical criterion of semi-stability of Geometric Invariant Theory:

$$x \in X \text{ semi-stable} \Leftrightarrow \forall g \in G, \forall \text{ one-parameter subgroups } \lambda \in \Gamma(T),$$

$$\mu(g \cdot x, \lambda) \geq 0.$$

If semi-stability is calculated by the embedding $\mathcal{O}_{\mathbf{P}}(k_1) \otimes \cdots \otimes \mathcal{O}_{\mathbf{P}}(k_m)$, then

$$\mu(g \cdot x, \lambda) = \max_{\begin{pmatrix} s_1 \in \mathcal{S}_1(gx) \\ \cdots \\ s_m \in \mathcal{S}_m(gx) \end{pmatrix}} \ell\Big(\sum_{\alpha=1}^{m} k_\alpha s_\alpha\Big)$$

where λ corresponds to $\ell \in \mathfrak{t}$. Therefore:

the gen. pt. $x \in X$ is $\mathcal{O}_{\mathbf{P}}(k_1) \otimes \cdots \otimes \mathcal{O}_{\mathbf{P}}(k_m)$-semi-stable $\Leftrightarrow$ $\forall$ generic T-states β, $\forall \ell \in \mathfrak{t}$

$$\max_{\begin{bmatrix} s_\alpha \in \mathcal{S}_{\alpha\beta} \\ 1 \le \alpha \le m \end{bmatrix}} \ell\left(\sum_{\alpha=1}^{m} k_\alpha \mathcal{S}_\alpha\right) \geq 0.$$

To analyze the right hand side, define

$$\mathcal{S}_\beta^* \subset \mathbf{R}^m \times \mathfrak{t}^*$$

by

$$\mathcal{S}_\beta^* = \{(k_1, \ldots, k_m, \Sigma\, k_\alpha s_\alpha) \mid k_i \in \mathbf{R},\ k_i \geq 0,\ s_\alpha \in \text{convex hull}\ (\mathcal{S}_{\alpha\beta})$$

Then $\mathcal{S}_\beta^*$ is rational convex polyhedral cone such that

$$\mathcal{S}_\beta^* \cap [(k_1, \ldots, k_m) \times \mathfrak{t}^*] = \sum_\alpha k_\alpha\ (\text{convexhull of } \mathcal{S}_{\alpha\beta}).$$

The condition on the right hand side is just that for all β,

$$0 \in \sum_\alpha k_\alpha (\text{convexhull of } \mathcal{S}_{\alpha\beta}).$$

But $\mathcal{S}_\beta^*$ may be defined by a finite set of inequalities

$$(k_1, \ldots, k_m, x) \in \mathcal{S}_\beta^* \Leftrightarrow \sum_{\alpha=1} e_{\alpha\beta\gamma} k_\alpha + \ell_{\beta\gamma}(x) \geq 0,$$

hence it follows that the right hand side is equivalent to $\Sigma_\alpha e_{\alpha\beta\gamma} k_\alpha \geq 0$, all β, γ which defines a rational convex polyhedral cone. Q.E.D.

Oscar Zariski

1899–1986

Oscar Zariski was born on April 24, 1899, in the town of Kobryn, which lies on the border of Poland and the U.S.S.R. It was Russian at the time of Zariski's birth, was Polish between the two world wars, and is now Russian again. He was the son of Bezalel and Chana Zaristky, and was given the name of Asher Zaristky, which he changed to Oscar Zariski when he came to Italy. Kobryn was a small town where his mother ran a general store, his father having died when he was two. In 1918, he went to the University of Kiev in the midst of the revolutionary struggle. He was seriously wounded in one leg when caught in a crowd that was fired at by troops, but recovered after two months in the hospital. As a student, he was attracted to the fields of algebra and number theory as well as to the revolutionary political ideas of the day. He supported himself partly by writing for a local Communist paper. This is most surprising for those of us who only knew him much later, but calls to mind the quip—A man has no heart if he is not a radical in his youth and no mind if he is not a conservative in his mature years.

Because of the limitations of the education available in the U.S.S.R. at the time, in 1921 Zariski went first to the University of Pisa and six months later to the University of Rome where the famous Italian school of algebraic geometry, Castelnuovo, Enriques, and Severi, was flourishing. He had no money and the fact that universities in Italy were free to foreign students was an important consideration. Zariski was especially attracted to Castelnuovo, who immediately recognized his talent. Castelnuovo took him on a three hour walk around Rome after which Zariski realized that he had been given an oral exam in every area of mathematics! Castelnuovo saw in Zariski a man who would not only push their subject further and deeper, but would find radically new ways to overcome its present limitations. Zariski was fond of quoting Castelnuovo as saying "Oscar, you are here with us, but are not one of us," referring to Zariski's doubts even then of how rigorous their proofs were. Zariski met his wife, Yole Cagli, while a student in Rome and they were married on September 11, 1924, in Kobryn.

He received his doctorate in the same year. His thesis ([1],[2]) classified all rational functions $y = P(x)/Q(x)$ of x such that 1) x can be solved for in terms of radicals starting with y, and 2) given any two solutions $x1$ and $x2$, all other solutions x are rational functions of $x1$ and $x2$. Already in his first work he strongly showed his ability to combine algebraic ideas (the Galois group), topological ideas (the fundamental group), and the "synthetic" ideas of classical geometry. The interplay of these different tools was to characterize his life's work.

He pursued these ideas with the support of a Rockefeller fellowship in Rome during the years 1925–1927. His son Raphael was born there on July 18, 1925. In 1927, he accepted a position at Johns Hopkins and in 1928 his family moved to the U.S.A. to join him. Here his daughter Vera was born on September 14, 1932.

A crucial paper in this phase of his career is his analysis [3] of an incomplete proof by Severi that the Jacobian of a generic curve of genus g has no nontrivial endomorphisms. Severi's paper reads as though the proof were complete. Zariski discovered the problem and found a very ingenious argument to remedy it, but neither were well received by Severi who published his own correction independently.

The effect of this discovery seems to have been to turn Zariski's interests to the study of the topology of algebraic varieties, especially of the fundamental group, where the rigor of the techniques was beyond question and the tools were clean and new. He travelled frequently to Princeton to discuss his ideas with Lefschetz. In this phase of his career, roughly from 1927 to 1935, he studied the fundamental group of a variety through the fundamental group of projective n-space minus a divisor. This work is characterized by the spirit of exploration and discovery and, in spite of much recent interest, it remains a largely uncharted area. One result will give the flavor of the new things he turned up: according to another incomplete paper of Severi, it was widely believed that all plane curves of fixed degree with a fixed number of nodes (ordinary double points) belonged to a single algebraic family. What Zariski found was that curves with a fixed degree and a fixed number of *cusps* (the next most complicated type of double point) could belong to several families. He exhibited curves $C1$ and $C2$ of degree 6 with 6 cusps such that the fundamental groups of their complements were not isomorphic!

In 1935, however, Zariski completed his monumental review of the central results of the Italian

D. Mumford, *Selected Papers*, Vol. II,

school, his Ergebnisse monograph *Algebraic Surfaces* [4]. His goal had been to disseminate more widely the ideas and results of their research, but the result for him was "the loss of the geometric paradise in which I so happily had been living"[1]. He saw only too clearly that the lack of rigor he had touched on was not a few isolated sores but a widespread disease. His goal now became the problem of restoring the main body of algebraic geometry to proper health. Algebra had been his early love and algebra was blooming, full of beautiful new ideas in the hands of Noether and Krull, and various applications to algebraic geometry had already been proposed by van der Waerden. Zariski threw himself into this new discipline. He spent the year 1935–1936 at the Institute for Advanced Study in Princeton, and met regularly with Noether, then at Bryn Mawr, learning the new field through first hand contact with the master.

The fifteen years or so that followed, 1938–1951, if you take the years between his paper [5] recasting the theory of plane curve singularities in terms of valuation theory and his monumental treatise [6] on his so-called "holomorphic functions" (sections of sheaves formed from completions of rings in I-adic topologies), saw the most incredible outpouring of original and creative ideas in which tool after tool was taken from the kit of algebra and applied to elucidate basic geometric ideas. Many mathematicians in their forties reap the benefits of their earlier more original work; but Zariski undoubtedly was at his most daring exactly in this decade. He corresponded extensively in this period with Andre Weil, who was also interested in rebuilding algebraic geometry and extending it to characteristic p with a view to its number-theoretic applications. Although they only rarely agreed, they found each other very stimulating, Weil saying later that Zariski was the only algebraic geometer whose work he trusted. They managed to get together in 1945 while both were visiting the University of Sao Paulo in Brazil.

At the same time, these were years of terrible personal tragedy. During the war, all his relatives in Poland were killed by the Nazis. Only his immediate family and those of two of his siblings who had moved to Israel escaped the holocaust. He told the story of how he and Yole were halfway across the U.S., driving back to the East Coast, the day Poland was invaded. They listened each hour to the news broadcasts on their car radio, their only link to the nightmare half a world away. There was nothing they could do.

In this period of his work, Zariski solved many problems with his algebraic ideas. Three themes in his work are particularly beautiful and deep and I want to describe them in some detail.

[1] Preface by Zariski to his Collected Works, MIT Press.

The first theme is the study of birational maps which lead him to the famous result universally known as "Zariski's Main Theorem". This was the final result in a foundational analysis of birational maps between varieties, "maps" which are one to one and onto outside of a finite set of subvarieties of the range and domain, but which "blow up" or "blow down" special points. Zariski showed that if there are points P and Q in the range and domain which are isolated corresponding points, i.e. the set of points corresponding to P contains Q but no curve through Q, and the set of points corresponding to Q contains P but no curve through P, and if, further, P and Q satisfy an algebraic restriction—they are **normal** points—then in fact Q is the only point corresponding to P and vice versa (slightly stronger: the map is **biregular** between P and Q). Zariski's proof of this was astonishingly subtle, yet short.

The second theme from this period is the resolution of singularities of algebraic varieties, which culminated in his proof that all algebraic varieties of dimension at most 3 (in characteristic zero) have "nonsingular models," i.e., are birational to nonsingular projective varieties. In dimension 3, this was a problem that had totally eluded the easy-going Italian approach. Even in dimension 2, although some classical proofs were essentially correct, many of the published treatments definitely were not. Zariski attacked this problem with a whole battery of techniques, pursuing it relentlessly over 6 papers and 200 pages. Perhaps the most striking new tool was the application of the theory of general valuations in function fields to give a birationally invariant way to describe the set of all places which must be desingularized. The result proved to the mathematical world the power of the new ideas. For many years, this work was also considered by everyone in the field to be technically the most difficult proof in all algebraic geometry. Only when the result was proven for surfaces in characteristic p by Abhyankar and later for varieties of arbitrary dimension in characteristic 0 by Hironaka[2] was this benchmark surpassed!

The third theme is his theory of abstract "holomorphic functions." The idea was to use the notion of formal completion of rings with respect to powers of an ideal as a substitute for the idea of convergent power series, and to put elements of the resulting complete rings to some of the same uses as classical holomorphic functions. The most striking application was to a stronger version of the "Main Theorem," known as the connectedness theorem. The connectedness theorem states that if a birational map from X to

[2] S.S. Abhyankar, *Local uniformization on algebraic surfaces of characteristic* $p \neq 0$, Annals of Math., **63** (1956); H. Hironaka, *Resolution of singularities of an algebraic variety of characteristic* 0, Annals of Math., **79** (1964).

Y is single-valued and if a point Q of Y is normal, then the inverse image of Q on X is connected (we are assuming X and Y are complete, e.g. projective). This result was later one of the inspirations for Grothendieck's immense work in rebuilding with yet newer tools the foundations of algebraic geometry[3].

This phenomenal string of papers caught the attention of the mathematical world. Zariski received the Cole Prize from the American Mathematical Society in 1944. In 1945, he moved to a Research Professorship at the University of Illinois. Early in the forties, his work had caught the attention of G.D. Birkhoff who decided he must come to Harvard[4] and, indeed, in 1947 he received and accepted an offer to come to Harvard University, where he remained for the rest of his life. He was a very strong influence on the mathematical environment at Harvard and he enjoyed the opportunity of luring the best people he could to Harvard and bringing out the best in each of his students. While he was chairman, the Dean, McGeorge Bundy, used to refer to him as that "Italian pirate," so shrewd was he in getting his way, inside or outside the usual channels. Whenever Harvard's baroque appointment rules, known as the Graustein Plan (after the earlier mathematician who invented it), jibed with his plans, he used them; but whenever they did not, he feigned ignorance of all that nonsense and insisted the case be considered on its own merits. Over the next thirty years, he made Harvard into the world center of algebraic geometry. His seminar welcomed Weil, Hodge, Nagata, Kodaira, Serre, Grothendieck, and many others. The stimulating evenings at his home and the warm welcome extended by Oscar and Yole were not easily forgotten.

His work of reconstruction of algebraic geometry had started with the writing of the monograph *Algebraic Surfaces*, and now that Zariski felt he had reliable and powerful general tools, it was natural for him to see if he could put all the main results of the theory of surfaces in order. He initiated the modern work on the duality theorems for cohomology (called by him the "lemma of Enriques-Severi" [7], before the topic was taken up by Serre and Grothendieck), the questions of the existence of minimal nonsingular models in each birational equivalence class of varieties [8], and on the classification of varieties following Enriques [9] (now known as the classification by Kodaira dimension). In each of these areas he spread before his students the vision of many possible areas to explore, many exciting prospects.

Although he himself had developed a fully worked out theory of the foundations of algebraic geometry, he welcomed the prospect of yet newer definitions and techniques being introduced because they would make the subject itself stronger. He embraced the new language of sheaf theory and cohomology, working through the basic ideas methodically as was his custom in the Summer Institute in Colorado in 1953 [10], although he never adopted this language as his own. When Grothendieck appeared in the field, he immediately invited him to Harvard. Grothendieck, for his part, welcomed the prospect of working with Zariski. Because Grothendieck's political beliefs did not allow him to swear the oaths of loyalty required in those unfortunate days, he even asked Zariski to investigate the feasibility of continuing his mathematical research from a Cambridge jail cell, i.e., how many books and visitors would be allowed!

The final phase of Zariski's mathematical career was a return to the problems of singularities. Zariski had absolutely no use for the concept of retirement and he dedicated his sixties and seventies and as much of his eighties as he could to a broad-based attack on the problem of "equisingularity". The goal was to find a natural decomposition of an arbitrary variety X into pieces Y_i, each one made up of a subvariety of X from which a finite set of lower dimensional subvarieties have been removed, such that *along* each subvariety Y_i, the big variety X had essentially the same type of singularity at each point. Zariski made major strides towards the achievement of this goal, but the problem has turned out to be quite difficult and is still unsolved.

Zariski's last years were disturbed by his fight with his hearing problem. Zariski was always very lively both in mathematical and in social interactions with his friends and colleagues, picking up every nuance. He was struck with tinnitus, which produced a steady ringing in his ears, a greater sensitivity to noise, and a gradual loss of hearing. This forced him into himself, into his research and kept him close to home. Only the boundless love of his family sustained him in his last years. He died at home on July 4, 1986.

Many honors flowed to Zariski in well-deserved appreciation of the truly extraordinary contribution he had made to the field of algebraic geometry. He received honorary degrees from Holy Cross in 1959, Brandeis in 1965, Purdue in 1974, and from Harvard in 1981. He received the National Medal of Science in 1965, and the Wolf Prize, awarded by the government of Israel, in 1982. His friends, his students, and his colleagues

[3] Grothendieck's style was the opposite of Zariski's. Whereas Zariski's proofs always had a punch-line, a subtle twist in the middle, Grothendieck would not rest until every step looked trivial. In the case of holomorphic functions, Grothendieck liked to claim that the result was so deep for Zariski because he was just proving it for the 0th cohomology group. The easy way, he said, was to prove it first for the top cohomology group, then use descending induction!

[4] The story, which I have from reliable sources, is that Birkhoff approached Zariski and said in his magisterial way: "Oscar, you will probably be at Harvard within the next five years."

will remember not only the beautiful theorems he found, but the forcefulness and the warmth of the man they knew and loved.

REFERENCES

[1] *Sulle equazioni algebriche contenenti linearmente un parametro e risolubili per radicali*, Atti Accad. Naz. Lincei Rend., Cl. Sci. Fis. Mat. Natur., serie V. vol. 33 (1924) pp. 80–82.

[2] *Sopra una classe di equazioni algebriche contenenti linearmente un parametro e risolubili per radicali*, Rend. Circolo Mat. Palermo, vol. 50 (1926) pp. 196–218.

[3] *On a theorem of Severi*, Amer. J. Math., vol. 50 (1928) pp. 87–92.

[4] *Algebraic Surfaces*, Ergebnisse der Mathematik, vol. 3, no. 5., Springer-Verlag, Berlin, 1935, 198 pp.; second supplemented edition, with appendices by S.S. Abhyankar, J. Lipman, and D. Mumford, Ergebnisse der Mathematik, vol. 61, Springer-Verlag, Berlin-Heidelberg-New York, 1971, 270 pp.

[5] *Polynominal ideals defined by infinitely near base points*, Amer. J. Math., vol. 60 (1938) pp. 151-204.

[6] *Theory and applications of holomorphic functions on algebraic varieties over arbitrary ground fields*, Mem. Amer. Math. Soc., no. 5 (1951) pp. 1-90.

[7] *Complete linear systems on normal varieties and a generalization of a lemma of Enriques-Severi*, Ann. of Math., vol. 55 (1952) pp. 552–592.

[8] *Introduction to the problem of minimal models in the theory of algebraic surfaces*, Publ. Math. Soc. Japan, no. 4 (1958) pp. 1–89.

[9] *On Castelnuovo's criterion of rationality $p_a = P_2 = 0$ of an algebraic surface*, Illinois J. Math., vol. 2 (1958) pp. 303–315.

[10] *Algebraic sheaf theory* (Scientific report on the second Summer Institute), Bull. Amer. Math. Soc., vol. 62 (1956) pp. 117–141.

A Foreword for Non-Mathematicians

When I first met Oscar Zariski, I was a lowly and invisible undergraduate, and he was a commanding figure preaching about a seductive world of which he was the master. Later I came to know him as a colleague, and as I gained confidence in my own strength as a mathematician, I could look at his work and see him as a fellow human being, struggling to shape half-glimpsed truths into tangible reality. Through the years he became a close friend and, as he declined physically, a friend in need of support in facing the losses that all people eventually face. It is very exciting for me to see how Carol Parikh has been able to bring to life the full development of Zariski as a person, from his youthful dreams, through his eager days as a student, to the central period in which he doubted his own teachers and found how to correct them and penetrate further into his beloved field of geometry.

I hope that this book will make the mathematical endeavor itself clearer to those readers who have always wondered what on earth mathematicians do. Zariski was a man caught up in many of the central conflicts of the twentieth century. He was torn between his early dedication to communism and his later, more sober, reflections on the success of capitalism. He was torn between an allegiance to an intellectual world that ignored the politics of race and his emotional need to find safety for those members of his family who escaped the Holocaust. Intellectually, he was torn between a love of the free-spirited, creative Italian vision of geometry and his appreciation of the need for strict logical rigor which he found in the Bauhaus-like school of the abstract German algebraists.

Unfortunately, like all working mathematicians, I have led my life with the realization that most of what I care about so passionately is nearly impossible to explain to the educated layman. "What do you mean," they say, "when you say this theorem is *beautiful* or that theorem is *deep*?" One cannot appreciate what drove Zariski and why his colleagues were so excited by his contributions without having some idea of the intellectual world in which he moved. Is it possible within the confines of this foreword to convey some idea of this world and why it is so vital for the dedicated group of mathematicians who pursue it? I won't try to explain all the terms needed to state Zariski's deepest theorems, but I think something of what draws people to his subject can actually be explained in two fairly easy illustrations.

Before I embark, I have to make one thing clear about the way mathematicians think about their world. Everyone knows that physicists are concerned with the laws

D. Mumford, *Selected Papers*, Vol. II,

of the universe and have the audacity sometimes to think they have discovered the choices God made when He created the universe in thus and such a pattern. Mathematicians are even more audacious. What they feel they discover are the laws that God Himself could not avoid having to follow. Now some would say all such laws must be obvious, that you can find nothing truly new beyond what you assumed in the beginning. But this isn't what mathematicians find. They find that by following the thread of logic, just as you would follow a river to its source, at every bend you find things that are totally unexpected. Because these things follow by logic, they have to be true in any world God creates, and yet there is no way in which they are evident on first sight. Or at least so it seems until some mathematician finds a way of rephrasing or recasting the facts; then, by some sleight of hand, they appear immediately evident. That's one of the things mathematicians mean by a beautiful proof. Yet other theorems continue to fascinate mathematicians because they have never been fully reduced to something intuitively obvious. Such theorems live on in a state of tension between seeming new and surprising and seeming clear and evident.

To be a mathematician is to be an out-and-out Platonist. The more you study mathematical constructions, the more you come to believe in their objective and prior existence. Mathematicians view themselves as explorers of a unique sort, explorers who seek to discover not just one accidental world into which they happen to be born, but the universal and unalterable truths of all worlds.

My first illustration will attempt to show in the simplest possible way how algebra and geometry come together in the field Oscar Zariski made his own, algebraic geometry. We want to go back to what was perhaps the first and arguably still one of the deepest mathematical truths—Pythagoras's theorem. We start with a right triangle A, B, C (see Figure 1), with a right angle at B, the side AC being the longest, the so-called hypotenuse. Pythagoras's theorem states that the square of the length AC equals the sum of the square of the length AB and the square of the length BC.

We're not going to prove this theorem; rather, we'll use it to build a fundamental link between algebra and geometry. To do this, we first need to use an idea of Descartes: we can name points in the plane by means of pairs of numbers, called their x and y coordinates. That is to say, to each point, we can assign two numbers,

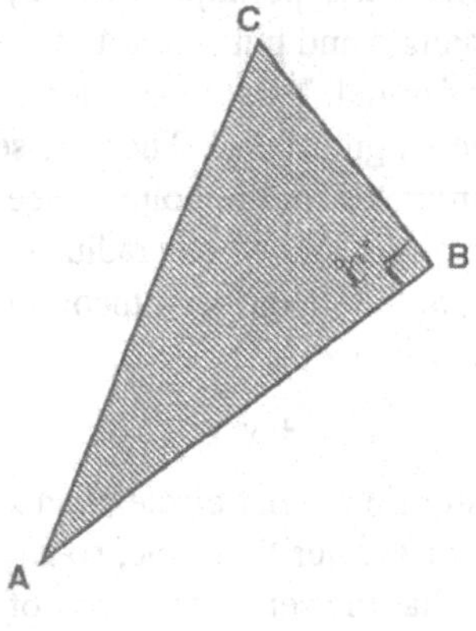

Figure 1. A right triangle

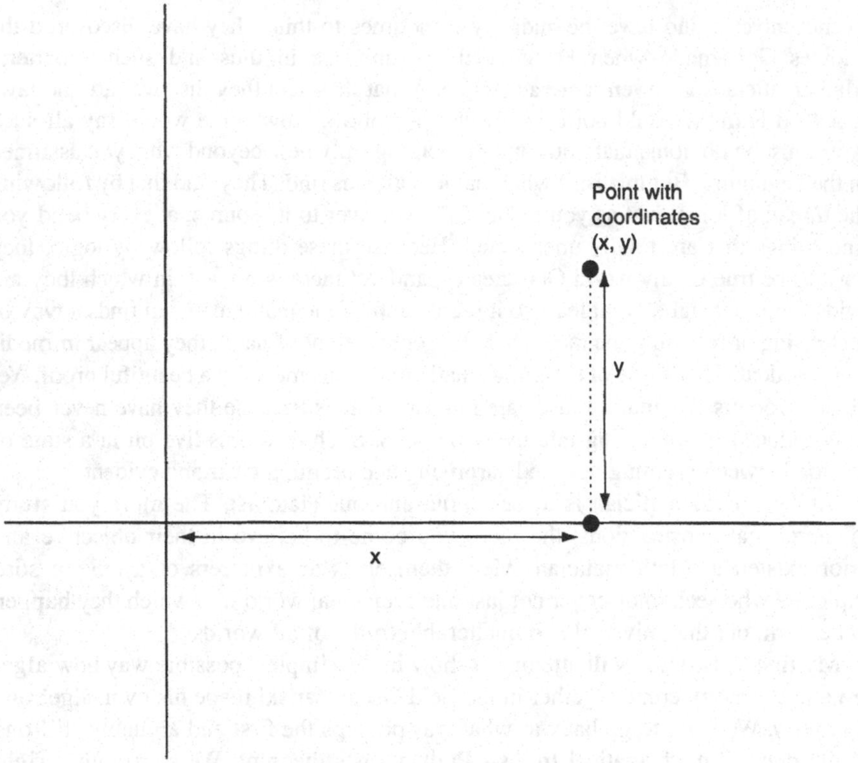

Figure 2. Cartesian coordinates

and conversely to any two numbers, we assign a single point (see Figure 2). This idea, although commonplace to anyone who has taken high school math, was an amazing step for Descartes; it was a step that the Greeks never took. In fact, the Greeks had terrible techniques for doing simple arithmetic, and they would never have thought of the reduction of geometry to arithmetic by means of coordinates as any sort of simplification (which was perhaps why they didn't think of it).

Now take Pythagoras's triangle and put point A at the origin of Descartes' coordinates and make side AB horizontal. This makes side BC vertical. Also let x be the length of AB and let y be the length of BC. Then we see that the pair of numbers x,y is simply Descartes's coordinates for the point C (see Figure 3). Finally, consider the circle whose center is the origin and whose radius is one. If C lies on that circle, then the length of AC is one, and Pythagoras's theorem tells us that the sum of the square of x and the square of y is one:

$$x^2 + y^2 = 1.$$

On the other hand, if C doesn't lie on that circle, then $x^2 + y^2$ is the square of some other number, less than one or greater than one, so $x^2 + y^2$ does not equal one. In other words, we have shown that the set of solutions of the equation

$$x^2 + y^2 = 1$$

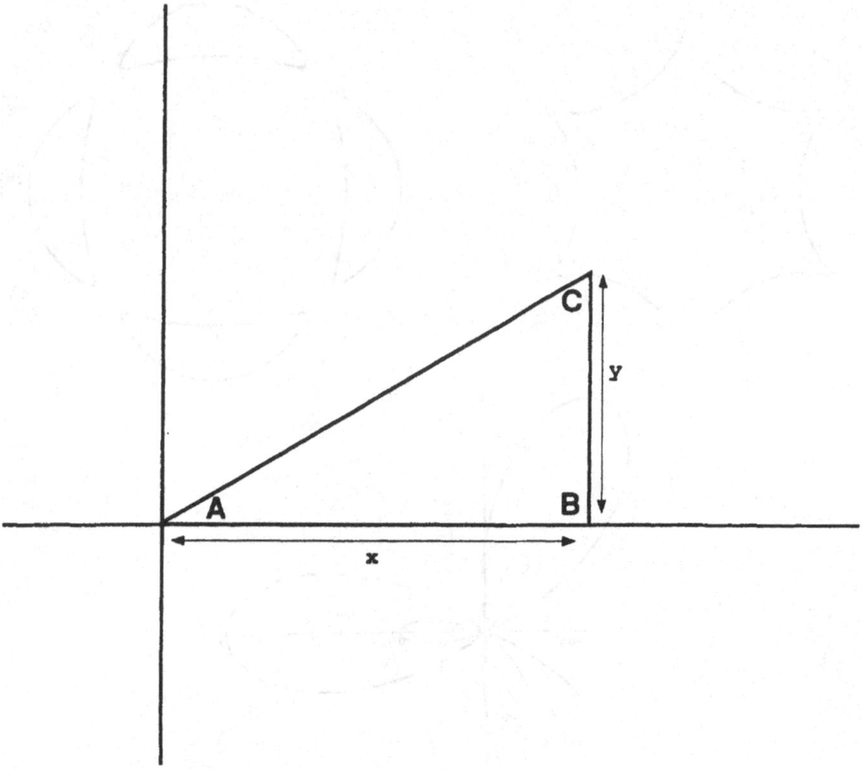

Figure 3. Pythagoras' theorem in Cartesian coordinates

is the same as the set of coordinates (x,y) of the points on our circle! We have an equation, and a simple one at that, for the most basic object of geometry. We have reduced the circle, one of the great building blocks of geometry, to a polynomial $x^2 + y^2$.

This idea, of taking equations of any kind and plotting their set of solutions using Cartesian coordinates, is the secret to the link between algebra and geometry, and the origin of algebraic geometry. What happens with other equations? We can take any equation made up by adding, subtracting, and multiplying x and y and ordinary numbers and out of it get a curve, which is called an algebraic curve. The curve is the set of points whose coordinates x,y solve the equation. In Figure 4, we have drawn three such curves to give you an idea what can happen. Clearly the algebra can produce a whole lot of geometry.

What sort of rules apply to this dictionary between equations and curves? We need some terminology. The equations are built by adding and multiplying the coordinates x,y by various numbers and by each other, and we call x and y the "variables" in the equation because they can be given any value. Some rules are easy: for instance, if the equation is linear (it doesn't multiply variables by each other, but only adds them

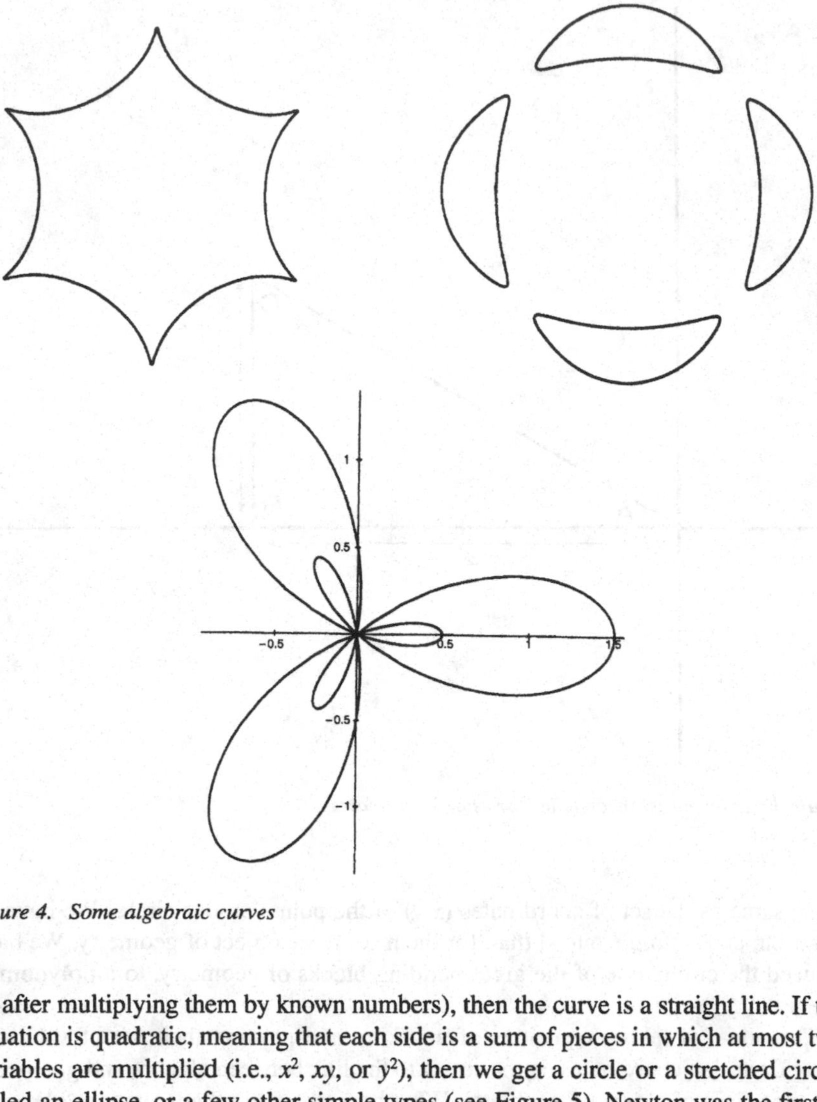

Figure 4. Some algebraic curves

up after multiplying them by known numbers), then the curve is a straight line. If the equation is quadratic, meaning that each side is a sum of pieces in which at most two variables are multiplied (i.e., x^2, xy, or y^2), then we get a circle or a stretched circle, called an ellipse, or a few other simple types (see Figure 5). Newton was the first to make a systematic study and to classify the curves obtained from cubic equations.

Now, here's our second illustration of the way mathematics works. We ask a simple question: If we start with two algebraic curves, is there a rule for predicting how many points they have in common, that is, how large is their intersection? Well, two lines always meet in exactly one point—unless they are parallel, a special case that we shall leave aside for the moment. A line and a circle can meet in two points, or in one point if they are tangent, or in no points if the line doesn't go near the circle at all (see Figure 6). Looks like a mess!

But here we can adopt another strategy that mathematicians love and that often leads to great surprises: if you find a mess in the world you start in, why not change

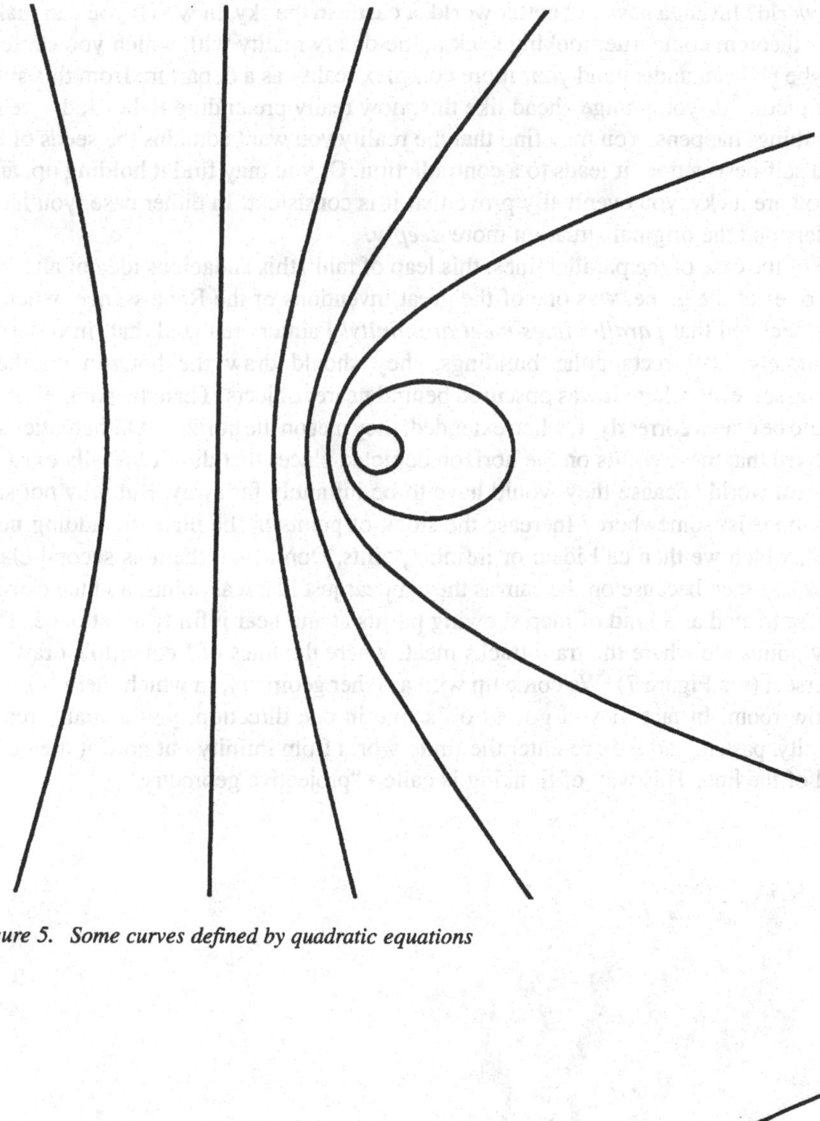

Figure 5. Some curves defined by quadratic equations

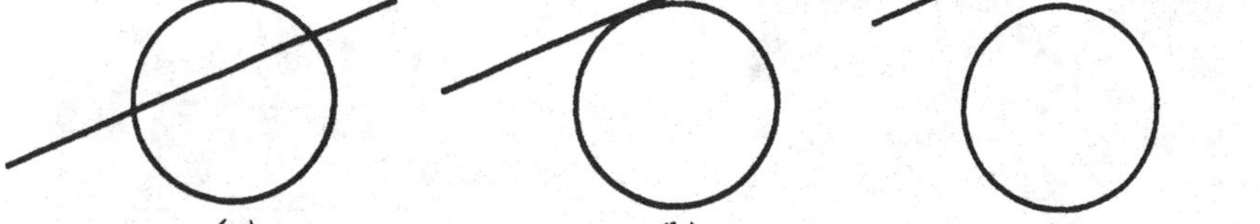

Figure 6. (a) A line and a circle that meet twice; (b) A line and a circle that meet once and are tangent; (c) A line and a circle that never meet

the world? Invent a new and better world, a castle in the sky, in which you can make your theorem come true; looking back at the dreary reality with which you started, maybe you can understand your more complex reality as a departure from this simpler picture. If you plunge ahead like this, now really pretending to be God, one of two things happens. You may find that the reality you want contains the seeds of its own self-destruction: it leads to a contradiction. Or you may find it holding up, and if you are lucky, you eventually prove that it is consistent. In either case, you have understood the original situation more *deeply*.

For the case of the parallel lines, this leap of faith, this audacious idea of altering the rules of the game, was one of the great inventions of the Renaissance, when it was declared that *parallel lines meet at infinity!* Painters realized that, in order to accurately draft rectangular buildings, they should draw the horizon on their canvasses, even where it was obscured behind nearer objects. Then the parallel lines would be drawn correctly if, when extended, they met on the horizon. Mathematicians realized that these points on the horizon depicted places that didn't literally exist in the real world because they would have to be infinitely far away. But why not say they do exist somewhere? Increase the stock of points in the plane by adding new ones, which we then call ideal or infinite points. Don't treat them as second-class citizens either, because on the canvas they appear just like real points, and the canvas can be treated as a kind of map showing points at and near infinity all at once. The new points are where the train tracks meet, where the lines of Leonardo's drawing intersect (see Figure 7). We come up with a richer geometry, in which there is more elbow room. In fact, if you go out on a line in one direction, you actually reach infinity, pass it, and then re-enter the finite world from infinity but now at the other end of the line. This way of thinking is called "projective geometry."

Figure 7. Leonardo da Vinci: Perspective study

Now how about the circle and the line? Ignore for a while the case where the circle is tangent to the line, as it is a special case. The two basic cases are where they meet twice and where they don't meet at all. We don't want to lose any points, so we are forced to add points again until a line totally outside a circle still "meets" it somewhere. Here is where some old ideas that originated in the Middle Ages come to our help: the square root of -1, called i, and the complex numbers built up from it, e.g., $2 + 3i$ or $-4.5 - 5i$. It had been known for a long time that solving polynomial equations seemed to work better if you allowed complex numbers in, either as the solutions themselves or, even if you only wanted the usual real roots, as intermediate steps in calculating the solutions. Such numbers had had an air of mystery and black magic about them at first, but gradually it was realized that there was nothing inconsistent about them; if you suspended your disbelief and admitted them for the sake of the game, you didn't reach any contradiction. A beautiful way of describing them by points in the plane, due to C. F. Gauss, the founder of the modern era of mathematics, made it totally clear that they were a perfectly consistent rigorous construction.

So where are the missing points, for example, where the line $x = 1.25$ and the circle $x^2 + y^2 = 1$ meet? One of them is the point $x = 1.25$, $y = 0.75i$, and the other is $x = 1.25$, $y = -0.75i$. (Just square 1.25 getting 1.5625, and square $.75i$ getting $-.5625$, which add up to 1.0.) With a little algebra, it's easy to see that this always works, so long as we let the coordinates x,y of the points in the plane be complex numbers. But what has this technique done to our geometry? In fact, it has made it much richer. Although we continue to treat it like a two-dimensional world, to specify a point requires two coordinates, and each of them, being complex, needs to have a real and an imaginary part (thus $2 + 3i$ has real part 2 and imaginary part 3). This means that we need in all four numbers of the ordinary sort to specify a point, so our geometry has now become four-dimensional. Moreover, we still have to add the line of points at infinity, including complex points at infinity. For instance, a circle, which in the ordinary sense doesn't go out to infinity at all, now can do so, provided the direction in which it goes has imaginary slope (the points at infinity on circles used to be called I and J and were nicknamed Isaac and Jacob by students in the college days of my colleague Lars Ahlfors). The whole affair is called the *complex projective plane* and is the place in which to "draw" algebraic curves and to do algebraic geometry.

To complete our story, what have we gained by these mental gymnastics? In fact, we have gained a tremendous amount, but to tell the story is to tell a large part of algebraic geometry. For this foreword, I'll only tell about Bezout's theorem—actually a theorem of Poncelet, I believe, but mathematicians are notorious for crediting things rather arbitrarily. Remember that any polynomial equation in x and y defines its curve of solutions. The degree of the equation is simply the largest number of times the variables are ever multiplied together (so 2 is the degree of $x^2 + y^2 = 1$, and 5 is the degree of $x^3 \bullet y^2 = -1$). Bezout's theorem states that two such curves, of degrees n and m, meet almost always in $n \bullet m$ points, and always in $n \bullet m$ points if special points of intersection, like a point where a line is tangent to a circle, are counted more than once in a careful way. (Finding techniques for counting these special points was, by the way, one of the principle technical accomplishments of Zariski's archrival Weil, *see* Ch. 12.) In other words, we have found a strong general link between the algebra of the polynomials on the one hand and the geometry of the curves on the other. Such links, many quite amazing on first sight, are the main concerns of algebraic geometry.

I want to touch on one more thing in this quick tour of the mathematician's world. The lay picture of the mathematician (as seen in *New Yorker* cartoons) shows a bespectacled, white-coated, rather unworldly man looking at a blackboard of bizarre equations. This man is probably dry and precise, following rules without fail; his failing to do so is cause for humor (see Figure 8). As discussed below, much of Zariski's life was devoted to seeking the right way to make precise a huge

Figure 8. © 1975 by Sidney Harris, American Scientist *magazine*

amount of writing and thinking produced by other mathematicians who were anything but precise. In fact, one of them was an out-and-out romantic and another a dictatorial dramatic man with a flair for wild driving. Let the truth be known: mathematicians are as subject to human error and emotion, as subject to the fashions of intellectual trends, and as often personifications of their national characteristics, as thinkers in any other field. They do strive, or claim, to be better and more detached, but their history reveals marvelous episodes in which they have driven right off the road in pursuit of their particular vision of truth.

This book deals with one of the most colorful episodes of this type. The Italian school of algebraic geometry was created in the late 19th century by a half dozen geniuses who were hugely gifted and who thought deeply and nearly always correctly about their field. They extended its ideas over a huge new area, especially what is called the theory of algebraic surfaces (we were discussing algebraic curves; surfaces come from equations in three variables, x, y, and z, instead of two). But they found the geometric ideas much more seductive than the formal details of proofs, especially when these proofs had to cover all the nasty special cases that so often crop up in geometry. So, in the twenties and thirties, they began to go astray. It was Zariski and, at about the same time, Weil who set about to tame their intuition, to find the principles and techniques that could truly express the geometry while embodying the rigor without which mathematics eventually must degenerate to fantasy.

The 20th century was, until its final decades, an era of "modern mathematics" in a sense quite parallel to "modern art" or "modern architecture" or "modern music." That is to say, it turned to an analysis of abstraction, it glorified purity and tried to simplify its results until the roots of each idea were manifest. These trends started in the work of Hilbert in Germany, were greatly extended in France by a secret mathematical club known as "Bourbaki," and found fertile soil in Texas, in the topological school of R. L. Moore. Eventually, they conquered essentially the entire world of mathematics, even trying to breach the walls of high school in the disastrous episode of the "new math." Now the trend has reversed: postmodern mathematics is quite different and has reintroduced the love of the baroque; it embraces the tool of the computer and seeks out rather than shunning the complexities of applications. The theory of chaos is the best-known example of this trend, but it extends from the vast number–theoretic speculations on modular forms to the paradoxically flat yet knotted "non-standard" four-dimensional spaces. Zariski's life is the story of a mathematician of this century, who lived with and loved and gave his soul to these struggles. He began his career with naive beliefs inherited from the nineteenth century; the middle part of his career was wholly devoted to "modern mathematics"; and in the last part, he began to look again at the richness and complexities of his material. But this is the story Carol Parikh has told so ably in the book that follows.

David Mumford

What Can Be Computed in Algebraic Geometry?

Dave Bayer * David Mumford

May 25, 1993

This paper evolved from a long series of discussions between the two authors, going back to around 1980, on the problems of making effective computations in algebraic geometry, and it took more definite shape in a survey talk given by the second author at a conference on Computer Algebra in 1984. The goal at that time was to bring together the perspectives of theoretical computer scientists and of working algebraic geometers, while laying out what we considered to be the main computational problems and bounds on their complexity. Only part of the talk was written down and since that time there has been a good deal of progress. However, the material that was written up may still serve as a useful introduction to some of the ideas and estimates used in this field (at least the editors of this volume think so), even though most of the results included here are either published elsewhere, or exist as "folk-theorems" by now.

The article has four sections. The first two parts are concerned with the theory of Gröbner bases; their construction provides the foundation for most computations, and their complexity dominates the complexity of most techniques in this area. The first part introduces Gröbner bases from a geometric point of view, relating them to a number of ideas which we take up in more detail in subsequent sections. The second part develops the theory of Gröbner bases more carefully, from an algebraic point of view. It could be read independently, and requires less background. The third part is an investigation into bounds in algebraic geometry of relevance to these computations. We focus on the *regularity* of an algebraic variety (see Definition 3.2), which, beyond its intrinsic interest to algebraic geometers, has emerged as a measure of the complexity of computing Gröbner bases (see [BS87a], [BS87b], [BS88]). A principal result in this part is a bound on the regularity of any smooth variety by the second author: Theorem 3.12(b). This bound has stimulated subsequent work, and has now been generalized

*Partially supported by NSF grant DMS-90-06116.

D. Mumford, *Selected Papers*, Vol. II,

by [BEL91]. Another result of interest is Proposition 3.13, which elucidates the scheme structure of the ideal membership problem. The fourth part is a short discussion of work on algorithms for performing some other key operations on varieties, some open problems about these operations and some general ideas about what works and what doesn't, reflecting the prejudices of the authors.

One of the difficulties in surveying this area of research is that mathematicians from so many specialties have gotten involved, and they tend both to publish in their own specialized journals and to have specific agendas corresponding to their area. Thus one group of researchers, the working algebraic geometers, are much more interested in actually computing examples than in worst-case complexity bounds. This group, including the first author, has put a great deal of work into building a functioning system, *Macaulay*, based on Gröbner bases, which has solved many problems and provided many examples to the algebraic geometry community [BS92a]. Another group comes from theoretical computer science and is much more interested in theoretical bounds than practical systems (cf. the provocative comments in Lenstra's survey [Len92]). It seems to us that more communication would be very helpful: On the one hand, the working algebraic geometer knows lots of facts about varieties that can be very relevant to finding fast algorithms. Conversely asymptotic and/or worst-case performance bounds are sometimes, at least, important indicators of real-time performance. These theoretical bounds may also reveal important distinctions between classes of procedures, and may pose new and deep problems in algebraic geometry. Thus we will see in Section 3 how regularity estimates flesh out a picture explaining why Gröbner basis computations can have such explosive worst case behavior, yet be so useful for the kinds of problems typically posed by mathematicians. Finally, to make this article more useful in bridging this gap, we have tried to include a substantial number of references in our discussions below.

1 A Geometric Introduction

Let X be a subvariety or a subscheme of projective n-space $\mathbf{P}^n$, over a field k. Let $\mathcal{F}$ be a vector bundle or a coherent sheaf supported on X. We would like to be able to manipulate such objects by computer. From algebra we get finite descriptions, amenable to such manipulations: Let $S = k[x_0, \ldots, x_n]$ be the homogeneous coordinate ring of $\mathbf{P}^n$. Then X can be taken to be the subscheme defined by a homogenous ideal $I \subset S$, and $\mathcal{F}$ can be taken to be the sheaf associated to a finitely generated S-module M. We can represent I by a list of generators $(f_1, \ldots, f_r)$, and M by a presentation matrix F, where

$$M_1 \xrightarrow{F} M_0 \longrightarrow M \longrightarrow 0$$

presents M as a quotient of finitely generated free S-modules M_0, M_1. We concentrate on the case of an ideal I; by working with the submodule $J = \mathrm{Im}(F) \subset M_0$, the module case follows similarly.

The heart of most computations in this setting is a deformation of the input data to simpler data, combinatorial in nature: We want to move through a family of linear transformations of $\mathbf{P}^n$ so that in the limit our objects are described by monomials. Via this family, we hope to pull back as much information as possible to the original objects of study.

Choose a one-parameter subgroup $\lambda(t) \subset GL(n+1)$ of the diagonal form

$$\lambda(t) = \begin{bmatrix} t^{w_0} & & & \\ & t^{w_1} & & \\ & & \cdots & \\ & & & t^{w_n} \end{bmatrix},$$

where $W = (w_0, \ldots, w_n)$ is a vector of integer weights. For each $t \neq 0$, $\lambda(t)$ acts on X via a linear change of coordinates of $\mathbf{P}^n$, to yield the subscheme $X_t = \lambda(t)X \cong X$. The limit

$$X_0 = \lim_{t \to 0} X_t$$

is usually a simpler object, preferable to X for many computational purposes.

Even if we start out by restricting X to be a subvariety rather than a subscheme of $\mathbf{P}^n$, it does not suffice to take the limit X_0 set-theoretically; often all we will get pointwise in the limit is a linear subspace $L \subset \mathbf{P}^n$, reflecting little besides the dimension of the original variety X. By instead allowing this limit to acquire embedded components and a nonreduced structure, we can obtain an X_0 which reflects much more closely the character of X itself.

We compute explicitly with the generators $f_1, \ldots, f_r$ of I: Let λ act on S by mapping each x_i to $t^{w_i}x_i$; λ maps each monomial $\mathbf{x}^A = x_0^{a_0} \cdots x_n^{a_n}$ to $t^{W \cdot A}\mathbf{x}^A = t^{w_0a_0 + \ldots + w_na_n} x_0^{a_0} \cdots x_n^{a_n}$. If $f = a\mathbf{x}^A + b\mathbf{x}^B + \ldots$, then $\lambda f = a\, t^{W \cdot A}\mathbf{x}^A + b\, t^{W \cdot B}\mathbf{x}^B + \ldots$. We take the projective limit $\mathrm{in}(f) = \lim_{t \to 0} \lambda f$ by collecting the terms of λf involving the least power of t; $\mathrm{in}(f)$ is then the sum of the terms $a\mathbf{x}^A$ of f so $W \cdot A$ is minimal. For a given f and most choices of λ, $\mathrm{in}(f)$ consists of a single term.

The limit X_0 we want is defined with all its scheme structure by the ideal $\mathrm{in}(I) = \lim_{t \to 0} \lambda I$, generated by the set $\{\, \mathrm{in}(f) \mid f \in I \,\}$. For a given I and most choices of λ, $\mathrm{in}(I)$ is generated by monomials. Unfortunately, this definition is computationally unworkable because I is an infinite set, and in(I) need not equal (in(f_1), ..., in(f_r)) for a given set of generators $f_1, \dots, f_r$ of I. To understand how to compute in(I), we need to look more closely at the family of schemes X_t defined by λ.

Let $S[t]$ be the polynomial ring $k[x_0, \dots, x_n, t]$; we view $S[t]$ as the coordinate ring of a one-parameter family of projective spaces $\mathbf{P}^n_t$ over the affine line with parameter t. For each generator f_j of I, rescale λf_j so the lowest power of t has exponent zero: Let $g_j = t^{-\ell}\lambda f_j$, where $\ell = W \cdot A$ is the least exponent of t in λf_j. Then $f_j = g_j|_{t=1}$ and in$(f_j) = g_j|_{t=0}$. Now, let $J \subset S[t]$ be the ideal generated by $(g_1, \dots, g_r)$; J defines a family Y over $\mathbf{A}^1$ whose central fiber is cut out by (in(f_1), ..., in(f_r)).

What is wrong with the family Y? Y can have extra components over $t = 0$, which bear no relation to its limiting behavior as $t \to 0$. Just as the set-theoretic limit $\lim_{t \to 0} X_t$ can be too small (we need the nonreduced structure), this algebraically defined limit can be too big; the natural limit lies somewhere in between.

The notion of a *flat* family captures exactly what we are looking for here. For example, if Y is flat, then there are no extra components over $t = 0$. While the various technical definitions of flatness can look daunting to the newcomer, intuitively flatness captures exactly the idea that every fiber of a family is the natural scheme-theoretic continuation of its neighboring fibers.

In our setting, all the X_t are isomorphic for $t \neq 0$, so we only need to consider flatness in a neighborhood of $t = 0$. Artin [Art76] gives a criterion for flatness applicable here: The *syzygies* of $g_1, \dots, g_r$ are the relations $h_1 g_1 + \dots + h_r g_r = 0$ for $h_1, \dots, h_r \in S[t]$. Syzygies correspond to elements $(h_1, \dots, h_r)$ of the $S[t]$-module $S[t]^r$; the set of all syzygies is a submodule of $S[t]^r$. Y is a flat family at $t = 0$ if and only if the restrictions $(h_1|_{t=0}, \dots, h_r|_{t=0})$ of these syzygies to the central fiber generate the S-module of syzygies of $g_1|_{t=0}, \dots, g_r|_{t=0}$.

When $g_1|_{t=0}, \dots, g_r|_{t=0}$ are single terms, their syzygies take on a very simple form: The module of syzygies of two terms $a\mathbf{x}^A$, $b\mathbf{x}^B$ is generated by the syzygy $b\mathbf{x}^C(a\mathbf{x}^A) - a\mathbf{x}^D(b\mathbf{x}^B) = 0$, where $\mathbf{x}^E = \mathbf{x}^C\mathbf{x}^A = \mathbf{x}^D\mathbf{x}^B$ is the least common multiple of $\mathbf{x}^A$ and $\mathbf{x}^B$. The module of syzygies of r such terms is generated (usually not minimally) by the syzygies on all such pairs.

We want to lift these syzygies to syzygies of $g_1, \dots, g_r$, working modulo increasing powers of t until each syzygy lifts completely. Whenever we get stuck, we will find ourselves staring at a new polynomial g_{r+1} so $t^\ell g_{r+1} \in J$ for some $\ell > 0$. Including g_{r+1} in the definition of a new $J' \supset J$ has no effect on the family defined away from $t = 0$, but will cut away unwanted portions of the central fiber; what we are doing is removing t-torsion. By iterating this process until every syzygy lifts, we obtain explicit generators $g_1, \dots, g_r, g_{r+1}, \dots, g_s$ for a flat family describing the degeneration of $X = X_1$ to a good central fiber X_0. The corresponding generators $g_1|_{t=1}, \ldots, g_s|_{t=1}$ of I are known as a *Gröbner basis* for I.

This process is best illustrated by an example. Let $S = k[w, x, y, z]$ be the coordinate ring of $\mathbf{P}^3$, and let $I = (f_1, f_2, f_3) \subset S$ for

$$f_1 = w^2 - xy,\ f_2 = wy - xz,\ f_3 = wz - y^2.$$

I defines a twisted cubic curve $X \subset \mathbf{P}^3$; X is the image of the map $(r, s) \mapsto (r^2 s, r^3, rs^2, s^3)$. Let

$$\lambda(t) = \begin{bmatrix} t^{-16} & & & \\ & t^{-4} & & \\ & & t^{-1} & \\ & & & t^0 \end{bmatrix}.$$

If $w^a x^b y^c z^d$ is a monomial of degree < 4, then $\lambda \cdot w^a x^b y^c z^d = t^{-\ell} w^a x^b y^c z^d$ where $\ell = 16a + 4b + c$. Thus, sorting the monomials of S of each degree < 4 by increasing powers of t with respect to the action of λ is equivalent to sorting the monomials of each degree in lexicographic order.

We have

$$\begin{aligned} g_1 &= t^{32}\lambda f_1 &= w^2 - t^{27}xy, \\ g_2 &= t^{17}\lambda f_2 &= wy - t^{13}xz, \\ g_3 &= t^{16}\lambda f_3 &= wz - t^{14}y^2. \end{aligned}$$

The module of syzygies on w^2, wy, wz is generated by the three possible pairwise syzygies; we start with the syzygy $y(w^2) - w(wy) = 0$. Substituting g_1, g_2 for the lead terms w^2, wy we get

$$y(w^2 - t^{27}xy) - w(wy - t^{13}xz) = t^{13}wxz - t^{27}xy^2$$

which is a multiple $t^{13}x$ of g_3. Thus, the syzygy

$$yg_1 - wg_2 - t^{13}xg_3 = 0$$

of g_1, g_2, g_3 restricts to the monomial syzygy $y(w^2) - w(wy) = 0$ when we substitute $t = 0$, as desired.

Similarly, the syzygy

$$zg_1 - t^{14}yg_2 - wg_3 = 0$$

restricts to the monomial syzygy $z(w^2) - w(wz) = 0$. When we attempt to lift $z(wy) - y(wz) = 0$, however, we find that

$$z(wy - t^{13}xz) - y(wz - t^{14}y^2) = -t^{13}xz^2 + t^{14}y^3.$$

xz^2 is not a multiple of w^2, wy, or wz, so we cannot continue; $J = (g_1, g_2, g_3)$ does not define a flat family. Setting $t = 1$, the troublesome remainder is $-xz^2 + y^3$. Making this monic, let $f_4 = xz^2 - y^3$; $f_4 \in I$ and

$$g_4 = t^4 \lambda f_4 = xz^2 - ty^3.$$

Adjoin g_4 to the ideal J, redefining the family Y. Now,

$$zg_2 - yg_3 + t^{13} g_4 = 0$$

restricts to $z(wy) - y(wz) = 0$ as desired.

The module of syzygies of w^2, wy, wz, and xz^2 is generated by the pairwise syzygies we have already considered, and by the syzygy $xz(wz) - w(xz^2) = 0$, which is the restriction of

$$-ty^2 g_2 + xzg_3 - wg_4 = 0.$$

Thus, $J = (g_1, g_2, g_3, g_4)$ defines a flat family Y, and

$$w^2 - xy,\ wy - xz,\ wz - y^2,\ xz^2 - y^3$$

is a Gröbner basis for I. The limit X_0 is cut out by the monomial ideal $\text{in}(I) = (w^2, wy, wz, xz^2)$, which we shall see shares many properties with the original ideal I. Note that $xz^2 - y^3 = 0$ defines the projection of X to the plane $\mathbf{P}^2$ in x, y, and z.

The scheme structure of X_0 is closely related to the combinatorial structure of the monomial k-basis for $S/\text{in}(I)$: For each degree d in our example, the monomials not belonging to $\text{in}(I)$ consist of three sets $\{x^d, x^{d-1}y, \ldots, y^d\}$, $\{x^{d-1}z, x^{d-2}yz, \ldots, y^{d-1}z\}$, $\{y^d, y^{d-1}z, \ldots, z^d\}$, and a lone extra monomial $x^{d-1}w$. The first two sets correspond to a double line supported on $w = z = 0$, the third set to the line $w = x = 0$, and the extra monomial to an embedded point supported at $w = y = z = 0$. Together, this describes the scheme structure of X_0. The first two sets consist of $d+1$ and d monomials, respectively; the third set adds $d-1$ new monomials, and overlaps two monomials we have already seen. With the extra monomial, we count $3d+1$ monomials in each degree, which agrees with the dimensions of the graded pieces of S/I. The embedded point is crucial; it makes this count come out right, and it alone keeps X_0 nonplanar like X.

The new monomial generator xz^2 of $\text{in}(I)$ excludes the line $w = y = 0$ from X_0; combinatorially, it excludes all but three monomials of the set $\{x^d, x^{d-1}z, \ldots, z^d\}$ from the monomial k-basis for each degree of the quotient $S/\text{in}(I)$. We can see that this line is unwanted as follows: Away from $t = 0$, Y is parametrized by $(r, s, t) \mapsto (t^{16}r^2s, t^4r^3, trs^2, s^3, t)$. Thus, fixing r and s, the curve $(r, ts, t) \mapsto (t^{17}r^2s, t^4r^3, t^3rs^2, t^3s^3, t)$, with projective limit $(0, 0, r, s, 0)$ as $t \to 0$. Similarly, the curve (r, t^3s, t^2) has as its limit $(0, r^2, s^2, 0, 0)$. These calculations show that the lines $w = z = 0$ and $w = x = 0$ indeed belong set-theoretically to the limit X_0. We can find no such curve whose limit is a general point on the line $w = y = 0$, for $(r, t^4 s, t^3)$ doesn't work. Thus, the line $w = y = 0$ sticks out of the good total space Y.

One usually computes Gröbner bases by working directly in the ring S, dispensing with the parameter t. The one-parameter subgroup λ is replaced by a total order on the monomials of each degree, satisfyhe *multiplicative* property $\mathbf{x}^A > \mathbf{x}^B \Rightarrow \mathbf{x}^C\mathbf{x}^A > \mathbf{x}^C\mathbf{x}^B$ for all $\mathbf{x}^C$. In fact, for our purposes these are equivalent concepts: The weight vector W associated with λ induces the order $\mathbf{x}^A > \mathbf{x}^B \iff W\cdot A < W\cdot B$, which is a total multiplicative order in low degrees as long as no two monomials have the same weight. Conversely, given any multiplicative order and a degree bound d, one can find many λ which induce this order on all monomials of degree $< d$. See [Bay82], [Rob85] for characterizations of such orders.

We shall be particularly interested in two multiplicative orders, the *lexicographic* order used in our example, and the *reverse lexicographic* order. The lexicographic order simply expands out the monomials of each degree into words, and sorts them alphabetically, i.e. $\mathbf{x}^A > \mathbf{x}^B$ iff the first nonzero entry in $A - B$ is positive. The reverse lexicographic order pushes highest powers of x_n in any expression back to the end, then within these groups pushes highest powers of x_{n-1} to the end, etc., i.e. $\mathbf{x}^A > \mathbf{x}^B$ iff the last nonzero entry of $A - B$ is negative.

What do these orders mean geometrically? The dominant effect of the lexicographic order is a projection from $\mathbf{P}^n$ to $\mathbf{P}^{n-1}$, eliminating x_0. A second order effect is a projection to $\mathbf{P}^{n-2}$, and so forth. We could compute the deformation from X to X_0 with respect to the lexicographic order in stages carrying out these projections, first applying a λ with $W = (-1, 0, \ldots, 0)$, then with $W = (-1, -1, 0, \ldots, 0)$, etc. Alternatively, for monomials of each degree $< d$, we can apply the single λ with $W = (-d^{n-1}, \ldots, -d, -1, 0)$, generalizing the λ used in our example. Use of the lexicographic order tends to muck up the family Y more than necessary in most applications, because projections tend to complicate varieties.

For the reverse lexicographic order, the dominant effect is a projection of $\mathbf{P}^n$ down to the last coordinate point $(0, \ldots, 0, 1)$. As a secondary effect, this order projects down to the last coordinate line, and so forth. In other words, this order first tries to make X into a cone over the last coordinate point, and only then tries to squash the result down to or cone it over the last coordinate line, etc. For monomials of each degree $< d$, this can be realized by applying λ with $W = (0, 1, d, \ldots, d^{n-1})$. Like such cones, the reverse lexicographic order enjoys special properties with respect to taking linear sections of X or X_0 by intersection with the spaces defined by the last variable(s) (see [BS87a]). The preferred status of the reverse lexicographic order can be attributed to this relationship, because generic linear sections do not complicate varieties.

For example, if we take X to be three general points in $\mathbf{P}^2$, then using the lexicographic order X_0 becomes a triple point on a line, because the first order effect is the projection of the three points to a line, and the second order limiting process keeps the points within this line. By contrast, if we use the reverse lexicographic order then X_0 becomes the complete first order neighborhood of a point (a point doubled in all directions). This is because the first order limiting process brings the three points together from distinct directions, tracing out a cone over the three points. The first order neighborhood of the vertex in this cone has multiplicity 3, and is the same as the complete first order neighborhood in the plane of this vertex.

For those familiar with the theory of valuations in birational geometry [ZS76, Vol. II, Ch. VI], the lexicographic and reverse lexicographic orders have simple interpretations. Recall that if X is a variety of dimension n, and

$$F : X = Z_0 \supset Z_1 \supset Z_2 \supset \ldots \supset Z_n$$

is a flag of subvarieties, $\operatorname{codim}_X(Z_i) = i$, with Z_i smooth at the generic point of Z_{i+1}, then we can define a rank n valuation v_F on X as follows: For each $i = 2, \ldots, n$, fix f_i to be a function on Z_{i-1} with a 1st order zero on Z_i. Then for any function f, we can define $e_1 = \operatorname{ord}_{Z_1}(f)$, $e_2 = \operatorname{ord}_{Z_2}((f/f_2^{e_1})\,|_{Z_1})$, etc., and $v_F(f) = (e_1, \ldots, e_n) \in \mathbf{Z}^n$, where the value group $\mathbf{Z}^n$ is ordered lexicographically. The arbitrarily chosen f_i are not needed to compare two functions f, g: We have $v_F(f) \succ v_F(g)$ if and only if $\operatorname{ord}_{Z_1}(f/g) > 0$, or if this order is zero and $\operatorname{ord}_{Z_2}((f/g)\,|_{Z_1}) > 0$, and so forth. Such a valuation also defines an order on each graded piece S_d of the homogeneous coordinate ring: take any $f_0 \in S_d$ and say $f > g$ if and only if $v_F(f/f_0) \succ v_F(g/f_0)$. More generally, one may take the Z_i to be subvarieties of a variety X' dominating X and pull back functions to X' before computing v_F.

The lexicographic order on monomials of each degree of $\mathbf{P}^n$ is now induced by the flag

$$\mathbf{P}^n \supset V(x_0) \supset V(x_0, x_1) \supset \ldots \supset V(x_0, \ldots, x_{n-1}).$$

For example, the first step in the comparison defining $v_F(\mathbf{x}^A/f_0) \succ v_F(\mathbf{x}^B/f_0)$ has the effect of asking if $a_0 - b_0 > 0$.

The reverse lexicographic order is induced by a flag on a blowup X of $\mathbf{P}^n$: First blow up $V(x_0, \ldots, x_{n-1})$ and let E_1 be the exceptional divisor. Next blow up the proper transform of $V(x_0, \ldots, x_{n-2})$, and let E_2 be this exceptional divisor. Iterating, we can define a flag

$$X \supset E_1 \supset E_1 \cap E_2 \supset \ldots \supset E_1 \cap \ldots \cap E_n$$

which induces the reverse lexicographic order on monomials in each degree. For example, looking at the affine piece of the first blow up obtained by substituting $x_0 = x_0' x_{n-1}, \ldots, x_{n-2} = x_{n-2}' x_{n-1}$, the power of x_{n-1} in the transform of $\mathbf{x}^A$ is $a_0 + \ldots + a_{n-1}$, which is the order of vanishing of this monomial on E_1. Thus, the first step in the comparison defining $v_F(\mathbf{x}^A/f_0) \succ v_F(\mathbf{x}^B/f_0)$ has the effect of asking if $a_0 + \ldots + a_{n-1} - b_0 - \ldots - b_{n-1} > 0$, which is what we want.

Taking into account the equivalence between multiplicative orders and one-parameter subgroups, the process we have described in $S[t]$ is exactly the usual algorithm for computing Gröbner bases. It is computationally advantageous to set $t = 1$ and dismiss our extra structure as unnecessary scaffolding, but it is conceptually advantageous to treat our viewpoint as what is "really" going on; many techniques of algebraic geometry become applicable to the family Y, and assist in analyzing the complexity of Gröbner bases. Moreover, this picture may help guide improvements to the basic algorithm. For example, for very large problems, it could be computationally more efficient to degenerate to X_0 in several stages; this has not been tried in practice.

The coarsest measure of the complexity of a Gröbner basis is its maximum degree, which is the highest degree of a generator of the ideal $\text{in}(I)$ defining X_0. This quantity is bounded by the better-behaved ***regularity*** of $\text{in}(I)$: The regularity of an ideal I is the maximum over all i of the degree minus i of any minimal i^{th} syzygy of I, treating generators as 0th syzygies. When I is the largest (the ***saturated***) ideal defining a scheme X, we call this the regularity of X. We take up regularity in detail in Section 3; here it suffices to know that regularity is ***upper semi-continuous*** on flat families, i.e. the regularity can only stay the same or go up at special fibers.

Let $\text{reg}(I)$ denote the regularity of I, and $\text{reg}_0(I)$ denote the highest degree of a generator of I. In our case, $t = 0$ is the only special fiber, and the above says that

$$\text{reg}_0(I) \leq \text{reg}(I) \leq \text{reg}(\text{in}(I)) \geq \text{reg}_0(\text{in}(I)),$$

where $\text{reg}_0(I)$ can be immediately determined from the input data, and $\text{reg}_0(\text{in}(I))$ is the degree-complexity of the Gröbner basis computation. In practice, each of these inequalities are often strict.

However when k is infinite, then for any set of coordinates for $\mathbf{P}^n$ chosen from a dense open set $U \subset GL(n+1)$ of possibilities, Galligo ([Gal74]; see also [BS87b]) has shown that the limiting ideal $\text{in}(I)$ takes on a very special form: $\text{in}(I)$ is invariant under the action of the Borel subgroup of upper triangular matrices in $GL(n+1)$. This imposes strong geometric conditions on X_0. In particular, the associated primes of $\text{in}(I)$ are also Borel-fixed, so they are all of the form $(x_0, \ldots, x_i)$ for various i. This means that the components of X_0 are supported on members of a flag.

In characteristic zero, it is shown in [BS87a] that the regularity of a Borel-fixed ideal is exactly the maximum of the degrees of its generators, or in our notation, that $\text{reg}(\text{in}(I)) = \text{reg}_0(\text{in}(I))$ when $\text{in}(I)$ is Borel-fixed. Thus, for generic coordinates in characteristic zero, the degree-complexity of computing Gröbner bases breaks down into two effects: the gap $\text{reg}_0(I) \leq \text{reg}(I)$ between the input degrees and the regularity of X, and the gap $\text{reg}(I) \leq \text{reg}(\text{in}(I))$ allowed by upper-semicontinuity.

A combination of theoretical results, hunches and experience guides the practitioner in assessing the first gap; what about the second? Does the regularity have to jump at all? One can easily find examples of ideals and total orders exhibiting such a jump, but in [BS87a], it is shown that for the reverse lexicographic order, in generic coordinates and any characteristic, there is no jump: $\text{reg}(I) = \text{reg}(\text{in}(I))$, so in characteristic zero we have

$$\text{reg}_0(\text{in}(I)) = \text{reg}(I).$$

In this sense, this order is an optimal choice: ***For the reverse lexicographic order, the degree-complexity of a Gröbner basis computation is exactly the regularity of the input data.*** This agrees with experience; computations made on the same inputs using the lexicographic order can climb to much higher degrees than the reverse lexicographic order, in practice.

For many applications, one is free to choose any order, but some problems restrict us to using orders satisfying combinatorial properties which the reverse lexicographic order fails to satisfy. An example, developed further in Section 2, is that of eliminating variables, or equivalently, of computing projections. To compute the intersection of I with a subring $R = k[x_i, \ldots, x_n]$, it is necessary to use an order which in each degree sorts all monomials not in R ahead of any monomial in R. The lexicographic order is an example of such an order, for each i simultaneously. This strength comes at a cost; we are paying in regularity gaps for properties we may not need in a particular problem. An optimal order if you need one specific projection (in the same sense as above) is constructed by sorting monomials by total degree in the variables to be eliminated, and then breaking ties using the reverse lexicographic order. See [BS87b] for this result, and a generalization to the problem of optimally refining any nonstrict order.

Using this elimination order, one finds that the inherent degree-complexity of a computation is given not by the regularity of X itself, but rather by the regularity of the ***flat projection*** X' of X, which is the central fiber of a flat family which animates the desired projection of X as $t \to 0$. The jump in regularity between X and X' is unavoidable; by choosing an optimal order, we avoid the penalty of a further jump in regularity between X' and X_0.

The regularity of algebraic varieties or schemes X is far from being well understood, but there is considerable interest in its study; this computational interpretation of regularity as the inherent degree-complexity of an ideal is but one more log on the fire.

From a theoretical computer science perspective, the full complexity of computing Gröbner bases is determined not merely by the highest degree $\text{reg}_0(I)$ in the basis, but by the total number of arithmetic operations in the field k required to compute this basis. This has not been analyzed in general, but for 0-dimensional ideals I, Lakshman and Lazard ([Lak91], [LL91]) have shown that the complexity of computing reduced Gröbner bases is bounded by a polynomial in d^n, where d is the maximum degree of the generators, and n is the number of variables.

2 Gröbner Bases

Let $S = k[x_0, \ldots, x_n]$ be a graded polynomial ring over the field k, and let $I \subset S$ be a homogeneous ideal.

Let S_d denote the finite vector space of all homogeneous, degree d polynomials in S, so $S = S_0 \oplus S_1 \oplus \ldots \oplus S_d \oplus \ldots$. Writing I in the same manner as $I = I_0 \oplus I_1 \oplus \ldots \oplus I_d \oplus \ldots$, we have $I_d \subset S_d$ for each d. Recall that the Hilbert function of I is defined to be the function $p(d) = \dim(I_d)$, for $d \geq 0$.

A total order $>$ on the monomials of S is said to be ***multiplicative*** if whenever $\mathbf{x}^A > \mathbf{x}^B$ for two monomials $\mathbf{x}^A$, $\mathbf{x}^B$, then $\mathbf{x}^C\mathbf{x}^A > \mathbf{x}^C\mathbf{x}^B$ for all monomials $\mathbf{x}^C$. This condition insures that if the terms of a polynomial are in order with respect to $>$, then they remain in order after multiplication

Definition 2.1 *Let $>$ be a multiplicative order. For a homogeneous polynomial $f = c_1\mathbf{x}^{A_1} + \ldots + c_m\mathbf{x}^{A_m}$ with $\mathbf{x}^{A_1} > \ldots > \mathbf{x}^{A_m}$, define the initial term* in(f) *to be the lead (that is, the largest) term $c_1\mathbf{x}^{A_1}$ of f. For a homogeneous ideal $I \subset S$, define the initial ideal* in(I) *to be the monomial ideal generated by the lead terms of all elements of I.*

Note that the definitions of in(f) and in(I) depend on the choice of multiplicative order $>$. See [BM88] and [MR88] for characterizations of the finite set of in(I) realized as the order $>$ varies.

Fix a multiplicative order $>$ on S.

Proposition 2.2 (Macaulay) *I and* in(I) *have the same Hilbert function.*

Proof. ([Mac27]) The lead terms of I_d span in$(I)_d$, because every monomial $\mathbf{x}^A \in$ in(I) is itself the lead term in(f) of some polynomial $f \in I$: Since

$\mathbf{x}^A = \mathbf{x}^C\mathbf{x}^B$ for some $\mathbf{x}^B = \mathrm{in}(g)$ with $g \in I$, we have $\mathbf{x}^A = \mathrm{in}(f)$ for $f = \mathbf{x}^C g$.

Choose a k-basis $B_d \subset I_d$ with distinct lead terms, and let $\mathrm{in}(B_d)$ be the set of lead terms of B_d; $\mathrm{in}(B_d)$ has cardinality $p(d) = \dim(I_d)$. Since any element of I_d is a linear combination of elements of B_d, any lead term of I_d is a scalar multiple of an element of $\mathrm{in}(B_d)$. Thus, $\mathrm{in}(B_d)$ is a basis for $\mathrm{in}(I)_d$, so $p(d) = \dim(\mathrm{in}(I)_d)$. ∎

One can compute the Hilbert function of I by finding $\mathrm{in}(I)$ and applying this result; see [MM83], [BCR91] and [BS92b].

Corollary 2.3 *The monomials of S which don't belong to* $\mathrm{in}(I)$ *form a k-basis for S/I.*

Proof. These monomials are linearly independent in S/I, because any linear relation among them is a polynomial belonging to I, and all such polynomials have lead terms belonging to $\mathrm{in}(I)$. These monomials can be seen to span S/I by a dimension count, applying Proposition 2.2. ∎

Two examples of multiplicative orders are the lexicographic order and the reverse lexicographic order. $\mathbf{x}^A > \mathbf{x}^B$ in the lexicographic order if the first nonzero coordinate of $A - B$ is positive. For example, if $S = k[w, x, y, z]$, then $w > x > y > z$ in S_1, and

$$w^2 > wx > wy > wz > x^2 > xy > xz > y^2 > yz > z^2$$

in S^2.

$\mathbf{x}^A > \mathbf{x}^B$ in the reverse lexicographic order if the last nonzero coordinate of $A - B$ is negative. For example, if $S = k[w, x, y, z]$, then $w > x > y > z$ in S_1, and

$$w^2 > wx > x^2 > wy > xy > y^2 > wz > xz > yz > z^2$$

in S^2. These two orders agree on S_1, but differ on the monomials of S of degree > 1 when $n \geq 2$.

The lexicographic order has the property that for each subring $k[x_i, \ldots, x_n] \subset S$ and each polynomial $f \in S$, $f \in k[x_i, \ldots, x_n]$ if and only if $\mathrm{in}(f) \in k[x_i, \ldots, x_n]$. The reverse lexicographic order has the property that for each $f \in k[x_0, \ldots, x_i]$, x_i divides f if and only if x_i divides $\mathrm{in}(f)$.

One can anticipate the applications of these properties by considering a k-basis $B_d \subset I_d$ with distinct lead terms, as in the proof of Proposition 2.2. With respect to the lexicographic order, $B_d \cap k[x_i, \ldots, x_n]$ is then a k-basis for $I_d \cap k[x_i, \ldots, x_n]$ for each i. With respect to the reverse lexicographic order, $B_d \cap (x_n)$ is then a k-basis for $I_d \cap (x_n)$. Thus, these orders enable us to find polynomials in an ideal which do not involve certain variables, or which are divisible by a certain variable. For a given degree d, one could construct such a basis B_d by applying Gaussian elimination to an arbitrary k-basis for I_d. However, this cannot be done for all d at once; such a computation would be infinite. We will finesse this difficulty by instead constructing a finite set of elements of I whose monomial multiples yield polynomials in I with every possible lead term.

Such sets can be described as follows:

Definition 2.4 *A list $F = [f_1, \ldots, f_r] \subset I$ is a (minimal) Gröbner basis for I if* $\text{in}(f_1), \ldots, \text{in}(f_r)$ *(minimally) generate* $\text{in}(I)$.

$\text{in}(I)$ is finitely generated because S is Noetherian, so Gröbner bases exist for any ideal I.

The order of the elements of F is immaterial to this definition, so F can be thought of as a set. We are using list notation for F because we are going to consider algorithms for which the order of the elements is significant. For convenience, we shall extend the notation of set intersections and containments to the lists F.

A minimal set of generators for an ideal I need not form a Gröbner basis for I. For example, if $S = k[x, y]$ and $I = (x^2 + y^2, xy)$, then with respect to the lexicographic order, $\text{in}(x^2 + y^2) = x^2$ and $\text{in}(xy) = xy$. Yet $y(x^2 + y^2) - x(xy) = y^3 \in I$, so $y^3 \in \text{in}(I)$. Thus, any Gröbner basis for I must include y^3; it can be shown that $\text{in}(I) = (x^2, xy, y^3)$ and $[x^2+y^2, xy, y^3]$ is a Gröbner basis for I.

On the other hand,

Lemma 2.5 *If $F = [f_1, \ldots, f_r]$ is a Gröbner basis for I, then $f_1, \ldots, f_r$ generate I.*

Proof. For each degree d, we can construct a k-basis $B_d \subset I_d$ with distinct lead terms, whose elements are monomial multiples of $f_1, \ldots, f_r$: For each $\mathbf{x}^A \in \text{in}(I)_d$, $\mathbf{x}^A$ is a scalar multiple of $\mathbf{x}^C \text{in}(f_i)$ for some $\mathbf{x}^C$ and some i; include $\mathbf{x}^C f_i$ in the set B_d. Thus, the monomial multiples of $f_1, \ldots, f_r$ span I. ∎

Proposition 2.6 (Spear, Trinks) *Let $R \subset S$ be the subring $R = k[x_i, \ldots, x_n]$. If $F = [f_1, \ldots, f_r]$ is a Gröbner basis for the ideal I with respect to the lexicographic order, then $F \cap R$ is a Gröbner basis for the ideal $I \cap R$. In particular, $F \cap R$ generates $I \cap R$.*

Proof. ([Spe77], [Zac78], [Tri78]) Let $f \in I \cap R$; $\text{in}(f)$ is a multiple of $\text{in}(f_i)$ for some i. Since $\text{in}(f) \in R$, $\text{in}(f_i) \in R$, so $f_i \in R$. Thus, $F \cap R$ is a Gröbner basis for $I \cap R$. By Lemma 2.5, $F \cap R$ generates $I \cap R$. ∎

Proposition 2.6 has the following geometric application: If I defines the subscheme $X \subset \mathbf{P}^n$, then $I \cap k[x_i, \ldots, x_n]$ defines the projection of X to $\mathbf{P}^{n-i} = \mathrm{Proj}(k[x_i, \ldots, x_n])$.

Recall that the saturation I^{sat} of I is defined to be the largest ideal defining the same subscheme $X \subset \mathbf{P}^n$ as I. I^{sat} can be obtained by taking an irredundant primary decomposition for I, and removing the primary ideal whose associated prime is the irrelevant ideal $(x_0, \ldots, x_n)$. I is saturated if $I = I^{\mathrm{sat}}$.

If the ideal I is saturated, and defines a finite set of points $X \subset \mathbf{P}^n$, then $I \cap k[x_{n-1}, x_n]$ is a principal ideal (f), where $\{f = 0\}$ is the image of the projection of X to $\mathbf{P}^1 = \mathrm{Proj}(k[x_{n-1}, x_n])$. Given a linear factor of f of the form $(bx_{n-1} - ax_n)$, we can make the substitution $x_{n-1} = az$, $x_n = bz$ for a new variable z, to obtain from I an ideal $J \subset k[x_0, \ldots, x_{n-2}, z]$ defining a finite set of points in $\mathbf{P}^{n-1}$. For each point $(c_0, \ldots, c_{n-2}, d)$ in the zero locus of J, $(c_0, \ldots, c_{n-2}, ad, bd)$ is a point in the zero locus of I.

If $X \subset \mathbf{P}^{n-1}$ is of dimension 1 or greater, then in general $I \cap k[x_{n-1}, x_n] = (0)$, because a generic projection of X to $\mathbf{P}^1$ is surjective. In this case, an arbitrary substitution $x_{n-1} = az$, $x_n = bz$ can be made, and the process of projecting to $\mathbf{P}^1$ iterated. Thus, the lexicographic order can be used to find solutions to systems of polynomial equations.

Recall that the ideal quotient $(I : f)$ is defined to be the ideal $\{g \in S \mid fg \in I\}$. Since S is Noetherian, the ascending chain of ideals $(I : f) \subset (I : f^2) \subset (I : f^3) \subset \ldots$ is stationary; call this stationary limit $(I : f^\infty) = \{g \in S \mid f^m g \in I \text{ for some } m\}$.

Proposition 2.7 *If $[x_n^{a_1} f_1, \ldots, x_n^{a_r} f_r]$ is a Gröbner basis for the ideal I with respect to the reverse lexicographic order, and if none of $f_1, \ldots, f_r$ are divisible by x_n, then $F = [f_1, \ldots, f_r]$ is a Gröbner basis for the ideal $(I : x_n^\infty)$. In particular, $f_1, \ldots, f_r$ generate $(I : x_n^\infty)$.*

Proof. ([Bay82], [BS87a]) We have $F \subset (I : x_n^\infty)$. Let $f \in (I : x_n^\infty)$; $x_n^m f \in I$ for some m, so $\mathrm{in}(x_n^m f)$ is a multiple of $\mathrm{in}(x_n^{a_i} f_i)$ for some i. Since f_i is not divisible by x_n, $\mathrm{in}(f_i)$ is not divisible by x_n, so $\mathrm{in}(f)$ is a multiple of $\mathrm{in}(f_i)$. Thus, F is a Gröbner basis for $(I : x_n^\infty)$. By Lemma 2.5, $f_1, \ldots, f_r$ generate $(I : x_n^\infty)$. ∎

If $I = \mathbf{q}_0 \cap \mathbf{q}_1 \cap \ldots \cap \mathbf{q}_t$ is a primary decomposition of I, then $(I : x_n^\infty) = (\cap \mathbf{q}_i : x_n^\infty) = \cap(\mathbf{q}_i : x_n^\infty)$. We have $(\mathbf{q}_i : x_n^\infty) = (1)$ if the associated prime $\mathbf{p}_i$ of $\mathbf{q}_i$ contains x_n, and $(\mathbf{q}_i : x_n^\infty) = \mathbf{q}_i$ otherwise. Thus, if I defines the subscheme $X \subset \mathbf{P}^n$, then $(I : x_n^\infty)$ defines the subscheme consisting of those primary components of X not supported on the hyperplane $\{x_n = 0\}$.

$(I : x_n^\infty)$ is saturated, because it cannot have $(x_0, \ldots, x_n)$ as an associated prime. If x_n belongs to none of the associated primes of I except

$(x_0, \ldots, x_n)$, or equivalently if $\{x_n = 0\}$ is a generic hyperplane section of $X \subset \mathbf{P}^n$, then $(I : x_n^\infty) = I^{\mathrm{sat}}$. Thus, the reverse lexicographic order can be used to find the saturation of I.

One of the most important uses of Gröbner bases is that they lead to canonical representations of polynomials modulo an ideal I, i.e. a division algorithm in which every $f \in S$ is written canonically as $f = \sum g_i f_i + h$, where $[f_1, \ldots, f_r]$ is a Gröbner basis for I, and h is the remainder after division.

Recall the division algorithm for inhomogeneous, univariate polynomials $f(x)$, $g(x) \in k[x]$: Let $\mathrm{in}(f)$ denote the highest degree term of f. The remainder of g under division by f can be recursively defined by

$$R_f(g) = R_f(g - cx^a f)$$

if $\mathrm{in}(f)$ divides $\mathrm{in}(g)$, where $cx^a = \mathrm{in}(g)/\mathrm{in}(f)$, and by

$$R_f(g) = g$$

otherwise.

Division can be generalized to homogeneous polynomials $f_1, \ldots, f_r, g \in S$, given a multiplicative order on S ([Hir64], [Bri73], [Gal74], [Sch80]): The remainder $R_F(g)$ of g under division by the list of polynomials $F = [f_1, \ldots, f_r]$ can be recursively defined by

$$R_F(g) = R_F(g - c\mathbf{x}^A f_i)$$

for the least i so $\mathrm{in}(g)$ is a multiple $c\mathbf{x}^A$ of $\mathrm{in}(f_i)$, and by

$$R_F(g) = \mathrm{in}(g) + R_F(g - \mathrm{in}(g))$$

if $\mathrm{in}(g)$ is not a multiple of any $\mathrm{in}(f_i)$. $R_F(g)$ is an element of S.

Thus, the fate of $\mathrm{in}(g)$ depends on whether or not $\mathrm{in}(g) \in (\mathrm{in}(f_1), \ldots, \mathrm{in}(f_r))$. Let I be the ideal generated by $f_1, \ldots, f_r$. If $F = [f_1, \ldots, f_r]$ fails to be a Gröbner basis for I, then the remainder is poorly behaved. For example, with respect to the lexicographic order on $k[x, y]$,

$$R_{[xy, x^2+y^2]}(x^2y) = x^2y - x(xy) = 0,$$

but

$$R_{[x^2+y^2, xy]}(x^2y) = x^2y - y(x^2 + y^2) = -y^3,$$

so the remainder $R_F(g)$ is dependent on the order of the list F. Note that $x^2y \in (x^2 + y^2, xy)$.

If on the other hand, F is a Gröbner basis for the ideal I, then $R_F(g)$ is a k-linear combination of monomials not belonging to $\mathrm{in}(I)$. By Corollary 2.3, these monomials form a k-basis for S/I, so each polynomial in S has a unique representation in terms of this k-basis, modulo the ideal I. The remainder gives this unique representation, and is independent of the order of F (but dependent on the multiplicative order chosen for the monomials of S). In particular, $R_F(g) = 0$ if and only if $g \in I$.

An algorithm for computing a Gröbner basis for I from a set of generators for I was first given by Buchberger ([Buc65], [Buc76]). This algorithm was discovered independently by Spear ([Spe77], [Zac78]), Bergman [Ber78], and Schreyer [Sch80]. It was termed the division algorithm by Schreyer, after the division theorem of Hironaka ([Hir64], [Bri73], [Gal74]).

Define $S(f_i, f_j)$ for $i < j$ by

$$S(f_i, f_j) = b\mathbf{x}^B f_i - c\mathbf{x}^C f_j,$$

where $\mathbf{x}^A = b\mathbf{x}^B \text{in}(f_i) = c\mathbf{x}^C \text{in}(f_j)$ is the least common multiple of $\text{in}(f_i)$ and $\text{in}(f_j)$. $b\mathbf{x}^B f_i$ and $c\mathbf{x}^C f_j$ each have $\mathbf{x}^A$ as lead term, so $\mathbf{x}^A$ cancels out in $S(f_i, f_j)$, and $\mathbf{x}^A > \text{in}(S(f_i, f_j))$.

If F is a Gröbner basis for the ideal I, then $R_F(S(f_i, f_j)) = 0$ for each $i < j$, since $S(f_i, f_j) \in I$. Conversely,

Proposition 2.8 (Buchberger) *If $R_F(S(f_i, f_j)) = 0$ for each $i < j$, then $F = [f_1, \ldots, f_r]$ is a Gröbner basis for the ideal $I = (f_1, \ldots, f_r)$.*

See [Buc65], [Buc76]. We postpone a proof until the theory has been extended to S-modules. This result can also be thought of as an explicit converse to the assertion that if F is a Gröbner basis, then division is independent of the order of F: Whenever we have a choice in division between subtracting off a multiple of f_i and a multiple of f_j, the difference is a multiple of $S(f_i, f_j)$. If division is independent of the order of F, then these differences must have remainder zero, so by Proposition 2.8, F is a Gröbner basis.

As sketched in Section 1, Proposition 2.8 can be used to compute a Gröbner basis from a set of generators $f_1, \ldots, f_r$ for the ideal I: For each $i < j$ so $f_{r+1} = R_F(S(f_i, f_j)) \neq 0$, adjoin f_{r+1} to the list $F = [f_1, \ldots, f_r]$. Note that $f_{r+1} \in I$. By iterating until no new polynomials are found, a Gröbner basis F is obtained for I. This process terminates because S is Noetherian, and each new basis element corresponds to a monomial not in the ideal generated by the preceding lead terms.

We now extend this theory to S-modules. Let M be a graded, finitely generated S-module, given by the exact sequence of graded S-modules

$$M_1 \xrightarrow{F} M_0 \longrightarrow M \longrightarrow 0,$$

where $M_0 = Se_{01} \oplus \ldots \oplus Se_{0q}$ and $M_1 = Se_{11} \oplus \ldots \oplus Se_{1r}$ are free S-modules with $\deg(e_{ij}) = d_{ij}$ for each i, j. We now think of F both as a list $[f_1, \ldots, f_r]$ of module elements, and as a map between free modules: Let $f_i = F(e_{1i}) \neq 0$ for $i = 1, \ldots, r$, and let $I \subset M_0$ be the homogeneous submodule generated by $f_1, \ldots, f_r$. Thus, $M = M_0/I$.

A monomial of M_0 is an element of the form $\mathbf{x}^A e_{0i}$; such an element has degree $\deg(\mathbf{x}^A) + d_{0i}$. An order on the monomials of M_0 is multiplicative if whenever $\mathbf{x}^A e_{0i} > \mathbf{x}^B e_{0j}$, then $\mathbf{x}^C \mathbf{x}^A e_{0i} > \mathbf{x}^C \mathbf{x}^B e_{0j}$ for all $\mathbf{x}^C \in S$. For some applications, such as developing a theory of Gröbner bases over quotients of S, one wants this order to be compatible with an order on S: If $\mathbf{x}^A > \mathbf{x}^B$, then one wants $\mathbf{x}^A e_{0i} > \mathbf{x}^B e_{0i}$ for $i = 1, \ldots, r$. The orders encountered in practice invariably satisfy this second condition, but it does not follow from the first, and we do not require it here.

One way to extend a multiplicative order on S to a compatible multiplicative order on M_0 is to declare $\mathbf{x}^A e_{0i} > \mathbf{x}^B e_{0j}$ if $i < j$, or if $i = j$ and $\mathbf{x}^A > \mathbf{x}^B$. Another way is to assign monomials $\mathbf{x}^{C_1}, \ldots, \mathbf{x}^{C_q}$ in S to the basis elements $e_{01}, \ldots, e_{0q}$ of M_0, and to declare $\mathbf{x}^A e_{0i} > \mathbf{x}^B e_{0j}$ if $\mathbf{x}^{A+C_i} > \mathbf{x}^{B+C_j}$, or if $A + C_i = B + C_j$ and $i < j$.

Fix a choice of a multiplicative order $>$ on M_0. The constructions developed for S carry over intact to M_0, with the same proofs ([Gal79], [Sch80], [Bay82]): Given an element $f \in M_0$, define $\text{in}(f)$ to be the lead term of f. Define $\text{in}(I)$ to be the submodule generated by the lead terms of all elements of $I \subset M_0$; $\text{in}(I)$ is a monomial submodule of M_0 with the same Hilbert function as I. Define $F = [f_1, \ldots, f_r] \subset I$ to be a Gröbner basis for I if $\text{in}(f_1), \ldots, \text{in}(f_r)$ generate $\text{in}(I)$; a set of generators for I need not be a Gröbner basis for I, but a Gröbner basis for I generates I. Given an element $g \in M_0$, define $R_F(g) \in M_0$ exactly as was done for the free module S. If F is a Gröbner basis for I, then $R_F(g) = 0$ if and only if $g \in I$.

The quotient of g under division by $f_1, \ldots, f_r$ can be recursively defined by

$$Q_F(g) = c\mathbf{x}^A e_{1i} + Q_F(g - c\mathbf{x}^A f_i)$$

for the least i so $\text{in}(g)$ is a multiple $c\mathbf{x}^A$ of $\text{in}(f_i)$, and by

$$Q_F(g) = Q_F(g - \text{in}(g))$$

if $\text{in}(g)$ is not a multiple of any $\text{in}(f_i)$. The quotient is an element of M_1.

Following the recursive definitions of the remainder and quotient, it can be inductively verified that

$$g = F(Q_F(g)) + R_F(g).$$

If F is a Gröbner basis for I, and $g \in I$, then $R_F(g) = 0$, so the quotient lifts g to M_1. In this case, the quotient can be thought of as expressing g in terms of $f_1, \ldots, f_r$.

Define $S(f_i, f_j)$ for $i < j$ by

$$S(f_i, f_j) = b\mathbf{x}^B f_i - c\mathbf{x}^C f_j,$$

if $\text{in}(f_i)$ and $\text{in}(f_j)$ have a least common multiple $\mathbf{x}^A e_{0k} = b\mathbf{x}^B \text{in}(f_i) = c\mathbf{x}^C \text{in}(f_j)$. Leave $S(f_i, f_j)$ undefined if $\text{in}(f_i)$ and $\text{in}(f_j)$ lie in different summands of M_0, and so don't have common multiples.

Recall that the module of syzygies of $f_1, \ldots, f_r$ is defined to be the kernel of the map F, which is the submodule of M_1 consisting of all $h \in M_1$ so $F(h) = 0$. Thus, if $h = h_1 e_{11} + \ldots + h_r e_{1r}$ is a syzygy, then $h_1 f_1 + \ldots + h_r f_r = 0$. Let $J \subset M_1$ denote the module of syzygies of $f_1, \ldots, f_r$, and let $K \subset M_1$ denote the module of syzygies of $\text{in}(f_1), \ldots, \text{in}(f_r)$.

Define the map $\text{in}(F) : M_1 \to M_0$ by $\text{in}(F)(e_{1i}) = \text{in}(f_i)$; K is the kernel of $\text{in}(F)$. For each $i < j$ so $S(f_i, f_j)$ is defined, define t_{ij} to be the element

$$t_{ij} = b\mathbf{x}^B e_{1i} - c\mathbf{x}^C e_{1j} \in M_1,$$

where $\mathbf{x}^A e_{0k} = b\mathbf{x}^B \text{in}(f_i) = c\mathbf{x}^C \text{in}(f_j)$ is the least common multiple of $\text{in}(f_i)$ and $\text{in}(f_j)$, as before. $\text{in}(F)(t_{ij}) = 0$, so each t_{ij} belongs to the syzygy module K. Observe that $F(t_{ij}) = S(f_i, f_j)$.

Assign the following multiplicative order on M_1, starting from the order on M_0 ([Sch80]; see also [MM86]): Let $\mathbf{x}^A e_{1i} > \mathbf{x}^B e_{1j}$ if $\mathbf{x}^A \text{in}(f_i) > \mathbf{x}^B \text{in}(f_j)$, or if these terms are k-multiples of each other and $i < j$. If the order on M_0 is compatible with an order on S, then this order on M_1 is compatible with the same order on S.

With respect to this order on M_1, we have

Lemma 2.9 *The list* $[\,t_{ij}\,]$ *is a Gröbner basis for the module* K *of syzygies of* $\text{in}(f_1), \ldots, \text{in}(f_r)$.

Proof. Let $h \in M_1$, so $\text{in}(F)(h) = 0$. Then $\text{in}(F)(\text{in}(h))$ is canceled by $\text{in}(F)(h - \text{in}(h))$ in M_0. Therefore, if $\text{in}(h) = \mathbf{x}^A e_{1i}$, then h has another term $\mathbf{x}^B e_{1j}$ so $\mathbf{x}^A \text{in}(f_i)$ and $\mathbf{x}^B \text{in}(f_j)$ are k-multiples of each other and $i < j$. Thus, t_{ij} is defined and $\text{in}(t_{ij})$ divides $\text{in}(h)$, so $[\,t_{ij}\,]$ is a Gröbner basis for K. ∎

Thus, the set $\{t_{ij}\}$ generates K. In general, the $[\,t_{ij}\,]$ are far from being a minimal Gröbner basis for K; we consider the effects of trimming this list in Proposition 2.10 below.

Define

$$s_{ij} = t_{ij} - Q_F(S(f_i, f_j))$$

whenever $R_F(S(f_i, f_j)) = 0$. Note that $\text{in}(s_{ij}) = \text{in}(t_{ij})$. Each s_{ij} is the difference of two distinct elements of M_1, each of which is mapped by F to $S(f_i, f_j)$, so $F(s_{ij}) = 0$. In other words, s_{ij} belongs to the syzygy module J. Conversely,

Proposition 2.10 (Richman, Spear, Schreyer) *Choose a set of pairs $T = \{(i,j)\}$ such that the set $\{t_{ij}\}_{(i,j)\in T}$ generates the module K of syzygies of $\text{in}(f_1), \ldots, \text{in}(f_r)$. If $R_F(S(f_i, f_j)) = 0$ for each $(i,j) \in T$, then*

(a) $F = [f_1, \ldots, f_r]$ is a Gröbner basis for I;

(b) the set $\{s_{ij}\}_{(i,j)\in T}$ generates the module J of syzygies of $f_1, \ldots, f_r$. Moreover,

(c) if $[\, t_{ij} \,]_{(i,j)\in T}$ is a Gröbner basis for K, then $[\, s_{ij} \,]_{(i,j)\in T}$ is a Gröbner basis for J.

Proof. ([Ric74], [Spe77], [Zac78], [Sch80]) First, suppose that $[\, t_{ij} \,]_{(i,j)\in T}$ is a Gröbner basis for K. Let $h \in J$, so $F(h) = 0$. By the same reasoning as in the proof of Lemma 2.9, we can find $(i,j) \in T$ so $\text{in}(t_{ij})$ divides $\text{in}(h)$. Since $\text{in}(s_{ij}) = \text{in}(t_{ij})$, $\text{in}(s_{ij})$ also divides $\text{in}(h)$, so $[\, s_{ij} \,]_{(i,j)\in T}$ is a Gröbner basis for J, proving (c).

Now, suppose that $\{t_{ij}\}_{(i,j)\in T}$ merely generates K. Let T' be a set of pairs so $[\, t_{\ell m} \,]_{(\ell,m)\in T'}$ is a Gröbner basis for K. It is enough to construct a list $[\, u_{\ell m} \,]_{(\ell,m)\in T'}$ of elements of J, generated by $\{s_{ij}\}_{(i,j)\in T}$, so $\text{in}(u_{\ell m}) = \text{in}(t_{\ell m})$ for all $(\ell, m) \in T'$. Then by the preceding argument, $[\, u_{\ell m} \,]_{(\ell,m)\in T'}$ is a Gröbner basis for J, so $\{s_{ij}\}_{(i,j)\in T}$ generates J.

Write each $t_{\ell m} = \sum g_{\ell m i j} t_{ij}$, for $(\ell, m) \in T'$ and $(i,j) \in T$, in such a way that the terms of $t_{\ell m}$ and each term of each product $g_{\ell m i j} t_{ij}$ map via $\text{in}(F)$ to multiples of the same monomial in M_0. In other words, find a minimal expression for each $t_{\ell m}$, which avoids unnecessary cancellation. Then define

$$u_{\ell m} = \sum g_{\ell m i j} s_{ij}.$$

We have $\text{in}(u_{\ell m}) = \text{in}(t_{\ell m})$, proving (b).

Let $f \in I$, and choose $g \in M_1$ so $f = F(g)$. Let $h \in M_1$ be the remainder of g under division by $[\, u_{\ell m} \,]_{(\ell,m)\in T'}$; $f = F(h)$. Since $\text{in}(h)$ is not a multiple of any $\text{in}(u_{\ell m}) = \text{in}(t_{\ell m})$, the lead term of $F(\text{in}(h))$ is not canceled by any term of $F(h - \text{in}(h))$. Therefore, if $\text{in}(h) = a\mathbf{x}^A e_{1i}$, then $\text{in}(f_i)$ divides $\text{in}(F)$. Thus, $F = [f_1, \ldots, f_r]$ is a Gröbner basis for I, proving (a). ∎

Proposition 2.8 follows as a special case of this result.

The above proof can be understood in terms of an intermediate initial form $\text{in}_0(h)$ for $h \in M_1$: Apply the map $\text{in}(F)$ separately to each term of h, and let $\mathbf{x}^A \in M_0$ be the greatest monomial that occurs in the set of image terms. Define $\text{in}_0(h)$ to be the sum of all terms of h which map via $\text{in}(F)$ to multiples of $\mathbf{x}^A$. Then in refines in_0, for according to the order we have defined on M_1, $\text{in}(h)$ is the term of $\text{in}_0(h)$ lying in the summand of M_1 whose basis element e_i has the smallest index i.

In this language, $t_{ij} = \text{in}_0(t_{ij}) = \text{in}_0(s_{ij})$. Our expressions for the $t_{\ell m}$ have the property that each $g_{\ell m i j} t_{ij} = \text{in}_0(g_{\ell m i j} t_{ij})$, with each term of each product for a given $t_{\ell m}$ mapping via $\text{in}(F)$ to multiples of the same monomial $\mathbf{x}^A$. Thus, each $\text{in}_0(g_{\ell m i j} s_{ij}) = g_{\ell m i j} t_{ij}$; the tails $g_{\ell m i j}(s_{ij} - \text{in}_0(s_{ij}))$ stay out of our way, mapping termwise via $\text{in}(F)$ to monomials which are less than $\mathbf{x}^A$ with respect to the order on M_0.

Observe that $Q_F(g)$ is a linear combination of monomials not belonging to $\text{in}(J)$, for any $g \in M_0$.

In [Buc79], Buchberger gives a criterion for selecting a set T of pairs (i, j) in the case where I is an ideal: If $(i_0, i_1), (i_1, i_2), \ldots, (i_{s-1}, i_s) \in T$, and the least common multiple of $\text{in}(f_{i_0}), \text{in}(f_{i_1}), \ldots, \text{in}(f_{i_s})$ is equal to the least common multiple of $\text{in}(f_{i_0})$ and $\text{in}(f_{i_s})$, then (i_0, i_s) need not belong to T. In other words, if $t_{i_0 i_s} \in (t_{i_0 i_1}, \ldots, t_{i_{s-1} i_s})$, then the pair (i_0, i_s) is unnecessary; this condition is equivalent to the condition of Proposition 2.10, for the case of an ideal.

Suppose that we wish to compute the syzygies of a given set of elements $g_1, \ldots, g_s$ of M_0. To do this, compute a Gröbner basis $f_1, \ldots, f_r$ for the submodule $I \subset M_0$ generated by $g_1, \ldots, g_s$. Keep track of how to write each f_i in terms of $g_1, \ldots, g_s$. Using these expressions, each syzygy of $f_1, \ldots, f_r$ can be mapped to a syzygy of $g_1, \ldots, g_s$. These images generate the module of syzygies of $g_1, \ldots, g_s$; the set of syzygies obtained in this way is not in general minimal.

Syzygies can be used to find a minimal set of generators for a submodule $I \subset M_0$ from a given set of generators $g_1, \ldots, g_s$: If $h_1 g_1 + \ldots + h_r g_r = 0$ is a syzygy of $g_1, \ldots, g_s$ with $h_1 \in k$, then $g_1 = (h_2 g_2 + \ldots + h_r g_r)/h_1$, so g_1 is not needed to generate I. All unnecessary generators can be removed in this way.

Alternatively, a careful implementation of Gröbner bases can directly find minimal sets of generators for submodules: Starting from an arbitrary set of generators, we can eliminate unnecessary generators degree by degree, by removing those which reduce to zero under division by a Gröbner basis for the ideal generated by the preceding generators.

Either way, we can trim the set of syzygies computed via Gröbner bases for a given set of generators $g_1, \ldots, g_s$ of I, to obtain a minimal set of generators for the syzygy module J. By starting with a minimal generating set for I, and iterating this method, a minimal free resolution can be found for I.

A beautiful application of these ideas yields a proof of the Hilbert syzygy theorem, that minimal free resolutions terminate (Schreyer [Sch80], [Sch91], for an exposition see also Eisenbud [Eis92]). At each stage of a resolution, order the Gröbner basis F for I in such a way that for each $i < j$, letting $\text{in}(f_i) = a\mathbf{x}^A e_{0k}$ and $\text{in}(f_j) = b\mathbf{x}^B e_{0\ell}$, we have $\mathbf{x}^A > \mathbf{x}^B$ in the lexicographic order. If the variables $x_1, \ldots, x_m$ are missing from the initial terms of the f_i, then the variables $x_1, \ldots, x_{m+1}$ will be missing from the initial terms of the syzygies s_{ij}. Iterating, we run out of variables, so the resolution terminates.

3 Bounds

How hard are the algorithms in algebraic geometry? We describe some key bounds. The best known example is the bound established by G. Hermann [Her26] for ideal membership:

Theorem 3.1 (G. Hermann) *Let k be any field, let $(f_1, ..., f_k) \subset k[x_1, ..., x_n]$ and let $d = \max(\deg(f_i))$. If $g \in (f_1, ..., f_k)$, then there is an expression*

$$g = \sum_{i=1}^{k} a_i f_i$$

where $\deg(a_i) \leq \deg(g) + 2(kd)^{2^{n-1}}$.

This type of bound is called "doubly exponential". However, with the advent of the concept of coherent sheaf cohomology [Ser55] and the systematic study of vanishing theorems, it has become apparent that the vanishing of these groups in high degrees is almost always the most fundamental bound. The concept of an ideal being "m-regular" or "regular in degrees $\geq m$" was introduced by one of us [Mum66] by generalizing ideas of Castelnuovo:

Definition 3.2 [1] *Let k be any field, let $I \subset k[x_0, \ldots, x_n]$ be an ideal generated by homogeneous polyomials, let I_d be the homogeneous elements in I of degree d, let $\mathcal{I}$ be the corresponding sheaf of ideals in $\mathcal{O}_{\mathbf{P}^n}$, and let $\mathcal{I}(d)$ be the dth twist of $\mathcal{I}$. Then the following properties are equivalent and define the term "m-regular":*

(a) the natural map $I_m \to H^0(\mathcal{I}(m))$ is an isomorphism and $H^i(\mathcal{I}(m-i)) = (0)$, $1 \leq i \leq n$

(b) the natural maps $I_d \to H^0(\mathcal{I}(d))$ are isomorphisms for all $d \geq m$ and $H^i(\mathcal{I}(d)) = (0)$ if $d + i \geq m$, $i \geq 1$.

(c) Take a minimal resolution of I by free graded $k[X]$-modules:

$$0 \to \bigoplus_{\alpha=1}^{r_n} k[\mathbf{x}] \cdot e_{\alpha,n} \xrightarrow{\phi_n} \ldots \xrightarrow{\phi_1} \bigoplus_{\alpha=1}^{r_0} k[\mathbf{x}] \cdot e_{\alpha,0} \xrightarrow{\phi_0} k[\mathbf{x}] \longrightarrow k[\mathbf{x}]/I \to 0.$$

Then $\deg(e_{\alpha,i}) \leq m + i$ for all α, i. (In particular, if $f_\alpha = \phi_0(e_{\alpha,0})$, then $f_1, \ldots, f_{r_0}$ are minimal generators of I, and $\deg(e_{\alpha,0}) = \deg(f_\alpha) \leq m$.)

[1] The definition has been slightly modified so as to apply to ideals I instead of the corresponding sheaf of ideals $\mathcal{I}$.

The intuitive idea is that past degree m, nothing tricky happens in the ideal I. Unfortunately, neither (a), (b) nor (c) can be verified by any obvious finite algorithm. This lack of a finitely verifiable criterion for m-regularity has been remedied by a joint result of the first author and M. Stillman [BS87a]:

Theorem 3.3 (Bayer-Stillman) *I is m-regular if and only if the degrees of the minimal set of generators of I are at most m, and there exists a set $y_0, \ldots, y_\ell$ of linear combinations of $x_0, \ldots, x_n$ such that for all homogeneous f of degree m,*

$$\begin{aligned} y_0 f \in I &\Rightarrow f \in I \\ y_1 f \in I &\Rightarrow f \in I + k[\mathbf{x}] \cdot y_0 \\ &\cdots \\ y_\ell f \in I &\Rightarrow f \in I + \sum_{i=0}^{\ell-1} k[\mathbf{x}] \cdot y_i \end{aligned}$$

and

$$f \in I + \sum_{i=0}^{\ell} k[\mathbf{x}] \cdot y_i.$$

Moreover, if this holds at all, it holds for $y_0, \ldots, y_\ell$ taken arbitrarily from a Zariski-open set in the space of $\ell + 1$ linear forms.

To see why m-regularity is a key bound, we want to show that it controls some of the geometric features of the ideal I. Let's introduce several refined notions of the "degree" of I:

Definition 3.4 *If $I = \mathbf{q}_0 \cap \mathbf{q}_1 \cap \ldots \cap \mathbf{q}_t$ is a primary decomposition of I, $\sqrt{\mathbf{q}_i} = \mathbf{p}_i$ is prime and $V(\mathbf{p}_i)$ is the subvariety Z_i of $\mathbf{P}^n$ for $i \geq 1$, while $\mathbf{p}_0 = (x_0, \ldots, x_n)$ (so that $V(\mathbf{p}_0) = \emptyset$), then first let $\mathbf{q}_1, \ldots, \mathbf{q}_s$ be the isolated components, (i.e., $Z_i \not\subset Z_j$ if $1 \leq i \leq s$, $1 \leq j \leq t$, $i \neq j$, or equivalently, $V(I) = Z_1 \cup \ldots \cup Z_s$ is set-theoretically the minimal decomposition of $V(I)$ into varieties). Then let*

$$\begin{aligned} \text{mult}(\mathbf{q}_i) \;=\;& \textit{length } \ell \textit{ of a maximal chain of } \mathbf{p}_i\textit{-primary ideals:} \\ & \mathbf{q}_i = J_\ell \subsetneq J_{\ell-1} \subsetneq \cdots \subsetneq J_1 = \mathbf{p}_i \end{aligned}$$

(Equivalently, this is the length of the local ring $k[\mathbf{x}]_{\mathbf{p}_i}/Ik[\mathbf{x}]_{\mathbf{p}_i}$, or, in the language of schemes, if η is the generic point of Z_i, then this is the length of $\mathcal{O}_{\eta,\mathbf{P}^n}$.)

$$\begin{aligned} \deg(Z_i) \;=\;& \textit{usual geometric degree of } Z_i\textit{:} \\ & \textit{the cardinality of } Z_i \cap L \textit{ for almost all} \end{aligned}$$

linear spaces L of complementary dimension.

$$\text{geom-deg}_r(I) = \sum_{\substack{i \text{ such that } \dim Z_i = r \\ 1 \le i \le s}} \text{mult}(\mathbf{q}_i)\deg(Z_i)$$

If $\mathbf{q}_i$ is one of the non-isolated, or embedded components, then we extend the concept of multiplicity more carefully: Let

$$I_i = \left\{\cap \mathbf{q}_j \middle| j \text{ such that } \mathbf{p}_j \subsetneqq \mathbf{p}_i \text{ or equivalently } Z_j \supsetneqq Z_i\right\} \cap \mathbf{p}_i$$

and

$$\begin{aligned} \text{mult}_I(\mathbf{q}_i) &= \text{length } \ell \text{ of a maximal chain of ideals:} \\ &\quad \mathbf{q}_i \cap I_i = J_\ell \subsetneqq J_{\ell-1} \subsetneqq \ldots \subsetneqq J_0 = I_i \\ &\quad \text{where each } J_k \text{ satisfies: } ab \in J_k, a \notin \mathbf{p}_i \Rightarrow b \in J_k. \end{aligned}$$

(Equivalently, J_k equals $\mathbf{q}_k \cap I_i$ for some $\mathbf{p}_i$-primary ideal $\mathbf{q}_k$.) In particular:

$$\begin{aligned} I_0 &= \bigcap_{j=1}^{t} \mathbf{q}_j \text{ is known as } I^{\text{sat}}, \text{ and} \\ \text{mult}_I(\mathbf{q}_0) &= \text{length } \ell \text{ of a maximal chain of ideals} \\ &\quad I = J_\ell \subsetneqq J_{\ell-1} \subsetneqq \ldots \subsetneqq J_0 = I^{\text{sat}} \\ &= \dim_k(I^{\text{sat}}/I). \end{aligned}$$

For $s+1 \le i \le t$, an equivalent way to define $\text{mult}_I(\mathbf{q}_i)$ is as the length of the module

$$I_i k[\mathbf{x}]_{\mathbf{p}_i} / I k[\mathbf{x}]_{\mathbf{p}_i}$$

or, in the language of schemes, the length of

$$I_i \mathcal{O}_{\eta,\mathbf{P}^n} / I \mathcal{O}_{\eta,\mathbf{P}^n}$$

where η is the generic point of Z_i.

Then write

$$\text{arith-deg}_r(I) = \sum_{\substack{i \text{ such that } \dim Z_i = r \\ 1 \le i \le s}} \text{mult}_I(\mathbf{q}_i)\deg(Z_i)$$

and

$$\text{arith-deg}_{-1}(I) = \text{mult}_I(\mathbf{q}_0).$$

The idea here is best illustrated by an example: let

$$I = (x_1^2, x_1x_2) \subset k[x_0, x_1, x_2].$$

Then

$$\begin{aligned} I &= \mathbf{q}_1 \cap \mathbf{q}_2 \\ q_1 &= (x_1),\ \mathbf{p}_1 = (x_1),\ Z_1 = \{\text{line } x_1 = 0\} \\ q_2 &= (x_1^2, x_1x_2, x_2^2),\ \mathbf{p}_2 = (x_1, x_2),\ Z_2 = \{\text{point } (1,0,0)\}. \end{aligned}$$

Then

$$\deg(Z_1) = 1,\ \text{mult}(q_1) = 1$$

so

$$\text{geom-deg}_1(I) = \text{arith-deg}_1(I) = 1.$$

One might be tempted to simply define

$$\text{mult}_I(\mathbf{q}_2) = \text{ length of chain of } \mathbf{p}_2\text{-primary ideals between } \mathbf{q}_2, \mathbf{p}_2$$

and since

$$k[\mathbf{x}]_{\mathbf{q}_2}/\mathbf{q}_2 k[\mathbf{x}]_{\mathbf{p}_2} \cong K \cdot 1 + K \cdot x_1 + K \cdot x_2, K = k(x_0)$$

this is 3. But embedded components are not unique! In fact,

$$\begin{aligned} I &= \mathbf{q}_1 \cap \mathbf{q}_2' \\ \mathbf{q}_2' &= (x_1^2 x_2) \text{ also,} \end{aligned}$$

which leads to

$$k[\mathbf{x}]_{\mathbf{p}_2}/\mathbf{q}_2' k[\mathbf{x}]_{\mathbf{p}_2} \cong K \cdot 1 + K \cdot x_2$$

which has length 2. The canonical object is not the local ring $k[\mathbf{x}]_{\mathbf{p}_2}/\mathbf{q}_2 k[\mathbf{x}]\mathbf{p}_2$ but the ideal

$$\text{Ker}\,(k[\mathbf{x}]_{\mathbf{p}_2}/I k[\mathbf{x}]_{\mathbf{p}_2} \rightarrow k[\mathbf{x}]_{\mathbf{p}_2}/\mathbf{p}_2 k[\mathbf{x}]_{\mathbf{p}_2}) \cong k \cdot x_1$$

which has length 1. Thus, the correct numbers are

$$\text{mult}_I(\mathbf{q}_2) = 1$$

and

$$\begin{aligned} \text{geom-deg}_0(I) &= 0 \\ \text{arith-deg}_0(I) &= 1. \end{aligned}$$

Now the question arises: find bounds on these degrees in terms of generators of I. For geometric degrees, a straightforward extension of Bezout's theorem gives:

Proposition 3.5 *Let $d(I)$ be the maximum of the degrees of a minimal set of generators of I. Then*

$$\text{geom-deg}_r(I) \le d(I)^{n-r}.$$

A proof can be found in [MW83]. The idea is clear from a simple case: Suppose $f, g, h \in K[x, y, z]$ and $f = g = h = 0$ consists of a curve C and ℓ points P_i off C. We can bound ℓ like this: Choose 2 generic combinations f', g' of f, g, h so that $f' = g' = 0$ does not contain a surface. It must be of the form $C \cup C'$, C' one-dimensional, containing all the P_i but not the generic point of C. Then by the usual Bezout theorem

$$\deg C' \le \deg f' \deg g' = d(I)^2.$$

Let h' be a 3rd generic combination of f, g, h. Then $C' \cap \{h' = 0\}$ consists of a finite set of points including the P_i's. Thus

$$\begin{aligned} \ell &= \#P_i \le \#(C' \cap \{h' = 0\}) \\ &\le \deg C' \cdot d(I) \text{ by Bezout's theorem} \\ &\le d(I)^3. \end{aligned}$$

Can arith-deg(I) be bounded in the same way? In fact, it cannot, as we will show below. Instead, we have

Proposition 3.6 *If $m(I)$ is the regularity of I, then for $-1 \le r \le n$,*

$$\text{arith-deg}_r(I) \le \binom{m(I) + n - r - 1}{n - r} \le m(I)^{n-r}$$

which replaces $d(I)$ by the regularity of I. A proof is given in the technical appendix.

We have introduced two measures of the complexity of a homogeneous ideal I. The first is $d(I)$, the maximum degree of a polynomial in a minimum set of generators of I. The second is $m(I)$, which bounds the degrees of generators and of all higher order syzygies in the resolution of I (Definition 3.2 (c)). Obviously,

$$d(I) \le m(I).$$

A very important question is how much bigger can $m(I)$ be than $d(I)$? The nature of the answer was conjectured by one of us in his thesis [Bay82] and this conjecture is being borne out by subsequent investigations. This conjecture is that in the worst case $m(I)$ is roughly the $(2^n)^{\text{th}}$ power of $d(I)$ – a bound like G. Hermann's. But that if $I = I(Z)$ where Z is geometrically nice, e.g. is a smooth irreducible variety, then $m(I)$ is much smaller, like the nth power of $d(I)$ or better. This conjecture then has three aspects:

(1) a doubly exponential bound for $m(I)$ in terms of $d(I)$, which is always valid,
(2) examples of I where the bound in (1) is best possible, or nearly so,
(3) much better bounds for $m(I)$ valid if $V(I)$ satisfies various conditions.

All three aspects are partially proven, but none are completely clarified yet. We will take them up one at a time.

A doubly exponential bound for $m(I)$ in terms of $d(I)$ may be deduced easily *in characteristic zero* from the work of M. Giusti [Giu84] and A. Galligo [Gal79]:

Theorem 3.7 *If* $\mathrm{char}(k) = 0$ *and* $I \subset k[x_0, \dots x_n]$ *is any homogeneous ideal, then*

$$m(I) \leq (2d(I))^{2^{n-1}}.$$

It seems likely that Theorem 3.7 holds in characteristic p, too. A weaker result can be derived quickly in any characteristic by straightforward cohomological methods:

Proposition 3.8 *If* $I \subset k[x_0, \dots x_n]$ *is any homogeneous ideal, then*

$$m(I) \leq (2d(I))^{n!}.$$

The proof is given in the technical appendix.

Next, we ask whether Theorem 3.7 is the best possible, or nearly so. The answer is yes, because of a very remarkable example due to E. Mayr and A. Meyer [MM82].

Example 3.9 Let I_n^A be the ideal in $10n$ variables $S^{(m)}$, $F^{(m)}$, $C_i^{(m)}$, $B_i^{(m)}, 1 \leq i \leq 4, 1 \leq m \leq n$ defined by the $10n - 6$ generators

$$2 \leq m \leq n \begin{cases} S^{(m)} - S^{(m-1)}C_1^{(m-1)} \\ F^{(m)} - S^{(m-1)}C_4^{(m-1)} \\ C_i^{(m)}F^{(m-1)}B_2^{(m-1)} - C_i^{(m)}B_i^{(m)}F^{(m-1)}B_3^{(m-1)} 1 \leq i \leq 4 \end{cases}$$

$$1 \leq m \leq n-1 \begin{cases} F^{(m)}C_1^{(m)}B_1^{(m)} - S^{(m)}C_2^{(m)} \\ F^{(m)}C_2^{(m)} - F^{(m)}C_3^{(m)} \\ S^{(m)}C_3^{(m)}B_1^{(m)} - S^{(m)}C_2^{(m)}B_4^{(m)} \\ S^{(m)}C_3^{(m)} - F^{(m)}C_4^{(m)}B_4^{(m)} \end{cases}$$

$$C_i^{(1)}S^{(1)} - C_i^{(1)}F^{(1)}(B_i^{(1)})^2, 1 \leq i \leq 4$$

Let I_n^H be the ideal gotten from I_n^A by homogenizing with an extra variable u. Then Mayr and Meyer [MM82, lemma 8, p. 318] prove:

Lemma 3.10 *Let $e_n = 2^{2^n}$. If M is any monomial in these variables, $S^{(n)}C_i^{(n)} - F^{(n)}M \in I_n^A$ if and only if*

$$M = C_i^{(n)}(B_i^{(n)})^{e_n},$$

and $S^{(n)}C_i^{(n)} - S^{(n)}M \in I_n^A$ if and only if

$$M = C_i^{(n)}.$$

Now note that the generators of I_n^A and I_n^H are all of the very simple type given by a difference of two monomials. Quite generally, if

$$\begin{aligned} J &\subset k[x_1, \ldots, x_n] \\ J &= (\ldots, \mathbf{x}^{\alpha_i} - \mathbf{x}^{\beta_i}, \ldots)_{1 \le i \le k} \end{aligned}$$

then the quotient ring $k[\mathbf{x}]/J$ has a very simple form. In fact, we get an equivalence relation between monomials generated by

$$\mathbf{x}^{\alpha_i + \gamma} \sim \mathbf{x}^{\beta_i + \gamma}, \text{ any } i, \gamma$$

and

$$k[\mathbf{x}]/J \cong \oplus_\delta\, k \cdot \mathbf{x}^\delta$$

where δ runs over a set of representatives of each equivalence class.

Bearing this in mind, let's look at the 1st order syzygies for the homogeneous ideal:

$$J_n^H = (S^{(n)}, F^{(n)}, I_n^H).$$

$S^{(n)}$ and $F^{(n)}$ are part of a minimal set of generators, and let $f_\alpha \in I_n^H$ complete them. Then syzygies are equations:

$$p\, S^{(n)} + q\, F^{(n)} + \sum r_\alpha\, f_\alpha = 0.$$

One such is given by:

$$\left[u^{e_n+e}\, C_i^{(n)}\right] S^{(n)} + \left[-u^e\, (B_i^{(n)})^{e_n}\, C_i^{(n)}\right] F^{(n)} + \sum R_\alpha\, f_\alpha = 0$$

for some R_α, and some $e \ge 0$ (the extra power u^e is necessary because some terms $R_\alpha f_\alpha$ have degree greater than $e_n + 2$) whose degree is $2 + e_n + e$. Now express this syzygy as a combination of a minimal set of syzygies. This gives us in particular:

$$\begin{aligned} u^{e_n+e}\, C_i^{(n)} &= \sum a_\lambda\, p_\lambda \\ -u^e\, (B_i^{(n)})^{e_n}\, C_i^{(n)} &= \sum a_\lambda\, q_\lambda \\ p_\lambda\, S^{(n)} + q_\lambda F^{(n)} + \sum R_{\alpha\lambda}\, f_\alpha &= 0\,. \end{aligned}$$

Then for some λ, p_λ must have a term of the form u^ℓ or $u^\ell C_i^{(n)}$, hence the monomial $u_\ell S^{(n)}$ or $u_\ell C_i^{(n)} S^{(n)}$ occurs in $p_\lambda S^{(n)}$. But by the general remark on quotient rings by such simple ideals, this means that this term must equal some second term $M S^{(n)}$ (M a monomial in p_λ) or $M F^{(n)}$ (M a monomial in q_λ) mod I_n^H. By the lemma, the first doesn't happen and the second only happens if the term $u^\ell C_i^{(n)} (B_i^{(n)})^{e_n}$ occurs in q_λ, in which case $e_n + 1 \le \deg q_\lambda = \deg(\mathrm{syzygy}(p_\lambda, q_\lambda, R_{\alpha\lambda})) - 1$. This proves:

Proposition 3.11 *J_n^H has for its bounds:*

$$\begin{aligned} d(J) &= 4 \\ m(J) &\ge 2^{2^n} + 1. \end{aligned}$$

Going on to the 3rd aspect of the conjecture, consider results giving better bounds for $m(I)$ under restrictive hypotheses on $V(I)$.

Theorem 3.12 *If $Z \subset \mathbf{P}^n$ is a reduced subscheme purely of dimension r, and $I = I(Z)$ is the full ideal of functions vanishing on Z, then*
(a) if $r \le 1$, or Z is smooth, char$(k) = 0$ and $r \le 3$, then:

$$m(I) \le \deg Z - n + r + 1$$

(b) if char$(k) = 0$ and Z is smooth,

$$m(I) \le (r+1)(\deg(Z) - 2) + 2.$$

Since $\deg(Z) \le d(I)^{n-r}$ (Proposition 3.5), these bound $m(I)$ in terms of $d(I)$.

Part (a) of this are due to Gruson-Lazarsfeld-Peskine [GLP83] for $r \le 1$, and to Pinkham [Pin86], Lazarsfeld [Laz87], and Ran [Ran90] for $r \le 3$. It is *conjectured* by Eisenbud and Goto [EG84], and others, that the bound in (a) holds for all reduced irreducible Z, and it might well hold even for reduced equidimensional Z which are connected in codimension 1. As this problem is now understood, the needed cohomological arguments follow formally, once one can control the singularities of a projection of the variety. These singularities become progressively harder to subdue as the dimension of the variety increases, and are what impedes definitive progress beyond dimension 3.

Part (b) is due to the second author and is proven in the technical appendix. It has been generalized by Bertram, Ein, and Lazarsfeld [BEL91] to show that any smooth characteristic 0 variety of codimension e defined as a subscheme of $\mathbf{P}^n$ by hypersurfaces of degrees $d_1 \ge \ldots \ge d_m$ is $(d_1 + \ldots d_e - e + 1)$-regular. Since we cannot decide the previous conjecture, this is a result of considerable practical importance, for it strongly bounds the complexity of computing Gröbner bases of smooth characterisitic 0 varieties in terms of the degrees of the input equations.

The biggest missing link in this story is a decent bound on $m(I)$ for any reduced equidimensional ideal I. We would conjecture that if a linear bound as in part (a) doesn't hold, at the least a so-called "single exponential" bound, i.e. $m(I) \leq d^{O(n)}$ ought to hold. This is an essential ingredient in analyzing the worst-case behavior of all algorithms based on Gröbner bases, and would complete the story about what causes the bad examples discussed above. At least in some cases Ravi [Rav90] has proven that the regularity of the radical of a scheme is no greater than the regularity of the scheme itself.

There is a direct link between the bounds that we have given so far and the G. Hermann bound with which we started the section. This results from the following:

Proposition 3.13 *Let $I^A \subset k[x_1, \ldots, x_n]$ have generators $f_1, \ldots, f_k$ and let $I^H \subset k[x_0, x_1, \ldots, x_n]$ be the ideal generated by homogenizations $f_1^h, \ldots, f_k^h$ of the f_i. Let $I^H = \mathbf{q}_0 \cap \ldots \cap \mathbf{q}_t$ be the primary decomposition of I^H, let $Z_i = V(\mathbf{q}_i)$ and let*

$$\mathrm{mult}_\infty(I^H) = \max\left[\mathrm{mult}_I(\mathbf{q}_{i_1}) + \ldots + \mathrm{mult}_I(\mathbf{q}_{i_k}) + \mathrm{mult}_I(\mathbf{q}_0)\right]$$

where the max *is taken over chains $V((x_0)) \supset Z_{i_1} \supsetneq \ldots \supsetneq Z_{i_k}$. If $g \in I^A$, then we can write:*

$$g = \sum_{i=1}^{k} a_i f_i$$

where

$$\deg a_i \leq \deg g + \mathrm{mult}_\infty(I^H).$$

The proof goes like this: Let g^h be the homogenization of g. Consider the least integer m such that $x_0^m g^h \in I^H$. Since $g \in I$, this m is finite. Moreover, if

$$x_0^m g^h = \sum x_0^{m_i} a_i^h f_i^h$$

then

$$g = \sum a_i f_i$$

and

$$\deg a_i = \deg(a^h) \leq \deg(x_0^m g^h) - \deg f_j \leq m + \deg(g).$$

Now in the primary decomposition of I^H, suppose that for some k,

$$x_0^k g^h \in \bigcap_{i \in S} q_i, \text{ and } x_0^k g^h \notin \mathbf{q}_j \text{ if } j \notin S.$$

Choose $\ell \notin S$ such that $V(q_\ell)$ is maximal. Since $g \in I^A$, we know $V(\mathbf{q}_\ell) \subset V((x_0))$, hence $x_0 \in \mathbf{p}_\ell$. Let

$$I_S = \bigcap_{i \in S} \mathbf{q}_i.$$

Then $\text{mult}_I(\mathbf{q}_\ell)$ is easily seen to be the length of a maximal chain of ideals between:

$$I \cdot k[\mathbf{x}]_{\mathbf{p}_\ell} \text{ and } I_S \cdot k[\mathbf{x}]_{\mathbf{p}_\ell}.$$

But look at the ideals J_p, for $p \geq 0$, defined by

$$I\, k[\mathbf{x}]_{\mathbf{p}_\ell} \subset \underbrace{(I, x_0^{k+p} g^h)\, k[\mathbf{x}]_{\mathbf{p}_\ell}}_{J_p} \subset I_S\, k[\mathbf{x}]_{\mathbf{p}_\ell}.$$

If $J_p = J_{p+1}$, then

$$\begin{aligned} x_0^{k+p} g^h &\in (I, x_0^{k+p+1} g^h) \\ \text{i.e., } x_0^{k+p} g^h &= a\, x_0^{k+p+1} g^h + b, \quad b \in I. \end{aligned}$$

But $1 - ax_0$ is a unit in $k[\mathbf{x}]_{\mathbf{p}_\ell}$, so

$$J_p = x_0^{k+p} g^h = (1 - ax_0)^{-1} b \in I\, k[\mathbf{x}]_{\mathbf{p}_\ell}.$$

This means that in any case

$$x_0^{k+\text{mult}_I(\mathbf{q}_\ell)} g^h \in I \cdot k[\mathbf{x}]_{\mathbf{p}_\ell}$$

hence, because q_ℓ is p_ℓ-primary:

$$x_0^{k+\text{mult}_I(\mathbf{q}_\ell)} g^h \in \mathbf{q}_\ell$$

Induction now shows that

$$x_0^{\text{mult}_\infty(I^H)} g^h \in I^H$$

Corollary 3.14 *Let I^A, I^H be as above. If $g \in I^A$, then*

$$g = \sum a_i f_i$$

where $\deg(a_i) \leq \deg(g) + \binom{m(I)+n+1}{n+1}$.

Proof. Combine Propositions 3.6 and 3.13.

If we further estimate $m(I)$ by Theorem 3.7 in characteristic 0 or by Proposition 3.8, we get somewhat weaker versions of Hermann's Theorem 3.1. But if $I = V(Z)$, Z a good variety, we may expect the Corollary to give much better bounds than Theorem 3.1.

Corollary 3.14 shows that any example which demonstrates the necessity of double exponential growth in Hermann's ideal membership bound (Theorem 3.1) also demonstrates the necessity of double exponential growth in the bounds on $m(I)$ given in Theorem 3.7 and Proposition 3.8. Thus we can make use of the general arguments for the existence of such examples given in [MM82], rather than depending on the single example of Proposition 3.11, to show that the bounds on $m(I)$ inevitably grow double exponentially: Since in Corollary 3.14, the degrees of the a_i are bounded by a single exponential function of $m(I)$, in all examples where the degrees of the a_i grow double exponentially, $m(I)$ also grows double exponentially.

This line of argument gives a geometric link between the ideal membership problem and $m(I)$: In Corollary 3.14, if I^A exhibits a_i of high degree, then I^H has primary components of high multiplicity. These components force $m(I)$ to be large, and distinguish I^H from good ideals considered in Theorem 3.12 and related conjectures.

A major step in understanding the gap between the double exponential examples and the strong linear bounds on the regularity of many smooth varieties was taken by Brownawell [Bro87] and Kollár [Kol88]. They discovered the beautiful and satisfying fact that if we replace membership in I by membership in $\sqrt{I}$, then there are single exponential bounds on the degrees of a_i:

Theorem 3.15 (Brownawell, Kollár) *Let k be any field, let $I = (f_1, ..., f_k) \subset k[x_1, ..., x_n]$ and let $d = \max(\deg(f_i), i = 1, \cdots, k; 3)$. If $n = 1$, replace d by $2d - 1$. If $g \in \sqrt{I}$, then there is an expression*

$$g^s = \sum_{i=1}^{k} a_i f_i$$

where $s \leq d^n$ and $\deg(a_i) \leq (1 + \deg(g))d^n$. In particular:

$$\left(\sqrt{I}\right)^{d^n} \subset I.$$

What this shows is that although the bad examples have to have primary components at infinity of high degree, nonetheless these primary ideals contain relatively small powers of $\sqrt{I^H}$. The picture you should have is that these embedded components at infinity are like strands of ivy that creep a long way out from the hyperplane at infinity, but only by clinging rather closely to the affine components.

Technical Appendix to Section 3

1. Proof of the equivalence of the conditions in Definition 3.2:

In [Mum66, pp. 99-101], it is proven that for any coherent sheaf $\mathcal{F}$ on $\mathbf{P}^n$, $H^i(\mathcal{F}(-i)) = (0)$, $i \geq 1$ implies that the same holds for $\mathcal{F}(d)$, all $d \geq 0$, and that $H^0(\mathcal{F}(d))$ is generated by $H^0(\mathcal{F}) \otimes H^0(\mathcal{O}(d))$. In particular, if you apply this to $\mathcal{F} = \mathcal{I}(m)$, the equivalence of (a) and (b) follows. (Note the diagram:

$$\begin{array}{ccc} I_d & \longrightarrow & H^0(\mathcal{I}(d)) \\ \cap & & \cap \\ k[\mathbf{x}]_d & \longrightarrow & H^0(\mathcal{O}_{\mathbf{P}^n}(d)) \end{array}$$

which shows that $I_m \to H^0(\mathcal{I}(m))$ is injective for every d). To show that (b) $\Rightarrow$ (c), first note that we may rephrase the reults in [Mum66] to say that if $H^i(\mathcal{F}(-i)) = (0)$, $i \geq 1$, then the degrees of the minimal generators of the $k[\mathbf{x}]$-module

$$\bigoplus_{d \in \mathbf{Z}} H^0(\mathcal{F}(d))$$

are all zero or less. So we may construct the resolution in (c) inductively: at the kth stage, say

$$\bigoplus_{\alpha=1}^{r_k+1} k[\mathbf{x}] \cdot e_{\alpha,k-1} \xrightarrow{\phi_{k-1}} \cdots \longrightarrow k[\mathbf{x}] \longrightarrow k[\mathbf{x}]/I \longrightarrow 0$$

has been constructed, let $M_k = \ker(\phi_{k-\ell})$ and let $\mathcal{F}_k$ be the corresponding sheaf of ideals. The induction hypothesis will say that $H^i(\mathcal{F}_k(m+k-1)) = (0)$, $i \geq 1$. Therefore M_k is generated by elements of degree $\leq m+k$, i.e., $d_\alpha = \deg e_{\alpha,k} \leq m+k$, all α. We get an exact sequence

$$0 \longrightarrow M_{k+1} \longrightarrow \bigoplus_{\alpha=1}^{r_k} (k[\mathbf{x}] \cdot e_{\alpha,k}) \longrightarrow M_k \longrightarrow 0$$

hence

$$0 \longrightarrow \mathcal{F}_{k+1} \longrightarrow \bigoplus_{\alpha=1}^{r_k} \mathcal{O}_{\mathbf{P}^n}(-d_\alpha) \longrightarrow \mathcal{F}_k \longrightarrow 0 \tag{1}$$

Therefore

$$\bigoplus_{\alpha=1}^{r_k} H^i(\mathcal{O}_{\mathbf{P}^n}(m+k-i-d_\alpha)) \longrightarrow k-i)) \longrightarrow \tag{2}$$

$$H^{i+1}(\mathcal{F}_{k+1}(m+(k+1)-(i+1))) \longrightarrow \bigoplus_{\alpha=1}^{r_k} H^{i+1}(\mathcal{O}_{\mathbf{P}^n}(m+k-i-d_\alpha))$$

is exact. But $m+k-i-d_\alpha \geq -i$ so $H^{i+1}(\mathcal{O}_{\mathbf{P}^n}(m+k-i-d_\alpha)) = (0)$. This shows that $\mathcal{F}_{k+1}$ satisfies the induction hypothesis and we can continue. Thus (c) holds. To see that (c) $\Rightarrow$ (a), we just use the same exact sequences (1) and prove now by descending induction on k that $H^i(\mathcal{F}_k(m+k-i)) =$

(0), $i \geq 1$. Since $I = \mathcal{F}_0$, this does it. The inductive step again uses (2), since $H^i(\mathcal{O}_{\mathbf{P}^n}(m+k-i-d_\alpha)) = (0)$ too.

2. Proof of Proposition 3.6:

Look first at the case $r = 0$. Let $\mathcal{I}$ be the sheaf of ideals defined by I and let $\mathcal{I}^* \supset \mathcal{I}$ be the sheaf defined by omitting all 0-dimensional primary components of I. Consider the exact sequence:

$$0 \longrightarrow \mathcal{I}(m-1) \longrightarrow \mathcal{I}^*(m-1) \longrightarrow (\mathcal{I}^*/\mathcal{I})(m-1) \longrightarrow 0$$

This gives us:

$$H^0(\mathcal{I}^*(m-1)) \longrightarrow H^0((\mathcal{I}^*/\mathcal{I})(m-1)) \longrightarrow H^1(\mathcal{I}(m-1))$$

Now $H^1(\mathcal{I}(m-1)) = (0)$ by m-regularity, and $h^0((\mathcal{I}^*/\mathcal{I})(m-1)) = h^0(\mathcal{I}^*/\mathcal{I}) = \text{length}(\mathcal{I}^*/\mathcal{I}) = \text{arith-deg}_0(I)$ since $\mathcal{I}^*/\mathcal{I}$ has 0-dimensional support. But $H^0(\mathcal{I}^*(m-1)) \subset H^0(\mathcal{O}_{\mathbf{P}^n}(m-1))$, so

$$\begin{aligned} \text{arith-deg}_0(I) &\leq h^0(\mathcal{I}^*(m-1)) \\ &\leq h^0(\mathcal{O}_{\mathbf{P}^n}(m-1)) \\ &= \binom{m+n-1}{n} \end{aligned}$$

If $r > 0$, we can prove the Proposition by induction on r. Let H be a generic hyperplance in $\mathbf{P}^n$, given by $h = 0$. Let $I_H = (I, h)/(h) \subset k[x_0, \ldots, x_n]/(h) \cong k[x'_0, \ldots, x'_{n-1}]$ for suitable linear combinations x'_i of x_i. Then it is easy to check that:

$$\text{arith-deg}_r(I) = \text{arith-deg}_{r-1}(I_H)$$

and that I_H is also m-regular, so by induction

$$\begin{aligned} \text{arith-deg}_{r-1}(I_H) &\leq \binom{m+(n-1)-(r-1)-1}{(n-1)-(r-1)} \\ &= \binom{m+n-r-1}{n-r} \end{aligned}$$

If $r = -1$, we use the fact that

$$0 \longrightarrow I_d \longrightarrow H^0(\mathcal{I}(d)) \xleftarrow{\approx} (I^{\text{sat}})_d$$

if $d \geq m$, hence

$$\dim(I^{\text{sat}}/I) \leq \dim k[\mathbf{x}]/(x_0, \ldots, x_n)^m = \binom{m+n}{n+1}.$$

3. Proof of Proposition 3.8:

Let $I \subset k[x_0, \dots x_n]$ and assume, after a linear change of coordinates, that x_n is not contained in any associated prime ideals of I. Let $\overline{I} \subset k[x_0, \dots x_{n-1}]$ be the image of I. Then $d(\overline{I}) = d(I)$ and by induction we may assume

$$m(\overline{I}) \leq (2d(I))^{(n-1)!}.$$

We will prove, in fact, that

$$m(I) \leq m(\overline{I}) + \binom{m(\overline{I}) - 1 + n}{n} \tag{3}$$

and then we will be done by virtue of the elementary estimate:

$$\text{if } m^* = (2d(I))^{(n-1)!}, \text{ and } d \geq 2, \text{ then } m^* + \binom{m^* - 1 + n}{n} \leq (2d(I))^{n!}$$

To prove (3), we use the long exact sequence

$$\begin{array}{ccccccccc} 0 & \longrightarrow & (I:(x_0))_{k-1} & \xrightarrow{x_0} & I_k & \longrightarrow & \overline{I}_k & \longrightarrow & 0 \\ & & \downarrow & & \downarrow & & \downarrow & & \\ 0 & \longrightarrow & H^0(\mathcal{I}(k-1)) & \longrightarrow & H^0(\mathcal{I}(k-1)) & \longrightarrow & H^0(\overline{\mathcal{I}}(k-1)) & \xrightarrow{\delta} & \\ & & & & & & & & \\ & \xrightarrow{\delta} & H^1(\mathcal{I}(k-1)) & \longrightarrow & H^1(\mathcal{I}(k)) & \longrightarrow & H^1(\overline{\mathcal{I}}(k)) & & \end{array}$$

where $(I:(x_0)) = \{ f \mid x_0 f \in I \}$. Let $\overline{m} = m(\overline{I})$. Note that $H^i(\overline{\mathcal{I}}(k-1)) = (0)$, $i \geq 1$, $k \geq \overline{m}$, hence

$$H^i(\mathcal{I}(k-1)) \to H^i(\mathcal{I}(k))$$

is an isomorphism if $k \geq \overline{m} - 1 + 1$ and $i \geq 2$. Since $H^i(\mathcal{I}(k)) = (0)$, $k \gg 0$, this shows that $H^i(\mathcal{I}(k)) = (0)$, $i \geq 2$, $k \geq \overline{m} - i$. Moreover $\overline{I}_k \to H^0(\overline{\mathcal{I}}(k))$ is an isomorphism if $k \geq \overline{m}$, hence $\delta = 0$ if $k \geq \overline{m}$, hence $H^1(\mathcal{I}(k)) = (0)$, $k \geq \overline{m} - 1$. But now look at the surjectivity of $I_k \to H^0(\mathcal{I}(k))$. For all k, let M_k be the cokernel. Then $\oplus M_k$ is a $k[\mathbf{x}]$-module of finite dimension. Multiplication by x_0 induces a sequence:

$$0 \longrightarrow \frac{(I:(x_0))_{k-1}}{I_{k-1}} \longrightarrow M_{k-1} \xrightarrow{x_0} M_k \longrightarrow 0$$

which is exact if $k \geq \overline{m}$. But if, for one value of $k \geq \overline{m}$,

$$(I:(x_0))_k = I_k \tag{4}$$

then by Theorem 3.3, I is k-regular and (4) continues to hold for larger k, and M_k must be (0). In other words,

$$\dim M_k, \;\; k \geq \overline{m} - 1$$

is non increasing and monotone decreasing to zero when $k \geq \overline{m}$. Therefore

$$\begin{aligned} m(I) &\leq \overline{m} + \dim M_{\overline{m}-1} \\ &\leq \overline{m} + \dim k[\mathbf{x}]_{\overline{m}-1} \\ &\leq \overline{m} + \binom{\overline{m}-1+n}{n} \end{aligned}$$

which proves (3).

4. Proof of Theorem 3.12(b):

Let Z be a smooth r-dimensional subvariety of $\mathbf{P}^n$ and $d =$ degree of Z. We first consider linear projections of Z to $\mathbf{P}^r$ and to $\mathbf{P}^{r-1}$. To get there, let $L_1 \subset \mathbf{P}^n$ be a linear subspace of dimension $n-r-1$ disjoint from Z and $L_2 \subset L_1$ a linear subspace of dimension $n-r-2$. Take these as centers of projection:

$$\begin{array}{ccc} \mathbf{P}^n - L_1 & \supset & Z \xrightarrow{p_2} \\ \downarrow & & \downarrow p_1 \\ \mathbf{P}^{r+1} - \{P\} & \supset & Z_1 \\ \downarrow & & \\ \xrightarrow{p_2} \mathbf{P}^r & & \end{array}$$

Let $x_0, \ldots x_{r+1}$ be coordinates on $\mathbf{P}^{r+1}$ so that $p = (0, \ldots, 0, 1)$, hence $x_0, \ldots x_r$ are coordinates on $\mathbf{P}^r$. Let $f(x_0, \ldots x_{r+1}) = 0$ be the equation of the hypersurface Z_1.

Now there are two ways of getting r-forms on Z: by pullback of r-forms on $\mathbf{P}^r$ and by residues of $(r+1)$-forms on $\mathbf{P}^{r-1}$ with simple poles along Z_1. The first gives us a sheaf map

$$p_2^* \, \Omega^r_{\mathbf{P}^r} \hookrightarrow \Omega^r_Z$$

whose image is $\Omega^r_Z(-B_1)$, B_1 the branch locus of p_2. Corresponding to this on divisor classes:

$$\begin{aligned} K_Z &\equiv p_2^*(K_{\mathbf{P}^r}) + B_1 \\ &\equiv -(r+1)H + B_1, \end{aligned} \tag{5}$$

where $H =$ hyperplane divisor class on Z. The second is defined by

$$a(\mathbf{x}) \cdot \frac{dx_1 \wedge \ldots \wedge dx_{r+1}}{f} \longmapsto p_1^* \left(a(\mathbf{x}) \cdot \frac{dx_1 \wedge \ldots \wedge dx_r}{\partial f / \partial x_{r+1}} \right) \tag{6}$$

and it gives us an isomorphism

$$p_1^*(\Omega^{r+1}_{\mathbf{P}^{r+1}}(Z_1) \,|_{Z_1}) \cong \Omega^r_Z(B_2)$$

B_2 is a divisor which can be interpreted as the *conductor* of the affine rings of Z over those of Z_1: i.e.,

$$f \in \mathcal{O}_Z(-B_2) \iff f \cdot (p_{1,*}\mathcal{O}_Z) \subset \mathcal{O}_{Z_1}.$$

In particular,

$$p_{1,*}(\mathcal{O}_Z(-B_2)) \cong \text{sheaf of } \mathcal{O}_{Z_1} - \text{ ideals } C \text{ in } \mathcal{O}_{Z_1}. \tag{7}$$

A classical reference for these basic facts is Zariski [Zar69], Prop. 12.13 and Theorem 15.3. A modern reference is Lipman [Lip84] (apply Def. (2.1)b to p_1 and apply Cor. (13.6) to $Z_1 \subset \mathbf{P}^{r+1}$). (4) gives us the divisor class identity:

$$\begin{aligned} K_Z + B_2 &\equiv p_1^*(K_{\mathbf{P}^{r+1}} + Z_1) \\ &\equiv (d-r-2)H. \end{aligned} \tag{8}$$

(5) and (8) together tell us that

$$B_1 + B_2 \equiv (d-1)H.$$

In fact, the explicit description (6) of the residue tells us more: namely that if $y_1, \ldots, y_r$ are local coordinates on Z, then

$$\frac{\partial(x_1, \ldots, x_r)}{\partial(y_1, \ldots, y_r)} \cdot \frac{1}{\partial f/\partial x_{r+1}}\, dy_1 \wedge \ldots \wedge dy_r$$

generates $\Omega^r_Z(B_2)$ locally. But $\frac{\partial(x_1,\ldots,x_r)}{\partial(y_1,\ldots,y_r)} = 0$ is a local equation for B_1, so this means that $\partial f/\partial x_{r+1} = 0$ is a local equation for $B_1 + B_2$. But $\partial f/\partial x_{r+1} = 0$ is a global hypersurface of degree $d-1$ in $\mathbf{P}^{r+1}$, hence globally:

$$B_1 + B_2 = p_1^*(V(\frac{\partial f}{\partial x_{r_1}}))$$

(equality of divisors, not merely divisor classes). All this is standard classical material.

(7) has an important cohomological consequence: let $C^* \subset \mathcal{O}_{\mathbf{P}^{r+1}}$ be the sheaf of ideals consisting of functions whose restriction to Z_1 lies in C. Then we get an exact sequence:

$$0 \to \mathcal{O}_{\mathbf{P}^{r+1}}(-Z_1) \to C^* \mathcal{O}_{\mathbf{P}^{r+1}} \to C\,\mathcal{O}_{Z_1} \to 0$$

hence an exact sequence

$$0 \to \mathcal{O}_{\mathbf{P}^{r+1}}(\ell - d) \to C^*\mathcal{O}_{\mathbf{P}^{r+1}}(\ell) \to p_{1,*}(\mathcal{O}_Z(\ell H - B_2)) \to 0$$

for all integers ℓ. But $H^1(\mathcal{O}_{\mathbf{P}^{r+1}}(\ell-d)) = (0)$, hence

$$H^0(C^*\mathcal{O}_{\mathbf{P}^{r+1}}(\ell)) \to H^0(\mathcal{O}_Z(\ell H - B_2))$$

is surjective, hence

$$H^0(\mathcal{O}_Z(\ell H - B_2)) \subset \operatorname{Im}\left[H^0(\mathcal{O}_{\mathbf{P}^{r+1}}(\ell)) \to H^0(\mathcal{O}_Z(\ell H))\right]. \qquad (9)$$

Now let us vary the projections p_1 and p_2. For each choice of L_1, we get a different B_1: call it $B_1(L_1)$, and for each choice of L_2, as different B_2: call it $B_2(L_2)$. By (5) and (8), all divisors $B_1(L_1)$ are linearly equivalent as are all divisors $B_2(L_2)$. Moreover:

$$\begin{aligned} \bigcap_{L_1} B_1(L_1) &= \emptyset \\ \bigcap_{L_2} B_2(L_2) &= \emptyset \end{aligned}$$

This is because, if $x \in Z$, then there is a choice of L_1 such that $p_1 : Z \to \mathbf{P}^r$ is unramified at y; and a choice of L_2 such that $p_2(x) \in Z_1$ is smooth, hence p_2 is an isomorphism near x. Thus

$$|B_1(L_1)| = |K_Z + (r+1)H|$$

and

$$|B_2(L_2)| = |K_Z + (d-r-2)H|$$

are base point free linear systems.

Next choose $(r+1)$ L_2's, called L_2^α, $1 \leq \alpha \leq r+1$, so that if $B_2^{(\alpha)} = B_2(L_2^{(\alpha)})$, then $\bigcap_\alpha B_2^{(\alpha)} = \emptyset$. Look at the Koszul complex:

$$\begin{aligned} 0 \to \mathcal{O}_Z(\ell H - \textstyle\sum B_2^{(\alpha)}) &\to \cdots \\ \to \sum_{\alpha,\beta} \mathcal{O}_Z(\ell H - B_2^{(\alpha)} - B_2^{(\beta)}) &\to \sum_\alpha \mathcal{O}_Z(\ell H - B_2^{(\alpha)}) \to \mathcal{O}_Z(\ell H) \to 0. \end{aligned}$$

This is exact and diagram chasing gives the conclusion:

$$H^i(\mathcal{O}_Z(\ell H - (i+1)B_2)) = (0), \text{ all } i \geq 1$$

$$\Rightarrow \sum_\alpha H^0(\mathcal{O}_Z(\ell H - B_2^{(\alpha)})) \to H^0(\mathcal{O}_Z(\ell H)) \text{ surjective}$$

hence by (9)

$$H^0(\mathcal{O}_{\mathbf{P}^n}(\ell)) \to H^0(\mathcal{O}_Z(\ell H)) \text{ surjective}$$

and

$$H^{i+j}(\mathcal{O}_Z(\ell H - (i+1)B_2)) = (0), \text{ all } i \geq 0$$
$$\Rightarrow H^j(\mathcal{O}_Z(\ell H) = (0).$$

Now $I(Z)$ is m-regular if and only if $H^i(\mathcal{I}_Z(m-i)) = (0)$, $i \geq 1$, hence if and only if

$$H^0(\mathcal{O}_{\mathbf{P}^n}(m-1)) \to H^0(\mathcal{O}_Z(m-1)) \text{ surjective}$$
$$H^i(\mathcal{O}_Z(m-i-1)) = (0), \; i \geq 1.$$

By the previous remark, this follows provided that

$$H^{i+j}(\mathcal{O}_Z((m-i-1)H - (j+1)B_2)) = (0), \text{ if } i,j \geq 0, \; i+j \geq 1.$$

But let us rewrite:

$$(m-i-1)H - (j+1)B_2 \equiv K_Z + jB_1 + (m-i-(j+1)(d-1)+r)H$$

using (5) and (8). Note that $jB_1 + \ell H$ is an ample divisor if $\ell \geq 1, j \geq 0$, because $|B_1|$ is base point free. Therefore by the Kodaira Vanishing Theorem,

$$H^i(\mathcal{O}_Z(K_Z + jB_1 + \ell H)) = (0), \; i,j \geq 1, \; j \geq 0$$

and provided $m = (r+1)(d-2)+2$, this gives the required vanishing.

4 Applications

From some points of view, the first main problem of algebraic geometry is to reduce the study of a general ideal I to that of prime ideals, or the study of arbitrary schemes to that of varieties. One way of doing this is to find a decomposition of the ideal into primary ideals: i.e. write it as an intersection of primary ideals. But even when non-redundancy is added, this is not unique, and usually one actually wants something less: to find its radical and perhaps write the radical as an intersection of prime ideals, or to find its top dimensional part, or to find its associated prime ideals and their multiplicities. There are really four computational problems involved here which should be treated separately: (i) eliminating the multiplicities in the ideal I, (ii) separating the pieces of different dimension, (iii) "factoring" the pieces of each dimension into irreducible components, and finally (iv) describing the original multiplicities, either numerically or by a primary ideal. Three of these four problems are the direct generalizations of the basic problems for factoring a single polynomial: we can eliminate multiple factors, getting a square-free polynomial, we can factor this into irreducible pieces and we can ask for the multiplicities with which each factor appeared in the original polynomial. There is a fifth question which arises when we work, as we always must do on a computer, over a non-algebarically closed field k: we can ask (v) for an extension field k' of k over which the irreducible components break up into absolutely irreducible components.

Classical algorithms for all of these of these rely heavily on making explicit projections of $V(I)$ to lower dimensional projective spaces. This can be done either by multi-variable resultants if you want only the set-theoretic projection, or by Gröbner bases with respect to the lexicographic order or an elimination order, to get the full ideal $I \cap k[X_0, \cdots, X_m]$. Recent treatments of multi-variable resultants can be found in [Can89], [Cha91], and a recent treatment of the basis method can be found in [GTZ88]. There is no evidence that either of these is an efficient method, however, and taking Gröbner bases in the lexicographical order or an elimination order is often quite slow, certainly slow in the worst case. The general experience is that taking projections can be very time consuming. One reason is that the degree of the generators may go up substantially and that sparse defining polynomials may be replaced by more or less generic polynomials. A specific example is given by principally polarized abelian varieties of dimension r: they are defined by quadratic polynomials in $(4^r - 1)$-space, but their degree here (hence the degree of their generic projection to $\mathbf{P}^{r+1}$) is $4^r r!$ [Mum70a]. In fact, any variety is defined purely by quadratic relations in a suitable embedding [Mum70b].

Instead of using real computational experience, the fundamental method in theoretical computer science for analyzing complexity of algorithms is to count operations. For algebraic algorithms, the natural measure of complexity is not the number of bit operations, but the number of field operations, addition, subtraction, multiplication and (possibly) division that are used. In this sense, any methods that involve taking Gröbner bases for any order on monomials will have a worst-case behavior whose complexity goes up with the regularity of the ideal hence will take "double exponential time". However, it appears that this worst-case behavior may in fact only concern problem (iv) – finding the primary ideals – and that problems (i), (ii) and (iii) may be solvable in "single exponential time". The idea that such algorithms should exist for finding $V(I)$ set-theoretically was proposed in the 1984 lecture on which this article is based, but turned out, in fact, to have been already proven by Chistov and Grigoriev, cf. their unpublished 1983 note [CG83]. Their line of research led, in some sense, to the work of Brownawell and Kollár, showing the single exponential bound $\left(\sqrt{I}\right)^m \subset I$ for $m = d^n$, where $d = \max($ degrees of generators of $I)$.

Based on this work, Giusti and Heintz [GH91] give a singly exponentially bounded algorithm for computing ideals $\mathbf{q}_i$ such that $V(\mathbf{q}_i)$ are the irreducible components of $V(I)$ (over the ground field k). The method depends on computing what is essentially the Chow form of each component, and leads to an ideal defining this variety but not its full ideal. In fact, their $\mathbf{q}_i$ may be guaranteed to be prime except for possible embedded components.

A direct approach to constructing both $\sqrt{I}$ and the intersection of the top-dimensional primary components of I, denoted $\mathrm{Top}(I)$, is given in a recent paper by Eisenbud, Huneke and Vasconcelos [EHV92]. Their construction of the radical uses the Jacobian ideals, i.e. the ideals of minors of various sizes of the Jacobian matrix of generators of I. This is certainly the most direct approach, but, again they have trouble with possible embedded components, and must resort to ideal quotients, hence they need a Gröbner basis of I in the reverse lexicographic order. They compute $\mathrm{Top}(I)$ as the annihilator of $\mathrm{Ext}^{\mathrm{codim}(I)}(k[X_0, \cdots, X_n]/I, k[X_0, \cdots, X_n])$, which is readily found from a full resolution using Gröbner bases. Their algorithm appears to be practical in some cases of interest, but still has double exponential time worst-case behavior.

It may turn out to be most effective in practice to combine these ideas. Often an ideal under study has regularity far smaller than the geometric degree of its top dimensional components; projecting these components to a hypersurface requires computing in degrees up to the geometric degree, which is wasteful. Other hand, methods such as those in [EHV92] work better in low codimensions, if only because there are fewer minors to consider in the Jacobian matrix. Thus, projecting an arbitary scheme down to low codimension and then switching to direct methods may work best of all.

This still does not settle the issue of the complexity of calculating $\sqrt{I}$, or, for that matter, calculating the full prime ideal of any subvariety of codimension greater than one. Chow form type methods give you an effective method of defining the set $V(I)$ but only of generating I up to possible embedded components. For this reason, the two schools of research, one based on the algebra of I, the other based on subsets of $\mathbf{P}^n$ have diverged. If we knew, as discussed in the previous section, that the regularity of a reduced ideal could be bounded singly exponentially, then we could bound the degrees of the generators of $\sqrt{I}$, and, using Brownawell-Kollár, we could determine $\sqrt{I}$ up to these degrees and get the whole ideal. But without such a bound, it is still not clear whether only $V(I)$ and not $\sqrt{I}$ can be found in worst-case single exponential time.

Let's look at problem (iii). Assume you have found a reduced equidimensional I. To study splitting it into irreducible or absolutely irreducible pieces, we shall assume initially it is a hypersurface, i.e. $I = (f)$. Computationally, there may often be advantages to not projecting a general I to a hypersurface, and we will discuss one such approach below. Geometrically, there is nothing very natural about irreducible but not absolutely irreducible varieties: from the standpoint of their properties, they behave like reducible varieties, except that, being conjugate over k, their components have very similar properties. If the ground field k gets bigger or smaller, the set of absolutely irreducible components gets partitioned in finer or coarser ways into the k-components. If one has never done any calculations, one would therefore be inclined to say – let's extend k as far as needed to split our algebraic set up into absolutely irreducible components. *This is a very bad idea!* Unless this extension k' happens to be something simple like a quadratic or cyclotomic extension of k, the splitting field k' is usually gigantic. This is what happens if one component of $V(I)$ is defined over an extension field k_1 of k of degree e, and the Galois group of k_1/k is the full symmetric group, a very common occurence. Then $V(I)$ only splits completely over the Galois closure of k_1/k and this has degree $e!$. The moral is: never factor unless you have to.

In fact, unless you need to deal simultaneously with more than one of its irreducible components, you can proceed as follows: the function field $K = k[X_0, \cdots, X_n]/(f)$ contains as a subfield an isomorphic copy of k_1: you find that field as an extension $k_1 = k[y]/(p(y))$ of k, and solve for the equation of one irreducible component $f_1 \in k_1[X_0, \cdots, X_n]$ by the formula $\mathrm{Norm}_{k_1/k}(f_1) = f$.

Pursuing this point, why should one even factor the defining equation f over k? Factoring, although it takes polynomial time [LLL82], is often very slow in real time, and, unless the geometry dictates that the components be treated separately, why not leave them alone. In some situations, for instance, [DD84] one may have an ideal, module or other algebraic structure defined by polynomials or matrices of polynomials over a *ground ring* $D = k[y]/(p(y))$, where p is a square-free polynomial. Thus D is a direct sum of extension fields, but there is no need to factor p or split up D until the calculations take different turns with the structures over different pieces of *Spec*(D).

The standard methods of factoring in computer algebra all depend on (i) writing the polynomial over a ring, finitely generated over $\mathbf{Z}$, and reducing modulo a maximal ideal $\mathbf{m}$ in that ring, obtaining a polynomial over a finite field; and (ii) restricting to a line L, i.e. substituting $X_i = a_i X_0 + b_i, i \geq 1$ for all but one variable, obtaining a polynomial in one variable over a finite field. This is then factored and then, using Hensel's lemma, one lifts factorization modulo higher powers of $\mathfrak{m}$ and of the linear space L. One then checks whether a coarsened version of this factorization works for f. This is all really the arithmetic of various small fields. Geometrically, every polynomial in one variable factors over a suitable extension field and the question of counting the absolutely irreducible components of a variety is really more elementary: it is fundamentally topological and not arithmetic. One should, therefore, expect there to be direct geometric ways of counting these components and separating them. Assuming I is a reduced, equi-r-dimensional ideal, the direct way should be to use Serre duality, computing the cohomology $H^r(\Omega^r_{V(I)})$, where $\Omega^r_{V(I)} \subset \omega_{V(I)}$ is the subsheaf of the top-dimensional dualizing sheaf of $V(I)$ of absolutely regular r-forms. Its dimension will be the number of absolutely irreducible components into which $V(I)$ splits. Calculating this cohomology involves two things: algebraically resolving the ideal I and geometrically resolving the singularities of $V(I)$ far enough to work out $\Omega^r_{V(I)}$. Classically, when $I = (f)$ was principal, $\Omega^r_{V(I)}$ was called its ideal of "subadjoint" polynomials.

There is one case where this is quite elementary and has been carried out: this is for plane curves. One can see immediately what is happening by remarking that a non-singular plane curve is automatically absolutely irreducible, hence one should expect that its singularities control its decomposition into absolutely irreducible pieces. Indeed, if $\mathbf{C} \subset k[X_0, X_1, X_2]/(f)$ is the conductor ideal, then $\Omega^1_{V(f)}$ is given by the homogeneous ideal $\mathbf{C}$, but with degree 0 being shifted to be polynomials of degree $d-3$, d the degree of f. To calculate H^1, assume X_0 is not zero at any singularity of $V(f)$ and look at the finite-dimensional vector space of all functions $k[X_1/X_0, X_2/X_0]/(\mathbf{C} + (f))$ modulo the restrictions $g/(X_0^{d-3})$ for all homogeneous polynomials g of degree $d-3$. This will be canonically the space of functions omponents of $V(f)$ with sum 0. In particular, it is (0) if and only if $V(f)$ is absolutely irreducible. This follows from standard exact sequences and duality theory. It was known classically as the Cayley-Bacharach theorem, for the special case where $V(f)$ was smooth except for a finite number of ordinary double points. It states that $V(f)$ is absolutely irreducible if and only if for every double point P, there is a curve of degree $d-3$ passing through all the double points except P.

This example gives one instance where a deeper computational analysis of varieties requires a computation of its resolution of singularities. We believe that there will be many instances where practical problems will require such an analysis. In many ways, resolution theorems look quite algorithmic, and, for instance, Abhyankar and his school have been approaching the problem in this way [Abh82], as have Bierstone and Milman [BM91]. However, the only case of resolution of singularities to be fully analyzed in the sense of computational complexity is that of plane curves. This has been done by Teitelbaum [Tei89], [Tei90]. His analysis is notable in various ways: he is extremely careful about not making unnecessary factorizations, let alone taking unnecesions, and uses the "D" formalism discussed above. He describes his algorithm so precisely that it would be trivial to convert it to code and, as a result, he gives excellent bounds on its complexity.

References

[Abh82] S. S. Abhyankar, *Weighted expansions for canonical desingularization*, Lecture Notes in Math., vol. 910, Springer-Verlag, 1982.

[Art76] M. Artin, *Lectures on deformations of singularities*, Tata Institute on Fundamental Research, Bombay, 1976.

[Bay82] Dave Bayer, *The division algorithm and the Hilbert scheme*, Ph.D. thesis, Harvard University, Department of Mathematics, June 1982, order number 82-22588, University Microfilms International, 300 N. Zeeb Rd., Ann Arbor, MI 48106.

[BEL91] Aaron Bertram, Lawrence Ein, and Robert Lazarsfeld, *Vanishing theorems, a theorem of Severi, and the equations defining projective varieties*, J. Amer. Math. Soc. **4** (1991), 587–602.

[Ber78] G. M. Bergman, *The diamond lemma for ring theory*, Adv. in Math. **29** (1978), 178–218.

[BM88] Dave Bayer and Ian Morrison, *Standard bases and geometric invariant theory I. Initial ideals and state polytopes*, J. Symb. Comput. **6** (1988), no. 2–3, 209–217, reprinted in [Rob89].

[BM91] E. Bierstone and P. Milman, *A simple constructive proof of canonical resolution of singularities*, Effective methods in algebraic geometry (Castiglioncello, 1990), Progr. Math., vol. 94, Birkhauser Boston, 1991, pp. 11–30.

[BCR91] A. M. Bigatti, M. Caboara, and L. Robbiano. *On the computation of Hilbert-Poincare series*, Applicable Algebra in Engineering, Communications, and Computing **2** (1991), 21–33.

[Bri73] J. Briancon, *Weierstrass prepare a la Hironaka*, Astérisque **7,8** (1973), 67–73.

[Bro87] W. D. Brownawell, *Bounds for the degrees in the Nullstellensatz*, Ann. of Math. (2) **126** (1987), 577–591.

[BS87a] Dave Bayer and Mike Stillman, *A criterion for detecting m-regularity*, Invent. Math. **87** (1987), 1–11.

[BS87b] Dave Bayer and Mike Stillman, *A theorem on refining division orders by the reverse lexicographic order*, Duke Math. J. **55** (1987), no. 2, 321–328.

[BS88] Dave Bayer and Mike Stillman, *On the complexity of computing syzygies*, J. Symb. Comput. **6** (1988), 135–147.

[BS92a] Dave Bayer and Mike Stillman, *Macaulay: A system for computation in algebraic geometry and commutative algebra*, 1982–1992, computer software available via anonymous ftp from zariski.harvard.edu.

[BS92b] Dave Bayer and Mike Stillman, *Computation of Hilbert functions*, J. Symb. Comput. **6** (1992), 31–50.

[Buc65] B. Buchberger, Ph.D. thesis, Univ. Innsbrück, 1965.

[Buc76] B. Buchberger, *A theoretical basis for the reduction of polynomials to canonical forms*, ACM SIGSAM Bull. **39** (1976), 19–29.

[Buc79] B. Buchberger, *A criterion for detecting unnecessary reductions in the construction of Gröbner bases*, Symbolic and Algebraic Computation (Proceedings of EUROSAM 79), Lecture Notes in Computer Science, vol. 72, Springer-Verlag, 1979, pp. 3–21.

[Can89] J. Canny, *Generalized characteristic polynomials*, Symbolic and Algebraic Computation (Proceedings of ISSAC 88), Lecture Notes in Computer Science, vol. 358, Springer-Verlag, 1989, pp. 293–299.

[CG83] A. L. Chistov and D. Yu. Grigoriev, *Subexponential-time solving systems of algebraic equations I, II*, Steklov Mathematical Institute, Leningrad department, LOMI Preprints E-9-93, 0E-10-c83, 1983.

[Cha91] Marc Chardin, *Un algorithme pour le calcu des résultants*, Effective methods in algebraic geometry (Castiglioncello, 1990), Progr. Math., vol. 94, Birkhauser Boston, 1991, pp. 47–62.

[DD84] C. Dicrescenzo and D. Duval, *Computations on curves*, Lecture Notes in Computer Science, vol. 174, Springer-Verlag, 1984.

[EG84] David Eisenbud and Shiro Goto, *Linear free resolutions and minimal multiplicity*, J. Algebra **88** (1984), no. 1, 89–133.

[EHV92] David Eisenbud, Craig Huneke, and Wolmer Vasconcelos, *Direct methods for primary decomposition*, Invent. Math. (1992), to appear.

[Eis92] David Eisenbud, *Commutative algebra with a view toward algebraic geometry*, 1992, in preparation.

[Gal74] A. Galligo, *A propos du theoreme de preparation de Weierstrass*, Fonctions de Plusieurs Variables Complexes, Lecture Notes in Math., vol. 409, Springer-Verlag, 1974, pp. 543–579.

[Gal79] A. Galligo, *Theoreme de division et stabilite en geometrie analytique locale*, Ann. Inst. Fourier (Grenoble) **29** (1979), 107–184.

[GH91] Marc Giusti and Joos Heintz, *Algorithmes—disons rapides—pour la decomposition d'une variete algebrique en composantes irreductibles et equidimensionnelles ["Fast" algorithms for the decomposition of an algebraic variety into irreducible and equidimensional components]*, Effective methods in algebraic geometry (Castiglioncello, 1990), Progr. Math., vol. 94, Birkhauser Boston, 1991, pp. 169–194.

[Giu84] Marc Giusti, *Some effectivity problems in polynomial ideal theory*, EUROSAM 84), Lecture Notes in Computer Science, vol. 204, Springer-Verlag, 1984, pp. 159–171.

[GLP83] L. Gruson, R. Lazarsfeld, and C. Peskine, *On a theorem of Castelnuovo, and the equations defining space curves*, Invent. Math. **72** (1983), 491–506.

[GTZ88] P. Gianni, B. Trager, and G. Zacharias, *Gröbner bases and primary decomposition of polynomial ideals*, J. Symb. Comput. **6** (1988), no. 2–3, 149–167, reprinted in [Rob89].

[Her26] Grete Hermann, *Die Frage der endlich vielen Schritte in der Theorie der Polynomideale*, Math. Ann. **95** (1926), 736–788.

[Hir64] H. Hironaka, *Resolution of singularities of an algebraic variety over a field of characteristic zero: I, II*, Ann. of Math. (2) **79** (1964), 109–326.

[Kol88] János Kollár, *Sharp effective Nullstellensatz*, J. Amer. Math. Soc. **1** (1988), no. 4, 963–975.

[Lak91] Y. N. Lakshman, *A simple exponential bound on the complexity of computing Gröbner bases of zero-dimensional ideals*, Effective methods in algebraic geometry (Castiglioncello, 1990), Progr. Math., vol. 94, Birkhauser Boston, 1991, pp. 227–234.

[Laz87] Robert Lazarsfeld, *A sharp Castelnuovo bound for smooth surfaces*, Duke Math. J. **55** (1987), 423–429.

[Len92] H. W. Lenstra, Jr., *Algorithms in algebraic number theory*, Bull. Amer. Math. Soc. (N.S.) **26** (1992), no. 2, 211–244.

[Lip84] Joseph Lipman, *Dualizing sheaves, differentials and residues on algebraic varieties*, Astérisque, vol. 117, 1984.

[LL91] Y. N. Lakshman and D. Lazard, *On the complexity of zero-dimensional algebraic systems*, Effective methods in algebraic geometry (Castiglioncello, 1990), Progr. Math., vol. 94, Birkhauser Boston, 1991, pp. 217–225.

[LLL82] A. K. Lenstra, H. W. Lenstra, Jr., and L. Lovász, *Factoring polynomials with rational coefficients*, Math. Ann. **261** (1982), 515–534.

[Mac27] F. S. Macaulay, *Some properties of enumeration in the theory of modular systems*, Proc. London Math. Soc. **26** (1927), 531–555.

[MM82] Ernst W. Mayr and Albert R. Meyer, *The complexity of the word problem for commutative semigroups and polynomial ideals*, Adv. in Math. **46** (1982), 305–329.

[MM83] H. Michael Möller and Ferdinando Mora, *Upper and lower bounds for the degree of Gröbner bases*, Computer Algebra (EUROCAL 83), Lecture Notes in Computer Science, vol. 162, Springer-Verlag, 1983, pp. 157–167.

[MM86] H. Michael Möller and Ferdinando Mora, *New constructive methods in classical ideal theory*, J. Algebra **100** (1986), no. 1, 138–178.

[MR88] T. Mora and L. Robbiano, *The Gröbner fan of an ideal*, J. Symb. Comput. **6** (1988), no. 2–3, 183–208, reprinted in [Rob89].

[Mum66] David Mumford, *Lectures on curves on an algebraic surface*, Princeton University Press, Princeton, New Jersey, 1966.

[Mum70a] David Mumford, ***Abelian varieties***, Oxford University Press, Oxford, 1970.

[Mum70b] David Mumford, ***Varieties defined by quadratic equations***, Questions on Algebraic Varieties, Centro Internationale Matematica Estivo, Cremonese, Rome, 1970, pp. 29–100.

[MW83] D. W. Masser and G. Wüstholz, ***Fields of large transcendence degree generated by values of elliptic functions***, Invent. Math. **72** (1983), 407–464.

[Pin86] Henry C. Pinkham, ***A Castelnuovo bound for smooth surfaces***, Invent. Math. **83** (1986), 491–506.

[Ran90] Ziv Ran, ***Local differential geometry and generic projections of threefolds***, J. Differential Geom. **32** (1990), 131–137.

[Rav90] M. S. Ravi, ***Regularity of ideals and their radicals***, Manuscripta Math. **68** (1990), 77–87.

[Ric74] F. Richman, ***Constructive aspects of Noetherian rings***, Proc. Amer. Math. Soc. **44** (1974), 436–441.

[Rob85] L. Robbiano, ***Term orderings on the polynomial ring***, Proceedings of EUROCAL '85 (Linz), Lecture Notes in Computer Science, vol. 204, Springer-Verlag, 1985, pp. 513–517.

[Rob89] Lorenzo Robbiano (ed.), ***Computational aspects of commutative algebra***, Academic Press, 1989, ISBN 0-12-589590-9.

[Sch80] Frank-Olaf Schreyer, ***Die Berechnung von Syzygien mit dem verallgemeinerten Weierstrass'schen Divisionssatz***, Diplomarbeit am Fachbereich Mathematik der Universität Hamburg, 1980.

[Sch91] Frank-Olaf Schreyer, ***A standard basis approach to syzygies of canonical curves***, J. Reine Angew. Math. **421** (1991), 83–123.

[Ser55] J.-P. Serre, ***Faisceaux algebrique coherents***, Ann. of Math. (2) **61** (1955), 197–278.

[Spe77] D. Spear, ***A constructive approach to commutative ring theory***, Proceedings of the 1977 MACSYMA Users' Conference, NASA CP-2012, 1977, pp. 369–376.

[Tei89] Jeremy Teitelbaum, *On the computational complexity of the resolution of plane curve singularities*, Symbolic and algebraic computation (Rome, 1988), Lecture Notes in Computer Science, vol. 358, Springer, 1989, pp. 285–292.

[Tei90] Jeremy Teitelbaum, *The computational complexity of the resolution of plane curve singularities*, Math. Comp. **54** (1990), no. 190, 797–837.

[Tri78] W. Trinks, *Über B. Buchberger's Verfahren, Systeme algebraischer Gleichungen zu lösen*, J. Number Theory **10** (1978), 475–488.

[Zac78] G. Zacharias, Bachelor's thesis, Mass. Inst. of Technology, 1978.

[Zar69] Oscar Zariski, *An introduction to the theory of algebraic surfaces*, Lecture Notes in Math., vol. 83, Springer-Verlag, 1969.

[ZS76] Oscar Zariski and Pierre Samuel, *Commutative algebra, Volumes I, II*, Graduate texts in mathematics, vol. 28–29, Springer-Verlag, 1975–1976.

IN MEMORIAM: GEORGE R. KEMPF 1944-2002

Photo courtesy of the Kempf family

I met George in 1970 when he burst on the algebraic geometry scene with a spectacular PhD thesis. His thesis gave a wonderful analysis of the singularities of the subvarieties W_r of the Jacobian of a curve C obtained by adding the curve to itself r times inside its Jacobian. This was one of the major themes that he pursued throughout his career: understanding the interaction of a curve with its Jacobian and especially to the map from the r-fold symmetric product of the curve to the Jacobian. In his thesis he gave a determinantal representation both of W_r and of its tangent cone at all its singular points, which gives you a complete understanding of the nature of these singularities. A major focus of his later work in this area were the Picard bundles: the vector bundles on the Jacobian whose projectivizations are r-fold symmetric powers of C, for $r \geq 2g - 1$. He unwound many of the mysteries of these bundles.

As George's research evolved, our work became closely intertwined in multiple ways. In particular, he worked on invariant theory and on abelian varieties,

D. Mumford, *Selected Papers*, Vol. II,

especially linear systems on abelian varieties. Maybe his result in invariant theory which I loved the most concerns the orbits in a representation of a reductive group that are "unstable," i.e. have 0 in their closure. He gave a beautiful construction of one canonical worst subgroup $\mathbf{G}_m$ in G carrying the point to 0. I had looked for this in awkward ways and found it in some cases, but he saw what was really going on. This result had many corollaries and completed the program in Geometric Invariant Theory in the best possible way. Later on, he studied extensively the singularities of orbit spaces, showing in many cases that they had only rational singularities; he also studied the effective construction of rings of invariants, and thus of orbit spaces.

Perhaps the area in which we were closest was his work on linear systems and the equations defining abelian varieties. I wrote three papers on this in 1966-67, much inspired by hearing Igusa's lectures on theta functions. But I used to joke that George was the only one in the world who actually read these papers. Again, he went deeper than I with more persistence and the deft touch by which I always recognized his work. He kept finding better and more satisfying reasons why abelian varieties are so wonderful. For example, there was his theorem that their homogeneous coordinate rings A were, in his terminology, exactly that: "wonderful." He defined "wonderful" to mean that all the modules $\mathrm{Tor}_i^A(k, k)$ are purely of degree i. This turns out to be the secret cohomological key to answer many questions. Another unexpected and lovely result was the one he dedicated to me for my 50th birthday: that multiplication gives an isomorphism between the tensor product of the vector space of rank 2 theta functions, generically twisted, and the vector space of rank 4 theta functions.

One of the things that distinguished his work was the total mastery with which he used higher cohomology. A paper which, I believe, every new student of algebraic geometry should read, is his elementary proof of the Riemann-Roch theorem on curves: "Algebraic Curves" in Crelle, 1977. That such an old result could be treated with new insight was the work of a master.

I won't discuss his work on the cohomology of homogeneous spaces or the representation theory of algebraic groups, which others know much better than I. Instead, I want to conclude by saying that this love of the simple and satisfying elegance which can be found in these abstract fields brought George and I together. One feels that, given the disease with which he struggled, this mathematics was a constant stable light to which he returned, that centered him when other things failed. We miss the light he shed for us.

David Mumford, September 2002

Article [u64a]
The Boundary of Moduli Schemes

David Mumford

1 Discussion

To begin with, what is a variety of moduli? Start with the set of all nonsingular complete varieties of dimension n and arithmetic genus p. For each isomorphism class of these, take one point: then try to put these points together in a variety. There are some more requirements: a "nearby" pair of varieties V_1, V_2 should correspond to a "nearby" pair of points: e.g.,

Let $\mathscr{S}$ = set of isomorphism classes of V's.
$U \subset \mathscr{S}$ is "open" if, for all families of varieties of the given type, varieties of type U occur over an open set in the parameter space.

Another requirement is that for all families

$$\pi\colon \mathscr{V} \longrightarrow S$$

suppose you map S to $\mathscr{S}$ by assigning to each $s \in S$ the class of the fibre $\pi^{-1}(s)$: then this map should be algebraic.

The problem, in this raw form, has been modified bit by bit so as to make it more plausible:

(I) Instead of classifying "bare" varieties V, one seeks to classify pairs $(V, \mathscr{D})$ where $\mathscr{D}$ is a numerical equivalence class of very ample divisors on V.

(II) Then break up the set $\mathscr{S}$ via the Hilbert polynomials of the divisors in $\mathscr{D}$: viz. for every P, let $\mathscr{S}^P$ = isomorphism classes of $(V, \mathscr{D})$ such that for all $D \in \mathscr{D}$

$$P(n) = \chi(\mathscr{O}_V(nD)).$$

Now we are close to a good problem:

for all $D \in \mathscr{D}$,
for all bases of $H^0(V, \mathscr{O}_V(D))$ you get a canonical immersion,

$$V \subset \mathbb{P}^n \qquad (n = \dim H^0(V, \mathscr{O}_V(D)) - 1)$$

such that hyperplane sections are linearly equivalent to D.

$$\text{i.e.,}\quad \mathcal{S}^P \cong \frac{\text{certain set of subvarieties } V \text{ of } \mathbb{P}^n}{\text{certain equivalence relation, especially projective equivalence}}$$

(III) Why insist that V be nonsingular? The only reason appears to be that over $\mathbb{C}$ families of nonsingular varieties are locally differentiably trivial: so one can view them as families of complex structures on a fixed differentiable manifold (or, as in the Bers–Ahlfors approach, on a fixed topological manifold). Algebraically, there is no point: let's let V be any complete variety at all, maybe even reducible, and assume that $\mathcal{D}$ is a class of *Cartier* divisors.

To go further, let's stop and ask what problems arise: first we should take a broad look at the topology which we are getting by throwing in all varieties—typically it will be very un-separated; second, we should try to find open subsets $U \subset \mathcal{S}^P$ such that, in their induced topology, they are separated, and "compact" if possible.

[This means that if U could be given the structure of a moduli variety, it would turn out complete; and it also means, directly, that if $(V, \mathcal{D}) \in U$, and we specialize the ground field, then we can find a specialization $(\overline{V}, \overline{\mathcal{D}})$ of $(V, \mathcal{D})$ also in U.]

Thirdly, we will finally have to find out if U can be made into a *variety*.

(IV) We understand the last problem better when we realize that, e.g., via Chow coordinates, *almost all* of U is bound to come out as a variety. We saw that $\mathcal{S}^P$ was a quotient of a piece $\mathcal{H}$ of the Chow variety by an algebraic equivalence relation. Such quotients always exist *birationally*, i.e., for a small enough Zariski-open subset $U^* \subset \mathcal{H}$,

[U^* modulo equivalence relation]

will be a good variety. So the 3rd problem is like the first two:

> *The only problem is to pick the "boundary" components shrewdly,* i.e., to decide which nongeneric varieties to allow.

There again, it would prejudice the issue to think that we should necessarily use all and/or only nonsingular varieties. And the choice should be made by a) checking the topology, and b) checking its "algebraizability".

(V) A final step in setting up the problem reasonably is to realize that all the same questions occur equally well for a much more general class of problems: viz. that of forming quotients of varieties by algebraic equivalence relations. Only by realizing this can we hope to find simple enough examples to study first so as to get the right feeling. Especially, the hard equivalence relations are the *noncompact* ones; and in the case of moduli, this occurs principally in forming:

$$\mathcal{H} / \{\text{Projective equivalence of } V\text{'s in } \mathbb{P}^n\}$$

i.e., in forming an orbit space by $PGL(n+1)$.

2 Present State of the Theory

very good (i) analogous problem in classifying vector bundles on a fixed curve
pretty good (ii) moduli of curves (canonically polarized)
half good (iii) moduli of polarized abelian varieties
no good (iv) moduli of surfaces of general type

3 An Example

Rather than analyze an actual moduli problem, I want to take one of the simplest nontrivial orbit space problems, in which all the features of the conjectured results occur:

$G = PGL(2)$ acting on $\mathbb{P}_n$, where $\mathbb{P}_n = n^{\text{th}}$ symmetric product of $\mathbb{P}^1$,
i.e., $PGL(2)$ acting on the set of 0-cycles of degree n.

(= theory of binary quantics).

a) **jump phenomenon:**
look at $\mathbb{P}_2/PGL(2)$. There are 2 orbits: $\{P+Q \mid P \neq Q\}$ and $\{2P\}$. Therefore, get 2 points x, y where x is open but not closed, y is closed but not open:

$$\bullet \longrightarrow \bullet$$

This occurs in all moduli problems, and one always must exclude some points to avoid this.
In $\mathbb{P}_n$, exclude the 0-cycles

$$kP + (n-k)Q$$

whose isotropy group is infinite.

b) **further nonseparation:**
take $n = 6$

generic cycle.

Let all points in group A come together; you get in the limit:

(*)

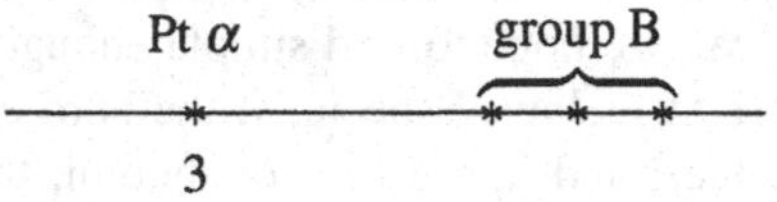

But suppose, as group A collapses to α, you apply a one-parameter subgroup $\mathbb{G}_m \subset PGL(2)$, moving points away from α to β. Then the following are projectively equivalent:

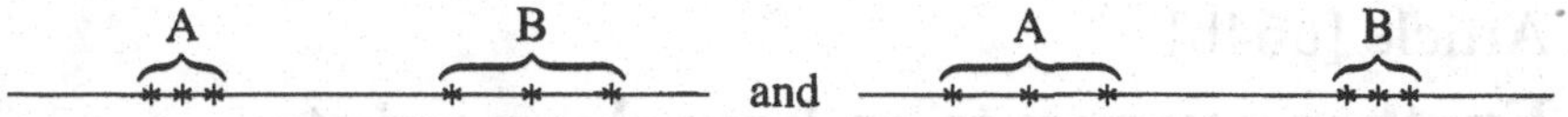

the latter approaches:

(**) 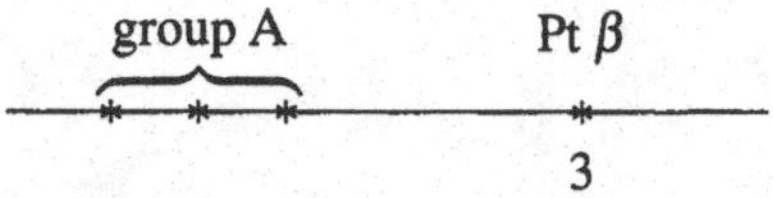

But the 0-cycles (*) and (**) are probably not projectively equivalent.

c) **the unitary retraction:** to avoid these bad things, define

$$\mathscr{K} \subset \mathbb{P}_n,$$

$\mathscr{K}$ = Set of 0-cycles $\sum_{i=1}^{n} P_i$, such that, putting the P_i on the Gauss sphere, and embedding the Gauss sphere in $\mathbb{R}^3$ as $x^2+y^2+z^2=1$, then the *vector* sum of the P_i in $\mathbb{R}^3$ is $(0,0,0)$.

One checks, if $x, y \in \mathscr{K}$, then x, y are equivalent under $PGL(2)$ if and only if they are equivalent under the maximal compact subgroup

$$K = SO(3;\mathbb{R}) \subset PGL(2,\mathbb{C}) = G.$$

But $\mathscr{K}$ is compact, therefore $\mathscr{K}/K$ is compact and separated. And

$$\mathscr{K} \cdot PGL(2) = \left\{ \mathfrak{a} \,\middle|\, \begin{array}{l} \text{no point } Q \text{ occurs in with multiplicity} > n/2; \text{ and} \\ \text{if } Q \text{ occurs with multiplicity } n/2, \text{ then } \mathfrak{a} = \frac{n}{2}(Q+Q') \end{array} \right\}.$$

d) **stability restriction:** $\mathscr{K} \cdot PGL(2)$ contains a Zariski-open set

$$U_{\text{stable}} = \left\{ \mathfrak{a} \,\middle|\, \begin{array}{l} \text{no point } Q \text{ occurs in } \mathfrak{a} \\ \text{with multiplicity} \geqq n/2 \end{array} \right\}.$$

So U_{stable}/G has separated topology, and is compact if n is odd. It is also a variety by virtue of a general theorem of mine.

e) **semi-stability:** When n is even, things are less clean.
$\mathscr{K}$ showed that there was a natural compactification of U_{stable}/G by adding a *single* point representing the cycles $(n/2)(Q+Q')$. In fact, there is a complete *variety* $\overline{V}_n$, with point ∞ and diagram of algebraic maps:

$$\begin{array}{ccc} U_{\text{semi-stable}} & \longrightarrow & \overline{V}_n \\ \cup & & \cup \\ U_{\text{stable}} & \longrightarrow & U_{\text{stable}}/G = \overline{V}_n - (\infty) \end{array}$$

where

$$U_{\text{semi-stable}} = \left\{ \mathfrak{a} \,\middle|\, \begin{array}{l} \text{no point } Q \text{ occurs in } \mathfrak{a} \\ \text{with multiplicity} > n/2 \end{array} \right\}.$$

Article [u64b]

Further comments on boundary points

David B. Mumford

In these notes I shall describe some joint work of A. Mayer and myself, as well as some related results, summarizing further comments made in my lecture and a 2nd lecture by Mayer.[1] During the institute lectures were also given by H. Rauch and L. Ehrenpreis, discussing various aspects of the Torelli and Teichmüller covering spaces of the moduli scheme for curves of genus g (cf. the notes of Ehrenpreis). The ground field will be assumed to be the complex numbers in our discussion. One word of apology: the full proofs of many of our results have not been written down, so, strictly speaking, much of what follows should be taken as conjectures not theorems.

1 Compact moduli spaces for vector bundles over curves.

This theory has been worked out by Seshadri, Narasimhan, and myself. Let $\mathbf{E}$ be a vector bundle of rank r over a curve C.

Definitions:

i) $\mathbf{E}$ is *regular* if the only endomorphisms of $\mathbf{E}$ are multiples of the identity,

ii) $\mathbf{E}$ is *stable* if, for all sub-bundles $\mathbf{F} \subset \mathbf{E}$, $\deg[c_1(\mathbf{F})] < \dfrac{\mathrm{rank}(\mathbf{F})}{\mathrm{rank}(\mathbf{E})} \cdot \deg[c_1(\mathbf{E})]$,

iii) $\mathbf{E}$ is *semi-stable* if, for all sub-bundles $\mathbf{F} \subset \mathbf{E}$, $\deg[c_1(\mathbf{F})] \leq \dfrac{\mathrm{rank}(\mathbf{F})}{\mathrm{rank}(\mathbf{E})} \cdot \deg[c_1(\mathbf{E})]$,

iv) $\mathbf{E}$ is *retractable* if it is a direct sum of stable bundles.

[1] Notes were prepared by A. Mayer for his talk in the 1964 Summer Institute at Woods Hole, but they do not seem to have been distributed during the Summer Institute. These notes were subsequently lost. A possibly expanded version of the notes exists, in pages 6–15 of *Seminar on Degeneration of Algebraic Varieties*, conducted by P.A. Griffiths, Lectures by C.H. Clemens, P.A. Griffiths, T.F. Jamois and A.L. Mayer. Institute for Advanced Study, Princeton, New Jersey, Fall Term, 1969-1970, 152 pp.

D. Mumford, *Selected Papers*, Vol. II,

If $\deg[c_1(\mathbf{E})] = 0$, $\mathbf{E}$ is retractable if and only if $\mathbf{E}$ admits a hermitian structure with curvature form 0.

To obtain a modulus space for vector bundles with given rank and $\deg(c_1)$, first one must throw out irregular bundles since they give rise to jump phenomena, i.e., constant families of bundles, which suddenly jump to another bundle (cf. my lecture notes, "Curves on an algebraic surface", Lecture 7, §4). In the remaining class of bundles the topology is still un-separated; but in the set of retractable bundles the topology is both compact and separated, since this set of bundles is isomorphic to the set of unitary representations of π_1 of the base curve (for $\deg[c_1(\mathbf{E})] = 0$; otherwise the argument can be modified). This set turns out to contain the open set of stable bundles, and to be contained in the open set of semi-stable bundles (it is not open itself). One finds that the stable bundles are classified by the points of a nonsingular variety V, and that V is an open subset of a *compact* variety $\overline{V}$. The set of points of $\overline{V}$ is isomorphic to the (nonalgebraic) set of retractable bundles, and there is even a natural map from the set of all semi-stable bundles to $\overline{V}$, but nonisomorphic bundles no longer correspond to distinct points:

$$\begin{array}{ccccc} \left\{\begin{matrix}\text{regular}\\ \text{bundles}\end{matrix}\right\} & \supset & \left\{\begin{matrix}\text{stable}\\ \text{bundles}\end{matrix}\right\} & \simeq & \left\{\begin{matrix}\text{points of}\\ V\end{matrix}\right\} \\ & & \cap & & \cap \\ & & \left\{\begin{matrix}\text{retractable}\\ \text{bundles}\end{matrix}\right\} & \simeq & \left\{\begin{matrix}\text{points of}\\ \overline{V}\end{matrix}\right\} \\ & & \cap & \nearrow & \\ & & \left\{\begin{matrix}\text{semi-stable}\\ \text{bundles}\end{matrix}\right\} & & \end{array}$$

2 Compact moduli spaces for abelian varieties: Satake

Let V_n denote the moduli scheme for principally polarized abelian varieties of dimension n. That is,

$$V_n \cong \mathfrak{H}_n/\Gamma_n \qquad \text{(as analytic space)},$$

where $\mathfrak{H}_n$ is the Siegel upper $\frac{1}{2}$-plane of type n, and Γ_n is the modular group acting on $\mathfrak{H}_n$. V_n has even a canonical structure of algebraic variety over $\mathbb{Q}$, due to its interpretation as a moduli scheme[2]. V_n carries a canonical class of ample invertible sheaves $\mathscr{L}(i)$ defined for all sufficiently large i,[3] and such that

[2] cf. Baily's work, or my "Geometric Invariant Theory". (original footnote by Mumford)

[3] More precisely for all sufficiently divisible $i \in \mathbb{N}$. We need this for the later statement that, as long as $n \geq 2$, the ring R_n is isomorphic to the graded ring of modular forms on $\mathfrak{H}_n$ with respect to Γ_n, that is $\Gamma(V_n, \mathscr{L}(i))$ corresponds to modular forms of weight i for (V_n, Γ_n). Without the divisibility requirement there are problems with elements of finite order in the modular group Γ_n. For instance the weight of every non-zero modular form for (V_n, Γ_n) is even if n is odd, because of the element $-\mathrm{Id}_n \in \Gamma_n$.

$$\mathscr{L}(i) \otimes \mathscr{L}(j) = \mathscr{L}(i+j)$$

when this makes sense. Therefore one has the graded ring

$$R_n = \bigoplus_{i \geq i_0} \Gamma(V_n, \mathscr{L}(i)),$$

which is known to be isomorphic to the ring of modular forms on $\mathfrak{H}_n$ with respect to Γ_n, if $n \geq 2$.

The Satake compactification of V_n is then the open immersion:

$$V_n \subset \mathrm{Proj}(R_n) = V_n^*.$$

It turns out that there is a canonical isomorphism of $V_n^* - V_n$ and V_{n-1}^*, so that set-theoretically:

$$V_n^* = V_n \cup V_{n-1} \cup \cdots \cup V_1 \cup V_0.$$

(V_0 is a single point). This amazing equation suggests that this compact variety, which is defined only as a kind of "minimal model", should have an interpretation as a moduli space. In fact, consider all commutative group schemes X connected and of finite type over $\mathbb{C}$.

Definition: X is *stable* if X is an abelian variety.

X is *semi-stable* if X is an extension of an abelian variety by multiplicative groups $(\mathbb{G}_m)^r$.

X is *retractable* if X is the product of an abelian variety by multiplicative groups.

Exactly as before, A. Mayer and I have proven:

$$\begin{array}{ccc}
\left\{\begin{array}{c}\text{stable } X \text{ with}\\ \text{polarization}\end{array}\right\} & \simeq & \left\{\begin{array}{c}\text{points of}\\ V_n\end{array}\right\} \\
\cap & & \cap \\
\left\{\begin{array}{c}\text{retractable } X\\ \text{with polarization}\end{array}\right\} & \simeq & \left\{\begin{array}{c}\text{points of}\\ V_n^*\end{array}\right\} \\
\cap & \nearrow & \\
\left\{\begin{array}{c}\text{semi-stable } X\\ \text{with polarization}\end{array}\right\} & &
\end{array}$$

Explanations

1° A polarization of X may be taken to mean a divisor D on X, determined up to algebraic equivalence, such that if

$$\pi : X \to X_0$$

is the projection of X onto its abelian part, and if $D = \pi^*(D_0)$ (recall that $\mathrm{Pic}(X_0) \twoheadrightarrow \mathrm{Pic}(X)$),[4] then D_0 is ample on X_0 and

$$\begin{cases} (D_0^{n_0}) = n_0!\,, \\ \quad n_0 = \dim X_0\,. \end{cases}$$

2° A family of these objects is a morphism

$$f\colon \mathfrak{X} \to S$$

with the structure of group scheme (i.e., a "multiplication" $\mu\colon \mathfrak{X} \times_S \mathfrak{X} \to \mathfrak{X}$, etc.) and a family of Cartier divisors $\mathscr{D}$ on $\mathfrak{X}$ determined up to algebraic equivalence, and replacements

$$\mathscr{D}' = \mathscr{D} + f^*(\mathscr{E})$$

for any Cartier divisors $\mathscr{E}$ on S, and inducing a polarization of each fibre $f^{-1}(s)$. With this definition, stable and semi-stable $\mathfrak{X}$'s form open sets, but retractable $\mathfrak{X}$'s do not.

3° The meaning of the arrows in the diagram is this: let $f\colon \mathfrak{X} \to S$ be a family of semi-stable objects where S is a *normal* algebraic variety. Map S to V_n^* by assigning to each $s \in S$ the point of V_{n_0} corresponding, in the classical way, to the abelian part of $f^{-1}(s)$ ($n_0 =$ dim of this abelian part). Then this is a *morphism*.

This last result is proven by reducing to the case where S is a curve. Then one passes to the corresponding analytic set-up, and replaces S by a disc $\{z \mid |z| < 1\}$ where all fibres of f are diffeomorphic except for $f^{-1}(0)$. Next one introduces the invariant and vanishing cycles on the general fibre, so as to put the period matrix $\Omega_{ij}(z)$ of the abelian part of $f^{-1}(z)$ in a normalized form. One then computes (using very helpful tricks of Kodaira):

$$\Omega_{ij}(z) = \frac{1}{2\pi i}\log z \left(\begin{array}{c|c} S & 0 \\ \hline 0 & 0 \end{array}\right) + \left(\begin{array}{c|c} A(z) & B(z) \\ \hline {}^tB(z) & C(z) \end{array}\right),$$

where S is integral, positive definite and symmetric, and is obtained from the monodromy substitution for the cycle $|z| = 1$; where A, B, C are holomorphic in z at $z = 0$; and where $C(0)$ is the period matrix of the abelian part of $f^{-1}(0)$. This implies that $\Omega_{ij}(z) \to C(0)$ in Satake's topology, when $z \to 0$.

4 In the mimeographed notes distributed during the 1964 Woods Hole Summer School, this passage reads "(recall that $\mathrm{Pic}(X) \cong \mathrm{Pic}(X_0)$)"; here we changed "$\cong$" to "$\twoheadrightarrow$". The natural map $\pi^*\colon \mathrm{Pic}(X_0) \to \mathrm{Pic}(X)$ is a surjection, but $\mathrm{Ker}(\pi^*)$ is nontrivial in general—it is subgroup of $\mathrm{Pic}(X_0)$ corresponding to all pushouts of the extension $0 \to T \to X \to X_0 \to 0$ by characters $T \to \mathbb{G}_m$, where $T = \mathrm{Ker}(\pi)$. See also the comments in the letter [1965Apr16] from Grothendieck, p. 706.

3 Compact moduli spaces for curves

Let $\mathcal{M}_g$ denote the moduli scheme for curves of genus g. Let

$$\Theta \colon \mathcal{M}_g \to V_g$$

be the morphism which assigns to a curve its jacobian variety with its theta-polarization. From the work of Baily, Matsusaka, and Hoyt, it is known that Θ is an isomorphism of $\mathcal{M}_g$ with a locally closed subvariety of V_g, which we also denote $\mathcal{M}_g$.[5] The simplest approach to compactifying $\mathcal{M}_g$ is to use its closure $\mathcal{M}_g^*$ in V_g^*. The boundary $\mathcal{M}_g^* - \mathcal{M}_g$ breaks up into two pieces

$$\mathcal{M}_g' = (\mathcal{M}_g^* \cap V_g) - \mathcal{M}_g,$$
$$\mathcal{M}_g'' = \mathcal{M}_g^* - (\mathcal{M}_g^* \cap V_g).$$

Matsusaka and Hoyt showed that $\mathcal{M}_g'$ is exactly the set of products of lower dimensional jacobian varieties. We have proven that $\mathcal{M}_g'' = \mathcal{M}_{g-1}^*$, so that

$$\mathcal{M}_g^* = \mathcal{M}_g \cup \mathcal{M}_g' \cup \mathcal{M}_{g-1} \cup \mathcal{M}_{g-1}' \cup \cdots \cup \mathcal{M}_0$$

($\mathcal{M}_0 = V_0$ is a single point).

The proof is based on two lemmas, and on the results of §2:

Lemma A *Let C be a curve and let $f \colon \mathfrak{X} \to C$ be a family of curves of arithmetic genus g [i.e., f is proper and flat and its fibres $f^{-1}(P)$ are connected curves of arithmetic genus g]. Let $P_0 \in C$ and assume that $f^{-1}(P)$ is nonsingular if $P \neq P_0$. Then there exists a diagram:*[6]

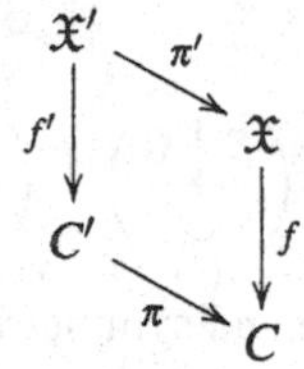

where

1) *C' is a curve and π is a finite morphism totally ramified over P_0: let $P_0' = \pi^{-1}(P_0)$,*
2) *f' is a family of curves over C',*
3) *$\mathfrak{X}' - f'^{-1}(P_0')$ is just the induced family of curves over $C' - P_0'$, i.e.,*

$$(C' - P_0') \times_C \mathfrak{X} = \mathfrak{X}' - f'^{-1}(P_0'),$$

4) *$f'^{-1}(P_0')$ is reduced and has only ordinary double points.*

---

[5] The symbol $\mathcal{M}_g$ in this and the next sentence are incorrectly typed as "$\mathcal{M}_g'$" in the original Woods Hole notes. This error was pointed out by Grothendieck in [1965Apr16].

[6] The proof of Lemma A of §3 was sketched on pages 7–8 in the notes of Mayer mentioned in footnote 1. For a vast generalization of Lemma A, the semistable reduction theorem for smooth varieties over a field of characteristic 0, see page 53 of [TE] (= *Toroidal Embeddings. I*). The proof spans Chapters II and III of [TE].

Lemma B *Let C be a curve and let*

$$f\colon \mathfrak{X} \to C$$

be a family of curves of arithmetic genus g such that each curve $f^{-1}(P)$ is reduced and has only ordinary double points. Then the set of generalized jacobian varieties of the curves $f^{-1}(P)$ forms a family of polarized semi-stable group varieties over C.

These lemmas give the inclusion $\mathscr{M}_g'' \subset \mathscr{M}_{g-1}^*$ directly; Lemma *B* and an easy construction of some actual families give the converse $\mathscr{M}_g'' \supset \mathscr{M}_{g-1}^*$.

Unfortunately, $\mathscr{M}_g^*$ is not a reasonable moduli space for curves: for example, let a point of $\mathscr{M}_g'$ correspond to

$$A_1 \times A_{g-1},$$

where A_1 is an elliptic curve, and A_{g-1} is the jacobian of a curve C of genus $g-1$. Let $x \in A_1$ and $y \in C$ be any points. Then $A_1 \times A_{g-1}$ is the generalized jacobian variety of the curve:

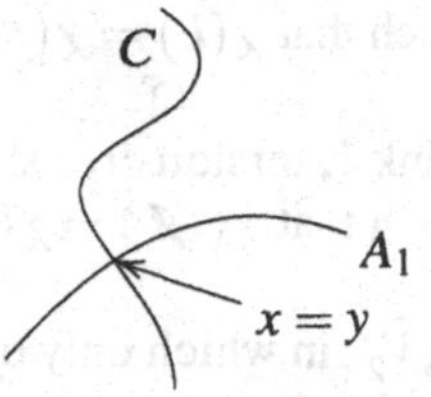

with an ordinary double point. In other words, the jacobian is independent of which y is chosen: i.e., Torelli's theorem is false for reducible curves. It is clearly necessary to blow up $\mathscr{M}_g'$. This phenomenon is closely related to the fact, discovered by Bers and Ehrenpreis, that the generic point of $\mathscr{M}_g'$ is not only singular on $\mathscr{M}_g^*$: it is not even "almost nonsingular" (= "Jungian" = "V-manifold"). In fact, Lemma *A* suggests

Definition: A curve C of arithmetic genus g is *stable* if C is reduced and connected, has only ordinary double points, and has only a finite group of automorphisms.

It appears that the set of all stable curves is open and compact and is naturally isomorphic to the set of points of a compact analytic space with almost nonsingular points: $\widetilde{\mathscr{M}_g^*}$. It is still unknown whether $\widetilde{\mathscr{M}_g^*}$ is a projective algebraic variety, although it is a $\mathbf{Q}$-variety. There is a proper holomorphic map

$$\widetilde{\mathscr{M}_g^*} \to \mathscr{M}_g^*,$$

which is an isomorphism over the open subset $\mathscr{M}_g$. One of the remarkable features of this case is that there are no semi-stable but not stable curves.

4 Compact moduli spaces for abelian varieties: blown up

The preceding construction suggests the possibility of blowing up V_n^* so as to obtain a $\widetilde{V}_n^*$ which corresponds to a moduli problem with a larger set of stable objects. We would like the stable points of $\widetilde{V}_n^*$ to correspond to polarized compactifications of commutative group schemes X. One approach is to compactify the generalized jacobian varieties of curves C. Say C is irreducible and reduced: let J be the generalized jacobian of C. Then one has an isomorphism

$$\left\{\begin{matrix}\text{points of}\\ J\end{matrix}\right\} \simeq \left\{\begin{matrix}\text{invertible sheaves } L \text{ on } C\\ \text{such that } \chi(L) = \chi(\mathscr{O}_C)\end{matrix}\right\}.$$

We can prove that there is a projective scheme J^* containing J as an open subset, and on which J acts, plus a natural isomorphism

$$\begin{matrix}\left\{\begin{matrix}\text{points of}\\ J\end{matrix}\right\} & \simeq & \left\{\begin{matrix}\text{invertible sheaves } L \text{ on } C\\ \text{such that } \chi(L) = \chi(\mathscr{O}_C)\end{matrix}\right\}\\ \cap & & \cap\\ \left\{\begin{matrix}\text{points of}\\ J^*\end{matrix}\right\} & \simeq & \left\{\begin{matrix}\text{rank 1, torsion-free sheaves } \mathscr{I} \text{ on } C\\ \text{such that } \chi(\mathscr{I}) = \chi(\mathscr{O}_C)\end{matrix}\right\}.\end{matrix}$$

Using this, we find an interesting $\widetilde{V}_2^*$, in which only one point is still mysterious: that is the point which is the image under Θ of the curve of genus 2 depicted below:[7]

[7] The picture Mumford drew on the board during the lecture in 1964 is "*a dollar sign lying on its side*" () according to A. Mayer. It represents two proper smooth rational curves meeting transversally at three distinct points. Grothendieck asked "Are there pages lacking, or were you making fun?" in [1965Apr16].

Article [u67a]

Abstract Theta Functions

David Mumford

Advanced Science Seminar in Algebraic Geometry
Sponsored by the National Science Foundation
Bowdoin College, Summer 1967
Notes by Harsh Pittie

1 Introduction: Let A be an abelian variety defined over k, an algebraically closed field complete with respect to a real[-valued] absolute value. Let R be the ring of integers in k, and $\overline{k}$ the residue field; suppose char $\overline{k} \neq 2$. Our aim is to show that A has a "good reduction" over R: i.e., that there is a fibre product diagram:

$$\begin{array}{ccccc} A & \hookrightarrow & \mathfrak{A} & \hookleftarrow & \overline{A} \\ \downarrow & & \downarrow & & \downarrow \\ \operatorname{Spec} k & \hookrightarrow & \operatorname{Spec} R & \hookleftarrow & \operatorname{Spec} \overline{k} \end{array}$$

where $\mathfrak{A}$ is a group scheme over R, and $\overline{A}$ is an extension of $\mathbb{G}_m^M$ by an abelian variety. The existence of such reductions provides an abstract analogue of the existence of the Satake compactification of the moduli scheme of A.

If $\overline{A} = (\mathbb{G}_m)^n$, we will say that A has totally-degenerate reduction: in this case one can get a p-adic[1] analytic uniformization $\pi\colon V \to A$, and hence the Tate–Morikawa–McCabe theory.

We will use an abstract theory of theta-functions to perform the reduction; and we begin by sketching such a theory.

2 Abstract Theta-functions: Classically, the theta-functions associated to an abelian variety A arise in the following way. Let A be defined over k, and suppose there is a surjective homomorphism $\pi\colon V \to A$; then $A \cong V/\operatorname{Ker}\pi$. For example, if $k = \mathbb{C}$, $V = \mathbb{C}^g$ ($g = \dim A$) and $\operatorname{Ker}\pi$ is a lattice; or in the Tate–Morikawa–McCabe theory, k is a local field, $V = (k^*)^g$ and $\operatorname{Ker}\pi$ is a "multiplicative (annular) lattice". Then the theta-functions on V associated to A are holomorphic functions on V which satisfy a certain functional equation with respect to $\operatorname{Ker}\pi$.

Quite generally, suppose we have a homomorphism $\pi\colon V \to A$; and an ample, invertible sheaf L on A so that if $\mathbb{L}$ is the induced line bundle, then $\pi^*\mathbb{L} \cong \mathbf{1}$ (the

[1] When the complete valued field k has residue characteristic $p > 0$.

D. Mumford, *Selected Papers*, Vol. II,

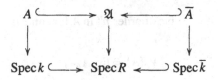

trivial line-bundle on V—i.e., induced from $\mathcal{O}_V$). Then sections $S \in \Gamma(A,\mathbb{L})$ pull back to sections $\pi^*(S) \in \Gamma(V,\mathbf{1})$, and these are naturally interpreted as k-valued functions on V which can be called theta-functions. These functions satisfy a kind of periodicity with respect to $\operatorname{Ker}\pi$, as the following argument shows. Let $\gamma \in \operatorname{Ker}\pi$, and interpret it as a translation map on V; then $\pi \circ \gamma = \pi$ so $\gamma^*\pi^* = \pi^*$. Therefore we have a commutative diagram

$$\begin{array}{ccc} \gamma^*\pi^*\mathbb{L} & = & \pi^*\mathbb{L} \\ \|\wr & & \|\wr \\ \gamma^*\mathbf{1} & \xrightarrow[\text{mult. by } c_\gamma]{} & \mathbf{1} \end{array}$$

where c_γ is a suitable nowhere-zero function on V. Thus, if f is the k-valued function $\pi^*(s)$, then we have

$$f(\gamma z) = c_\gamma(z) f(z), \quad z \in V.$$

We apply this formulation as follows. Let p be a prime,

$$A_{p^\infty} = \text{pts of order } p^n \text{ in } A \text{ for some } n,$$

and[2]

$$V_p(A) = \varprojlim A_{p^\infty},$$

the Tate-module of A at p. Then there is an exact sequence

$$0 \to \Lambda_p \to V_p(A) \xrightarrow{\pi} A_{p^\infty} \to 0,$$

where π is given by $\pi(a_0, a_1, \ldots) = a_0$. Recall that there are isomorphisms $V_p(A) \cong (\mathbb{Q}_p)^{2g}$ and $\Lambda_p \cong (\mathbb{Z}_p)^{2g}$ ($g = \dim A$). Thus we can discuss "local" theta-functions corresponding to the uniformization $\pi\colon V_p(A) \to A_{p^\infty}$. There is an analogous theory of global theta-functions in which $V_p(A)$ is replaced by the adèle group $\prod_p V_p(A)$. However, there seem to be difficulties in the local case for $p \neq 2$; hence we shall restrict our attention, from now on, to the case $p = 2$.

3 Construction of Theta-functions

Let L be an ample invertible sheaf on A, $\ell\colon \mathbb{L} \to A$ the corresponding line bundle. For $y \in A$, let $\mathbb{L}_y = \ell^{-1}(y)$, the fibre over y. Assume that

i) There is an isomorphism $\rho\colon i^*\mathbb{L} \to \mathbb{L}$ where $i\colon A \to A$ is the map $i(x) = -x$; i.e., that $\mathbb{L}$ is symmetric.
ii) We are given a specific isomorphism $\varphi_0\colon \mathbb{L}_0 \xrightarrow{\sim} k$.

2 The projective system which defines $V_p(A)$ is indexed by $\mathbb{N}$, each term is A_{p^∞}, and the transition maps are induced by $p\cdot \mathrm{Id}_A$.

Now we can trivialize $\pi^*\mathbb{L}$ if we can find isomorphisms $\mathbb{L}_x \xrightarrow{\sim} k$ for all $x \in A_{2^\infty}$. We proceed to do so as follows.

Let $t_x: A \to A$ be the translation $t_x(y) = x + y$, and suppose that for some particular x we are given an isomorphism $\tau_x: t_x^*\mathbb{L} \xrightarrow{\sim} \mathbb{L}$. Consider the diagram:

$$\begin{array}{ccc}
i^*t_x^*\mathbb{L} & \xrightarrow{i^*\tau_x} & i^*\mathbb{L} \\
\| & & \big\downarrow \rho \\
(t_x \circ i)^*\mathbb{L} & & \\
\| & & \mathbb{L} \\
(i \circ t_{-x})^*\mathbb{L} & & \big\downarrow t_{-x}^*\tau_x \\
\| & & \\
t_{-x}^*i^*\mathbb{L} & \xrightarrow{t_{-x}^*\rho} & t_{-x}^*\mathbb{L}
\end{array}$$

As it stands, there is no reason to expect this diagram to commute. However, if we modify τ_x by a suitable automorphism of $\mathbb{L}$ (which is just an element of k^*—since $\mathbb{L}$ is a line bundle over a projective variety) we can force the diagram to commute. Now suppose that $\alpha, \beta \in k^*$ are automorphisms such that $\alpha \cdot \tau_x$ and $\beta \cdot \tau_x$ make the diagram commutative, then an easy chase shows that $\alpha^2 = \beta^2$, or $\alpha = \pm\beta$. Thus $\alpha \cdot \tau_x$ and $-\alpha \cdot \tau_x$ are the only isomorphisms of $t_x^*\mathbb{L}$ with $\mathbb{L}$ which make the diagram commutative, and if we stipulate that this should be so, then an isomorphism $\tau_x': t_x^*\mathbb{L} \xrightarrow{\sim} \mathbb{L}$ is determined canonically up to ± 1.

We can define a *completely* canonical isomorphism σ_{2x} from $t_{2x}^*\mathbb{L}$ to $\mathbb{L}$ as follows,

$$\sigma_{2x}: t_{2x}^*\mathbb{L} = t_x^*(t_x^*\mathbb{L}) \xrightarrow{t_x^*\tau_x'} t_x^*\mathbb{L} \xrightarrow{\tau_x'} \mathbb{L},$$

since $-\tau_x'$ and τ_x' give the same σ_{2x}. Thus we can get canonical isomorphisms

$$(\sigma_{2x})_0: (t_{2x}^*\mathbb{L})_0 \to \mathbb{L}_0,$$

and from this $\varphi_0 \circ (\sigma_{2x})_0: (t_{2x}^*\mathbb{L})_0 \to k$. But $(t_{2x}^*\mathbb{L})_0 = \mathbb{L}_{2x}$. Therefore we have a canonical isomorphism

$$\varphi_0 \circ (\sigma_{2x})_0: \mathbb{L}_{2x} \to k.$$

Therefore we can trivialize $\pi^*\mathbb{L}$ along those fibers $\mathbb{L}_x$ such that $t_y^*\mathbb{L} \xrightarrow{\sim} \mathbb{L}$ for some y solving $2y = x$. But this isomorphism exists for only a few points in A. We use the following lemma to enable us to obtain isomorphisms $\mathbb{L}_x \xrightarrow{\sim} k$ for all $x \in A_{2^\infty}$. Put $H(L) = \{x \in A \mid t_x^*\mathbb{L} \xrightarrow{\sim} \mathbb{L}\}$.

Lemma. *($n^2 > n$). Let $n\delta: A \to A$ be the isogeny $n\delta(x) = nx$. For all $x \in A$ of finite order, $(n\delta)^*t_x^*\mathbb{L}$ is isomorphic to $(n\delta)^*\mathbb{L}$ for some n ($n =$* order of x *will do).*

(The proof is easy: let $x = ny$, so y has order n^2. Then $(n\delta)^*\mathbb{L} \xrightarrow{\sim} \mathbb{L}^{n^2}$. But for any $\mathbb{M}$ and m, $H(\mathbb{M}^m) \supseteq A_m$. So $(n\delta)^*t_x^*\mathbb{L} \xrightarrow{\sim} t_y^*(n\delta)^*\mathbb{L} \to t_y^*\mathbb{L}^{n^2} \simeq (n\delta)^*\mathbb{L}$.)

Now let $x_0 \in A_{2^\infty}$ be some fixed but arbitrary element. Then x_0 sits in at least one sequence $(x_0, x_1, \ldots) \in V_2(A)$. We will not in general have an isomorphism $t_{x_1}^* \mathbb{L} \xrightarrow{\sim} \mathbb{L}$; however, for large enough m, $(2^m\delta)^*(t_{x_1}^*\mathbb{L}) \xrightarrow{\sim} (2^m\delta)^*\mathbb{L}$ by the lemma. Since $(2^m\delta)^* t_{x_1}^* \mathbb{L} \to t_{x_{m+1}}^*(2^m\delta)^*\mathbb{L}$ we get a canonical (up to ± 1) isomorphism $\tau'_{x_{m+1}} : t_{x_{m+1}}^*(2^m\delta)^*\mathbb{L} \to (2^m\delta)^*\mathbb{L}$ and thus a completely canonical σ_{x_m}, and therefore isomorphisms $\varphi_0 \circ (\sigma_{x_m})_0 : ((2^m\delta)^*\mathbb{L})_{x_m} \to k$. But $((2^m\delta)^*\mathbb{L})_{x_m} = \mathbb{L}_{x_0}$. Thus we get the desired isomorphism of the fiber $\mathbb{L}_{x_0}$ with k.

Glossing over the development, the final theory comes out something like this. We begin with a symmetric, ample, invertible sheaf L on A of degree 1, as above. L determines

i) a bimultiplicative, skew-symmetric form

$$e\colon V_2(A) \times V_2(A) \to \{2^n\text{th roots of 1 in } k \text{ for some } n\}$$

ii) a "quadratic character" $e_*\colon \frac{1}{2}\Lambda/\Lambda \to \{\pm 1\}$ satisfying

$$e(\alpha,\beta)^2 = e_*(\overline{\alpha}+\overline{\beta})\, e_*(\overline{\alpha})\, e_*(\overline{\beta})$$

for $\alpha, \beta \in \dfrac{1}{2}\Lambda$.

We can assume that the Arf invariant of e_* is zero by replacing L by some $t_x^* L$, $x \in A_2$, if necessary.
(The quadratic form[3] e is classical: see for example Lang, *Abelian Varieties*).

In terms of this data we obtain theta-functions $\theta_{[s]}\colon V_2(A) \to k$ for all $s \in \Gamma(A, \mathbb{L}^n)$, satisfying

$$(*) \qquad \theta_{[s]}(\alpha+\beta) = [e_*(\beta/2)\cdot e(\beta/2,\alpha)]^n\, \theta_{[s]}(\alpha) \qquad \text{for all } \alpha \in V_2(A),\ \beta \in \Lambda.$$

This gives a homomorphism of k-algebras

$$\theta\colon \bigoplus_{n=1}^{\infty} \Gamma(A,\mathbb{L}^n) \to \{k\text{-valued functions on } V_2(A) \text{ satisfying } (*)\}$$

where multiplication of sections s_1, s_2 is given by $s_1 \otimes s_2$. Further, θ is injective; in fact, if $a = (a_0, a_1, \ldots) \in V_2(A)$, $\theta_{[t]}(a) = 0$ if and only if $t(a_0) = 0$. If s_0 denotes the canonical section of $\mathbb{L}$ then we put $\theta_{[s_0]} = \Theta$ the Riemann theta-function. It satisfies

$$(**) \qquad \Theta(-\alpha) = \Theta(\alpha)$$

$$(***) \qquad \prod_{i=1}^{4}\Theta(\alpha_i) = 2^{-g} \sum_{\eta \in \frac{1}{2}\Lambda/\Lambda} e(\gamma,\eta) \prod_{i=1}^{4}\Theta(\alpha_i+\gamma+\eta)$$

$$\text{where } \gamma = -\frac{1}{2}\sum \alpha_i$$

$$(****) \qquad \text{For every } \alpha \in V_2(A)\ \exists \beta \in \frac{1}{2}\Lambda \text{ so that } \Theta(\alpha+\beta) \neq 0.$$

[3] The skew symmetric form actually.

What is remarkable about these theta-functions is that beginning with just Θ we can recover the pair $(A,\mathbb{L})$. Suppose we start with a vector space V isomorphic to $(\mathbb{Q}_2)^{2g}$, Λ a maximal isotropic lattice in V, the form e and the quadratic character e_*. Then we can define a theta-function Θ on V as a k-valued function satisfying (**), (***) and (****). We then put

$$M = k\text{-vector space spanned by } e(\alpha,\beta)\Theta(2\alpha-\beta), \text{ where } \alpha \in V,\ \beta \in \frac{1}{2}\Lambda.$$
$$\text{(This will equal the space of } \theta_{[s]}\text{'s,}\quad s \in \Gamma(A,\mathbb{L}^4).)$$
$$S_0(M) = k,$$
$$S_1(M) = M,$$
$$S_n(M) = \text{space spanned by } n\text{-fold products of elements from } M.$$

Then $A = \mathrm{Proj}(\bigoplus S_n(M))$ is the abelian variety sought for, and $\mathbb{L}$ is easily recovered from A and $\bigoplus S_n(M)$.

Finally, let us note an important correspondence between theta-functions on V and finitely-additive measures on a certain subspace of V. These measures arise from the Fourier transforms of the theta-functions, and are examples of Schwartz–Bruhat distributions. Explicitly, we can describe them as follows: decompose V as $V_1 \oplus V_2$, where V_1, V_2 are isotropic with respect to the pairing e, $\Lambda = (\Lambda \cap V_1) + (\Lambda \cap V_2)$ and $e_*(\alpha) = 1$ for $\alpha \in \frac{1}{2}\Lambda \cap V_i$. A finitely-additive measure μ on the Boolean ring of compact open subsets of V_1 is called *Gaussian* if and only if

i) $\mu(U) = \mu(-U)$
ii) Given the map $\xi: V_1 \times V_1 \to V_1 \times V_1$, $\xi(x,y) = (x+y, x-y)$, then $(\mu\times\mu)(\xi U) = (\nu\times\nu)(U)$, where ν is some other measure on the same ring.

The correspondence between theta-functions on V and Gaussian measures μ on V_1 is given thus:

$$\mu(\alpha_1 + 2^n\Lambda_1) = 2^{-ng} \sum_{\alpha_2 \in 2^{-n}\Lambda_2/\Lambda_2} e(\alpha_1, \alpha_2/2)\Theta(\alpha_1+\alpha_2)$$

and

$$\Theta(\alpha_1+\alpha_2) = e(\alpha_1,\alpha_2/2) \int_{\alpha_1+\Lambda_1} e(\alpha_2,\beta)\, d\mu(\beta),$$

where $\Lambda_i = \Lambda \cap V_i$ $(i = 1,2)$ and $\alpha_i \in \Lambda_i$.

4 The Reduction of A over R

We now analyze the relation of Θ to the integers R in k (R, k as in section 1). Let $|\ |$ denote the real absolute value of k; $V_2(A)$, Λ and Θ as before.

Proposition. $\max_{\alpha\in V} |\Theta(\alpha)|$ *is finite and is taken on for some* $\alpha \in \frac{1}{2}\Lambda$.

Proof. The Riemann theta-relation

$$\prod_{i=1}^{4}\Theta(\alpha_i)=2^{-g}\sum_{\eta\in\frac{1}{2}\Lambda/\Lambda} e(\gamma,\eta)\prod_{i=1}^{4}\Theta(\alpha_i+\gamma+\eta)$$

gives

$$\prod_{i=1}^{4}|\Theta(\alpha_i)|\leq \max_{\eta\in\frac{1}{2}\Lambda/\Lambda}\prod_{i=1}^{4}|\Theta(\alpha_i+\gamma+\eta)|$$

since $|\Theta(\alpha)|$ is constant on cosets of Λ.

If we put $\alpha_1=\alpha_2=\alpha_3=\alpha=-\alpha_4$, then

$$(\dagger)\qquad |\Theta(\alpha)|^4\leq \max_{\eta\in\frac{1}{2}\Lambda/\Lambda}|\Theta(\eta)|^3\cdot|\Theta(2\alpha-\eta)|;$$

since (†) is valid for all $\alpha\in V$, applying it successively to $2\alpha-\eta, 4\alpha-3\eta,\ldots,$ $2^n\alpha-(2^n-1)\eta,\ldots,$ and substituting back in (†) we get

$$|\Theta(\alpha)|^4\leq \max_{\eta\in\frac{1}{2}\Lambda/\Lambda}|\Theta(\eta)|^{r_n}\cdot|\Theta(2^n\alpha-(2^n-1)\eta)|^{s_n},$$

where[4] $r_n=\sum_{i=0}^{n}3/4^i$, $s_n=\dfrac{1}{2^{2n}}$.

Now in the 2-adic topology $2^n\alpha-(2^n-1)\eta$ converges to η. We know r_n converges to 4. Therefore we get

$$|\Theta(\alpha)|^4\leq \max_{\eta=\frac{1}{2}\Lambda/\Lambda}|\Theta(\eta)|^4,$$

whence

$$\max_{\alpha\in V}|\Theta(\alpha)|^4\leq \max_{\eta\in\frac{1}{2}\Lambda/\Lambda}|\Theta(\eta)|^4\leq\max_{\alpha\in V}|\Theta(\alpha)|^4,$$

which yields the result.

Using this proposition we can normalize Θ so that its values lie in R, but not all lie in the maximal ideal M; that is, if $\overline{\Theta}$ denotes the induced function to $\overline{k}$, $\overline{\Theta}(\alpha)\neq 0$ for some $\alpha\in V$.

We now invoke the main result used for the Satake compactification (see [1]).

Theorem 1. *For every theta-function Θ on V, (i.e., a function satisfying $(*)$, $(**)$, $(***)$ but* not *necessarily $(****)$) there is a subspace $W\subseteq V$ with $W^{\perp}\subseteq W$ ($\perp$ with respect to e) and a nondegenerate theta-function Φ on $W/W^{\perp}$ such that*

$$\operatorname{supp}\Theta\subseteq W+\Lambda+\eta_0,\ \eta_0\in\frac{1}{2}\Lambda$$

and

$$\Theta(\eta_0+\eta_1+\alpha)=e_*(\eta_1/2)e(\eta_1/2,\alpha)e\Big(\frac{\eta_0+\eta_1}{2},\alpha\Big)\Phi(\overline{\alpha})\quad(\eta_1\in\Lambda,\ \alpha\in W).$$

---

[4] A misprint "$s_n=\dfrac{1}{2n}$" in the original is corrected here.

The theta-function Φ is used to construct an abelian variety B over $\bar{k}$ of dimension h (where $\dim W/W^\perp = 2h$) in the same way that A was constructed from Θ (see section 3). Then the special fibre $\bar{A}$ of the sought-for group scheme $\mathfrak{A}$ should be in an extension

$$0 \to \mathbb{G}_m^h \to \bar{A} \to B \to 0.$$

Notice that if the reduction is to be totally-degenerate, then we must have $W = W^\perp$, and hence $B = \{0\}$.

To construct $\mathfrak{A}$ however, we must first study how many R-valued theta-functions come from a given $\bar{k}$-valued nondegenerate theta-function Φ on a vector space $W/W^\perp$ of smaller dimension. For this question the measure-theoretic point of view (outlined at the end of section 3) is much better. For ease of exposition we confine ourselves to the case of totally-degenerate reduction, i.e., $W = W^\perp$. In this case the R-valued measure μ corresponding to Θ reduces to a $\bar{k}$-valued measure $\bar{\mu}$, where $\bar{\mu}$ is just the point mass at 0, δ_0 (the so-called "Dirac delta-function"). The main result concerning these measures is this.

Theorem 2. *Let μ be a nondegenerate R-valued Gaussian measure on $\mathbb{Q}_2^g$ such that $\bar{\mu} = \delta_0$. Then there is a unique subgroup M' in $\mathbb{Q}_2^g$ isomorphic to $\mathbb{Z}\left[\frac{1}{2}\right]^g$ (and equal to it after a suitable change of co-ordinates) and a unique quadratic character $c' \colon M' \to R - \{0\}$ such that*

$$\mu = \sum_{x \in M'} c'(x)\, \delta_x .$$

Moreover, if we tensor M' with $\mathbb{R}$, then there is a positive-definite quadratic form $Q \colon M' \otimes \mathbb{R} \to \mathbb{R}$ so that $|c'(x)| = e^{-Q(x)}$.

Ideally, at this point we should write down $\mathfrak{A}$ explicitly in terms of Θ. However, this presents certain complications, and it is faster to construct $\mathfrak{A}$ by means of the theory of the Néron model and to check that its special fibre $\bar{A}$ is $\mathbb{G}_m^h$ by means of Galois theory (following a suggestion of Grothendieck).

Choose a subfield $k_0 \subseteq k$ with a discrete absolute value so that A is defined over k_0, and let $\mathfrak{A}_0$ be the Néron model of A over R_0 = integers in k_0. Let

$$\mathfrak{A} = \mathfrak{A}_0 \text{ (minus the components of its special fibre not containing zero)}.$$

Then $G = \mathrm{Gal}(k/k_0)$ acts on $V_2(A)$ preserving Λ, e, W, M', and Q. On V/W the action of G is determined by its action on $M' \cap (\Lambda/\Lambda \cap W) \cong \mathbb{Z}^g$. Hence we have a representation

$$G \to O(Q)_{\mathbb{Z}}$$

into an integral orthogonal group (corresponding to the quadratic form (Q) on $\mathbb{Z}^g$).[5] But this group is finite! Hence replacing k_0 by a finite extension k_1 if necessary, we see that $G_1 = \mathrm{Gal}(k/k_1)$ acts trivially on V/W. Since the action of G_1 preserves e, it acts trivially on W too; thus the representation takes the form

[5] A misprint "corresponding to the quadratic form (Q) on $\mathbb{Q}_2^{2g}$" is corrected here.

$$\sigma \longrightarrow \left(\begin{array}{c|c} I & * \\ \hline O & I \end{array}\right).$$

Thus A_{2^∞} contains a subgroup H which is k_1-rational and is a maximal isotropic subgroup of points of order 2^n, isomorphic to $(\mathbb{Q}_2/\mathbb{Z}_2)^g$. Now by one of the key properties of Néron models, all k_1-rational points of A extend to R_1-rational points of $\mathfrak{A}_1$ ($\mathfrak{A}_1 =$ Néron model of A over R_1). Since H is divisible, all points of H give R_1-rational points of $\mathfrak{A}_1$ hence H induces a subgroup $\overline{H} \subseteq \overline{A}$, isomorphic to $(\mathbb{Q}_2/\mathbb{Z}_2)^g$.

Now from quite general structure theorems on group schemes, we have an exact sequence

$$0 \to L \to \overline{A} \xrightarrow{\pi} B \to 0,$$

when L is a linear group of dimension r, and B is an abelian variety. It can be shown that $\pi(\overline{H})$ is still isotropic in B and since B has dimension $g-r$, $\pi(\overline{H}) \cong (\mathbb{Q}_2/\mathbb{Z}_2)^k$, $k \leq g-r$. Therefore $\overline{H} \cap L$ has a subgroup $(\mathbb{Q}_2/\mathbb{Z}_2)^r$, whence $L = \mathbb{G}_m^r$. Using the total-degeneracy of the theta function we can then show that $\overline{A} = \mathbb{G}_m^g$—i.e., that $B = \{0\}$.

5 Analytic Theta-Functions.

In this section we will show how our theta-functions with totally-degenerate reduction are essentially *equal* to suitable holomorphic theta-functions of Tate–Morikawa–McCabe, and hence that the abelian varieties uniformized by the Tate theory are exactly those with totally-degenerate reduction.

In the algebraic theory we have outlined, the exact sequence

$$0 \to \Lambda \to V_2(A) \to A_{2^\infty} \to 0$$

is the analogue of the sequence

$$0 \to M \xrightarrow{q} V(M) \to A(k) \xrightarrow{?} 0$$ [6]

(where $V(M)$ is the g-dimensional torus with character group M) of the holomorphic theory. See Tate's Bowdoin Colloquium talks for details. Now using the theorem of the previous section we can express every theta-function with totally degenerate reduction, Θ_a (the subscript a emphasizes that it is the *algebraic* theta-function) as

$$\Theta_a(\alpha+\beta) = e(\beta/2,\alpha) \sum_{x \in M''} e(\beta,x)c'(\alpha+x),$$

where $V = V_1 \oplus V_2$ is a suitable decomposition, $\alpha \in V_1$, $\beta \in V_2$, $M'' = M' \cap \Lambda$, $M' \subset V_1$, V_2 is the W of the previous section, c' and M' as in Theorem 2.

[6] The map $V(M)(k) \to A(k)$ is surjective when the abelian variety A over k is the quotient in the category of rigid analytic spaces of a *split* torus $V(M)$ over k by the period subgroup $q(M)$.

5 Analytic Theta-Functions.

In the holomorphic theory there is a quadratic character $c\colon M \to R$ which determines q via the identity

$$(q^x)^y = \frac{c(x+y)}{c(x)c(y)}.$$

The unique holomorphic theta-function Θ_h is equal to

$$\Theta_h(u) = \sum_{x \in M} c(x) u^x.$$

It is now easy to relate Θ_h and Θ_a. Explicitly, we construct a map

$$\begin{array}{ccc} f\colon M' \oplus (V_2/V_2 \cap \Lambda) & \longrightarrow & \{x \in V(M) \mid x^{2^m} \in q^M \text{ for some } m \in \mathbb{N}\} \\ \cap\!\!| & & \cap\!\!| \\ V_2(A)/V_2 \cap \Lambda & & V(M) \end{array}$$

so that if $\alpha \in V_2$, $x \in M$, then

$$f(\alpha)^x = e(\alpha, x),$$

and if $\alpha \in M'$, $x \in M$, then

$$f(\alpha)^x = \frac{c(\alpha+x)}{c(\alpha)c(x)}.$$

Note that $f(V_2/V_2 \cap \Lambda) \xrightarrow{\sim} \{\text{points of order } 2^n \text{ in } V(M)\}$ and $f(M')$ is a "2-divisible hull" of q^M in $V(M)$.
Define:

$$\gamma\colon M' \oplus (V_2/V_2 \cap \Lambda) \to R \quad \text{by}$$
$$\gamma(\alpha+\beta) = c(\alpha)^{-1} e(\alpha, \beta/2).$$

Then a simple verification yields

$$\Theta_h(f(x)) = \gamma(x) \cdot \Theta_a(X).$$

Since we have essentially the same theta functions in the algebraic and holomorphic cases, it is easy to deduce that the two theories provide uniformizations of the same abelian variety.

Reference

[1] Mumford, D., *On Equations Defining Abelian Varieties II*, to appear.[7]

[7] *Invent. Math.* **3**, 1967, 75–135.

Article [u67b]

Degeneration of algebraic theta functions[1]

David Mumford

1 2-adic theta functions, values in a complete valued field

Problem:

Given K: complete algebraically closed valued field; integers $\mathscr{O}$, residue field $k = \mathscr{O}/\mathfrak{m}$, absolute value $|\ |: K^\times \to \mathbb{R}_{>0}$, char.$(k) \neq 2$.

Given V: $2g$-dimensional vector space over $\mathbb{Q}_2$, plus e, e_*, Λ.[2]

Given $\Theta : V \to K$, a *theta function* w.r.t. e, e_*, coarse support$(\Theta) = V$.[3]

Analyze structure of Θ

---

[1] This is a slightly edited version of a set of handwritten notes by Mumford in the summer of 1967. It contains an essentially complete proof of the results in the letter [1967undated] to Grothendieck, despite a disclaimer in the letter (see footnote 114, p. 722). Mumford lectured on these results in the 1967 Summer School at Bowdoin, see the previous article, [u67a]. Appendix II in the 1984 Ph.D. dissertation of C.-L. Chai, London Math. Soc. Lecture Notes Series 107, 1985, pp. 237–286 is a modified version based on the same set of notes.

The notes come in two batches, reproduced as two sections. The first section contains the key results on the structure of 2-adic theta functions associated to abelian varieties over a local field. This structure theory is applied in §2 to the 2-adic monodromy of abelian varieties over local fields. Two pages of the original notes are essentially the same as the last section of the Bowdoin lecture notes [u67a]; they are not reproduced here.

[2] The notations and results in [66a], [67a] and [67b], *Equations defining abelian varieties I, II, III*, referred to as [Eq I, II, III] in the footnotes, are used extensively in this set of notes. In particular $e: V \times V \longrightarrow \mu_{2^\infty}(K)$ is a skew-symmetric bi-multiplicative nondegenerate pairing from $V \times V$ to the group of all roots of unity whose order is a power of 2, Λ is a maximal isotropic $\mathscr{O}$-lattice in V, and $e_*: \frac{1}{2}\Lambda/\Lambda \longrightarrow \{\pm 1\}$ is a quadratic character such that

$$e_*(\alpha+\beta)e_*(\alpha)e_*(\beta) = e(\alpha,\beta)^2 \qquad \forall \alpha,\beta \in \frac{1}{2}\Lambda.$$

[3] That $\Theta : V \longrightarrow K$ is a theta function for (V,Λ,e,e_*) means that it satisfies
theta transformation law: $\Theta(\alpha+\beta) = e_*(\beta/2)e(\beta/2,\alpha)\Theta(\alpha)\ \ \forall \alpha \in V, \forall \beta \in \Lambda$.
symmetry: $\Theta(-\alpha) = \Theta(\alpha)\ \ \forall \alpha \in V$.
Riemann theta relation: For all $\alpha_1,\alpha_2,\alpha_3,\alpha_4 \in V$ we have

$$\prod_{1\le i\le 4} \Theta(\alpha_i) = 2^{-g}\cdot \sum_{\eta\in\frac{1}{2}\Lambda/\Lambda} e(\gamma,\eta) \prod_{1\le i\le 4} \Theta(\alpha_i+\gamma+\eta), \text{ where } \gamma = -\tfrac{1}{2}(\alpha_1+\alpha_2+\alpha_3+\alpha_4).$$

See [Eq III], p. 216. The *coarse support* of an algebraic theta function $\Theta: V \to K$ is the set of all $\alpha \in V$, for which there exists $\eta \in \frac{1}{2}\Lambda$ such that $\Theta(\alpha+\eta) \neq 0$.

D. Mumford, *Selected Papers*, Vol. II,

(I) All values $\Theta(\alpha)$ are integrally dependent on $\{\Theta(\beta) \mid \beta \in \frac{1}{2}\Lambda\}$; hence $\max|\Theta(\alpha)|$ exists and is taken on for some $\alpha \in \frac{1}{2}\Lambda$.[4] So multiply Θ by a constant s.t.

(a) $\Theta(\alpha) \in \mathcal{O}$ for all $\alpha \in \frac{1}{2}\Lambda$,
(b) $\exists \alpha \in \frac{1}{2}\Lambda$ with $\Theta(\alpha) \notin \mathfrak{m}$, or equivalently $|\Theta(\alpha)| = 1$.

$\therefore$ Get a nonzero theta function $\overline{\Theta}(\alpha) := [\Theta(\alpha) \bmod \mathfrak{m}] \in k$.

(II) Say coarse support$(\overline{\Theta}) = W + \frac{1}{2}\Lambda$, $W \subseteq V$ a *cusp*.[5]

$(*)$ Choose a symplectic transformation T of V s.t. $T(\Lambda) = \Lambda$, $e_* \equiv 1$ on $T(W^\perp) \cap \frac{1}{2}\Lambda$. Change Λ by this: Then 0 is an origin[6] for W. Later, will have to apply T in reverse to the structure Th. we get for Θ.

$\rightsquigarrow$ OK:

$$\begin{aligned} \overline{\Theta}(\alpha) &= e_*(\eta/2) \cdot e(\eta/2, \alpha) \cdot \overline{\Theta}^*(\alpha_0^*) \\ \text{if} \quad & \alpha = \eta + \alpha_0,\ \eta \in \Lambda,\ \alpha_0 \in W, \\ & \alpha_0^* = \text{image of } \alpha_0 \text{ in } W/W^\perp, \\ & \overline{\Theta}^* = k\text{-valued nondegen. theta fcn. on } W/W^\perp. \end{aligned}$$

Choose:

$$\left. \begin{aligned} & V = W_1 \oplus W_2 \\ & \Lambda = \Lambda_1 \oplus \Lambda_2, \qquad \Lambda_i = \Lambda \cap W_i \\ & e_* = 1 \quad \text{on } \tfrac{1}{2}\Lambda_i \end{aligned} \right\} \text{ standard decomp. of } V$$

s.t. $(0) \subset W^\perp \subset W_1 \subset W \subset V$, so $W = W_1 \oplus \widetilde{W}_2$, $\widetilde{W}_2 \subset W_2$.

Given $V = W_1 \oplus W_2$ and $W = W_1 \oplus \widetilde{W}_2$ as above:
$\exists$ 1-1 correspondence between

(a) $\mathcal{O}$-valued theta fcns. Θ on V s.t. coarse supp$(\overline{\Theta}) = W + \frac{1}{2}\Lambda$,
(b) $\mathcal{O}$-valued *Gaussian measures*[7] μ on W_2 s.t. supp$(\overline{\mu}) = \widetilde{W}_2$.

---

[4] See Prop. 1 of *Abstract theta functions*. paper [u67a] in this volume.

[5] See Theorem on p. 230 of [Eq III] for this assertion. A *cusp* is a vector subspace $W \subseteq V$ such that $W^\perp \subseteq W$; see p. 229 *loc. cit.*

[6] An *origin* of a cusp W is an element $\eta_0 \in \frac{1}{2}\Lambda$ such that $e_*(\eta_0) = 1$ and $e_*(\alpha) = e(\alpha, \eta_0)^2$ for all $\alpha \in W^\perp \cap \frac{1}{2}\Lambda$; see p. 229 *loc. cit.*

[7] A k-valued *even* measure on W_2 is a *Gaussian measure* if there exists a k-valued measure ν on W_2 such that $(\mu \times \mu)(U) = (\nu \times \nu)(\xi(U))$ for all compact open subsets U in $W_2 \times W_2$, where $\xi\colon W_2 \times W_2 \longrightarrow W_2 \times W_2$ is defined by $\xi(x,y) = (x+y, x-y)$; see p. 118 of [Eq II].

In fact[8]

$$\boxed{\begin{array}{ll} \mu(a_2+2^n\Lambda_2)=2^{-ng}\displaystyle\sum_{a_1\in 2^{-n}\Lambda_1/\Lambda_1} e(a_2,a_1/2)\cdot\Theta(a_1+a_2) & \forall a_2\in W_2,\ \forall n\in\mathbb{N} \\ \Theta(a_1+a_2)=e(a_2,a_1/2)\displaystyle\int_{a_2+\Lambda_2} e(a_1,\beta)\cdot d\mu(\beta) & \forall a_1\in W_1,\ \forall a_2\in W_2 \end{array}}$$

Esp:

$$\sup\left\{|\mu(a_2'+2^n\Lambda_2)|\,\middle|\,\begin{array}{l}a_2'\in a_2+\widetilde{W}_2+\frac{1}{2}\Lambda_2\\ n\geq 0\end{array}\right\}=\sup\left\{|\mu(U)|\,\middle|\,\begin{array}{l}U\subset a_2+\widetilde{W}_2+\frac{1}{2}\Lambda_2\\ U \text{ compact open}\end{array}\right\}$$

$$\|$$

$$\sup\left\{|\Theta(a_1+a_2')|\,\middle|\,\begin{array}{l}a_1\in W_1\\ a_2'\in a_2+\widetilde{W}_2+\frac{1}{2}\Lambda_2\end{array}\right\}=\sup\left\{|\Theta(b)|\,\middle|\,b\in a_2+W+\frac{1}{2}\Lambda\right\}$$

(III) *Next step: Show that* $\forall\Theta$ *or* μ, *and* $\forall a_2$, *this* sup *is a* max.

Proof. Associate Φ to Θ s.t.

$$\Phi(\alpha)\Phi(\beta)=\sum_{\zeta\in\frac{1}{2}\Lambda_1/\Lambda_1} e(\alpha,\zeta)\cdot\Theta(\alpha+\beta+\zeta)\cdot\Theta(\alpha-\beta+\zeta)$$

$$2^g\,\Theta(2\alpha)\Theta(2\beta)=\sum_{\zeta\in\frac{1}{2}\Lambda_2/\Lambda_2} e(\alpha,\zeta)^2\cdot\Phi(\alpha+\beta+\zeta)\cdot\Phi(\alpha-\beta+\zeta)$$

$$\therefore |\Phi(\alpha)|\cdot|\Phi(\beta)|\leq\max_{\zeta\in\frac{1}{2}\Lambda_1}|\Theta(\alpha+\beta+\zeta)|\cdot|\Theta(\alpha-\beta+\zeta)|$$

$$|\Theta(\alpha+\beta)|\cdot|\Theta(\alpha-\beta)|\leq\max_{\zeta\in\frac{1}{2}\Lambda_2}|\Phi(\alpha+\zeta)|\cdot|\Phi(\beta+\zeta)|$$

$$\therefore\ \max_{\zeta\in\frac{1}{2}\Lambda_1}|\Theta(\alpha+\beta+\zeta)|\cdot|\Theta(\alpha-\beta+\zeta)|=\max_{\zeta\in\frac{1}{2}\Lambda_2}|\Phi(\alpha+\zeta)|\cdot|\Phi(\beta+\zeta)|.$$

So

$$\max_{\zeta\in\frac{1}{2}\Lambda_1}|\Theta(\alpha+\beta+\zeta)|\cdot\max_{\zeta\in\frac{1}{2}\Lambda_1}|\Theta(\alpha-\beta+\zeta)|=\max_{\zeta\in\frac{1}{2}\Lambda_2+\frac{1}{4}\Lambda_1}|\Phi(\alpha+\zeta)|\cdot|\Phi(\beta+\zeta)|.$$

Now assume $\beta\in W$. Use $\forall x\in W+\Lambda_2\ \exists\eta\in\frac{1}{4}\Lambda_1$ s.t. $|\Phi(x+\eta)|=1$.
Let $\tau(\gamma)=\max_{\zeta\in\frac{1}{4}\Lambda_1}|\Theta(\gamma+\zeta)|$.

$$\therefore \tau(\alpha+\beta)\tau(\alpha-\beta)=\max_{\substack{\zeta_1,\zeta_2\in\frac{1}{2}\Lambda_2+\frac{1}{8}\Lambda_1,\\ \zeta_1+\zeta_2\in\frac{1}{4}\Lambda_1}}|\Phi(\alpha+\zeta_1)|\cdot|\Phi(\beta+\zeta_2)|.$$

Def. $\alpha\in V$ is *normal* if $\max_{\zeta\in\frac{1}{8}\Lambda_1}|\Phi(\alpha+\zeta)|=\max_{\zeta\in\frac{1}{2}\Lambda_2+\frac{1}{8}\Lambda_1}|\Phi(\alpha+\zeta)|$.

[8] See pp. 116–117 of [Eq II].

$$\left[\forall\alpha\ \exists\eta\in\frac{1}{2}\Lambda_2 \text{ s.t. } \alpha+\eta \text{ is normal.}\right]$$

So if α normal, $\beta\in W$, then

$$\tau(\alpha+\beta)\tau(\alpha-\beta) = \max_{\zeta\in\frac{1}{8}\Lambda_1+\frac{1}{2}\Lambda_2} |\Phi(\alpha+\zeta)| =: \rho(\alpha).$$

Esp.

$$\tau(\alpha+\beta)\tau(\alpha-\beta)=\tau(\alpha)^2.$$

Note: *If $\eta\in\frac{1}{2}\Lambda_2$, $\alpha+\eta$ normal, then $\tau(\alpha+\eta)\geq\tau(\alpha)$.*

Proof of Note.

$$\begin{aligned}\tau(\alpha)^2 &= \max_{\substack{\zeta_1,\zeta_2\in\frac{1}{2}\Lambda_2+\frac{1}{8}\Lambda_1,\\ \zeta_1+\zeta_2\in\frac{1}{4}\Lambda_1}} |\Phi(\alpha+\zeta_1)|\cdot|\Phi(\zeta_2)|\\ &\leq \max_{\zeta\in\frac{1}{2}\Lambda_2+\frac{1}{8}\Lambda_1} |\Phi(\alpha+\zeta)| = \rho(\alpha);\end{aligned}$$

$\alpha+\eta$ normal $\Longrightarrow \tau(\alpha+\eta)^2=\rho(\alpha+\eta)=\rho(\alpha)\geq\tau(\alpha)^2.$ Q.E.D.

Now suppose $\alpha_n\in a+W+\frac{1}{2}\Lambda$ s.t.

$$|\Theta(\alpha_n)| \longrightarrow \sup\left\{|\Theta(\beta)|\,\middle|\,\beta\in a+W+\frac{1}{2}\Lambda\right\} =: s.$$

W.l.o.g. can assume $|\Theta(\alpha_n)|=\tau(\alpha_n)$ & α normal (in view of Note above).
OK: Pass to subsequence s.t.

$$\alpha_n-\alpha_m\in W+\Lambda \quad (\text{all } n,m).$$

W.l.o.g. may assume $\alpha_n-\alpha_m\in W$ for all n, m. Now if

$$\tau(\alpha_n)=|\Theta(\alpha_n)|>\sqrt{s\cdot|\Theta(\alpha_1)|}=\text{geom. mean of } s \text{ and } |\Theta(\alpha_1)|,$$

then

$$s\cdot\tau(\alpha_1)<\tau(\alpha_n)^2=\tau(\alpha_n+(\alpha_1-\alpha_n))\cdot\tau(\alpha_n-(\alpha_1-\alpha_n))<\tau(\alpha_1)\cdot s,$$

contradiction.

$$\therefore s = \tau(\alpha_n) \quad \text{for all } n.$$

Step (III) is proved.

We conclude

Proposition 1. *For any $\mathcal{O}$-valued Gaussian measure μ on W_2 such that the k-valued measure $\overline{\mu}$ is not zero, let $\widetilde{W}_2=\operatorname{supp}(\overline{\mu})$. Then $\forall$ compact open subgroup $\Lambda_2'\subset V_2$ and $\forall a\in W_2$,*

$$\sup\left\{|\mu(U)|\,\middle|\,U\subseteq\widetilde{W}_2+\Lambda_2'+a\right\}$$

is attained.[9]

[9] Prop. 1 has been proved for $\Lambda'=\frac{1}{2}\Lambda_2$. Apply an automorphism A of V_2 such that $A(\frac{1}{2}\Lambda_2)\subset\Lambda_2'$.

(IV) Let $\mathscr{V}$ be a finite-dimensional vector space over $\mathbb{Q}_2$, let W be a vector subspace of $\mathscr{V}$, and let $\Lambda \subset \mathscr{V}$ be a compact open subgroup.[10]

Theorem 2. *Let μ be a Gaussian measure on $\mathscr{V}$ with values in $\mathscr{O}$. Let ν be the dual Gaussian measure of μ, i.e., $\xi_*(\mu \times \mu) = \nu \times \nu$.*[11]
Assume

(1) *$\overline{\mu}, \overline{\nu}$ have support $W \subset \mathscr{V}$,*
(2) $\forall w \in \mathscr{V}$

$$\max\{\,|\mu(V)|\,|\,V \subset w+\Lambda+W\,\} =: \sigma(w),$$
$$\max\{\,|\nu(V)|\,|\,V \subset w+\Lambda+W\,\} =: \tau(w)$$

exist.

Then for all $w \in \mathscr{V}$, if $c \in \mathscr{O}$ s.t. $|c| = \sigma(w) = \max_{\eta \in \frac{1}{2}\Lambda} (\sigma(w+\eta))$, then

$$\operatorname{supp}\left\{\overline{\frac{\mu}{c}\bigg|_{w+\Lambda+W}}\right\} = w+\eta_0+W \qquad \text{for some } \eta_0 \in \Lambda.$$

Proof. **Claim 1:** $\tau(w)^2 = \sigma(w)$.

(1) $\exists U \subset w+\Lambda+W$ s.t. $|\mu(U)| = \sigma(w)$.[12]

$$\therefore\ |(\mu \times \mu)(U \times \Lambda)| = \sigma(w), \qquad \therefore\ |(\nu \times \nu)(\xi(U \times \Lambda))| = \sigma(w).$$

But $\xi(U \times \Lambda) \subset (w+\Lambda+W) \times (w+\Lambda+W)$,

$$\therefore\ \exists U_1,\, U_2 \subset w+\Lambda+W \quad \text{s.t.} \quad |(\nu \times \nu)(U_1 \times U_2)| = \sigma(w)$$
$$\|$$
$$|\nu(U_1)| \cdot |\nu(U_2)|$$

$$\therefore\ \tau(w) \geq \max\{|\nu(U_1)|, |\nu(U_2)|\} \geq \sqrt{\sigma(w)}.$$

(2) $\exists U \subset w+\Lambda+W$ s.t. $|\nu(U)| = \tau(w)$.

$$\therefore\ |(\nu \times \nu)(U \times U)| = \tau(w)^2.$$

But

$$(w+\Lambda+W) \times (w+\Lambda+W) = \bigcup_{\text{disjoint}} \xi((w+\Lambda+W+\eta) \times (\Lambda+W+\eta)) \quad \eta \in \tfrac{1}{2}\Lambda,$$

$\therefore\ \exists \eta,\ \exists U_1 \subset w+\Lambda+W+\eta,\ \exists U_2 \subset \Lambda+W+\eta$, s.t.

$$\tau(w)^2 = |(\nu \times \nu)(\xi(U_1 \times U_2))| = |\mu(U_1)| \cdot |\mu(U_2)|$$
$$\leq |\mu(U_1)| \leq \sigma(w+\eta) \leq \sigma(w).$$

We have proved Claim 1.

[10] The general notation for §1 is suspended in Steps (IV), (V) and the first statement of Thm. 1 in (VI). In application the triple $(\mathscr{V}, \Lambda, W)$ in this theorem will be $(W_2, \Lambda_2, \widetilde{W})$. Also the meaning of the function τ here is *different* from that in the proof of Step (III).

[11] As before, $\xi\colon \mathscr{V} \times \mathscr{V} \longrightarrow \mathscr{V} \times \mathscr{V},\ \xi\colon (x,y) \mapsto (x+y, x-y)$.

[12] Here U is compact open; the same for the U_1, U_2 and U below.

E

Theorem: Let μ be a Gaussian measure on $\mathcal{V}$, values in ϑ. Let ν = dual Gaussian measure. $\Lambda \subset \mathcal{V}$ a compact open subgp.

Assume

1) $\bar\mu, \bar\nu$ have support $W_0 \subset \mathcal{V}$

2) ~~$\forall \Lambda \subset \mathcal{V}$ compact open subgps~~,

$\forall\ w \in \mathcal{V}$

$$\max \{ |\mu(V)| \mid V \subset w + \Lambda + W \} = \sigma(w)$$
$$\max \{ |\nu(V)| \mid V \subset w + \Lambda + W \} = \tau(w).$$

exist.

Then for all $w \in \mathcal{V}$, if $|c| = \sigma(w) = \max_{\eta \in \frac{1}{2}\Lambda} (\sigma(w+\eta))$

$$\mathrm{Supp}\left\{ \overline{\frac{\mu}{c}\Big|_{w+\Lambda+W}} \right\} = w + \eta_0 + W, \quad \text{some } \eta_0 \in \Lambda.$$

Pf. We have $\xi(\mu\times\mu) = \nu\times\nu$

Claim $\tau(w)^2 = \sigma(w)$.

(1) $\exists\, U \subset w + \Lambda + W$ s.t. $|\mu(U)| = \sigma(w)$.

$\therefore |\mu\times\mu(U\times\Lambda)| = \sigma(w)$

$\therefore |\nu\times\nu(\xi(U\times\Lambda))| = \sigma(w)$

But $\xi(U\times\Lambda) \subset (w+\Lambda+W)\times(w+\Lambda+W)$

$\therefore \exists\, U_1, U_2 \subset w + \Lambda + W$

$$|\nu\times\nu(U_1\times U_2)| = \sigma(w)$$
$$\parallel$$
$$|\nu(U_1)|\cdot|\nu(U_2)|$$

$\therefore \tau(w) \geq \max |\nu(U_i)| \geq \sqrt{\sigma(w)}$

(2) $\exists\, U \subset w + \Lambda + W$ s.t. $|\nu(U)| = \tau(w)$

$\therefore |\nu\times\nu(U\times U)| = \tau(w)^2$

But $(w+\Lambda+W)\times(w+\Lambda+W) = \bigcup_{\text{disj}} \{ (w+\Lambda+W+\eta)\times(\Lambda+W+\eta) \}$ $\eta \in \frac{1}{2}\Lambda$

$\therefore \exists\, \eta,\ U_1 \subset w + \Lambda + W + \eta$

$U_2 \subset \Lambda + W + \eta$ s.t.

Look at measures

$$\overline{\frac{\mu}{c}\Big|_{w+\Lambda+W}} =: \mu_w, \qquad \overline{\frac{\nu}{\sqrt{c}}\Big|_{w+\Lambda+W}} =: \nu_w .$$

Claim 2. (a) $\xi_*(\mu_w \times \overline{\mu}) = (\nu_w \times \nu_w)|_{\xi((w+\Lambda+W)\times(\Lambda+W))}$.
(b) *The restriction of the measure* $\nu \times \nu$ *to*

$$(w+\Lambda+W)\times(w+\Lambda+W) - \xi((w+\Lambda+W)\times(\Lambda+W))$$

has absolute values strictly less than $\sigma(w)$.
(c) $\xi_*(\mu_w \times \overline{\mu}) = \nu_w \times \nu_w$ *as measures on* $(w+\Lambda+W)\times(w+\Lambda+W)$.

Clearly (a) holds, and (b) implies (c). To see (b), suppose that

$$\begin{aligned} &U_1 \subset w+\Lambda+W+\eta, \\ &U_2 \subset \Lambda+W+\eta, \qquad\qquad \eta\in\frac{1}{2}\Lambda,\ \eta\notin\Lambda+W. \\ &\quad U_1 \text{ and } U_2 \text{ compact open,} \end{aligned}$$

Then

$$\begin{aligned} |\nu\times\nu(\xi(U_1\times U_2)| &= |\mu(U_1)|\cdot|\mu(U_2)| \\ &\le \sigma(w+\eta)\cdot\sigma(\eta) \\ &< \sigma(w+\eta) \qquad (\because\ \eta\notin\Lambda+W) \\ &\le \sigma(w) \qquad (\because \text{ assumption on } w) \end{aligned}$$

Claim 2 (b) is proved.[13]

Theorem 2 is a formal consequence of (c):

$$\xi_*(\mu_w\times\overline{\mu}) = \nu_w\times\nu_w \quad \text{implies} \quad \xi(\mathrm{supp}(\mu_w)\times W) = \overbrace{\mathrm{supp}(\nu_w)}^{\text{call this } T}\times\mathrm{supp}(\nu_w),$$
$$\| $$
$$\{(a+w,a-w)\mid a\in \overbrace{\mathrm{supp}(\mu_w)}^{\text{call this } S},\ w\in W\}.$$

Start with $a\in S$, $u\in W$. Then $a\pm u\in T$, so $(a+u,a+u)\in T\times T$, and $a+u\in S$ too because $\xi^{-1}(a+u,a+u)=(a+u,0)$. We have shown that $a+W\subset S$ for all $a\in S$. If $b\in S$ also, then $b\pm u\in T$ as before, and $(a+u,b+u)\in T\times T$.

$$\therefore\ \frac{1}{2}(a-b)\in W \quad \because \xi^{-1}(a+u,b+u)\in S\times W.$$

So $a-b\in W$ for all a, $b\in S$. Theorem 2 is proved. Q.E.D.

[13] Because each compact open subset of $((w+\Lambda+W)\times(w+\Lambda+W))-\xi((x+\Lambda+W)\times(\Lambda+W))$ is a finite disjoint union of subsets of the form $\xi(U_1\times U_2)$ satisfying the above conditions.

(V) We now reformulate what has been proved so far, and what is expected.
Let $\mathscr{V}$ be a vector space over $\mathbb{Q}_2$, $W \subset \mathscr{V}$ a vector subspace, $\pi: \mathscr{V} \twoheadrightarrow \mathscr{V}/W$, $\dim \mathscr{V} = g$, $\dim W = g-r$. Let μ be an $\mathscr{O}$-valued Gaussian measure on $\mathscr{V}$ s.t. $\mathrm{supp}(\overline{\mu}) = W$.
We have proved:

(1) *For any compact open subset* $U \subset \mathscr{V}/W$,

$$\sigma_U := \sup\{\,|\mu(U')|\,\big|\, U' \subset \pi^{-1}(U),\, U' \text{ compact open}\,\}$$

is reached by some compact open subset U'.

(2) *Let* U *and* σ_U *be as in (1) above, let* $c_U \in K$ *be s.t.* $|c_U| = \sigma_U$, *and let*

$$\mu_U = \overline{\left[\frac{\mu}{c_U}\Big|_{\pi^{-1}(U)}\right]}.$$

Then $\mathrm{supp}(\mu_U)$ *is a finite union of cosets of* W.

Expectation 3: $\exists S \subset \mathscr{V}/W$,

$$\begin{array}{ccc} S & \subset & \mathscr{V}/W \\ \wr\| & & \wr\| \\ \mathbb{Z}[1/2]^r & \subset & \mathbb{Q}_2^r \end{array}$$

and $\exists$ *a function*[14] $\sigma = \sigma_{ss}: S \longrightarrow \mathbb{R}$ *of the form*

$$\sigma(x) = e^{-Q(x,x)},\ Q \text{ a pos. def. quad. form on } S,$$

s.t. $\forall U \subset \mathscr{V}/W$ *compact open, if*[15]

$$\sigma_U := \max_{x \in U \cap S} \sigma(x), \quad \text{and} \quad c_U \in K \text{ s.t. } |c_U| = \sigma_U,$$

then[16]

(a) $|\mu(U')| \le \sigma_U$ *for all compact open* $U' \subset \pi^{-1}(U)$.

(b) $\mu_U = \overline{\left[\frac{\mu}{c_U}\Big|_{\pi^{-1}(U)}\right]}$ *is a k-valued measure whose support is exactly*

$$\bigcup_{y \in S \cap U,\ \sigma(y) = \sigma_U} \pi^{-1}(y).$$

Def. The *singular set* $S = S(\mu)$ of μ is defined by

$$S = S(\mu) := \left\{ x \in \mathscr{V}/W \;\middle|\; \begin{array}{l} \exists \text{open neighborhood } U \text{ of } x \text{ in } \mathscr{V}/W \\ \text{s.t. } \mathrm{supp}(\mu_U) = \pi^{-1}(x) \end{array} \right\}.$$

Def. Define the sup. map $\sigma: S \longrightarrow \mathbb{R}$ for the Gaussian measure μ by

$$\sigma(x) = \max\{\,|\mu(U')|\,\big|\, U' \subset \pi^{-1}(U)\,\}$$

---

[14] The function σ here on the singular set S of μ is *different* from the function σ on W in Theorem 2.

[15] The use of the notation σ_U here is compatible with the notation in (1) at the beginning of (V); cf. the *definition* of the function $\sigma: S \to \mathbb{R}$ below and Theorem 4 (1).

[16] In terms of the function $\sigma: S \longrightarrow \mathbb{R}$ here, for $w \in \mathscr{V}$, the positive number $\sigma(w)$ in Theorem 1 is equal to $\max_{x \in S \cap \pi(w+\Lambda+W)} \sigma(x)$.

for $x \in S$, where U is as above, i.e., an open neighborhood of x in $\mathscr{V}/W$ s.t. $\operatorname{supp}(\mu_U) = \pi^{-1}(x)$. This definition is independent of the choice of U.

It remains to show that

$$\begin{array}{ccc} S & \subset & \mathscr{V}/W \\ \wr\| & & \wr\| \\ \mathbb{Z}[1/2]^r & \subset & \mathbb{Q}_2^r \end{array}$$

and

$$\sigma(x) = e^{-Q(x,x)}, \quad Q \text{ pos. def.}$$

Proof of Expectation 3. Let ν be the $\mathscr{O}$-valued Gaussian measure dual to μ, i.e., $\xi_*(\mu \times \mu) = \nu \times \nu$. Let $T = S(\nu) \subset \mathscr{V}/W$ be the singular set of ν, and let[17]

$$\tau \colon T \longrightarrow \mathbb{R}$$

be the sup. map for the Gaussian measure ν on $\mathscr{V}$. Then

$$\xi(S \times S) = T \times T$$

and

$$\sigma(x) \cdot \sigma(y) = \tau(x+y) \cdot \tau(x-y) \qquad \text{for all } x, y \in S.$$

From these we deduce

(a) S is a subgroup of $\mathscr{V}/W$ and $2S = S$.
(b) $Q := -\log \sigma$ is a quadratic form from S to nonnegative real numbers.[18]

Let $\Lambda \subset \mathscr{V}/W$ be a neighborhood of 0, and let $S_0 := \Lambda \cap S$, a subgroup of S s.t.

$$\bigcup_{n \in \mathbb{N}} 2^{-n} S_0 = S.$$

(c) Let $x_1, \dots, x_n \in S_0$ be $\mathbb{Z}$-linearly independent elements in S_0. Look at the maximal H s.t. $\exists$

$$\begin{array}{ccc} \mathbb{Q}^n & & \\ \cup & & \\ H & \overset{\phi}{\searrow} & \\ \cup & & \\ \mathbb{Z}^n & \longrightarrow & S_0 \end{array} \qquad \phi((a_1, \dots, a_n)) = \sum a_i x_i.$$

Let Q' be the quadratic form on $\mathbb{Q}^n$ s.t.

$$Q'(a,a) := -\log \sigma(\phi(a_1, \dots, a_n)) \qquad \text{for all } a = (a_1, \dots, a_n) \in H.$$

Note that[19]

---

[17] The function τ here on the singular set T of ν should not be confused with the function $\tau(w)$ on $W \subset \mathscr{V}$ in Theorem 2, nor with the function τ on V in the proof of (III).

[18] For $s \in S$, we have $\sigma(s) = 1 \Leftrightarrow s \in W = \operatorname{supp}(\overline{\mu})$.

[19] The quadratic form Q' on $\mathbb{Q}^n$ satisfies $Q'(a) = Q(\phi(a)) > 0 \ \forall a \in \mathbb{Z}^n$. This implies that the extension of Q' to $\mathbb{R}^n$ is positive semidefinite. Otherwise the open cone $C \subset \mathbb{R}^n$ where Q' is negative is nonempty, and $C \cap \mathbb{Q}^n$ is a dense subset of C, a contradiction.

$$\left.\begin{array}{l} Q' \text{ is a pos. semi-definite quad. form.} \\ Q'(a) = 0, \ a \in \mathbb{Q}^n \implies a = 0. \end{array}\right\}$$

(c_1) $[H : \mathbb{Z}^n] < \infty$.

If not, $\exists$ $\mathbb{Q}$-vector subspace $L \subset \mathbb{Q}^n$ s.t. $H \cap L$ is dense in L in classical topology. But $\forall a \in \mathbb{Z}^n$, $\phi(a) \in S_0$,

$$\therefore \text{ in } \phi(a) + 2^m\Lambda,\ \sigma(\phi(a)) \geq \sigma(b') \text{ for all } b' \in \phi(a) + 2^m\Lambda$$

if m is large enough. Thus

$$Q'(a,a) \leq Q'(b,b) \qquad \text{for all } b \in a + 2^m H.$$

Take $a \in L \cap \mathbb{Z}^n$ and $b \in (a + 2^m H) \cap L$ in particular: then the possible b's are dense in L. So there are some b's for which $Q'(b,b) <$ any given ε, and get a contradiction.

Corollary. *H is a finitely generated abelian group: w.l.o.g. $H = \mathbb{Z}^n$.*

(c_2) *Q' is positive definite.*[20]

If not, get

$$\begin{array}{ccc} \mathbb{R}^n & \xrightarrow[\pi]{\text{proj.}} & \mathbb{R}^m \qquad (m < n) \\ \cup & & \\ H & & \end{array}$$

and a quadratic form Q'' on $\mathbb{R}^m$ s.t. $Q'(a) = Q''(\pi(a))$ for all $a \in \mathbb{R}^n$, and $\pi(H) \subset \mathbb{R}^m$ is *not* discrete. i.e., $\exists$ $\mathbb{R}$-vector subspace $L \subset \mathbb{R}^m$ s.t. $\pi(H) \cap L$ is dense in L.[21] Get the same contradiction as above.

(d) *S_0 is a free abelian group of rank $r = \dim(\mathscr{V}/W)$.*

Proof. Define $r := \dim(\mathscr{V}/W)$, $d := \dim_{\mathbb{Q}}(S_0 \otimes \mathbb{Q})$. Then

$$d \text{ finite} \iff S_0 \text{ fin. gen. by } (c_1) \text{ and } (c_2).$$

If $d < r$, then S_0 is too small to be dense[22] in Λ, OUT.
If $d > r$, well

$$S_0/2S_0 \subset \Lambda/2\Lambda \cong (\mathbb{Z}/2\mathbb{Z})^r. \qquad \text{Q.E.D.}$$

$$\therefore \quad \begin{array}{ccc} & & \mathbb{R}^r \\ & & \cup \\ S & \cong & \mathbb{Z}[1/2]^r \\ \cup & & \cup \\ S_0 & \cong & \mathbb{Z}^r \end{array}$$

---

[20] We know that the extension of Q' to $\mathbb{R}^n$ is a positive semidefinite; the assertion here is that Q' is positive definite on $\mathbb{R}^n$.

[21] $\because$ the closure of $\pi(H)$ in $\mathbb{R}^m$ is a Lie subgroup of $\mathbb{R}^m$ of positive dimension.

[22] S is dense in $\mathscr{V}/W$ because $\operatorname{supp}(\mu) = \mathscr{V}$.

and $\sigma = e^{-Q(a,a)}$, Q a pos. def. quad. form on $\mathbb{R}^r$. Expectation 3 is proved.

(VI)

Theorem 4 (1) *Every Gaussian measure* μ *on* $\mathscr{V}$ *(as above) can be written as*

$$\mu = \sum_{x\in S} \mu_x,$$

where each μ_x [23] [24] *is an* $\mathscr{O}$*-valued measure on* $\mathscr{V}$ *with*

$$\left.\begin{array}{ll} \mathrm{supp}(\mu_x) = \pi^{-1}(x) & \forall x \in S \\ \sup\{|\mu_x(U)| \mid U \subset \mathscr{V} \textit{ compact open}\} = \sigma(x) & \forall x \in S \end{array}\right\}.$$

Similarly, the dual measure ν *can be written as*

$$\nu = \sum_{x\in S} \nu_x$$

with similar properties as above. Moreover

$$\xi_*(\mu_x \times \mu_y) = \nu_{x+y} \times \nu_{x-y} \qquad \forall x, y \in S.$$

(2) *Correspondingly, suppose that* Θ *is an* $\mathscr{O}$*-valued theta function w.r.t.* (V, e, e_*, Λ) *related to an* $\mathscr{O}$*-valued Gaussian measure* μ *on* W_2 *as in* (II), *and let*

$$(0) \subset W^\perp \subset W \subset V, \quad V = W_1 \oplus W_2 \supset W_1 \oplus \widetilde{W} = W \quad \textit{and} \quad \Lambda = \Lambda_1 \oplus \Lambda_2$$

be as in (II). *Let* $\pi : V \twoheadrightarrow V/W = W_2/\widetilde{W}$ *be the projection map, and let* $S \subset W_2/\widetilde{W} = V/W$ *be the singular set of* μ *given by* (1) *above. Then*

$$\Theta(\alpha) = \sum_{x\in S} \Theta_x(\alpha),$$

where each Θ_x [25] *is a function on* V *such that*

a) $\Theta_x(\alpha+\beta) = e_*(\beta/2)\,e(\beta/2,\alpha)\,\Theta_x(\alpha) \qquad \forall \beta \in \Lambda,$
b) $\mathrm{supp}(\Theta_x) \subset \pi^{-1}(x) + \Lambda,$
c) $\Theta_x(\alpha+\beta) = e(\beta, \gamma_x - \alpha/2)\,\Theta_x(\alpha) \qquad \forall \beta \in W^\perp$ *if* $\gamma_x \in V$ *satisfies* $\pi(\gamma_x) = x$.

---

[23] For $x \in S$, the measure μ_x is the push-forward to $\mathscr{V}$ of a measure μ'_x on $\pi^{-1}(x)$, defined as follows. For any compact open subset U' of $\pi^{-1}(x)$, let $\{U_i\}_{i\in\mathbb{N}}$ be a decreasing family of compact open subsets of $\mathscr{V}$ such that $\bigcap_{i\in\mathbb{N}} U_i = U'$. Then $\mu'_x(U') = \lim_{i\to\infty} \mu(U_i)$.

[24] The $\mathscr{O}$-valued measures μ_x should not be confused with the k-valued measure μ_w in Claim 2 in the proof of Theorem 2.

[25] The functions Θ_x is related to the measures μ_x as in (II):

$$\mu_x(a_2 + 2^n\Lambda_2) = 2^{-ng} \sum_{a_1 \in 2^{-n}\Lambda_1/\Lambda_1} e(a_2, a_1/2)\,\Theta_x(a_1 + a_2),$$

$$\Theta_x(a_1 + a_2) = e(a_2, a_1/2) \int_{a_2+\Lambda_2} e(a_1, \beta)\cdot d\mu_x(\beta), \qquad a_1 \in V_1, a_2 \in V_2.$$

Defining an associated tower of toroidal groups

Θ on V gives

$$\begin{cases} (0) \subset W^{\perp} \subset W \subset V,\ \pi\colon V \twoheadrightarrow (V/W) \\ S \subset V/W \\ \Theta_x \text{ on } \pi^{-1}(x) + \Lambda\,. \end{cases} \qquad \begin{pmatrix} \text{we assume } e_*(\alpha/2) = 1, \\ \forall \alpha \in W^{\perp} \cap (1/2)\Lambda \end{pmatrix}$$

(1) Θ_0 on $W/W^{\perp}$ defines a tower of abelian varieties B_α, indexed by compact open subsets $U_\alpha \subset W/W^{\perp}$.

(2) If $U \subset V$ is a compact open subgroup, get

(a) $U_\alpha = (U \cap W)/(U \cap W^{\perp}) \subset W/W^{\perp}$, hence B_α.

(b) $\pi(U) \cap S = S_0$, a lattice in S.
$\forall x \in S_0$, choose $\gamma_x \in U \cap \pi^{-1}(x)$. Set

$$\Phi_x(\beta) = e(\gamma_x/2, \beta) \cdot \Theta_x(\beta + \gamma_x) \quad \beta \in W,$$

a function on $W/W^{\perp}$ "related" to Θ_0.

$\therefore$ Φ_x defines a point $P_\alpha(x) \in B_\alpha$.

[If $\gamma_x' = \gamma_x + \eta$, $\eta \in U + W$, then $\Phi_x'(\beta) = \text{const} \cdot e(\eta/2, \beta)\, \Phi_x(\beta + \eta)$, so $P_\alpha(x)$ doesn't change.]

Get a homomorphism $S_0 \xrightarrow{P_\alpha} B_\alpha$

$$x \longmapsto P_\alpha(x)\,.$$

(c) $G_\alpha = \mathbf{Spec}_{B_\alpha}\left(\bigoplus_{x \in S_0} \{ T^*_{P_\alpha(x)} L_\alpha \otimes L_\alpha^{-1} \} \right)$.

A class of rigid analytic maps

Given: $K =$ complete valued field, $C = \widehat{\overline{K}}$,
Given: G, a comm. alg. grp. over K of type

$$\begin{array}{c} G \\ {\scriptstyle \mathbb{G}_m^r}\big\downarrow{\scriptstyle \pi} \\ A \end{array} \quad \text{abelian var.,}$$

L, ample inv. sheaf on A, all rational over K.

Now

$$G \cong \mathbf{Spec}_A \left\{ \bigoplus_{n \in \mathbb{Z}^r} (K_1^{n_1} \otimes \cdots \otimes K_r^{n_r}) \right\},$$

where $K_1, \ldots, K_r$ are invertible sheaves on A, alg. equiv. to $\mathcal{O}_A$.

To define a rigid analytic map $\phi: G_C \longrightarrow \mathbb{P}^m_C$,
need $m+1$ analytic sections of $\pi^*(L)$ over G_C,

$$\downarrow$$

$$m+1 \text{ Laurent-type expressions } L_i = \sum_{n\in\mathbb{Z}^r} s(n,i), \qquad 0 \le i \le m,$$
$$s(n,i) \in \Gamma(A, L\otimes K_1^{n_1}\otimes\cdots\otimes K_r^{n_r}).$$

CONVERGENCE: $\forall x \in G_C$, get $\pi(x) = y \in A_C$, plus $K_i(y) \xrightarrow{\sim} C$ for $i = 0, 1, \dots, m$. Then *evaluate*:

$$s(n,i) \longmapsto \mathrm{Val}_x[s(n,i)] \in L(y).$$

Ask that

$$\sum_n \mathrm{Val}_x[s(n,i)] \qquad \begin{cases} \text{exists in } L(y) \text{ for all } i, \\ \text{\& not be } 0 \text{ for all } i. \end{cases}$$

Hence ϕ comes out.

2 Application to monodromy: method of theta functions

Given:

(a) an abelian variety X over K $\longrightarrow$ get $T_2(X)$, a module over $\mathbb{Z}_2[\mathrm{Gal}(\overline{K}/K)]$,

(b) a principal polarization on X plus an even symmetric theta-divisor D_Θ representing it $\longrightarrow$ get a theta function $\Theta: V_2(X) \to \overline{K}$ s.t.
$$\Theta(\sigma x) = \Theta(x)^\sigma \quad \forall \sigma \in \mathrm{Gal}(\overline{K}/K).$$

[State converse: all such $(V, \Lambda, e, e_*, \Theta)$ come from (X, D_Θ), with $V_2(X) = V$, $T_2(X) = \Lambda$, e and e_* induced by D_Θ.]

Problem is to show:

if K = local field, alg. cl. res. field k, $\mathrm{char}(k) \neq 2$,
& if $\Gamma := \mathrm{Gal}(\overline{K}/K)$ acts on $T_2(X)$ via a homomorphism ρ_X from its tamely ramified quotient Γ_{tame} to the 2-adic symplectic group $\mathrm{Sp}(V, \Lambda, e) \cong \mathrm{Sp}_{2g}(\mathbb{Z}_2)$,[26]

then

$\exists$ an open subgroup $U \subset \Gamma_{\text{tame}}$ s.t.
$\forall \gamma \in U$ $\rho_X(\gamma)$ operates unipotently on $T_2(X)$.

Method: a complete description of the solutions to the theta functional equations over a local field; viz. $\exists$

(i) $(0) \subset W^\perp \subset W \subset V$ subspaces, $\pi: V \to V/W$

[26] $\Gamma_{\text{tame}} \cong \prod_\ell \mathbb{Z}_\ell(1)$, where ℓ runs through all prime numbers which are invertible in k.

(ii)
$$\begin{array}{ccc} S & \subset & V/W \\ \wr\Vert & & \wr\Vert \\ \mathbb{Z}[1/2]^r & \subset & \mathbb{Q}_2^r \end{array}$$

(iii) $Q\colon S \longrightarrow \mathbb{R}$ pos. def. quad. form

s.t.

$$\Theta = \sum_{s \in S} \Theta_x$$

(a) $\operatorname{supp}(\Theta_x) \subset \pi^{-1}(x) + \Lambda$,
(b) $\max_y |\Theta_x(y)| = e^{-Q(x,x)}$,
(c) $\Theta_x(\alpha+\beta) = e_*(\beta/2) \cdot e(\beta/2, \alpha) \cdot \Theta_x(\alpha)$ for all $\beta \in \Lambda$,
(d) $\Theta_x(\alpha+\beta) = e(\beta, \gamma_x - \alpha/2) \cdot \Theta_x(\alpha)$ for all $\beta \in W^\perp$ if $\gamma_x \in \pi^{-1}(x)$.

Claim: *It follows that*

$$\gamma = \mathrm{id} \ \textit{on } W \textit{ and on } V/W^\perp \qquad \forall \gamma \in U,$$

i.e., the matrix representation of γ has the form

$$\begin{pmatrix} I & O & * \\ O & I & O \\ O & O & I \end{pmatrix}.$$

This Claim will be proved in two steps below.

Step 1. *Assume that $\Theta\colon V \to k$ is an algebraic theta function for (V, Λ, e, e_*), and σ is an element of* $\mathrm{Sp}(V, \Lambda, e)$ *such that*

- $\forall x \in V, \exists \eta \in \frac{1}{2}\Lambda$ *s.t.* $\Theta(x+\eta) \neq 0$,
- $\Theta(\sigma x) = \Theta(x)$ *for all* $x \in V$.

Then σ is of finite order.

Proof of Step 1. Replace σ by a suitable power so that $(\sigma - 1)\Lambda \subseteq 4\Lambda$. We will show that[27] for any $n \geq 2$,

$$(\sigma - 1)\Lambda \subseteq 2^n \Lambda \implies (\sigma - 1)\Lambda \subseteq 2^{2n-1}\Lambda.$$

For any $x \in 2^{-n}\Lambda$, we have

$$\begin{aligned} \Theta(x) &= \Theta(\sigma x) \\ &= \Theta(x + (\sigma x - x)) \qquad \because \sigma x - x \in \Lambda \\ &= e_*\Big(\frac{\sigma x - x}{2}\Big) \cdot e\Big(\frac{\sigma x - x}{2}, x\Big) \cdot \Theta(x) \end{aligned}$$

[27] This statement was proved only in the case $n = 2$ in the original notes; the original left it to the reader to do this for $n > 2$.

$\therefore\ e(\sigma x - x, x) = 1$ if $\Theta(x) \neq 0$. Pick an $\eta \in \frac{1}{2}\Lambda$ such that $\Theta(x+\eta) \neq 0$. Then

$$\begin{aligned}1 &= e((\sigma-1)(x+\eta),(x+\eta)) \\ &= e((\sigma-1)x,x)\cdot e((\sigma-1)\eta,x)\cdot e((\sigma-1)x,\eta)\cdot e((\sigma-1)\eta,\eta).\end{aligned}$$

The last factor is 1 because $(\sigma-1)\eta \in 2^{n-1}\Lambda \subset 2\Lambda$. The product of the two middle factors is

$$\begin{aligned}e((\sigma-1)\eta,x)\cdot e((\sigma-1)x,\eta) = e(\sigma\eta,x)\cdot e(x,\sigma^{-1}\eta) &= e((\sigma^2-1)\eta,\sigma x) \\ &= e((\sigma-1)^2\eta + 2(\sigma-1)\eta,\sigma x) = 1\end{aligned}$$

because $(\sigma-1)^2\eta \in 2^{2n-1}\Lambda \subseteq 2^n\Lambda$, $2(\sigma-1)\eta \in 2^n\Lambda$ and $\sigma x \in 2^{-n}\Lambda$. So

$$q(x) := e((\sigma-1)x,x) = 1 \qquad \forall x \in 2^{-n}\Lambda.$$

Now we have

$$\begin{aligned}1 = \frac{q(x+y)}{q(x)\cdot q(y)} = e((\sigma-1)x,y)\cdot e((\sigma-1)y,x) &= e(x,\sigma^{-1}y)\cdot e(\sigma y,x) \\ &= e(x,\sigma^{-1}y-\sigma y)\end{aligned}$$

for all $x,y \in 2^{-n}\Lambda$, therefore $\sigma^{-1}y - \sigma y \in 2^n\Lambda$ for all $y \in 2^{-n}\Lambda$. Write $\sigma = 1+\tau$, we have

$$2^n\Lambda \ni \sigma^2 y - y = 2\tau y + \tau^2 y \qquad \forall y \in 2^{-n}\Lambda.$$

But $\tau^2 y \in 2^n\Lambda$, therefore $\tau y \in 2^{n-1}\Lambda$ for all $y \in 2^{-n}\Lambda$, i.e., $(\sigma-1)\Lambda \subseteq 2^{2n-1}\Lambda$.
Q.E.D.

We go back to the algebraic theta function Θ for $(V = V_2(X), \Lambda = T_2(X), e, e_*)$ attached to (X, D_θ). Let $W \subset V$ be the associated cusp, $W^\perp \subset W$. Every element $\gamma \in \Gamma_{\mathrm{tame}}$ operates on V via an element of $\mathrm{Sp}(V,\Lambda,e)$ s.t. $\gamma(W) \subseteq W$, $\gamma(W^\perp) \subseteq W^\perp$. By Step 1, there exists an open subgroup $U \subset \Gamma_{\mathrm{tame}}$ such that the action $\rho_X(\gamma)$ on (V,Λ) has the block form

$$\begin{pmatrix} A & B & * \\ O & I & C \\ O & O & {}^tA^{-1}\end{pmatrix}$$

for all $\gamma \in U$, i.e., γ operates on $W/W^\perp$ as the identity.

Step 2. Claim: $\left.\begin{matrix} A = I \\ B = 0\end{matrix}\right\}$ *i.e., look at* $\gamma|_W$
know γ *on* $W/W^\perp$ *is* $\mathrm{id}_{W/W^\perp}$
want $\gamma|_W = \mathrm{id}_W$.

It then follows that $C = 0$ too, i.e., $(\gamma-1)V \subset W^\perp$, or equivalently

$$e(\gamma x - x, y) = 1 \qquad \forall x \in V,\ \forall y \in W,$$

since

$$\begin{aligned} e(\gamma x - x, y) = e(x, \gamma^{-1} y) \cdot e(x, -y) &= e(x, \gamma^{-1} y - y) \\ &= e(x, 0) \qquad \because \gamma|_W = \mathrm{id}_W \\ &= 1. \end{aligned}$$

So the action $\rho_X(\gamma)$ on (V, Λ) has the form

$$\begin{pmatrix} I & O & D \\ O & I & O \\ O & O & I \end{pmatrix}, \qquad {}^t D = D.$$

We need the following facts for the proof of Step 2; they are consequences of the results in §1, summarized at the beginning of this section.

Fact (a). $\kappa(x) := \sup_{\eta \in \frac{1}{2}\Lambda} |\Theta(x + \eta)|$ depends only on the image of x in $V/(W + \frac{1}{2}\Lambda)$.[28]

Fact (b). $\forall x \in V, \exists \xi_x \in V$, depending only on the image of x in V/W, s.t.

$$|\Theta(x + u) - e(\xi_x, u)\Theta(x)| < \kappa(x) \qquad \forall u \in W^{\perp}.$$

We know that $\Theta(\gamma x) = \Theta(x)^{\gamma} \quad \forall x \in V, \gamma(\Lambda) \subseteq \Lambda$ and $(\gamma - 1)W \subseteq W^{\perp}$. Replacing U by an open subgroup, we may assume[29] that $\rho(U) \cong \mathbb{Z}_2$ and

$$(\gamma - 1)(\Lambda) \subseteq 8\Lambda + W^{\perp}, \quad (\gamma - 1)(\Lambda \cap W) \subseteq 8\Lambda \quad \forall \gamma \in U,$$

i.e., $\mathbf{P}(3)$ holds, where $\mathbf{P}(n)$ stands for the statement

$$\mathbf{P}(n): \qquad (\gamma - 1)(\Lambda) \subseteq 2^n\Lambda + W^{\perp} \text{ and } (\gamma - 1)(\Lambda \cap W) \subseteq 2^n\Lambda \quad \forall \gamma \in U.$$

It is clear that Step 2 follows from Claim 3 below.

Claim 3. *Suppose $n \geq 3$ and $\mathbf{P}(n)$ holds, then $\mathbf{P}(2n - 1)$ holds.*[30]

The first part of $\mathbf{P}(2n - 1)$ implies that if $x \in \Lambda \cap W$, $\xi = 2^{-2n+1}\lambda + w$, $\lambda \in \Lambda$, $w \in W$, then $(\gamma^{-1} - 1)\xi \in \Lambda + W^{\perp}$, so

$$e((\gamma - 1)x, \xi) = e(x, (\gamma^{-1} - 1)\xi) = 1.$$

i.e., the second part of $\mathbf{P}(2n - 1)$ follows from the first part.

Let $x \in 2^{-n}\Lambda$, $n \geq 3$. Write $(\gamma - 1)x = \eta + u$, $\eta \in \Lambda$, $u \in W^{\perp}$. Then

$$\Theta(x)^{\gamma} = \Theta(\gamma x) = \Theta(x + \eta + u) = e_*(\eta/2) \cdot e(\eta/2, x + u) \cdot e(\xi_x, u) \cdot \Theta(x).$$

---

28 This sup was denoted by $\sigma(x)$ in the original. Since the notation "$\sigma(x)$" was already overused in the previous section, it is replaced by "$\kappa(x)$" here.

29 Because $\mathrm{Sp}(V, \Lambda, e)$ is an extension of a finite group by a pro-2 group.

30 Claim 3 was extracted from the original notes; the argument there amounts to: if $n \geq 3$, then $\mathbf{P}(n)$ implies $\mathbf{P}(n')$ for some $n' > n$, therefore $\mathbf{P}(3) \Longrightarrow \gamma|_W = \mathrm{id}_W$.

Changing x by an element of $\frac{1}{2}\Lambda$, we may assume that $|\Theta(x)| = \kappa(x)$. Change x to x' with $w := x' - x \in 2^{-n}\Lambda \cap W$. Then $|\Theta(x')| = \kappa(x') = \kappa(x)$ too by Facts (a), (b) above. We know[31] that

$$\frac{\Theta(x')^\gamma}{\Theta(x')} = \frac{\Theta(x)^\gamma}{\Theta(x)}.$$

$$\therefore\ e_*(\eta/2)e(\eta/2, x+u)e(\xi_x, u) = e_*(\eta/2)e(\eta'/2, x'+u)e(\xi_x, u)$$

$$\therefore\ e(\eta, x+u) = e(\eta', x'+u).$$

We have $(\gamma-1)x = \eta + u$, $(\gamma-1)x' = \eta' + u$, $\eta' - \eta = (\gamma-1)w \in W^\perp \cap \Lambda$ by $\mathbf{P}(n)$.

$$\therefore e(\eta, x) = e(\eta', x') = e(\eta + (\gamma-1)w, x+w)$$

$$\begin{aligned}\therefore 1 = e(\eta, w)e((\gamma-1)w, x) &= e(w, -\eta)e(w, (\gamma^{-1}-1)x)\\ &= e(w, -\eta + (\gamma^{-1}-1)x)\end{aligned}$$

for all $w \in W \cap 2^{-n}\Lambda$.

$$\therefore -\eta + (\gamma^{-1}-1)x \in 2^n\Lambda + W^\perp$$

$$\therefore \gamma\eta + (\gamma-1)x \in 2^n\Lambda + W^\perp$$

$$\therefore \gamma\eta + \eta \in 2^n\Lambda + W^\perp \quad \because u \in W^\perp.$$

Hence

$$2\eta \in 2^n\Lambda + W^\perp$$

by $\mathbf{P}(n)$, i.e.,

$$\eta \in 2^{n-1}\Lambda + W^\perp.$$

We have shown that $\forall y \in 2^{-n}\Lambda$, $\exists x \in y + \frac{1}{2}\Lambda$ s.t. $(\gamma-1)x \in 2^{n-1}\Lambda + W^\perp$ for all $\gamma \in U$. So

$$(\gamma-1)\Lambda \subset 2^{2n-1}\Lambda + W^\perp.$$

Claim 3 and Step 2 are proved. Q.E.D.

[31] From the structure of tamely ramified extensions of local fields.

Part II
Correspondence

Correspondence 1958–1986

Grothendieck to Oscar Zariski, 6 August, 1958

Paris, August 6, 1958

Dear Professor Zariski,

I am very sorry to be obliged to tell you about a most silly misadventure which has happened to me: a letter with documents concerning my visa application, sent to the American embassy on July 19, has got lost in the mail, devil knows how. Therefore I would need again two papers which were already sent me by the University: my certificate of eligibility to an exchange program (which, according to instructions of the embassy, I should have in *two* copies); a statement that I will have a sufficient salary paid to cover all costs. I am very sorry to bother you once again with these trivial details, and I am convinced it will be the last time before coming to Harvard at last.

Writing down the theory of schemas, I got what seems to me now the definitive form of your theorem on holomorphic functions[1] proved by the same standard arguments (implying, in particular, a *decreasing* induction on the dimension on cohomology, so that the statement implying H^0 is proved last!) as the general finiteness theorem for proper maps. The statement is as follows: if $f: X \to Y$ is a proper morphism of noetherian schemas (think for instance of a proper morphism of algebraic varieties), $\mathscr{F}$ an algebraic coherent sheaf on X, and the $R^n f_*(\mathscr{F})$ the "higher direct images" of $\mathscr{F}$ by f (the sections of $R^n f_*(\mathscr{F})$ on an affine open set U of Y being the group $H^n(f^{-1}(U), \mathscr{F})$); if y is a point of Y which for convenience of statement we assume closed, and if we consider the sheaf of ideals $\mathscr{J}$ in $\mathscr{O}_X$ defined by the fibre $f^{-1}(y)$, then (i) the $R^n f_*(\mathscr{F})$ are coherent sheaves, (ii) the completion of $R^n f_*(\mathscr{F})_y$ for the $\mathfrak{m}_y$-adic topology is isomorphic to $\varprojlim_k H^n(f^{-1}(y), \mathscr{F} \otimes_{\mathscr{O}_X} \mathscr{O}_X/\mathscr{J}^k)$. Taking

[1] Grothendieck eventually published a different proof, of a fancier theorem, in EGA III 4. Something like the simple-minded proof outlined in this letter, of the less general theorem, may be found in section III.11 of R. Hartshorne, *Algebraic geometry*, Grad. Texts in Math. 52, Springer-Verlag, 1977.

D. Mumford, *Selected Papers*, Vol. II,

for instance $\mathscr{F} = \mathscr{O}_X$, $n = 0$ we see that $f_*(\mathscr{O}_X)$ is a coherent sheaf of commutative $\mathscr{O}_Y$-algebras and the set of maximal ideals of $f_*(\mathscr{O}_X)_y$ is in one to one correspondence with the set of connected components in the fiber $f^{-1}(y)$. This gives a reinforcement of your connectedness theorem; namely if X, Y are both irreducible, their sheaves without nilpotent elements, f surjective, and the field of Y quasi-algebraically closed in the field of X, then $f_*(\mathscr{O}_X)_y$ is contained in the integral closure of $\mathscr{O}_y$ in $k(X)$; if we know that $\mathscr{O}_y$ is "unibranch", that is if there is only one maximal ideal in the integral closure of $\mathscr{O}_y$ in its quotient field, then there can be only one maximal ideal in $f_*(\mathscr{O}_X)_y$ as well: this gives the connectedness theorem, without analytic irreducibility needed. Besides, using the connectedness theorems and the same standard techniques, one gets as a consequence your "main theorem" for *arbitrary* noetherian rings. So a "local" result is proved by global means.

— As an application of the general connectedness theorem, I give the following example (which was given to me as a problem by Serre). If $\mathscr{O}$ is a noetherian local ring, S the associated graded ring, X the projective algebraic set (over the residue field) defined by S, then X is connected provided $\mathscr{O}$ is unibranch.—

I have some hope also of solving your two open problems on holomorphic functions by these methods, using perhaps a general duality theorem, which I am now developing, and which holds for arbitrary complete schemas (the singularities do not matter). I will tell more about it in the seminar at Harvard.

Sincerely yours
(signed) A Grothendieck

Grothendieck to Mumford, 5 October, 1960

Paris Oct 5, 1960

My dear Mumford,

I beg very much your pardon for not having replied to your last letter nor acknowledged receipt of your manuscript. Unfortunately, it is too late now to publish it this very year; besides the referee was prevented from rereading the manuscript, and I will have to give it to another one. I appreciate your effort to give complete proofs and hope that you will let us publish your paper in spring 1961—which is certainly possible if there are no other gaps of importance in the proofs.

I am glad to hear you are interested in the existence theorem for Picard schemata,[2] yet I did not prove it with the generality you believe. I have to assume:

1) X flat and *projective* over S.
2) The fibers $f^{-1}(s)$ are *"absolutely irreducible"* and *without embedded primes.*
3) For every fiber, $H^0(f^{-1}(s), \mathscr{O}_{f^{-1}(s)}) \xleftarrow{\sim} k(s)$.

[2] See Footnote 21 in p. 650.

To go much further will presumably demand a considerably greater effort, which I do not intend to go into myself, but which I expect to be very much worthwhile.

Sincerely yours,
(signed) A Grothendieck

P.S. I will send you a copy of Chapter I of the Elements. For other issues of the "Publications IHÉS" you should write to the publisher, I believe.

Grothendieck to Mumford, 20 April, 1961

A. Grothendieck
23 Boul. de Levallois
Neuilly (Seine)

Paris April 20, 1961

Dear Mumford,

I am much interested by a result of yours on passage to the quotient by semi-simple algebraic groups, which Zariski has reported to me. Would it be possible for you to send me an outline of the proof? Even for the group $PGL(2)$, or the reductive group $\mathbb{G}_m$, I am not able to solve the problem, and I begin to have doubts even if such general results (as yours, which looks very like the conjecture 8.1 in the Bourbaki talk III on construction techniques)[3] really exist. I just found various counterexamples to my conjecture as it was formulated. For instance (as I wrote Tate) even for very standard operations on $PGL(n)$, the graph may not be closed and then even a nonseparated quotient may not exist; thus in 8.1. 1° one has at least to assume a closed graph (as indeed people generally do). Moreover conjecture 8.1. 2° seems hopelessly false, take for instance $Y = \mathbb{P}^1$, and X the principal projective bundle on Y associated to the standard ample line bundle $\mathcal{O}_Y(1)$ and the natural homomorphism $\mathbb{G}_m \to PGL(2)$; it is easy to see that X is "quasi-affine" i.e., an open subset of an affine space, i.e., $\mathcal{O}_X$ is ample, yet the sheaf $\mathcal{O}_Y$ of which it is the inverse image is of course not ample! Thus seems to escape the hope of any "general" construction of an ample sheaf on a quotient X/G, knowing one on X.

I knew the existence of varieties of moduli (for curves, or polarized abelian varieties) over $\mathbb{Q}$, for all "levels" (Stufe), but using Baily's transcendental results,[4] guaranteeing the existence over $\mathbb{C}$ and hence over $\mathbb{Q}$ of the quotient variety you know. I am now able to perform directly the same constructions *over* $\mathbb{Z}$, *but only for high levels* n. As I do not know by now if the corresponding schemes are quasi-projective over $\mathbb{Z}$, and hence if the groups $\Gamma = \mathrm{Sp}(2g, \mathbb{Z}/n\mathbb{Z})$ have orbits contained

[3] TDTE III: Préschémas quotients, in *Séminaire Bourbaki* 1960/61, n° 212.

[4] W.L. Baily, Jr., On the theory of θ-functions, the moduli of abelian varieties, and the moduli of curves, *Ann. Math.* **75** (1962) 342–381.

in affine sets, I cannot yet pass to the quotient by Γ to construct $\mathscr{M}_n$ for smaller levels. If your arguments are correct, they may yield the lacking proof for the quasi-projectivity(?).

Sincerely yours,
(signed) A Grothendieck

Grothendieck to Mumford, 25 April, 1961

April 25, 1961

Dear Mumford,

I thank you very much for your letter, and would like to congratulate you on your results. Still I would appreciate very much getting a sketch of your key-theorem (theorem 1 of your letter). It is of course obvious that the quotient U/G exists and U is a locally trivial principal bundle over U/G, even if U means the bigger open set of all $(x_1, \ldots, x_d)$ such that, for at least one choice of $n+2$ distinct indices, we get a projective basis of $\mathbb{P}^n$. What is not clear to me is what you define to be the ample sheaf on U/G, or probably rather how you prove that the obvious sheaf you get on U/G (say by descending the inverse of the sheaf of highest differentials on $(\mathbb{P}^n)^d$) is ample. Is your hypothesis on U really necessary?

I had obtained in the meanwhile the same counterexamples as you based on Hironaka's construction,[5] and was all the more afraid your proof was erroneous, as the theorem Zariski read to me from your letter resembled very much to my false conjecture. I am glad to know you are as skeptical as I about general criteria for passing to the quotient by the projective group, and feel more confident now. Besides, my construction of schemata of moduli for high levels (as defined axiomatically in my Cartan Seminar talks[6] or in an older letter to Tate) resembles very much to yours, except that I did not observe that the suitably embedded polarized abelian varieties are completely determined by their sets of points of order n, (n big enough), which then leads you to a rather specific situation for passing to the quotient.

It seems to me that, because of your lack of some technical background on schemata, some proofs are rather awkward and unnatural, and the statements you give not as simple and strong as they should be. Therefore I suggest you to wait for writing a detailed paper till August, where I would appreciate very much discussing these matters with you. It is much to be desired, at last, to have on these questions a paper having the conceptual clarity in statements and proofs they deserve (especially after work like that of Igusa, which is most discouraging to read!). As for the

[5] For this counterexample see Chap. 4, §3 of [GIT]; H. Hironaka, An example of a non-Kählerian complex-analytic deformation of Kählerian complex structures, *Ann. Math.* **75** (1962) 190–208.

[6] *Séminaire Henri Cartan* 13, 1960/61, Exposé 9–16.

Grothendieck to Mumford, 25 April, 1961

April 25, 1961

Dear Mumford,

I thank you very much for your letter, and like to congratulate you on your results. Still I would appreciate very much getting a sketch of your key-theorem (theorem 1 of your letter). It is of course obvious that the quotient U/G exists and U is a locally trivial principal bundle over U/G , even if U means the bigger open set of all $(x_1,..,x_d)$ such that at least one choice of n+2 distinct indices, we get a projective base of P^n . What is not clear to me is what you define to be the ample sheaf on U/G , or probably rather how you prove that the obvious sheaf you get on U/G (say by descending the inverse of the sheaf of highest differentials on $(P^n)^d$) is ample. Is your hypothesis on U really necessary ?

I had obtained in the meanwhile the same counterexamples as you based on Hironaka's construction, and was all the more affraid your proof was erroneous, as the theorem Zariski read to me from your letter resembled very much to my false conjecture. I am glad to know you are as sceptical as I about general criteria of passing to the quotient by the projective group, and feel more confident now. Besides, my construction of schemata of moduli for high levels (as defined axiomatically in my Cartan Seminar talks, or in an older letter to Tate) resembles very much to yours, except that I did not observe that the suitably imbedded polarized abelian varieties are completely determined by their sets of points of order n, (n big enough), which then leads you to a rather specific situation for passing to the quotient.

It seems to me that because of your lack of some technical background on schemata, some proofs are rather awkward and unnatural, and the statements you give not as simple and strong as they should be. Therefore I suggest you to wait for writing a detailed paper till August, where I would appreciate very much discussing these matters with you. It is much to be desired, at last, to have on these questions a paper having the conceptual clarity in statements and proofs they deserve (especially after work like that of Igusa, which is most discouraging statement of the results, I believe my Cartan talk is a good model, and scarcely anything needs to be changed. This of course would not prevent you to announce your results at once in a random way, before writing your detailed paper.

Can you prove, as is plausible from the transcendental approach, that for modular spaces of level $n \geq 3$ (over which therefore the modular family of curves and jacobians is defined), the invertible sheaf on the modular scheme defined by the highest degree differentials on the jacobians is ample? (Indeed, this really should stem from the corresponding result for schemata of moduli for abelian polarized varieties.) In fact, there are quite a few candidates for ample sheaves on $\mathcal{M}_{g,n}$, and it would be interesting to know about their relations.

In your "appendix", you refer to a result of Matsusaka I did not hear of before, namely the connectedness or irreducibility of the variety of moduli for curves of genus g, in any characteristic. I did not know there was any algebraic proof for this (whatever way you state it). Yet I have some hope to prove the connectedness of the $\mathcal{M}_{g,n}$ (arbitrary levels) using the transcendental result in char. 0 and the connectedness theorem; but first one should get a natural "compactification" of $\mathcal{M}_{g,n}$ which should be simple over $\mathbb{Z}$.[7] I would like to know what is known to you concerning connectedness. I insist once more that the most interesting objects are not the classical $\mathcal{M}_g$'s, but the schemata with operators $\mathcal{M}_{g,n}$, which have much nicer properties and achieve much preciser aims than $\mathcal{M}_g$ alone. For instance $\mathcal{M}_{g,n}$ ($n \geq 3$) is simple over $\mathbb{Z}$. Of course the strongest connectedness theorems will be concerned with the $\mathcal{M}_{g,n}$'s, big n, or (still better) with their Teichmüller analogues.

I indeed wrote a precise theory of the so-called "Hilbert schemata" which are to replace Chow coordinates, but are in fact rather different. They contain as open subsets the nonmultiple parts of symmetric products, but the points corresponding to multiple cycles are blown up there, because an ideal at a point, primary for the maximal ideal, is not known by telling the multiplicity (except on a nonsingular curve!). I will give a rather detailed account in my next Bourbaki talk,[8] alluded to in talk III.

Sincerely yours
(signed) A Grothendieck

---

[7] A proof of the connectedness, along these lines, was eventually published by Deligne and Mumford; see [69c].

[8] A. Grothendieck, TDTE IV: Les schémas de Hilbert, *Séminaire Bourbaki* 1960/61, Fasc 3, Exposé 221. A better and more elegant treatment of Hilbert schemes was found later, by Mumford; see Chapters 8 and 14 of [CAS] (= *Lectures on Curves on an Algebraic Surface*). Chap. 8 deals with flattening stratifications, Chap. 14 with the boundedness of the scheme. See also [1964Aug31].

Grothendieck to Mumford, 10 May, 1961

Paris May 10, 1961

Dear Mumford,

I thank you very much for your letter and the proof of your key theorem. I think I will be able in the next days to read it thoroughly.

Please excuse me if I omitted to write you some time ago that I received the revised version of your MS on the blowing down of a surface, which has been given to the printer. You will probably get the proofs from the printer during this month. I hope you will not be too dissatisfied with the delay of publication!

It occurred to me that I had sent you only a copy of my Bourbaki talk on quotients,[9] but none of my Cartan talk.[10] This is done now; I will send you the following ones within the next weeks.

The formation of the modular schemas $\mathcal{M}_{g,n}$ $(n \geq 3)$, representing contravariant functors, is obviously compatible with base extension. But I doubt the same be true for $\mathcal{M}_{g,1}$, which is the sub-product of $\mathcal{M}_{g,n}$ obtained by dividing by the finite group of automorphisms $G = \mathrm{Sp}(2g, \mathbb{Z}/n\mathbb{Z})$ (at least when restricting to the part of $\mathcal{M}_{g,1}$ lying over the open subset of $\mathrm{Spec}(\mathbb{Z})$ complement of the set of primes dividing n), and does not represent any reasonable contravariant functor (but, as you remarked, a covariant one). Such a commutation would mean that for every open affine set of $\mathcal{M}_{g,n}$, stable under G and with affine ring A, $H^1(G,A) = 0$. This can be expressed equivalently by introducing for every $x \in \mathcal{M}_{g,n}$ the inertia group G_x of x (which is the group of automorphism of the corresponding algebraic curve), and demanding that $H^1(G_x, \mathcal{O}_x) = 0$, where $\mathcal{O}_x$ is the local ring of x in $\mathcal{M}_{g,n}$. You can also replace the latter by its completion, which is nothing else but the local ring describing the "formal variety of moduli" of the given curve, in the sense of my Bourbaki talk II,[11] and the consideration of which is independent of the global theory. Although I did not make any effective computation, I do not see why such a relation should hold (even for genus $g = 2$); it does however for $g = 1$, because an equivalent formulation of the question is whether the fibers of $\mathcal{M}_{g,1}$ over the different points of $\mathrm{Spec}(\mathbb{Z})$ are normal, which is indeed true for genus 1. In the same direction, there is the question whether the natural morphism from $\mathcal{M}_{g,1}$ into the corresponding modular space for polarized abelian varieties is really an embedding; a priori one can say only that $\mathcal{M}_{g,1}$ is the normalisation of a (nonclosed) subschema of the latter, which may not be normal.

Sincerely yours
(signed) A Grothendieck

---

[9] *Séminaire Bourbaki* 1960/61, n° 212.

[10] *Séminaire Henri Cartan* 13, 1960/61, Exposé 9–16.

[11] *Séminaire Bourbaki* 1960/61, n° 195.

Grothendieck to Mumford, 29 January, 1962 (a)

Paris Jan 29, 1962

Dear Mumford,

Thanks for your letter, and best wishes to you and your wife for your son!

Your ampleness criterion looks nice indeed. I would appreciate to have an outline of the proof some time.

I am afraid you will not convince me of the usefulness of Chow coordinates, in fact your example shows again that the wrong method will lead to prove statements under unnatural assumptions (such as normality). Although I did not check it, I am convinced that the method I used for the theorems of passage to the quotient in my Bourbaki talk III[12] will yield:

Let X quasi-projective over S loc. noeth., $\mathscr{R} \subset X \times_S X$ a closed subscheme such that

(i) $\mathscr{R}$ is "set-theoretically an equivalence relation",
(ii) $\mathrm{pr}_1 : \mathscr{R} \to X$ is proper (hence projective) and universally open.

Then $Y = X/\mathscr{R}$ exists, $X \to Y$ is proper (and universally open), $\mathscr{R} \to X \times_Y X$ is a bijective closed immersion. [I checked this long ago when X is finite over S; no openness conditions are then required.]

If such a statement should be of use somewhere, I can include it in Chap. V. However, I never needed it, as it is much too coarse for the kind of problems I was considering. In fact, it should be considered rather as a statement of a theory of schemes "modulo $\mathscr{M}$", where $\mathscr{M}$ is the set of all morphisms which are "universal homeomorphisms", which one wants to consider as isomorphisms in the new category (obtained by adjoining formally their inverses). N.B. under the usual finiteness assumptions, "universal hom." = "finite surjective radicial morphism". If one sticks to, say, algebraic groups, one gets Serre's "quasi-algebraic groups" = groups mod purely inseparable isogeny.

Tate wrote me you are talking in your seminar on your theorem about passage to the quotient. I would appreciate to know when you obtain results on existence of Picard schemes (I am giving a Bourbaki talk on Picard on Feb 18).

Sincerely yours,
(signed) A Grothendieck

---

12 *Séminaire Bourbaki* 1960/61, n° 212. Note that, in Théorème 6.1 on p. 212-14, the hypothesis of universal openness is replaced by the hypothesis of flatness, and the proof uses quasi-sections. In a later Exposé, *Séminaire Bourbaki* 1961/62, n° 232, Grothendieck mentions a second proof, which uses the Hilbert scheme; see Remarque 5.1 on p. 232-13. The details were explained by Mumford in a conversation with Altman and Kleiman, who published them on p. 70 of their article Compactifying the Picard scheme, *Adv. in Math.* **35** (1980) 50–112.

Grothendieck to Mumford, 29 January, 1962 (b)

Paris Jan 29, 1962

Dear Mumford,

I have been too rash in my reply to your last letter: in effect, I was thinking of a reduction of the general case (concerning passage to the quotient under the conditions you know) to the case where $\mathscr{R} \to X$ is *finite*. However, in the latter, I have no means of attacking the problem, which in fact meets with a few unsolved problems on equivalence relations I still had in store. Therefore I grant you that, for the time being, in your example Chow coordinates do give mathematical information about existence of quotients which is not obtained by other means. I do not expect this situation to hold for long still! Besides, Chevalley had nontrivial unpublished results on quotients, of course never using Chow coordinates, which may well cover the cases we have in mind. Unfortunately he is very sick at the moment, with a so-called "pancreatitis" and there is no asking him about anything now. One more comment: it seems that, for the application of Chow coordinates, your regularity assumptions: X normal, $\mathscr{R} \to X$ univ. open, are not the right thing exactly, unless you assume $\mathscr{R}$ irreducible. What is needed, in effect, seems that all components of all fibers of $\mathscr{R} \to X$ have the same dimension (which *does not* follow from the assumptions as you stated them).

It also appears to me that in my Bourbaki talk III,[13] on quotients, in th. 6.1. (i), the assertion that $X/\mathscr{R} = Y$ is *quasi-projective* is proved only if $\mathscr{R}$ is really an *equivalence* relation (or $\mathscr{R} \to X$ *finite*), and not only a *pre*equivalence (this excludes the case of groups operating with fixed points!). I do not know at present if there may be a counterexample in the general case. I guess you are right to say that "the last word has not been said" at all in the theory of quotient schemes!

Best regards
(signed) A Grothendieck

Grothendieck to John Tate, 5 February, 1962

Paris Feb 5, 1962

Dear John,

In connection with my Bourbaki talk,[14] I pondered again on Picard schemes. For instance, as I told Mumford, I proved that if X/S is projective and simple,[15] then

[13] TDTE III: Préschémas quotients, *Séminaire Bourbaki* 1960/61, n° 212.

[14] TDTE V. Les schémas de Picard: Théorèmes d'existence, *Séminaire Bourbaki* 1961/62, n° 232, and TDTE VI. Les schémas de Picard: Propriétés générales, *ibid.*, n° 236.

[15] The standard terminology has changed from "simple" to "smooth".

$\mathscr{P}ic^{\tau}_{X/S}$ is of finite type over S.[16] More generally, the decomposition of $\mathscr{P}ic_{X/S}$ according to the Hilbert polynomials (in fact, the first two nontrivial coefficients of the polynomial suffice) consists of pieces which are of finite type, hence projective over S. Another way of stating this is to say that a family of divisors D_i on the geometric fibers of X/S is "limited" iff the projective degrees of the D_i and D_i^2 are bounded.

Another result, of interest in connection with your seminar, is a proof of the fact that, for an abelian scheme A/k, k a perfect field, the absolute formal scheme of moduli over $\mathrm{W}_\infty(k)$ is simple over k. This comes from the following general fact: Let X_0/S_0 be simple, X_0'/X_0 étale, S_0 subscheme of S defined by an ideal $\mathscr{I}$ of square 0. Let $\xi_0 \in H^2(X_0, \mathfrak{G}_{X_0/S_0} \otimes_{\mathscr{O}_{S_0}} \mathscr{I})$ and[17] $\xi_0' \in H^2(X_0', \mathfrak{G}_{X_0'/S_0} \otimes_{\mathscr{O}_{S_0}} \mathscr{I})$ be the obstruction for lifting. Then ξ_0' is the inverse image of ξ_0 under the obvious map. As a consequence, if X_0/S_0 is abelian, taking $X_0' = X_0$, $X_0' \to X_0$ multiplication by n prime to the residue characteristic, we get $\xi_0 = n^*(\xi_0)$. If $S = \operatorname{Spec} \Lambda$, Λ local artin, and $\mathfrak{m}\mathscr{I} = 0$, then we are reduced to an obstruction in the H^2 of the reduced $X_0 \otimes_{\Lambda_0} k = A$, satisfying $\xi = n^*(\xi)$ for n prime to p. Using the structure

$$H^*(A, \mathfrak{G}_{A/k}) \simeq \overset{*}{\bigwedge} H^1(A, \mathscr{O}_A) \otimes t_A,$$

we get $n^*(\xi) = n^3 \xi$, hence $(n^3 - 1)\xi = 0$. Taking $n = -1$ we get $2\xi = 0$, hence $\xi = 0$, and we win!

I just noticed[18] the proof does not give any information for residue char. $= 2$! Here is a simple proof valid in any char.: Consider the obstruction η_0 for lifting $X_0 \times_{S_0} X_0$, then $\eta_0 = \xi_0 \otimes 1 + 1 \otimes \xi_0$, and η_0 is invariant under the *automorphism* $(x,y) \rightsquigarrow (x, y+x)$ of $X_0 \times_{S_0} X_0$. Thus we get an element $\xi = \sum_{i,j} \lambda_{i,j} e_i \wedge e_j$ in $H^2(A, \mathscr{O}_A) = \bigwedge^2 t$, s.th. $\eta = \sum_{i,j} \lambda_{i,j} e_i' \wedge e_j' + \sum_{i,j} \lambda_{i,j} e_i'' \wedge e_j''$ in $\bigwedge^2 (t \oplus t)$ is *invariant* under $(x,y) \rightsquigarrow (x, y+x)$, carrying $e_i' \rightsquigarrow e_i' + e_i''$ and $e_i'' \rightsquigarrow e_i''$, hence trivially $\xi = 0$!

As a consequence, we get that the scheme of moduli for the *polarized* abelian schemes, with polarization degree d, is simple over $\mathbb{Z}$ at all those primes p which do not divide d. This comes from the fact that the obstruction to polarized lifting lies in a module $H^2(A, \mathscr{E})$, where $\mathscr{E}$ is an extension (the"Atiyah extension")

$$(*) \qquad 0 \to \mathscr{O}_A \to \mathscr{E} \to \mathfrak{G}_{A/k} \to 0$$

whose class c in $H^1(A, \Omega^1_{A/k})$ is just the Chern class $\frac{d\mathscr{L}}{\mathscr{L}}$ of the invertible sheaf $\mathscr{L}$ on A defining the polarization. Now in the exact sequence of cohomology for $(*)$, the map

$$\begin{array}{ccc} H^i(\mathfrak{G}_{A/k}) & \xrightarrow{\partial^{(i)}} & H^{i+1}(\mathscr{O}_A) \\ \wr\Vert & & \wr\Vert \\ \bigwedge^i t' \otimes t & & \bigwedge^{i+1} t' \end{array} \qquad t = t_A,\ t' = t_{\hat{A}}$$

[16] See Footnote 25.

[17] Here $\mathfrak{G}_{X_0/S_0}$ and $\mathfrak{G}_{X_0'/S_0}$ denote the relative tangent sheaves for X_0/S_0 and X_0'/S_0 respectively.

[18] This paragraph was penned in vertically, in the left margin.

Grothendieck to Tate, 5 February, 1962

Paris. Feb 5, 1962

Dear John,

In connection with my Bourbaki talk, I pondered again on Picard schemes. For instance, as I told Mumford, I proved that if X/S is projective + simple, then $Pic^{\tau}_{X/S}$ is of finite type over S. More generally, the decomposition of $Pic_{X/S}$ according to Hilbert polynomials (in fact, the first two non trivial coefficients of the polynomial suffice) consists of pieces which are of finite type, hence projective, over S. Another way of stating this is to say that a family of divisors D_i on the geometric fibers of X/S is "limited" iff the projective degrees of the D_i and D_i^2 are bounded.

Another result, of interest in connection with your seminar, is a proof of the fact that for an abelian scheme A/k, k a perfect field, the absolute formal scheme of moduli over $W_\infty(k)$ is simple over it. This comes from the following general fact: Let X_0/S_0 be simple, X'_0/X_0 étale, S_0 subscheme of S defined by an ideal J of square 0. Let $\xi_0 \in H^2(X_0, \mathcal{G}_{X_0/S_0} \otimes_{\mathcal{O}_{S_0}} J)$ and $\xi'_0 \in H^2(X'_0, \mathcal{G}_{X'_0/S_0} \otimes_{\mathcal{O}_{S_0}} J)$ be the

$$\begin{array}{ccc} H^i(\mho_{A/k}) & \xrightarrow{\gamma^{(i)}} & H^{i+1}(\mathcal{O}_A) \\ \wr\| & & \wr\| \\ \overset{i}{\Lambda} t' \otimes t & & \overset{i+1}{\Lambda} t' \end{array} \qquad t = t_A, \ t' = t_{\hat{A}}$$

is trivially described in terms of

$$c \in H^1(A, \Omega^1_{A/k}) \simeq \mathrm{Hom}(t, t'),$$

~~and~~ where the homom. $c : t \to t'$ is just the tangent map for $\varphi : A \to \hat{A}$ defined by the polarization. This map being surjective by assumption, $\gamma^{(i)}$ is surjective, hence $H^i(\mathcal{E}) \to H^i(\mho_{A/k})$ is injective, in particular

$$H^2(\mathcal{E}) \to H^2(\mho_{A/k})$$

is injective. As the obstructions obtained in $H^2(\mho_{A/k})$ are zero, the same holds for the polarized obstructions in $H^2(\mathcal{E})$, hence the assertion of simplicity. (If however $p \mid d$, simplicity does not hold at any point of M over p!)

Using the simplicity for formal schemes of moduli of abelian varieties, I can prove the following:

Let X/Λ be flat, proper, $H^0(X_0, \mathcal{O}_{X_0}) \simeq k$, where Λ local artin with residue field k. Assume $\underline{\mathrm{Pic}}_{X_0/k}$ exists, and is simple $/k$ (always true in ~~dim~~ car 0), i.e. $\dim \underline{\mathrm{Pic}}_{X_0/k} = \dim H^1(X_0, \mathcal{O}_{X_0})$. Then a) $\underline{\mathrm{Pic}}^{\circ}_{X/\Lambda}$ exists and is an abelian scheme over Λ

is trivially described in terms of

$$c \in H^1(A, \Omega^1_{A/k}) \simeq \mathrm{Hom}(t, t'),$$

where the homomorphism $c\colon t \to t'$ is just the tangent map for $\varphi\colon A \to \hat{A}$ defined by the polarization. This map being surjective by assumption, $\partial^{(i)}$ is surjective, hence $H^i(\mathscr{E}) \to H^i(\mathfrak{G}_{A/k})$ is injective, in particular

$$H^2(\mathscr{E}) \to H^2(\mathfrak{G}_{A/k})$$

is *injective*. As the obstructions obtained in $H^2(\mathfrak{G}_{A/k})$ are zero, the same holds for the polarized obstructions in $H^2(\mathscr{E})$, hence the assertion of the simplicity. (If however $p \mid d$, simplicity *does not hold* at *any* point of $\mathscr{M}$ over p!)

Using the simplicity for the formal scheme of moduli of abelian varieties, I can prove the following:

Let X/Λ be flat, proper, $H^0(X_0, \mathscr{O}_0) \xleftarrow{\sim} k$, where Λ is local artin with residue field k. Assume $\mathscr{P}ic_{X_0/k}$ exists, and is *simple* over k, i.e., $\dim \mathscr{P}ic_{X_0/k} = \dim H^1(X_0, \mathscr{O}_{X_0})$ (always true in char 0). Then

a) $\mathscr{P}ic^0_{X/\Lambda}$ exists and is an *abelian* scheme over Λ.
b) The "base extension property" holds for $R^i f_*(\mathscr{O}_X)$ in dimension 1, and more generally in any dimension i such that

$$\bigwedge^i H^1(X_0, \mathscr{O}_{X_0}) \to H^i(X_0, \mathscr{O}_{X_0})$$

is *surjective*, and $H^1(X, \mathscr{O}_X)$ is free over Λ.

Idea of proof:

a) $\mathscr{P}ic^0_{X/k}$ is constructed stepwise. Having $\mathscr{P}ic^0_{X_{n-1}/k} = A_{n-1}$, to get A_n we first lift *arbitrarily* A_{n-1} to an abelian scheme A'_n. We then try to construct the can. invertible "Weil sheaf" on $X_n \times_{\Lambda_n} A'_n$, extending the given Weil sheaf on $X_{n-1} \times_{\Lambda_{n-1}} A_{n-1}$. The obstruction lies in

$$H^2(X_0 \times A_0, \mathscr{O}_{X_0 \times A_0}) \simeq H^2(\mathscr{O}_{X_0}) \times H^2(\mathscr{O}_{A_0}) \times H^1(\mathscr{O}_{X_0}) \otimes H^1(\mathscr{O}_{A_0}),$$

and in fact, as easily seen, in the last factor $H^1(X_0, \mathscr{O}_{X_0}) \otimes H^1(A_0, \mathscr{O}_{A_0}) \simeq t_{A_0} \otimes H^1(A_0, \mathscr{O}_{A_0}) \simeq H^1(A_0, \mathfrak{G}_{A_0/k})$. This space is exactly the group operating in a simply transitive way on the set of all extensions of A_{n-1}. Thus we can *correct* A'_n in just one way to get an A_n with a "Weil sheaf" on it! This does it.
b) Let ω be the conormal sheaf to the unit section of $A = \mathscr{P}ic^0_{X/S}$, thus ω is *free* because A/S is simple, and by the definition of $\mathscr{P}ic^0_{X/S}$ we have

$$H^1(X, \mathscr{O}_A) \simeq \mathrm{Hom}(\omega, \mathscr{O}_S).$$

This description holds also after any base extension, hence the fact that $H^1(X, \mathscr{O}_X)$ is free over Λ and its formation commutes with base extension. This implies also

$H^1(X,\mathscr{O}_X) \to H^1(X_0,\mathscr{O}_{X_0})$ surjective, hence $H^i(X,\mathscr{O}_X) \to H^i(X_0,\mathscr{O}_{X_0})$ is surjective for the i's as in the theorem, ok.

Corollary. Let A/S be any abelian scheme, then the modules $R^if_*(\mathscr{O}_A)$ on S are locally free and in fact $\simeq \bigwedge^i R^1f_*(\mathscr{O}_A)$. If $\mathscr{P}ic_{A/S}$ exists, then $\mathscr{P}ic^0_{A/S}$ is open and is an abelian scheme over S.

(Moreover, biduality holds, as follows easily from the statement over a field ...).

Corollary. Let $f\colon X \to S$ be flat, proper, $k(s) \xrightarrow{\sim} H^0(X_s,\mathscr{O}_{X_s})$ for every s, let $s \in S$ be such that $\dim H^1(X_s,\mathscr{O}_{X_s}) = \dim \mathscr{P}ic_{X_s/k(s)}$, (the latter defined, if $\mathscr{P}ic_{X_s/k(s)}$ is not known to exist, in terms of the formal Picard scheme). Then $R^1f_*(\mathscr{O}_X)$ is free at s.

This is always applicable if char $k = 0$.

I do not know if, in the case considered, the $R^if_*(\mathscr{O}_X)$ or even $R^if_*(\Omega^j_{X/S})$ are also free at s, even in char 0. It is true for $f_*(\Omega^1_{X/S})$ whenever we know that $\dim H^1(X_s,\mathscr{O}_{X_s}) = \dim H^0(X_s,\Omega^1_{X_s})$, for instance if char $k(s) = 0$ and $f\colon X \to S$ is projective and simple. (If *moreover* S is reduced, Hodge theory implies *all* $R^if_*(\Omega^j_{X/S})$ are free at s; but if S is artin, I have no idea!)

I now doubt very much that it be true in general that $\mathscr{P}ic^\tau_{X/S}$ is flat over S, or even only universally open over S, when X/S is simple. Here is an idea of an example, inspired by Igusa's surface. Let A/S be an abelian scheme, G a finite group of automorphisms of A. If G operates without fixed points on B/S projective and simple over S, with $\mathscr{O}_S \xrightarrow{\sim} g_*(\mathscr{O}_B)$, we construct $X = B \times_G \hat{A}$ which is an abelian scheme over $Y = B/G$, and one checks

$$\mathscr{P}ic_{X/S} \simeq \mathscr{P}ic_{Y/S} \times_S (\mathscr{P}ic_{\hat{A}/S})^G$$

(where upper G denotes the subscheme of invariants), hence

$$\boxed{\mathscr{P}ic^\tau_{X/S} \simeq \mathscr{P}ic^\tau_{Y/S} \times_S A^G\,.}$$

Hence for getting examples of bad $\mathscr{P}ic^\tau_{X/S}$, we are led to study schemes of the type A^G, with S say spectrum of a discrete valuation ring V. Thus we are led to the questions:

a) Can it occur that there are components of $C = A^G$ which do not dominate S? For instance, $A_1^G =$ unit subgroup (set theoretically, or even scheme-theoretically) and $A_0^G \neq$ unit subgroup set theoretically—where A_0, A_1 are the special and the generic fibers.

b) If $C_1 = A_1^G$ is connected (for instance is the unit subgroup), and hence $C^\circ = C_0^\circ \cup C_1^\circ$ is open, can it occur that C° is nonflat over S [for instance $C_1 = \{e\}, C_0^\circ \neq \{e\}$]?

c) Same questions for $H^1(A,\mathscr{O}_{A/S})^G = t_{\hat{A}}{}^G$ and $H^0(A,\Omega^1_{A/S})^G = t_A{}^G$ (in order to get examples where the dimensions $h^{0,1}$ and $h^{1,0}$ for the fibers make a jump in the case of *equal characteristics*).

The trouble is I have no idea how to get nontrivial ways of letting a finite group operate on an abelian variety. It seems that starting with products of elliptic curves and using only endomorphisms of the factors, for instance letting a finite subgroup of $\mathrm{GL}(n,R)$ operate on E^n, where R is the ring of endomorphisms of the elliptic curve E, won't give a counterexample (I more or less proved this latter statement). If p is the residue characteristic, one sees easily that the only trouble against flatness can come from a Sylow p-subgroup of G. For instance, in a) the question is equivalent to getting an example where $T_p(\overline{A_0}) \to T_p(\overline{A_1})$ (where T_p is the contravariant Tate functor, $T_p(M) = \mathrm{Hom}({}_{p^\infty}M, \mathbb{Q}_p/\mathbb{Z}_p)$, and $\overline{A_0}$ and $\overline{A_1}$ are the *geometric* fibers) induces

$$\hat{H}^{-1}(G, T_p(\overline{A_0})) \to \hat{H}^{-1}(G, T_p(\overline{A_1}))$$

which is *not injective*. I am convinced such things can happen. Perhaps you or Mumford are cleverer than I and find a counterexample? What I did get easily was an example of an *abelian* scheme X/S [product of two elliptic curves over S] such that multiplication p: $\mathscr{P}ic_{X/S} \to \mathscr{P}ic_{X/S}$ is *not* universally open, i.e., such that there exists an irreducible component C of $\mathscr{P}ic_{X/S}$ not dominating S, but such that pC is contained in a component dominating S. [N.B. if n prime to all residue char., multiplication by n in any $\mathscr{P}ic_{X/S}$ is *étale*.]

Best regards to Karin, kids etc.
(signed) Schurik

P.S. I just proved: If $X \to S$ is *simple* and *projective*, then $\mathscr{P}ic^{\tau}_{X/S}$ is *projective* over S. Method:

a) From the fact that the fibers of $\mathscr{P}ic^{0}_{X/S}$ are proper, follows that $\mathscr{P}ic^{0}_{X/S}$ is proper over S, hence closed in $\mathscr{P}ic_{X/S}$, hence easily that $\mathscr{P}ic^{\tau}_{X/S}$ is *closed* in $\mathscr{P}ic_{X/S}$. It remains to prove it is of *finite type* over S—hence proper over S, and quasi-projective over S, hence projective.

b) For every $n > 0$, the kernel of $\mathscr{P}ic_{X/S} \xrightarrow{n} \mathscr{P}ic_{X/S}$ is of finite type over S [and even more: the multiplication μ by n is of finite type, hence finite]. If n is prime to the residue characteristics, this follows from the fact that μ is *étale* and has finite fibers. This reduces to the case S of char $p > 0$, $n = p$. Then I use a technique of descent involving the "relative p-power scheme" $(X/S)^{(p)}$, following a suggestion of Serre.

c) For variable $s \in S$ (S noetherian), the Néron–Severi torsion group of X_s remains of bounded order. This can be shown using the method of Matsusaka's proof for the finiteness of the "torsion group". From a), b), c), the theorem follows.

Remark: Using the Picard–Igusa inequality for ρ = rank of Néron–Severi, and Lefschetz type theorems I told you about, one gets also that $\rho(X_s)$ remains bounded for $s \in S$ (S noetherian).

Question: Is $\mathscr{P}ic^{\tau}_{X/S}$ always of finite type over S, under merely the usual assumptions for existence of $\mathscr{P}ic_{X/S}$? I have no proof even if $X \to S$ is normal! Same question for ρ. This seems related to the question of uniform majorization of the Mordell–Weil–Néron–Lang finiteness theorem, for a *variable* abelian variety.

Grothendieck to Mumford, 31 March, 1962

31.3.1962

My dear Mumford,

I was quite interested by your letter. Concerning your example of a nonflat $\mathcal{P}ic^{\tau}$,[19] I am convinced there should be still stronger counterexamples, insofar as 1°) $\mathcal{P}ic^{\tau}$ need not be flat over S even at points of the connected component of the identity 2°) $\mathcal{P}ic^{\tau}$ need not even be universally open over S, except at points corresponding to the part of the torsion of Néron–Severi prime to the characteristic; i.e., in the case where S is the spectrum of a valuation ring, there may be components of $\mathcal{P}ic$ which do not dominate S. Even in case 1°; examples with S the spectrum of a valuation ring (discrete of course) should exist. I wrote to Tate about the matter about two months ago, telling him how one could adapt Igusa's example so as to reduce oneself to producing suitable examples of finite groups of automorphisms on abelian schemes (which should give also examples where the $H^{1,0}$ and $H^{0,1}$ of fibers of simple morphism make jumps...) and begging him for help, but that unnice chap never answered a word. Besides, did the same tell you that I proved the simplicity of moduli for abelian schemes (either formal moduli, or polarized moduli with polarization degree prime to the char)? Using this, I can prove that $H^{0,1}$ behaves decently whenever the $\mathcal{P}ic$ of the special fiber is simple....

I am particularly happy with your simple example of a nonexisting $\mathcal{P}ic$.[20] I still naively surmised descent of Picard schemes would not cause any difficulty,

---

[19] Grothendieck described this example in TDTE VI, *Séminaire Bourbaki* 1961/62, n° 236, Remarque 2.9 as a deformation of an Igusa surface over an Artinian local ring. Here is an attempt to reconstruct this example: Let E_1, E_2 be ordinary elliptic curves over an algebraically closed field k of char. 2, and let $a \in E_2(k)$ be a nontrivial 2-torsion point of E_2. Let $X = (E_1 \times E_2)/(x,y) \sim (-x, y+a)$. The Hodge numbers of such an Igusa surface X is the same as those of an abelian surface, and the Hodge-to-de Rham spectral sequence degenerates. Let I be the k-linear dual of $H^1(X, \Theta_X)$, and let $R := k \oplus I$ be the Artinian local k-algebra with $I^2 = (0)$. Let $X_1 \to \mathrm{Spec}(R)$ be the universal first order equi-characteristic deformation of X, i.e., its Kodaira–Spencer class $\gamma \in I \otimes H^1(X, \Theta_X)$ is the identity map for $H^1(X, \Theta_X)$. Then $\mathcal{P}ic^{\tau}(X_1/S)$ has two connected components, the neutral component $\mathcal{P}ic^0(X_1/S)$ and another component $\mathcal{P}'$. The structural morphism $\mathcal{P}' \to S$ factors through a closed subscheme of S defined by a nonzero ideal of R. So $\mathcal{P}ic^{\tau}(X_1/S)$ is not flat over S at points of $\mathcal{P}'$. The key facts are:

(a) For any nontrivial line bundle $\mathcal{L}$ on X with $\mathcal{L}^{\otimes 2} \cong \mathcal{O}_X$, the Chern class $c_1^{\mathrm{dR}}(\mathcal{L})$ is an element of $\mathrm{Fil}^1_{\mathrm{hodge}} H^2(X, \Omega_X^{\bullet})$ whose image $c_1(\mathcal{L})$ in $\mathrm{gr}^1_{\mathrm{hodge}} = H^1(X, \Omega_X^1)$ is nonzero.

(b) The natural map $H^1(X, \Omega_X^1) \times H^1(X, \Theta_X) \longrightarrow H^2(X, \mathcal{O}_X) \cong k$ is a nondegenerate pairing.

An example of a nonflat $\mathcal{P}ic^{\tau}$, where the base scheme is the spectrum of a discrete valuation ring with mixed characteristics $(0, p)$, was published in Prop. 4.2.4 on p. 138 of M. Raynaud, "p-torsion" du schéma de Picard, *Astérisque* 64 (1979) 87–148. In Thm. 4.1.2 on p. 132 *loc. cit.*, it is shown that $\mathcal{P}ic^{\tau}_{X/R}$ is flat if X proper flat over a discrete valuation ring R with mixed characteristics $(0, p)$ and absolute ramification degree $e < p-1$. By the counterexample mentioned in the previous paragraph this statement is false when $e \geq p-1$; it is also false for the equicharacteristic p case.

[20] See TDTE VI *Séminaire Bourbaki* 1961/62, n° 236, 0.a. This example is discussed on p.210 of the book *Néron Models* by S. Bosch, W. Lütkebohmert and M. Raynaud, Springer-Verlag, 1990, and in greater detail in 9.4.14 on p.267 of S. Kleiman: The Picard scheme, in *Fundamental Algebraic Geometry,* Math. Surveys Monogr. 123, Amer. Math. Soc. 2005, pp. 235–321.

Grothendieck to Mumford, 31 March, 1962

31. 3. 1962

My dear Mumford,

I was quite interested by your letter. Concerning your example of a non flat <u>Pic</u>$^{\tau}$, I am convinced there should be still stronger counter-examples, insofar as 1º) <u>Pic</u>$^{\tau}$ needs not be flat over S even at points of the connected component of the identity 2º) <u>Pic</u>$^{\tau}$ needs not even be universally open over S, except at points corresponding to the part of the Torsion of Neron-Severi prime to the caracteristic; i.e. in the case S is the spectrum of a valuation ring, there may becomponents of <u>Pic</u> which do not dominate S. Even in case 1º; examples with S the spectrum of a valuation ring (discrete of course) should exist. I wrote to Tate about the matter about two months ago, telling him how one could adapt Igusa's example so as to reduce oneself to produce suitable examples of finite groups of automorphisms on abelian schemes (which should give also examples where the H^{10} and H^{01} of fibers of simple morphism make jumps ...) and begging him for help, but that unnice chap never answered a word. Besides, did same tell you that I proved simplicity of moduli for abelian schemes (either formal moduli, or polarized moduli with polarization degree prime to the car) ? Using this, I can prove that H^{01} behaves decently whenever the <u>Pic</u> of the special fiber is simple ...

I am particularly happy with your simple example of a non existing <u>Pic</u> , I still naively surmised descent of Picard schemes would not cause any difficulty, and had felt satisfied in your very example with proving the existence of <u>Pic</u> in the case the irréducible components of the special fiber are geometrically irreducible. Still there remains the hope that <u>Pic</u>$^{\tau}$ exists in great generality, or at least (in case the Pic groups of the fibers are simple) <u>Pic</u>^{o} , obtained by taking the sub-fonctor of the Pic functor corresponding to invertible sheaves inducing on the special fibers sheaves that are alg equiv to 0;(in case <u>Pic</u> exists and its fibers are simple, I proved that <u>Pic</u>^{o} is open).

It is quite mysterious to me how from your general remarks on Severi-Brauer schemes you will deduce the existence of <u>Pic</u> in the case you claim, fibers separable and irred components being geom irred. Whatever way you present technicalities, it seems to me you will need a theorem of the following type: f:X→S being as before (of course, also projectiv and had felt satisfied in your very example with proving the existence of $\mathcal{P}ic$ in the case where the irreducible components of the special fiber are geometrically irreducible.[21] Still there remains the hope that $\mathcal{P}ic^{\tau}$ exists in great generality, or at least (in case the Pic groups of the fibers are simple) $\mathcal{P}ic^{0}$, obtained by taking the subfunctor of the Pic functor corresponding to invertible sheaves inducing on the special fibers sheaves that are algebraically equivalent to 0;[22] (in case $\mathcal{P}ic$ exists and its fibers are simple, I proved that $\mathcal{P}ic^{0}$ is open).

It is quite mysterious to me how from your general remarks on Severi–Brauer schemes you will deduce the existence of $\mathcal{P}ic$ in the case you claim, fibers separable and irreducible components being geometrically irreducible. Whatever way you present technicalities, it seems to me you will need a theorem of the following type: $f\colon X \to S$ being as before (of course, also projective and flat), there should exist a family (U_i, S_i) of finite étale multisections of X over open subsets $U_i \subset S$, such that, for every S' over S and every $\xi \in \mathrm{Pic}(X'/S')$ "sufficiently ample", corresponding to some immersion of X' into a $\mathbb{P}^N_{S'}$ as usual (we can in fact suppose without loss that the Brauer–Severi schema corresponding to ξ is trivial), and every $s' \in S'$, there exists a U_i below s' such that the multisection S'_i of X' over U'_i deduced from S_i, viewed as a family of zero cycles on the fibers of $\mathbb{P}^N_{S'}$, consists only of zero-cycles in your open set (where passage to the quotient by the projective group *and* the symmetric group in $(\mathbb{P}^N)^m$ is possible). Did you prove anything such? I wonder how you will use the hypothesis on the irreducible components of the fibers of X/S!

You make an allusion to results of yours "over $\mathbb{Q}$" for vector bundles over non-singular curves. Do you just mean "in char. 0"—as just afterwards you assume the ground field algebraically closed. I confess the little you say about it does not suggest much to me! The reference in my notes on properness criteria which you did not understand was to III 5.5.1. (I guess you will get Chap III very soon, as it has appeared by now; I had copies sent to Hartshorne and Lichtenbaum too). This states that if X is separated of finite type over say a complete local noetherian ring, and if Z_0 is an open and proper subset of the special fiber X_0, then there exists an open and closed subset Z of X, proper over S, whose special fiber is Z_0 (and Z will in fact be the biggest closed subset of X proper over S).

Lubkin's result seems very unlikely to me too, but although I had a little thought of constructing a counterexample over $S = \mathbb{P}^1$ (keeping in mind $\pi_2(S) = \mathbb{Z}$), I did not succeed. I will keep the question in mind, and discuss it with Serre when he comes back from Bourbaki next week.

Sincerely yours
(signed) A Grothendieck

---

[21] The representability theorems for the Picard functor come in two flavors, as schemes or as algebraic spaces. See 8.1 and 8.2 of S. Bosch, W. Lütkebohmert & M. Raynaud, *Néron Models*, Springer-Verlag, 1990. See also 19.4 of S. Kleiman, The Picard scheme, cited in Footnote 20. Mumford's existence theorem of the Picard scheme, stated on the first page of TDTE VI, *Séminaire Bourbaki* 1961/62, n° 236, and also on p. viii of [CAS] (= *Lectures on Curves on an Algebraic Surface*) in a slightly weaker form, is still unpublished; see also Remark 19.4.18 of Kleiman's article *loc. cit.*

[22] See the last paragraph of [1965Jan23] for a counterexample based on a remark of M. Raynaud.

Grothendieck to Mumford, 23 June, 1962

Neuilly June 23, 1962

Dear Mumford,

I was of course quite interested by the results you stated in your last letter. If you want to explain the ideas of the proof to me when I come to Harvard this will take a while, as it turns out that I won't come this year, due to health troubles for my wife and children. If you have time to give me an idea of the proof for finite type of $\mathcal{P}ic^P_{X/S}$ by letter, I would appreciate it. I wonder if you can prove the slightly stronger result I had in case $f\colon X \to S$ is simple, and the fibers of pure dimension d, namely that it is sufficient that, in $P(n) = a_0n^d + a_1n^{d-1} + a_2n^{d-2} + \cdots$, the coefficients a_1 and a_2 remain bounded [in terms of divisors, D and D^2 have bounded projective degrees] in order for the invertible sheaves considered to belong to a quasi-compact subset of $\mathcal{P}ic_{X/S}$? My proof, following Matsusaka, uses equivalence criteria and Riemann–Roch for surfaces, and is technically rather involved.

Sincerely yours,
(signed) A Grothendieck

P.S. Do you know if $\mathcal{P}ic^{\tau}_{X/S}$ is of finite type over S, when $f\colon X \to S$ is separable (= flat with reduced geom. fibers)?[23]

Grothendieck to Mumford, 6 July, 1962

23 Boul. de Levallois
Neuilly (Seine)

Neuilly July 6, 1962

Dear Mumford,

I enjoyed very much the proofs you gave me in the last letter, your proof of finite type for $\mathcal{P}ic^f$ is certainly much simpler than mine (which gives a more precise result, in a less general case). The main step in my proof (besides Matsusaka's method using Riemann–Roch, to deal with the case of nonsingular surfaces—a method now superseded by your proof) is the following:

Theorem. *Let $f\colon X \to S$ be a projective and flat morphism whose fibers are of depth ≥ 3 at closed points, Y a Cartier Divisor on X, transversal to the fibers i.e., flat/S, and ample relative to S, assume $\mathcal{P}ic_{X/S}$ and $\mathcal{P}ic_{Y/S}$ exist, then the morphism*

$$\mathcal{P}ic_{X/S} \to \mathcal{P}ic_{Y/S}$$

is of finite type.

[23] See Footnote 25.

Idea of proof. Let $Y_0 = Y$, define Y_m $(m \geq 0)$ as usual, then using exact sequences of cohomology, it is not hard to show that $\mathcal{P}ic_{Y_m/S} \to \mathcal{P}ic_{Y_0/S}$ is of finite type. (NB if $S \neq \emptyset$, there exists a nonempty open subset U of S such that the restriction of the previous morphism over U is affine—a fortiori of finite type. This proves finite type for the morphism by noetherian induction on S). This permits us to replace Y_0 by Y_m, large m. Using the depth ≥ 2 assumption and the (easy part of) equivalence criteria as developed in my IHÉS Seminar 1962, one gets that, for large m, $\mathcal{P}ic_{X/S} \to \mathcal{P}ic_{Y_m/S}$ is a monomorphism. One is reduced to proving that, for any section of $\mathcal{P}ic_{Y_m/S}$ over S, its inverse image in $\mathcal{P}ic_{X/S}$ is of finite type over S, and using that its projection to S is a monomorphism, one is reduced to proving the following: Assume S irreducible with generic point s, let $\mathcal{M}$ be an invertible sheaf on Y such that $\mathcal{M}_s$ does not come from an invertible sheaf on X_s, then there is an open neighbourhood U of s such that $t \in U$ implies that $\mathcal{M}_t$ does not come from an invertible sheaf on X_t. To prove this, using depth ≥ 3 and the "existence" part of the equivalence criteria, one gets that the assumption on $\mathcal{M}_s$ means that either (i) there exists $m' \geq m$ such that $\mathcal{M}_t$ does not come from an invertible sheaf on $Y_{m'}$ or (ii) there exists a coherent sheaf $\mathcal{L}_s$ on X_s, invertible in a neighbourhood of Y_s but not invertible on the whole of X_s, having depth ≥ 2 at all closed points, and inducing $\mathcal{M}_s$. It is now easy to see that either property (i) or (ii) will still hold in a neighbourhood of s.

Unfortunately this proof involves a considerable technical background. The theorem just stated, together with your finiteness theorem, proves the following:

Theorem. *Let $f: X \to S$ be flat projective with geometrically integral fibers, assume the fibers are of depth $\geq d$ at closed points, and let $d' = \mathrm{Sup}(0, d-2)$. Then, in order to ensure quasi-compactness for a subset of $\mathcal{P}ic_{X/S}$ in terms of the coefficients of the Hilbert polynomials, one can neglect the d' last coefficients.*

If for instance the fibers of f are Cohen Macaulay and of dimension n, this means that one needs to look only at the first three coefficients (the first one being inessential anyhow, being the projective degree of the fibers). Does this statement become false if the fibers of f are not Cohen–Macaulay, and (say) normal of dim 3?

I doubt if there will be an occasion for me to expound the theory of formal moduli in a seminar, before Chap V is published. (Next year I will run a seminar together with Demazure on semi-simple group schemes, whereas my main interest will lie in developing (at last) Weil cohomology for schemes). I had noticed also, at the very start of my ponderings on the subject, that, in the case where "all obstructions vanish", i.e., the functor one wants to represent is "simple", the existence of formal moduli is immediate; the main point of the theory is of course to construct also *singular* formal modular varieties.

I include a copy of a letter to Hironaka, containing various questions. I would appreciate any comments you would make; Mike Artin has perhaps an idea on some of them. Besides, is Mike still in Cambridge? I wrote him lately to ask him to write us now a firm answer if he wants to come to Paris in 63/64, but did not get any answer. Perhaps you could give him a call about it, if he is still there.

Sincerely yours
(signed) A Grothendieck

P.S. I have a few comments on Picard[24] of a projective scheme over a field k, (which we may assume algebraically closed). First, if X is any scheme, denote by $K(X)$ the usual group constructed with locally free sheaves on X, this is augmented into $H^0(X,\mathbb{Z}) = \mathbb{Z}^{\pi_0(X)}$ (by rank), let $I(X)$ be the kernel. Serre proved (in a very elementary way) that when X is quasi-compact and has an ample sheaf, and $\dim X = d < +\infty$, then $I(X)^{d+1} = 0$. This has various applications, for instance: let P be the group of invertible elements of $K(X)$, these are of the type $1+y$, $y \in I(X)$, consider the inclusion map $P \to K(X)$, this as a map of $\mathbb{Z}$-modules is a *polynomial map of degree* $\leq d$. Therefore the natural map $\mathrm{Pic}(X) \to K(X)$ has the same property, and of course keeps it if we follow it by any linear map. Thus, if again X is projective over a field k, and if $\mathscr{F}$ on X is coherent, then $\mathscr{L} \rightsquigarrow \chi(\mathscr{F} \otimes \mathscr{L})$ is a polynomial map on the group $\mathrm{Pic}(X)$. As Serre remarked some time ago (before Riemann–Roch was proved), from this follows that $\chi(\mathscr{F} \otimes \mathscr{L})$ does not change if we replace $\mathscr{L}$ by a sheaf which is congruent to it mod $\mathrm{Pic}^\tau(X)$, in other words, for any $\mathscr{F}$ and $\mathscr{L} \in \mathrm{Pic}^\tau(X)$, we have $\chi(\mathscr{F} \otimes \mathscr{L}) = \chi(\mathscr{F})$. (Use the fact that $\mathrm{Pic}^\tau(X)$ has a composition series where the factors are divisible, or torsion groups. One can also prove this invariance of χ by a direct argument, using still $I^{d+1} = 0$ but not the polynomial type of χ). This proves for instance that under the conditions of your finiteness theorem for $\mathscr{P}ic^P_{X/S}$ for a projective flat morphism $f\colon X \to S$ (integral geometric fibers), $\mathscr{P}ic^\tau_{X/S}$ is contained in one $\mathscr{P}ic^P_{X/S}$ (provided S connected), and therefore of finite type over S, and moreover the pieces $\mathscr{P}ic^P_{X/S}$ are stable under translation by $\mathscr{P}ic^\tau_{X/S}$.

Besides, there is a converse to the previous result, to the effect that "τ-equivalence" is in fact *equivalent* to "numerical equivalence", namely if $\mathscr{L}$ invertible is such that for every coherent $\mathscr{F}$, $\chi(\mathscr{F} \otimes \mathscr{L}) = \chi(\mathscr{F})$ i.e., $\chi(\mathscr{F}(\mathscr{L} - 1)) = 0$, then $\mathscr{L} \in \mathrm{Pic}^\tau(X)$. Indeed, it is sufficient ($\mathscr{O}(1)$ denoting as usual an ample sheaf on X relative to k) to assume $\chi(\mathscr{O}(n) \otimes \mathscr{L}^{\otimes m}) = \chi(\mathscr{O}(n))$ for any integers n, m. This means in fact that the sheaves $\mathscr{L}^{\otimes m}$ have the same Hilbert polynomial, hence remain in a quasi-compact subset of $\mathscr{P}ic_{X/S}$, which by definition means $\mathscr{L} \in \mathrm{Pic}^\tau(X)$. (NB the argument supposes X integral, but it is easy to get rid of this assumption in the original statement). One interesting consequence of the last criterion, for a projective morphism $f\colon X \to S$ as above: the subscheme $\mathscr{P}ic^\tau_{X/S}$ is not only open, but also closed!

There seems to be another characterisation of τ-equivalence to 0 for $\mathscr{L}$ on a projective X/k. With the previous notations, note first that if $\mathscr{L}$ is τ-equivalent to 0, then for any ample sheaf $\mathscr{M}$ on X, $\mathscr{L} \otimes \mathscr{M}$ is again ample, therefore $\mathscr{L}(1)$ and more generally the sheaves $\mathscr{L}^{\otimes n}(1)$ must be ample. (This fact is well known and an easy consequence of the fact that for every neighbourhood U of 0 in $\mathscr{P}ic^0 = G$, $U \cdot U = G$; a still simpler proof—in fact a trivial one—is obtained using your

[24] See Footnote 25.

ampleness criterion, and the fact that $\mathscr{L}$ is numerically equivalent to 0). I believe the converse should be true. Let V be the Néron–Severi group of X tensored by the reals, which is a finite dimensional vector space over $\mathbb{R}$, endowed with an open convex cone P (generated by ample sheaves), let $\overline{P}$ be its closure. The previous conjecture would follow from the fact that $\overline{P}$ does not contain any line (i.e., P does not come from a cone in a smaller quotient space...). Another way of stating this is that for any $x \in P$, the set $P \cap (x - P)$ is relatively compact (which would yield an interesting finite-type criterion in $\mathscr{P}ic$). Generally speaking, what facts are known to you concerning the shape of P?

A last question about finiteness criteria. Consider the map $\chi : V \to \mathbb{R}$ (polynomial of degree $\leq d$) deduced from χ: $\mathrm{Pic}(X) \to \mathbb{Z}$ by ring extension. Select an $a \in P$ (corresponding to the choice of an ample sheaf $\mathscr{O}(1)$, for instance), then for any $\xi \in V$, $\chi(a + n\xi)$ is a polynomial with respect to n, say $P_\xi(n)$ (the Hilbert polynomial of ξ with respect to a). Let $c_i(\xi)$ be its coefficients, which are polynomial functions in ξ. If ξ varies in V in such a way that the coefficients $c_i(\xi)$ remain bounded, does ξ remain bounded (we now assume X irreducible)? Perhaps this is just a formal consequence of your finiteness result, (which corresponds to taking a, ξ in the original lattice of V); this should be considered as a generalisation of the known fact that on the Néron–Severi space of a nonsingular surface, the intersection form has just one positive square. Of course, under suitable assumptions on the depth of X at closed points, one should be able to disregard some of the last coefficients $c_i(\xi)$, in the criterion of boundedness.

Grothendieck to Mumford, 12 July, 1962

12 July, 1962

Dear Mumford,

I had a little more thought about finiteness questions for Pic, and have finally come to a solution of about all the questions I had met with.[25] The key facts I will state in

Theorem 1. *Let S be a noetherian prescheme.*

(i) *Let* $f: X \to Y$ *be a surjective morphism of proper S-schemes, suppose* $\mathscr{P}ic_{X/S}$ *and* $\mathscr{P}ic_{Y/S}$ *both exist, then* $f^*: \mathscr{P}ic_{Y/S} \to \mathscr{P}ic_{X/S}$ *is of finite type.*

[25] An account of the finiteness theorems, along the lines of this letter and the previous letter [1962Jul06], appeared in two exposés in SGA6, LNM 225, Springer-Verlag, 1971: Exposé XII, M. Raynaud, *Un théorème de représentabilité relative sur le foncteur de Picard,* pp. 595–615; Exposé XIII, S.L. Kleiman, *Les théorèmes de finitude pour le foncteur de Picard,* pp. 616–666. Results on the Picard functors are explained in Chap. 8 of S. Bosch, W. Lütkebohmert & M. Raynaud, *Néron Models*, Springer-Verlag, 1990, and also in S. Kleiman, The Picard scheme, in *Fundamental Algebraic Geometry*, Math. Surveys Monogr. 123, Amer. Math. Soc. 2005, pp. 235–321.

(ii) *Let Y be a projective S-scheme, X a "hyperplane section" i.e., the sub-prescheme of zeros of a section of an invertible sheaf $\mathscr{L}$ on Y ample relative to S. Assume again $\mathscr{P}ic_{Y/S}$ and $\mathscr{P}ic_{X/S}$ exist, then $f^* : \mathscr{P}ic_{Y/S} \to \mathscr{P}ic_{X/S}$ is of finite type, provided all irreducible components of the geometric fibers of X/S are of dimension ≥ 3.*

(NB in other words, in both statements, a subset of $\mathscr{P}ic_{Y/S}$ is quasi-compact iff its image in $\mathscr{P}ic_{X/S}$ is. It is evident how to state these theorems so that they make sense without the assumption of existence for the Picard schemes, and the proofs work as well. The same remark holds for all other statements which seem to make use of the existence of certain Picard preschemes. The proof shows also that in cases (i) and (ii), if S is the spectrum of a field, the morphism f^* is even affine).

(iii) *Let X be a projective S-scheme, with integral geometric fibers all of dimension n, endowed with a sheaf $\mathscr{O}_X(1)$ very ample over S. In order for a subset M of $\mathscr{P}ic_{X/S}$ to be quasi-compact it is necessary and sufficient that, in the Hilbert polynomials $a_0x^n + a_1x^{n-1} + \cdots$ of the elements of M, the coefficients a_1 and a_2 remain bounded.*
(NB It can be shown also that if we express the invertible sheaves on geometric fibers stemming from M in terms of Cartier divisors D, the condition is also equivalent with asking that D and D^2 should have bounded projective degrees—this statement makes sense even when the fibers are singular, because if D is a Cartier divisor one can give a meaning to D^k and $\deg D^k$ *for every k....)*

(iv) *Let X be a proper S-scheme such that $\mathscr{P}ic_{X/S}$ exists. Then, for every integer $n \neq 0$, multiplication by n in this group prescheme is a morphism of finite type.*

As a corollary of (i) and (iii) we get the following

Corollary 1. *Let X be proper over S such that $\mathscr{P}ic_{X/S}$ exists, then $\mathscr{P}ic^{\tau}_{X/S}$ is of finite type over S.*

Also, as a trivial consequence of (i) and (ii):

Corollary 2. *Under conditions* (i) *or* (ii), *if $\mathscr{L}$ is an invertible sheaf on Y, then $\mathscr{L}$ is τ-equivalent to 0 iff its inverse image on X is. In other words, if k is an algebraically closed field and S its spectrum, denoting by $LN(Y)$ the Néron–Severi group of Y mod torsion, $LN(Y) \to LN(X)$ is injective, a fortiori for the Picard numbers $\rho(Y) \leq \rho(X)$.*

In the same way, using (i) to reduce to the projective case, (ii) to cut down the dimension of fibers to be ≤ 2, then again (i) and resolution of singularities for a surface (over an algebraically closed field) to reduce to the case of a simple morphism, and lastly Néron's theorem, the Igusa–Picard inequality, and corollary 1 we get:

Corollary 3. *Let X be proper over S. Then the Néron–Severi groups of the geometric fibers of X/S are of finite type, and of bounded rank and bounded order for the torsion subgroups.*

I will give the idea of the proof of theorem 1. Logically, (i) comes first, (ii) uses a weaker version of (iii) and is needed itself to prove (iii) in full strength, (iv) uses (i) and corollary 1 (in the case X/k normal, to ensure that the kernel of multiplication by n, n prime to the residue characteristic, is of finite type over S if S is the spectrum of a field k...) hence to a certain extent (ii) and (iii) or some other known information as Néron's theorem, or finite generation of fundamental group.

The proof of (i) relies heavily on the ideas of nonflat descent (expounded roughly in my Bourbaki talks), it is pretty natural although cumbersome in details. At first sight, there seems to be a drawback because of the lack of criteria for effectiveness of descent data in the case of a nonflat morphism (assumed to be of descent with respect to locally free sheaves say); if we had always effectivity we would be able to conclude that, if $S \neq \emptyset$, there exists in S a nonempty open set U such that over U the morphism $\mathcal{P}ic_{Y/S} \to \mathcal{P}ic_{X/S}$ is affine. I do not know if this is a true statement in general (as the schemes involved are not of finite type over S, it does *not* follow from the corresponding known fact over a field, when applied to the generic fiber...). However, having only in mind the finite type property, one gets along by remarking (for the simple types of morphisms f one can reduce to) that $\mathcal{P}ic_Y \to \mathcal{P}ic_X$ can be factored through an S-prescheme Q, with $Q \to \mathcal{P}ic_X$ an *affine* morphism, and $\mathcal{P}ic_Y \to Q$ a *monomorphism.* Indeed, Q expresses the classification of invertible sheaves on X with descent data relative to f, (to give such descent data on a given $\mathcal{L}$ on X is expressed in taking a section of a suitable scheme *affine* over S), the fact that $\mathcal{P}ic_Y \to Q$ is a monomorphism comes from the fact that we assume f a morphism of descent for invertible sheaves, (universally with respect to base changes $S' \to S$). Now we are reduced to proving that $\mathcal{P}ic_Y \to Q$ is of finite type, which amounts to verifying that if a descent datum on a given invertible sheaf $\mathcal{L}$ on X induces on the generic fiber a noneffective one, it is noneffective on the neighbouring fibers as well—a very easy fact indeed.

For (ii) I use your finiteness theorem. However, it seems to me that your proof is incomplete at one point, namely when you conclude that (granting H^0 of dim > 1 for all invertible sheaves considered) the effective divisors D yielding the sheaves $\mathcal{L}$ remain in a quasi-compact subset of $\mathcal{D}iv$. In fact, we know only that the Hilbert polynomial for D is $P_{\mathcal{O}_X} - P_{\mathcal{L}^{-1}}$, now it does not seem obvious to me that, from the assumption that the Hilbert polynomials $P_{\mathcal{L}}$ remain bounded, the same is true for the polynomials $P_{\mathcal{L}^{-1}}$. Therefore it seems that your argument applies only, a priori, if you know that the sheaves $\mathcal{O}_D$ can be chosen in a way so as not to have embedded primes, (at least no embedded primes of dimension 0); indeed, from the induction assumption it follows at least that *except for the constant terms*, the coefficients of the Hilbert polynomials for the divisors D remain bounded, (and the case where the relative dimension of X/S is 0 or 1 does not offer any difficulty). If however the fibers of X/S satisfy Serre's property (S_2), for instance are normal (the only case I will use in the proof of (ii)), then the $\mathcal{O}_D$ are without any embedded primes, and we get through. Once (ii) is proved, one can recover your original statement without restriction, and in the stronger form of the theorem 1 (iii), as follows. By criterion (ii) one reduces easily to the case when X/S is of relative dimension 2 (in which case your version and mine agree). Of course, we can always reduce to proving quasi-compactness when restricting over some nonempty open subset of S, S integral. But with this restriction in mind, it is easily seen that we can find a finite morphism $f\colon X' \to X$, such that X'/S satisfies to the same conditions as X/S, but has moreover fibers satisfying (S_2), and such that f induces on every geometric fiber an isomorphism except at isolated points. (NB of course, the verification of this fact reduces to the case when S is the spectrum of a field; then the set Z of points where X is not S_2, i.e., not Cohen–Macaulay, is finite because X is integral of dim 2, and, denoting by i the inclusion $U = X - Z \to X$, we take $X' = \mathrm{Spec}(i_*(\mathcal{O}_U)))$. Moreover, an invertible sheaf on a fiber of X and its inverse image have the same Hilbert polynomial, except for a *fixed constant* (namely length of $\mathcal{O}_{X'_s}/\mathcal{O}_{X_s}$). This way we are reduced to proving your criterion for X' instead of X, for which it is already known.

To prove (iv) we can assume n a prime, and are reduced to proving that, for every section of $\mathcal{P}ic_X$ over S, its inverse image by n is of finite type over S. Now this inverse image is a formally homogeneous principal space over ${}_n\mathcal{P}ic$, which is of finite type over S by corollary 1. This reduces us to proving the following: if the fiber of this prescheme at the generic point s of S (assumed integral) is empty, so are the neighbouring ones. If n is distinct from the characteristic of $k(s)$, we can assume it is prime to all residue characteristics, then the scheme considered is étale over S, hence easily the conclusion. (NB in fact, a universally open morphism which is locally of finite type and has finite fibers is of finite type—thus we need only the part of corollary 1 stating that, over a field, the torsion of Néron–Severi killed by n prime to the characteristic is finite). If n is equal to char $k(s)$, we can assume S to be of characteristic $n = p > 0$, and then, using the Frobenius functor relative to S, we get a canonical factorisation of multiplication by p as

$$(+)\qquad \mathcal{P}ic_{X/S} \xrightarrow{g} \mathcal{P}ic_{X^{(p)}/S} = (\mathcal{P}ic_{X/S})^{(p)} \xrightarrow{f^*} \mathcal{P}ic_{X/S},$$

where

$$f\colon X \to X^{(p)}$$

is the Frobenius morphism, and the first map in (+) is the Frobenius morphism for the prescheme $P = \mathcal{P}ic_{X/S}$ over S. As the latter is locally of finite type over S, it follows that g is finite. Moreover f is finite and *surjective*, and therefore, by (i), f^* is of finite type. Hence f^*g is of finite type and we are through. (NB I proved first (iv) for a *simple* morphism, a few months ago, in this case $f\colon X \to X^{(p)}$ is flat, and the theory of flat descent implies easily that f^* is *affine*, without using the more delicate theorem 1 (i)).

I did not solve in full generality the following problem: Let X/S be projective over S, such that $\mathcal{P}ic_{X/S}$ exists, let M be a subset of $\mathcal{P}ic_{X/S}$, then prove M is quasi-compact iff there exists n such that

$$\mathcal{O}_X(-n) \leq M \leq \mathcal{O}_X(n)$$

(inequality with respect to the order relation on all fibers defined by the cone of ample sheaves). Using (i) and (ii), one can reduce to the case where X/S is of relative dimension 2 and with normal irreducible fibers. However, if S is the spectrum of a field (which we may assume alg. closed) the answer is affirmative, as results at once from the more general:

Theorem 2. *Let X/k be proper, k alg. closed. There exists a finite number of integral curves C_i in X, with normalisations C_i', such that, for a subset M of $\mathscr{P}ic_{X/k}$ to be quasi-compact, it is necessary and sufficient that the numbers* $\deg \mathscr{L}_{C_i'}$ $(\mathscr{L} \in M)$ *remain bounded.*

Proof: using (i) and (ii) we are reduced to the case where X is a normal irreducible surface. Using resolution of singularities for a surface, we can assume X nonsingular. In this case, the fact is known and results from (a) Néron's finiteness theorem (b) the fact that the fundamental bilinear form on $LN(X)$ is nondegenerate. [The latter results formally, besides, from the weak RR for surfaces and your finiteness theorem (which yields also, besides, the fact on the signature of the quadratic form): using RR, your criterion is equivalent with: putting $D' = 2D + K$, (K the canonical divisor), if D'^2 and $D'E$ remain bounded, D remains in a finite subset of $LN(X)$. This excludes the possibility of the bilinear form being degenerate, because the set of $D \in LN(X)$ in the kernel of the bilinear form satisfies the finiteness criterion. Also (b) implies directly that $LN(X)$ is free. Then Igusa's argument applies to yield the Igusa–Picard inequality, without using Néron's result. Thus corollary 3 of Theorem 1 (a common generalisation of Néron's and Igusa's result) is now proved without reference to Néron's result, (using heights etc.). I wonder if you are able to give a direct proof of Néron's theorem from your finiteness criterion, without using Igusa's involved argument (using the structure of the fundamental group of a curve), and to get rid in the proof of Theorem 2 of resolution of singularities of a surface. I would expect that this is possible, using the following argument. Let X be any complete surface over k (not necessarily normal), then using Serre's remark that $I(X)^3 = 0$, we get a canonical *bilinear* form in $LN(X)$, (X need not be projective, see below proof of (i) $\Longrightarrow$ (ii) in corollary 1), by setting

$$B(\mathscr{L},\mathscr{L}') = \chi(\mathscr{O}_X) - \chi(\mathscr{L}) - \chi(\mathscr{L}') + \chi(\mathscr{L}\otimes\mathscr{L}')$$

(this definition, and the whole of intersection theory, generalizes to varieties of arbitrary dimension...). Now, *if X is integral and proj., this form is nondegenerate, and has just one positive square.* (This statement of nondegeneracy + Néron of course implies theorem 2). This is an easy consequence of resolution of the singularities of X and of theorem 1 (i), taking into account that the canonical bilinear form is compatible with the maps $LN(X) \to LN(X')$ stemming from morphisms of degree 1, $f\colon X' \to X$. Do you have any idea of how to get rid, in the proof of nondegeneracy, of the resolution of singularities? What happens if X is not projective?[26] (I used projectivity through the fact that there is at least one positive square for the bilinear form, and that any subspace, in a quadratic space of signature $(1,s)$, which contains one positive square, is nondegenerate).]

---

[26] Both questions were addressed in two articles by Kleiman cited in Footnote 25: in section 7, pp. 662–666 of Exposé XIII in SGA6 (LNM 225), and in Appendix B, pp. 319–321 in Math. Surveys Monogr. 123.

your finiteness criterion, without using Igusa's involved argument (using the structure of the fundamental group of a curve), and to get rid in the proof of Theorem 2 of resolution of singularities of a surface. I would expect that this is possible, using the following argument. Let X be any (complete) surface over k (not necessarily normal), then using Serre's remark that $I(X)^3=0$, we get a canonical bilinear form in LN(X), by setting

$$B(L,L') = \chi(\underline{O}_X) - \chi(\underline{L}) - \chi(\underline{L}') + \chi(\underline{L}\otimes\underline{L}')$$

(this definition, and the whole of intersection theory, generalizes to varieties of arbitrary dimension ...). Now, if X is integral and proj. this form is non degenerate, and has just one positive square. (That statement of non degeneracy of course implies theorem 2). This is an easy consequence of resolution of the singularities of X and of theorem 1 (i), taking into account that the canonical bilinear form is compatible with maps $LN(X) \rightarrow LN(X')$ stemming from morphisms of degrée 1 $f: X' \rightarrow X$. Do you have any idea of how to get rid, in the proof of non degeneracy, of the resolution of singularities ? What happens if X is not projective ? (I used projectivity through the fact that there is at least one positive square for the bilinear form, and that any subspace, in a quadratic space of signature (1,s), which contains one positive square, is non degenerate).

As an easy consequence of Theorem 2, we get:

Corollary 1 Let X/k be proper, $\underline{L}$ an invertible sheaf on X. The following conditions are equivalent:

(i) $\underline{L}$ is τ-equivalent to zero.

(ii) For every coherent F on X, $\chi(F\underline{L}) = \chi(F)$

As an easy consequence of Theorem 2, we get:

Corollary 1. *Let X/k be proper, $\mathscr{L}$ an invertible sheaf on X. The following conditions are equivalent:*

(i) *$\mathscr{L}$ is τ-equivalent to zero.*
(ii) *For every coherent $\mathscr{F}$ on X, $\chi(\mathscr{F}\otimes\mathscr{L})=\chi(\mathscr{F})$.*
(ii *bis*) *As before, with $\mathscr{F}=\mathscr{O}_Y$, Y an integral curve contained in X.*
(iii) *For every integral curve contained in X, letting Y' be its normalisation, $\deg\mathscr{L}_{Y'}=0$ (NB with notations of theorem 2, it is enough to take for Y one of the C_i).*

And for completeness, if X/k is projective, I state the following equivalent conditions:

(iv) *$\mathscr{L}^{\otimes m}(1)$ ample for every integer m.*
(v) *(If X is integral) $\chi(\mathscr{L}^{\otimes m}(n))=\chi(\mathscr{O}(n))$ for every m, n, i.e., the sheaves $\mathscr{L}^{\otimes m}$ all have the same Hilbert polynomial.*

Proof: (i) $\Rightarrow$ (ii). By a devissage argument and Serre's result $I(X)^{d+1}=0$ for a quasi projective X of dimension d, one proves (without projectivity assumption, for any prescheme of finite type over k) $I(X)^{d+1}K_\bullet(X)=0$, where $K_\bullet(X)$ is the Grothendieck group for the category of all coherent sheaves on X (not only locally free ones as in the definition of $K^\bullet(X)$; $K_\bullet$ behaves covariantly for proper morphisms, $K^\bullet$ contravariantly for arbitrary morphisms). It follows again that the map $\mathscr{L}\rightsquigarrow\mathscr{L}\otimes\mathscr{F}$ from $\mathrm{Pic}(X)$ into $K_\bullet(X)$ is polynomial of degree $\leq d$ if $\mathscr{F}$ is a coherent sheaf on X; hence, if X is complete, $\mathscr{L}\rightsquigarrow\chi(\mathscr{F}\otimes\mathscr{L})$ has the same property. From this, by Serre's remark, follows that the function is constant on classes modulo $\mathrm{Pic}^\tau(X)$.

(ii) $\Rightarrow$ (ii bis) $\Rightarrow$ (iii) is trivial, (iii) $\Rightarrow$ (ii) follows trivially from theorem 2, (i) $\Rightarrow$ (iv) is known and (ii) $\Rightarrow$ (v) trivial, (iv) $\Rightarrow$ (i) results trivially from theorem 2, and (v) $\Rightarrow$ (i) from your finiteness theorem. As I remarked in my previous letter, the criterion (v) is useful in order to prove the

Corollary 2. *Let $f\colon X\to S$ be flat projective with integral geometric fibers, then $\mathscr{P}ic^\tau_{X/S}$ is open* and closed *in $\mathscr{P}ic_{X/S}$.*

This raises some questions: does the result remain true if we drop the projectivity assumption? Of course one is reduced to the case where S is the spectrum of a valuation ring, and one would like to apply the corollary 2 to theorem 1, and Chow's lemma; but there is a difficulty, as in Chow's lemma X'/S will not have integral geometric fibers, therefore the conclusion to be proved for X/S may be false for X'/S (example: a conic degenerating into two lines). In the previous corollary, is it enough to assume the geometric fibers of X/S irreducible (not necessarily reduced)? If X/S is normal i.e., flat with normal geometric fibers, is it true that $\mathscr{P}ic^\tau_{X/S}$ is proper over S (as is $\mathscr{P}ic^0_{X/S}$)? This is equivalent with asking that ${}_n\mathscr{P}ic_{X/S}$ (kernel of nth power) should be proper over S, and I doubt it is true. In characteristic 0, this is equivalent with stating that the Néron–Severi torsion groups of the geometric fibers are of the same order (in fact, isomorphic) if S is connected; I doubt very much that this is true. (Of course, the point is that I do not assume X/S simple).

I now become aware I forgot to give indications for the proof of theorem 1 (ii). First, using (i), one can assume Y/S to be normal relative to S, with irreducible geometric fibers, and X equally flat over S, and distinct from Y, therefore a relative Cartier divisor. Moreover, replacing X by a suitable multiple and using (i), we can assume the ample sheaf $\mathscr{L}$ (whose section gives X) to be *very ample*, i.e., X is really a hyperplane section. Moreover, we can now assume (again by suitable base change) that there exists another hyperplane section X'/S which is normal over S, and is a relative Cartier divisor. Let M be a subset of $\mathscr{P}ic_{Y/S}$ whose image M_X in $\mathscr{P}ic_{X/S}$ is quasi-compact, then the Hilbert polynomials of the elements of M_X remain bounded, therefore the same is true for the Hilbert polynomials of the elements of $M_{X'}$. By your criterion *in the normal case*, it follows that $M_{X'}$ is also quasi-compact. Then we can replace in the argument X by X'. Now as the fibers of Y/S are assumed of dimension ≥ 3, these of X/S are of dimension ≥ 2, and as the fibers of X/S and Y/S are normal, they are of depth ≥ 2 at their closed points. This is enough to use the "equivalence criteria" I alluded to in my last letter (the assumption that the geometric fibers of X be of depth ≥ 3 at their closed points being stronger than actually needed!), and to carry through the argument I indicated there. Ouf!

I hope my sketchy indications are clear enough to convince you, modulo the IHÉS seminar of this year. Of course I will send you a copy of the seminar as soon as everything is written up.

Sincerely yours
(signed) A Grothendieck

Grothendieck to Murre, 18 July, 1962

July 18, 1962

My dear Murre,

I recently had some thoughts on finiteness conditions for Picard preschemes, and substantially improved on the results stated in the last section of my last Bourbaki talk.[27] The main result stated there, for a simple projective morphism with connected geometric fibers (namely that the pieces $\mathscr{P}ic^{P}_{X/S}$ are of finite type over S), has been extended by Mumford to the case where instead of f simple we assume only f flat with integral geometric fibers (at least if these are normal). Using his result (the proof of which is quite simple and beautiful) I could get rid of the normality assumption, and even (as in theorem 4.1. of my talk) restrict to the consideration of the two first nontrivial coefficients of the Hilbert polynomials. The key results for the reduction are the following (the proofs being very technical, and rather different for (i) and (ii), except that (ii) uses (i) to reduce to the normal case; moreover (ii)

[27] Referring to TDTE VI, *Séminaire Bourbaki* 1961/62, n° 236.

uses Mumford's result and the equivalence criteria as developed in my last Seminar):

(i) Let X, Y be proper over S noetherian, let $f: X \to Y$ be a *surjective* S-morphism, assume, for simplicity of the statement, that the Picard preschemes exist, then $f^*: \mathscr{P}ic_{Y/S} \to \mathscr{P}ic_{X/S}$ is of finite type (and in fact affine if S is the spectrum of a field), i.e., a subset M of $\mathscr{P}ic_{Y/S}$ is quasi-compact iff its image in $\mathscr{P}ic_{X/S}$ is.
(ii) The same conclusion holds for a canonical immersion $X \to Y$ if Y/S is projective, with fibers all components of which are of dimension ≥ 3, and if X is the subscheme of zeros of a section over Y of an invertible sheaf $\mathscr{L}$ ample with respect to S.

A connected result is that, for any X/S proper and integer $n \neq 0$, the nth power homomorphism in the Picard prescheme is of finite type.

I tell you about this, namely (i), because of the method of proof, involving of course considerations of nonflat descent. The fact that I do not have any good effectivity criterion does not hamper, by just recalling what the effectivity of a given descent datum means. Now it turns out that, by a slightly more careful analysis of the situation, one can prove the following theorem, of a type very close to the one you have proved recently, and to some you still want to prove as I understand it.

Theorem. *Let S be an integral noetherian scheme, X and X' proper over S, and $f: X' \to X$ a surjective S-morphism. Look at the corresponding homomorphism for the Picard functors $f^*: \mathscr{P}ic_{X/S} \to \mathscr{P}ic_{X'/S}$. Assume:*

a) *the existence problem A defined below for X/S has always a solution (this is certainly true when X/S is projective).*
b) *the morphism $f_s: X'_s \to X_s$ induced on the generic fiber is a morphism of descent, i.e., $\mathcal{O}_{X_s} \to f(\mathcal{O}_{X'_s}) \rightrightarrows h(\mathcal{O}_{X''_s})$ is exact.*

Then, provided we replace S by a suitable nonempty open set, the homomorphism f^* is representable by a quasi-affine morphism, *more specifically in the factorisation of f^* via the functor representing suitable descent data, $f^* = vu$ with u affine and v a monomorphism (as you well know), v is in fact representable by a finite direct sum of immersions.*

Corollary. *Without assuming b), but instead in a) allowing X/S to be replaced by suitable other schemes X_i finite over X, the same conclusion holds, namely f^* is representable by quasi-affine morphisms.*

This follows from the theorem, using a suitable factorisation of f. For instance, using Chow's lemma and the main existence theorem in my first talk on Picard schemes,[28] one gets:

Corollary 2. *Assume X/S proper satisfies the condition:*

a′) *for every X' finite over X, there exists a nonempty open subset S_1 of S such that problem A for $X'|S_1$ has always solution*

[28] Referring to TDTE V, *Séminaire Bourbaki* 1961/62, n° 232.

(this condition is satisfied if X/S is projective). Then, provided we replace S by a suitable S_1 nonempty and open, $\mathscr{P}ic_{X/S}$ exists, is separated, and its connected components are of finite type over S.

N.B. The proof does not give any evidence towards the fact that, in the theorem, one could replace "quasi-affine" by "affine". This is true however over a field, because a quasi-affine algebraic group is affine! It would be interesting to have a counterexample, say, over a ring of dimension 1 such as $k[t]$, X and X' projective and simple over S and $X' \to X$ birational, or, alternatively, X and X' projective and normal over S and $f: X' \to X$ finite. A counterexample in the latter case would of course provide a counterexample to the effectivity problem for a finite morphism raised in my first talk on descent....

"Problem A" is the following: given X/S and a module $\mathscr{F}$ on X, to represent the functor on the category of S-preschemes taking any S'/S into a one-element or into the empty set, according as to whether $\mathscr{F}'$ on X'/S' is flat with respect to S' or not, where $X' = X \times_S S'$, $\mathscr{F}' = \mathscr{F} \times_S S'$.

Given X/S, we say that "Problem A for X/S has always a solution" if, for every coherent $\mathscr{F}'$ on some X'/S', the previous functor on (Sch)$/S'$ is representable by an S'-scheme of finite type. The main step in my proof of existence of Hilbert schemes shows that this condition is satisfied when X/S is projective; in the proof, essential use is made of the Hilbert polynomial, in fact we get a solution as a disjoint sum of subschemes of S corresponding to various Hilbert polynomials. Still I would expect that the functor is representable as soon as X/S is proper. In view of the application we have in mind here, it would be sufficient (for any integral S) to find in S a nonempty open set S_1 such that Problem A has always a solution for $X_1 = X \times_S S_1$ over S_1. To prove this weaker existence result, it is well possible that a reduction to the projective case is possible, using Chow's lemma and some induction on the relative dimension perhaps. I also would expect that a proof will be easier when working over a complete noetherian local ring, hence the case of a general noetherian local ring by flat descent. And it is well possible that, putting together two such partial results, a proof of the existence in general could be obtained. (I met with such difficulties already some time ago in a very analogous nonprojective existence problem, which besides I have not solved so far!) This problem A has been met also by Hartshorne (a Harvard student), but I doubt he will work seriously on it. Thus I now write to you in the hope you may be interested to have a try at this problem. As a general fact, our knowledge of nonprojective existence theorems is exceedingly poor, and I hope this will change eventually.

Sincerely yours,
A. GROTHENDIECK

Grothendieck to Robin Hartshorne, 17 September, 1962

Bures Sept. 17, 1962

My dear Hartshorne,

I thank you very much for the notes on your work on Hilbert schemes. They strike me as very ingenious. The main result is striking, the methods of proof illuminating, the technical difficulties to be overcome quite serious—I am sure it will be a very good thesis indeed.[29] I did not check enough the hard part, namely Chap IV, but I am confident your constructions are all right. Moreover, I think your method should enable you to make a still closer analysis of the structure of Hilb^P, for instance to determine the irreducible components, their dimensions, and their mutual incidence relations. For instance, for every set of integers $m_* = (m_1, \dots, m_r)$, consider the locally closed subset M_{m_*} of $M = \mathrm{Hilb}^P$ of points having that invariant, then the irreducible components of M are among the closures of the irreducible components of the M_{m_*}'s. The first question one might try to solve is whether the M_{m_*}'s are irreducible, also to determine their dimension and the incidence relations between their closures, etc. [For given P, m, the results will probably be different according to the characteristic, for (according to Serre) there are components of $\mathrm{Hilb}_{\mathbb{Z}}$ lying over single primes.] Quite a few pieces of information along these lines seem already contained in your proof of the connectedness of M (cf. my P.S.). Perhaps such an analysis will lead you to solve (in the context of Hilbert schemes, replacing Chow varieties) Weil's problem whether the geometric irreducible components are already defined over the prime field; this would follow of course if you could prove that the M_{m_*}'s are geometrically irreducible. I recall the following remarkable implication of a positive answer to Weil's problem: if X is a nonsingular projective variety defined over the field $\mathbb{C}$ of complex numbers, u an automorphism of $\mathbb{C}$, and X^u the variety over $\mathbb{C}$ deduced from X by u (via the base change $\mathbb{C} \xrightarrow{u} \mathbb{C}$, or, equivalently, by applying u to the coefficients of the equations describing X) then X and X^u are homeomorphic, hence have same homology and homotopy invariants. (At this moment, it is not known whether this statement is true.[30])

Another type of problem to be investigated: consider the open subset M' of M corresponding to simple subvarieties of $\mathbb{P}^r$; determine the irreducible components of M' (i.e., those of M that meet M'), in particular determine for which Hilbert polynomial P we have $M' \neq \emptyset$ i.e., there exists a *nonsingular* subvariety of X admitting this P. Here Borel's theorem will not help much, as M' is not complete.

One suggestion in case the invariants n_i are not enough to get hold of the components of M: there are various other invariants that may help, and have the same semi-continuity property, for instance the values of the Hilbert *function*, or even the integers $\dim H^i(X, \mathcal{O}_X(n))$, any i, n. I do not know anything about these, (except the semi-continuity and the fact that the alternating sum is continuous and a polynomial

---

[29] This 1963 Princeton University thesis was published in R. Hartshorne, Connectedness of the Hilbert scheme, *Publ. Math. IHÉS* **29** (1966) 5–48.

[30] See *C. R. Acad. Sci. Paris* **258** (1964) 4194–4196, for a counterexample by Serre.

in n), any other general information about this set of integers, for variable X, would be welcome.

Another problem, of a very different type, is to determine the category of locally free sheaves over $M = \mathrm{Hilb}_S^P$, where S is for instance the spectrum of $\mathbb{Z}$, or of a prime field, or an algebraically closed field. This problem is trivially equivalent with finding all functors, associating to an S' over S and a subscheme X of $\mathbb{P}^r_{S'}$ with Hilbert polynomial P, a locally free sheaf $\mathscr{E}$ on S', in a manner compatible with base change. One way to get such functors is to take $\mathscr{E} = f_*(\mathscr{O}_X(n))$, with large n, (and also $R^i f_*(\mathscr{O}_X(n))$ for suitable n, depending on P and i), and those obtained from such sheaves $\mathscr{E}$ by the usual tensor operations. For instance, taking exterior powers of maximal order, one gets various *invertible* sheaves $\mathscr{E}_n$ (n large). A first question is to determine the relations between these $\mathscr{E}_n$ (viewed as elements of the Picard group $\mathrm{Pic}(M)$ say), and to see if these generate the latter. Of course, in the above construction of $\mathscr{E}$ by means of direct images of $\mathscr{O}_X$ twisted by n, we could as well replace $\mathscr{O}_X$ by any other sheaf $\mathscr{F}$, flat with respect to S', and depending functorially on (S',X') (or what amounts to the same, a sheaf $\mathscr{F}$ on the universal $X = X_M$ over the modular scheme $S' = M$, $\mathscr{F}$ flat over M)—as one would get for instance starting with a locally free sheaf of $\mathbb{P}^r_S$, and inducing it on the subschemes X of $\mathbb{P}^r_{S'}$. In other words, one is led to investigate equally the category of locally free sheaves on X_M, and on $\mathbb{P}^r_S$, and their various interrelations by means of direct and inverse images (and of course tensor operations). A complete picture (even for the category of locally free sheaves on projective space only) is probably quite out of reach for the time being. However if, instead of the full category of locally free sheaves, one is content to work with the ring $K(M)$ generated by their elements (as studied in connection with the Riemann–Roch theorem in Serre–Borel's paper), it would be possible perhaps to achieve complete results. These would allow to determine at least the group $\mathrm{Pic}(M)$, and presumably $\mathrm{Pic}(X_M)$, in terms of $K(M)$ and $K(X_M)$. Moreover, once one knows $\mathrm{Pic}(M)$, one should determine for every $\mathscr{L} \in \mathrm{Pic}(M)$ the group $H^0(M,\mathscr{L})$ (having an evident functorial interpretation, in terms of the functor corresponding to $\mathscr{E}$), and of course the tensor operations $H^0(\mathscr{L}) \times H^0(\mathscr{L}') \to H^0(\mathscr{L} \otimes \mathscr{L}')$, in particular one should know the algebras $\coprod_{n \geq 0} H^0(\mathscr{L}^{\otimes n})$, and have thus a complete insight into all possible projective embeddings of M. For the time being even $H^0(M, \mathscr{O}_M)$ is not known, because, although M is geometrically connected, it is generally not reduced (even for curves in $\mathbb{P}^3$ over a field of char. 0, according to Mumford), therefore it is not clear whether $H^0(M, \mathscr{O}_M)$ as a ring has nilpotent elements or not! Of course, the knowledge of this ring alone implies your connectedness theorem; thus the questions raised here, which are concerned with M as a scheme and not only as a topological space, may turn out to be rather tough. It is not even clear whether or not $\mathrm{Pic}(M)$ is discrete. If S is the spectrum of an algebraically closed field, so that $\mathrm{Pic}(M)$ is the set of points rational over k of the Picard group-scheme $\mathscr{P}ic_{M/k}$, the question in char 0 amounts to the question if $H^1(M, \mathscr{O}_M) = 0$ (in char $p > 0$, one can only say that the latter relation implies that $\mathscr{P}ic_{M/k}$ is discrete, hence $\mathrm{Pic}(M)$ finitely generated). An argument in support of this conjecture (discreteness) would be that those invertible sheaves on M one gets from locally free sheaves on $\mathbb{P}^r$ (by twisting with *large* n, inducing on X_M, taking the direct image, and highest exterior power) form a finitely generated group, as one sees using the fact that $K(\mathbb{P}^r)$ is generated as a ring by the class of $\mathscr{O}(1)$. Thus at first sight I do not see a way of constructing a nonconstant continuous family of invertible sheaves on M!

I include in this letter some trivial comments on your notes. I am not sure I will find time very soon to work through the details of Chap 4, hoping that in your final version it will simplify a little? Is it possible to keep your manuscript?

Sincerely yours
(signed) A Grothendieck

P.S. It is known that Borel's fixed point theorem extends to an arbitrary ground field, provided "solvable" is understood as "solvable over k"; this applies to the triangular group, in particular.

Grothendieck to Mumford, 2 October, 1962

Bures Oct. 2, 1962

Dear Mumford,

Thanks for your letter which has just arrived. I did not completely understand what you are after by looking at Chow points. However I can certainly help you in defining your map $\mathscr{H}ilb \to \mathscr{D}iv\mathscr{G}rass$.[31] As usual, I like to give a general setting. I confess I did not systematically write down all that I am going to state, but enough bits of it here and there, some time ago, to be sure it can be done without much effort.

Let X be a quasi-compact prescheme having an ample sheaf, so as to allow locally free resolutions of coherent sheaves. Let's denote by $K(X)$ the ring of classes of locally free sheaves on X; as well known this is also, as a group, the group of classes of coherent sheaves on X having finite projective dimension. Taking highest exterior powers of locally free sheaves, one gets a natural homomorphism:

$$\text{dét}\colon K(X) \to \text{Pic}(X)$$

which you called rightly the first Chern class. Thus $\text{dét}(\mathscr{F})$ is also defined for any $\mathscr{F}$ of finite cohomological dimension, and behaves multiplicatively with respect to exact sequences of such $\mathscr{F}$'s. To define it, take a locally free resolution of $\mathscr{F}$ by

[31] The proof sketched in this letter, with the details completed, appeared in J. Fogarty, Truncated Hilbert functors, *J. Reine Angew. Math.* **234** (1969) 65–88.

$\mathscr{L}_i$'s, and take the alternating product of highest exterior powers. Besides, looking closer it turns out one can even define *functorially* (with respect to isomorphisms) in $\mathscr{F}$ an invertible sheaf dét($\mathscr{F}$) on X this way, for instance an automorphism of $\mathscr{F}$ defines one of dét($\mathscr{F}$), i.e., a section of $\mathscr{O}_X^*$! As a consequence of this remark, the definition of dét($\mathscr{F}$) does not really require global resolutions, and is valid on any locally ringed space whatever!

Next let $f\colon X \to Y$ be a quasi-projective morphism, for simplicity choose a Y-immersion $X \overset{j}{\hookrightarrow} \mathbb{P}_Y^r$. A coherent sheaf $\mathscr{F}$ on X is called "of finite projective dimension relative to Y" if it is of finite projective dimension on $\mathbb{P}_Y^r$. This is easily seen to be a purely local property on $\mathscr{F}$ ($\mathbb{P}_Y^r$ can be replaced by $\mathbb{A}_Y^r$), independent of the chosen immersion. It is always satisfied if $\mathscr{F}$ is flat with respect to Y. Assume supp $\mathscr{F}$ proper over Y. Then one can define an element $f_!(\mathscr{F})$ the following way,

$$f_!(\mathscr{F}) \in K(Y):$$

The assumption implies that we can resolve $\mathscr{F}$ on $\mathbb{P}_Y^r$ by sheaves of the type $g^*(\mathscr{E}_i)(n)$, or rather sums of such, where $\mathscr{E}_i$ on Y is locally free. (Indeed, we may assume $\mathscr{F}$ free on $\mathbb{P}_Y^r$, and for n big, represent $\mathscr{F}$ as a quotient of $g^*(g_*(\mathscr{F}(n)))$, but for such n, $g_*(\mathscr{F}(n))$ is locally free on Y; going on this way, one shows one eventually gets a resolution of $\mathscr{F}$ of the desired type). One then defines

$$f_!(\mathscr{F}) = \sum_i (-1)^i \mathscr{E}_i \cdot g_!(\mathscr{O}_{\mathbb{P}_Y^r}(n))$$

with

$$g_!(\mathscr{O}_{\mathbb{P}_Y^r}(n)) = \sum_i (-1)^i R^i g_*(\mathscr{O}_{\mathbb{P}_Y^r}(n)).$$

(NB g denotes the projection of $\mathbb{P}_Y^r$ on Y). Of course one verifies the independence of this definition from all choices performed. Besides $f_!$ is characterized by the following properties:

a) Additivity for exact sequences.

b) Transitivity, if one has $X \xrightarrow{f} X' \xrightarrow{g} Y$, with g flat.

c) If $\mathscr{F}$ is flat with respect to Y, and all $R^i f_*(\mathscr{F})$ are locally free, then

$$f_!(\mathscr{F}) = \sum_i (-1)^i R^i f_*(\mathscr{F}).$$

Moreover, we have the following way to get $f_!(\mathscr{F})$ up to torsion if $\mathscr{F}$ is flat over Y:

d) For big n, the locally free sheaf $f_!(\mathscr{F}(n))$ on Y, as an element of the abelian group $K(Y)$, is a polynomial in n, whose constant coefficient is precisely $f_!(\mathscr{F})$. (In fact, $n \mapsto f_!(\mathscr{F}(n))$ is a polynomial in n, coinciding for big n with the previous function).

NB I convinced myself that the general Riemann–Roch theorem can be stated and proved for a morphism $f\colon X \to Y$ which is quasi-projective and "a complete is precisely $f_!(F)$. (In fact, $n \mapsto f_!(F(n))$ is a polynomial in n, coinciding for big n with the previous function).

NB I conviced myself that the general Riemann-Roch theorem can be stated and proved for a morphism $f:X \to Y$ which is quasi-projective and [illegible] "a complete intersection" (i.e. such that X is a complete intersection in P^r_Y), for any sheaf F which has the properties stated, allowing to define $f_!(F)$.

On the other hand, you have in mind how intersection theory can be phrased in terms of the ring operations of K(X), which allow besides to get rid to a large extent of all regularity assumptions, provided we do not try to intersect any two cycles (because the sheaves they define will not be of finite cohomological dimension in general), but rather classes of sheaves instead.

Now your definition ! Let $X \subset P^r_Y$ be proper and flat over Y, with relative dimension $\leq$ d, we want to associate to it a section of $\underline{\text{Div}}\ \underline{\text{Grass}}_{r-d-1}(P^r_Y)$, in a functorial way with respect to base change, so as to have $\underline{\text{Hilb}} \to \underline{\text{Div}}\ (\underline{\text{Grass}})$. [illegible] We make the base change $Y' \to Y$, with $Y' = \underline{\text{Grass}}_{r-d-1}(P^r_Y)$. Now in $P^r_{Y'}$ we have canonically a projective sub-bundle $M^{r-d-1} = M'$, [illegible] The structure sheaves of X' and M' can be viewed as coherent sheaves on $P^r_{Y'}$, we can take the product of the elements they define in $K(P^r_{Y'})$, and take the image under $g_!$: $K(P^r_{Y'}) \to K(Y')$, which is defined as $P^r_{Y'}$ is projective and flat over Y'. Now take the dét :

$$\underline{L} = \text{dét}\ g_!(\underline{O}_{X'} \cdot \underline{O}_{M'})$$

to get an invertible sheaf on Y'. Let $T = g'(X' \cap M')$, $Y'_0 = Y' - T$, then intersection" (i.e., such that X is a complete intersection in $\mathbb{P}^r_Y$), for any sheaf $\mathscr{F}$ which has the properties stated, allowing to define $f_!(\mathscr{F})$.

On the other hand, you have in mind how intersection theory can be phrased in terms of the ring operations of $K(X)$, which allow besides to get rid to a large extent of all regularity assumptions, provided we do not try to intersect any two cycles (because the sheaves they define will not be of finite cohomological dimension in general), but rather classes of sheaves instead.

Now your definition! Let $X \hookrightarrow \mathbb{P}^r_Y$ be proper and flat over Y, with relative dimension $\leq d$, we want to associate to it a section of $\mathscr{D}iv\mathscr{G}rass_{r-d-1}(\mathbb{P}^r_Y)$, in a functorial way with respect to base change, so as to have $\mathscr{H}ilb \longrightarrow \mathscr{D}iv\mathscr{G}rass$. We make the base change $Y' \longrightarrow Y$, with $Y' = \mathscr{G}rass_{r-d-1}(\mathbb{P}^r_Y)$. Now in $\mathbb{P}^r_{Y'}$ we have canonically a projective subbundle $M^{r-d-1} = M'$, the structure sheaves of X' and M' can be viewed as coherent sheaves on $\mathbb{P}^r_{Y'}$, we can take the product of the elements they define in $K(\mathbb{P}^r_{Y'})$, and take the image under $g'_!\colon K(\mathbb{P}^r_{Y'}) \longrightarrow K(Y')$, which is defined as $\mathbb{P}^r_{Y'}$ is projective and flat over Y'. Now take the dét:

$$\mathscr{L} = \text{dét}\, g'_!(\mathscr{O}_{X'} \cdot \mathscr{O}_{M'})$$

to get an invertible sheaf on Y'. Let $T = g'(X' \cap M')$, $Y'_0 = Y' - T$, then $\mathscr{O}_{X'} \cdot \mathscr{O}_{M'}$ restricted to $g'^{-1}(Y'_0)$, is of course 0, therefore (as the definition of $g'_!$ is local on the base), the sheaf $\mathscr{L}|Y'_0$ is trivial. Looking at it more closely, one even finds a *canonical* trivialisation of $\mathscr{L}|Y'_0$, $\mathscr{L}$ being itself canonically defined as a sheaf (not only as an isomorphism class of sheaves). To make this precise, I should have defined more precisely $g_!$ as associating, to an $\mathscr{F}$ having the stated conditions, not only an element of $K(Y)$, but an object of the category of finite complexes of locally free sheaves on Y, with morphisms being the hyperext. Such an object of course defines an element of $K(Y)$ by taking alternating sums of the components (which is the same if we replace the object by an isomorphic one), and thus taking the dét we get again an element of $\text{Pic}(Y)$; but more precisely we have directly a functor $\text{dét}\colon C \longrightarrow \text{Inv}$ of the aforesaid category into the category of invertible sheaves (generalizing my remark on dét in the beginning). Moreover, the definition should be extended of course, assuming for simplicity X flat over Y, from a single $\mathscr{F}$ to the category $C(X)$ of finite complexes of locally free sheaves on X, getting thus *functors* $C(X) \xrightarrow{f_!} C(Y) \xrightarrow{\text{dét}} \text{Inv}(Y)$. Now in $C(X)$ the tensor product is functorially defined, and going back to the situation with $\mathscr{H}ilb$ etc., viewing $\mathscr{O}_{X'}$ and $\mathscr{O}_{M'}$ as defining objects of $C(\mathbb{P}^r_{Y'})$ (via resolutions), written $\mathscr{O}_{X'}$ and $\mathscr{O}_{M'}$ for simplicity, and taking their product and applying the functor $\text{dét}\, g_!$, we get $\mathscr{L}$ in a functorial way. (Thus, X could be replaced by any sheaf $\mathscr{F}$ on $\mathbb{P}^r_Y$ flat with respect to Y, and $\mathscr{L}$ depends functorially on such an $\mathscr{F}$ ...). This then makes clear that we have a canonical section of $\mathscr{L}|Y'_0$, coming from a canonical isomorphism $\mathscr{O} \xrightarrow{\sim} (\mathscr{O}_{X'} \cdot \mathscr{O}_{M'})|g'^{-1}(Y'_0)$ and applying the previous functor to this. Now it is easy to verify (I hope) that this rational section of $\mathscr{L}$ over $Y' = \mathscr{G}rass$ is in fact everywhere defined, due to the fact that Y' is simple over Y and that Y'_0 contains the generic points of the fibers of Y' over Y, and that one shows that, on every fiber of Y', the section is regular and defines the usual Chow divisor (which of course I did not check). Of course, instead of taking

$\mathscr{G}rass$, one could also take multiprojective space over Y, as does Chow, I do not know if the theory of Chow coordinates works the same using $d+1$ hyperplanes, or a linear subspace of codimension $d+1$ as you suggest, except that it is of course still true that a pure cycle of dimension d is determined by a Chow point in your version. Sticking to Chow's definition, M' would be an intersection of $d+1$ hyperplanes which might at some points intersect excessively, therefore we better keep the system of $d+1$ divisors D_i and directly look at the product $\mathscr{O}_{X'} \cdot \mathscr{O}_{D_1} \cdots \mathscr{O}_{D_{d+1}}$ instead of $\mathscr{O}_{X'} \cdot \mathscr{O}_{M'}$.

To compute $\mathscr{L}$, one can as well first induce $\mathscr{O}_{M'}$ on X' (in the sense of course of the categories $C(\mathbb{P}^r_{Y'})$, $C(X')$, i.e., first taking a locally free resolution of $\mathscr{O}_{M'}$), then project on Y' by $f'_!$, where $f' \colon X' \to Y'$ is the projection, and take the dét. Working, for simplicity, with Y' the multiprojective scheme, the induced object in $C(X')$ is a complex having as underlying graded module the exterior algebra of $\left(\coprod_{0 \le i \le d} M \otimes M_i\right)$, where M [resp. M_i] is the inverse image of $\mathscr{O}_{\mathbb{P}^r_Y}(1)$ by the projection of $\mathbb{P}^r_{Y'} \to \mathbb{P}^r_Y$ [resp. $\mathbb{P}^r_{Y'} \to Y' \to$ (ith factor $\mathbb{P}^r_Y$ of $Y' = (\mathbb{P}^r_Y)^{d+1}$)]. Perhaps this may help to identify the sheaf you constructed on Y (as highest coefficient in the Hilbert polynomial expressing the function $n \rightsquigarrow \text{dét} f_* \mathscr{O}_X(n)$ for big n, or equivalently $n \rightsquigarrow \text{dét} f_! \mathscr{O}_X(n)$ for any n), as the inverse image of a suitable ample sheaf of $\mathscr{D}iv(Y')$ under the section we just defined of $\mathscr{D}iv(Y')$, as you suggest. I have no feeling whether such an interpretation is possible. These questions are of course related to the problem I proposed to Hartshorne a few weeks ago, namely to determine (over various ground schemes S such as Spec($\mathbb{Z}$), the spectrum of a field or others) the complete structure of $K(M)$ and $K(X_M)$, where M is a component of the $\mathscr{H}ilb$ scheme, and $X_M \subset \mathbb{P}^r_M$ the universal flat subscheme of $\mathbb{P}^r$—together of course with the operations $f^!$ and $f_!$ coming from the projection $f \colon X_M \to M$. I made one or two wishful conjectures, including that $K(M)$ is generated by invertible sheaves and that Pic(M) is "discrete", but I grant I have no serious support for such conjecture. I wonder if you are able to compute $H^0(M, \mathscr{O}_M)$ and $H^1(M, \mathscr{O}_M)$ over a field, say $k = \mathbb{C}$. Is the first k, the second 0? Even the first question has no obvious answer, because of the existence of nilpotent elements in M. A good knowledge of M should even contain, not only Pic(M), but also knowledge of the $H^0(M, \mathscr{L})$ for $\mathscr{L} \in \text{Pic}(M)$, the tensor operations etc.

For the computations you have in mind it may be enough to know that, for a projective bundle $\mathbb{P}$ associated to a locally free $\mathscr{E}$ of rank $r+1$ over a base Y admitting an ample sheaf, $K(\mathbb{P})$ is completely determined (as a λ ring) by the fact that, as a module over $K(Y)$, it has a basis consisting of the classes of $\mathscr{O}_{\mathbb{P}}(n) = \mathscr{L}^n$ with $n_0 \le n \le n_0 + r$. The most convenient is to take $n_0 = -r$, i.e., write uniquely an element x of $K(\mathbb{P})$ as $\sum_i c_i \mathscr{L}^i$ $(-r \le i \le 0)$, then $f_!(x) = c_0$. As for the ring structure of $K(X)$, known when $\mathscr{L}^{-r-1}$ is known as a linear combination of the basis elements $\mathscr{L}^i$ $(-r \le i \le 0)$, it is obtained by writing simply $\lambda_{-1}(\mathscr{E} - \mathscr{L}) = 0$, i.e., $\lambda_t(\mathscr{E})$ vanishes when substituting t by $-\mathscr{L}^{-1}$. From this, the K of various flag fiber spaces associated to $\mathscr{E}$, including grassmannians, can be determined in terms of $K(Y)$, $\mathscr{E} \in K(Y)$ in a purely formal way, as in the talks I gave in Chevalley's Seminar. As $\mathscr{D}iv\mathscr{G}rass$ is essentially a projective fiber bundle over $\mathscr{G}rass$ (due to the fact that the Picard scheme of $\mathscr{G}rass$ is étale over the base, and any invertible sheaf on $\mathscr{G}rass$ is "cohomologically flat" over the base), the K of this scheme is easily determined too. Of course, when you have an $X \subset \mathbb{P}$ flat over Y, the invertible sheaf it defines on Y by our previous construction is easily computed in terms of the element $\sum c_i \mathscr{L}^{-i}$ of $K(\mathbb{P})$ defined by X. The fact that X has relative dimension $\leq d$ over Y is then expressed by the fact that, if we take for $K(\mathbb{P})$ the basis formed by the elements H^i $(0 \leq i \leq r)$, with $H = 1 - \mathscr{L}$, (H "hyperplane section"), then the coefficients c_i with $i > r-d$ have augmentation 0. From this I guess your assertion about the Δ^{k+1} should follow formally. Anyhow, I guess the story will be clearer if instead of an X you take a coherent sheaf on $\mathbb{P}^r_Y$ flat with respect to Y, or complexes on $\mathbb{P}^r_Y \ldots.$

To come back to your initial problem of moduli for projective invariants of varieties, I do not see the point in what you call "my best result so far", concerning moduli for projective curves. Once you know the existence of a modular scheme for jacobi curves of level n, does it not follow trivially that there is a modular scheme for curves in $\mathbb{P}^3$, by taking the previous modular scheme, a suitable open subset of the Picard scheme of the modular curve over it, and a suitable open subset of some grassmannian scheme over the latter? I did not figure this out, therefore I wonder if there are some difficulties, or if you stick to your approach via "stable Chow forms" only because of the hope that some time that method might yield results on higher dimensional varieties?

I did not understand at all your suggestion concerning Murre's theorem.[32] In fact, Murre has two theorems, one (the easier, the first he got) concerns a group functor which is embedded in a representable one; then his criterion does not need any Rosenlicht type condition. In this case your suggestion falls short, as Murre's criterion applies also when $H \to G$ is *not* a closed immersion, for instance is a monomorphism $\mathbb{Z} \to G$, where G is a group scheme of finite type over k such as $\mathbb{G}_m$, say. On the other hand in his second criterion, concerning a functor which is not embedded in another, (this condition being replaced by the Rosenlicht condition), I do not see what your condition 4) could possibly mean. In any case, besides, the ground field need not be algebraically closed.

I guess you heard that Mike proved that, over $\mathbb{C}$, the Weil cohomology = usual cohomology. Pondering over his proofs this now appears almost trivial, moreover his method yields some basic results in arbitrary characteristic. The development of a large part of Weil cohomology now seems to me a mere routine matter, and I feel the complete equivalent of the classical theory, including Weil's conjectures, should be obtained within the next one or two years. Just the typical Kähler–Hodge–Lefschetz type things will perhaps offer some serious difficulty.

Sincerely yours
(signed) A Grothendieck

[32] The results were published in J.P. Murre, On contravariant functors from the category of preschemes over a field into the category of abelian groups (with an application to the Picard functor), *Publ. Math. IHÉS* **23** (1964) 5–43.

P.S. I had an afterthought on the relative Cartier divisor you request on $Y' = \mathscr{G}rass$; I gave you one, but no proof that it is a *positive* divisor. However as Y' is simple, hence *flat* over Y, it is enough to prove it is positive at points of Y' which are *divisorial* on the fiber of $Y' \to Y$ (simple reasons of "depth"), and besides reduce to the case where Y is local artinian (and points $y \in Y$ such that $\dim \mathscr{O}_{Y,y} = 1$). This implies it is enough to look at what happens at generic points x' of the intersection $\operatorname{supp} \mathscr{F}' \cap M'$. Now such a point will project onto a generic point of $\operatorname{supp} \mathscr{F}$, hence $\mathscr{F}'$ will have projective dimension $r-d$ at it (i.e., will be Cohen–Macaulay). Moreover the elements of $\mathscr{O}_{P',x'}$ defining M' as a complete intersection will form an $\mathscr{F}'$-sequence, so that the higher $\operatorname{Tor}_i^{\mathscr{O}_{P'}}(\mathscr{F}', \mathscr{O}_{M'})$ vanish, and $\mathscr{F}' \otimes \mathscr{O}_{M'}$ is of projective dimension $(r-d)+(d+1) = r+1$. Moreover, $\operatorname{supp}(\mathscr{F}' \otimes \mathscr{O}_{M'})$ will be finite over Y' when localizing at such a point y'. From this it follows easily that $f'_!(\mathscr{F}' \otimes \mathscr{O}_{M'})$ is (on this neighbourhood of y') just the usual $f'_*(\mathscr{F}' \otimes \mathscr{O}_{M'})$, and the latter is at y' of cohomological dimension $(r+1)-r = 1$, where r is the rel. dimension of P'/Y'. But, as you already noticed, such a sheaf on Y' *does* define a positive Cartier divisor at y'. This concludes the proof! I wonder if there is something simpler to do it?

Grothendieck to Mumford, 18 October, 1962

Bures 18.10.1962

Dear Mumford,

Thanks for your letter. Unfortunately, I have no idea how to rigidify in general polarized varieties with no infinitesimal automorphisms in a discrete way. Anyhow, didn't you give me once an example with no automorphisms whatsoever (infinitesimal or finite) for which there was no reasonable local modular family?[33] In this

[33] This sentence is somewhat mystifying. Suppose that X is an algebraic variety proper over a field k with "no automorphisms whatsoever", then the deformation functor $\operatorname{Def}(X)$ of X is pro-representable by Schlessinger's criterion; see M. Schlessinger, Functors of Artin rings, *Trans. Amer. Math. Soc.* **130** (1968) 208–222. In other words, over the complete local ring R which pro-represents $\operatorname{Def}(X)$, there exists a flat proper formal scheme $\mathscr{X}$ over $\operatorname{Spf}(R)$ such that every deformation of X over an Artinian ring S is the pull-back of $\mathscr{X}$ by a unique homomorphism $R \to S$ of local rings. So the formal scheme $\mathscr{X}$ over $\operatorname{Spf}(R)$ can be regarded as a "reasonable local modular family" for X. On the other hand the formal scheme $\mathscr{X}$ over R may not be algebraic (i.e., there may not exist a proper scheme $\mathbb{X}$ over R whose formal completion is $\mathscr{X} \to \operatorname{Spf}(R)$), which might be what Grothendieck had in mind. The following example, due to P. Deligne, illustrates this possibility. Let A be an abelian variety over an algebraically closed field k of dimension $g \geq 2$. Choose a finite set of distinct closed points $x_1, \ldots, x_n \in A$ and let $X = \operatorname{Bl}(A, x_1, \ldots, x_n)$ be the blowing up of A with center $\{x_1, \ldots, x_n\}$. If n is large and $x_1, \ldots, x_n$ are in general position, then the variety X will have no non-trivial automorphisms whatsoever. On the other hand one can show that the universal deformation $\mathscr{X}$ of X is not a scheme over R—because the universal deformation of A is not algebraic.

L. Illusie suggested that Grothendieck might have meant the following: an example of a proper algebraic variety X with no automorphisms whatsoever whose local deformation space is trivial. Fake projective spaces, e.g., those constructed by Mumford in [79a], have the above properties.

extension by prime to p homogeneous spaces under an algebraic group of automorphisms.

Your result of finiteness on polarized non singular surfaces is quite interesting. I would appreciate very much to have an idea of the proof.

It seemed to me, looking at Mike's arguments, that his lemma on divisor classes (existing variation) can be completely eliminated. Unfortunately, for the time being, everything is tied to your constructions (in fact, even to <u>algebraic</u> schemes). The key lemma is the following one:

If X^n is non singular over a field k alg. closed, then every point has an open neighborhood U having the context, I never really developed (in the spirit of my talk on formal moduli) the question of a *modular field* for say an algebraic scheme X over some field k, s.t. $H^0(X, \mathcal{O}_X) = 0$, and, for simplicity, the group of automorphisms Γ of X being finite (say, by imposing if needed a polarization on X, or some other extra structure). To such an X there should be associated something like

a) A field k_0, finitely generated over the prime field
b) A galois extension k_1 of k_0, with group Γ
c) A scheme X_1/k_1, such that for $g \in \Gamma$, then $\exists$ k_1-isomorphism $X_1^g \simeq X_1$ (but of course, no descent data to k_0!)
d) An isomorphism $X_{1\Omega} \simeq X_\Omega$, where Ω is some common (big) extension of k_1 and k.

Data a), b), c) in terms of X should be canonically definable, independent from field extension or k, k_0 should be contained in any "field of definition" k of X, and, if k is algebraically closed, k_0 should be something like the field of invariants of all $\sigma \in \operatorname{Aut} k$ such that $X^\sigma \simeq X$, at least up to inseparability. In the rigid case, $\Gamma = e$, k_0 will be just the *smallest* field of definition for X, and X comes in a unique way from an X_0 over k_0. Did you ever try to work out these things? If you do not assume $H^0(X, \mathcal{O}_X) = 0$, something should still be feasible, replacing Galois extensions by principal homogeneous spaces under an algebraic group of automorphisms.

Your result of finiteness on polarized nonsingular surfaces[34] is quite interesting. I would appreciate very much to have an idea of the proof.

It seemed to me, looking at Mike's arguments, that his lemma on divisor classes (using resolutions) can be completely eliminated. Unfortunately, for the time being, everything is tied to equal characteristics (in fact, even to *algebraic* schemes). The key lemma[35] is the following one:

If X^n is nonsingular over a field k, alg. closed, then every point has an open neighbourhood $\mathcal{U}$ having the following structure: There exists

$$\mathcal{U} = \mathcal{U}_n \xrightarrow{f_n} \mathcal{U}_{n-1} \longrightarrow \cdots \longrightarrow \mathcal{U}_1 \xrightarrow{f_1} \mathcal{U}_0 = \operatorname{Spec}(k),$$

where every f_i is an "elementary" morphism; namely obtained from a simple proper morphism $g\colon V \longrightarrow W$ with geometric fibers connected of dim 1 (V a nice relative curve over W) by removing an étale multisection Z. Thus, from the point of view of topology, $\mathcal{U}$ is remarkably simple, its universal covering being *contractible* and its π_1 a successive extension of free groups [hence $\mathcal{U} \simeq B\pi_1$, the classifying space of such a group π_1!].

Sincerely yours,
(signed) A Grothendieck

34 This refers to the second main theorem in [64].

35 A proof of this key lemma was published as Prop. 3.3 in SGA4 Exposé XI, on p. 69 of LNM 305, Springer-Verlag, 1973.

Grothendieck to Michael Artin, 14 February, 1963

Bures Feb 14, 1963

My dear Mike,

I was just going to write you when I got your letter. First I want to ask you if you feel like refereeing Néron's big manuscript on minimal models for abelian varieties[36] (it has over 300 pages). I wrote to Mumford in this matter, who says he will have no time in the next months, do you think you would? Otherwise I will publish it as it is, as it seems difficult to find a referee, and the stuff is doubtlessly to be published, even if it is not completely OK in the details.

I started thinking on the cohomology of schemes, after reading your notes which I find quite useful. (As for comments of detail, we will discuss about it when you are here and we are organizing the seminar). I got a few results:

1) Let f be a *proper* morphism of locally noetherian schemes, then for any torsion sheaf $\mathcal{F}$ on X, formation of $R^i f_*(\mathcal{F})$ commutes with arbitrary base-extension. (It is equivalent to state that, for Y strictly local, i.e., the spectrum of a local hensel ring with separably closed residue field, the maps $H^i(X,\mathcal{F}) \to H^i(X_0,\mathcal{F}_0)$, where X_0 is the special fiber, are isomorphisms.)
2) Let f be as above, assume $\mathcal{F}$ a constructible torsion sheaf (constructible means that, for any $x \in X$, the restriction of $\mathcal{F}$ to the closure Z of x is given by an étale group-scheme over a nonempty open set of Z. It is equivalent to say when X is noetherian that $\mathcal{F}$ is a noetherian object of the category of sheaves on $X\ldots$). Then the sheaves $R^i f_*(\mathcal{F})$ are constructible.

The same should hold if f is only assumed to be of finite type, provided $\mathcal{F}$ is prime to the residue characteristics. By virtue of 2), it is enough to show it for an open immersion $U \to Y$ and $\mathcal{F} = \mu_n$. Whenever resolution of singularities is available, one is even reduced to the case where Y is regular, and U the complement of a divisor having only normal crossings, and then it would follow from the conjectural statement about the $R^i f_*(\mu_n)$ when U is the complement of a regular divisor. Thus, using your local result, we get:

3) Let f be a morphism of finite type of locally algebraic preschemes over a field of char 0, $\mathcal{F}$ a constructible torsion sheaf on X, then the sheaves $R^i f_*(\mathcal{F})$ are constructible.

The same technique yields the comparison theorem:

4) Under the conditions of 3), assume the ground field is $\mathbb{C}$, the field of complex numbers. Then formation of $R^i f_*(\mathcal{F})$ is compatible with passing to the underlying "usual" topological spaces and sheaves.

The result 1) on base extension should be true if we drop the properness assumption, assuming instead that $\mathcal{F}$ is prime to the residue characteristics, and that the base change $Y' \to Y$ is "regular", namely flat with geometrically regular fibers. This would be applicable to situations like Y a "good" local ring, and Y' its completion,

[36] Published in *Publ. Math. IHÉS* **21** (1964) 361–484.

or situations deduced from this one by base extension on Y, which would be a nice thing to have in order to know once for all that, for instance, for cohomological purposes, a "good" hensel ring can always be replaced by its completion. For the time being I cannot prove that general result, even in characteristic 0 (NB when resolution of singularities is available, it can be shown to be equivalent with the statement about the regular divisor in a regular scheme...). However, using the local Lefschetz techniques, I proved your "key lemma"[37] without resolution of singularities, and from this:

5) The conclusion of 1) remains valid when dropping the properness assumption, assuming f of finite type, $\mathscr{F}$ prime to the residue characteristics, and the base change morphism $Y' \to Y$ *simple* (which means regular and locally of finite type).

This implies the usual result on the cohomological structure of a regular scheme and a regular divisor in it in various "relative" cases. Using this, and 1), one gets in a pretty formal way:

6) Let $f: X \to Y$ be proper and simple, G a commutative group scheme over X, finite and étale over X, (we say that the sheaf defined by G is "locally constant"), prime to the residue characteristics. Then the sheaves $R^i f_*(G)$ on Y are equally locally constant. The same holds true if we replace X by $X - Z$, where Z is a closed subscheme of X simple over Y (but of course G has to be defined on the whole of X).

Truth to tell, I checked this only when Y is the spectrum of a discrete valuation ring, but I think from this and 2) the general result should follow. I think also that 1) will yield the Künneth formula for a product, over a field, of two preschemes one of which is proper; of course, the same should hold true without properness, sticking to coefficients prime to the characteristic. Of course, the main interest of 6) is to allow computations of cohomology in characteristic $p > 0$ from transcendental results in characteristic 0, just as for the fundamental group. Besides, the main steps in the key results 1) and 5) are the analogous statements on fundamental groups. The main techniques I developed so far in algebraic geometry have to be used: the existence theorem on coherent algebraic sheaves, nonflat descent, Hilbert and Picard schemes (the latter for nice relative curves only), Lefschetz techniques. Thus it was not so silly after all to postpone Weil cohomology after all this.

Here is what I can say about the Brauer group $\mathrm{Br}(X)$ of a prescheme (more generally, a ringed space). We define it as the group of classes of Azumaya algebras over X, two such algebras A and B being considered equivalent if there exist locally free sheaves $\mathscr{E}$, $\mathscr{F}$ on X and an isomorphism

$$A \otimes \mathrm{End}(\mathscr{E}) \simeq B \otimes \mathrm{End}(\mathscr{F})$$

or, what amounts to the same, if there is a locally free $\mathscr{E}$ and an isomorphism

$$B^{\circ} \otimes A \simeq \mathrm{End}(\mathscr{E}).$$

[37] A version of the "key lemma" here, also referred to as *Mike's "key lemma"* in the last paragraph of [1963Feb21], was published as Lemma 2.6 in SGA4 Exposé XV, on p. 196 of LNM 305, Springer-Verlag, 1973.

Viewing an Azumaya algebra of rank n^2 as being defined by an element of $H^1(X, PGL(n))$ (NB étale "locally finite" topology[38]), and using the obstruction (coboundary) map corresponding to the exact sequence

$$e \to \mathbb{G}_m \to GL(n) \to PGL(n) \to e,$$

one obtains a homomorphism[39]

$$c\colon \mathrm{Br}(X) \to H^2(X, \mathbb{G}_m)$$

which is always injective (as results formally from the fact that the vanishing of the obstruction means the possibility of lifting the structure sheaf to $GL(n)$). Denoting by X_Z the prescheme X with the Zariski topology, and using the map $f\colon X \to X_Z$ and the Leray spectral sequence, we get

$$0 \to H^2(X_Z, \mathbb{G}_m) \to H^2(X, \mathbb{G}_m) \to H^0(X_Z, R^2 f_*(\mathbb{G}_m)) \to H^3(X_Z, \mathbb{G}_m)$$

(using $R^1 f_*(\mathbb{G}_m) = 0$); if X is regular this shows (using $H^i(X_Z, \mathbb{G}_m) = 0$ for $i \geq 2$):

$$H^2(X, \mathbb{G}_m) = H^0(X_Z, \mathscr{B}r_X) \qquad (X \text{ regular}),$$

where $\mathscr{B}r_X = R^2 f_*(\mathbb{G}_m)$ is the sheaf on X whose fibers are the groups

$$H^2(\mathrm{Spec}(\mathscr{O}_{X,x}), \mathbb{G}_m) = \mathrm{Br}(\mathrm{Spec}(\mathscr{O}_{X,x}))\,.$$

The last equality comes from

$$\mathrm{Br}(X) \simeq H^2(X, \mathbb{G}_m) \quad \text{if } X \text{ local},$$

as you noticed, since, by the choice of the topology, we have

$$(*) \qquad H^i(X, \mathscr{F}) \simeq H^i(\pi, H^0(\bar{X}, \mathscr{F}))$$

for any sheaf $\mathscr{F}$, $\bar{X}$ being the universal covering. (NB I do not know if we would obtain the same result taking the étale topology you like best; this amounts to the question whether, for X local, $H^2(\bar{X}_{\mathrm{Mike}}, \mathbb{G}_m) = 0$.[40] Did you check this result, at least for X regular?). In fact, now that I am writing about it, I get aware that (lacking the foundations on the étale locally finite topology, which I grant is not too nice), I do not even know if $(*)$ above is true, therefore I do not know even for X local if $\mathrm{Br}(X) = H^2(X, \mathbb{G}_m)$.[41] I wonder even if by chance the category of sheaves for the two topologies (the étale and étale locally finite) are not equivalent, at least for X normal say, so that the cohomological theory is the same for both; remember that the covering families I take are by no means closed under composition, and that

---

[38] This is the topology (etf), "topologie étale finie"; see SGA3 Exposé IV 6.3.

[39] The typed formula reads $\mathrm{Br}(X) \to H^2(X, \mathscr{O}_X)$. The $\mathscr{O}_X$ may have meant $\mathscr{O}_X^*$.

[40] Here "the étale topology you like best" is the standard étale topology, denoted (et) in SGA3 Exposé IV 6.3. So $H^2(\bar{X}_{\mathrm{Mike}}, \mathbb{G}_m) = H^2(\bar{X}_{\mathrm{et}}, \mathbb{G}_m)$.

[41] The 2nd argument of this H^2 was handwritten and looks like either $\mathscr{O}_X$ or $\mathbb{G}_m$.

by saturating we may well come very close to the good étale topology of yours. Anyhow, all I stated before is good taking any topology between the étale locally finite and the flat quasi-compact one, the latter gives the "largest" $H^2(X,\mathbb{G}_m)$ (NB they are included ones in the others), and, I hope, all $H^i(X,\mathbb{G}_m)$ should be the same in all these topologies, at least for the étale and the quasi-finite and flat one. The flat topologies have the advantage that, for any n, we have the exact sequence

$$e \to \mu_n \to SL(n) \to PGL(n) \to e;$$

this shows for instance that the part of $H^2(X,\mathbb{G}_m)$ (and hence of $\mathrm{Br}(X)$) coming from $H^1(X,PGL(n))$ comes in fact from $H^2(X,\mu_n)$ and hence is annihilated by n. If you write this as meaning that $A\otimes A\otimes\cdots\otimes A \simeq \mathrm{End}(\mathscr{E})$, some $\mathscr{E}$, for any Azumaya algebra A of rank n^2, the tensor product being n-fold, I do not see any direct geometric description of the $\mathscr{E}$ in terms of A!

As for your question whether $\mathrm{Br}(X) = H^2(X,\mathbb{G}_m)$ in general, I very much doubt it is true, even if X is regular.[42] Granting it is true if X is local and regular, this would mean that for variable U on a regular prescheme X, $U \rightsquigarrow \mathrm{Br}(U)$ is a sheaf on X, or also that for any two open sets U, V and elements of $\mathrm{Br}(U)$, $\mathrm{Br}(V)$ that match in the intersection, there is an element of $\mathrm{Br}(U\cup V)$ inducing them. Granting the standard local results (which are proved in algebraic geometry over a field), this would imply that whenever Y is a closed subset of codimension at least 2 in X, and u an element of $\mathrm{Br}(X-Y)$, it comes from an element of $\mathrm{Br}(X)$.[43] All I could do along these lines is remark that any element of $H^2(X,\mathbb{G}_m)$ can be represented by an element of $\mathrm{Br}(U)$, where $U = X - Y$ with Y of codimension ≥ 3 (using the fact that a reflexive module over a regular local ring of dimension 2 is free). Thus

[42] The cohomology group $H^2(X,\mathbb{G}_m)$ in this letter is the cohomology for the étale finite topology (etf). In the rest of this footnote we use the étale topology.

The Brauer group $\mathrm{Br}(X)$ is a torsion group. The cohomology group $H^2(X_{\mathrm{et}},\mathbb{G}_m)$ is torsion if X is regular, but not so in general. So any scheme X such that $H^2(X_{\mathrm{et}},\mathbb{G}_m)$ is not torsion (so X is not regular) will have $\mathrm{Br}(X) \neq H^2(X_{\mathrm{et}},\mathbb{G}_m)$; see 7) P.S. of [1963Feb23] (p. 685) for such an example.

Denote by $\mathrm{Br}'(X)$ the torsion subgroup of $H^2(X_{\mathrm{et}},\mathbb{G}_m)$, called the cohomological Brauer group of X. One can ask whether the canonical map $\mathrm{Br}(X) \to \mathrm{Br}'(X)$ is an isomorphism for X quasi-compact and separated. (If either the quasicompact or the separated assumption is dropped one can produce counterexamples with X locally noetherian and integral.) It is a theorem of O. Gabber that this "conjecture" $\mathrm{Br} = \mathrm{Br}'$ is true if there exists an ample invertible sheaf on X. Gabber's proof is unfortunately not recorded. A different proof is in a preprint by A.J. de Jong, *A result of Gabber*, available from `http://www.math.columbia.edu/~dejong/`

In the case when X is affine, this statement is proved in Gabber's dissertation, Some theorems on Azumaya algebras, published in *The Brauer group* (Sem. Les Plans-sur-Bex, 1980), pp. 129–209, LNM 844, Springer-Verlag, 1981. Some other cases, including abelian schemes, toric varieties and surfaces were known before Gabber's general theorem. The reader can consult the survey article by R.T. Hoobler, When is $\mathrm{Br}(X) = \mathrm{Br}'(X)$? in *Brauer Groups in Ring Theory and Algebraic Geometry* (Wilrijk, 1981), LNM 917, pp. 231–244, Springer-Verlag, 1982. More references can be found in the Math. Reviews articles MR2017247 (2005b:14032), MR1970808 (2004b:14029), MR1866495 (2002i:14023).

[43] In the case when Y is regular, the purity result discussed here for torsion prime to the residue characteristic can be found in Thm. 6.1 (p. 134) of A. Grothendieck, GB III : Exemples et compléments, in *Dix Exposés sur la Cohomologie des Schémas*, North-Holland, 1968, pp. 88–188. (In the statement of Thm. 6.1 in *op. cit.* one should assume that Y is regular.)

the first counterexample should be looked for in dimension 3. Here is a suggestion for a "universal" counterexample[44] over $\mathbb{C}$: take the Eilenberg–Mac Lane space $K(\mathbb{Z}/n\mathbb{Z},2)$, approximate it homotopically up to dimension $d \geq 2$ by a nonsingular variety, (this is possible, if I remember well, by a construction of Atiyah), and take the canonical class in $H^2(X,\mathbb{Z}/n\mathbb{Z})$ and its image in $H^2(X,\mathbb{G}_m)$. The idea is that perhaps one can give some nonempty necessary topological conditions on an element of $H^2(X,\mathbb{Z}/n\mathbb{Z})$ to come from a projective (topological) bundle, or rather for an element of $H^3(X,\mathbb{Z}) \simeq H^2(X_{\text{top}},\mathbb{G}_m)$ to be the Bockstein coboundary of such an element of $H^2(X,\mathbb{Z}/n\mathbb{Z})$. To be precise, ask Bott (say) if the Bockstein of the canonical $H^2(K(\mathbb{Z}/n\mathbb{Z},2),\mathbb{Z}/n\mathbb{Z})$ can be defined by a projective bundle on $K(\mathbb{Z}/n\mathbb{Z},2)$, as an obstruction to lifting to $GL(n)$, i.e., as the inverse image of a certain obvious canonical class in $H^2(B_{PGL(n)},\mathbb{Z}/n\mathbb{Z})$. If not, we get the expected counterexample....[45]

Grothendieck to Mumford, 21 February, 1963

Bures 21.2.1963

My dear Mumford,

I think I can give an affirmative answer to your question. First note that, to give a functor of the kind you say, is equivalent to giving a functor $F\colon (\mathrm{Sch}/S)^\circ \to (\mathrm{Ens})$ endowed with a structure of $\mathscr{O}$-module, where $\mathscr{O}$ is the functor $T \rightsquigarrow \Gamma(T,\mathscr{O}_T)$, and such that F be of "local type", i.e., for every argument T, $U \rightsquigarrow F(U)$ for U an open set of T is a sheaf (of modules) on T, and that moreover the previous sheaf be coherent. Your problem then is whether F is representable (in the usual sense) by a vector bundle, in a way to respect the module structures. Of course, the question is local on S, so we may assume S affine for simplicity, say $S = \mathrm{Spec}(A)$. It turns out that, in practice, a functor F as above is in fact always defined via a functor $M \rightsquigarrow G(M)$ from arbitrary A-modules to abelian groups (or A-modules, this amounts to the same), by putting $F(T) = G(B)$ if $T = \mathrm{Spec}(B)$, and deducing $F(T)$ in general by recollement. I guess there should be a simple way of expressing in general equivalence between giving an F or a G, via "Nagata's trick" for instance, using, to define $G(M)$ in terms of F, the algebra $D_A(M) = A \oplus M$ (M ideal of square zero)... but in fact I do not care too much, as in practice one has a direct hold of G. The question therefore becomes to characterize (given a noetherian ring A) the covariant functors $C_A \to (\mathrm{Ens})$ (C_A = category of A-modules) which are representable by a module of finite type. (To say that this functor comes from a functor with values in the category (Ab) of abelian groups just means G is additive, i.e., transforms finite products into products). Here is a set of necessary and sufficient conditions, which give the answer in those cases I have needed so far:

1) G commutes with filtering direct limits.

---

[44] Cf. 5) in [1963Feb23] (p. 684).

[45] What we have of the letter ends here, without a signature.

2) G is left exact (which means also: additive, and left exact in the sense of additive functors of abelian categories). Of course, due to 1), it is enough to check left exactness for arguments of finite type over A.
3) For every noetherian algebra B over A, separated and complete for some J-adic topology, and every module of finite type M over B, the map

$$G(M) \to \varprojlim G(M/J^{n+1}M)$$

is an isomorphism.
4) For every noetherian algebra B over A, and M a module of finite type over B, $G(M)$ is a module of finite type over B. (In fact, it is enough to check it for $B = A$).
5) For every ideal J such that $A/J \neq 0$, there exists a nonnilpotent element f in A/J such that, if we put $B = (A/J)_f$, the "induced functor" $G_B \colon C_B \to (\mathrm{Ens})$ is representable by a B-module of finite type.

Conditions 1), 2), 3) are "exactness conditions", in 3) it is enough to take the case where B is either the completion of a local ring of A for the usual topology, or the completion of A itself for some ideal J. 4) is a simple finiteness condition. In practice 3) and 4) are verified by the standard theorems of the type "finiteness" "comparison" "existence" of EGA III, whereas 1) and 2) are about trivial. Condition 5), of "generic representability", is more delicate to verify in the applications. Restricted to prime ideals J, this condition is equivalent to

5a) The function $\mathfrak{p} \rightsquigarrow \mathrm{rank}_{k(\mathfrak{p})}\, G(k(\mathfrak{p}))$ on $\mathrm{Spec}(A)$ is constructible.

However, this does not imply 5) in general. If A is quotient of a regular ring (harmless condition, by standard reduction steps to algebras of finite type over $\mathbb{Z}$!), 5) is equivalent to the following:

5b) For every ideal J in A such that $B = A/J \neq 0$, and every module Ω of finite type over A/J, there exists $f \in B$, nonnilpotent, with the following property: for every prime $\mathfrak{p} \not\ni f$ of B and every regular sequence (f_i) of parameters of $B_{\mathfrak{p}}$, the canonical homomorphism

$$G(\Omega_{\mathfrak{p}}) \otimes_B \Big(B_{\mathfrak{p}} \Big/ \Big(\sum_i f_i B_{\mathfrak{p}}\Big)\Big) \longrightarrow G\Big(\Omega_{\mathfrak{p}} \otimes_B \Big(B_{\mathfrak{p}} \Big/ \sum_i f_i B_{\mathfrak{p}}\Big)\Big)$$

is an isomorphism.

(NB This condition is anyhow necessary, even if A is not a quotient of a regular ring, and in the stronger form where one does not assume the f_i to be a whole system of parameters).

The main application I had in mind was in the following situation: let $f \colon X \to Y$ be a proper morphism (Y locally noetherian), $\mathscr{E}$, $\mathscr{F}$ two coherent modules on X, with $\mathscr{F}$ flat with respect to Y, and consider the functor

$$\mathscr{M} \rightsquigarrow \mathrm{Hom}_{\mathscr{O}_X}(\mathscr{E}, \mathscr{F} \otimes_Y \mathscr{M}).$$

This functor is representable by a coherent sheaf $\mathscr{P}$ over Y ($\mathscr{M}$ is a variable quasi-coherent sheaf on Y). Same is true for $\mathrm{Ext}^i_{\mathscr{O}_X}(X; \mathscr{E}, \mathscr{F} \otimes \mathscr{M})$, assuming for simplicity Y affine, provided $\mathrm{Ext}^{i-1}_{\mathcal{O}_X}(X;\mathcal{E},\mathcal{F}\otimes\mathcal{M})$ is identically zero (to ensure left exactness). In the Harvard seminar I gave another simpler proof, valid only if $\mathcal{E}$ was a cokernel of a homomorphism of locally free sheaves, for instance for f projective; I was unable then to deal with the general case.

As a consequence of the previous statement it follows that, for $f\colon X \to Y$ proper and flat and $g\colon Z \to X$ affine, the prescheme $\prod_{X/Y} Z/X$ exists and is affine over Y, and of finite type over Y if g is of finite type. An important particular case is the one when g is a closed immersion, i.e., Z is a closed subscheme of X, then we get a closed subscheme of Y, whose points are the points of Y the fibers of which are majorized by Z. In the general application I stated the proof of condition 5b) is not quite trivial, and (apart from standard constructibility considerations) uses some local duality theory!

By a very analogous technique, just a little more delicate, I was able also to give a general characterization of those functors $(\mathrm{Sch}/S)^\circ \to (\mathrm{Ens})$ which are representable by S-preschemes X which are *locally quasi-finite* and *separated* over S. This criterion becomes especially handy in the important case when we want X to be not only locally quasi-finite, but locally nonramified (for instance a monomorphism). In all cases which I have looked at, when I expected to find such a representability, I have been able to prove it by this general criterion, of course independently of any projectiveness assumption. For instance for correspondence classes, Néron–Severi schemes when they are likely to exist etc. In particular, for any abelian scheme I can construct the Néron–Severi scheme, for any two abelian schemes also $\mathcal{H}om_{S\text{-gr}}(A,B)$ and the scheme $\mathcal{C}orr_S(A,B)$ of correspondence classes, etc. As an application, if A is any abelian scheme over S locally noetherian and *geometrically unibranch*, then A is globally projective over S. Another application is to the "flattening functor" you discussed once about with Hartshorne, corresponding to a given proper morphism $f\colon X \to Y$ and a coherent sheaf $\mathcal{F}$ on X, which we want to "make flat over Y". As a consequence, as I once mentioned to you, we get that, for any proper scheme X over an integral noetherian S, there exists a nonempty open set U in S such that the Picard scheme of $(X|_U)/U$ exists, and has various good extra properties such as $\mathcal{P}ic^\tau$ being both open and closed in $\mathcal{P}ic$, flat over the base etc.[46] However, this technique does not seem to give the case when S is not integral, even assuming the Picard scheme over the local ring of the generic point exists. Anyhow I do not intend to investigate this any further, as I have started at last working on Weil cohomology and this keeps me busy enough. I got some satisfactory results, including good behaviour of cohomology under specialisation, and I am quite optimistic about cohomology being ready-to-use within the next one or two years.

I got a few byproducts about birational transformations, and I wonder if these are known. Let $f\colon X \to Y$ be proper birational, X and Y regular schemes. Then $H^2(Y,\mathbb{G}_m) \to H^2(X,\mathbb{G}_m)$ is bijective (birational invariance of the extended Brauer group)—at least, for the time being, if everything is of finite type over a field, or simple over the integers (these restrictions are certainly superfluous, and will be eliminated with the solution of some pending local question in the case of a regular scheme and a regular divisor in it...). Of course $\pi_1(X) \to \pi_1(Y)$ is bijective

[46] Mumford's comment in the margin: "NOT assuming $f_*(\mathcal{O}_X) = \mathcal{O}_S$??" See also Footnote 25.

by purity, and from this I can deduce[47] that all geometric fibers Z are simply connected. Hence $\mathscr{P}ic^{\tau}(Z)$ is unipotent, and I believe is zero (this I checked if Z is of dimension 1). Using resolution of singularities in the very strong form of Hironaka, namely the fact that X can be dominated by X' deduced from Y by nice quadratic transformations, (available for "good" preschemes of characteristic 0), it follows also that $R^i f_*(A) = 0$ if i is odd, A any coefficient group prime to the residue characteristics, hence $H^i(Z) = 0$ for i odd for such coefficients. I wonder if you can check (I won't think, counter-examplify!) such things in characteristic $p > 0$? Besides, $H^2(Z, \mathbb{G}_m) = 0$ (perhaps assuming Z regular, I don't remember exactly), in any case $H^2(Z, \mu_n)$ is "algebraic" i.e., equal to the image of $H^1(Z, \mathbb{G}_m)$[48] (if n prime to the characteristic of the ground field for Z). —I begin to realize it would be extremely handy to have resolution for all "good" rings, as now seems reasonable; there are still various things I cannot prove without. However I got Mike's "key lemma"[49] about $A\{t\}$ without assumption on A, using local Lefschetz theory as expounded in my seminar of last year. In fact, about every technique I worked out so far seems to be needed to get the basic properties of cohomology of schemes in sufficient generality (and apparently, more will be needed still!).

Sincerely yours
(signed) A Grothendieck

Grothendieck to Michael Artin, 23 February, 1963

Bures 23.2.1963

My dear Mike,

I want to ask you a few questions and give some complements to my last letter.

1) What about Néron's Manuscript?

2) What about Lichtenbaum's notes of Grothendieck's and Mumford–Tate's seminars?[50]

---

47 A proof of the hoped-for purity result, for torsion prime to the residue characteristic, is due to O. Gabber, using methods in earlier work of R.W. Thomason, Absolute cohomological purity, *Bull. Soc. Math. France*, **112** (1984) 397–406. Gabber's proof can be found in K. Fujiwara, A proof of the absolute purity conjecture (after Gabber), *Algebraic Geometry 2000, Azumino (Hotaka)*, Adv. Stud. Pure Math. 36, Math. Soc. Japan, 2002, pp. 153–183. More information about the purity conjecture can be found in Gabber's 2004 Oberwolfach Report, *On purity for the Brauer group,* Report No. 37/2004, Mathematisches Forschungsinstitut Oberwolfach, Oberwolfach Reports 1 (2004), 1971–1973, European Math. Soc.

48 A typo in the original, "equal to the image of n", is corrected here.

49 Lemma 2.6 in SGA4 Exposé XV, on p. 196 of LNM 305.

50 This refers to the Mumford–Tate seminar in the spring of 1962. Lichtenbaum's notes on the lectures by Grothendieck, Mumford and Tate have not been published. See also Remark 9.4.18 in S.L. Kleiman, The Picard scheme, in *Fundamental Algebraic Geometry*, Amer. Math. Soc. 2005, pp. 235–321, where the contents of Mumford's personal folder for this seminar are described.

3) I feel very silly lately, as I am wondering if the following is not always true: Let X be a prescheme, X_0 a closed subscheme of X, $\mathscr{F}$ an injective torsion sheaf on X, then is $\mathscr{F}_0 = \mathscr{F}|_{X_0}$ injective, or at least is it true that $H^i(X_0,\mathscr{F}_0)=0$, $i>0$? The analogous statement for Zariski topologies is *false* anyhow, but the étale topology may resemble more to paracompact topologies! This would imply that, whenever $H^0(X,\mathscr{F}) \xrightarrow{\sim} H^0(X_0,\mathscr{F}_0)$ for every torsion-sheaf $\mathscr{F}$ (which, for X noetherian, simply means that for every X' finite over X and connected, nonempty, X'_0 is connected, nonempty), then $H^i(X,\mathscr{F}) \xrightarrow{\sim} H^i(X_0,\mathscr{F}_0)$, every i! Thus the 1°) of my previous letter would become evident (whereas my proof is a simple but nontrivial one, and uses far more than just the "connectedness theorem" as would be the case if the "conjecture" above were true). Moreover, it would give the analogous comparison theorem for the spectrum X of any noetherian ring A separated and complete for some I-adic topology, with $X_0 = \operatorname{Spec} A/I$, (which I have not proved as yet!)

4) I got a result on cohomological dimension of *affine* schemes, in a rather formal way from the statement 1) of my previous letter (in fact I need only that, for $X = \mathbb{P}^1_Y$, Y strictly local, and any torsion sheaf $\mathscr{F}$ on X, $H^i(X,\mathscr{F})=0$ for $i \geq 2$).

Theorem *Let $\mathscr{F}$ be a torsion sheaf on $\mathbb{A}^m_Y$ (Y strictly local, noeth. of dim. n), which is "zero in codimension $< d$," i.e., if $\bar{x}$ is a geom. point in codim. $< d$, then $\mathscr{F}_{\bar{x}}=0$. Then $H^i(\mathbb{A}^m_Y,\mathscr{F})=0$ if $i>m+n-d$,* provided *at least Y comes from a scheme of finite type over a noeth. ring of dim. ≤ 1 by "strict localization."*

Corollary 1 *Let X be a closed subscheme of $\mathbb{A}^m_Y$, of codimension $\geq d$. Then*

$$\operatorname{cd}(X) \leq n+m-d.$$

Corollary 2 *Let X be any affine scheme of finite type over Y, let a be the closed point of Y and $Y' = Y - \{a\}$, $X' = X|_{Y'}$, $X_0 = X_a = X - X'$, and $\nu = \sup(\dim X_0, \dim X'+1)$. Then*

$$\operatorname{cd}(X) \leq \nu.$$

Corollary 3 *Let X be an affine scheme of finite type over a field k, k sep. closed. Then*

$$\operatorname{cd}(X) \leq \dim X$$

(of course, in fact equality holds).

Corollary 4 *Let Y be as in the theorem, and U an affine open subset of Y (for instance $U = Y_f$, some f), then $\operatorname{cd}(U) \leq n$. (Take $m=0$ in the theorem).*

From Corollary 3 follows the

Lefschetz Theorem *Let X be projective over k sep. closed, Y a hyperplane section, Y and X regular. Then the natural map*

$$H^{i-2}(Y,\mathscr{F}_Y \otimes \check{\mathscr{T}}) \longrightarrow H^i(X,\mathscr{F})$$

is surjective if $i \geq n+1$, bijective if $i \geq n+2$, where $n = \dim X$. Here $\mathscr{F}$ is a locally free torsion sheaf of order prime to the char, and $\mathscr{T}$ the "Tate sheaf" prime to char. k.

This gives the result (which I understand from Tate you know already)

$$H^n(X, \mu_N^{\otimes n}) \simeq \mathbb{Z}/N\mathbb{Z}$$

(X projective, simple over k, connected, N prime to the char, $n = \dim X$), and once we have duality, by transposition, the usual statement

$$H^j(X, \mathscr{F}) \longrightarrow H^j(Y, \mathscr{F}_Y)$$

is a monomorphism if $j \leq n-1$, bijective if $j \leq n-2$.

N.B. I doubt not that the theorem is true for any Y, at least if Y is "good," for instance complete. I checked cor. 4 for Y_f if $\dim Y \leq 2$, any Y. Whenever cor. 4 is true, it implies the following for the field of fractions $K(Y)$ of Y, when Y is integral:

$$\operatorname{cd} K(Y) \leq n.$$

I wonder if this result can be proved directly in all cases.

5) The suggestion in the previous letter for a counterexample concerning the Brauer group is somewhat inaccurate, in various ways. Anyhow, Serre checked there is no hope to get a counterexample through topological obstructions, namely, for any finite complex X and any torsion element ξ in $H^3(X, \mathbb{Z})$, there exists a projective bundle on X (some n), whose obstruction is ξ. Thus $H^3(X, \mathbb{Z}) = H^2(X, \mathbb{C}^*)$ is really the "topological Brauer group" of X![51]

Besides, did you notice that the extended Brauer group $H^2(X, \mathbb{G}_m)$, for regular X, is invariant under proper birational morphisms, at least whenever the standard local theorem $\mathscr{H}^i_Y(\mathbb{G}_m) = \ldots$ is true (for instance X of finite type over a field...). Thus, at least for surfaces over a perfect field (when resolution is available), $\operatorname{Br}(X)$ is an invariant for the function field K (X a complete regular model).[52]

6) I tried again to prove that, for a "good" strictly local ring A, the fibers of $\operatorname{Spec}\widehat{A} \to \operatorname{Spec} A$ are acyclic (for coefficients prime to the residue char. of A), and simply connected (with same restriction on Galois groups). For the statement "simply connected" I can reduce, by local Lefschetz theory, to the case $\dim A = 2$, A normal, and to prove that thus any Galois covering of $\hat{A}$, unramified outside the origin, comes from a Galois covering of A. Thus we would be through if one could resolve singularities at least in dim. 2! I have the feeling that the dim. 2 case is really irreducible in a way, and demands some other methods than those I know.... I begin to respect dim. 2!

Sincerely yours,
(signed) A Grothendieck
i.e., Schurik

---

51 See Thm. 1.6 of Grothendieck, GB I, *Séminaire Bourbaki* 1964/65, n° 290.

52 Proofs of results in this paragraph appeared in section 7 of GB III. They can now be complemented using Gabber's purity theorem mentioned in footnote 47 for [1963Feb21] (p. 682).

7) P.S. It is not always true that $\mathrm{Br}(X) = H^2(X, \mathbb{G}_m)$, even if X is local and normal of dim. 2. In fact, if X is not regular, $H^2(X, \mathbb{G}_m)$ is not necessarily a torsion group. To see this, look at the resolution

$$0 \longrightarrow \mathbb{G}_{mX} \longrightarrow \mathscr{R}_X^* \longrightarrow \mathscr{D}_X \longrightarrow \mathscr{P}_X \longrightarrow 0$$

with $\mathscr{R}_X^*$ the sheaf of rational invertible functions, $\mathscr{D}_X$ the sheaf of Weil divisors, $\mathscr{P}_X$ the cokernel of $\mathscr{R}_X^* \to \mathscr{D}_X$, which can be called the "sheaf of (strictly) local divisor class groups." The cohomology of X in dim $\neq 0$ with coefficients in $\mathscr{R}_X^*$, $\mathscr{D}_X$ is torsion, hence *mod torsion* we have

$$H^i(X, \mathbb{G}_{mX}) \equiv H^{i-2}(X, \mathscr{P}_X), \qquad i \geq 3,$$
$$H^2(X, \mathbb{G}_{mX}) \equiv H^0(X, \mathscr{P}_X) / \mathrm{Im} H^0(X, \mathscr{D}_X).$$

Assume, for instance, X has an isolated singularity x, let $\mathscr{O}_x$ ($\mathscr{O}_{\bar{x}}$) be the local ring (resp. its strict henselization), thus

$$H^2(X, \mathbb{G}_m) = \mathrm{Cl}(\mathscr{O}_{\bar{x}}) / \mathrm{Im}\,\mathrm{Cl}(\mathscr{O}_x)$$

when Cl is the divisor class group. Now there are I believe examples of Mumford's where $\mathrm{Cl}(\mathscr{O}_x) = 0$, whereas $\mathrm{Cl}(\mathscr{O}_{\bar{x}}) \neq 0$, and even $\mathrm{Cl}(\mathscr{O}_{\bar{x}})$ a nontorsion group (you should check this point).[53] Hence the counterexample.

One last remark: for a complete, nonsingular surface X over an alg. closed field, we get an interpretation of $b_2 - \rho$ as the rank of the module of points of order n of $\mathrm{Br}(X)$, which is a free module over $\mathbb{Z}/n\mathbb{Z}$, where n is prime to the characteristic and to the torsion of the Néron–Severi group.

Best regards to Jean and the kids,

(signed) Your Schurik

Grothendieck to Mumford, 11 June, 1963

Bures, June 11, 1963

Dear Mumford,

I understand you published something (e.g., so-called "pathology"[54]), please send me a reprint; I hope you will put me on your general mailing list, if you have one.

Matsumura lately told me he proved representability of $\mathscr{A}ut_k(X)$, X proper over a field k, using Murre's method.[55] I then tested my criterion for representability of a

---

53 See [61a].

54 [61b] and [62a].

55 This result was published in H. Matsumura and F. Oort, Representability of group functors, and automorphisms of algebraic schemes, *Invent. Math.* 4 (1967) 1–25.

functor F/S by a scheme separated and unramified over S, I told you about, and got more general results. For instance if X and Y are locally of finite type over a field k, X proper over k, then there exists a finite separable extension k' of k, depending only on X, such that $\mathcal{H}om_{k'}(X_{k'}, Y_{k'})$ is representable; thus, if k is separably closed, $\mathcal{H}om_k(X,Y)$ is representable. The analogous result holds if X, Y are locally of finite type over S noetherian and integral, X being proper and flat over S, provided we restrict to some open nonempty subset of S. Besides, without assuming S integral, but say X with integral geometric fibers and admitting a section along which X is simple over S, $\mathcal{H}om_S(X,Y)$ is representable. The general result from which these can be easily deduced is as follows: Let X, Y, Z be locally of finite type over S, X and Y proper and flat, let $\phi : Z \to X$ be given (for instance, Z is a flat finite multisection of X/S), hence a homomorphism of functors

$$F = \mathcal{H}om_S(X,Y) \longrightarrow G = \mathcal{H}om_S(Z,Y);$$

consider the subfunctor $U = F_\phi$ of F where $F \to G$ is unramified, more precisely, its points with values in S'/S consist of those $u' : X' \to Y'$ such that, for every $s' \in S'$, the following map be injective:

$$\mathrm{Hom}_{\mathcal{O}_{X'_{s'}}}(u'^*_{s'}(\Omega^1_{Y'_{s'}}), \mathcal{O}_{X'_{s'}}) \longrightarrow \mathrm{Hom}_{\mathcal{O}_{Z'_{s'}}}(v'^*_{s'}(\Omega^1_{Y'_{s'}}), \mathcal{O}_{Z'_{s'}}),$$

where $v' = u'\phi'$. This functor U is an *open* subfunctor of F (and the idea is to exhaust F by such open subfunctors F_ϕ, with suitably large Z's). Look at the induced homomorphism

$$U = F_\phi \longrightarrow G;$$

the result is that *this homomorphism is representable by unramified separated morphisms*. As a consequence, if $G = \mathcal{H}om_S(Z,Y)$ is representable, so is F_ϕ.

It is possible that, for any X, Y over S as above, $\mathcal{H}om_S(X,Y)$ is representable *locally* for the flat quasi-finite topology; this can be checked (even for the étale topology) when X is simple over S, or only with separable fibers. A question which is not solved by the method is (in the case of a ground field) whether the connected components of $\mathcal{H}om_k(X,Y)$ are of finite type over k; I suspect not.

Murre is in Bures for one month now. He is trying to prove his general criterion of representability for group functors, dropping the commutativity condition, and the last two conditions (Rosenlicht's condition, and the separation axiom), by relying still more on the techniques of formal moduli and descent. The idea is to construct first the local ring of the generic point of the connected component of e, by the smallest ring of definition for the canonical point of the functor G with values in the function field of the formal group prorepresenting G at e, and then use an easy generalisation of Weil's theorem on group varieties to construct the whole connected component. But there are considerable technical difficulties involved, such as effectivity criteria for "birational" equivalence relations or "birational" descent data (everything in nonflat, nonfinite, cases). It would be quite a progress for the nonprojective construction techniques if Murre could overcome these difficulties, even if the final theorem on representation of group functors should not be of frequent use. Perhaps Murre will run a seminar on formal moduli in 64/65.

Anything new going on there? Here Bass proved a beautiful theorem using his K^1 functor of rings, namely for $SL(n,\mathbb{Z})$, $n \geq 3$, the topology of subgroups of finite index equals the topology of congruence subgroups.[56]

Sincerely yours
(signed) A Grothendieck

Grothendieck to Mumford, 5 August, 1963

5.8.1963

My dear Mumford,

Thank you for your letter. I am glad you had such an interesting time in Japan. However, for the time being I do not think of traveling myself, and probably my first long trip abroad will be to Harvard, in perhaps two years or three. I am sorry to hear Hironaka accepted a position at Columbia and will not be around in Cambridge any longer. Couldn't he get a comparable salary at Brandeis?

I knew in effect that it was possible to check Serre's conjecture on the tangent space of $\mathrm{Pic}^{\mathrm{red}}$ by duality, when I wrote the last formula of SGA1 XI; but I never actually did it.

Let X/k be proper, and for every $x_i \in \mathrm{Ass}\, \mathscr{O}_X$, let $y_i \in \bar{x}_i$, and let k_i be the separable algebraic closure of k in $k(y_i)$. Let k' be a field extension of k that splits all of the extensions k_i/k. Let Y be any locally algebraic scheme over k, then $\mathscr{H}om_{k'}(X_{k'}, Y_{k'})$ exists. As you see, there is no need for the geometric irreducible components of X/k to be defined over k'. The proof is easy by the general result I stated in my last letter. Besides, in all this, a more general and more convenient point of view is to abide with the functor $\prod_{X/k} Y/X$ when X proper over k, and Y/X given; everything I did applies to this situation. Besides, instead of assuming that k'/k splits the k_i/k, it would be enough to assume that $\prod_{k'_i/k'}(Y_{y_i})_{k'}$ are representable (where $k'_i = k_i \otimes_k k'$). For instance if the fibers Y_{y_i} are quasi-projective, then $\prod_{X/k} Y/X$ exists, and is in fact an increasing union of a sequence of quasi-projective open subsets.

I have been pretty busy for the last two weeks writing an outline for Hartshorne's seminar on residues and duality. It takes me longer than I thought it would to put things in a decent order, but I think I can begin typing in a few days and he will have most of it by the end of August.[57] Verdier has promised me to write an outline of

---

[56] Published in H. Bass, M. Lazard & J.-P. Serre, Sous-groupes d'indice fini dans $SL(n,\mathbb{Z})$, *Bull. Amer. Math. Soc.* **70** (1964) 385–392.

[57] This appeared in R. Hartshorne, *Residues and Duality*, LNM 20, Springer-Verlag, 1966. A compendium, in which some of the tricky points are worked out in greater detail, was published in B. Conrad, *Grothendieck Duality and Base Change*, LNM 1750, Springer-Verlag, 2000. An extensive account of the developments in the theory since Grothendieck's manuscript, written by J. Lipman and entitled *Notes on derived functors and Grothendieck duality*, will appear in: *Foundations of Grothendieck Duality for Diagrams of Schemes*, LNM 1960, Springer-Verlag, 2009, pp. 1–261.

those results on the foundations of homological algebra I told him I would need. I think, in order to get started, Hartshorne should take these foundations as granted, at least as far as proofs are concerned. Anyhow, as you know, it is planned that Verdier comes to Harvard in 64/65, and he will give probably some course or seminar on homological algebra and duality for topological spaces (the results being formally exactly parallel to duality for coherent modules on schemes).

As for me, I am running two joint-seminars next year, one with Demazure on group schemes, continuing the one of last year (the writing down of my own talks is far from finished and takes a long time!), one with Mike on étale cohomology. I hope we will have time to include duality and the application to *L*-functions over finite fields.

I will be interested to know what is going on in Harvard, and especially what you are doing yourself. Is there no resolution of singularities for good schemes in view?

Yours
(signed) A Grothendieck

Grothendieck to Mumford, 16 September, 1963

Bures Sept. 16, 1963

Dear Mumford,

Artin transmitted your question concerning passage to quotient in analytic spaces, to construct a Picard modular space as stated in my talk in Cartan's Seminar.[58] Looking back at it, I see I used without proof the following fact, (which I hope is not false!): if X/S is a projective flat scheme such that $\mathscr{O}_S \xrightarrow{\sim} f_*(\mathscr{O}_X)$ universally, looking at the modular space $\mathscr{F}$ for immersions $X \to \mathbb{P}^n_S$ of the special type considered *loc. cit.*, is it true that $PGL(n+1) \times \mathscr{F} \to \mathscr{F} \times \mathscr{F}$ is an *immersion*? (Or at least, when S is of finite type over $\mathbb{C}$, a homeomorphism into for the *usual* topologies). To prove it is an immersion is equivalent to the following: Assume $S = \operatorname{Spec} V$, V discrete valuation ring, and assume $i_1, i_2 \colon X \rightrightarrows \mathbb{P}^n_S$ are given, such that on the *two fibers*, i_{1s} and $i_{2s} \colon X_s \to \mathbb{P}^n_s$ $(s \in S)$ are conjugate, to prove i_1 and i_2 are conjugate under an element of $PGL(n+1)(V)$. Can you say something about this problem?[59]

Yours
(signed) A Grothendieck

---

[58] See *Séminaire Henri Cartan* 13, 1960/62, Exposé 16, Thm. 3.1 and its proof.

[59] Mumford's comment at the bottom margin:

$$\left.\begin{array}{ll}\text{Have} & \lambda\colon \mathbb{G}_m \longrightarrow PGL_{n+1},\ Z_i \subset \mathbb{P}^n,\ Z_i \longrightarrow Z. \\ \text{Assume} & \alpha_i \in \mathbb{G}_m \text{ s.t. } Z_i \hookrightarrow \mathbb{P}^n \xrightarrow{\lambda(\alpha_i)} \mathbb{P}^n \text{ approaches } Z \hookrightarrow \mathbb{P}^n. \\ \text{Say} & H^0(\mathscr{O}_Z) = k,\ H^1(\mathscr{O}_Z(1)) = (0),\ H^0(\mathscr{O}_{\mathbb{P}^n}(1)) \xrightarrow{\sim} H^0(\mathscr{O}_Z(1)). \end{array}\right\} \overset{?}{\Longrightarrow} \alpha_i \text{ have a limit}$$

Grothendieck to Mumford, 1963/4 undated[60]

Dear Mumford,

Thanks for your letter, and also for your manuscript on "Geometric Invariant Theory", which I have not quite read through yet. I noticed in Chap I some inaccuracies (for instance concerning questions of openness of morphisms), a detailed list would perhaps be tedious. I hope in the final draft the cross-references will be easier to find than in the one I read, where I did not always succeed to get the right reference. One remark on terminology: your use of the word "reductive" seems to me misleading, as it conflicts with the terminology generally adopted (which Demazure and I follow also in our seminar), couldn't you invent some other word? Apart from terminology, you seem always implicitly to assume your groups (at least the reductive ones) smooth over the ground field, without ever stating this. Strictly speaking, your definition of "reductive groups", in char $p > 0$, yields exactly the "multiplicative type groups" i.e., the duals of usual discrete comm. groups (if k alg. closed), including such groups as μ_p. Besides, I give a rather detailed study of these groups (from a point of view of course very different from yours, and over arbitrary ground schemes) in talks VIII to X of SGA3 (and you should get pretty soon talks VIII to XIV, which are being bound).

I was surprised to find the corollary you missed in EGA III 7 was not there.[61] In a way, we should have repeated as corollaries, in sections 7.7 and 7.8., whatever we did in the previous sections! But I agree the one you state is particularly useful, and should not have been forgotten.

The references you ask: [62]

a) $f\colon X \longrightarrow Y$ *open* finite type, X, Y irred., Y noeth.
$\Longrightarrow f$ equidimensional, IV 14.2.2.;
converse if Y is normal (or geom. unibranch) IV 14.4.4. (Chevalley)

b) $f\colon X \longrightarrow Y$ finite type, then $x \rightsquigarrow \dim_x f^{-1}f(x)$ is upper semicontinuous
IV 13.1.3. (Chevalley)

c) $f\colon X \longrightarrow Y$ finite type, X Cohen–Macaulay, Y regular, then f open $\Longleftrightarrow f$ flat
IV 15.4.2.

d) $X \xrightarrow{f} Y$, $g\colon X \to S$, $h\colon Y \to S$. Everything of finite pres., g, h flat, f_s flat $\Longrightarrow f$ flat on X_s
IV 17.... (I will have it more precise when Dieudonné is back from Japan with the manuscript!)

e) X reduced of finite type over $\mathbb{Z} \Longrightarrow \widetilde{X}$ finite over X
IV 7.7. (same remark as above). (Nagata)

---

[60] We are placing this letter according to its position in Mumford's file. [GIT] was written during the academic year 1962/63 at the IAS.

[61] The corollary in question was published by Mumford in [AV]; see the theorem towards the bottom of p. 46. The same corollary is found in R. Hartshorne, *Algebraic Geometry*, Springer-Verlag, 1977, Prop. III.12.2. Generalizations were studied in SGA6, and for applications to K-theory see R.W. Thomason and T. Trobaugh, Higher algebraic K-theory of schemes and of derived categories, *The Grothendieck Festschrift*, Vol. III, Progr. Math. 88, Birkhäuser, 1990, pp. 247–435.

[62] In this paragraph IV = EGA IV and III = EGA III.

f) $PGL(m+1)_S$ represents $\mathscr{A}ut_S(\mathbb{P}^m_S)$, will be in one of the later paragraphs of III which have not been written up. A sketch of the proof, using available references of Chap III, should take no more than half a page. The reason why I did not include it in III 4 was that I have to use the fact that $\mathrm{Pic}(\mathbb{P}^m_k)$ is the group generated by $\mathscr{O}(1)$, for which one needs that the local rings are UFD..., which we had not available there.

Besides, part 1 (out of 4) of Chap IV has just appeared, and you will have a reprint pretty soon. It contains only IV and IV 1, part 2 (already at the printer) contains paragraphs 2–7 (including the theory of "excellent local rings"), most of the references you need are in part 3, (paragraphs 8 to 15), which will be given to the printer within a month or two. Part one contains also the list of all paragraphs 1 to 21 (of which only the two last are still to be written in a publishable shape).

About your functor $M_g(\mathscr{S})$, I wonder what you mean by "ordinary *double* point", namely how are you to prevent (if you want the valuative criterion of properness) two double points to collapse to a triple point? It seemed to me one should allow multiple points of any order, but of "loose" type (as coordinate axes in n-space) so that $\mathrm{Pic}(C)$ does not acquire a unipotent component. But I confess I did not think this over seriously. By the way, Igusa seems to have a really beautiful *nonsingular projective* model in char 0 for compactifying the usual modular varieties with levels, which he has completely worked out for $g = 2$, but which according to him should generalize to all g, for principally polarized abelian varieties.[63]

I am sorry not to have heard anything before on Schlessinger's thesis,[64] which sounds interesting; but what you say about it is somewhat short for me to understand, especially what you mumble about the case $H^0(\mathfrak{g}) \neq 0$ and the smooth topology. If some day there is anything mimeographed or printed available I would appreciate getting a copy!

I did not prove anything noticeable in the last year, although I lately spent one month or two trying to prove Weil's conjectures. I have found lots of conjectures on algebraic cycles,[65] which I expect will keep me (and others perhaps) busy for quite a while. Mike will tell you about it next month I guess. I'll try again during the vacation to prove something along these lines, it seems time at last to know something at least on algebraic cycles which are not divisors.

Yours sincerely
(signed) A Grothendieck

---

[63] The generalization of Igusa's result mentioned here is known as the *toroidal compactification*; see [SC] (= *Smooth Compactification of Locally Symmetric Varieties*).

[64] Part of M. Schlessinger's 1964 Harvard Ph.D. thesis was published as M. Schlessinger, Functors of Artin rings, *Trans. Amer. Math. Soc.* **130** (1968) 208–222.

[65] See A. Grothendieck, Standard conjectures on algebraic cylces, in *Algebraic Geometry (Internat. Colloq., Tata Inst. Fund. Res., Bombay, 1968)*, Oxford Univ. Press, 1969, pp. 193–199.

Grothendieck to Mumford, 31 August, 1964

Bures 31.8.64

Dear Mumford,

I am just through reading Dieudonné's final version of EGA IV paragraphs 11 to 15. You may be interested in our final version of Chevalley's openness criterion, which reads as follows

Theorem 14.4.1 *Let $f\colon X \to Y$ be locally of finite presentation, $y \in Y$, x maximal in $X_y = f^{-1}(y)$, assume y geometrically unibranch on Y. The following conditions are equivalent:*

(a) *f is universally open at x (or, what trivially amounts to the same, at every point in the closure of x in X_y).*

(b) *If z is the maximal generisation of y, i.e., the generic point of the unique irreducible component Y_0 of Y through y, there exists an irreducible component Z of X, containing x and equidimensional over Y_0 at x, i.e., such that* $\dim_x Z_y = \dim Z_z$.

(b′) *For every open neighbourhood U of x in X, and every generisation y' of y,* $\dim U_{y'} \geq \dim_x U_y$ *holds.*

If Y is locally noetherian, these conditions are also equivalent to the following:

(c) *f is open at x.*

NB The equivalence of (b) and (b′) is about trivial, the essential part of the theorem being (b) ⇒ (a). This theorem gives as a corollary 14.4.2, equivalent conditions for f to be universally open at all points of X_y, namely through conditions (b), (b′) or (c) at the maximal points of X_y; however, it is easily seen that we may state these conditions as well at *all* points of X_y, and also state equivalently

(b″) $\dim U_z \geq \dim U_y$ for every open subset U of X.

As another consequence 14.4.8, we get a necessary and sufficient condition for a morphism locally of finite presentation $f\colon X \to Y$ to be universally open. (In fact, the criterion obtained is really a pointwise criterion on maximal points of fibers): if Y' is the normalisation of Y_{red}, it is necessary and sufficient that $X' = X \times_Y Y' \to Y'$ be open (universally so if Y is not supposed locally noetherian), which is equivalent also to either of the conditions in terms of dimensions seen above. It is very likely besides that everything holds without any reference to a noetherian condition, but we could not settle this point (and I guess you do not care anyhow).

I included also a proposition of yours as follows:

Proposition 14.5.10 *Let Y be noetherian, $f\colon X \to Y$ locally of finite type, surjective and universally open. Then there exists a finite surjective morphism $Y' \to Y$ such that $X' = X \times_Y Y'$ admits sections locally over Y'.*

As a corollary 14.5.11, we state the conclusion you had in mind, namely that if a morphism $Y_1 \to Y$, locally of finite type, becomes affine after the base change

$X \to Y$, then it was affine before. We also give the analogous descent statement 14.5.12 for ampleness of an invertible sheaf on Y_1, relative to Y.

Thank you very much for your notes on surfaces,[66] which I looked through with pleasure. Here are a few comments and questions.

1° Page 8.1. you state the problem of the existence of a flattening stratification for a proper morphism $X \to S$ and a coherent sheaf $\mathscr{F}$ on X, S locally noetherian. Now this problem is practically solved, namely as I indicated to you in an old letter of mine, the flattening functor in this case is indeed representable by a prescheme S' of finite type over S, and a monomorphism $S' \to S$. The only question which remains (and I would rather guess the answer to be negative) is whether $S' \to S$ is a stratification; but this is, I believe, rather inessential for all applications. (NB as $S' \to S$ is a monomorphism it is automatically quasi-affine, a fortiori quasi-projective).

2° I appreciated very much your Chapter 14, particularly the theorem on page 14.4. I did not check through the details of your proof, but I guess your result holds true if, instead of taking sheaves of ideals, you take subsheaves, or equivalently and preferably for my taste, quotient sheaves, of a fixed coherent sheaf? In this form, your theorem is a significant amelioration of a finiteness theorem in my Bourbaki talk on Hilbert schemes, namely *loc. cit.* 2.1. (where instead of "il faut" one must of course read "il faut et suffit"). Now it would seem very likely that an analogous quantitative version should equally exist for *loc. cit.* 2.2. (where instead of $\leq s-1$ one should read $\geq s$, and instead of $s-2$ one should read $s-1$; in the reformulation 2.3 read s instead of $s-1$).[67] Namely that the limitedness of quotients $\mathscr{F}/\mathscr{H}_i = \mathscr{G}_i$, as expressed, say, by the twisting n_0 needed so that for $n \geq n_0$, both Serre's statements hold for $\mathscr{H}_i(n)$, can be estimated by a polynomial with respect to the coefficients of degree $\geq s$, if we restrict to quotients $\mathscr{G}_i$ such that the associated cycles are all of dimension $\geq s$ (the polynomial depending only on X, $\mathscr{F}$, $\mathscr{O}_X(1)$ and s). I wonder if you checked this variant of your theorem, which I would like to consider as a starting point for a systematic "quantitative" version of the standard finiteness theorems (as once Mike told me about, à propos making quantitative Noether's theorem of finiteness of integral closure).[68]

3° I liked also Bergman's Chap 26–27,[69] and especially his universal Witt scheme, realized as a formal power series functor. This meets with some old ponderings of mine on power series beginning with 1, on which I make some comments in my little paper on Chern classes (the appendix to the Serre–Borel paper). As I point out there this is not only a ring, but a λ-ring (and even a "special" λ-ring), on the other hand since Gabriel's seminar on formal groups I had the feeling that the Witt rings must also have a λ-structure (or something very close to it). Namely, according to Dieudonné–Cartier–Gabriel, certain algebras over $W_\infty(k)$ (the Witt vector ring over the perfect field k) allow to classify, either commutative formal groups without

[66] Published in [CAS] (= *Lectures on Curves on an Algebraic Surface*).

[67] Mumford wrote a question mark (?) in the left margin.

[68] Mumford wrote a question mark (?) in the left margin.

[69] "Bergman's Chap 26–27" appeared as Chap 26 in the published version.

toroidal part (one might call them ind-unipotent), or ordinary unipotent algebraic groups (the two classifications being in fact dual), in terms of modules over these algebras. Now, in the categories in question, one has not only a structure of abelian category, but also the notion of tensor product and consequently of exterior power. Now this extra structure should be reflected in some extra structure of the mentioned classifying algebra, and presumably, en dernière analyse, by W itself. This question should certainly be investigated some day, and perhaps Bergman has a good starting point. I guess, besides, you noticed that, analogously, the classifying space of the infinite unitary group of K-theory is not only a group in the hot-category, but actually a λ-ring, and the same remark applies to the orthogonal case, these facts reflecting simply the λ-ring structure of the K-functors.

4° The proof of the fundamental theorem, Chap 23, via Kodaira–Spencer, is not really different from the proof in Chap 25. This is still more striking if one has in mind Cartier's own proof of his theorem on smoothness of algebraic groups in char 0, or rather formal groups, which is precisely using the exponential, so I suspect that what Kodaira–Spencer do is just giving Cartier's proof. Besides, the proof you give of Cartier's theorem is also the one I intend to include in EGA; it goes further for it proves also that if G is a group scheme locally of finite type over a noetherian ground scheme S say, such that S is of char 0 i.e., lies over Spec($\mathbb{Q}$), and that the sheaf $\omega_{G/S} = \mathscr{I}/\mathscr{I}^2$ ($\mathscr{I}$ is the augmentation ideal on G coming from the unit section) is locally free on S, then G is smooth over S along the unit section (and therefore on the connected components of the identity of the fibers). This can be applied for instance to Picard schemes, where we have a direct construction of $\omega_{G/S}$ and where the assumption of local freeness of the latter just means that X/S is "cohomologically flat in dimension 1" i.e., what amounts here to the same, satisfies the base change property for $R^1f_*(\mathscr{O}_X)$ (f is assumed flat, proper, with $f_*(\mathscr{O}_X) = \mathscr{O}_Y$ universally). One remark which could have been made already in Chap 23 is that Kodaira–Spencer's theorem is valid as stated there, simply replacing the char 0 assumption by the assumption that $\mathscr{P}ic_{X/k}$ is smooth. It is not clear why in Chap 25 you feel obliged to give a weaker statement of the theorem, assuming $H^1(X,\mathscr{L}) = 0$.

The result and proof in Chap 28[70] is really very nice and elegant. Suggestion for a thesis: give a version of this theorem as a criterion for smoothness along the unit section of the Picard scheme (or Picard proscheme) over an arbitrary base (or an artin base, which amounts to the same). The right cohomological operations, replacing Serre's Bockstein operations when the base is not of a given char $p > 0$, will clearly be the ones arising from the formal powers series scheme, and they would seem to deserve more study.

5° I was interested by your numerical result on page 17.7. Do you have an analogous result for higher dimensional varieties? This reminds me also of some positivity questions I once discussed with Mike, which he promised he would tell you about, (but I am not sure he kept his promise!). First take a surface (say projective nonsingular), and the vector space over $\mathbb{Q}$ defined through numerical equivalence of divisors, in this space we have the closed cone Q defined by the quadratic form,

[70] Chap 27 in the published version.

proscheme) over an arbitrary base (or an artin base, which amounts to the same). The right cohomological operations, replacing Serre's Bockstein operations when the base is not of a given car $p > 0$, will clearly be the ones arising from the formal powers series scheme, and they would seem to deserve more study.

5º) I was interested by your numerical results on page 17.7. Do you have analoguous results for higher dimensional varieties? This reminds me also of some positivity questions I once discussed with Mike, which he promised he would tell you about, (but I am not sure he kept promise !). First take a surface (say projective non singular), and the vector space over $\mathbb{Q}$ defined through numerical equivalence of divisors, in this space we have the closed cone Q defined by the quadratic form, restricting to the part of degree $\geqslant 0$, the closed cone $P \subset Q$ generated by ample sheaves, whose interior consist exactly of elements having a positive integral multiple defined by an ample sheaf, and the closed cone $\overline{R}$ defined by postive divisors, genererating the cone R. By prop. on page 18.1. $\overset{\circ}{Q} \subset R$ hence $Q \subset \overline{R}$, moreover $P \subset Q$. Using this and Nakai's criterion for ampleness, one finds that $P = R^{o}$, the polar of R or $\overline{R}$, hence $\overline{R} = P^{o}$. This however gives not Nakai's result, it suggests the following: if a divisor D is such that for every divisor $C > 0$, onde has $C.D > 0$, is it true that D is ample? A priori, we know only that it follows $D^2 \geqslant 0$, and by Nakai this is almost what is needed to imply ampleness, namely we need $D^2 > 0$. Do you know the answer ? More generally, if X is projective smooth of any dimension, D a divisor, and if $D.C > 0$ for any curve on X, is it true that D is ample ? Is it true at least that D is in the closure P of the cone defined by ample divisors, which would then

and, restricting to the part of degree ≥ 0, the closed cone $P \subset Q$ generated by ample sheaves, whose interior consist exactly of elements having a positive integral multiple defined by an ample sheaf, and the closed cone $\overline{R}$ defined by positive divisors, generating the cone R. By the prop. on page 18.1 $Q^\circ \subset R$, hence $Q \subset \overline{R}$, moreover $P \subset Q$. Using this and Nakai's criterion for ampleness, one finds that $P = R^\circ$, the polar of R or $\overline{R}$, hence $\overline{R} = P^\circ$. This however gives not Nakai's result, but it suggests the following: if a divisor D is such that, for every divisor $C > 0$, one has $C \cdot D > 0$, is it true that D is ample? A priori, we know only that it follows $D^2 \geq 0$, and by Nakai this is almost what is needed to imply ampleness, namely we need $D^2 > 0$. Do you know the answer?[71] More generally, if X is projective smooth of any dimension, D a divisor, and if $D \cdot C > 0$ for any curve on X, is it true that D is ample? Is it true at least that D is in the closure P of the cone defined by ample divisors, which would then imply $D^p Z_p \geq 0$, if Z is a subvariety of X of dimension p. This weaker statement is certainly true on a surface, what about a threefold?[72]

I would like even a lot more to be true, namely the existence of a numerical theory of ampleness for cycles of any dimension. Assume for simplicity X projective nonsingular connected of dim. n, let $A^i(X)$ be the vector space over $\mathbb{Q}$ deduced from numerical equivalence for cycles of codimension i (presumably this is of finite dimension over $\mathbb{Q}$), and $A_i(X) = A^{n-i}(X)$ defined by cycles of dimension i, presumably A_i and A^i are dual to each other. Let A_i^+ be the closed cone generated by positive cycles, and let $P^i \subset A^i$ be the polar cone. The elements of P^i might be called pseudo-ample, those in the interior of P^i ample (which for $i = 1$ would check with the notion of ample divisor, if for instance the strengthening of Mumford–Nakai's conjecture considered above is valid). The strongest in this direction I would like to conjecture is that the intersection of pseudo-ample (resp. ample) cycles is again pseudo-ample (ample), thus the intersection defines

$$P^i \times P^j \to P^{i+j}.$$

If i and j are complementary, $i + j = n$, this also means that the natural map $u_i : A^i \to A_{n-i}$ maps P^i into A_{n-i}^+ (and one certainly expects an ample cycle to be at least

[71] Mumford wrote a "NO" in the margin. A counterexample is in the next letter; see Footnote 74.

[72] Mumford wrote a question mark (?) in the margin. This conjecture turned out to be true; a generalization to all dimensions (not just threefolds) was proved by S.L. Kleiman, Toward a numerical theory of ampleness, *Ann. Math.* **84** (1966) 293–344. Given an irreducible projective variety one says, in current terminology, that a divisor D on X is *nef* (or *numerically effective*) if $(D \cdot C) \geq 0$ for every irreducible curve C on X. These generate a closed cone $\mathrm{Nef}(X) \subseteq N^1(X)_{\mathbb{R}}$ in the finite-dimensional real vector space of numberical equivalence classes of divisors. Kleiman's theorem is that the interior of $\mathrm{Nef}(X)$ is exactly the open cone of ample divisor classes. In particular, as Grothendieck suggests, if $(D \cdot C) > 0$ for all curves C—or even if $(D \cdot C) \geq 0$ for all C—then D lies in the closure of the ample cone of X. (However Mumford's example shows that these do not imply that D is ample.) For a detailed discussion of these matters see Chapter 1.4 of R.K. Lazarsfeld, *Positivity in Algebraic Geometry. I*, Springer-Verlag, 2004; Kleiman's theorem appears there as Theorem 1.4.8 on page 44. It is equivalent to the assertion that the cone of nef divisors is dual to the closed cone $\overline{NE}(X) \subseteq N_1(X)_{\mathbb{R}}$ generated by all effective curves on X. Mori discovered that parts of this cone of curves have a remarkable special structure, and this plays a central role in the minimal model program. See Chapters 1–3 of J. Kollár and S. Mori, *Birational geometry of algebraic varieties*, Cambridge Tracts in Mathematics 134, Cambridge University Press, 1998.

equivalent to a positive one!). For i and j arbitrary, the above inclusion can also be interpreted as meaning that the intersection of an ample cycle with a positive cycle is again (equivalent to) a positive cycle. Of course, one would expect an ample positive cycle to move a lot within its equivalence class, allowing to consider proper intersections with another given positive cycle. I wonder if you have any material against, or in favor of, these conjectures?

I am busy right now, granting Weil's conjectures (via the Lefschetz and Hodge type statements for algebraic cycles) plus Tate's, to get the right feel for what should replace the *rational* cohomology of schemes (there is certainly also something like an integral cohomology, but this is too sharp for the time being), namely to define the right category of "sheaves" and their basic properties.[73] One striking fact, which is certainly true, is that for a scheme X of finite type over the integers, taking l-adic sheaves over X (l prime to the residue char.) arising through any simple "geometric" construction (as higher direct images of $\mathbb{Q}_l$ etc), say any of the Tate sheaves $\mathbb{Q}_l(n)$, the cohomology modules $H^i(X,\mathbb{Q}_l(n))$ for variable l are canonically isomorphic to some $H^i(X,\mathbb{Q}(n)) \otimes_{\mathbb{Q}} \mathbb{Q}_l$, where $H^i(X,\mathbb{Q}(n))$ is a certain vector space of finite dimension over $\mathbb{Q}$. Vaguely speaking, this is the (common) subspace of the elements of $H^i(X,\mathbb{Q}_l(n))$ which can be constructed "in terms of algebraic cycles".... The philosophy is here that in a way, for a scheme of finite type over $\operatorname{Spec}\mathbb{Z}$, the whole of its cohomology is "algebraic" i.e., has direct arithmetic significance. For the time being unfortunately, nothing new concerning the proofs of the basic conjectures. All I did was to construct "intermediate jacobians" in terms of cycles algebraically equivalent to zero, the necessary majorization for the construction coming from the l-adic Betti-numbers. But except for the definition, Ind duality for complementary (to $\dim X - 1$) dimensions, which is practically part of the definition, I have no result concerning these abelian varieties. (NB in the classical case, they correspond just to a small piece of Weil's intermediate jacobians).

Sincerely yours
(signed) A Grothendieck

Mumford to Grothendieck, 1964 undated

Dear Grothendieck,

Thanks for the long and *very* interesting letter. I've been thinking off and on for the last 2 weeks on some of your questions.

(I)[74] There is a surface F, with divisor D such that

$(D^2) = 0$
$(D \cdot C) > 0$, all positive divisors C.

---

[73] Mumford wrote a question mark (?) in the margin.

[74] A counterexample to a question raised by Grothendieck in the previous letter. See Footnote 71.

Proof. The idea is to take char $= 0$ and F to be a "generic" ruled surface over a curve Γ of sufficiently high genus g, of *even* type (i.e., $F = \mathbb{P}(\mathscr{E})$ where $\deg c_1(\mathscr{E})$ is even). Then, mod algebraic equivalence, $\mathrm{Pic}(F)$ is generated by

f—the fibre of the ruling
E—any "unisecant", i.e., cross-section of the ruling.

Then one may as well replace E by $E - kf$ so that $(E^2) = 0$. This E is the example.

One has to check $(E \cdot C) > 0$, all positive *irreducible* C.
The idea is this: say C is algebraically equivalent to $aE + bf$, hence $(E \cdot C) = b$; also either $a = 0 \Longrightarrow C = f$ or $a > 0 \Longrightarrow C$ an a-fold covering of Γ.

1° K (the canonical class) is algebraically equivalent to $-2E + (2g-2)f$. Then

$$2p_a(C) - 2 = (C + K) \cdot C \geq a(2g-2)$$

since C is an a-fold covering of Γ. This implies immediately that $b \geq 0$ or $a = 1$.
2° If $a = 1$, $b \leq 0$, then you have my "un-stable" ruled surfaces which depend on only $2g - 1$ parameters (as opposed to the $3g - 3$ moduli for generic ruled surfaces over Γ).
3° If $a > 1$, $b = 0$, then F contains a curve C which is an *unramified* a-fold covering of Γ.
3° (i) $a = 2$. Then we get a diagram

$$\begin{array}{ccc} & & F \times_\Gamma C \\ & \swarrow & \downarrow \\ F & & C \\ \downarrow & \swarrow & \\ \Gamma & & \end{array}$$

and $F \times_\Gamma C \cong \mathbb{P}(\mathscr{L}_1 \oplus \mathscr{L}_2)$ over C and $F = F \times_\Gamma C/\mathbb{Z}_2$. It is easy to check that there are very few of these ruled surfaces.
3° (ii) $a > 2$. Let $\widetilde{C}/C$ be a further unramified covering s.t. $\widetilde{C}/\Gamma$ is Galois with group π. Let $\widetilde{F} = F \times_\Gamma \widetilde{C}$. Then $\widetilde{F}$ has ≥ 3 disjoint sections, so $\widetilde{F} \cong \mathbb{P}^1 \times \widetilde{C}$, so

$$F = \mathbb{P}^1 \times \widetilde{C}/\pi.$$

Also the inverse image of $C \subset F$ in $\widetilde{F}$ is a set of ≥ 3 sections $\{a_i\} \times \widetilde{C} \subset \mathbb{P}^1 \times \widetilde{C}$ permuted by π. Therefore π acts on $\mathbb{P}^1 \times \widetilde{C}$ by a product of some action on $\mathbb{P}^1$ with the given action on $\widetilde{C}$; there are only a few such, of course.

(II) **Re** limited families of sheaves of ideals $\mathscr{I} \subset \mathscr{O}_{\mathbb{P}^n}$ (the generalization to subsheaves of any $\mathscr{F}$ is easy by the way). Let $\mathscr{O}/\mathscr{I} = \mathscr{F}$. Let

$$P(m) = \chi(\mathscr{F}(m)) = \sum_{i=0}^{n} a_i \binom{m}{i}.$$

Then we know that determining $a_0, \ldots, a_n$ puts $\mathscr{I}$ in a limited family. Suppose you go further and assume $\mathscr{F}$ has no 0-dimensional associated cycles. Fixing $a_1, \ldots, a_n$

does $\boxed{not}$ put $\mathscr{I}$ in a limited family. The example is below. I notice that you don't prove that in Exposé 221 either (so there is no contradiction). (For $\leq s-1$ in 2.2, write $\geq s-1$ and that seems to be what you proved). *BUT*:

If $\mathscr{F}$ has neither 0 nor 1-dimensional associated cycles, then we're ok. In fact:

Theorem. *Look at the set of all coherent sheaves $\mathscr{F}$ on $\mathbb{P}^n$ satisfying*

i) $\chi(\mathscr{F}(m)) - \chi(\mathscr{F}(0)) =$ *given* $P(m)$,

ii) *$\mathscr{F}$ has no 0 or 1-dimensional assoc. cycles,*

iii) $H^i(\mathscr{F}(m)) = (0)$, $i > 0$, $m \geq n_0$.

Then there is a polynomial f_n in n_0 and the coefficients $a_1, \ldots, a_n$ of P such that

$$|\chi(\mathscr{F})| \leq f_n(n_0; a_1, \ldots, a_n).$$

Cor. *If $\mathscr{F} = \mathscr{O}_{\mathbb{P}^n}/\mathscr{I}$, then by my arguments in Ch. 14, one verifies (iii) for an n_0 depending polynomially on $a_1, \ldots, a_n$* (n.b. *control of H^2 for $\mathscr{I}$ is the same as control of H^1 for $\mathscr{F}$). Hence one gets a polynomial $g_n(|a_1|, \ldots, |a_n|)$ s.t. $m \geq g_n(|a_1|, \ldots, |a_n|) \Longrightarrow \mathscr{I}$ is m-regular.*

Sketch of Proof:

Lemma 1. *Let the coherent sheaf $\mathscr{F}$ on $\mathbb{P}^n$ be n_0-regular, and let*

$$\chi(\mathscr{F}(m)) = \sum_{i=0}^{n} a_i \binom{m}{i}.$$

Then

$$\dim H^i(\mathscr{F}(n_0 - \ell)) \leq \binom{\ell - 1}{i} \cdot \sum_{j=i}^{n} a_j \binom{n_0 - i - 1}{j - i}$$
$$= g_{i,n}(\ell, n_0, a_i, \ldots, a_n)$$

if $\ell \geq 1$, $0 \leq i \leq n$.

Now, for any $\mathscr{F}$, put $\mathscr{D}^i(\mathscr{F}) = \mathscr{E}xt^i_{\mathscr{O}_{\mathbb{P}^n}}(\mathscr{F}, \Omega^n_{\mathbb{P}^n})$. One checks that if H is a "good" hyperplane, then

$$\mathscr{D}^i(\mathscr{F}) \otimes \mathscr{O}_H = \mathscr{E}xt^i_{\mathscr{O}_H}(\underbrace{\mathscr{F} \otimes \mathscr{O}_H}_{\mathscr{F}_H}, \Omega^{n-1}_H)(-1).$$

Call this $\mathscr{D}^i(\mathscr{F}_H)(-1)$.

GENERALITIES:

1° $W_i = \operatorname{supp} \mathscr{D}^i(\mathscr{F})$ has codim [at least] i, dim [at most] $n-i$.

2° For $m \gg 0$, $\dim H^0(\mathscr{D}^i(\mathscr{F})(m)) = \dim H^{n-i}(\mathscr{F}(-m))$.

3° $(n-i)$-dimensional components of W_i are exactly the $(n-i)$-dimensional associated prime cycles of $\mathscr{F}$.

(∗) *Now say $\mathscr{D}^i(\mathscr{F}_H)$ is n_1-regular, $i = 0, 1, \ldots, i_0$.*

Then $\mathscr{D}^i(\mathscr{F}) \otimes \mathscr{O}_H$ is $(n_1 + 1)$-regular, $0 \leq i \leq i_0$.

Sketch of Proof:

<u>Lemma 1</u> Let the coherent sheaf $\mathcal{J}$ on $\mathbb{P}_n$ be n_0-regular, and let

$$\chi(\mathcal{J}(m)) = \sum_{i=0}^{n} a_i \binom{m}{i}.$$

Then

$$\dim H^i(\mathcal{J}(n_0-\ell)) \leq \binom{\ell-1}{i} \cdot \underbrace{\sum_{j=i}^{n} a_j \binom{n_0-i-1}{j-i}}_{g_{i,n}(\ell, n_0, a_i, \dots, a_n)}$$

[if $\underline{\ell \geq 1}$, $0 \leq i \leq n$.]

Now, for any $\mathcal{J}$, put $\mathcal{D}^i(\mathcal{J}) = \mathcal{E}xt^i_{\mathcal{O}_{\mathbb{P}_n}}(\mathcal{J}, \Omega^n_{\mathbb{P}_n})$.
One checks that if H is a "good" hyperplane, then

$$\mathcal{D}^i(\mathcal{J}) \otimes \mathcal{O}_H = \mathcal{E}xt^i_{\mathcal{O}_H}(\underbrace{\mathcal{J}\otimes\mathcal{O}_H}_{\mathcal{J}_H}, \Omega^{n-1}_H)(-1)$$

$\uparrow$ call this $\mathcal{D}^i(\mathcal{J}_H)(-1)$.

(*) <u>Now say $\mathcal{D}^i(\mathcal{J}_H)$ is n_1-regular, $i = 0, 1, \dots, i_0$</u>

Then $\mathcal{D}^i(\mathcal{J}) \otimes \mathcal{O}_H$ is n_1+1-regular, $0 \leq i \leq i_0$.

$\therefore$ (as in Ch. 14)

$$H^j(\mathcal{D}^i(\mathcal{J})(m)) = (0), \quad j \geq 2, \quad m+j \geq n_1+1$$

Look at the Sp-Seq. of duality:

*	*	*	*	...
$H^0(\mathcal{D}^{i_0+1})$	$H^1(\mathcal{D}^{i_0+1})$	$H^2(\mathcal{D}^{i_0+1})$	*	...
$H^0(\mathcal{D}^{i_0})$	$H^1(\mathcal{D}^{i_0})$	0	0	...
$\vdots$	$\vdots$			
$H^0(\mathcal{D}^0)$	$H^1(\mathcal{D}^0)$	0	0	...

$m \geq n_1 - 1$

all transgr. are 0 here, not killed either.

$\therefore$ (as in Ch. 14)

$$H^j(\mathscr{D}^i(\mathscr{F})(m)) = (0), \qquad j \geq 2,\ m + j \geq n_1 + 1.$$

Look at the spectral sequence of duality [here $\mathscr{D}^i = \mathscr{D}^i(\mathscr{F})(m)$]:

$$\begin{array}{|cccc}
* & * & * & * \cdots \\
H^0(\mathscr{D}^{i_0+1}) & H^1(\mathscr{D}^{i_0+1}) & H^2(\mathscr{D}^{i_0+1}) & * \cdots \\
H^0(\mathscr{D}^{i_0}) & \boxed{H^1(\mathscr{D}^{i_0})} & 0 & 0 \cdots \\
\vdots & \vdots & \vdots & \vdots \\
H^0(\mathscr{D}^0) & \boxed{H^1(\mathscr{D}^0)} & 0 & 0 \cdots \\
\hline
 & \uparrow & &
\end{array} \qquad m \geq n_1 - 1$$

all transgressions are 0 here
not killed either.

Therefore

$$\begin{aligned}
\dim H^1(\mathscr{D}^i(\mathscr{F})(m)) &\leq \text{dim of } (i+1)^{\text{st}} \text{ term of abutment} \\
&= \dim H^{n-i-1}(\mathscr{F}(-m)) \\
&\leq g_{n-i-1,n}(n_0 + m, n_0, a_{n-i-1}, \ldots, a_n)
\end{aligned}$$

by lemma 1. Hence as in Ch. 14, we get estimate:

$\mathscr{D}^i(\mathscr{F})$ is n_2-regular, where
$n_2 = n_1 + 1 + g_{n-i-1,n}(n_0 + n_1, n_0, a_{n-i-1}, \ldots, a_n)$.

Prop. *Using induction, we prove that there are polynomials $G_{n,i}$ such that:*

$\mathscr{D}^0(\mathscr{F})$ *is $G_{n,0}(n_0, a_{n-1}, a_n)$-regular,*
$\mathscr{D}^1(\mathscr{F})$ *is $G_{n,1}(n_0, a_{n-2}, a_{n-1}, a_n)$-regular,*
....
$\mathscr{D}^{n-2}(\mathscr{F})$ *is $G_{n,n-2}(n_0, a_1, \ldots, a_n)$-regular,*
$\mathscr{D}^{n-1}(\mathscr{F})$ *is $G_{n,n-1}(n_0, a_0, a_1, \ldots, a_n)$-regular,*
$\mathscr{D}^{n}(\mathscr{F})$ *is $G_{n,n}(n_0, a_0, a_1, \ldots, a_n)$-regular.*

Now apply this to our case: by hypothesis (ii), get

$$\mathscr{D}^n = (0), \text{ and } \mathscr{D}^{n-1} \text{ has 0-dimensional support.}$$

Therefore, $\exists\, G(n_0, a_1, \ldots, a_n)$ s.t. *all* $\mathscr{D}^i$ are $G(n_0, a_1, \ldots, a_n)$-regular. Hence, if $m \geq G(n_0, a_1, \ldots, a_n)$

$$\dim H^0(\mathscr{F}(-m)) \leq \sum_{i=0}^{n} \dim H^{n-i}(\mathscr{D}^i(\mathscr{F})(m)) = 0.$$

$$\therefore\ |\chi(\mathscr{F}(-m))| \leq \sum_{i=1}^{n} \dim H^i(\mathscr{F}(-m)) \leq h(m, n_0, a_1, \ldots, a_n). \qquad \text{QED}$$

Example:

Family of 1-dimensional subschemes $X_n \subset \mathbb{P}^3$ s.t.

(a) X_n has no 0-dimensional component
(b) $\deg(X_n) = 2$, all n
(c) $\{X_n\}$ not limited, esp. $\dim H^0(\mathcal{O}_{X_n}) \to +\infty$ in n.

Proof: X_n is to be double line: 2ℓ.

Let line ℓ be $x = y = 0$, where x, y, u, v are homogeneous coordinates. Let F_n be a surface $xf_n(u,v) + yg_n(u,v) = 0$, where f_n, g_n homog. of degree $n-1$, no common linear factor. Then F_n has degree n, $F_n \supset \ell$, and F_n is nonsingular along ℓ. Let X_n be the Cartier divisor on F_n, 2ℓ; i.e., ideal is $(x^2, xy, y^2, xf_n + yg_n)$. Then in fact:

$$0 \longrightarrow \mathcal{I} \longrightarrow \mathcal{O}_{X_n} \longrightarrow \mathcal{O}_\ell \longrightarrow 0,$$

where $\mathcal{I}$ is the invertible sheaf $\mathcal{O}_\ell(n-2)$.

$$\therefore \dim H^0(\mathcal{O}_{X_n}) = n.$$

(III) Apropos of:[75]

$$\left.\begin{array}{c} V \text{ a n.s. projective 3-fold} \\ D \text{ a divisor on } V \\ (D\cdot\gamma) \geq 0 \text{ for all curve } \gamma \subset V \\ \Downarrow ? \\ (D^3) \geq 0 \end{array}\right\} \quad (1)$$

or of the *stronger* conjecture:

$$\left.\begin{array}{c} V \text{ a n.s. projective 3-fold} \\ D \text{ a divisor on } V \\ \forall \text{ surfaces } F \subset V, \text{ assume that the} \\ \text{divisor class } (D\cdot F) \text{ on } F \text{ is ample} \\ \Downarrow ? \\ D \text{ ample on } V. \end{array}\right\} \quad (2)$$

☞ I don't know anything about their validity. However, they don't look impossibly difficult. Maybe the fellow Kleiman here will have an idea.

---

75 Of the questions raised below, we now know that III (1) is true; see Footnote 72. Aside from III (1) and III (2) Mumford also wrote the following in the margin near this line in his copy of this letter:

$$\left.\begin{array}{l} (D^3) > 0 \\ (D^2\cdot H) > 0 \\ (D\cdot H^2) > 0 \end{array}\right. \underset{?}{\Longrightarrow} \left.\begin{array}{c} |nD| \neq 0 \\ \text{for some } n \end{array}\right\}$$

I've been working out Néron's latest height paper for applications to Mordell's conjecture. Personally, Néron seems to me to make a real mess of his theory; but I am finally seeing what it amounts to. It looks as if

a) there is an intersection theory on absolute surfaces (i.e., regular 2-dimensional F, proper over $\mathrm{Spec}(\mathbb{Z})$), and the index theorem is still valid,
b) as a consequence, if $F \to \mathrm{Spec} R \to \mathrm{Spec}(\mathbb{Z})$ is Stein factorization, g (= genus of generic fibre over R) ≥ 2, and if there is an infinite sequence $x_1, x_2, \ldots$ of rational points (i.e., sections of $F/\mathrm{Spec}(R)$) then $\exists\, a > 0$, b real s.t.

$$\mathrm{ht}(x_i) \geq 2^{ai+b} \qquad \text{(additive height)}$$

or

$$\mathrm{ht}(x_i) \geq 2^{(2^{ai+b})} \qquad \text{(multiplicative height).}$$

Best wishes,
(signed) David Mumford

Grothendieck to Mumford, 19 December, 1964

Bures 19.12.1964

Dear Mumford,

Can you please tell me if you have a counterexample to the conjecture, say, that in the category of complex analytic spaces (or schemes over $\mathbb{C}\ldots$) the functor corresponding to the classification of projective nonsingular surfaces with polarization, without automorphisms, is representable. What if we do not put the polarization in the structure, working in $(\mathrm{Sch})/\mathbb{C}$ say, and still restricting to structures without automorphisms? All I definitely remember is that you made an example showing the modular space is not separated. Did you ever publish examples of that kind?

I take the opportunity to ask you if you know an example of an algebraic surface (proj. nonsingular) over $\mathbb{C}$, whose H^2 is spanned by algebraic cycles, which is not ruled? Or where moreover $H^1 = 0$, and which is not rational?[76] In fact, there are analogous question for varieties of arbitrary dimension....

Sincerely yours
(signed) A Grothendieck

[76] If we interpret H^2 and H^1 as the Betti cohomology with coefficients $\mathbb{Q}$, then any Enriques surface X gives an example Grothendieck asked for. Indeed we have $h^{2,0}(X) = h^{0,2}(X) = 0$, $h^{1,1}(X) = 10$, so $H^2(X(\mathbb{C}), \mathbb{Q})$ is spanned by fundamental classes of divisors. Moreover $\pi_1(X(\mathbb{C})) \cong \mathbb{Z}/2\mathbb{Z}$, and $H^1(X(\mathbb{C}), \mathbb{Z}) = (0)$.

Grothendieck to Mumford, 17 January, 1965

17.1.1965

Dear Mumford,

Thank you very much for your letter answering my questions concerning moduli. I am not sure I understand well what you mean though, and would appreciate some more precise information, if possible.

a) Let F be some open subfunctor of the following one G: $G(S)$ = set of all classes (up to isomorphism) of X projective and flat over S, satisfying $H^0(X_s, \mathcal{O}_{X_s}) \xleftarrow{\sim} k(s)$, endowed with polarization (i.e., section of the Pic functor over S) which is *very* ample on each fiber and satisfies $H^1(X_{\bar{s}}, \mathcal{L}_{\bar{s}}) = 0$, $\dim H^0(X_{\bar{s}}, \mathcal{L}_{\bar{s}}) = N$ for every s (these last restrictions for mere convenience), and every $X_{\bar{s}}$ "without automorphisms respecting the polarization". By a standard argument using Hilbert schemes, if one assumes that F corresponds moreover to a fixed Hilbert polynomial, F is just a quotient M/G, where M is a projective scheme over $\mathrm{Spec}(\mathbb{Z})$ and G the projective group operating freely on M. As far as I understand from your letter, you have no example to the effect that F, i.e., M/G, is nonrepresentable by a *pre*scheme (of finite type over $\mathrm{Spec}(\mathbb{Z})$ necessarily!); the phenomenon you allude to when speaking of "birationally ruled surfaces" is just the fact that if you do not exclude these, your functor F will be nonseparated and more precisely the image of $G \times M$ in $M \times M$ will be nonclosed? On the other hand, when working in the category of analytic spaces rather than preschemes, you state (at least in the case of surfaces) you are pretty sure the quotient is representable—at least when you exclude the above-mentioned surfaces. Now I once verified a general theorem of passage to quotient, in the context of analytic spaces, by a flat equivalence relation $R \hookrightarrow M \times M$, the conclusion being that the quotient is representable *if and only if* $R \hookrightarrow M \times M$ is an immersion (or what amounts to the same, iff the topology of R is induced by the one of $M \times M$); of course, if this condition (trivially necessary) is satisfied, then M/R is separated if and only if the immersion $R \longrightarrow M \times M$ is *closed*. Now in the case when the analytic situation comes from an algebraic one over $\mathbb{C}$, immersion, respectively, closed immersion in the algebraic or in the analytic sense is the same. On the other hand, in the algebraic case, we have the valuative criterion in order to check immersions, respectively, closed immersions. Thus we are led exactly to the following question: Let X/S be a polarized scheme as above, corresponding to an element in $F(S)$, S being the spectrum of a discrete valuation ring, let X'/S be another one, let s_0 (s_1) be the special (generic) point of S, assume that X_{s_1} and X'_{s_1} are isomorphic, *and* X_{s_0} and X'_{s_0} isomorphic (by "isomorphism" we mean one respecting polarization; it must be necessarily unique). Is it true that X and X' are isomorphic? An affirmative answer is equivalent with the statement that $R \hookrightarrow M \times M$ is an immersion, and if we restrict to S lying over $\mathrm{Spec}(\mathbb{C})$, or simply "of char 0", it is equivalent with the possibility of passing to the quotient analytically. Do you have any result concerning this question? Did you actually check that, when restricting to surfaces which are not birationally ruled, you even have the stronger result about $R \hookrightarrow M \times M$ being a closed immersion, i.e., X_{s_1} and X'_{s_1} isomorphic implies X and X' are isomorphic? The ideal would be to be able to state some geometric conditions on the fibers of X/S characterizing those F for which either of the two valuative conditions are satisfied.

b) I did not understand your motivation for the feeling that, when dealing with variation of structure for nonpolarized *analytic* varieties without automorphisms, the functors you get will generally not be representable. The case of infinite discrete groups operating badly (as in the case of complex tori) seemed to me to occur precisely because of the existence of automorphisms, which I have excluded. On the other hand, your Blow_F functor is not an open subfunctor of the big functor I was considering, and the big one might be representable without the small one being so. In other words, it may happen that you have a family X/S of surfaces "without automorphisms", such that the condition making the fibers birationally equivalent to a given fiber (a rather screwy condition by the way, of which I would expect nothing good anyhow) is *not* representable. What would be more convincing would be a case when, on the *local* variety of moduli M for a given analytic compact nonsingular variety X_0 without automorphisms, there are points s arbitrarily near to s_0 where the family X/M is no longer modular, for instance the Zariski tangent space to M at s has not dimension equal to $\dim H^1(X_s, \mathscr{T}_{X_s})$. I wonder if such kind of phenomena are actually known to you.

c) Serre told me about your remark on Siegel's remark,[77] which is extremely nice indeed. Do you think one can recover also the case of arbitrary polarizations (not necessarily separable ones)? I confess I am afraid that the additive type part of the kernel of the polarization might make trouble, as introducing a continuous set of indeterminacies.... Besides, have you got any results on the actual dimension of the modular variety in char $p > 0$ for polarized abelian varieties in the case of inseparable polarizations? You remember perhaps that the Zariski tangent space becomes bigger than usual, which implies that the modular variety must either have bigger dimension (which would imply that there are polarized abelian varieties which do not lift to char 0) or else be nonreduced everywhere. Did Serre tell you about his candidate for an abelian variety in char p which should not lift (an inseparable quotient of the product of two elliptic curves with Hasse invariant 0)?[78]

Sincerely yours
(signed) A Grothendieck

---

[77] Neither Mumford nor Serre could remember exactly what this was about. Mumford suggested that, maybe, it concerned reducing the moduli spaces of arbitrary separably polarized abelian varieties (of a fixed degree) to the principally polarized case by isogenies.

[78] Mumford announced in [69b] that every abelian variety in char p can be lifted to an abelian variety in char 0. P. Norman and F. Oort, Moduli of abelian varieties, *Ann. Math.* **112** (1980) 413–439, completed this program, and also showed that the moduli space $\mathscr{A}_{g,d}$ of g-dimensional abelian varieties with a polarization of degree d in char p has dimension $g(g+1)/2$. A theorem of Mumford asserts that the local deformation spaces for $\mathscr{A}_{g,d}$ are of the form $\mathrm{Spf}(k[[t_1,\ldots,t_{g^2}]]/I)$, where I is an ideal generated by $g(g-1)/2$ elements; see Thm. 2.3.3 of F. Oort, Finite group

Grothendieck to Mumford, 23 January, 1965

23.1.1965

Dear Mumford,

Thanks for your letter, I finally looked at your example with Blow_2 and got your idea,[79] namely that the valuative criterion I was asking for becomes definitely false if one does not include polarization into the statement—a strange fact in a way! Do you know if for surfaces as the ones of your example, there are *strict* local varieties of moduli, namely that the local variety of moduli (of Kuranishi ...) is still modular at points near the center? If so, one gets an "open" subfunctor of the big functor I told you about (except one is forgetting about polarization) which can be written as $\mathscr{M}/\mathscr{R}$, $\mathscr{M}$ the little Kuranishi modular variety, $\mathscr{R}$ an analytic equivalence relation which is *étale* (i.e., $\mathscr{R} \xrightarrow{\mathrm{pr}_1} \mathscr{M}$ is étale) but $\mathscr{R} \to \mathscr{M} \times \mathscr{M}$ not being an immersion. It still would remain possible that one can represent the general functor corresponding to varieties of (say) analytic compact spaces without automorphisms, as $\mathscr{M}/\mathscr{R}$, $\mathscr{M}$ an analytic space and $\mathscr{R}$ an étale equivalence relation—a problem equivalent, I guess, to whether the Kuranishi local modular variety is "strictly" so.

I have no particular use for an answer to the valuative question I asked you, except that one certainly should know one day what is going on! Besides, I once had to ask you the analogous question for the Picard functor, in order to prove theorem 3.1, page 16-13 of Cartan's seminar 60/61; I begin to believe that this theorem is probably false, as I do not see any reason why the corresponding valuative criterion should be valid. You would probably be able to get a counterexample out of your shirt's sleeve, if you tried.

By the way, Raynaud remarked that if $G \to S$ is a group prescheme over S, flat, of finite presentation, with *connected fibers*, then $G \to S$ is *necessarily separated.*[80] This yields lots of cases where $\mathscr{P}ic^0_{X/S}$ and $\mathscr{P}ic_{X/S}$ are not representable, although the standard conditions (implying formal representability) are satisfied. Take for instance $X \to S$ with S the spectrum of a discrete valuation ring, f proper, X regular, $[\dim(X) = 2]$ generic fiber X_1 smooth and geom. connected, special fiber $X_0 = \sum \nu_i C_i$ (C_i prime divisors, $\nu_i > 0$), let $d = \gcd((\nu_i))$. Then if $d > 1$, $\mathscr{P}ic^0_{X/S}$ is nonseparated and hence *nonrepresentable*, therefore $\mathscr{P}ic_{X/S}$ is not representable either. Besides, if $d = 1$ one gets a canonical morphism of functors

$$\varphi\colon \mathscr{J}^0 \to \mathscr{P}ic^0_{X/S}$$

---

scheme, local moduli for abelian varieties, and lifting problems, *Algebraic geometry, Oslo 1970* (Proc. Fifth Nordic Summer-School in Math.), Wolters-Noordhoff, 1972, pp. 223–254. The two results imply that $\mathscr{A}_{g,d,n}$, classifying g-dimensional abelian varieties in char p with a polarization of degree d and a level-n structure, where $n \geq 3$, $\gcd(n,p) = 1$, is a local complete intersection.

[79] Blow_2 stands for the moduli space of 2 times blown up surfaces—subtleties arise when the points come together, since the order of blowing up must be specified.

[80] For a proof of Raynaud's remark see SGA3 $\mathrm{VI_B}$, Propriétés générales de schemas en groupes, Cor. 6.5 on page 351 of LNM 151, Springer-Verlag, 1970.

where $\mathscr{J}$ is the Néron model of the Jacobian of X_1, and $\mathscr{P}ic^0_{X/S}$ representable $\Longleftrightarrow$ φ is an isomorphism $\Longleftrightarrow$ φ_0 is an isomorphism ($\varphi_0 = \varphi \otimes k(S)$). I do not know if these conditions are always satisfied when $d = 1$.[81]

Sincerely yours
(signed) A Grothendieck

Grothendieck to Mumford, 16 April, 1965

Bures April 16, 1965

Dear Mumford,

I read through your nice notes from Woods Hole: "Further comments on boundary points".[82] Unfortunately, my copy stops on page 7 with the words: "the curve of genus 2 depicted below:". Are there pages lacking, or were you making fun?[83] I have just two mathematical comments: Page 4, lines 1 and 2,[84] the equality $\mathrm{Pic}(X) = \mathrm{Pic}(X_0)$ holds only in the stable case, otherwise you have to replace the Pic groups by the Néron–Severi groups to have a correct statement. Page 5, line 10,[85] it does not seem clear to me (but rather unlikely!) that Θ is really an isomorphism of $\mathscr{M}_g$ with a locally closed subvariety of V_g, say N,[86] all I know is that the image N of $\mathscr{M}_g$ is indeed locally closed, and $\mathscr{M}_g \to N$ is finite and radicial, and in fact makes $\mathscr{M}_g$ a normalization of N (N turns out to be geometrically unibranch). Your statement would mean that N is normal, which I doubt to be true. Besides, line 11 [87] reads $\mathscr{M}_g$ instead of $\mathscr{M}'_g$. Also, page 7, I did not quite see what you mean

---

[81] The condition "$\dim(X) = 2$" was inserted by the editors. The proof of the assertions about $\mathscr{P}ic^0_{X/S}$ and further information can be found in M. Raynaud, Spécialisation du foncteur de Picard, *Publ. Math. IHÉS* **38** (1970) 27–76. The assertion that $\mathscr{P}ic^0_{X/S}$ is not representable if $d > 1$ is part of Thm. 2.1 on p. 66. The statement about $\varphi\colon \mathscr{J}^0 \to \mathscr{P}ic^0_{X/S}$ is proved in Thm. 8.2.1 on page 66. The answer to the question in the last sentence is *yes* if for instance the residue field of the closed point of S is perfect; see Thm. 8.1.4 on page 65.

[82] [u64b] in this volume.

[83] The missing picture depicting a stable curve of genus 2 was "a dollar sign lying on its side"; see p. 604, Footnote 7.

[84] p. 601, lines 1 and 2 in this volume; see also Footnote 4 in the same page.

[85] P. 602, line 6 in this volume.

[86] The Torelli map in [u64b] is indeed an immersion of $\mathscr{M}_g$ to V_g; this is true over a field of characteristic 0. However, the same statement may not be true over a field of characteristic $p > 0$, and it *is* false when $p = 2$. See F. Oort and J. Steenbrink, The local Torelli problem for algebraic curves, in *Journées de géométrie algébrique d'Angers (juillet 1979) / Algebraic Geometry Angers 1979*, Sijthoff & Noordhoff, 1980, pp. 157–204

[87] This misprint in [u64b] is corrected in the retyped version in this volume; see p. 602, Footnote 5.

on line 12,[88] do you just mean to say that there is no natural definition, by analogy with the previous discussions, of a notion of semi-stable curves?

One question about your result on page 4, giving as it seems an axiomatic description of the Satake compactification, analogous to the axiomatic description of V_g itself: a) Is there a hope to get such a description over Spec($\mathbb{Z}$)? b) What about throwing in "levels"? How should one define a level n rigidification on your group schemes?

Sincerely yours

Grothendieck to Mumford, 4 October, 1965

Bures 4.10.1965

Dear Mumford,

Thank you very much for your notes on theta functions (pages 1 to 77),[89] and your book on geometric invariant theory.[90] I was certainly pleased by the advertising you are doing there for schemes, and flattered by the opinion you express in your introduction on my own work. However, I am sorry to state I have not really read it as yet, although this is in my program and the short size and vivid style will make it not too hard for me to stick to it, I hope! I just had a quick reading of your notes on θ-functions, which look very nice indeed; I hope, when you will have written up the whole, you'll send a copy of the remainder too. Maybe you could include too a description of the group of automorphisms of the extension $\mathfrak{G}(\mathscr{L})$ (is the first letter supposed to be a gothic G?),[91] which will eventually act on the modular scheme, and of what happens when you replace $\mathscr{L}$ by $\mathscr{L}^{\otimes n}$, variable n, as one would like to know how the corresponding modular schemes match together. A paragraph giving the connection with the transcendental construction would be nice too, for ignorant people like myself,—or is this what you intend to do in your paragraph 6? How do you intend to publish the theory? Another book would not seem a bad idea! In this case, or for any other book you would care to write (e.g., theory of surfaces), I would like to mention to you that Kuiper and I (and maybe a third man who is still not well determined) are starting to publish a new series of advanced books on pure mathematics, in North Holland Publishing Company, and we would certainly appreciate to have you in the series. The first book in our program will be one of Giraud on noncommutative homological algebra,[92] essentially his thesis in fact. He

88 P. 603, line −1 in this volume.

89 [66a].

90 [GIT].

91 Grothendieck probably referred to a handwritten symbol in the original manuscript of [66a]; in the published version it appeared as $\mathscr{G}(L)$.

92 Published as J. Giraud, *Cohomologie non abélienne*, Grundlehren der Math. Wiss. 179, Springer-Verlag, 1971.

did quite a good job, although to a great extent expository, and it will certainly become a standard reference within a few years. For instance, I will have to use his formalism intensively to formulate a Galois theory for motives. Besides, I am more or less decided to write a book myself on the theory of motives, despite the fact that it is likely that the whole theory will remain conjectural for a long time, so it should be called a program of a theory rather than a theory.

By the way, I have the feeling that one main point in realizing something like Kronecker's "Jugendtraum" is a theory of moduli for *motives*, and it is already clear to me how in this optic to generalize the Siegel generalized half-plane and the Siegel modular group acting on it (these corresponding just to polarized motives of weight 1). The deeper point is how to put an algebraic structure on the quotients one obtains, and how to interpret geometrically these quotients as modular schemes for polarized motives. Now this question should in essence be identical with the following one, for which your present work perhaps may give you some idea. By a *Hodge structure*, I will mean a free module of finite rank M over $\mathbb{Z}$, together with a bigrading of the complex vector space $M \otimes_{\mathbb{Z}} \mathbb{C}$, having positive partial degrees, and total degree n if we want a Hodge structure of weight n, and such that there should exist a bilinear form $\phi : M \times M \to \mathbb{Z}$, alternating or symmetric according as n is odd or even, such that $\phi_{\mathbb{C}}$ is compatible with the bigradings when $\mathbb{C}$ is considered of degree (n,n), and that $\phi_{\mathbb{C}}(\eta x, \bar{x})(-1)^n$ should be a positive definite Hermitian form on $M_{\mathbb{C}}$, where η is multiplication by $(-1)^p$ on the component of first degree p. Such a form will be called a *polarization of the Hodge lattice M*. If X is a projective nonsingular variety defined over $\mathbb{C}$, then by Hodge theory the lattices $H^n(X,\mathbb{Z})$ mod torsion can be viewed as Hodge lattices, any polarization of X (in the classical sense) defining a polarization of that Hodge lattice. In this way, one gets a functor from the category of (semi-simple effective) motives defined over $\mathbb{C}$ (never mind what that means for the moment!) into the category of Hodge lattices. Hodge's conjecture just amounts to saying that this is a fully faithful functor, at least when working modulo isogeny, and I feel the whole story in this respect should be that it is even an *equivalence* of categories. In a down to earth way, this essentially amounts to saying that any Hodge lattice is isomorphic to a sub-Hodge lattice (in fact, a direct factor) of some Hodge lattice $H^n(X,\mathbb{Z})$, always up to isogeny. By general principles, one should be able to take $\dim X = n$, (although this would not be a good idea when starting with a Hodge structure of weight 1, as then the natural X to take would be the associated abelian variety, not some generating curve on it), and to restrict to consideration of the primitive part (in the sense of Hodge–Lefschetz theory, namely the part that, restricted to a hyperplane section, vanishes). The problem then is to give, in terms of the "transcendental data": a polarized Hodge structure M, an explicit construction of some canonical X, presumably with a definite projective embedding, realizing that Hodge lattice, for instance as being isomorphic (or isogenous) just to the primitive part of $H^n(X,\mathbb{Z})$. To do things quite canonically, probably something like your θ-level structure will be required as an extra-structure on M. More important, the construction should be so canonical as to carry over to continuous, and more specifically, to complex-analytic families of Hodge structures (these being defined in a rather evident way, but taking care that an "integrability condition" has to be satisfied in the complex analytic case). Also, in view of Kronecker's dream, the finite modules M/nM over $\mathbb{Z}/n\mathbb{Z}$, variable n, should be recoverable in some way on the model X, say as finite subsets, as in the case $n = 1$ the points of finite order of an abelian variety. This latter point, I confess, is still very vague in my mind; it will have to be tied up with some extra structures on X, replacing the additive structure of an abelian variety (maybe a "hypergroup" structure, as inherited for instance by the quotient of a group by any finite group of automorphisms acting).... Of course, this would not be of any arithmetic use unless it also behaves properly with respect to families. Besides, I feel much less positive about the possibility of this latter element of structure, perhaps the analogy with abelian varieties is fallacious in this direction. After all, even without it, it will turn out that the transcendental functions of passage to quotient, from the Siegel-type modular spaces (which can be defined as certain homogeneous spaces of algebraic groups defined over $\mathbb{Q}$) to their algebraic quotient spaces (obtained by passing to the quotient by various discrete subgroups, commensurable with the group of integral points) will allow a description of the various "motivic" classes of infinite Galois extensions, as generated by values of transcendental functions constructed some way or other from the previous ones.

One question about your theory of θ-functions. You always make that assumption of separable polarization. But you have certainly noticed that for any abelian scheme X over a base S, endowed with a polarization, defining a morphism $X \to X'$ with kernel the finite flat group scheme K, your definition yields a canonical extension

$$1 \to \mathbb{G}_{mS} \to E \to K \to 1,$$

and the group-scheme E operates on $f_*(\mathscr{L})$ whenever your polarization is given by an actual invertible sheaf $\mathscr{L}$ (not a great restriction as you know). Maybe it's nicer to view K as acting on the Brauer–Severi bundle defined by the polarization! How far does your theory extend to this case? For instance, do you know, when S is the spectrum of an algebraically closed field, if this representation is still irreducible? Because of the variety of possibilities of structure of K (as a nonseparable group scheme), you cannot reduce all possible extensions E to a discrete set of standard types, by which you then rigidify. However, nothing would prevent you from starting with one extension E_0, and looking at those polarized abelian schemes whose extension E is locally isomorphic (for the fpqc topology, say) to E_0, and looking at modular schemes for these. Certainly Cartier's theorem will tell you anyhow that the alternating form $K \times K \to \mathbb{G}_m$ is nondegenerate (i.e., K is autodual in the sense of Cartier), thus the main point seems first to pick out a K with such an autoduality. For instance, you can start with any flat commutative K', and take $K = K' \times K''$, with $K'' = D(K')$ (Cartier dual). The case when you take precisely for K' an *étale* group, and K'' its dual, is essentially the one you consider in your notes, but one does not see any reason why to restrict to residue characteristics prime to rank K'; in a way, you should still get then the more general polarizations in any characteristics (namely those for which you have the maximum number of geometric points in K)—they deserve not to get lost on your way! Besides, it looks an interesting question to determine, over a perfect field say, all finite commutative group schemes over k endowed with an alternating autoduality—there should not be any difficulty to get the complete picture and see whether any such K can be written as $K' \times D(K')$ and the obvious form on it (one will have to use Dieudonné–Gabriel's structure theory in terms of modules over the Witt vectors, together with the operations $V, F\ldots$).[93] Maybe, even, you will be able to get in your system *smooth* modular schemes, which might allow you to solve the problem of lifting an abelian variety to char. 0.[94]

Best wishes
(signed) A Grothendieck

Grothendieck to Mumford, 1 November, 1965

Algiers Nov. 1, 1965

Dear Mumford,

I have become aware that I have been a bit rash with the conjectures I told you about Hodge structures, as in the form I stated them they contradict Tate's conjecture, which implies the following: if X is a smooth, connected *simply connected* scheme over $\mathbb{C}$, and $\mathscr{V}$ a polarized complex analytic family of Hodge structures parametrized by X^{hol}, then $\mathscr{V}$ is "algebraic" only if it is a *constant* family. Now it is easy to get examples of $\mathscr{V}$'s, (with X say *any* homogeneous space under an affine group over $\mathbb{C}$) which are *non*constant, therefore (one hopes!) not "algebraic" (as one would like Tate's conjectures to hold). However, one should try to test if one really *cannot* get an algebraic $\mathscr{V}$ that way. To start, I wonder if a single Hodge structure $\mathscr{V}$ with *even partial degree* (therefore giving rise to a "Hodge-group" G s.th. $G_{\mathbb{R}}$ is *compact*) can be algebraic (except when G is commutative). For instance, do you know an algebraic smooth projective surface Y over $\mathbb{C}$, such that $H^{1,1}(Y,\mathbb{C})$ is spanned by algebraic cycles, and $\dim H^{0,2}(Y,\mathbb{C}) \geq 2$??[95] I would appreciate your comments!

Yours,
(signed) A Grothendieck

P.S. What about your θ-functions, are you going to write a book?

---

93 Not every such K has this property: Let K be the group $E[p]$ of p-torsion points on a supersingular elliptic curve E over $\overline{\mathbb{F}_p}$. Then K is indecomposable, while the K is autodual via the Weil pairing.

94 See Footnote 78, at the end of [1965Jan17].

95 In [1965Dec03], Grothendieck said that Mumford solved this question affirmatively.

Grothendieck to Mumford, 1 November, 1965

Algiers Nov. 1, 1965.

Dear Mumford,

I have become aware that I have been a bit rash with the conjectures I told you about Hodge structures, as in the form I stated them they contradict Tate's conjecture, which implies the following: if X is a smooth, connected simply connected scheme over $\mathbb{C}$, and V a polarized complex analytic family of Hodge structures parametrized by X^{hol}, then V is "algebraic" only if it is a constant family. Now it is easy to get examples of V's, with X say any homogeneous space under an affine group over $\mathbb{C}$?) which are non constant, therefore (one hopes!) not "algebraic" (as one would like Tate's conjectures to hold). However, one should try to test if one really cannot get an algebraic V that way. To start, I wonder if a single Hodge-structure V with even partial degrees (therefore giving rise to a "Hodge-group" G s.th. $G_{\mathbb{R}}$ is compact) (except if the G is comm) can be algebraic. For instance, do you know an algebraic smooth projective surface Y over $\mathbb{C}$, such that $H^{1,1}(Y, \mathbb{C})$ is spanned by algebraic cycles, and $\dim H^{0,2}(Y, \mathbb{C}) \geq 2$?? I would appreciate your comments! Yours

A Grothendieck

P.S. What about your Θ-fns, are you going to write a book?

P.S. Here what are the modular variation one gets from consideration of motive-theoretic galois theory.

$G_{\mathbb{Q}}$ = reductive connected alg. group defined over $\mathbb{Q}$

$U_{\mathbb{R}}$ = dimension one twisted real Torus, with $U_{\mathbb{R}}(\mathbb{R})$ identified with the group of complex units of module 1

$i_0 : U_{\mathbb{R}} \longrightarrow G_{\mathbb{R}}$ homomorphism of algebraic groups over $\mathbb{R}$, s.th. $i_0(-1)$ is central, and s.th. the centralizator $K_{\mathbb{R}}$ of $i_0(\sqrt{-1})$ is such that $K_{\mathbb{R}}(\mathbb{R})$ is <u>maximal</u> <u>compact</u> in $G_{\mathbb{R}}(\mathbb{R})^{\circ}$. Moreover, for every $\mathbb{R}$-simple component of $G_{\mathbb{R}}$, the corresponding component of i_0 is not trivial.

$C_{\mathbb{R}} = \underline{\mathrm{Centr}}_{G_{\mathbb{R}}}(i_0) \subset K_{\mathbb{R}}$

$\mathcal{Q} = G_{\mathbb{R}} / C_{\mathbb{R}}$

"Siegel-space" = $\mathcal{Y} = G(\mathbb{R})^{\circ} / G(\mathbb{R})^{\circ} \cap C_{\mathbb{R}}(\mathbb{R}) \xrightarrow{\text{(a connected component)}} Q(\mathbb{R})$

N.B. $\mathcal{Y}$ has a natural complex structure (inherited from near $Q(\mathbb{R})$ in fact) invariant by operation of $G(\mathbb{R})$. Every linear representation (over $\mathbb{Q}$) of G into a vector space V (over $\mathbb{Q}$), together with a lattice V_0 of V, defines a '<u>complex</u> <u>analytic</u> family of Hodge structures' on $\mathcal{Y}$, at least when one fixes a "total degree" n

[Margin note, inserted after "family":] ((~~[illegible]~~ necessarily? globally polarizable ~~[illegible]~~ to get a polarization of the family, ~~[illegible]~~ one picks a "form of polarization" φ on V, invariant by G).

P.S. Here is what are the modular varieties one gets from considerations of motive-theoretic Galois theory.

$G =$ reductive connected alg. group defined over $\mathbb{Q}$

$\mathscr{U}_{\mathbb{R}} =$ dimension one twisted real torus, with $\mathscr{U}_{\mathbb{R}}(\mathbb{R})$ identified with the group of complex numbers of modulus 1

$i_0 \colon \mathscr{U}_{\mathbb{R}} \to G_{\mathbb{R}}$ homomorphism of algebraic groups over $\mathbb{R}$, s.th. $i_0(-1)$ is central, and s.th. the centralizer $K_{\mathbb{R}}$ of $i_0(\sqrt{-1})$ is such that $K_{\mathbb{R}}(\mathbb{R})^0$ is *maximal compact* in $G_{\mathbb{R}}(\mathbb{R})^0$. Moreover, for every $\mathbb{R}$-simple component of $G_{\mathbb{R}}$, the corresponding component of i_0 is not trivial.

$C_{\mathbb{R}} = \underline{\mathrm{Centr}_{G_{\mathbb{R}}}}(i_0) \subset K_{\mathbb{R}}$

$Q = G_{\mathbb{R}}/C_{\mathbb{R}}$

$$\text{“Siegel-space”} = \mathfrak{S} = \frac{G(\mathbb{R})^0}{G(\mathbb{R})^0 \cap C_{\mathbb{R}}(\mathbb{R})} \xrightarrow[\qquad]{\text{(a connected component)}} Q(\mathbb{R})$$

N.B. $\mathfrak{S}$ has a natural complex structure (inherited from the one on $Q(\mathbb{R})$ in fact) invariant by the operation of $G(\mathbb{R})$. Every linear representation (over $\mathbb{Q}$) of G in a vector space V (over $\mathbb{Q}$), together with a lattice V_0 of V, defines a *complex analytic* family $\mathscr{V}$ of Hodge structures (necessarily globally polarizable—to get a polarization of the family, one picks a “form of polarization” φ *on* V invariant by G) on $\mathfrak{S}$, at least when one fixes a “total degree” n and assumes that $i_0(-1)$ operates on V as the homothety $(-1)^n$. (It is not hard to describe $\mathfrak{S}$ by a universal property as representing some functor $F(X)$, X complex analytic, explaining that every V as above should define, in an additive and multiplicative way for varying V, a Hodge-structure $\mathscr{V}$ over X.) If $\Gamma \subset G(\mathbb{R})$ is the group leaving V_0 fixed, then Γ operates on $(\mathscr{V}, \mathfrak{S})$.

Note that in general $\mathfrak{S}$ is *not* Riemannian symm., but is a fiber space over a R.S. space, with fibers which are projective complex homogeneous spaces under $K_{\mathbb{C}}$. This implies at least that $\mathfrak{S}$ is *simply connected*, hence is the univ. covering of $\mathscr{M} = \mathfrak{S}/\Gamma$ if Γ operates freely.[96]

The modular varieties I expect to carry algebraic structure *over* $\mathbb{Q}$ (and morally, $\mathbb{Z}$) are the varieties $\boxed{\mathfrak{S}/\Gamma}$, where $\Gamma \subset G(\mathbb{Q}) \cap G(\mathbb{R})^0$ is commensurable with $G(\mathbb{Z})$ (for some matrix representation of G, giving a meaning to $G(\mathbb{Z})$). However, it is possible that one will have to impose on the data some extra conditions, either arithmetic (involving existence of “Frobenius elements” in $I(\mathbb{Q})$, $I = G/\text{flat.aut.}$),[97] or geometric on $G_{\mathbb{R}}$ (such as: $G_{\mathbb{R}}$ has *no compact factor*). A typical case which fits in the general description, but for which I have no evidence so far if $\mathfrak{S}/\Gamma$ has an algebro-geometric interpretation in terms of moduli for motives, is the following: start with a vector space of finite dimension V over $\mathbb{Q}$, with a fundamental bilinear form φ given, symmetric or alternating, take $G = SO(\varphi)$, take any $i_0 \colon \mathscr{U}_{\mathbb{R}} \to G_{\mathbb{R}}$ s.th. $i_0(-1) = \mathrm{id}$ (if φ is symmetric), $i_0(-1) = -\mathrm{id}$ (if φ is skew-symmetric), and

[96] This paragraph was written vertically in the left-hand margin.

[97] The original is hard to read, e.g., “flat.” can be “int.”, but I is meant to be an inner form of G.

a positivity condition of φ relative to the bigrading of $V_{\mathbb{C}}$ defined by i_0, which I will not write down. Then $\mathfrak{S}$ is a direct generalization of Siegel's half-space, and classifies all Hodge-structures with underlying $\mathbb{Q}$-vector space V, i.e., all bigradings of $V_{\mathbb{C}}$, satisfying the usual symmetry condition for complex conjugation, which are compatible with φ (i.e., φ is a polarization), and for which the dim $V_{\mathbb{C}}^{pq}$ have given values (depending on the weights of $\mathscr{U}_{\mathbb{C}}$ operating on $V_{\mathbb{C}} \ldots$). The truth is that, as these special types of $\mathfrak{S}$, $\mathfrak{S}/\Gamma$ are in a way "universal" for all others (namely the $\mathscr{V}$'s considered above are induced from ones on such special type of $\mathfrak{S}$), if the "modular family" $\mathscr{V}$ of Hodge-structures on such a $\mathfrak{S}/\Gamma$ was "algebraic," i.e., came from a relative motive over the complex analytic space $\mathfrak{S}/\Gamma$, (or even on some underlying structure of alg. var. over $\mathbb{C}$), the same would be true for *every* modular family on some $\mathfrak{S}/\Gamma$. Taking the case $G(\mathbb{R})$ compact, and thus $\mathfrak{S}/\Gamma = \mathfrak{S}$ = complex projective homogeneous space X under $G_{\mathbb{C}}$ by Borel, we would get a polarized motive over the analytic space $\mathfrak{S}$ corresponding to X. If we admit the GAGA yoga (coming from polarized abelian varieties) that such a family must also be a motive over X, (or the above strengthening of the original assumption), we get a contradiction with Tate's conjectures (because X is simply connected, as well known). So I really do not know what to believe![98]

To come back to the general case, maybe I should add that a *necessary* condition for $\mathfrak{S}/\Gamma$ to be algebraic and the Hodge-structure $\mathscr{V}/\Gamma$ over $\mathfrak{S}/\Gamma$ algebraic too, is (*granting* Tate's and Hodge's conjectures): for every $g \in G(\mathbb{R})^0$, and every Γ_0 of finite index in $G(\mathbb{Z})$, the smallest alg. subgroup H of G such that $H_{\mathbb{R}}$ contains $g[i_0(\mathscr{U}_{\mathbb{R}})]g^{-1}$ *and* Γ_0 is G itself.[99] Question: If $G_{\mathbb{R}}$ is semi-simple without compact factor, is $G(\mathbb{Z})$ always Zariski-dense in G??[100]

By the way: did you make out if, for the modular varieties $\mathfrak{S}/\Gamma$ of your Boulder talk (which I have just read), the image of $\mathfrak{S}/\Gamma$ in the usual Siegel modular variety is an algebraic variety, or at least constructible?[101]

Grothendieck to Mumford, 3 December, 1965

Bures Dec. 3, 1965

Dear Mumford,

Thanks for your letter. I am sorry my own from Algiers has not reached you. Before writing it all over again I'll wait to see if by chance it has not gone by sea mail.

---

98 Mumford's answer "*PROBABLE ANS.* $\sim\exists$ alg. families of $\mathfrak{S}/\Gamma$ (since $\mathfrak{S}/\Gamma$ not alg. itself)" is written vertically in the left-hand margin.

99 The following sentence was crossed out in the hand-written letter. "I did not try to test directly this condition, even in the case of the *usual* Siegel modular space, if it should be false in that case, as we know that $\mathfrak{S}/\Gamma$ *is* algebraic and that so are the $\mathscr{V}/\Gamma$ over it, it would follow that either Hodge's or Tate's conjecture is wrong."

100 An arrow is drawn from this question to Mumford's answer "Yes??" in the left-hand margin.

101 Mumford's answer "I believe so, but haven't written it down." is in the left-hand margin.

Please tell me if it should still arrive. Indeed a few days after my over-optimistic letter of October I got more realistic on the matter of deducing more or less arbitrary families of Hodge structures from motives, as this was going to conflict with Tate's conjectures. In my Sahara letter, I was expounding for you in some detail my perplexities, and gave also a detailed description of the complex analytic modular varieties I had been interested in, generalizing the Siegel–Griffiths ones.[102] I well know they are not bounded domains in general, but fiber spaces over such bounded domains, with fibers homogeneous spaces under complex linear groups.... This in itself would not bother me. Nor do I understand what are the results of Griffiths, and if they are really conclusive to the effect that there are no algebraic structures where I first expected some. With the notations of your letter, did you mean to say that for *every* Γ-linearized invertible bundle on D, $H^0(D/\Gamma, \mathscr{L}^{\otimes n})$ is zero (and this still when replacing Γ by any subgroup of finite index, so as to achieve for instance that Γ should operate freely)? Besides, I never got any reprint from Griffiths. If his result concerns only the "canonical" bundle of highest differential forms, this alone does not look so convincing. From my point of view, the main trouble is that it is very easy, among the modular varieties I alluded to for varying Hodge structures, to get any type of compact algebraic homogeneous space under a complex algebraic linear group, for instance $X = \mathbb{P}^1$, but such families cannot stem from a motive (say over the field of functions of X, or its algebraic closure...), without contradicting Tate's conjectures, which imply the following: let k be a field, (here $k = \mathbb{C}$), K a "regular" extension of finite type, M a motive over K such that the operations of $\mathrm{Gal}(\overline{K}/K)$ on $T_l(M) = l$-adic cohomology realization of M (some l prime to the char) is trivial, then M comes from a motive over k (and in the case $k = \mathbb{C}$, K the function field of some X over $\mathbb{C}$, this implies that the corresponding family of Hodge structures on X, or some Zariski-open subset of X to be more accurate, is constant). Now $\mathbb{P}^1$ is simply connected, and any family of Hodge structures parametrized by $\mathbb{P}^1$ must give rise to a trivial T_l! Notice that I do not demand that the "family" of smooth projective algebraic varieties, whose Hodge cohomology should contain the given Hodge structure over X as part of it, needs to be defined on the whole of X; it is enough that it be defined on a Zariski-open subset $\neq \emptyset$ (or only on some $X' \neq \emptyset$ étale over X), in order to contradict Tate's conjectures. As you surmised, what matters is not that the given family of projective varieties should be nonconstant, but that its Hodge cohomology (or rather, the piece of it we are looking at...) should be so. The question I was asking, about the existence of individual surfaces with $\rho = h^{1,1}$ and $h^{2,0} > 2$, and which you solved affirmatively,[103] arises when one wants to get

---

[102] The "Siegel–Griffiths ones" and "results of Griffiths" below refer to the period spaces announced by P. Griffiths in *Proc. Nat. Acad. Sci. USA* **55** (1966) 1303–1309 and 1392–1395, *ibid.* **56** (1966) 413–416; later published in *Amer. J. Math.* **90** (1968) 568–626.

[103] The question in [1965Nov01] asks for surfaces with $\rho = h^{1,1}$ and $h^{2,0} \geq 2$. The example given by Mumford is not recorded, but is likely to have been an elliptic surface. The following example, courtesy of P. Deligne, is closely related to modular forms. Let $p \geq 5$ be a prime number, and let $f\colon \mathscr{E} \to X(p)$ be the universal elliptic curve over the compactified modular curve $X(p)$ attached to the principal congruence subgroup $\Gamma(p) \subset SL(2,\mathbb{Z})$ of level p. The fibers of $\mathscr{E}$ over the cusps of $X(p)$ are cyclic chains of p copies of $\mathbb{P}^1$'s. By the Leray spectral sequence for f and the Eichler–Shimura isomorphism, the Hodge structure for $H^2(\mathscr{E}(\mathbb{C}), \mathbb{Q})$ is a direct sum of a Hodge structure

such a compact simply connected modular variety for Hodge structures of degree 2, by looking at the "Hodge-group" for it and asking that it be compact. By the way, do you know any actual example of a Hodge structure which cannot be embedded in the Hodge cohomology of an algebraic variety?

Sincerely yours
(signed) A Grothendieck

P.S.: Did you discuss a bit with Artin on the notion of motive?

Grothendieck to Mumford, 9 December, 1965

Dec 9.12.1965

Dear David,

I am glad to know you got my Algiers letter after all. Maybe you can say me a few words more about Griffiths, as I asked already in my previous letter, which I wrote Monday.

The fact that motives over *any* field form an abelian category follows from conjectures A and B of the letter to Serre of which I sent you a copy.[104] These, I feel, are considerably less remote than Tate's and Hodge's conjectures, at least I hope so. By the way, why don't you discuss a bit with Mike on motives—I spent about one day telling him about the yoga. Also, in a letter from Algiers to him, I raised a few questions, in connection with "p-adic" cohomology in char. p and Hodge cohomology, with the hope that if some answers are negative, he or you would know and tell me right away.

Motchane tells me you are planning to come to IHÉS in 1967/68. That would be great—but please tell me if it is not just an extrapolation by M. of what you really stated to him!

Yours,
(signed) Grothendieck

---

of type $(1,1)$ and a Hodge structure of type $\{(2,0),(0,2)\}$. Moreover the type $(2,0)$ Hodge component is naturally isomorphic to the space of all cusp forms of weight 3 with respect to $\Gamma(p)$. The Hodge number $h^{2,0}(\mathscr{E})$ is given by a standard formula for the space of cusp forms for $\Gamma(p)$; it grows like $p^3/12$ as $p \longrightarrow \infty$.

[104] This 27 August 1965 letter to Serre was published in *Correspondance Grothendieck–Serre,* Soc. Math. France, 2001, pp. 232–235.

Grothendieck to Mumford, 9 May, 1966

Pisa 9.5.1966

Dear Mumford,

Thanks a lot for your preprint "Abelian quotients of the Teichmüller modular group".[105] I have a few questions and comments.

P. 1 I think the formulation you give of Riemann's existence theorem is due not to Artin but to Serre, who gave a Bourbaki talk on this about ten years ago.[106]

P. 2 it seems to me that you prove th. 2 only in a weaker form, replacing "rational maps" by "morphisms". If th. 2 is true as stated, maybe you should show why.

P. 6 the lemma is due, I believe, to Matsumura (in his thesis). By the way, your proof of b) using the assumption of blowing down, whereas the statement works on any normal base, is somewhat misleading.

P. 7 cor. 1, do you know if "nonsingular" can be replaced by "normal"?

P.10 when you pretend you can compactify the modular variety for curves of genus g, adding only pieces of codimension at least 2, should you not assume $g \geq 3$? It seems to me Igusa has proved that for $g = 2$, the modular variety is affine. In any case, I think it would be useful for the reader that you state somewhat more explicitly what your and Mayer's result says; (or is it really in your Woods Hole talk?[107]).

P.15 the statement of Dehn's main result, by a simple reference to the figure 2, is not very clear. Maybe you could say what the generating curves are? One thing I found very misleading when reading your drawings was the double sense of the word "holes", which you use in a certain sense p.14, whereas on figure 3 it seems to mean a "handle".

P.21 can you send me a reprint of [6]?

As a general impression, I found it kind of astonishing that you should be obliged to dive so deep and so far in order to prove a theorem whose statement looks so simple-minded. For instance, using linear pencils of plane curves, could one not prove that any two sufficiently general curves of genus g can be connected by a linear family of curves?

I recently got a general result on modular varieties for rigidified abelian varieties, which I believe should be shared by all or at least many of the nonsingular algebraic varieties you get from arithmetic type discrete groups operating on suitable homogeneous spaces—namely the following: if X is connected, reduced and locally of finite type over the ground field k, with given geometric point x, then a morphism $X \to M$ is known when you know the geometric point image of x, and the action on the fundamental groups $\pi_1(X,x) \to \pi_1(M,f(x))$. This is a corollary of the following: if A,

[105] Published in [67d].

[106] See Thm. 1 of J.-P. Serre, Revêtement ramifié du plan projectif, *Séminaire Bourbaki* 1959–60, n° 204.

[107] Referring to [u64b].

B are two abelian schemes over X, l a prime number, $u_l: T_l(A) \to T_l(B)$ a homomorphism, and if the restriction of u_l to x comes from a homomorphism $A_x \to B_x$, then u_l comes from a homomorphism $u: A \to B$ (of course unique).[108] These results I can prove only if k is of char 0, and the proof uses quite sophisticated means, such as some very recent result of Tate's on his "p-divisible groups" on local fields with unequal characteristics, and Serre and Tate's lifting theory for abelian varieties in char $p > 0$. I wonder if you could think of any purely transcendental proof of the same result? I am finishing writing up the story and will send you a preprint of the proof pretty soon.

Best wishes

(signed) A Grothendieck

P.S. I am in Pisa till May 30.

Grothendieck to Mumford, 18 May, 1966

Manua di Pisa 18.5.66

Dear David,

I wonder if I did not misunderstand the last question of your letter (under the name, devil knows why, of "a Mordell–Weil theorem"), as the affirmative answer seems so trivial. Namely let A, B be abelian preschemes over any connected S of char. 0, $s \in S$, $u_s: A_s \to B_s$ a homomorphism, claim: there exists a largest abelian subscheme Z of A, such that $u_s|_{Z_s}$ lifts to $Z \to B$, namely: if Z, Z' are such, they are majorized by a third one Z'' having the same property. To see this, let $v: Z \times_S Z' \to A$ be the natural morphism, N its kernel. N is smooth over S (true for the kernel of any homomorphism of proper and smooth group preschemes over an S of char 0, as you will easily check), and of course proper over S. Take $Z'' = (Z \times_S Z')/N$. (NB This exists even without projectivity assumption on $Z \times_S Z'$ over S, by the way, using a general theorem of passage to quotient which, I believe, is stated in Murre's talk on unramified functors.)[109] This Z'' can be identified with an abelian sub-group-prescheme of A. Of course, if u_s is a monomorphism, then for every Z as above, $Z \to B$ is a monomorphism, as the kernel is smooth and proper over the connected S, and the fiber of said kernel at s is zero, hence also the whole kernel.

In char $p > 0$ or unequal characteristics, these results are no longer true, as follows from Koizumi's example. It will be true however if S is regular of dimension 1 if, instead of insisting on sub-group-schemes Z of A, you look at morphisms $Z \to A$ whose kernel at every fiber is radicial, and on the maximal fibers reduced to 0.

---

[108] This result was published in A. Grothendieck, Un théorème sur les homomorphismes des schémas abéliens, *Invent. Math.* **2** (1966) 59–78.

[109] *Séminaire Bourbaki* 1964/65, n° 294.

Why do you conjecture the Albanese varieties of the level modular varieties are zero? It's false for genus 1! What about genus two? What about the subgroups of finite index, made abelian, of the integral symplectic group $\mathrm{Sp}(2g,\mathbb{Z})$, do you know if they are finite?[110] It seems clear anyhow that they cannot contain a factor $\mathbb{Z}$, because then the p-adic analytic group $\mathrm{Sp}(2g,\mathbb{Z}_p)$ would have an open subgroup having a quotient isomorphic to $\mathbb{Z}_p$ (if $g \geq 2$, and using the nontrivial theorem of Bass–Lazard–Serre[111]), which it cannot. But on the other hand, is it not known by Borel–Harish-Chandra that the subgroups of finite index of $\mathrm{Sp}(2g,\mathbb{Z})$ are finitely generated? If so, this would prove that made abelian they become finite. By the way, thinking of the geometric interpretation of those subgroups as fundamental groups, it occurs to me we know beforehand they are finitely generated, without using Borel–HC, so it seems I myself answered the question I asked you. This solves also your $g=2$ case, which you thought you had to solve separately, as for $g=2$, the two types of modular varieties, for curves or abelian varieties, are essentially birationally equivalent; or does that silly Galois-group $\mathbb{Z}/2\mathbb{Z}$ cause serious trouble? Of course it might.... By the way do not the Teichmüller groups, just as the $\mathrm{Sp}(2g,\mathbb{Z})$, correspond to some algebraic groups, which could allow to apply the p-adic argument above? I confess I do not have any feeling so far for these Teichmüller groups. It would be nice to have them fit in the general yoga of arithmetic type discrete groups!

I am sorry for this somewhat chaotic letter. I came just back from Pisa where I spent two hours trying without success to overshout the tremendous noise coming from the street, while giving some introductory talk on l-adic cohomology. The noise here is just killing, otherwise everything is quite nice.

Yours
(signed) A Grothendieck

Grothendieck to Mumford, 4 November, 1966

Massy 4.11.1966

Dear Mumford,

I do not see why I should take offense when you tell me your opinions on some mathematical matters, all the less when I begged you to do so! As for your fears

[110] The answer is "yes": the abelianization of any subgroup of finite index in $Sp(2g,\mathbb{Z})$ is finite if $g \geq 2$, for instance because $Sp(2g,\mathbb{R})$ and $Sp(2g,\mathbb{Z})$ both satisfy Kazhdan's property (T). For a more general statement on the finiteness of the maximal abelian quotient of an arithmetic subgroup, see Chap. VIII, Corollary 2.8 on page 266 of G.A. Margulis, *Discrete Subgroups of Semisimple Lie Groups*, Springer-Verlag, 1991.

[111] Referring to H. Bass, M. Lazard and J.-P. Serre, Sous-groupes d'indice fini dans $SL(n,\mathbb{Z})$, *Bull. Amer. Math. Soc.* **70** (1964) 385–392. The congruence subgroup problem for the group $\mathrm{Sp}(2g,\mathbb{Z})$ here is solved in H. Bass, J. Milnor and J.-P. Serre, Solution of the congruence subgroup problem for SL_n ($n \geq 3$) and Sp_{2n} ($n \geq 2$), *Publ. Math. IHÉS* **33** (1967) 59–137.

concerning the eventual inclusion of the SGA seminars into EGA go, I can assure you that in every special instance, we (essentially Deligne and I) will decide to do this work only if the duplication in substance is not excessive, if the treatment of the known material can be made considerably simpler and more satisfactorily than in the SGA texts. This question will not arise before chapter IX, anyhow, as nearly all the material contemplated for the Chapters V to VIII (with the exception of part of what might be included in VII) is not available in the literature even in imperfect form. (Thus nearly everybody—and maybe you are the only exception—considers that my Bourbaki talks on construction techniques are too condensed to be of another use than one of preliminary information on what may be done, and a few indications of proofs). If we go on with a comparable speed as in the past, these chapters alone will keep us busy for another eight years or so, and by then we will have a clearer picture of what would be most useful to do next—and maybe to decide whether we should push the treatise any further at all. Of course in Chap VIII we will include about all we will know by then on Picard schemes, including the existence questions you allude to, about which I do not think I really know much more than you; Raynaud is working on the question for pencils of curves though, and there are a few precise results available in this case, including when the Picard functor is *not* representable. I guess that within the next few years the things one would really like to know along these lines will be clarified, and EGA VIII will be the equivalent of a book, giving an account of a well-understood subject!

As for the very spirit of a treatise like EGA, similar to Bourbaki, I have experienced so far that to write a really systematic treatment, even when including topics which are considered well known, is in the long run the more economical thing to do (on the one hand), and, by forcing you to more care and overthought on even familiar matters, an incentive to new progress as well. Of course, it does not serve exactly the same purpose, and should not be read the same way, as a paper or a moderately sized book on a more limited topic. With this restriction in mind I still believe (why should I not defend myself a bit!) that it is quite helpful to those who want to work in the field, by relieving them many times from tedious tasks by the possibility of reference to ready-to-use statements (not always is the situation as the one you complained of for Künneth-type relations in EGA III!), and providing some ready-to-use techniques.

I told Mlle Rolland to send you SGA3 3 and 4, and I hope you will finally get it!

Yours
(signed) A Grothendieck

P.S. I am sending you back in this same cover the paycheck from Harvard, as I do not consider to have done any work there, and hence do not think it proper to take any pay for it. I feel bad enough that I was obliged so abruptly and disappoint a few nice people, including myself!

Grothendieck to Mumford, 7 March, 1967

7.3.1967.

Dear David,

You once stated that you may be able to come to the IHÉS for something like two months in spring 1968. I wonder if your plans have grown more definite now and if you could tell us about it. There would be of course many people around here very interested if you could come for some time.

I would like to ask you a mathematical question, related with some results of Bott. One states that if there is, on a nonsingular analytic compact variety V, a holomorphic vector field with only isolated zeros, then the Chern numbers can be computed in terms of the behaviour of the vector field at these zeros, and hence vanish if the vector field does not vanish—a somewhat surprising result I found. The other states that if a finite group G acts holomorphically on V, so that for $g \neq e$, g acts with only isolated fixed points, then the Chern numbers mod $N = \operatorname{card} G$ can be computed in terms of the local action of the isotropy groups at the various points of V, and hence if G operates freely, then the Chern numbers are congruent to 0 mod N (the latter result being rather trivial directly, by the way). Using a somewhat different proof from Bott's, by using Segal's techniques, Illusie can extend the case of a finite group acting to the case of a smooth proper algebraic scheme over any field, it seems.[112] It is plausible too that by using the Lefschetz–Verdier formula for coherent sheaves, one should be able to work out an analogue of Bott's result for a vector field in the abstract case, however it seems that the method would yield information only for the Chern numbers mod p. Now, the yoga is that in char. $p > 0$, to give a vector field is not really better than to give an action of a p-group (for instance, if $A^p = 0$ resp. $A^p = A$, giving A amounts to giving an action of α_p resp. μ_p), and therefore I would not really expect anything better than a result mod p to hold. But on the other hand, of course, the last Chern class on V *can* be computed as the number of zeros (with multiplicities) of the vector field A (as an integer, not only mod p). So an example should decide what to expect. Thus, I would be pleased if you could find a surface in char. $p > 0$ with a vector field that vanishes nowhere, and still such that $c_1^2 = K^2$ (K the canonical bundle) be $\neq 0$. Or let's say it the following way: assume a surface (projective, nonsingular) V be given in char. 0, with the group $\mathbb{Z}/p\mathbb{Z}$ operating freely on it, and assume that we can find a nondegenerate reduction of V into char. p, in such a way that the given action extends to a free action of the group-scheme μ_p say; now why should this imply that $K^2 = 0$ (instead of $K^2 \equiv 0 \bmod p$ which we know before-hand)? Of course, we must have $c_2(V) = 0$, because the c_2 is invariant under specialization, and the specialized V will have a nowhere vanishing vector field; so there does in fact exist a nontrivial cohomological necessary condition for being able to reduce as stated the situation into characteristic p.

[112] Comments by L. Illusie: "The result which Grothendieck alludes to is in my paper, Nombres de Chern et groupes finis, *Topology* 7 (1968), 255–269. However, I worked in the context of almost complex varieties, with Atiyah–Segal K-theory. I didn't discuss the case of smooth proper schemes over an arbitrary field, and I haven't written any other paper on this topic."

By the way, I would appreciate having your comments on my comments to your proof of the Tate–Serre conjecture on the Néron model; do you agree with my criticism? Did you think again about it? I did not, and I doubt I will have time to include it into this year's seminar.

Did I tell you that Raynaud and I decided to write a book on Picard together? Also, Raynaud is willing to join us to finish writing up EGA (Chapters VI to VIII). So maybe Picard will be part of EGA after all, but the point is not too important of course. Raynaud is developing extremely nice representability theorems concerning Pic in his seminar.

Yours
(signed) Schurik

Mumford to Grothendieck, 1967 undated[113]

Dear Schurik,

I thought about your question of finding a surface with μ_p or α_p acting freely, yet $(K^2) \neq 0$, but I couldn't find one. There may be a better chance with higher dimensional varieties, since, using Kodaira's classification and some old Italian theorems (that I don't trust), there would appear to be very few surfaces in char. 0 for which $c_2 = 0, (K^2) \neq 0$.

About the Néron model: yes, your comments were quite correct and the "proof" that you indicated in your letters to Serre does indeed use essentially local uniformization and my "proof" is quite false. However, I have applied my theta functions to the problem, and, if K is a complete discrete valued field, residue char. $\neq 2$, I think I can now prove both (a) the result on the monodromy and (b) the result on the "stable" Néron model. I say "think" because I haven't written down the details systematically. In fact, one should get a rather complete "structure theorem" for these abelian varieties (I hope).[114]

$$\begin{cases} K = \text{complete discrete valued field, alg. cl. residue field } k,\ \text{char}(k) \neq 2. \\ C = \widehat{\overline{K}} = \text{completion of alg. cl. of } K. \\ \text{Let } X/K \text{ be an abelian variety.} \end{cases}$$

Then, after replacing K by a finite algebraic extension, one constructs

a) an algebraic group Y/K of "toroidal" type, i.e.,

$$0 \longrightarrow \mathbb{G}_m^r \longrightarrow Y \longrightarrow Y^* \longrightarrow 0,$$

Y^* an abelian variety,

b) $f\colon Y_C \to X_C$ a rigid analytic homomorphism defined over K,

---

[113] This letter was written sometime between March and August of 1967—Mumford wrote [u67b] in Maine during that summer before going to India. We are placing it here because the first part of the letter answers a question in [1967Mar02].

[114] See [u67b] in this volume, which, however, was written in the same summer.

Dear Schurik,

I thought about your question of finding . surface with μ_p or α_p acting freely, yet $(K^2)\neq 0$, but couldn't find one. There may be a better chance ith higher dimensional varieties, since, using Kodaira's lassification and some old Italian theorems (that I n't trust), there would appear to be very few surfac char. 0 for which $c_2 = 0$, $(K^2)\neq 0$.

About the Néron model: yes, your comments were ite correct and the "proof" that you indicated your letters to Serre does indeed ~~[illegible]~~ use essentially
is quite false
local uniformization/. However, ~~after~~ I have applied my theta functions to the problem, and, if K is a complete, discrete valued field, residue char. $\neq 2$, I think I can now prove both (a) the result on the monodromy and (b) the result on the "stable" Néron model. I say "think" because I haven't written down the details systematically. In fact, one should gets a rather complete "structure theorem" for these abelian varieties (I hope)

$$\begin{cases} K = \text{complete discrete valued field, alg. cl. residue field } \operatorname{char}(k)\neq 2. \\ C = \hat{\bar{K}} = \text{compl. of alg. cl. of } K \\ \text{Let } X/K \text{ be an abelian variety.} \end{cases}$$

Then, after replacing K by a finite algebraic extension, one constructs

a) an algebraic group Y/K of "toroidal" type, i.e.

$$0 \to \mathbb{G}_m^n \to Y \to Y^* \to 0$$

Y^* abelian variety

b) $f : Y_C \to X_C$ a rigid analytic homomorphism defined$/K$

such that :

such that:

(I) Y^* has nondegenerate reduction, hence there exists a scheme $\mathscr{Y}/\mathrm{Spec}(R)$ (smooth, connected fibres) whose generic fibre is Y, whose special fibre $\overline{Y}$ is again an algebraic group of toroidal type.

(II) f induces an algebraic isomorphism

$$\bar{f}: \overline{Y} \longrightarrow [\text{connected component of special fibre of Néron model of } X],$$

where the latter group is stable under finite extensions of K.

(III) If

$$\begin{aligned} Y'_C &= \{x \in Y_C \mid \text{the set of powers } \{x^n\} \text{ is bounded}\}^{115} \\ &= \{x \in Y_C \mid \text{closure of } x \text{ in } \mathscr{Y} \text{ meets } \overline{Y}\}, \end{aligned}$$

then Y_C splits canonically:

$$0 \longrightarrow Y'_C \longrightarrow Y_C \longrightarrow \Gamma^r \longrightarrow 0$$

(Γ = value group of C) [i.e., there is a canonical reduction of the structure group $(C^*)^r$ of Y_C/Y^*_C to (integers of C)^{*r}]. Then

$$\mathrm{Ker}(f) \text{ is } \begin{cases} \text{torsion-free, finitely generated} \\ \text{all its elements are rational over } K \\ \mathrm{Ker}(f) \cap Y'_C = \{0\}\ . \end{cases}$$

(IV) For all finite algebraic extensions L/K, $f(Y_L) = X_L$.

About visiting Paris in 1968: yes, I would like to do this and I had meant to write about this for some time. I would like to come for the month of May if this is alright. I will be coming (from India) with wife, 2 kids, and a Norwegian girl who helps with the kids. Perhaps you could ask your secretary to write me about the kind of accommodation that is available? I understand the Institute maintains some apartments? Thanks for your help in this.

Best Wishes,
(signed) David

[115] The part "$\mathscr{Y}$ meets $\overline{Y}$" in the second description of Y'_C was inserted by the editors. The original words near the margin were cut off during the photocopying process.

Grothendieck to Mumford, 2 May, 1967

Massy 2.5.1967

Dear David,

Thanks for your interesting letter on connections and stratifications. I had already wondered if by chance it is not always true for, say, a quotient of a power series ring over a field of char. 0, that the formal de Rham complex is a resolution of k, and if the de Rham complex for an algebraic variety, i.e., a scheme of finite type over k, does not always yield the correct cohomology (this would be a consequence, when X is complete, of a complex analytic variant of Poincaré's lemma for X, using Serre's GAGA). I had noticed with some surprise that it works for quite a few singularities of curves (like ordinary double points and cusps), and also a few Artin rings, and did not succeed to construct an example of an Artin ring giving the wrong H^0 of de Rham. So I am happy you got that silly question out of the way. I wonder though if by chance it does not work right for sufficiently "simple" singularities? I will also appreciate very much getting details on your ideas on compactification of the modular varieties and the connectedness theorem, as soon as you have some notes available. Today I sent you by air mail a photocopy of the notes on de Rham cohomology and crystals; they are still extremely sketchy, despite the floods of sweat they took to the redactors, who I am afraid did not understand too well so far what they were writing. I hope though that you will find the definition of the connection on de Rham cohomology explicit enough; I did not check though that the curvature tensor was zero, but do not doubt this is so.[116] Also you will find, when X/S is smooth and *moreover* S of char. zero, a definition of the absolute stratification of $\mathbb{R}f_*(\Omega^*_{X/S})$. As in the case of the connection, this does not

---

[116] The fact that the Gauss–Manin connection has curvature zero was proved in N.M. Katz & T. Oda, On the differentiation of de Rham cohomology classes with respect to parameters, *J. Math. Kyoto Univ.* **8** (1968) 199–213.

Here are two additional pieces of information; they may be both correct:

(a) W. Messing informed us that he "believe[s] Grothendieck had checked [the same fact] also, by the Autumn of 1967 (which was when the Katz–Oda paper was written)."

(b) In a letter to T. Oda, dated December 12, 1967, Grothendieck said:

"Thanks to you and Katz for your letter* on [the] integrability of the Gauss–Manin connection. I believe the same result should come out more conceptually, as well as the Leray spectral sequence,† using the 'crystalline site', which allows an interpretation of de Rham cohomology without using differentials. [...] I did not know the direct proof you propose for integrability, which looks interesting."

* A letter from N.M. Katz and T. Oda to Grothendieck on November 20, 1967, announcing and describing their algebraic proof of the integrability of the Gauss–Manin connection.

† Katz and Oda's letter also said, "A similar idea provides, in special cases, the Leray spectral sequence of de Rham cohomology which is conjectured at the end of your paper on de Rham cohomology."

does not à priori imply a corresponding stratification on the cohomology sheaves $R^i f_*(\underline{\Omega}^{\cdot}_{X/S})$, except in the case when we know that their formation commutes with arbitrary base change. Now when f is proper, this condition (by the standard Künneth type arguments) also just means that the $R^i f_*(\underline{\Omega}^{\cdot}_{X/S})$ are locally free, and this condition turns out to be a formal consequence of the fact that these sheaves are coherent (assuming S noetherian) and that ~~Rfx~~ the complexe $K^{\cdot} = Rf_*(\underline{\Omega}^{\cdot}_{X/S})$ from which they stem is bounded from above and endowed with an absolute connection, i.e. here a connection over a ground field k (here $k=\underline{Q}$). To see this, we are reduced to the case when S is of finite type over k by a standard limit argument, ~~then to the case~~ S ~~reduced to a point~~ (~~and~~ finite) over ~~k by another such argument, and by descent to the case~~ k alg closed. On the other hand, we now get the result by descending induction on i, using the fact that when the $\underline{H}^i(\underline{K})$ are locally free for $i > i_o$, then the formation of $\underline{H}^{i_o}(\underline{K})$ commutes with base change, hence $\underline{H}^{i_o}(\underline{K})$ has a stratification, hence is locally free. – One can prove the same result by a transcendental argument, not using the canonical stratification, by reducing to the case when k is the field $\underline{C}$ of complex numbers, S is reduced to a point (hence artinian), using Gaga to reduce to the corresponding complex analytic (analytic) statement, and using Poincarés lemma for the De Rham complex of X/S. I hope the same transcendental argument, using a suitable "relative" Hodge theory (over an Artin ring over $\underline{C}$) should prove that the $R^q f_*(\underline{\Omega}^p_{X/S})$ are equally locally free, which will imply that the spectral sequence beginning with these as $E_1^{p,q}$ and ending up with $R^n f_*(\underline{\Omega}^*_{X/S})$ degenerates, i.e. that the relative Hodge cohomology is just Gr $(R^n f_*(\underline{\Omega}^*_{X/S}))$, the graded sheaf associated to relative De Rham cohomology. The argument applies in any case (as was pointed

a priori imply a corresponding stratification on the cohomology sheaves $R^if_*(\Omega^*_{X/S})$, except in the case when we know that their formation commutes with arbitrary base change. Now when f is proper, this condition (by the standard Künneth type arguments) also just means that the $R^if_*(\Omega^*_{X/S})$ are locally free, and this condition turns out to be a formal consequence of the fact that these sheaves are coherent (assuming S noetherian) and that the complex $K^\bullet = \mathbb{R}f_*(\Omega^*_{X/S})$ from which they stem is bounded from above and endowed with an absolute connection, i.e., here a connection over a ground *field* k (here $k = \mathbb{Q}$). To see this, we are reduced to the case when S is of finite type over k by a standard limit argument. On the other hand, we now get the result by *descending induction on i*, using the fact that when the $\mathscr{H}^i(\mathscr{K}^\bullet)$ are locally free for $i > i_0$, then formation of $\mathscr{H}^{i_0}(\mathscr{K}^\bullet)$ commutes with base change, hence $\mathscr{H}^{i_0}(\mathscr{K}^\bullet)$ has a stratification, hence is locally free. One can prove the same result by a transcendental argument, not using the canonical stratification, by reducing to the case when k is the field $\mathbb{C}$ of complex numbers and S is reduced to a point (hence artinian), using GAGA to reduce to the corresponding complex analytic statement, and using Poincaré's lemma for the (analytic) de Rham complex of X/S. I hope the same transcendental argument, using a suitable "relative" Hodge theory (over an Artin ring over $\mathbb{C}$) should prove when f is projective that the $R^qf_*(\Omega^p_{X/S})$ are equally locally free, which will imply that the spectral sequence beginning with these as $E_1^{p,q}$ and ending up with $R^nf_*(\Omega^*_{X/S})$ degenerates, i.e., that the relative Hodge cohomology is just $\mathrm{Gr}(R^nf_*(\Omega^*_{X/S}))$, the graded sheaf associated to relative de Rham cohomology.[117] The argument applies in any case (as was pointed out to me long ago by Hironaka) when S is reduced. I could not get it out when S is just artinian, by a purely algebraic proof, using the corresponding result over the residue field and the fact that the de Rham cohomology is free, by some general argument of spectral sequences; maybe I just did not try hard enough, as the information available on the spectral sequence seems already rather strong. Notice also that the fact that $R^1f_*(\mathscr{O}_X)$ is locally free is proved in my second talk on Picard schemes (corollaire 3.6); when f is of relative dimension 2, what about the prospective dual $R^1f_*(\Omega^2_{X/S})$? (In fact, the former is a priori the dual of the latter, by the global duality theorem....)

Of course, the theory of crystals gives considerably more on the $R^if_*(\Omega^*_{X/S})$ than just a stratification, namely it endows them with a canonical structure of an absolute crystal; i.e., these sheaves extend automatically to sheaves over any infinitesimal neighbourhood of S. Although this point of view is not worked out in the notes, it must come out rather formally, I am convinced, by interpreting these sheaves as coming from $R^if_{\mathrm{cris}}(\mathscr{O}_{X_{\mathrm{cris}}})$, where $f_{\mathrm{cris}}\colon X_{\mathrm{cris}} \to S_{\mathrm{cris}}$ is the morphism of the *absolute* crystalline topoi associated to the morphism of schemes $f\colon X \to S$. The same remark should hold in arbitrary characteristics too, using now "crystal" in the sense of the IHÉS notes, namely involving divided powers; this is something pretty more precise than just a connection on the sheaves $R^if_*(\mathscr{O}_{X_{\mathrm{cris}}})$.... For the applications of

[117] The degeneration at $E_1^{p,q}$ of the relative Hodge-to-de Rham spectral sequence was proved in Thm. 5.5 of P. Deligne, Théorème de Lefschetz et critères de dégénérence de suites spectrales, *Publ. Math. IHÉS* **35** (1968) 107–126.

this to varieties or formal groups in char. $p > 0$, in particular to the interpretation of Dieudonné module and infinitesimal variations for p-divisible groups, as suggested in my Pisa letter to Tate, I want just to point out that if A is a ring, and p a prime number which is nilpotent in A, (for instance $A = W_n(k)$, k a perfect field of char. $p > 0$) then the ideal pA admits a canonical structure of divided powers. Thus I think that my interpretation of infinitesimal variations of an abelian variety or a p-divisible group works correctly when one takes, as infinitesimal parameter varieties, varieties endowed with an extra-structure of divided powers in the augmentation ideal. This (I think) is enough for various applications, such as a nice description of the Dieudonné module (using the previous remark on the $W_n(k)$), and at the same time rules out the unpleasant counterexample in my Pisa letter, which was concerned precisely with "vertical" (relative to $\mathrm{Spec}(\mathbb{Z})$ or $\mathrm{Spec}(W(k))$ variations of structure, namely those remaining in char. $p > 0$—the explanation now being that in an ideal $\mathfrak{m}$ of a ring of char. $p > 0$ there can be no divided power structure unless $\mathfrak{m}^{(p)} = 0$![118] Of course, in char. 0, the divided power structure always exists and is unique, so does not add anything, which explains why things worked so smoothly in char. 0.

Yours
(signed) Schurik

Grothendieck to Mumford, 1 August, 1967

Letter from Groth. 8/1/67 [119]

Coates and Jussila are very busy writing down the notes of my talks on de Rham cohomology,[120] which should be ready within a few weeks. I will send you however in this letter a copy of the two pages containing a direct construction of the Gauss–Manin connections. By the way, I had some extra thought about the definition of p-adic cohomology in char. p, and believe I have the right definition at last, using still another site, the so-called "de Rham site" of a relative scheme X/S, whose objects are (Zariski) open subsets U of X, together with a "thickening" U' of U i.e., a nilpotent immersion $U \to U'$ over S, and moreover a "divided powers structure" on the augmentation ideal for $\mathcal{O}_{U'} \to \mathcal{O}_U$. When working in char. 0, this extra structure is uniquely determined and we get the usual "site cristallin", whose cohomology

[118] Here $\mathfrak{m}^{(p)}$ is the ideal generated by elements of the form x^p with $x \in \mathfrak{m}$. In the original the formula looks more like $\mathfrak{m}^p$ than $\mathfrak{m}^{(p)}$. Apparently Grothendieck forgot to put in the pair of parentheses by hand—otherwise the statement is false as he surely knew. A counterexample due to Koblitz and Ogus is in 4.4.1 of P. Berthelot and W. Messing, Théorie de Dieudonné cristalline III: théorèmes d'equivalence et de pleine fidelité, *The Grothendieck Festschrift*, Vol. I, Progr. Math. 86, Birkhäuser, 1990, pp. 173–247.

[119] Handwriting of Mumford; so August 1 may be the date of receipt.

[120] Published as: Crystals and the de Rham cohomology of schemes, in *Dix Exposés sur la Cohomologie des Schémas*, North-Holland, 1968, pp. 306–358.

with coefficients in the structure sheaf of local rings is just (when X smooth over S) the relative de Rham cohomology $\mathbb{H}^*(X_{\text{zar}}, \Omega^*_{X/S})$. It now seems to me that essentially the same proof will show the same result in arbitrary characteristics, when working with the de Rham site, involving divided powers. These divided powers seem in a most subtle way to rule out the troubles I had come upon in my Italian letter to Tate on crystals[121] (in connection with elliptic curves of Hasse invariant zero). It seems to me that, at least for those smooth proper schemes in char. $p > 0$ which lift to char. 0, all the usual properties for the p-adic cohomology will then follow readily from the known results in char. 0, the ground ring for the cohomology theory being actually the *ring* of Witt vectors $W(k)$ (not its field of fractions, as in Washnitzer–Monsky's theory). This should then rule out the possible presence of denominators p^r in Weil's conjectures, which I alluded to in my letter to Serre on the standard conjectures for algebraic cycles.[122] I also got finally quite an interesting letter from Griffiths. As far as I could understand, though, he does not know about any new "finite" relations for the "geometric" Hodge structures (those embeddable in the Hodge structure associated to a projective smooth variety over $\mathbb{C}$), but *infinitesimal* relations concerning variation of Hodge structure coming from a variation of an algebraic variety. The result he states is enough in any case to take care about my worries concerning Tate's conjectures, as you pointed out to me yourself. Maybe after all there are no such finite relations as we contemplated, but only infinitesimal ones. It would be of course highly interesting if the necessary conditions he obtains for an infinitesimal variation of Hodge structure to be "geometric" are also sufficient, and to get a corresponding result for finite variations. His results anyhow show the a priori possibility of a completely different picture for moduli of motives from the one I originally had in mind, with an essential part being played by differential equations. This checks very well with my de Rham yoga, where differential operators of arbitrarily high orders are involved in quite an essential fashion.

Yours
(signed) Schurik

Grothendieck to Mumford, 18 March, 1968

Massy, March 18, 1968

Dear David,

Thank you very much for your letter, which has crossed with Deligne's telling me very much the same thing as you told me in yours! I believe your proof is concep-

[121] A scan of this letter is available from
http://www.math.jussieu.fr/~leila/grothendieckcircle/mathtexts.php

[122] This letter on the standard conjectures, dated 27 August, 1965, is in *Correspondance Grothendieck–Serre*, Soc. Math. France, 2001; *Grothendieck–Serre Correspondence* (bilingual edition), Amer. Math. Soc., 2004.

tually less sophisticated and therefore simpler than Deligne's (although in essence the same)[123], but that Deligne's systematic use of "étale topoi" as generalized varieties will eventually provide the better insight into the geometry of these questions. I am convinced this is really an important generalization of the notion of scheme, and we will have to deal with it systematically, alongside with Mike's intermediate generalization[124], starting with EGA VI.

The proof in my letter to Serre about semi-stable reduction of abelian schemes (via a monodromy theorem $(1-g^N)^2 = 0$ for l-adic H^1) works all right in all cases, and yields even in arbitrary dimension (when written out with care) $(1-g^N)^{i+1} = 0$ in H^i, as was known to Griffiths by transcendental methods in the complex case. Thus I have two essentially different proofs of the monodromy theorem, one arithmetic (which works over any discrete valuation ring R which has residue field of finite type, or which is localized from an algebra of finite type over a field), the other geometric (which works without restriction on R, provided $i = 1$).[125] Both work in characteristic zero (all i, any R); both use resolution some way, and both would work for any R, any i if resolution was known for schemes of finite type over $\mathbb{Z}$, except that the geometric proof uses moreover purity. En revanche, it gives a little more precise information in the smooth case (namely the exact exponent $i+1$); for $i = 1$, we need resolution and purity only in a range of dimensions where both are known (by Abhyankar, and Zariski–Nagata's purity theorem). The arithmetic proof on the other hand applies also to nonconstant coefficients—all that is needed is that the l-adic sheaf whose cohomology we are taking comes from a situation of finite type over the integers; morally, this means that it comes from a motive over your scheme X. I did not try to make the geometric proof yield the same kind of generalization; it may turn out notably more difficult. All these things will be explained at length, including applications to Néron models and the like, in my own exposés in SGA7 "Groupes de Monodromie Locale", which is a joint seminar by Deligne and myself, and has started this month. Later Deligne will give an algebraic proof of Lefschetz's theorem[126] about the Pic of the "general" surface of degree ≥ 4 in $\mathbb{P}^3$, which works in any characteristics, after some general facts about vanishing cycle theory. By the way, as you will have gathered, Deligne is extremely bright; I believe, brighter than anybody else I know in mathematics.

Now to your comments about the use of the word "variety". I was a bit surprised to see that this question nearly upsets you, and still more by what you say on algebraic geometry becoming "a still more unpleasant subject" with its "rival

[123] Referring to the proof of the irreducibility of M_g, paper [69e] in this volume.

[124] The notion of *algebraic spaces.*

[125] Grothendieck's arithmetic proof of the monodromy theorem was published in the appendix to J.-P. Serre and J. Tate, Good reductions of abelian varieties, *Ann. Math.* **88** (1968) 492–517, and also in the appendix of SGA7, Exposé I, LNM 288, Springer-Verlag, 1972. The geometric proof was published in §3 of SGA7, Exposé I. See SGA7, Exposé IX, *Modèles de Néron et monodromie,* for further discussion of the monodromy theorem. For a more elementary proof of the stable reduction theorem for curves see M. Artin and G. Winters, Degenerate fibres and stable reduction of curves, *Topology* **10** (1971) 373–383.

[126] The proof appeared in Exposé XIX of SGA7, LNM 340, 1973, pp. 328–340.

schools".... I never realized algebraic geometry was an unpleasant subject, nor that there was any rivalry among algebraic geometers; I have lived so far in the belief that all of us, although tastes and yogas may largely differ from one to another, are working towards a common goal of better insight into geometry, and that every one among us is glad about any good result any of his colleagues would get, and eager to make use of it, whatever the methods and the spirit in which this result may have been obtained or exposed. Does your own experience really tell you anything to the contrary? Also, I wish to assure you that I would never think of suspecting you or anybody else of being "personal", as you feared apparently I would, when discussing about any mathematical question, including questions of terminology. I would like you, on the other hand, to be as sure that I don't take it personally either, and that, when working out a terminology, my aim is not to tease, annoy or hurt anybody, and you less than anybody else. In fact, I learned with Bourbaki (and through my own experience) how much a good terminology is important for an easier understanding of mathematics and smoother working, and to be very painstaking in these matters, more than the average mathematician I would think, and to spend a nonnegligible amount of time on it. It is after serious consideration that I decided myself with Dieudonné, for instance, to change "simple" into "smooth" after SGA3, "non ramifié" into "net" more recently, "locally free sheaf of rank one" into "invertible sheaf" according to Tate's suggestion, and "prescheme" into "scheme" resp. "scheme" into "separated scheme". I do not pretend the result to be perfect, but I guess it is coherent and reasonably suggestive, and does not seem to offer difficulties to young people who have no "mental blocks" by too strong a habit of one of the other existing terminologies. But even if you believe the result to be bad, discussion will be easier if you don't assume it is so by purpose and in order to annoy anybody. Now let me come to the points you make about the word "variety", and to my own points.

1) You contend the word has already a precise and generally accepted meaning, I accept that it has for various mathematicians, but that the meaning is now pretty much different from one to another. You wrote me which is yours: an integral separated algebraic scheme. Mike on the other hand uses the word to mean just any algebraic scheme, and, according to context, he will understand implicitly that the scheme is separated, or that it is only locally of finite type (instead of finite type as an algebraic scheme should be); by the way, he uses much the same way analytic variety to mean any analytic space (in the sense of my exposés in Cartan's seminar). Weil's use of "variety" is closer to yours, but still different, as he assumes it to be geometrically integral. I am not too sure what is Zariski's terminology, but I guess it will be still different, something like a subset of projective space defined by a set of equations; I would have asked him if he were around, and maybe he would have been a bit embarrassed really to tell me what he means by "variety"!
2) I do not think there is a strong tendency in French to imply nonsingularity by the use of the word "variété". In algebraic geometry at least such an implication has never existed. In topology, most times nowadays when topologists speak about "varieties", they admit varieties with boundaries, and as soon as one starts tak-

ing products, even the boundaries acquire singularities. In the Cartan seminars, "variété analytique" does imply nonsingularity, but this terminology is by no means universally accepted. See Mike's use above. Also, if people like Thom or Whitney speak about "analytic varieties", they are mostly interested in their singularities, which they want to stratify in various ways!

3) Quite generally, I think that the natural trend now in the use of the word "variety" is to make its meaning ever wider—so much so as to include even functional spaces, when allowing the varieties to be infinite dimensional! This is in a way not better nor worse than viewing an arbitrary scheme (not only noetherian ones, or those of finite type over a field) as being "varieties". I believe that the only 'a priori' natural limitation to the use of this word should be that it should extend as far as does the specific geometric intuition, and some of the main technical features, of the objects which were initially considered. Objects which have local rings, tangent spaces and higher order differential invariants, for which all or most of the most important geometric constructions (projective bundles and other fibrations, Picard varieties and the like, normalisation etc) can be performed, seem to me to be eligible for the term "variety". I think however, like Deligne, that objects like the "étale topoi" present some essentially new features, which demands for a more sophisticated kind of geometric intuition than the usual one, mainly through the fact that morphisms of such objects (and in particular, their geometric points) may have nontrivial automorphisms—and that for this reason it would be unwise to subsume them also under the name of "variety".

4) As you admit yourself in your letter, it seems rather likely that the most natural "compact" moduli objects, for curves or abelian varieties, are not schemes, but just what we would like to call "varieties". Now these modular objects, you will agree, are among the most basic and important ones geometers would like to study, and I am convinced that their importance will still increase both for geometry and number theory in the next fifty years or more. So, just because of a taboo coming from some particular training of yours, you would forbid yourself forever to call these remarkable beings "modular varieties", as everybody has done so far since Riemann, I believe? I think you have just missed that point, that these "varieties" are precisely the good kind of objects, just varieties! Not really any different from what one has considered so far, and providing just a closer and better link with the usual analytic varieties (or "analytic spaces")—as various operations which could so far be performed only in the complex analytic context acquire a meaning also in algebraic geometry.

5) Quite generally, it is becoming rather clear now that the new "varieties" are the more "natural" objects when compared say with schemes, because the category of these varieties seems to have a remarkable stability with respect to those geometric constructions which seem the most important, and which sometimes get us out of the category of classical minded varieties, or schemes: contractions and other types of passage to quotient, representation of functors of Hilbert and Picard type, modular spaces of all kinds.... Therefore these objects do deserve a simple name, and possibly one which has already a rich intuitive content through the use which has been made of it before? I believe this is by no means an insult

to the classical people, but will *eventually* turn out to be an homage to them—as *at present* it is intended to stress for the "usager" the geometric significance of this comparatively new notion.

I discussed the matter with Deligne, who essentially shares my opinion on this matter. However, he told me that he did not wish to use this word consistently at the cost of upsetting you as it seems it does. We have little hope to convince you that the usage we want to make of the word "variety" is at present the best, but we do hope at least that you will let yourself be convinced that there is nothing offending to anybody in this use, that each of us came to the conclusion that this is best by objective motives, and not personal ones wishing to hurt anybody. After all, it is really not catastrophic if you go on using the word variety according to your own taste, and certainly quite a few others will do the same. At present, the only motive which keeps us from using the terminology which objectively seems to us the best, and to go on working with our minds in peace, is a personal one, as neither of us has a wish to hurt your feelings! Therefore, please consider the matter again and write us if your feelings need really be hurt, or if you believe that it is reasonable that *we* should adopt a terminology which, after careful consideration, seems best to both of us, and for which neither of us is able to find a satisfactory substitute.

With my best wishes to you and your family

(signed) Schurik

Please give my best regards to the Seshadris, to Ramanujam and to Ramanathan.

Grothendieck to Mumford, 2 August, 1968

Massy 2.8.1968

Dear David,

I looked through Cartier's notes on formal groups, as I am interested in a description of p-divisible groups over a scheme of char $p > 0$, or more generally over a scheme all of whose residue characteristics are p or zero. His description does not look too handy directly, especially the filtration he has to use is rather annoying. One would like something which directly generalizes Dieudonné's description over a perfect field: free modules of finite type over the ring W of Witt vectors, together with F and V satisfying the three known relations (in fact, V following from $F\ldots$). Did you work out any such description using Cartier's work? If so, I would be very grateful to you to write me what you know. I have been trying a bit to make more precise what I mumbled to you about crystals and p-divisible groups; that is why I need Cartier's stuff. By the way, I convinced myself that the description I suggested for p-divisible groups over unequal characteristic discrete valuation rings works only if the maximal ideal has topologically nilpotent divided powers structure. But this restriction should be unnecessary when dealing with p-divisible groups *up to isogeny* over V: such a structure should correspond exactly to a Dieudonné space M over the field of fractions K of W, and a filtration of $M \otimes_K L$ ($L = K \otimes_W V$) subject to the only condition that the dimensions of the two occurring factor spaces should be the correct ones (namely the dimensions of the group and its dual in char. p). I more or less checked this when the group in char p is "ordinary", i.e., extension of any ind-étale by a multiplicative type p-divisible group.

Best regards
(signed) Schurik

Grothendieck to Mumford, 9 August, 1968

Massy, August 9, 1968

Dear David,

Thanks a lot for your letter of July 24, which I got yesterday. I am sorry you cannot come for a whole year in 70/71, but am glad that you think you can come for about two months. Thanks for suggesting that I should come to Warwick for a while that same year. I guess it could be done, if you do not expect me to be there for longer than a week or ten days. I am ready to tell other people about that symposium to suggest participation, but maybe it would be useful if you could tell me a few words more what such a "low-pressure symposium" will be supposed to look like. Also thanks for your invitation to join the panel of invitations for the next international Congress; as I am not too convinced of the usefulness of such Congresses, I believe however you better leave me out!

I got Griffiths' preprints and "disclaimer"[127] at the same time as your letters. It looks quite startling indeed, but I had no time to look at it seriously as yet. And I managed to lose the preprints, and had to ask Gr. for another copy! By the way, his results (about which he is himself dubious) do not affect what I really call the standard conjectures, on which the theory of motives relies; these do not assert anything about τ-equivalence. But in order to come to a coherent picture concerning intermediate jacobians, and the tie they provide between Hodge's index theorem and the Néron–Tate form (by interpreting the intersection form on primitive cycles as a Néron–Tate form on a suitable intermediate jacobian), it has been extremely tempting to surmise that τ-equivalence equals numerical equivalence. I will have to reconsider the matter anew if really this assumption should turn out to be false. These questions were on my holidays' program, but I did not start so far, as I was still busy trying to come to some understanding on crystals and their relations to p-divisible

[127] The papers were published as On the periods of certain rational integrals, I, II, *Ann. Math.* **90** (1969) 460–495 and 496–541. However the published version contains no "disclaimer". Note also that Grothendieck gave a purely algebraic proof of Griffiths' theorem, which was written up by N. Katz in Éxposé XX of SGA7, LNM 340, Springer-Verlag, 1973, pp. 341–362.

groups (which should really be called Barsotti–Tate groups[128], as p-divisible should just mean that multiplication by p is an epimorphism, and not more). I hope you got my last letter asking you questions in this connection on Cartier's theory, and that you will be able to give me some information I need.

Best wishes
(signed) Schurik

Grothendieck to Mumford, 4 September, 1968

Sept. 4, 1968

Dear David,

Thanks for your letter. If you think I can give you advice on selecting speakers for the next congress, I will certainly not refuse giving it to you; but I guess for this there is no need for me to join your panel. A rather evident thing would be to ask that Griffiths should give a one hour talk. I am particularly impressed by theorem (**) stated in his "disclaimer", opening completely new perspectives. But have you been able to discover where, in Gr.'s paper, this theorem is proved, or even to convince yourself that the proof is OK? In any case, his theorem E (5.6) completely convinced me that my feelings on the relations between Hodge's index theorem and the Néron–Tate form were erroneous, so that I have no reluctance any longer to admit that τ-equivalence is indeed distinct from homological equivalence.[129] As for the explanations you give me in your letter, they seem to me to concern rather Gr.'s

[128] Grothendieck's footnote: Tell me if you agree, please.

[129] Comments by Phillip Griffiths, concerning [1968Aug09] and [1968Sep04]):

"Thank you for sending me the email with Grothendieck's letters to David. I believe what they are referring to (the part about my stuff) is

$$\text{homological} \neq \text{algebraic}$$

equivalence for higher codimensional cycles, even modulo torsion. Later, Clemens showed the quotient

$$\text{LHS}/\text{RHS}$$

is a countably but not finitely generated abelian group. My methods (and intuition) were classical analytic/geometric and it seems that David—to none of our surprise—was able to understand the argument and convince Grothendieck. Together with David's example of $\dim CH^2(X) = \infty$ on a surface X with $p_g(X) \neq 0$ this opened up an era (maybe "can of worms") of stuff regarding cycles with, at least to me, the main real progress since being the conjectures of Bloch–Beilinson which at least bring some order into $CH^*(X)$ and explain why $\dim CH^2(X) = \infty$ occurs (cf. the paper by Mark Green and myself in IMRN, 2003 that treats this).

Of course the central question—the Hodge conjecture together with its generalizations by Grothendieck—has seen no real progress in 50+ years (except that it has so far been consistent with other known / conjectural things).

What comes through also in Grothendieck's letters, and this was my personal experience as well, is how direct and to the point he is. Oh that he had written EGA."

theorem E, whose proof I understand (as I knew Lefschetz's proof of Noether's theorem); but I do not see why this should directly give an example, say, of a curve on a three-fold, homologous to zero and not τ-equivalent to zero. With your notations (V a general hypersurface section of high degree of W), the question remains why a primitive cycle of middle dimension on X, whose restriction to W gives the zero element in Weil's intermediate jacobian $J(V)$, should be itself homologous to zero? Gr. himself refers for this to the rather technical §'s 11 to 15!

I made some progress with the relations between Tate–Barsotti groups and crystals since I last wrote you, but was waiting to get your answer to my questions before starting some final checking for the crystal interpretation of the Dieudonné module, and also in order to check that if S is any (?) scheme of char. $p > 0$, then a Tate–Barsotti group on S is "the same thing" as a crystal M of locally free modules over S (crystal in the absolute sense, i.e., over $\operatorname{Spec}\mathbb{Z}$, or $\operatorname{Spec}\mathbb{Z}_p$, and in a sense slightly more sophisticated than in my notes, by asking that the divided power structures in the definition of the crystalline site should be compatible with the one we have on the maximal ideal of $\mathbb{Z}_p$), *plus* the maps F and V between M and $M^{(p)}$ satisfying $FV = p \cdot \mathrm{id}$, $VF = p \cdot \mathrm{id}$. (In case $p = 2$ one will have to be more careful, but I believe that an analogous statement will still make good sense). Then the analogue of Tate's theorem for the equicharacteristic case should follow from the general crystal-theoretic fact (which I did not try to check either so far) that if M, M' are crystals of locally free modules over the noetherian normal connected scheme S (no F and V here, and indeed I like to think about crystals as coming from cohomology groups of higher dimension as well), then any morphism between the generic fibers of these crystals is induced from a morphism $M \to M'$. I also see along which lines to look for a generalization of Tate's theorem to crystals in the unequal characteristic case, via the definition of a functor from filtered crystals with "Frobenius" F to Galois modules over the generic point; the description of this functor remains however the most mysterious point, which I will have to elucidate first in the case of Tate–Barsotti groups, with the help of Cartier's theory and Tate's ideas. The theorem should be that this functor is fully faithful working mod isogeny. Granting this functor, I see also what should replace Serre–Tate's theorem (cor. 1 to th. A in my Inventiones paper)[130] in higher dimensions, so as to get a principle of proof of conjecture 1.4 of that paper in arbitrary dimensions: namely for a projective smooth scheme X over the unequal characteristic discrete complete valuation ring V, a de Rham cohomology class should be algebraic if it is algebraic when interpreted as a crystalline cohomology class *of the special fiber* X_0, and if moreover it has the correct filtration. In other words, the functor from semi-simple motives over V, to pairs of a motive M_0 over the residue field k, together with a filtration of its crystalline realization $T_{\mathrm{cr}}(M)$ (a finite dimension vector space over the quotient field K of V), is fully faithful. (Analogue of Hodge's conjecture!) Maybe these statements will even turn out to be provable!

In connection with these questions, I wonder if for a projective smooth variety X_0 over k, one can foresee the value of the $h^{p,q}$ of any lifted variety from the structure

---

[130] *Invent. Math.* **2** (1966) 59–78; the corollary is in p. 63.

of the crystalline $H^n(X_0)$ $(n = p+q)$ together with the F-structure on it, namely the semi-linear map $H^n \to H^n$ stemming from the Frobenius map $X_0 \to X_0^{(p)}$. For instance, are the $h^{p,q}$ independent of the lifting? I would appreciate to know if you have any idea on this.

Best wishes
(signed) Schurik

My personal address:
2 Av. de Verrières, Massy (Essonne) France

NB. I take my mail at the IHÉS only once a week.

Re P.S. I am puzzled about Gr.'s 10.12, which looks false: take a family of subvarieties W of $\mathbb{P}^r$ with variable periods, and blow them up! Therefore I am dubious about the proof of 10.13 as well.

Grothendieck to Mumford, October 10, 1968

Massy 10.10.1968

Dear David,

Thanks for your letter. In the meantime I have thought some more about Griffiths' result, and come to exactly the kind of proof you outline. It seems to me (although I did not check carefully enough) that the same argument carries over to char. $p > 0$, whenever we know that Lefschetz's hard theorem holds true (no trouble for complete intersections for instance!) and provided we know moreover that, for a *general* hypersurface section of X^{2m} of some high enough degree, the vanishing cohomology of Y^{2m-1} is not of *level* 1; and for this, if we take the ground field to be finite, it is enough that we can find *some* hypersurface section Y of that degree, nonsingular, defined over a finite field with q elements, such that the proper values of Frobenius acting on $E(Y^{2m-1})$ (the vanishing part of the cohomology of Y), divided by q^{m-1}, are not all algebraic integers; or what amounts to the same, that the coefficients c_i of the corresponding polynomial $f(t) = \prod(1+\alpha_i t)$ are not divisible by $(q^{m-1})^i$. Of course, the transcendental situation suggests that we should even get maximum level $2m-1$, i.e., we should not be able to divide by q^i even. Now Katz told me that this can be effectively checked for various *complete intersections* in $\mathbb{P}^r$: indeed, it is enough that Y has no point rational over its field of definition k, because the number of such points (if Y is of dimension n, even or odd, it does not matter of course) is

$$1+q+\cdots+q^n+(-1)^n\sum\alpha_i,$$

which implies that not all algebraic integers α_i are divisible by q if this sum is to be zero! On the other hand, you can find a hypersurface of given degree, multiple of $q-1$, rational over the field with q elements and which has no point over that field, by taking $\sum a_i X_i^{d(q-1)} = 0$, the $a_i \in F_q$ being such that no partial nonempty sum of them is equal to zero; this works at least if q is > number of variables, by taking all a_i equal to 1. If the intersection of that hypersurface with X is nonsingular, we win. This works for instance for Griffiths' quadric in $\mathbb{P}^5$ (if char. $\neq 2$, at least).

A weird fact is that Griffiths' construction gives examples only over function fields, no field algebraic over the prime field. It seems quite hard to deduce an example over a number field, say, although there should certainly be such an example! I feel less secure over a finite field, and would not be surprised if it turned out that in the case where the ground field is the algebraic closure of a finite field, then numerical equivalence implies τ-equivalence. One heuristic reason is that the points of abelian varieties over such a field are of finite order. Another key invariant we can associate to a cycle Z on X/k which is cohomologically trivial on $X_{\bar{k}}$, namely the element of $H^1(k, R^{2i-1}f_*(\mathbb{Q}_l(i)))$ stemming from the Leray spectral sequence, should vanish (by virtue of the Weil conjectures) when k is a finite field. I would suspect that for k of finite type in the absolute sense, the vanishing of that class implies τ-equivalence to zero, and more precisely should be characteristic of some more refined equivalence, something like Picard-equivalence up to torsion, which in the transcendental case would be expressible by the fact that the image of the class in Griffiths' torus is a torsion element.

I did not prove what I surmised about the relations of Barsotti–Tate groups in char. p to Dieudonné crystals, and have not been thinking about these things for some time. Cartier says he checked the statement I proposed about classification of B–T groups over unequal characteristic discrete valuation rings with divided powers in the maximum ideal, in terms of a filtration of the extended Dieudonné module. I do not think he looked at the corresponding statement (without divided powers) for classification up to isogeny.

Yours
(signed) Schurik

Grothendieck to Mumford, 20 November, 1968

20.11.1968

Dear Mumford,

Thanks for your letter and your very nice paper on rational equivalence, which I just read.[131] Some trivial comments: in par. 1 the nonsingularity of X is not used. On page 4, line -6, the relation $S = \widetilde{S}/G$ would need a word of explanation, as it uses normality of S and char. 0. Page 5, instead of "*any*" you mean "commutative" I guess; it took me a while to understand what you meant to say before lemma 1, till I realized that $f\colon S \to Y$ and η_f were as before, but that you just allowed $\widetilde{S}$ and p and

[131] [69d] (= article 28 (pp. 753–762) in [SP1]).

$\widetilde{f}$ to change. (NB I often have trouble when reading you with such trivial matters, whereas otherwise your informal style makes understanding rather easier.) Page 6, line 10 the meaning of "too" is mysterious. Page 9, line −2, I guess m depends on i too. I think you may add as a corollary to your result that Samuel's conjecture (or question) that $((x)-(y)) \times ((x')-(y'))$ on a product variety is rationally equivalent to zero is false already for the product of two elliptic curves (as this would imply in this case that rational equivalence of 0-cycles is the same as Albanese equivalence).

Your objection to carrying over Griffiths' construction to char. p is met with by observing that the monodromy representation of $\pi_1(\mathbb{P}^1 - \{\text{critical points}\})$ is irreducible on the vanishing cycles space, just as in the transcendental setup; a fortiori, the corresponding "motive" cannot split. This is proved by Lefschetz's argument using the Picard–Lefschetz formula and the fact that the vanishing cycles corresponding to various critical points are still conjugate to each other. The first fact will be proved by Deligne in our seminar this year, by an argument of reduction from char. 0 to char. p, which requires care but is of rather standard nature; in char zero the transcendental theory can be used. (I do not know if a purely algebraic proof can be found;[132] I suppose yes for a pencil of curves, and you should know best....) The conjugacy statement should come out just the same way as in the classical case, using the irreducibility of the variety of critical hyperplanes, and will certainly be done too by Deligne in this seminar. He needs it for proving Noether–Lefschetz's theorem in char. p. I confess I never wrote out full proofs myself, but am quite confident it will come out all right. I would expect Deligne to talk on this in January or February, and if you are interested he may send you a Xerox of his notes then (or even now if they exist already in readable shape).

Katz made an interesting suggestion[133] towards proving the conjecture suggested by what happens in char zero, namely that the level of the vanishing cycles space of the general hypersurface section in a pencil, of a given high degree, is maximal, by [reducing to] the conjecture that the pth power map

$$H^{n-1}(Y_t, \mathcal{O}_{Y_t})^{(p)} \longrightarrow H^{n-1}(Y_t, \mathcal{O}_{Y_t})$$

(or rather on the "vanishing part") is not nilpotent (possibly even semi-simple). The same would hold then for any sufficiently general t, and, if we are working over a *finite* ground field, this would imply that, for most specialisations of Y_t to a finite ground field, the Frobenius acting on the vanishing part of $H^{n-1}(Y_s, \mathcal{O}_{Y_s})$ has (some) nonvanishing proper values. *If* the proper values of Frobenius acting on Hodge cohomology were just the reductions mod p of the proper values of Frobenius acting on l-adic (or crystalline) cohomology, we would be through, and we would have

132 An algebraic proof of the Lefschetz theorem (due to M. Noether) is in P. Deligne, Le théorème de Noether, SGA7 Exposé XIX, LNM 340, Springer-Verlag, 1973, 328–340.

133 Katz carried out his own proposal; see Exposés XX and XXI of SGA7, especially [Exposé XX, part 4.1.2 of Theorem 4.1] and [Exposé XXI, Theorems 1.4 and 4.2].

Relevant later work can be found in N. Koblitz, p-adic variation of the zeta-function over families of varieties defined over finite fields, *Compositio Math.* **31** (1975) 119–218, as well as L. Illusie, Ordinarité des intersections complètes générales, *The Grothendieck Festschrift*, Vol. II, Progr. Math. 87, Birkhäuser, 1990, pp. 376–405.

the extremely precise statement: the number of proper values of Frobenius (acting on some H^i or a piece thereof) which are units is exactly equal to the semi-simple rank of Frobenius acting on the corresponding $H^i(X, \mathcal{O}_X)$ resp. a piece thereof (the contribution of the other pieces of the Hodge cohomology does not count for obvious reasons). Now things in general won't be that simple, because of p-torsion phenomena for Y, which will make the Hodge cohomology a little too big as far as rank, i.e., number of proper values, is concerned. But Lefschetz's theorems as stated in his Borel tract suggest that the torsion of Y should be just the torsion of X, i.e., independent of the degree of the hypersurface section, so I guess asymptotically (for high degrees) it should not count—provided we prove that the semi-simple rank of Frobenius acting on $H^{n-1}(Y_t, \mathcal{O}_{Y_t})$ becomes large. So this is really what should be proved. The trouble is that, although this is again a purely geometric question, it does not seem at all a trivial one, even restricting to the case X a surface (say even $X = \mathbb{P}^2$!) and taking a general hypersurface section (not restricted to belong to a pencil). The question then is whether the general curve thus obtained has (many) étale coverings of order p not coming from coverings of X. Do you have any feeling about this question?[134]

Do you have any idea how to get Griffiths' example over a number field? And no illuminating examples concerning the Hodge conjecture?

I wonder if you, Mike and Hironaka would object in principle to including a paper by Atiyah in Zariski's blue volume[135], in case a paper should be ready before the fixed deadline. He promised us a paper already a while ago from his index series, and wants to keep his promise now, but would not like to have too long publication delays. He says he would be quite willing to dedicate something to Zariski, if he got a suitable paper ready in time.

Yours
(signed) Schurik

PS Maybe it will interest you that I worked out something like a formal substitute, for a smooth morphism $f\colon X \longrightarrow S$ with S of char 0, and a cycle on X which is cohomologically equivalent to zero on the fibers with respect to de Rham cohomology, of the corresponding section of the system of Weil–Griffiths jacobians of the fibers (which make sense only transcendentally). Namely, the tangent space $\mathfrak{t}$ along the zero section of this system makes sense purely algebraically in terms of relative de Rham cohomology, and heuristically Griffiths' section, when expressed locally as an exponential, defines a jet of infinite order of that vector bundle over S, at least up to translation by the image of a horizontal section of the de Rham cohomology sheaf ω on S. Now this jet (more precisely a certain section of $\mathscr{P}^{\infty}_{S/\mathbb{Q}}(\mathfrak{t})$ modulo the image of ω) can be given a purely algebraic definition, and even a pretty simple one. As a consequence, the images of the Griffiths section, à la Manin, corresponding to

[134] False for $\mathbb{P}^2$; $\mathbb{P}^2$ is simply connected, but the general curve of high degree on it has many cyclic covers of order p.

[135] Volume dédié au Professeur Oscar Zariski à l'occasion de son 70e anniversaire, *Publ. Math. IHÉS* **36**, 1969.

"Picard–Fuchs equations", can be given also a purely algebraic description. I think analogous constructions can be made in unequal characteristics, but I did not clear up my mind on this as yet.

Grothendieck to Mumford, 8 April, 1969

Massy 8.4.1969

Dear David,

I am establishing a bibliography of your papers and would like you to help me, as there are a few of your papers which I do not have (including some of which you sent me preprints or reprints, which I gave away without keeping track to whom). Could you either give me precise bibliographical indication or send me a reprint of the following of your works:

1° Your paper on Teichmüller groups (don't have it any longer).
2° Your book on abelian varieties, after your Tata course.
3° Your theorem about liftings of abelian varieties to char. 0,[136] which you explained at Tata. Reference to that Tata talk would do (title?).
4° Extension to char. $p > 0$ of the Italian theorems characterizing rational and ruled surfaces.[137] (Never got a preprint.)
5° Your recent counterexample to Severi's "theorem" on 0-cycles on surfaces (don't have any preprint left).[138]
6° Is there any better reference for your work on abstract θ-functions, compactifications of Néron models etc, than the Bowdoin notes by H. Pittie[139] (which I found pretty poor)?

Sorry to write you such an uninteresting letter! Nothing very interesting to report upon. I guess Katz sent you notes of his nice theorem about L-functions mod p, which will be used in our seminar to do Griffiths' example for complete intersections. Berthelot found a better definition for the crystalline site, dropping the nilpotency condition for the divided powers and replacing it by the condition that p is nilpotent on the objects of the site. The trouble with char. $p = 2$ disappears, and the construction of Dieudonné modules via crystals (for Barsotti–Tate groups) comes out beautifully for all p. Exponentials no longer exist but logarithms do, and this is OK for defining Chern classes for instance. No doubt left that the definition of Berthelot is good.[140] But there is an immense amount of work to be done on crystalline cohomology! Also, to tie it up with Deligne's beautiful generalized Hodge

---

[136] [69b] in this volume.

[137] [69a] (= article 22 in [SP1]).

[138] [69d] (= article 28 in [SP1]).

[139] [u67a] in this volume.

[140] Berthelot's thesis, containing the work Grothendieck mentions, was published in *Cohomologie Cristalline des Schémas de Caractéristique* $p > 0$, LNM 407, Springer-Verlag, 1974.

theory for arbitrary algebraic varieties over $\mathbb{C}$....[141] I guess I will spend the next time trying to understand a little better crystalline cohomology in char. $p > 0$, and just leave Griffiths and Deligne to find out how things look like over $\mathbb{C}$—and Deligne to explain everything to us in the coming year's seminars!

I hope Mike, Hei and you were not too annoyed at the blue journal not coming out when expected, and that by now Oscar got at least the title pages with the dedications.

Best regards
(signed) Schurik

Grothendieck to Mumford, 14 April, 1969

14.4.1969

Dear David,

Thanks a lot for your letter. I appreciated very much your proof of the fundamental intersection formula, and (with your permission) would like to include your proof in SGA5 VII, as an appendix to the exposé (to be ready soon) of Jouanolou, where he proves the same formula, but in the relative case over any base (X, Y smooth over S), no quasi-projectivity assumption, and working in the l-adic cohomology ring.[142] Indeed, Deligne remarked about one year ago that $f^*f_*(y)$ is multiplicative in y, i.e., $= yf^*f_*(1)$, in the cohomological context, which comes out rather trivially (such types of results will be in his exposé SGA4 XVIII of Poincaré duality for étale cohomology, which he is supposed to finish writing this summer), and using this Jouanolou proves formula (4.6) of SGA6 XIV[143] by reducing to the case of codimension 1, using the blowing up, in a rather simple way (less sophisticated than yours). I find it amazing how nicely things come out finally, just introducing these blown-up schemes which at first may seem extraneous to the situation! By the way, did you try also to get a proof of the formula (4.8) of SGA6 XIV (p.11)[144] by analogous arguments?[145] If so, I would appreciate knowing your proof, and reproducing it alongside with your other proof. In the case of l-adic cohomology, Jouanolou again did the work (not neglecting torsion), and in SGA5 VII this will

141 P. Deligne, Théorie de Hodge I, in *Actes du Congrès International des Mathématiciens (Nice, 1970)*, Tome 1, Gauthier-Villars, Paris, 1971, 425–430; Théorie de Hodge II, *Publ. Math. IHÉS* **40** (1971) 5–57; Théorie de Hodge III, *Publ. Math. IHÉS* **44** (1974) 5–77.

142 Mumford's proof of the "self intersection formula" appeared in SGA5 VII, Thm. 9.2 on p. 337 of LNM 589, Springer-Verlag, 1977. The proof of the "key formula" appeared in SGA5 VII, Prop. 9.6 on p. 343 of LNM 589.

143 LNM 225, Springer-Verlag, 1971, p. 676.

144 *Ibid.*, p. 677.

145 Another question is written on the left margin: "And, do you have a proof of RR without denominators for an immersion, in the *Chow ring*?"

figure, together with the corresponding structure theorem for the cohomology of the blown up scheme (which previously I could handle only up to torsion, and only using Deligne's preliminary result!). By the way, in this connection I would like to point out to you a nice foundational paper of Manin on motives (not using any conjectures!) whose main result is the computation of the *motive* of a blown-up variety. It is "Correspondences, motives and monoidal transforms", Mat. Sbornik, T.77 (119), n° 4, p.475–507. I hope it will be translated into English.

I did not understand your allusion to Schottky–Wirtinger, as I am ashamed to confess that I never heard about them. I will ask Serre about it, and would of course be interested to get any notes of yours.

Incidentally, I have already your θ-functions I to III[146] in Inventiones, and am at present only out of part I, which I lent to Raynaud. If you have parts I left, please send one to Raynaud (you sent him only II and III) and one more to me, to make a complete set I to III which I can give away to some chap here; otherwise I send you back II and III.

I spent a few days in Romania. There are two or three bright chaps there, and some more pretty good young people, and I enjoyed discussing with them, including on nonmathematical topics; but life as a whole looks pretty grim there, it gives the impression of a devastated country from the very start. People hate the Russians a great deal, and their own police still more, but they will say the first aloud (although never in print), but the second they won't. There is still a seminar going on applications of matérialisme dialectique to mathematics, an inheritance from times which remain pretty little removed and quite fresh in everybody's memory—but I am pretty sure there is not a single person in Romania who really gives a damn for communism, at least statistically speaking (because nuts you will find anywhere if you look out for them), and excepting the police of course.

Yours
(signed) Schurik

Grothendieck to Mumford, 8 August, 1969

Massy, 8.8.1969

Dear David,

Thanks for your letter. I have no comment to make to your tentative list of invited speakers—except that Lubkin [...].[147] As for the generalization you suggest for EGA IV 17.5.5 (and a variant of 17.15.15) I agree I should have included it, and will do so in the next edition, if Dieudonné lives old enough. (By the way, there

[146] [66a], [67a] and [67b] (= articles 10–12 in [SP1]).

[147] One phrase each in this and two other letters, [1986Jan09] and [1987Feb11], are deleted per instructions from Mumford. Grothendieck and Mumford wrote their convictions—they did not believe in everything Lubkin claimed.

is quite a bit already I would like to change in EGA IV!). However I believe you forget one assumption on $f\colon X \to Y$ (besides f loc. of finite type, Y integral, all components of X dominate Y and have generic fiber of dim $\geq n$, $\Omega^1_{X/Y}$ is locally free of rank $\leq n$), namely: Y geometrically unibranch. (Otherwise take Y to be a curve with an ordinary double point, and X the normalisation.) It is enough then that, instead of assuming $\Omega^1_{X/Y}$ to be locally free of rank $\leq n$, we assume it generated locally by n elements. This implies that we have locally a factorization of f as $X \to X' = Y[t_1, \ldots, t_n] \to Y$, with $X \to X'$ neat (= unramified); but X' is integral and geom. unibranch since Y is, and it is easily seen that every component of X dominates X', hence by EGA IV 18.10.2, $X \to X'$ is étale, hence $X \to Y$ is smooth.

Best regards
(signed) Schurik

Grothendieck to Mumford, 5 January, 1970

Massy, 5.1.1970

Dear David,

A month ago I got an official invitation to give a talk at the Congress of Nice, to my surprise, as I knew the panel on alg. geom. had not proposed me as a speaker. After asking Serre about it, he explained that the organizing committee had proposed me directly, and that by giving a 50 minutes talk I would not prevent any other geometer who may have more interesting things to say, i.e., by refusing no one else would be invited instead. So I accepted, with the idea of giving an outline of my ideas (or what will have become out of them by September) on relations between Barsotti–Tate groups and crystals—as it is my intention to devote most of my research time during the next months to these questions; I hope it is OK with you.

Jouanolou has finally worked out in all details your nice proof of the self-intersection formula in the Chow ring (without neglecting torsion), which will be part of SGA5 VII.9. Using still your idea, he was able to prove also the "key-formula" for a blown-up scheme in the Chow ring, still without neglecting torsion (same reference), and the Riemann–Roch formula without denominators for an immersion of quasi-projective smooth schemes over a field (also in the Chow ring). He also is able to prove the correct formula for $\lambda^n(i_*(x))$ as an element of $K(X)$, as given in my 1957 RR report in the case of char. zero.[148] I think I already wrote you that he proved some time ago the correct l-adic formula for the cohomology of a blown up variety, again without neglecting torsion, and even as a formula in a

[148] The results mentioned here appeared in J.P. Jouanolou, Riemann–Roch sans dénominateurs, *Invent. Math.* **11** (1970) 15–26.

derived category.... Thus a number of questions raised in the last exposé[149] of the Riemann–Roch Seminar SGA6 are settled.

During the month I was in Italy, I worked out various foundational questions on Barsotti–Tate groups over a more or less arbitrary base, assuming only (most times) p to be nilpotent. The fact that the base is no longer assumed to be artinian demands considerable extra care, and practically everything one wants to prove for the BT group $G = \varinjlim_n G(n)$ has to be refined to statements about truncated BT groups $G(n)$. Moreover, for this, a general deformation theory of flat group schemes is needed, for which I have quite clear-cut *statements* ready, but which has still to be proved as part of Illusie's thesis. Granting this, one of the striking byproducts is the following, which for simplicity I will state over a field k of char. p: consider the formal variety of moduli M of the BT group G_0 over k, hence a universal deformation G_M of G_0 over M. Consider $G_M(1) = \mathrm{Ker}(p \cdot \mathrm{id}_{G_M})$, which is a flat deformation of $G_0(1)$ over M. Then M is *versal* for $G_0(1)$ (viewed as a group *killed by p*) in the sense of Schlessinger, i.e., "G_0 and $G_0(1)$ have the same variety of (formal) moduli".

In this connection, I wonder if the following might be true: assume k alg. closed, let G and H be BT groups, and assume that $G(1)$ and $H(1)$ are isomorphic. Are G and H isomorphic? This is true, according to Lazard, if G is a formal group of dimension 1. Another question is: what are the finite groups Γ which are isomorphic to a $G(1)$, for G a BT group? A necessary condition is that Γ be killed by p and that the sequence

$$G \xrightarrow{F} G^{(p)} \xrightarrow{V} G$$

be exact. Is this condition sufficient?

Thanks for your notes on "varieties defined by quadratic equations",[150] which I had no time to look through as yet. Since I set back to do some research, I considerably had to cut down reading! Still, I keep on my table whatever I think I should read sooner or later, so please do not stop sending me reprints!

Best wishes to your and your family for the new year!

Yours
(signed) Schurik

P.S. Please send your mail to my *personal address*, as I stopped working at IHÉS (as Deligne probably told you).

[149] Exposé XIV.

[150] [70] in this volume.

Mumford to Grothendieck, January, 1970[151]

About your questions — over an algebraically closed field $k \supset \mathbb{F}_p$.

1. *If G is finite, $p=0$ in G and*

$$G \xrightarrow{F} G^{(p)} \xrightarrow{V} G$$

is exact, then indeed G can be embedded in a BT-group.

Pf. Use Dieudonné modules. The question becomes: $\forall k$-vector spaces M with p-linear (resp. p^{-1}-linear) endomorphisms F, V such that

$$\mathrm{Ker}(F) = \mathrm{Im}(V), \qquad \mathrm{Ker}(V) = \mathrm{Im}(F),$$

does there exist a free $W(k)$-module N with σ-linear (σ^{-1}-linear) endomorphisms F, V such that $FV = VF = p$ and $(N \otimes_{W(k)} k, F \otimes 1, V \otimes 1) \cong (M, F, V)$. One can check that all such M's have the following type of bases:

$$M \cong \left\{ \text{span of } e_1, \ldots, e_m, \ldots, \underset{1 \le i \le r_j}{F^i e_j}, \ \ldots, \underset{1 \le \ell \le s_j,\ 1 \le j \le m}{V^\ell e_j}, \ldots\ldots \right\}$$

$$\bmod F^{r_j} e_j = \sum_{t=1}^{m} b_{jt} \cdot V^{s_t}(e_t), \ 1 \le j \le m, \quad \det(b_{jt}) \neq 0,$$

where $FV^\ell e_j = 0 \ \forall \ell \ge 1; VF^i e_j = 0 \ \forall i \ge 1$.

Let N be the identical module over $W(k)$, with b_{jt} any nonsingular matrix lifting the b_{jt} above, *except*

$$FV^\ell e_j = p \cdot V^{\ell-1} e_j \ \forall \ell \ge 1, \qquad VF^\ell e_j = p \cdot F^{\ell-1} e_j \ \forall \ell \ge 1.$$

2. *In general, if G_1, G_2 are two BT-groups,*

$$G_1(1) \cong G_2(1) \not\Longrightarrow G_1 \cong G_2.$$

Pf. Just look at Manin's long paper,[152] classifying all BT-gps over k & this is pretty clear. For instance, take the case of 2-dimensional G's (i.e., the associated formal gp is 2-dimensional). The $G(1)$'s, as described in **1**, depend on at most 4 parameters, while Manin's types depends on arbitrarily many. This is not precise but it "clearly" could be made so.[153]

---

[151] Part of a letter from Mumford to Grothendieck, written between Jan. 5 1970 and Jan. 15 1970. This fragment of Mumford's response is all we have.

[152] Y. Manin, The theory of commutative formal groups over fields of finite characteristic, *Usp. Math.* **18** (1963) 3–90; *Russ. Math. Surveys* **18** (1963) 1–80.

[153] However $G_1(1) \cong G_2(1) \Longrightarrow G_1 \cong G_2$ if one of the two BT-groups G_1, G_2 is *minimal*; see Thm. 1.2 of F. Oort, Minimal p-divisible groups, *Ann. of Math.* **161** (2005) 1021–1036. See also 4.1 and 4.2 *loc. cit.* for examples of the negated implication in **2**.

Grothendieck to Mumford, 15 January, 1970

15.1.1970

Dear David,

I agree that the answer to the question of characterizing groups $G(1)$ for G a BT group, over an algebraically closed (or more generally, perfect) field is rather trivial, using Dieudonné's theory. By the way, a similar argument shows that if Γ is a finite group over k which is a flat module over $\mathbb{Z}/p^n\mathbb{Z}$ (or, what amounts to the same, killed by p^n and such that $\mathrm{Ker}(p \cdot \mathrm{id}_\Gamma) = \mathrm{Im}(p^{n-1} \cdot \mathrm{id}_\Gamma)$), then, if $n \neq 1$, there exists a BT group G over k (perfect field) such that $\Gamma \simeq G(n)$ if k algebraically closed or Γ is radicial unipotent. On the other hand, using the (not as yet proved) deformation theory of Illusie for flat groups I alluded to in my last letter,[154] one can prove that if S is a local complete noetherian scheme with residue field k of char. p, and Γ a finite flat group scheme over S which is killed by p^n, then, if Γ_0 is isomorphic to a group $G_0(n)$, Γ is isomorphic to a group $G(n)$ (G_0, G being BT groups over k, S). Thus, if k is perfect, we get a nice characterisation of the groups $G(n)$, which presumably should hold also without any restriction on k.

On the other hand, I could not make any sense out of the indications you gave me for constructing an example (over alg. closed k) where $G(1) \simeq G'(1)$ but $G \not\simeq G'$. You say that for two-dimensional formal p-divisible groups G, the moduli for $G(1)$ form a variety of dimension at most four; but this seems nonsense, because if you fix not only the dimension d of the Zariski tangent space to $G(1)$, but also the corresponding number d^* for the Cartier dual $G(1)^*$ (so that $d + d^*$ is the "height"), then the moduli space for $G(1)$ is of dimension dd^*, which for variable d^* gets arbitrarily high! Maybe there has been some misunderstanding on my part.

Thanks for your comments concerning my troubles with IHÉS. Fortunately things got arranged, as I was backed by my colleagues from IHÉS for demanding that no military funds should be used for the budget. Finally Motchane told us that no such funds were being used in 1970, and that he gave us "une assurance morale" (not being qualified to give us a formal commitment in this respect) that no such funds were to be used in the future. Thus I have taken up my job at IHÉS again, which of course is also the best solution in personal respect, as the position at IHÉS is quite satisfactory in various respects. Maybe you can inform Deligne about this outcome, as I will probably not write him before a week or so.

Best wishes to you and your family

(signed) Schurik

---

[154] Published as L. Illusie, *Complexe Cotangent et Déformations* I and II, LNM 239 (1971) and LNM 283 (1972).

Mumford to Grothendieck, 24 May, 1984

May 24, 1984

Professor Alexander Grothendieck
Mathématiques
University of Montpellier
2, Place Eugène Bataillon
34060 Montpellier
France

Dear Grothendieck,

Thank you very much for sending me your "Esquisse" which I have shared with half a dozen others here. I felt thrilled to hear what you are thinking about and very excited by your ideas.

You asked me in the margin about whether $T_{g,\nu}$ was known to be the same as π_1("$M_{g,\nu}$ at ∞"). I had wondered about this, too—a long time ago I had asked Tits about it. It seems to me that it follows now from the Thurston–Hatcher paper in Topology.[155] But it also follows from the much more powerful and amazing result of Harer[156] (conjectured by Mosher). He constructs a complex C_g of dimension $6g-4$ on which $T_{g,1}$ operates (the idea will apply directly if $\nu = 1$ and adapts, I guess, to other ν), plus a subcomplex D_g *containing the* $(2g-1)$*-skeleton of* C_g, plus a $T_{g,1}$-equivariant homeomorphism

$$(C_g - D_g) \xrightarrow{\approx} \text{Teich space of type } (g,1).$$

In fact, C_g is a union of k-simplices, one for each $(k+1)$-tuple $\sigma_0, \ldots, \sigma_k$ of disjoint arcs in a reference surface S_g with base point P_g, such that all σ_i begin and end at P_g, σ_i are not homotopically trivial and no two are isotopic, all mod isotopy of (S_g, P_g). And $C_g - D_g$ is the set of simplices for which $S_g - \bigcup \sigma_i$ is union of cells. The homeomorphism can be constructed elegantly using the theory of Strebel differentials, or the theory of measured foliations. In any case, I think it proves what you want—and more! Enclosed is a xerox of a letter[157] of mine with more details.

I would like to ask you, on another level, would you consider coming here for some period—it could be a short period, or it could be much longer—to pursue your research? We can offer you travel and support at $850 per week; or a full salary if you can come for a longer period. I realize you are deeply attached to your retreat in the south of France, but perhaps you'd like to think over the possibility of coming here. We can offer you time to do research and contact with lots of people with common interests. It would be wonderful to welcome you here again.

---

[155] Referring to A. Hatcher & W. Thurston, A presentation for the mapping class group of a closed orientable surface, *Topology* **19** (1980) 221–237.

[156] This result was published in J. Harer, The virtual cohomological dimension of the mapping class group of an orientable surface, *Invent. Math.* **84** (1986), 157–176.

[157] Mumford's note on the margin: "I can't find this right now. I'll send it later."

On a more personal note, let me tell you that I have been working in artificial intelligence recently—specifically computer vision. There are some fascinating ideas and problems here. I'd love to discuss these with you if we ever get a chance.

I hope to hear from you soon,

Warmly,
(signed) David
David Mumford

Grothendieck to Mumford, 29 June, 1984

Les Aumettes 29.6.84

Dear David,

I was very pleased to get your warm and enthusiastic letter in response to my "Esquisse". This is the first time since 1978 (the first glimpse upon the anabelian iceberg!) that one of my former friends in the mathematical world shows a sign of interest in these things, which my own instinct very evidently tells me are basic and exciting indeed. It is very strange—and I do feel like a stranger today among those people I used to like a lot. I should add that it seems kind of natural to me that you should be the one exception, which is in accordance with some lively former impression I've got about you.

I must apologize for being late in replying. One reason is that I've been sick for a few weeks now, from overwork I'm sorry to say—something really stupid! During the last four months I've been "just about to finish" some work, which had started as the "introduction" to Pursuing Stacks, and which has grown into a 500 page retrospective of my life as a mathematician, and of the predicaments which struck some of my work after I left the mathematical milieu. It was a very interesting and fruitful reflection, which is going to be the main part of vol. 1 of Reflexions Mathématiques. But with this feeling, of just having to finish up the stuff and be through, I overstrained, and it turns out I'll have to take a complete rest for a few weeks, maybe months. Still, I think I'll send you a copy of that retrospective (in French) sometime in September, and hope it will interest you. I've learned a lot writing it....

For the reason just said I'll not, at present, dive into mathematical matters in connection with your letter, except to tell you I'm glad the property that I needed for $\pi_1(\mathscr{M}_{g,n})$ is ok.[158]

Thanks also a lot for your suggestion to spend some time at Harvard. If any place should be congenial to me, apart from my home, for discussing some of the things in math which have been interesting me in the last years, it is the Harvard area indeed.

[158] In this case $\mathscr{M}_{g,n}$ means the moduli *stack* of curves of genus g with n marked points, and the π_1 is a mapping class group. In the letter Grothendieck refers to the stacks $\mathscr{M}_{g,n}$ as "Teichmüller multiplicities".

It is not clear though that I'll feel the desire or need within the next years to leave my home for the sake of doing maths more efficiently. If it should happen I'll contact you again. In the meanwhile, if you've a chance to drop by at my place, you'll be very welcome (I am able to accommodate you, if you feel like staying a few days). In any case I'll write you if I've mathematical questions which I feel you may know about (or simply be interested in). But for the time being I've got to rest!

Affectionately, (signed) Alexander

PS Rather write to my *personal* address: Les Aumettes, 84570 Mormoiron, France.

Mumford to Grothendieck, 26 December, 1985

Dec. 26, 1985

Dear Grothendieck,

I've spent quite a bit of time in the last week trying to read your long testimony—something I was very interested in as you have always been a very important and vivid figure in my life. I have been impeded by my inadequate knowledge of French—it is very easy for me to read French mathematics, but not at all easy to read more elaborate things. But I have been very moved by many of the things you say and very upset by others. I hesitated for some days trying to imagine how I might reply. I want to do so but I don't know what I can say which is helpful.

One thing I want to write you about is a rather specific suggestion. For at least 10 years I have had the hope that at some point the right occasion would arise to propose the publication, in a suitably edited form, of a large number of your mathematical *letters* to your friends. For me, the letters that you wrote me are by far the most important things which explained your ideas and insights. The letters are vivid and clear and unencumbered by the customary style of formal French publications. I assume that the letters you wrote to others are similar. They express succinctly the essential ideas and motivations and often give quite complete ideas about how to overcome the main technical problems. I have been very conscious of the difficulties that the younger generation has in getting a clear idea of your theories. This may be blunt and insensitive, but I should say that I find the style of the finished works, esp. EGA, to be difficult and sometimes unreadable because of its attempt to reach a superhuman level of completeness. But, for myself, I never liked Bourbaki either! This is a personal thing, but the point is that your letters would offer a clear alternative for students who wished to gain access rapidly to the core of your ideas. My proposal would be to approach someone with a broad knowledge of your theories, such as Artin or Mazur, and give them permission to approach the others to whom you wrote at length to send them copies of your letters (with personal details removed). They could then examine the whole corpus and glue and paste and provide orienting remarks, producing a first draft of a publication for you to review. I feel sure that such a collection would be extremely useful to the younger generation,

Dear Grothendieck,

I've spent quite a bit of time in the last week trying to read your long testimony — something I was very interested in as you have always been a very important and vivid figure in my life. I have been impeded by my inadequate knowlege of french — it is very easy for me to read french mathematics, but not at all easy to read more elaborate things. But I have been very moved by many of the things you say and very upset by others. I hesitated for some days trying to imagine how I might reply. I want to do so but I don't know what I can say which is helpful.

One thing I want to write you about is a rather specific suggestion. For at least 10 years I have had the hope that at some point the right occasion would arise to propose the publication, in a suitably edited form, of a large number of your mathematical <u>letters</u> to your friends. For me, the letters that you wrote me are by far the most important things which explained your ideas and insights. The letters are vivid and clear and unencumbered by the customary style of formal french publications. I assume that the letters you wrote to others are similar. They express succinctly the essential ideas and motivations and often give

many of whom don't have a good appreciation of your ideas at all. I'm not just thinking of *motives* or *D-modules* $\sim$ *crystals* of which you write at length. I find equally distressing the lack of understanding of *duality* (no-one reads Hartshorne's book on duality because they find it too long), of *topoi/stacks* (or the related general existence theorem of Artin's), or (earliest of all) the comparison theorem for π_1 (never published outside SGA1 I believe). There are hordes of smaller but crucial insights such as your letter to me about the yoga of Koszul ($\phi \circ \psi = (-1)^{ab}\, \psi \circ \phi, \ldots$) which showed me what was what with determinant bundles.[159] What do you think? The collection of letters would also serve as a key to the other published things—SGA, EGA.

On another level, I wanted to make some other remarks vis-a-vis intellectual "burial" and the influence of other people. It is very impressive that very few truly innovative ideas ever become current through the straightforward direct route of simply being published, read, understood and used by one's contemporaries. Quite often others "rediscover" them—which is a euphemism for the idea coming to them directly or indirectly at a time when they are not prepared to understand it—and then, when they are prepared, they think it is their idea. Other times, the focus of the international community shifts and a beautiful insight is forgotten for 30–40–50 years, until a new eddy carries people back to renew the old. Then again, people's idiosyncrasies sometimes prevent them from publishing their ideas in an accessible way: I think of Thurston especially who has published almost nothing of his basic theory of 3-manifolds and the associated developments in the theory of surfaces (e.g., geodesic laminations, "train-tracks", etc.). I think that it is rare that a beautiful idea is really forgotten, but it is probably fairly common that ideas get mis-attributed. I was quite disturbed by the merely passing recognitions that is given the work of Shafarevich, Arakelov and Parshin, when Faltings provided the last push that achieved the Mordell conjecture. (I don't mean that he is not a very strong and deep thinker too, whose faith and belief in the method was very very crucial—but just that the Russian school had provided both the basic tools and the motivating ideas of the proof.) Anyway, I do *not* think your ideas are buried, at least not buried too deeply! I believe there is a world-wide reaction today, a trend, towards mathematics that is more concrete and even computational, as opposed to extremely abstract ideas epitomized by *categories/functors/"Tohoku"*. This is not universal. Mathematics (as opposed to, e.g., particle physics) has the luxury of meandering and dividing like a river in a delta (Shafarevich's metaphor was the amorphous growth of an amoeba). But my perspective is that the intellectual center of gravity is shifting *right now* back to rather concrete problems, and this makes it very hard for people to read SGA and EGA. There is clearly a dialectic in mathematics between the concrete and the abstract, and nothing short of the death of mathematics could prevent the center of gravity from shifting back, however, at some point. My feeling therefore

[159] The letter in question, from Grothendieck to F.F. Knudsen, was published as Appendix B to F.F. Knudsen, Determinant functors on exact categories and their extensions to categories of bounded complexes, *Michigan Math. J.* **50** (2002) 407–444; the appendix appears in pages 441–444. Knudsen's article is based on Grothendieck's letter and develops its ideas, in particular of having determinants with values in Picard categories.

is that certain themes in your work, esp. the definition itself of a scheme and the theory of étale and crystalline cohomology, will continue to play a central role, while some of the perspectives will be less appreciated for a while. But this is all facile generalization: the reality is always more complicated.

On a more personal note, in a minor way I can share some of your feelings. What has happened in my life is that I came to a point where I felt that my life was passing quickly and that there were other ideas, other questions that I had once wanted to think about but had totally forgotten while I was immersed in my career as a pure mathematician. These questions were those about the nature of intelligence, about how one thinks and what "thought" consists of. I felt as a student, and I feel now, that the *computational* perspective offers one a fantastic tool to help disentangle these questions and wanted to pursue these ideas before it was too late. Anyway, for three years now I have done essentially no algebraic geometry. What is most disconcerting for me is the feeling of being, on a professional level, in a limbo: before, I had a clear acknowledged position in a clear limited field. When I dropped that I was a complete unknown, sometimes a mistrusted outsider, often a confused student faced with a diverse array of unfamiliar specialties. Anyway, it has been a bit difficult.

Thank you for sending me your testimony, and let me wish you, in the conventional phrase, a "very happy new year". I do indeed wish you a fruitful and rewarding New Year.

Sincerely, with Best Wishes,
(signed) David Mumford

Grothendieck to Mumford, 9 January, 1986

Les Aumettes Jan. 9, 1986

Dear David,

Thank you very much for your letter, and for your sympathy and concern. It is all the more precious and welcome to me as, among the host of my former friends and students from before my departure, there have been extremely few who so much as took the trouble to reply to ReS,[160] and (believe it or not) altogether only three including you, who would express the feeling that there was maybe something wrong somewhere. The other two are Samuel and (remarkably enough) Illusie, one of the three main artisans of the "Burial". Besides these three, I got a number of letters expressing interest, sympathy and concern from more recent friends from the mathematical milieu, and still more from people wholly outside and (presumably for this

[160] A. Grothendieck, *Récoltes et Semailles*. Available from
`http://www.math.jussieu.fr/~leila/grothendieckcircle/recoltesetc.php`

very reason) less reluctant to accept a certain disturbing picture of this milieu, which progressively comes into the fore as the reflection in my "testimony" proceeds.

I greatly appreciate too the effort you made to dig through the French, which certainly isn't easy, as I have been using often rather colloquial expressions which you wouldn't find in any dictionary. I was contacted by a New York publisher who wants to have the book translated and distributed in the US. I hope the project is realized, and will be glad to have you get a copy as soon as it is available.

Your letter strikes me as friendly and thoughtful, eager to be "helpful" one way or another—which in itself is comforting, and I am grateful for your concern. Let me be outspoken though, David, and tell you that I feel the *emphasis* in your letter and in your concern is misplaced. Namely, you need not worry (any more than I did and do) about *me*, or about my work and the recognition given to it. This is not the problem. As far as my person goes, my life is a very happy and fulfilled one, and the episode of writing "Reaping and Sowing", and discovering the "Burial" in its various aspects and impressive proportions, has been part of it (including the moments which have been kind of hard, and which all the less I would wish not to have gone though…). If I don't get a renewal of nomination in CNRS (which has become still more hazardous through the publication of ReS), I'm entitled anyhow to retirement after two more years, which will then allow me to devote myself entirely to those things which fascinate me most, foremost among which, meditation. But even before this, I am to a large extent independent of the good or bad will of my colleagues in the mathematical milieu. As for my work, including the part of it which is still buried or dismantled or made fun of, it is quite clear to me that, if mathematics (and mankind) goes on for a while still (which I don't really feel any more sure about than fifteen years ago…), people couldn't possibly prevent themselves from exhuming my ideas and a certain overall vision, or else rediscovering it. And as far as paternities go, I am not sure really many people seriously believed (even if they pretended to) that étale cohomology, motives and motive-theoretic Galois groups are due to Deligne, étale duality and derived categories to Verdier, the key ideas and the very notion of crystal to Berthelot—and for those who decided to forget what the score was, I guess that (whether they like it or not) the publication of ReS is going to call it back to their minds. And this will be all the more so if (as I now plan) I do spend a few years still, giving a sketch of the vision which was buried, and developing a little, in an informal way, one or the other of its tenets: six operations, crystals, motives, "stacks" and a certain approach to homotopy theory. (And presumably, leaving out the big program of Teichmüller–Galois theory and of anabelian algebraic geometry.)

If something deserves thought and concern it isn't me and my work, but the air in the mathematical milieu you are part of[161], as I once was part of it. You are

[161] Grothendieck's footnote: (Jan. 10) The expression "air" (in the mathematical community) strikes me as inadequate. My perception of reality would be better expressed by saying that I see this community as a gangrened body, of someone who doesn't care to take notice. In such a case, whatever you may say to him about what you are seeing is lost—the words, however plain, have just lost their meaning. This impression has been very strong lately, with the response to ReS from within the community…. And I wonder: what is the sense of doing mathematics, with ears and eyes shut, within such a context?

explaining to me, but perhaps rather explaining to yourself as a way to reassure yourself, about the pendulum of fashion or moods swinging back and forth between the "concrete" and the "abstract" (which is surely familiar to me, be it only from my own work...), and about people rediscovering things which had become forgotten (which is just as familiar to me). However, all this has nothing to do with the Burial and the spirit of the Burial. You know it yourself, and I need not explain it to you. What I do not as yet understand at all is what my particular person, and my particular impact on the mathematics of my time, or on the friends and students sharing with me the same milieu (and the same passion for mathematics), has to do with the present deep degradation of the professional ethics and of the quality of relations between mathematicians; and notably, the relation between those in position of prestige and power, and the others. What is clear to me though, from various echoes I got from here and there, is that this degradation is by no means restricted to the super-fraud and the derision around my work (and the work of Mebkhout) and the derision around my person. More still than by your suggestions of how to "remedy" the oblivion of, or the difficulties in approaching my work, I would be interested in your personal testimony, echoing mine, about your own contacts with the Burial during the last sixteen years, and beyond this, possibly, about occasional glimpses of the general degradation I have been alluding to; and be it only to tell me that you haven't noticed anything at any time, if this should be so. This degradation need not express itself, necessarily, by outright dishonesty and cynicism, it may just as well show through the gradual thickening or fading-out of liveliness, mutual concern and delicacy in relations between people. To speak of just two close colleagues of yours with whom I felt ties of close friendship and sympathy; my last letters to Barry Mazur and to John Tate go back to 1976 and to 1981 and never got a reply, and none of the two (nor Raoul Bott, who I was quite fond of too) took the trouble to reply to Récoltes et Semailles, which I sent with a personal dedication to each. I mention them because they happen to be at Harvard—but similar cases have been countless for the last ten years, and still more so with the sending out of personal copies of Récoltes et Semailles.

But let me get back to cases of outright fraud and ruthless cynicism, as exemplified by the remarkable volume LN 900 on motives (one of the most cited books in the literature), or by the very *name* "SGA4$\frac{1}{2}$" (same remark), or, more shameless than all, the Colloque Pervers (same remark again for the Proceedings of the Colloquium, in two volumes, published in Astérisque). The fraud in these cases is evident and glaringly clear to all those who are in touch with the topics dealt with—in the third case, it should add up to at least fifty world-wide known specialists, including such "stars" as Deligne, MacPherson, Beilinson, Malgrange, Verdier, and many others. Here it is not some well-known "ancestor", who used to be in a position of power himself but who isn't around any more, who is being plundered—or, if he is indeed, this isn't really the crux of the matter. The whole Colloquium (exhibiting for the first time a substantial portion of the panoply of the unnamed ancestor...) took place through the solitary and obstinate work of an unknown pioneer, who (drawing inspiration from the ancestor) succeeded to do the work that Deligne had been unable to conceive of and to do, ten years before. Through the connivance of the participants in this Colloquium this "unknown soldier", who did the work which none of these brilliant people had ever dreamed of, isn't named at all in the first volume of the Proceedings, and in the second only quite incidentally and never with reference to the main result which was the very spring of the Colloquium. I said enough about this affair in ReS, so that I need not dwell upon any more details. In connection with your letter, the association came to me with the Lubkin affair—I believe you never really looked too closely at what happened, and (as far as I remember) I never took the trouble to discuss the matter with you, which appeared to me as [...][162] —the only case I was confronted with in the period till 1970 (when I left). I was surprised that it should go through, unnoticed, but I was too busy then doing mathematics to worry too much. Maybe, in those clement times, what protected Lubkin was that his status was a modest one, so people would feel it wasn't nice in such a case to be too fussy. And I am not sure there is a direct link between this particular isolated case of fraud, and the "new times" which set in about ten years or so later, when fraud gradually has become the "new look" in mathematics. What strikes me is that the situation is now exactly reversed—nowadays it is a whole bunch of some among the most prestigious mathematicians, who will (by common agreement) shamelessly rob an obscure "assistant" in a provincial university, that nobody (except the bunch in the know) has ever heard of....

The point I want to make is this one. This kind of "new look" is alien to your own ways, sure enough, and getting to come into touch with it some way or other is embarrassing and painful, so it is understandable you prefer to turn away your eyes and forget about it. And I am afraid that most of the mathematicians and maybe all, who have not been won over to the "new look" (and I doubt not there are many still), just react that very same way. Maybe they'll say "poor Grothendieck" or (à la rigueur) "poor Mebkhout" (as far as the Burial goes) or "poor such and such"—and then turn to more pleasant thoughts. It may be more beneficial though to have a closer look to what is going on (however painful), and, when this is done, not to hesitate to call a fraud a fraud, and a crook a crook, even if this should be disagreeable to those who like to carry on their swindle, and to trample on the defenseless. Just turning your head away will indeed benefit the new style, and contribute your own share (in the passive mode) to the proliferation of such things as the Colloque Pervers, which already now is considered as something perfectly normal and honorable by the entire mathematical community. At any rate I got precious few responses, from within the math community, who would clearly imply they do *not* consider it normal nor honorable. Your response at any rate, however thoughtful, friendly and sympathetic, does not.

Just one more word. The situation of Mebkhout, since ReS was sent out, has become more difficult than ever—he has to face the hostility of nearly all colleagues, who will let off their own embarrassment (at the inequity of what was done to him through some of the most brilliant members) by holding him responsible for ReS, as the one holding the strings etc. He is at present at IAS, for three months, and I suppose he must be very isolated and ill at ease at this place, with people like

[162] Short phrase deleted.

Deligne, Langlands, Faltings (all very much "partie prenante" of the Burial). His friends (none of whom has the courage or weight to speak out publicly) foretell him that he may well be fired from CNRS (where he finally got admitted a few years ago), and that his future will no doubt be a dark one. And this prediction cannot but turn true, if the climate of cynicism, and of indifference to it I have been trying to describe should prevail. I am not sure that any single action by anyone, however prominent, can really remedy this state of affairs. But I do know that (on a wholly different level) just *one* act of decency and respect, of just anyone, however humble his status, means a lot....

This letter has become prohibitively long, David, and still I am far from having responded to all you touched upon in your long and thoughtful letter. Maybe some other time—this time, I responded first to what was strongest on my mind, I hope you won't mind (indeed, I'm sure you will not!). I look forward to hearing from you again. Please give my regards to our common friends, and above all to Oscar and Yole, if you have a chance. (I heard from Yole, and later from Mike Artin,[163] that Oscar isn't at all in good shape.)

Affectionately
(signed) Alexander

All the best for the New Year too!

Personal address (much speedier)
Les Aumettes
84570 Mormoiron
France

Mumford to Grothendieck, 11 February, 1986

February 11, 1986

Alexander Grothendieck
Les Aumettes
84570 Mormoiron
France

Dear Alexander,

Thank you for your long and moving letter which I have thought about quite a bit. I also discussed the questions you raise with various friends to see if this would give me a different point of view.

But after all this, I really must say I don't agree with you that there has been a general degradation in the manners and customs of the mathematical community.

[163] Grothendieck's footnote: Mike had the courtesy to answer and acknowledge receipt of ReS, in a few embarrassed lines before passing to news about Oscar, and then to mathematical matters which to me have just no meaning any more...

By moving into different fields, what I have found is that on the contrary pure mathematics has better manners and is much more gentle than any of the other fields I have touched. So many other fields have a standard custom of not acknowledging your rivals' work if you can avoid it (and, in fact, everyone has "rivals" to begin with) and being brutal in your criticism of others whenever you have an opportunity. I feel that there are lapses in the mathematical world, but they are rare, the people involved are usually guilty more due to oversight than to intent, and almost everyone tries to rectify the errors. On the other hand there is more of a general tendency in mathematics to forget whatever the previous generation did!

I can't comment on the specific cases of fraud you talk about because I haven't been involved and I don't want to second guess who did what without asking them. I know that others besides yourself have been very upset by particular incidents (Siggy Helgason or Gabriel Stolzenberg for instance). As for Lubkin, as I recall it, he was indeed [...],[164] but I also believe he had some good ideas of his own (his version of étale Cech cohomology for instance) and he got his sad reward by being wholly ignored.

So that's my perception of this isolated corner of the world: still rather blissfully lucky. I'm sorry that this perception is so different from yours. I feel this letter is a very inadequate way of communicating. It would be very nice if we could meet in person again some time. I hope you know how vivid and influential a figure you were in my life and my development at one time. Let me extend my very best wishes to you now.

Sincerely,
(signed) David
David Mumford

---

[164] Short phrase deleted.

Notes

In these notes, "page *x* (*y*)" means page *x* in the original publication and page *y* in this volume; "page *x* (vol. I, p. *y*)" means page *x* in the original and page *y* in [SP1].

[61a] The topology of normal singularities of an algebraic surface and a criterion for simplicity.

1. Page 5 (3), line −2 (a note concerning the Theorem): The generalization of the Theorem for (V^n, P) becomes false for every $n > 2$. Brieskorn has produced examples in which P is a normal isolated hypersurface singularity of type $z_0^{a_0} + \ldots + z_n^{a_n} = 0$. See *Proc. Nat. Acad. Sc. USA* **55** (1966) 1395–1397, and *Invent. Math.* **2** (1966) 1–14.
2. Page 6 (4), line 3 of the 4th paragraph:

   "note (a) $S_{ij} \geq 0$" should read "note (a) $S'_{ij} \geq 0$".

   Of course these two statements are equivalent, but next sentence refers to properties of the matrix S'.
3. Page 15 (13), line −5: "minimal prime ideal" should read "height one prime ideal".
4. Page 16 (14), lines −5 and −4: The first conjecture in the last paragraph on this page, that the ideal class groups of the holomorphic local ring $\mathfrak{o}$ and its completion $\mathfrak{o}^*$ are canonically isomorphic, is true. In fact the obvious generalization to a henselian pair (A, I) is also true; see the Corollary at the bottom of p. 573 of R. Elkik, Solutions d'équations à coefficients dans un anneau hensélien, *Ann. scient. Éc. Norm. Sup.*, Ser. 4, **6** (1973) 553–603.
   Notice that the holomorphic local ring $\mathfrak{o}$ is strictly henselian. The case where $A = \mathfrak{o}$ is a two-dimensional henselian local ring was proved by Boutot in 1971.[1]

[1] J.-F. Boutot, Groupe de Picard local d'un anneau hensélien, *C. R. Acad. Sc. Paris* **272** (1971) A1248–A1250.

5. Page 16 (14), lines −3 and −2: A modified version of the second conjecture in the last paragraph is true. Denote by $\hat{X}$ the formal completion of the reduced exceptional divisor $E := E_1 + \cdots + E_n$ in F'. Then there exists an exceptional divisor $Z \geq E$ supported in E such that the natural map $\mathrm{Pic}(\hat{X}) \to \mathrm{Pic}(D)$ is an isomorphism for every divisor $D \geq Z$ supported in E. This follows from Lemma 2.10 on p. 494 of M. Artin, Some numerical criteria for contractibility of curves on an algebraic surface, *Amer. J. Math.* **84** (1962) 485–496, and the standard GFGA theorems. The (strict interpretation of the) original conjecture, that the natural map $\mathrm{Pic}(\hat{X}) \to \mathrm{Pic}(E)$ is an isomorphism, is false. An example is provided by Example 6.3 (i) on p. 193 of H.C. Pinkham, Normal surface singularities with $\mathbb{C}^*$ action, *Math. Ann.* **227** (1977) 183–193. In this example, $E = E_1 + E_2 + E_3$, where E_1 is a smooth elliptic curve, E_2 and E_3 are isomorphic to $\mathbb{P}^1$, with the following intersection numbers: $(E_1 \cdot E_1) = -1$, $(E_2 \cdot E_2) = (E_3 \cdot E_3) = -2$, $(E_1 \cdot E_2) = (E_2 \cdot E_3) = 1$, $(E_1 \cdot E_3) = 0$. Moreover the restriction $\mathrm{Tr}_{E_1}(E) := \mathscr{O}(E) \otimes \mathscr{O}_{E_1}$ of $\mathscr{O}(E)$ to the elliptic curve E_1 is trivial. The short exact sequence $0 \to \mathrm{H}^1(E_1, \mathrm{Tr}_{E_1}(-E)) \to \mathrm{Pic}(E + E_1) \to \mathrm{Pic}(E) \to 0$ shows that the strict version of the second conjecture fails.

[61b] Pathologies of modular algebraic surfaces.

1. The displayed formula on p. 342 (vol. I, p. 734) should read

$$\phi_1^*(dt/t) = \phi_2^*(dt/t) = \phi_3^*(dt/t) = d(xyz)/(xyz).$$

[62b] The canonical ring of an algebraic surface.

1. The canonical ring of a complex algebraic variety of general type has since been proved to be finitely generated in any dimension. An analytic proof was given by Y.-T. Siu, Finite generation of canonical ring by analytic method, *Sci. China* Ser. A 51 (2008) 481–502. An algebraic proof exists in preprint form in C. Birkar, P. Cascini, C. Hacon and J. McKernan, *Existence of minimal models for varieties of log general type.*
2. Page 613 (22): In the first displayed formula, "$(K^2) - \deg(c_2)$" should read

$$(K^2) + \deg(c_2).$$

3. Page 614 (23), footnote: The cited paper by M. Artin appeared in *Amer. J. Math.* **84** (1962) 485–496.

[65a] A remark on Mordell's conjecture.

1. Page 1008 (47), line 15: "V_k" should be "X_k".

2. Page 1009 (48), line −2: In the definition of a *divisorial correspondence*, instead of requiring $\delta \in \mathrm{Pic}(X \times Y)$ to be trivial when restricted to both $X \times \{p_Y\}$ and $\{p_X\} \times Y$, one should impose the first (resp. the second) condition only when $\dim(Y) > 0$ (resp. $\dim(X) > 0$). (The universal mapping property (*) on p. 1010 (49) then gives a k-rational point $\tilde{\eta} \in J_k$ for any element $\eta \in \mathrm{Pic}(C)$ of degree 0 in the Picard group of C as in the proof of Proposition 2 on p. 1013 (52).)
3. Page 1010 (49), line 17: "J is characterized" should read "$\hat{J}$ is characterized".
4. Page 1010 (49), line −4: "with $X = J$" should read "with $X = \hat{J}$".
5. Page 1014 (53), line 7: "$p_1^*(x\,)$" should read "$p_1^*(x_0)$".

[65b] Picard groups of moduli problems.

1. Page 33 (56), line 8: There has been enormous progress on the question of the rationality of M_g, which remains a subject of active research. In characteristic 0 the following are some highlights.

   - $\mathcal{M}_g$ is of general type if $g \geq 24$. See [83]; J. Harris, Invent. Math. 75 (1984) 437–466; J. Harris and D. Eisenbud, Invent. Math. 90 (1987) 359–387.
   - $\mathcal{M}_g$ is rational for $g = 1,2,4,6$. See I. Dolgachev, Proc. Symp. Pure Math. 46, Part 2, 1987, pp. 3–16 for more information.
   - $\mathcal{M}_g$ is unirational for $g \leq 13$. For $g \leq 10$ this is classical and was known to Severi in 1915. For $g = 12$ see E. Sernesi, *Ann. Sc. Ec. Norm. Sup. Pisa* **8** (1981) 405–439; for $g = 11$ or 13 see M.C. Chang and Z. Ran, *Invent. Math.* **76** (1984) 41–54.

2. Page 47 (70), 3rd line of §3: Insert "complete" after "reduced".
3. Page 49 (72), in the displayed formula (b): The term on the far right, i.e., the expression "$[(S_2 \times \mathscr{X}_2) \times_{(S_1 \times S_2)} T]$", should read

$$[(S_1 \times \mathscr{X}_2) \times_{(S_1 \times S_2)} T].$$

4. Page 53 (76), line 15: In the displayed formula of line −14, that is right after the phrase "We get a diagram", add g below the arrow $T \longrightarrow S$, to become $T \underset{g}{\longrightarrow} S$.
5. Page 57 (80), line 14: "$\frac{\lambda-1}{\gamma}$" should read "$\frac{\lambda-1}{\lambda}$".

[67d] Abelian quotients of the Teichmüller modular group.

1. Page 227 (105): See the extensive comments on a preprint version of this paper by Grothendieck in his 1966 May 9 letter to Mumford (p. 717 in this volume).
2. Page 228 (106), line −4: It is clear from the context that the precise conjecture here is that the rank of $\mathrm{H}^2(\Gamma_g, \mathbb{Z})$ is one for $g \geq 3$: as mentioned on p. 243 (121), $H^2(\Gamma_2, \mathbb{Z}) \cong \mathbb{Z}/10\mathbb{Z}$. The first significant breakthrough on this question was

made by J. Harer, who showed in 1983 that $\mathrm{H}^2(\Gamma_g, \mathbb{Z}) \cong \mathbb{Z}$ for $g \geq 5$; see *Invent. Math.* **72** (1983) 221–239. Note that the claim in the paper that the torsion part of $\mathrm{H}^2(\Gamma_g, \mathbb{Z})$ is isomorphic to $\mathbb{Z}/(2g-2)\mathbb{Z}$ was incorrect. Since then there has been enormous progress on the cohomology of the mapping class group and the cohomology of $\mathscr{M}_{g,n}$ and $\overline{\mathscr{M}}_{g,n}$. In particular, $\mathrm{H}^2(\Gamma_g, \mathbb{Z})$ has been computed: it is isomorphic to $\mathbb{Z}$ for $g > 2$, and is isomorphic to $\mathbb{Z}/10\mathbb{Z}$ for $g = 2$.

3. Page 233 (111), line -3: replace "$x - x^{\gamma}$" by "$x - x^{y}$"
4. Page 244 (122): Paper [8] in Bibliography is [u64b].

[69a] Enriques' classification of surfaces in char p.

1. Page 329 (vol. I, p. 664), lines 6, 9, 10, 11: "$\mathscr{O}_{E_i}(-E_i^2)$" should read "$\mathscr{O}_{E_i}(-E_i)$"; similarly in line 8, "$H^1(\mathscr{O}_{E_i}(-E_i^2))$" should read "$H^1(\mathscr{O}_{E_i}(-E_i))$"

[69b] Biextension of formal groups.

1. Page 307 (141), lines 2 and 3: In a letter to P. Cartier in the winter of 1967/8, Mumford explained the definition of a multiplicative biextension of commutative formal groups and a related notion of a "polarized canonical module" over a commutative ring R of characteristic p and outlined a proof that generic abelian varieties in characteristic p are ordinary, in (a) and (b) below, similar to (ii)–(iv) on p. 307: (a) when R is a complete noetherian local ring, a polarized canonical module ovr R determines a deformation of polarized p-divisible groups over R; (b) a "generic" polarized canonical module deforming that of an abelian variety X over $R/\mathfrak{m}_R$ can be shown to be ordinary, by reducing to the case when the original abelian variety X satisfies $\alpha(X) = 1$.

   A version of the proof sketched in steps (i)—(iv) was published by P. Norman and F. Oort, *Ann. Math.* **112** (1980) 413–439. The proof uses a version of the method mentioned in (iii) below, published by P. Norman in *Ann. Math.* **101** (1975) 499–509. The notion of a *displayed Dieudonné module* in Norman's paper was generalized by T. Zink into a new Dieudonné theory for p-divisible groups, called the *theory of displays*. See T. Zink, The display of a formal p-divisible group, in *Cohomologies p-adiques et applications arithmétiques I*, Astérisque 278 (2002), 127–248; T. Zink, A Dieudonné theory for p-divisible groups, in *Class field theory—its centenary and prospect (Tokyo, 1998)*, Adv. Stud. Pure Math. 30, Math. Soc. Japan, Tokyo, 2001, 139–160; W. Messing, Travaux de Zink, *Séminaire Bourbaki* 2005/2006, Exp. 964, Astérisque 311 (2007), ix, 341–364.
2. Page 307 (141), line -6: Cartier's results were published by M. Lazard in LNM 443, Springer-Verlag, 1975.
3. Page 308 (142), line 5: The following should be added as a footnote for the definition of the ring A_R:

'The ring A_R is the completion of the noncommutative ring $W(R)[V][F]$ with respect to the right ideals $V^n W(R)[V][F]$. Explicitly, the ring A_R consists of all infinite series of the form

$$\sum_{m,n\geq 0} V^m [a_{mn}] F^n \qquad \text{with } a_{mn} \in R \;\forall\, m,n \geq 0$$

with the property that for every $m \geq 0$, there exists an integer C_m such that $a_{mn} = 0$ for all $n \geq C_m$. Here $[a_{mn}]$ denotes the Witt vector $(a_{mn}, 0, 0, \ldots)$, the "Teichmüller representative" of a_{mn}.'

4. Page 310 (144), §2: The notion of biextension was further developed in Exposés VII, VIII of SGA7I, LNM 288, Springer-Verlag, 1972. The set of isomorphism classes of biextensions of $G \times H$ by F is $\mathrm{Ext}^1(G \otimes^{\mathbb{L}} H, F)$, when G, H, F are abelian groups in a topos. In LNM 980, L. Breen developed the notion of *cubical structures* and related it to biextensions.
5. Page 313 (147), line -10: "Φ, Ψ are the group laws of G and H respectively" should read "Φ, Ψ are the group laws of H and G respectively".
6. Page 319 (153), line 3: "$\beta(Pm, Qn) = P.(m,n).Q^*$" should read "$\beta(Pm, Qn) = P \cdot \beta(m,n) \cdot Q^*$".

[69d] Rational equivalence of 0-cycles on surfaces.

1. Page 197 (vol. I, p. 755), 3 lines above the displayed commutative square: "induced 2-form" should read "induced q-form".

[70] Varieties defined by quadratic equations.

1. Page 70 (232), line 4 of the second paragraph: The word "natural" should be eliminated. In fact the homomorphism $\rho_\alpha : K(L) \to \mathcal{G}(P_\alpha)$ which splits the extension
$$1 \to k^* \to \mathcal{G}(P_\alpha) \to K(L) \to 1$$
is unique up to $\mathrm{Hom}(K(L), k^*)$.
2. Page 75 (237), line 12: W denotes the image of
$$\sum_{\alpha \in \hat{X}} \Gamma(L \otimes P_\alpha) \otimes \Gamma(M \otimes P_{-\alpha}) \to \Gamma(L \otimes M)$$
as in the proof of the Lemma on p. 68 (230).
3. Page 79 (241), line 2 from bottom: The displayed formula should read
$$P(x,y) = \chi(L) \prod_{i=1}^{g} (x - \alpha_i y).$$

4. Page 80 (242): The displayed formula on line 8 should read

$$P(x,y) = (\text{constant}) \cdot \prod_{i=1}^{r} (x - \alpha_i y) \cdot y^{g-r}.$$

5. Page 99 (261), line 4: Insert "abelian" between "complementary" and "subvariety".

[71a] Theta characteristics of an algebraic curve.

1. Page 189 (vol. I, p. 488), line 12: "E is a $\pi^* \mathscr{O}_{X'}$-algebra" should read "E is a $\pi_* \mathscr{O}_{X'}$-module".

[71b] A remark on Mahler's compactness theorem.

1. Page 291 (265): The last displayed formula (second formula in the statement of Theorem 2) should read

$$\{\Gamma \in \mathfrak{M}_G^C \mid \Gamma \cap U_\varepsilon = \{e\},\ \text{measure}(G/\Gamma) \le D\}$$

(in other words, the superscript "C" should be moved from Γ^C to $\mathfrak{M}_G^C$).

[72d] Introduction to the theory of moduli.

1. Page 176 (274), 4 lines before Corollary: "ψ: $M \to N$" should read "Ψ: $M \to N$".

[73c] A remark on the paper of M. Schlessinger.

1. Page 117 (326), the last line: The paper by H. Pinkham was published in *J. Algebra* **30** (1974) 92–102.

[75b] Matsusaka's big theorem.

1. Page 513 (327): A refinement of Theorem 1 by J. Kollár and T. Matsusaka states that k_0 depends only on the first two coefficients of the Hilbert polynomial $P(k)$; see *Amer. J. Math.* **105** (1983) 229–252. An effective estimate of k_0 is provided in Y.-T. Siu, *Ann. Inst. Fourier* **43** (1993) 1387–1405. An exposition of the effective version of Matsusaka's big theorem is given in 10.2 of the book *Positivity in Algebraic Geometry. II: Positivity for Vector Bundles and Multiplier Ideals* by R. Lazarsfeld, Springer-Verlag, 2004.
2. Page 529 (343), line 7: Insert "by" between "parametrized" and "a suitable countably infinite set".

[76a] Hilbert's fourteenth problem—the finite generation of subrings such as rings of invariants.

1. Page 438 (359): In the displayed formula in line −8, "$1, f_1, \dots, f_{k-1}$" should read
$$1, \mathscr{L}(bH+aE), \mathscr{L}(2bH+2aH), \dots, \mathscr{L}((k-1)bH+(k-1)aE).$$
2. Page 443 (364): The paper by E. Formanek and C. Procesi mentioned in Added in proof was published in *Advances in Math.* **19** (1976) 292–305.
3. Page 444 (365): Entry [25] of REFERENCES: W.J. Haboush, *Reductive groups are geometrically reductive*, was published in *Ann. Math.* **102** (1975) 67–83.

[76b] The projectivity of the moduli space of stable curves I: Preliminaries on "det" and "Div".

1. Page 20 (367), line −14: This display formula should be
$$\psi(l \otimes m) = (-1)^{\alpha(x)\beta(x)}(m \otimes l).$$

[78a] An algebro-geometric construction of commuting operators and of solutions to the Toda lattice equation, Korteweg de Vries equation and related nonlinear equations.

1. Page 116 (404), line 12: "$[a_1, a_2]$, $[b_1, b_2]$" should be "$[-a_1, a_2]$, $[-b_1, b_2]$". In §1, it is also assumed that a_1, a_2, b_1, $b_2 > 0$.
2. Page 116 (404), line −9: "$T_{x,p}$" should be "$T_{X,P}$".
3. Page 119 (407), 1st line of Proposition: "$\{X, P, Q, R\}$" should be "$(X, P, Q, \mathscr{F})$".
4. Page 120 (408), lines 5–8: "$|x_{n+1}| \geq C|x_n|$" (line 8) should read "$|x_{n-1}| \geq C|x_n|$". For conditions "$|x_n| \geq C|x_{n-1}|$" or "$|x_{n-1}| \geq C|x_n|$" to define neighborhoods of the two points at infinity, we should assume $\Sigma \cap \Theta = \emptyset$ as in p. 127 (415), and take a suitably normalized R among the equivalent rings, e.g., one for which A satisfies $A_{i,i+a_2} = 1$.
5. Page 120 (408), line −8: "$BC = CA$" should be "$BC = CB$".
6. Page 121 (409), line −13 [4th displayed formula]: "$\frac{a_2-1}{a_1} \geq \frac{\lambda}{\mu} \geq \frac{b_2}{b_1-1}$" should be "$\frac{a_2}{a_1} > \frac{\lambda}{\mu} > \frac{b_2}{b_1}$" as in the first displayed formula in the page, and subsequent arguments in p. 122 (410) should be changed appropriately, e.g., in line 5: "$R_{\lambda(a_2+b_1-1)} \cap R^{\mu(a_2+b_1-1)}$" should read "$R_{\lambda(a_2+b_1)-1} \cap R^{\mu(a_2+b_1)-1}$", etc. (Otherwise we need to also assume $(a_2-1)/a_1 \geq b_2/(b_1-1)$; this can be achieved, e.g., by replacing A and B by suitable powers of them.)
7. Page 121 (409), line −7: "$A^\lambda \in \mathscr{R}_{a_1}$" should read "$A^\lambda \in \mathscr{R}_{a_2}$".

8. Page 127 (415), line 6: "$(n,a_1) = (n,a_2) = 0$" should be "$(n,a_1) = (n,a_2) = 1$". This may be regarded as a special case of the condition $(a_1,b_1) = (a_2,b_2) = 1$ in Data B. When we consider $[A,S^n] = 0$ without the restriction on (n,a_1) and (n,a_2), the resulting picture looks simpler than the general Data A′–Data B′ correspondence: see [79b].
9. Page 133 (421), line 2 of Proof of Lemma: "$\deg A = \alpha$" should be "$\deg A = a$", and "$a+n+1$" should be "$a+n-1$".
10. Page 136 (424), line −6 and Page 137 (425), line 1: $\bigoplus_{i=1}^{r-1}$ should be $\bigoplus_{i=0}^{r-1}$.
11. Page 153 (441), line 13: H. McKean's article appeared in *Partial differential equations and geometry (Proc. Conf., Park City, Utah, 1977)*, Lecture Notes in Pure and Appl. Math., 48, Dekker, New York, 1979, pp. 237–254.
12. Page 153 (441), line 16: Airault *et al.*'s article appeared in *Comm. Pure Appl. Math.* **30** (1977), no. 1, 95–148. For further development in this direction, see I. Krichever, Elliptic solutions of Kadomtsev-Petviashvili equations and integrable systems of particles, *Funct. Anal. Appl.*, **14** (1980), no. 1, 45–54 (In Russian), 282–290 (Translation), A. Treibich and J.-L. Verdier, Solitons elliptique, in *The Grothendieck Festschrift*, Vol. III, Progr. Math. 88, Birkhäuser, 1990, pp. 437–480, and the literature cited therein.

[78c] Some footnotes to the work of C.P. Ramanujam.

1. Page 249 (447), line 3 of the footnote: "y^t" should read "Y^t".
2. Page 250 (448), bottom of the page: The displayed statement (1) is part of the Theorem on page 121 of C.P. Ramanujam, Supplement to the article "Remarks on the Kodaira vanishing theorem", *J. Indian Math. Soc.* **38** (1974), 121–124.
3. Page 251 (449), line 8: The paper by D. Gieseker is On a theorem of Bogomolov on Chern classes of stable bundles, *Amer. J. Math.* **101** (1979) 77–85.
4. Page 251 (449): Mumford's proof of Kodaira vanishing theorem (the proof (2)$\Longrightarrow$(1), starting on line 9) also appeared in Reid's article [13] as an appendix.
5. Page 254 (452), line 5: The morphism "resp" is $p|_E$, the restriction to E of the morphism $p: F \to C$.
6. Page 254 (452): The displayed formula before "*Q.E.D.*" should read
$$\mathrm{H}^1(F_0, \mathscr{O}(-H)) \neq (0).$$
7. Page 255 (453), line 3 of Theorem: "$\pi: X^* \to \operatorname{Spec} \mathscr{O}$" should read "$\pi: X \to \operatorname{Spec} \widetilde{\mathscr{O}}$".
8. Page 257 (455), line −4: The result of H. Matsumura referred to is in the article Geometric structure of the cohomology rings in abstract algebraic geometry, *Mem. Coll. Sci. Univ. Kyoto*, Ser. A Math. **32** (1959) 33–84.
9. Page 258 (456): The beginning of the two displayed lines in the statement of the Lemma should be "$H^i_{\{x\},\mathrm{int}}(\mathscr{O})$".

10. Page 258 (456): The local cohomology groups in the displayed lines −8 and −6 should be $H^i_{\pi^{-1}x}(\mathscr{F})$ and $H^i_{\pi^{-1}x}(\mathscr{O}_X)$, respectively.
Line −2: "$R_i\pi_*(\Omega^n_X)$" should read "$R^i\pi_*(\Omega^n_X)$".
11. Page 261 (459), line 11: The referenced Corollary of Proposition 2.6 in Chapter V of article [17] appeared on page 144 of LNM 632, Springer-Verlag, 1978.
12. Article [2] by F. Bogomolov appeared as Holomorphic tensors and vector bundles on projective manifolds, *Izv. Akad. Nauk SSSR* Ser. Mat. **42** (1978) 1227–1287, 1439.
13. Article [3] by D. Buchsbaum and D. Eisenbud appeared in *Amer. J. Math.* **99** (1977) 447–485.
14. Article [13] by M. Reid appeared in *Proceedings of the International Symposium on Algebraic Geometry (Kyoto, 1977)*, Kinokuniya Book Store, Tokyo, 1978, pp. 623–642.
15. Article [17] by J.-F. Boutot appeared as LNM 632, Springer-Verlag, 1978.

[82] On the Kodaira dimension of the moduli space of curves.

1. A gap in the proof of Theorem 4 on p. 58 (vol. I, p. 206) was found by S. Mochizuki; see the Remarks on p. 372 and p. 392 of The geometry of the compactification of the Hurwitz scheme, *Publ. RIMS Kyoto Univ.* **31** (1995) 355–441. In §3 *loc. cit.*, one finds an exposition of the notion of *log admissible covering* and a proof of the existence of an algebraic stack with a log structure which classifies log admissible coverings; see Thm. 3.22 on p. 389 *loc. cit.* This supplies a proof of Theorem 4 on p. 58 (vol. I, p. 206); see §3.1 and the Remark on p. 377 *loc. cit.*

Printed in the United States
By Bookmasters